CLIMATE CHANGED

A PERSONAL JOURNEY THROUGH THE SCIENCE

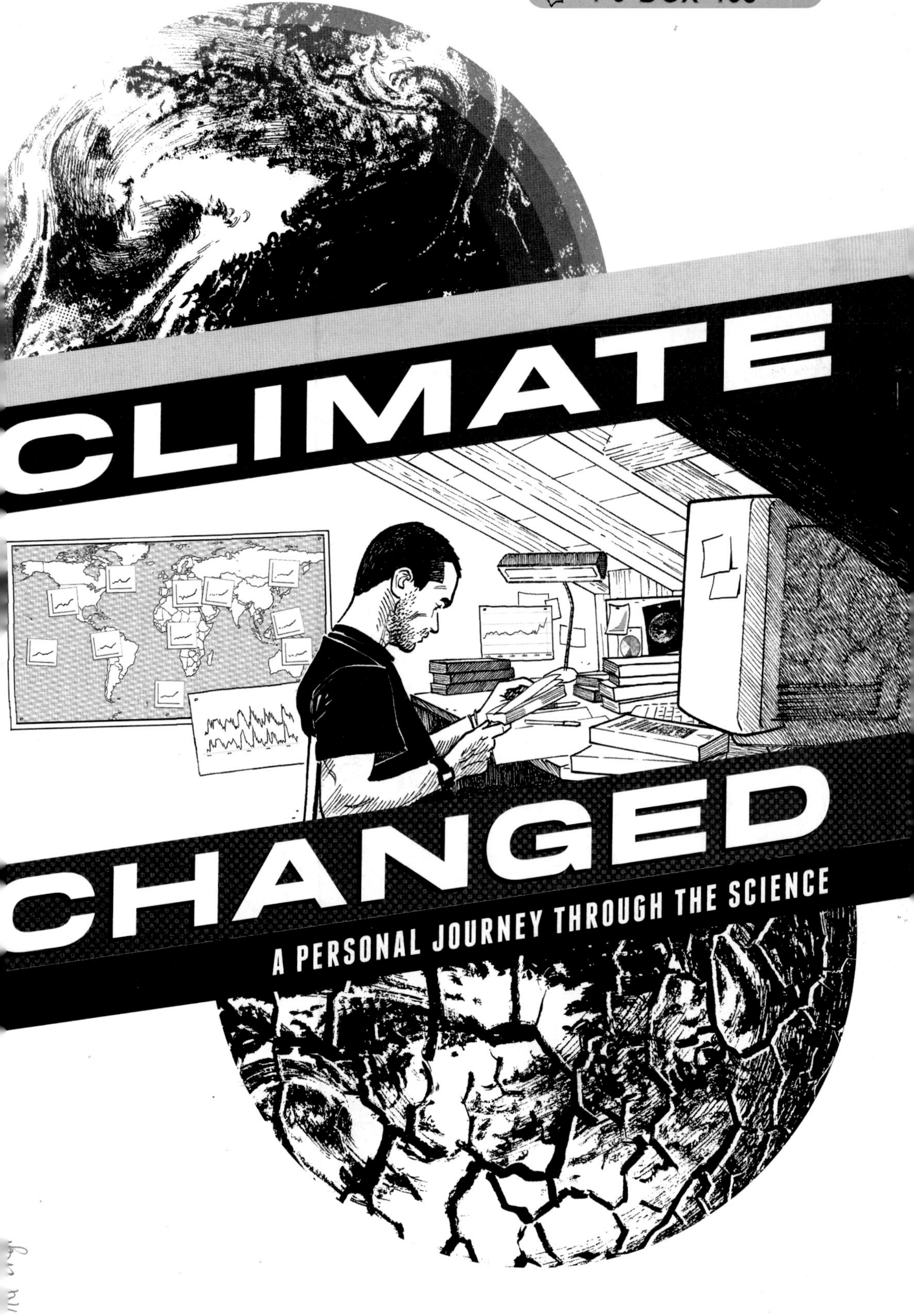

PHILIPPE SQUARZONI

CLIMATE

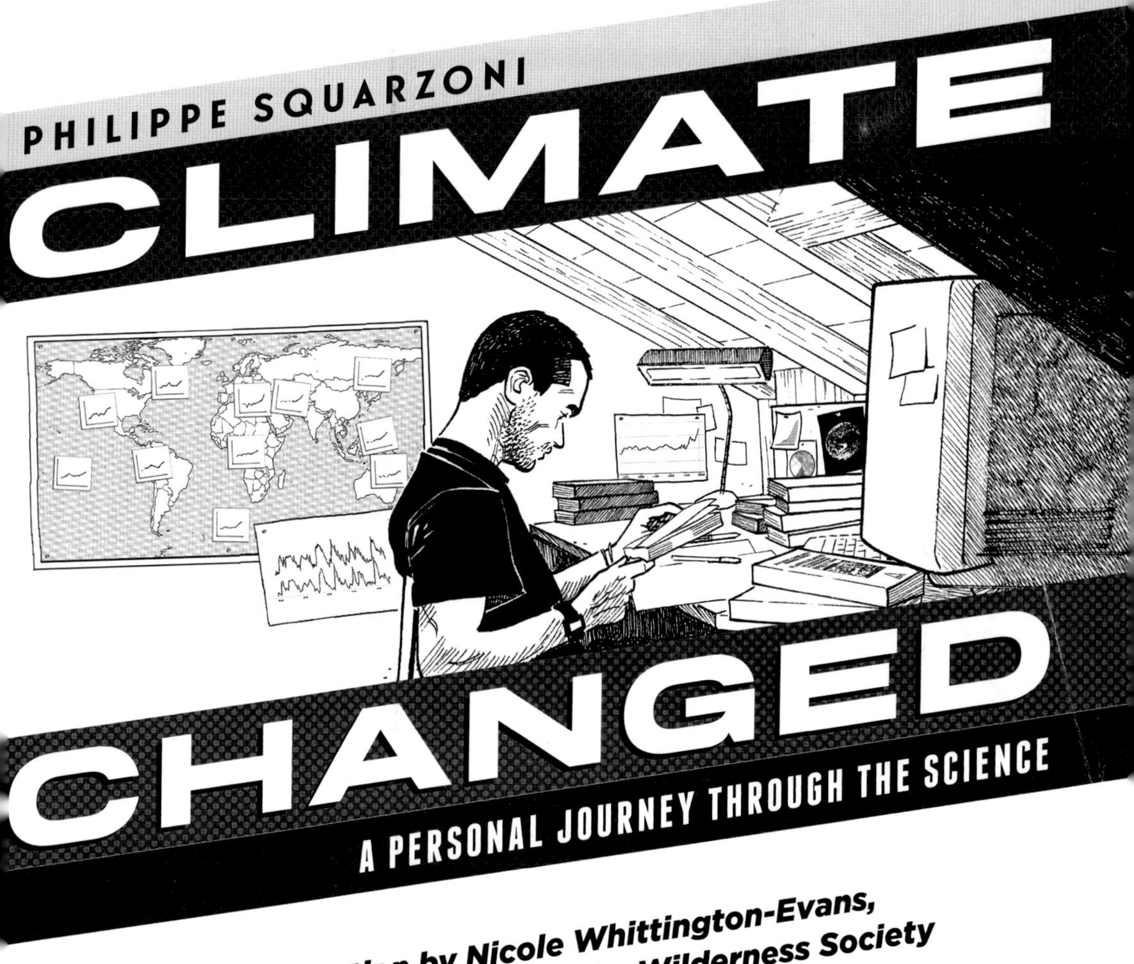

CHANGED

A PERSONAL JOURNEY THROUGH THE SCIENCE

Introduction by Nicole Whittington-Evans, Alaska Regional Director, The Wilderness Society

Translated by Ivanka Hahnenberger

Abrams ComicArts • New York

Editor: Carol M. Burrell
Designer: Meagan Bennett
Production Manager: Alison Gervais
Managing Editor: Jen Graham

Cataloging-in-Publication Data has been applied for
and may be obtained from the Library of Congress.
ISBN: 978-1-4197-1255-5

Saison Brune by Philippe Squarzoni
© Éditions Delcourt—2012
Text and illustrations by Philippe Squarzoni
Translation by Ivanka Hahnenberger
English translation copyright © 2014 Harry N. Abrams, Inc.
Lettering by Grace Lu

Originally published in French in 2012 under the title
Saison Brune by Éditions Delcourt.

Printed and bound in U.S.A.
10 9 8 7 6 5 4 3 2 1

Abrams ComicArts books are available at special discounts when purchased
in quantity for premiums and promotions as well as fundraising or educational use.
Special editions can also be created to specification. For details,
contact specialsales@abramsbooks.com or the address below.

ABRAMS
THE ART OF BOOKS SINCE 1949

115 West 18th Street
New York, NY 10011
www.abramsbooks.com

The views expressed by the author in this book do not necessarily represent those of the interviewees, the institutions for which they work, or the organizations to which they belong. Errors may have crept into the manuscript, despite the vigilance of the author and editors. Finally, this book is also a story, depicting events as they took place. Some factual information and figures cited herein were current when the events in the chapters unfolded, and some more recent events are not discussed.

A note on measurements: Greenhouse gases, CO_2 equivalents, and carbon equivalents are given in metric tons, in keeping with the standard used in the IPCC reports.

1 metric ton = 1.102 US tons ("short tons").

INTRODUCTION

In Alaska where I live and in the Arctic where I work, where the climate is warming at approximately twice the rate of other parts of the globe, it's obvious that climate change is real and pervasive. Alaskans are witnessing climate change across a wide spectrum of experiences and at an alarming pace. Entire communities are in need of relocation because of dramatic coastal erosion from violent sea storms. As sea ice declines in the Arctic Ocean, polar bears are swimming long distances—and sometimes drowning—in search of seals and pack ice, while a lack of ice is forcing walrus and seals to haul out and rest on land. In turn, indigenous Inupiat hunters are traveling farther from home, at great risk, in search of food. The natural ice cellars where these hunters store meat are collapsing under thawing. Even south of the Arctic, Alaska is experiencing visible changes, such as the invasion of alpine tundra habitats by increased growth of shrubs and alders, disappearing lakes and wetlands, greater and more frequent fires, and glaciers receding at an astounding pace. These are only snapshots, but every day, people are experiencing the effects of climate change here in Alaska.

In *Climate Changed*, Philippe Squarzoni has written a very important account that reveals the significance of the changing climate and its effects on humans and the environment. He explains in simple and compelling terms the complexities of climate change, so that we all can understand why Earth's atmosphere is warming and how a changing climate will affect the globe and its inhabitants. He describes his own struggles and personal choices while confronting this issue. Only when people understand the urgency and importance of the impacts of climate change will they begin to take action and alter their behavior.

It is vital for each of us to recognize that we have the ability, opportunity, and power to make decisions that can slow and minimize our impacts on the planet's climate. We *can* make choices through individual and collective action, such as making smart consumer decisions or speaking with elected officials and industry leaders. Only through these actions will we be able to steer our collective future to a safer outcome.

At The Wilderness Society, we work to protect wilderness throughout the United States. This year, 2014, marks the fiftieth anniversary of the 1964 Wilderness Act, a law that enables Congress to designate specific areas as wilderness to maintain them in their natural state. The founders of The Wilderness Society—Robert Sterling Yard, Robert Marshall, Harvey Broome, Benton Mackaye, and Aldo Leopold—were the architects and advocates of this law, and since its passage, it has been used to establish 757 wilderness areas totaling nearly 110 million acres in 44 states. In Alaska there are 48 wilderness areas totaling approximately 57.4 million acres, representing 52 percent of the nation's designated wilderness acreage. Because wilderness areas provide large, intact landscape-level natural habitat, they will continue to play a significant role in providing species the time and space to adapt to a changing climate.

America's Arctic still contains vast, wild landscapes and seascapes that are essentially unmarked by industrial development but threatened by those who want to extract oil and gas resources. These areas include globally significant habitat as well as thriving indigenous communities whose inhabitants continue to live off the land and water and practice cultural traditions they have pursued for thousands of years. The Wilderness Society in Alaska is focused primarily on protecting lands and waters north of the Arctic Circle that are wild, ecologically significant, and important for Alaska Native communities and all Americans. Specifically, The Wilderness Society is working to protect, permanently, the coastal plain and other portions of the Arctic National Wildlife Refuge and special areas in the western Arctic. These areas contain, for example, some of the most important onshore denning habitat for polar bears in America; the calving grounds for Alaska's largest caribou herds; and the largest wetlands complex in all of the circumpolar Arctic, which hosts some of the greatest known densities of nesting shorebirds and molting waterfowl in this region of the world. Protecting these critical wildlands will be an important step in establishing refugia for a variety of species to adapt to climate change, including polar bears, whales, walrus, seals, birds, and several of America's largest caribou herds, as well as a step toward curbing carbon emissions.

As Philippe Squarzoni tells us through this book, industrial development is the leading cause of the carbon emissions responsible for climate change. The Wilderness Society works to halt oil and gas development in the Arctic Ocean because of the impacts industrial development would have on this sensitive marine environment, on the special areas onshore, and to primary food sources for coastal villages. We also work to halt oil and gas development in the Arctic Ocean because the oil and gas reserves in this relatively untouched seascape would add significantly to global carbon emissions. Such drilling would be premature, because oil companies are not technically prepared to develop the Arctic Ocean safely, and the federal government has not yet written appropriate regulations for Arctic drilling operations, much less designated which areas are so sensitive that they should be protected from any drilling at all. Government estimates for undiscovered, technically recoverable oil and gas in the Arctic Ocean—from both the Chukchi and Beaufort Seas—are approximately 15.8 billion tons of carbon. If extracted and burned, the carbon produced would be equivalent to more than nine years of emissions from all U.S. transportation modes calculated at 2011 levels. Keeping this carbon source in the ground is one of the most significant actions our nation can take at this critical juncture to curb future carbon emissions and slow worldwide temperature increases so that they don't exceed 2 degrees Celsius (3.6 degrees Fahrenheit) by the end of this century—the internationally accepted goal for limiting global temperature increase. At The Wilderness Society, we are working with local, national, and indigenous partners to ensure that the Arctic Ocean and special onshore areas in America's Arctic remain free of industrial development for decades to come. These efforts will go a long way toward curbing America's carbon emissions. There is much more to be done, however.

My plea to you is to read this book, listen, learn, and continue to educate yourself about the effects of climate change—and take a personal journey, as Philippe Squarzoni has, to determine what your future actions will be, in what must become a collective effort to curb carbon emissions and slow Earth's warming.

Nicole Whittington-Evans, Alaska Regional Director
The Wilderness Society

Nicole Whittington-Evans works with members of the public, tribes, and federal land-managing agencies to advance conservation measures in the state of Alaska. She has worked in environmental policy and advocacy for nearly two decades and currently leads a team of staff focusing on issues including the Arctic National Wildlife Refuge, the western Arctic's National Petroleum Reserve, and the Arctic Ocean.

THERE ARE MANY WAYS TO START A BOOK...

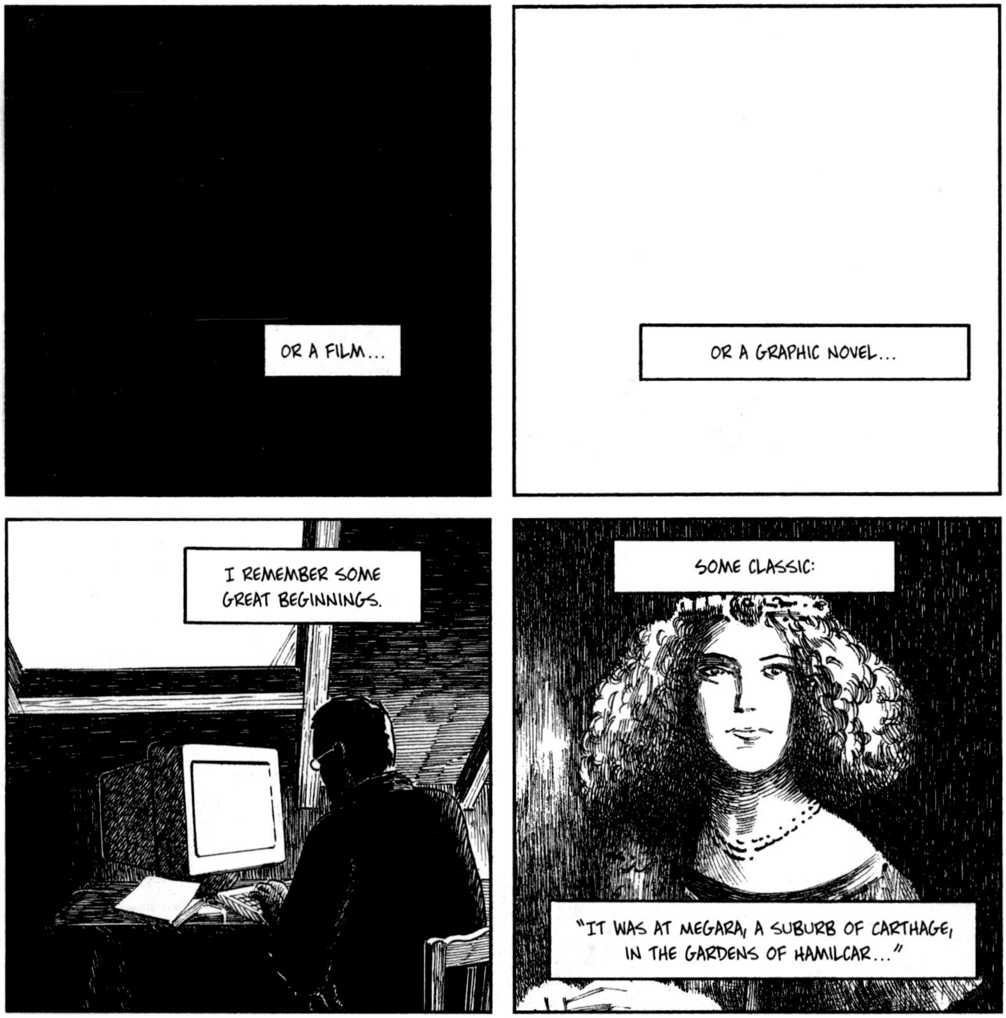
OR A FILM...
OR A GRAPHIC NOVEL...
I REMEMBER SOME GREAT BEGINNINGS.
SOME CLASSIC:
"IT WAS AT MEGARA, A SUBURB OF CARTHAGE, IN THE GARDENS OF HAMILCAR..."

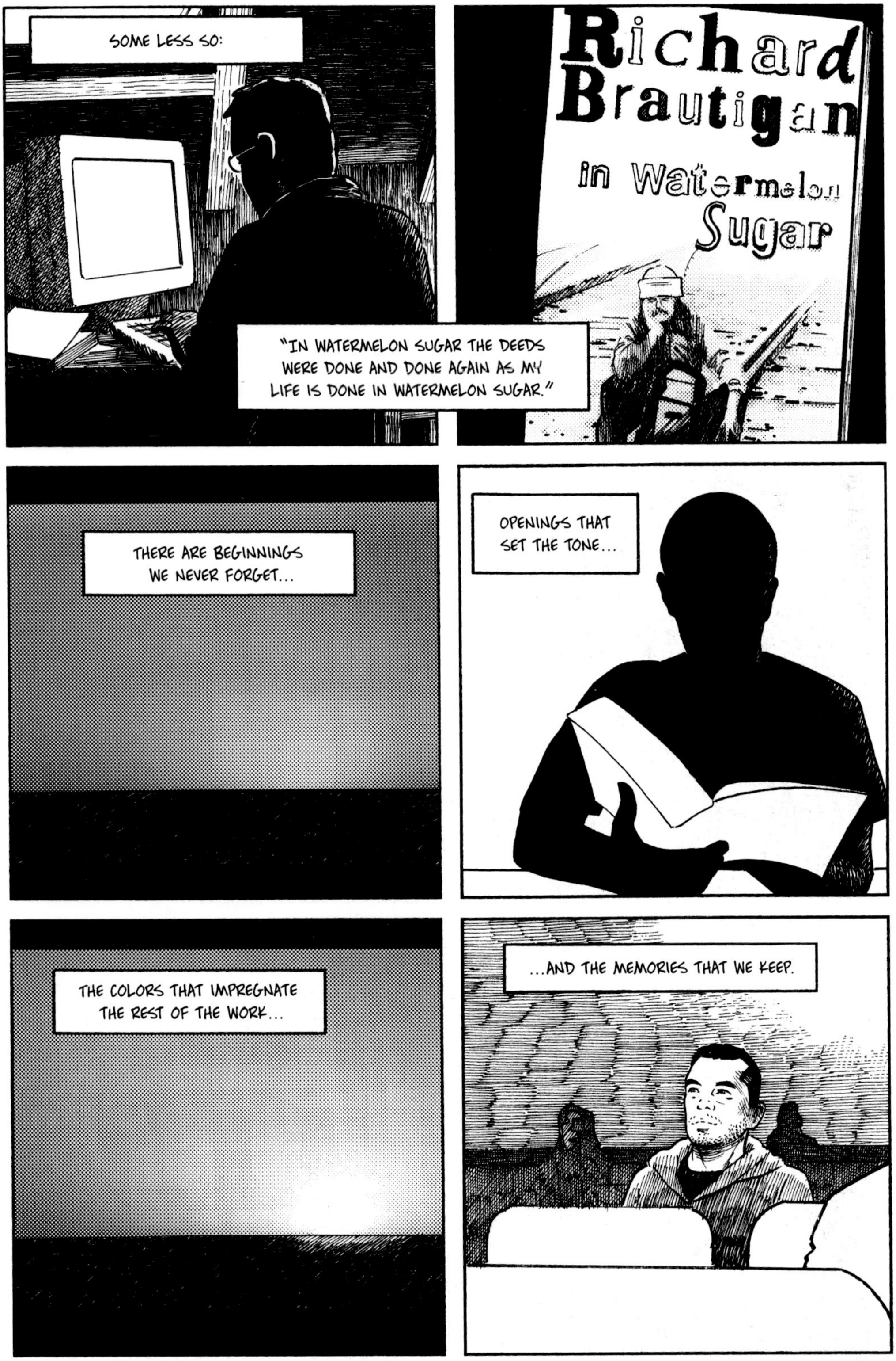
SOME LESS SO:
Richard Brautigan
in Watermelon Sugar
"IN WATERMELON SUGAR THE DEEDS WERE DONE AND DONE AGAIN AS MY LIFE IS DONE IN WATERMELON SUGAR."
THERE ARE BEGINNINGS WE NEVER FORGET...
OPENINGS THAT SET THE TONE...
THE COLORS THAT IMPREGNATE THE REST OF THE WORK...
...AND THE MEMORIES THAT WE KEEP.

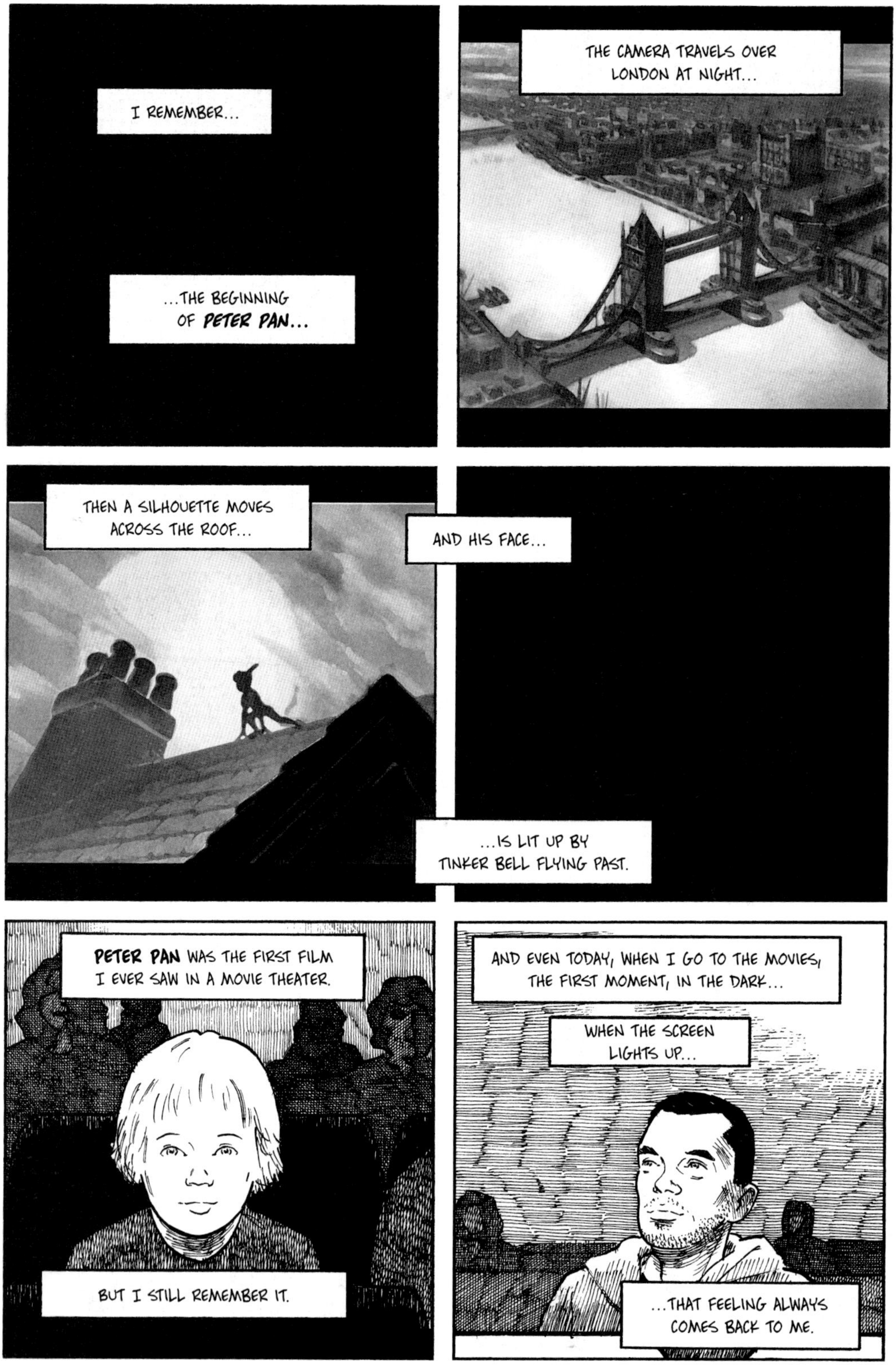

I REMEMBER...
...THE BEGINNING OF **PETER PAN**...
THE CAMERA TRAVELS OVER LONDON AT NIGHT...
THEN A SILHOUETTE MOVES ACROSS THE ROOF...
AND HIS FACE...
...IS LIT UP BY TINKER BELL FLYING PAST.
PETER PAN WAS THE FIRST FILM I EVER SAW IN A MOVIE THEATER.
BUT I STILL REMEMBER IT.
AND EVEN TODAY, WHEN I GO TO THE MOVIES, THE FIRST MOMENT, IN THE DARK...
WHEN THE SCREEN LIGHTS UP...
...THAT FEELING ALWAYS COMES BACK TO ME.

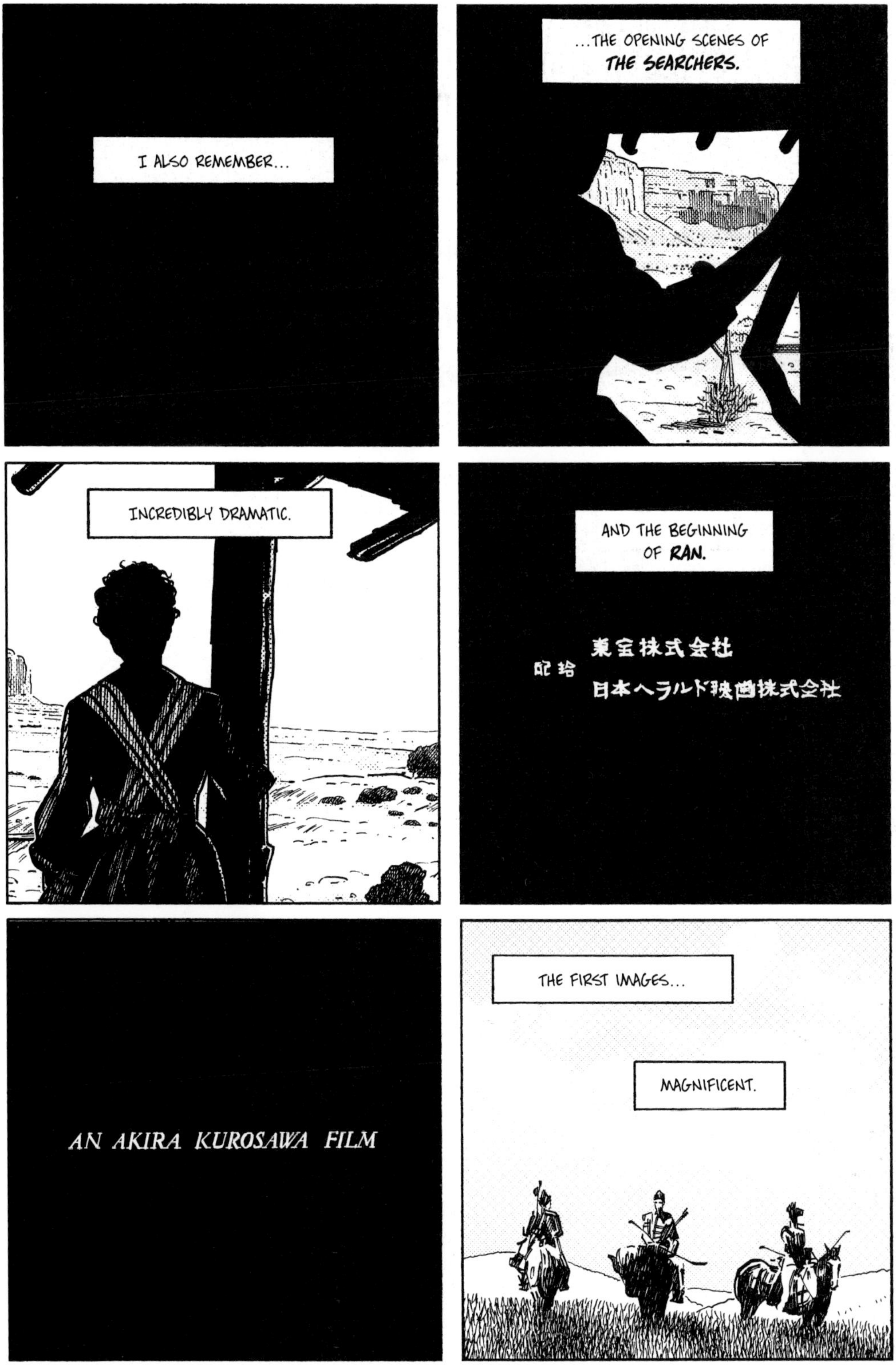
I ALSO REMEMBER...
...THE OPENING SCENES OF **THE SEARCHERS.**
INCREDIBLY DRAMATIC.
AND THE BEGINNING OF **RAN.**
東宝株式会社
配給
日本ヘラルド映画株式会社
AN AKIRA KUROSAWA FILM
THE FIRST IMAGES...
MAGNIFICENT.

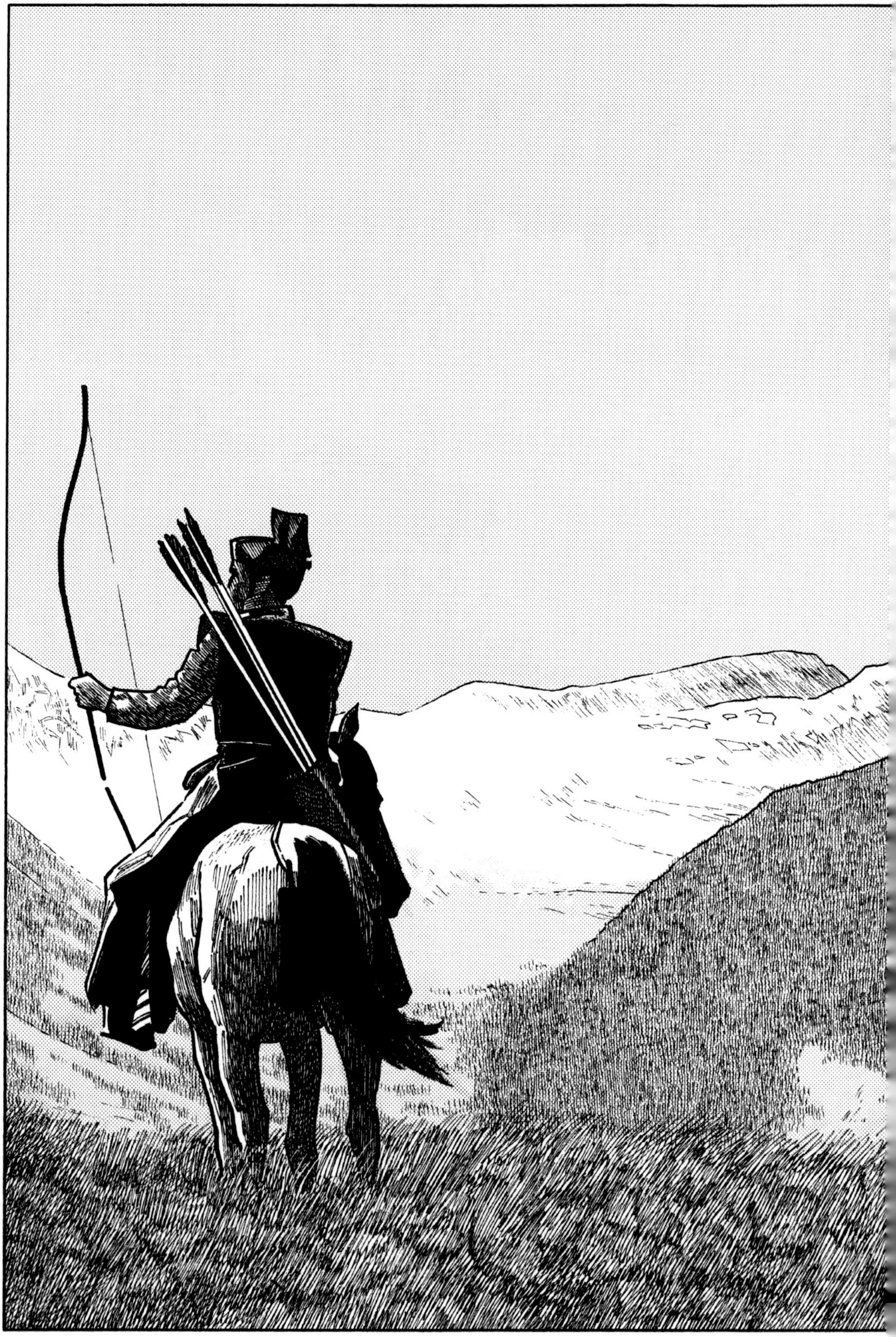

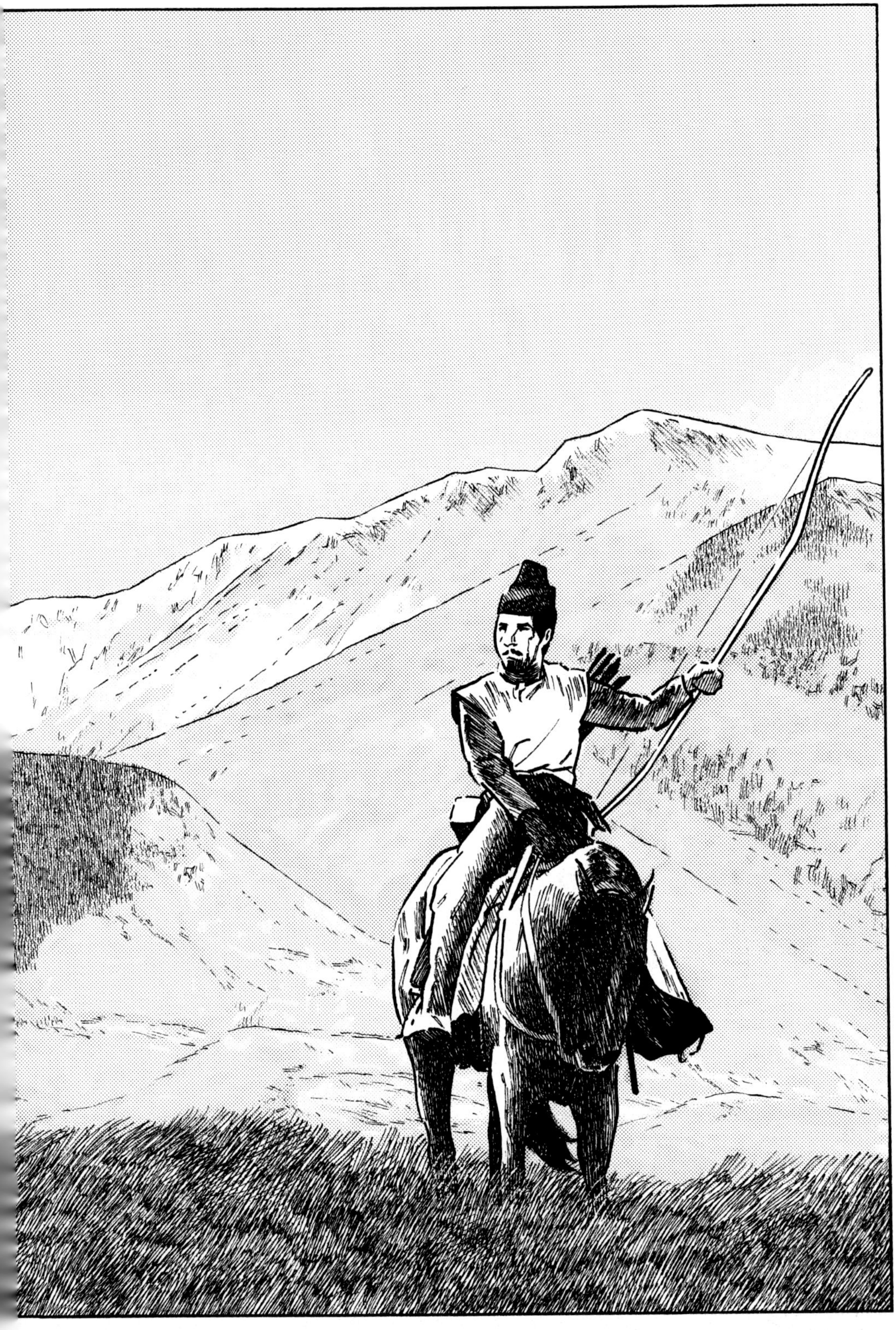

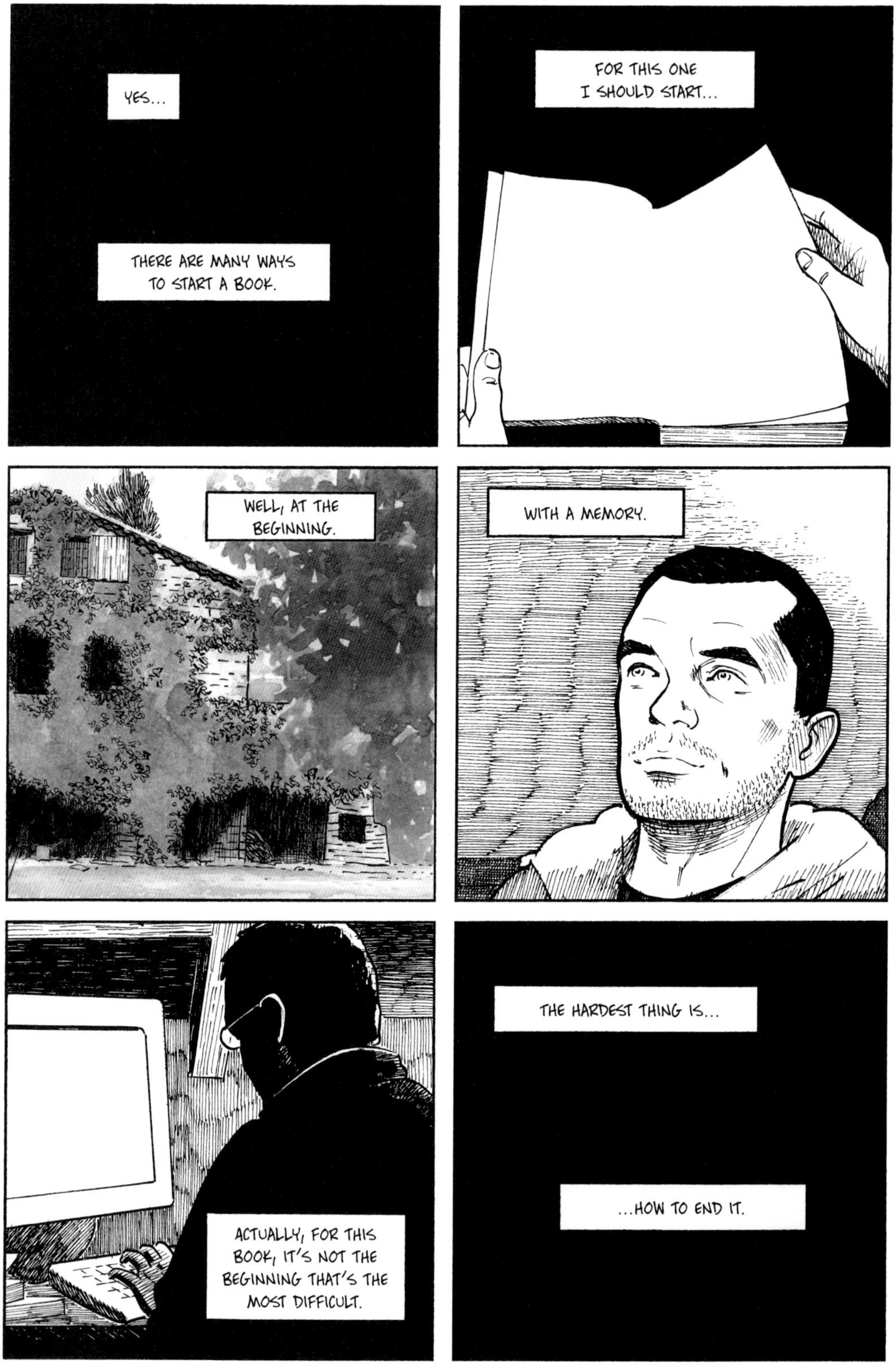
YES...
THERE ARE MANY WAYS TO START A BOOK.
FOR THIS ONE I SHOULD START...
WELL, AT THE BEGINNING.
WITH A MEMORY.
ACTUALLY, FOR THIS BOOK, IT'S NOT THE BEGINNING THAT'S THE MOST DIFFICULT.
THE HARDEST THING IS...
...HOW TO END IT.

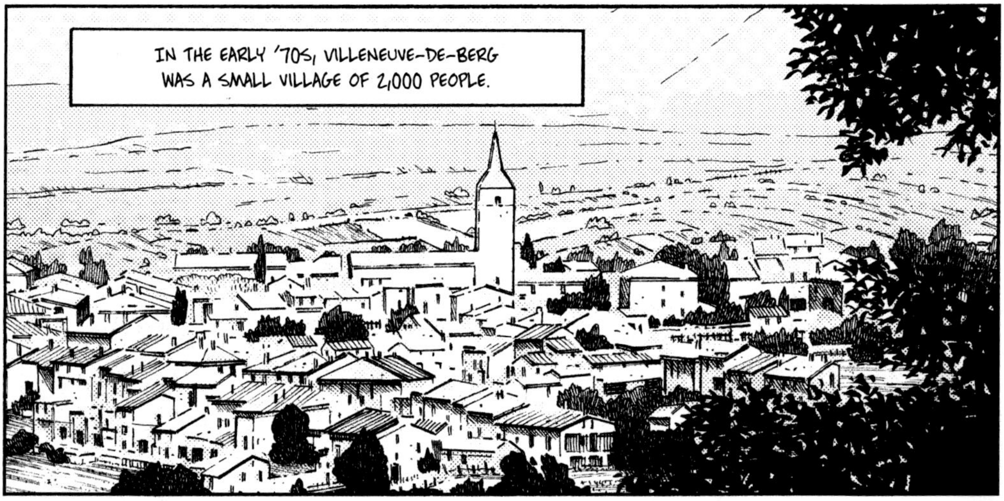
IN THE EARLY '70S, VILLENEUVE-DE-BERG WAS A SMALL VILLAGE OF 2,000 PEOPLE.

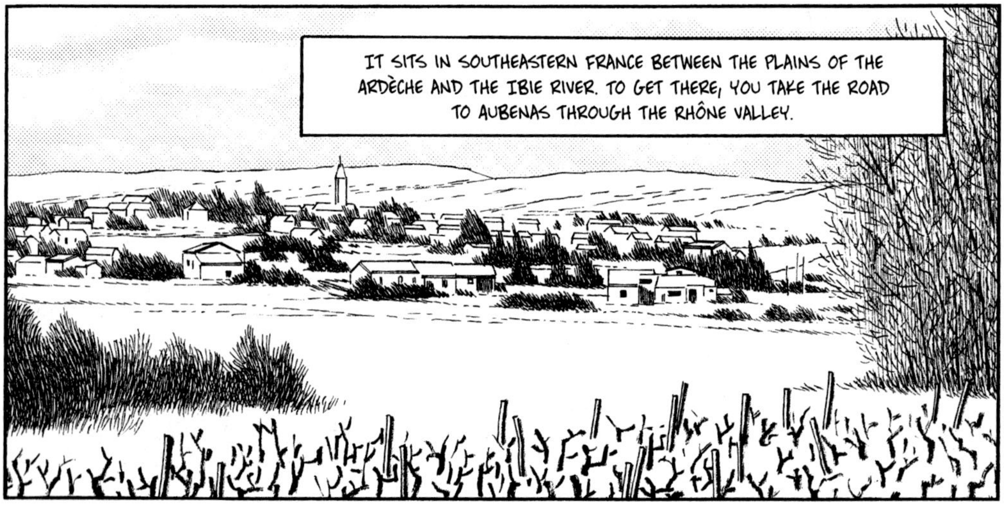
IT SITS IN SOUTHEASTERN FRANCE BETWEEN THE PLAINS OF THE ARDÈCHE AND THE IBIE RIVER. TO GET THERE, YOU TAKE THE ROAD TO AUBENAS THROUGH THE RHÔNE VALLEY.

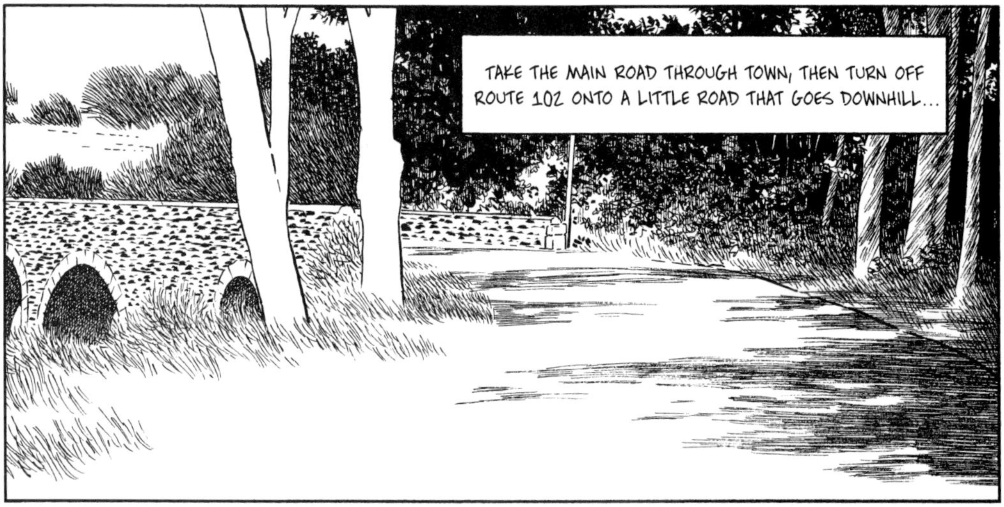
TAKE THE MAIN ROAD THROUGH TOWN, THEN TURN OFF ROUTE 102 ONTO A LITTLE ROAD THAT GOES DOWNHILL...

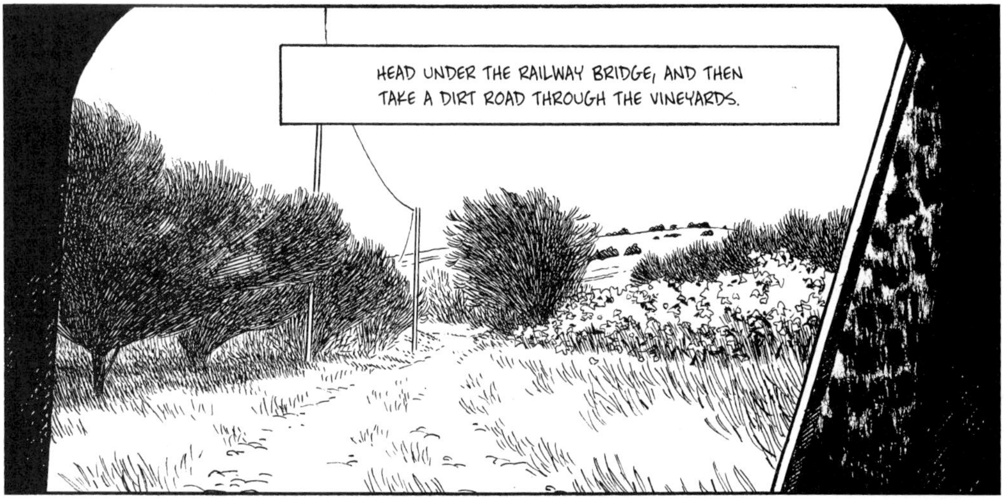
HEAD UNDER THE RAILWAY BRIDGE, AND THEN TAKE A DIRT ROAD THROUGH THE VINEYARDS.

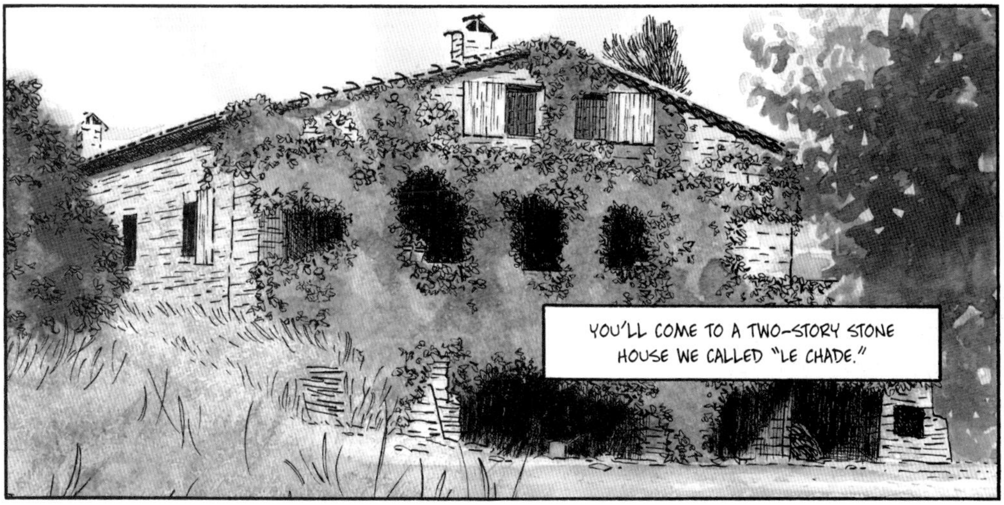
YOU'LL COME TO A TWO-STORY STONE HOUSE WE CALLED "LE CHADE."

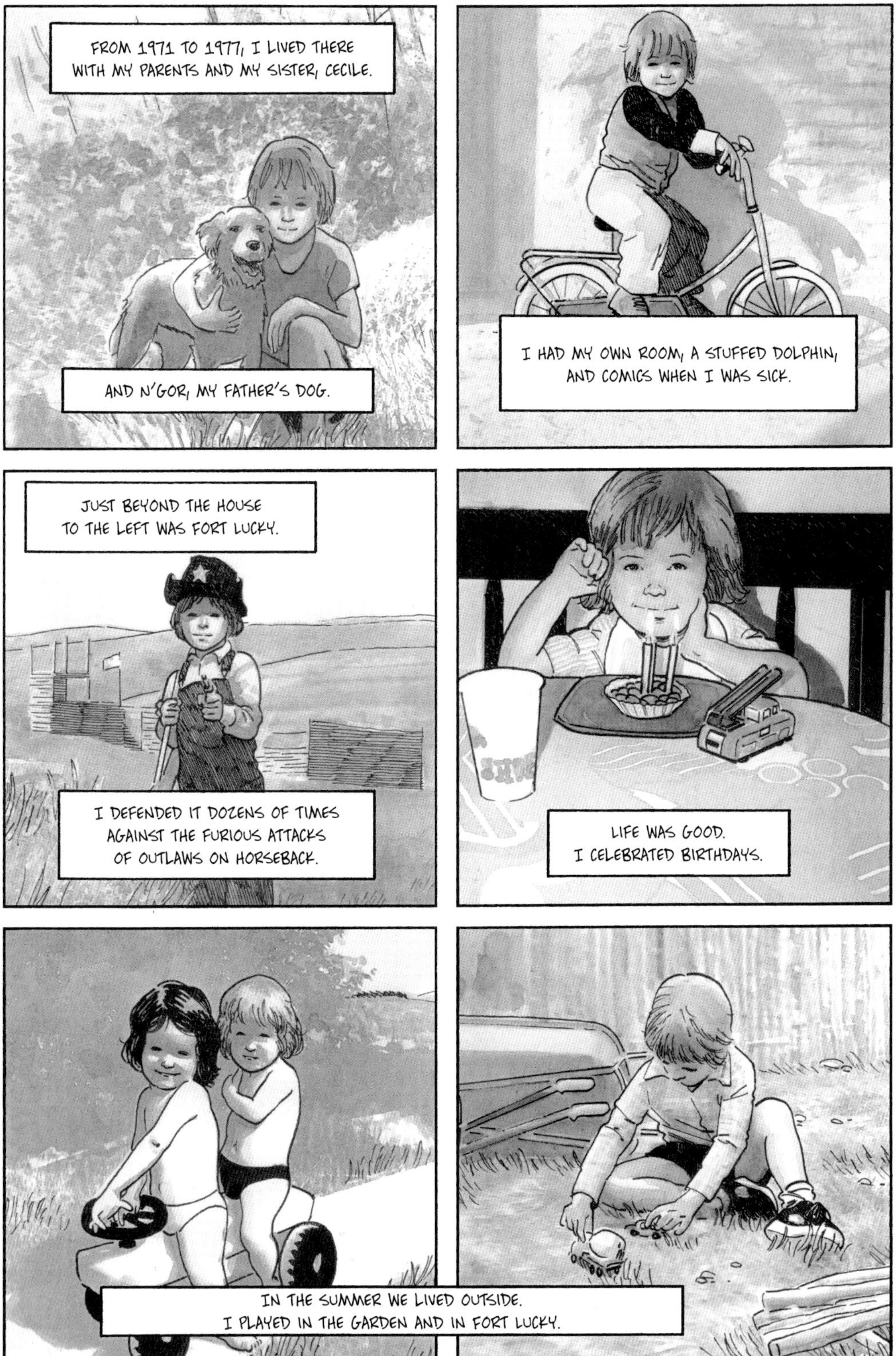
FROM 1971 TO 1977, I LIVED THERE WITH MY PARENTS AND MY SISTER, CECILE.
AND N'GOR, MY FATHER'S DOG.
I HAD MY OWN ROOM, A STUFFED DOLPHIN, AND COMICS WHEN I WAS SICK.
JUST BEYOND THE HOUSE TO THE LEFT WAS FORT LUCKY.
I DEFENDED IT DOZENS OF TIMES AGAINST THE FURIOUS ATTACKS OF OUTLAWS ON HORSEBACK.
LIFE WAS GOOD. I CELEBRATED BIRTHDAYS.
IN THE SUMMER WE LIVED OUTSIDE. I PLAYED IN THE GARDEN AND IN FORT LUCKY.

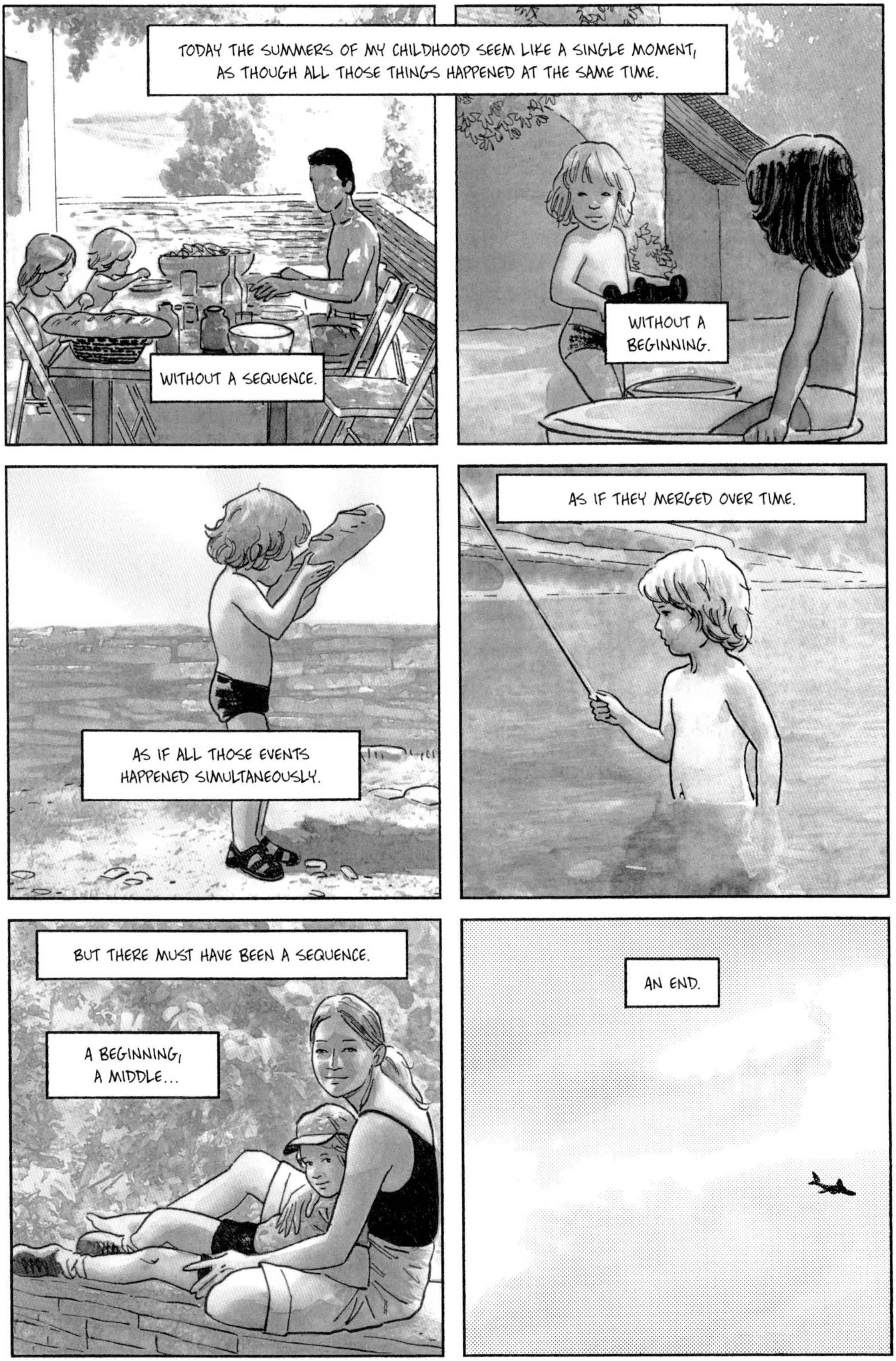
TODAY THE SUMMERS OF MY CHILDHOOD SEEM LIKE A SINGLE MOMENT, AS THOUGH ALL THOSE THINGS HAPPENED AT THE SAME TIME.
WITHOUT A SEQUENCE.
WITHOUT A BEGINNING.
AS IF ALL THOSE EVENTS HAPPENED SIMULTANEOUSLY.
AS IF THEY MERGED OVER TIME.
BUT THERE MUST HAVE BEEN A SEQUENCE.
A BEGINNING, A MIDDLE...
AN END.

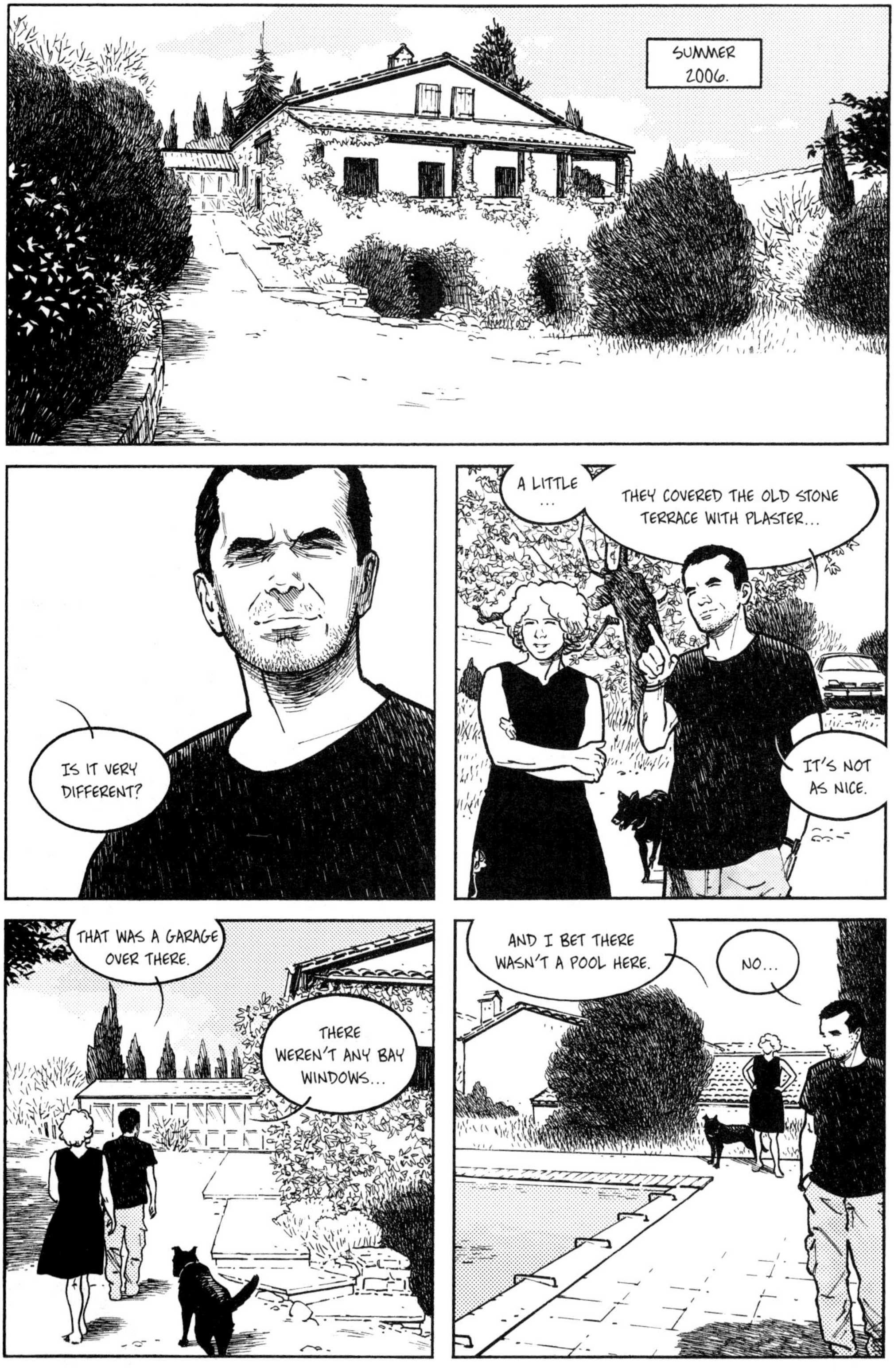
SUMMER 2006.
IS IT VERY DIFFERENT?
A LITTLE ...
THEY COVERED THE OLD STONE TERRACE WITH PLASTER...
IT'S NOT AS NICE.
THAT WAS A GARAGE OVER THERE.
THERE WEREN'T ANY BAY WINDOWS...
AND I BET THERE WASN'T A POOL HERE.
NO...

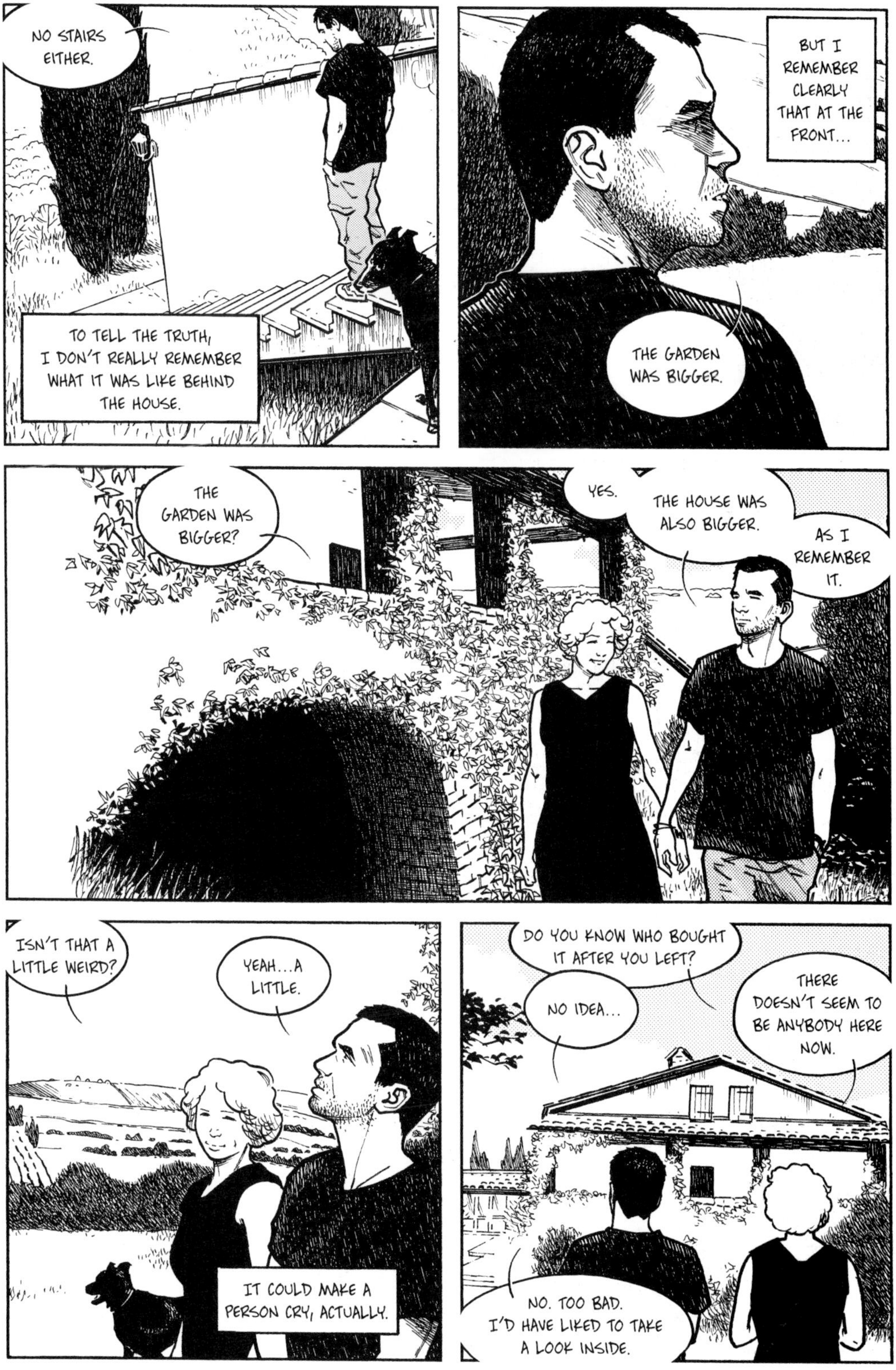
NO STAIRS EITHER.
TO TELL THE TRUTH, I DON'T REALLY REMEMBER WHAT IT WAS LIKE BEHIND THE HOUSE.
BUT I REMEMBER CLEARLY THAT AT THE FRONT...
THE GARDEN WAS BIGGER.
THE GARDEN WAS BIGGER?
YES.
THE HOUSE WAS ALSO BIGGER.
AS I REMEMBER IT.
ISN'T THAT A LITTLE WEIRD?
YEAH...A LITTLE.
IT COULD MAKE A PERSON CRY, ACTUALLY.
DO YOU KNOW WHO BOUGHT IT AFTER YOU LEFT?
NO IDEA...
THERE DOESN'T SEEM TO BE ANYBODY HERE NOW.
NO. TOO BAD. I'D HAVE LIKED TO TAKE A LOOK INSIDE.

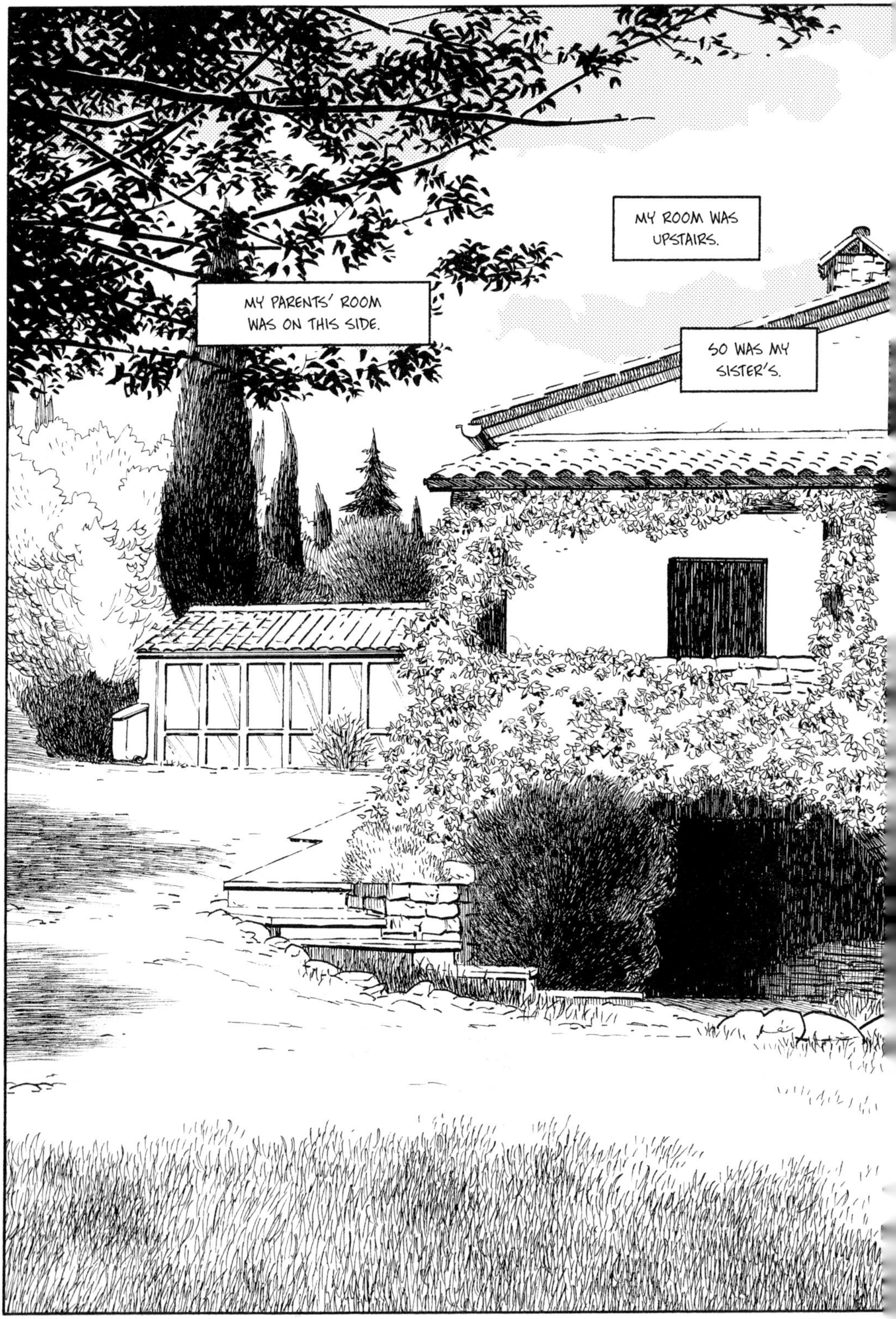
MY PARENTS' ROOM WAS ON THIS SIDE.
MY ROOM WAS UPSTAIRS.
SO WAS MY SISTER'S.

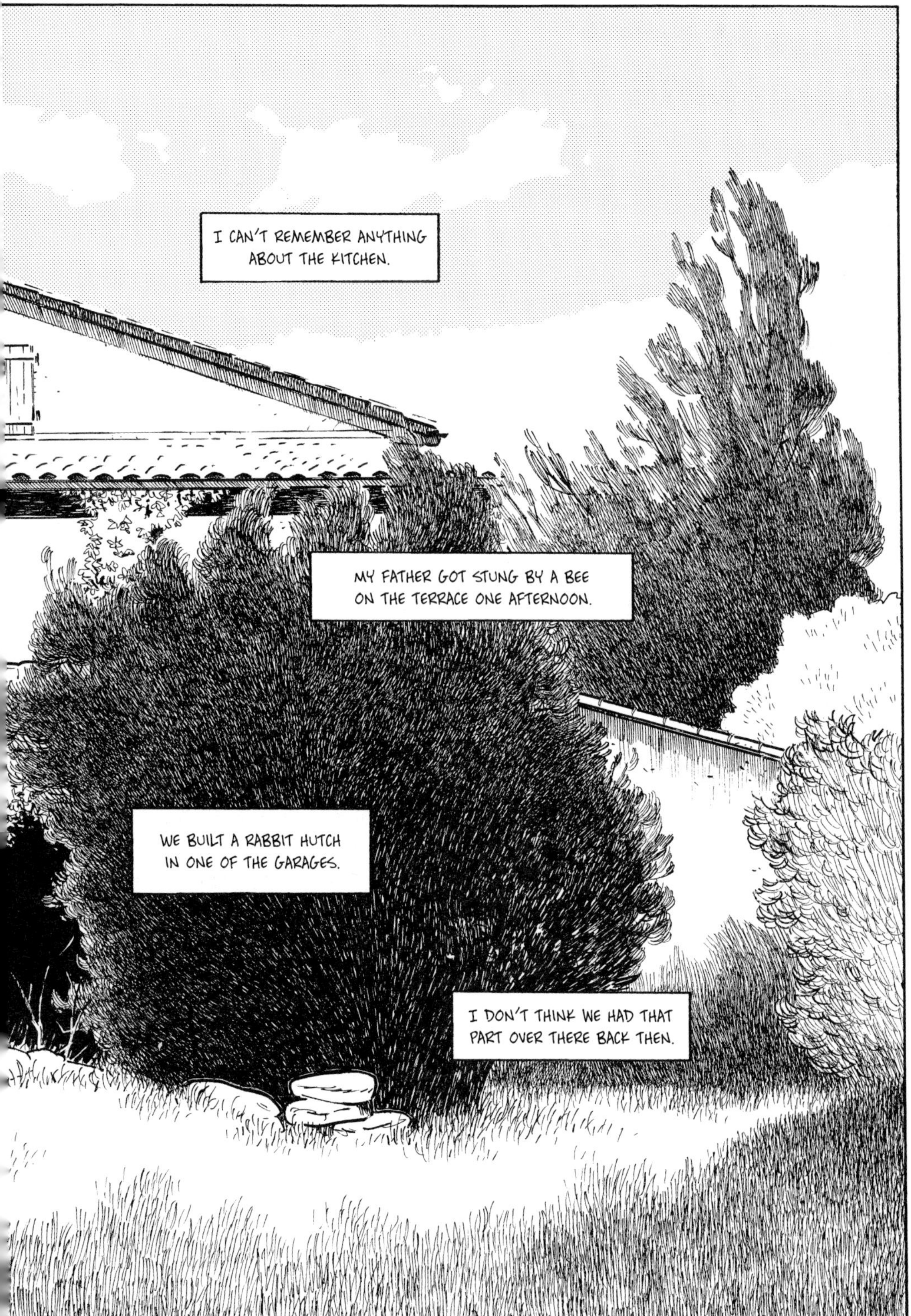
I CAN'T REMEMBER ANYTHING ABOUT THE KITCHEN.
MY FATHER GOT STUNG BY A BEE ON THE TERRACE ONE AFTERNOON.
WE BUILT A RABBIT HUTCH IN ONE OF THE GARAGES.
I DON'T THINK WE HAD THAT PART OVER THERE BACK THEN.

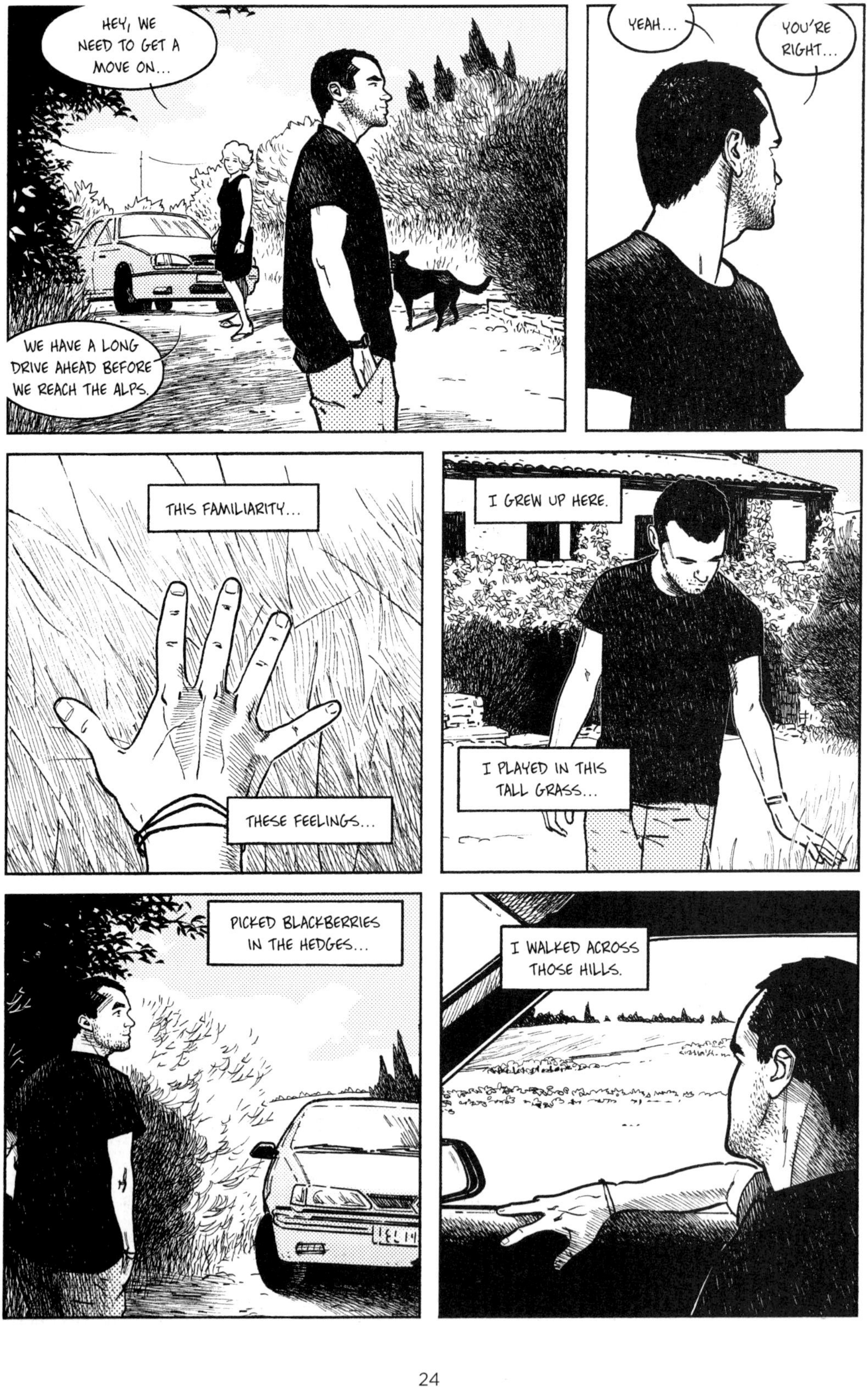
HEY, WE NEED TO GET A MOVE ON...
WE HAVE A LONG DRIVE AHEAD BEFORE WE REACH THE ALPS.
YEAH...
YOU'RE RIGHT...
THIS FAMILIARITY...
THESE FEELINGS...
I GREW UP HERE.
I PLAYED IN THIS TALL GRASS...
PICKED BLACKBERRIES IN THE HEDGES...
I WALKED ACROSS THOSE HILLS.

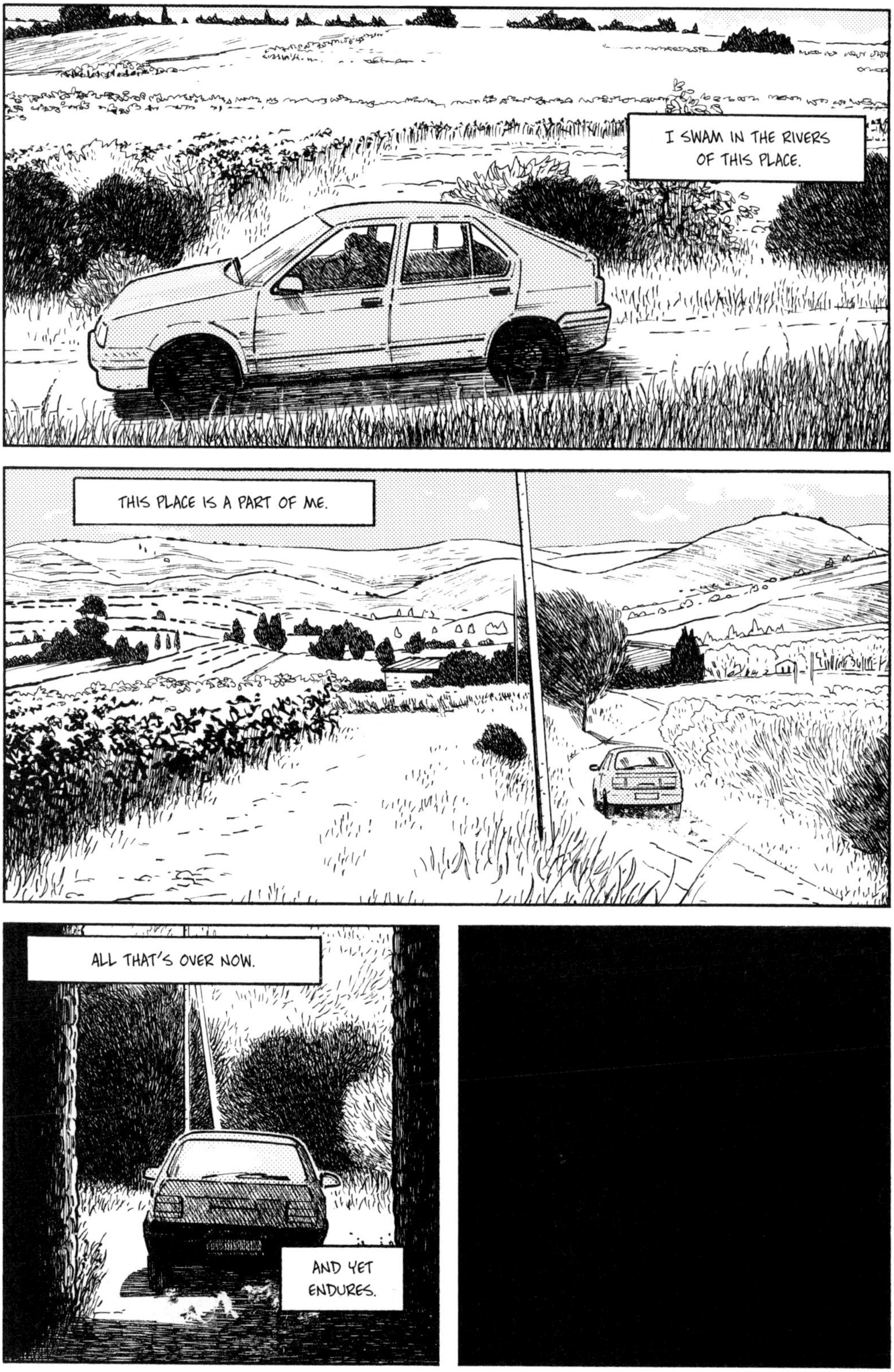
I SWAM IN THE RIVERS OF THIS PLACE.
THIS PLACE IS A PART OF ME.
ALL THAT'S OVER NOW.
AND YET ENDURES.

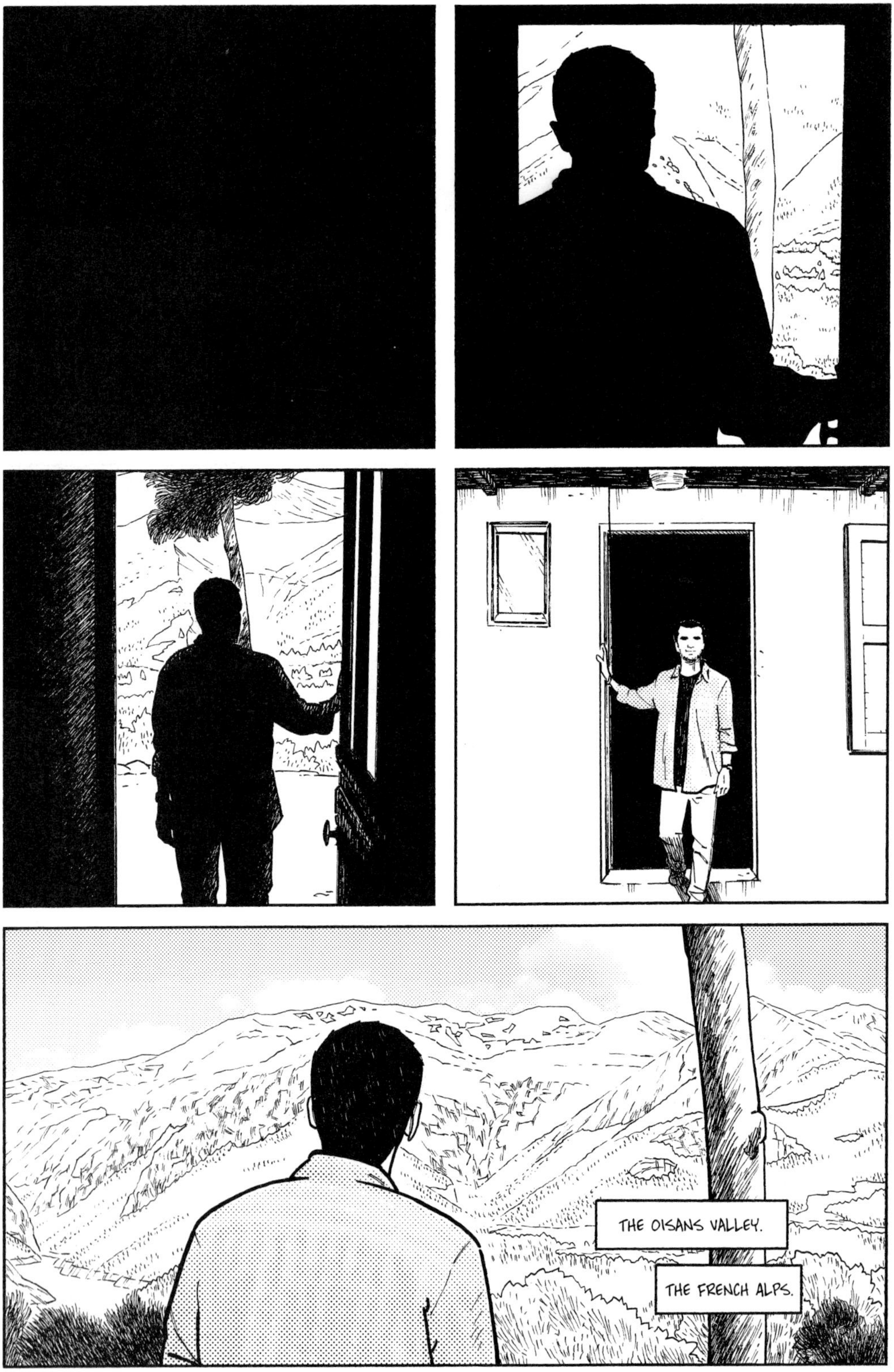
THE OISANS VALLEY.
THE FRENCH ALPS.

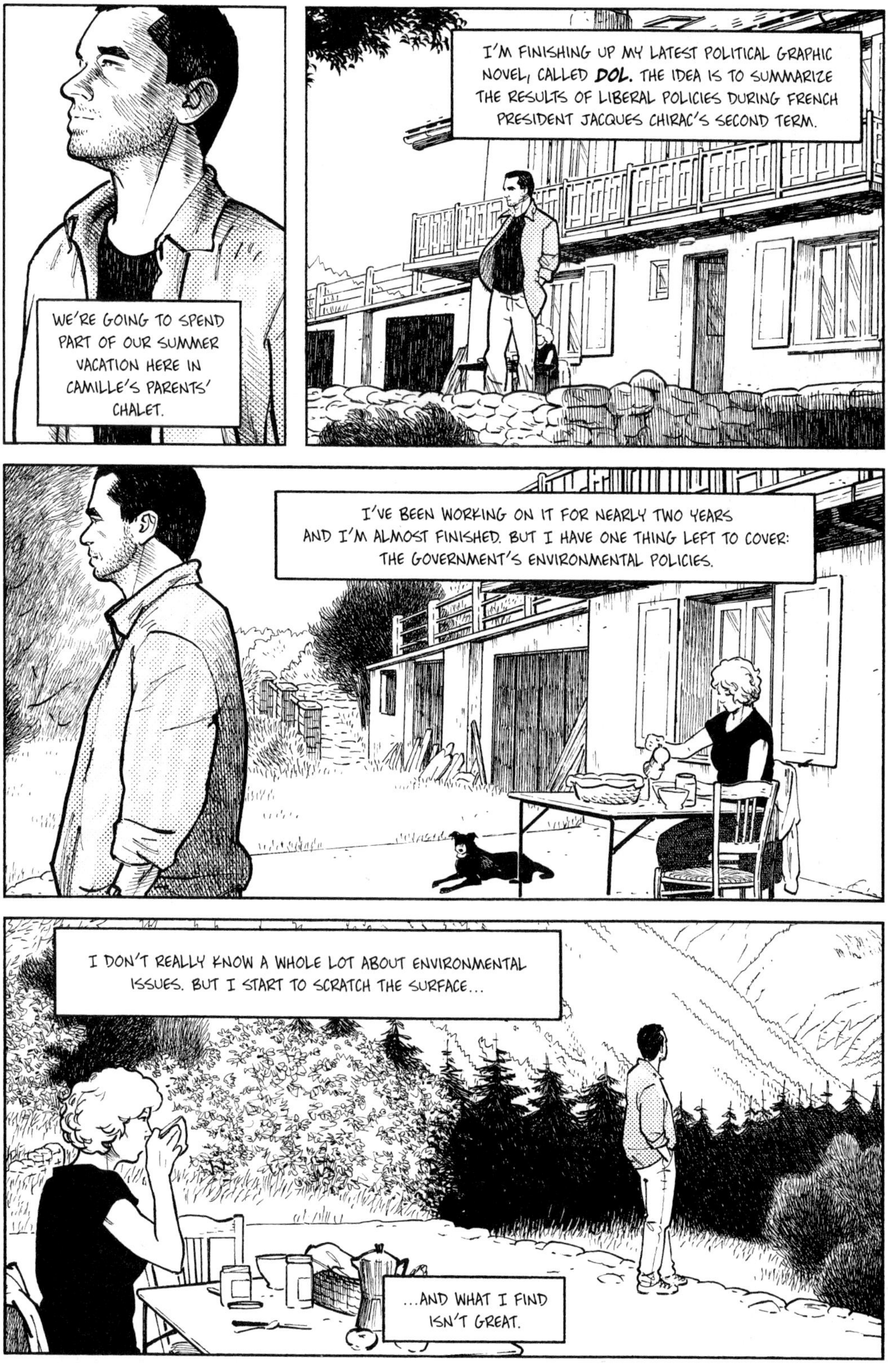
WE'RE GOING TO SPEND PART OF OUR SUMMER VACATION HERE IN CAMILLE'S PARENTS' CHALET.
I'M FINISHING UP MY LATEST POLITICAL GRAPHIC NOVEL, CALLED **DOL.** THE IDEA IS TO SUMMARIZE THE RESULTS OF LIBERAL POLICIES DURING FRENCH PRESIDENT JACQUES CHIRAC'S SECOND TERM.
I'VE BEEN WORKING ON IT FOR NEARLY TWO YEARS AND I'M ALMOST FINISHED. BUT I HAVE ONE THING LEFT TO COVER: THE GOVERNMENT'S ENVIRONMENTAL POLICIES.
I DON'T REALLY KNOW A WHOLE LOT ABOUT ENVIRONMENTAL ISSUES. BUT I START TO SCRATCH THE SURFACE...
...AND WHAT I FIND ISN'T GREAT.

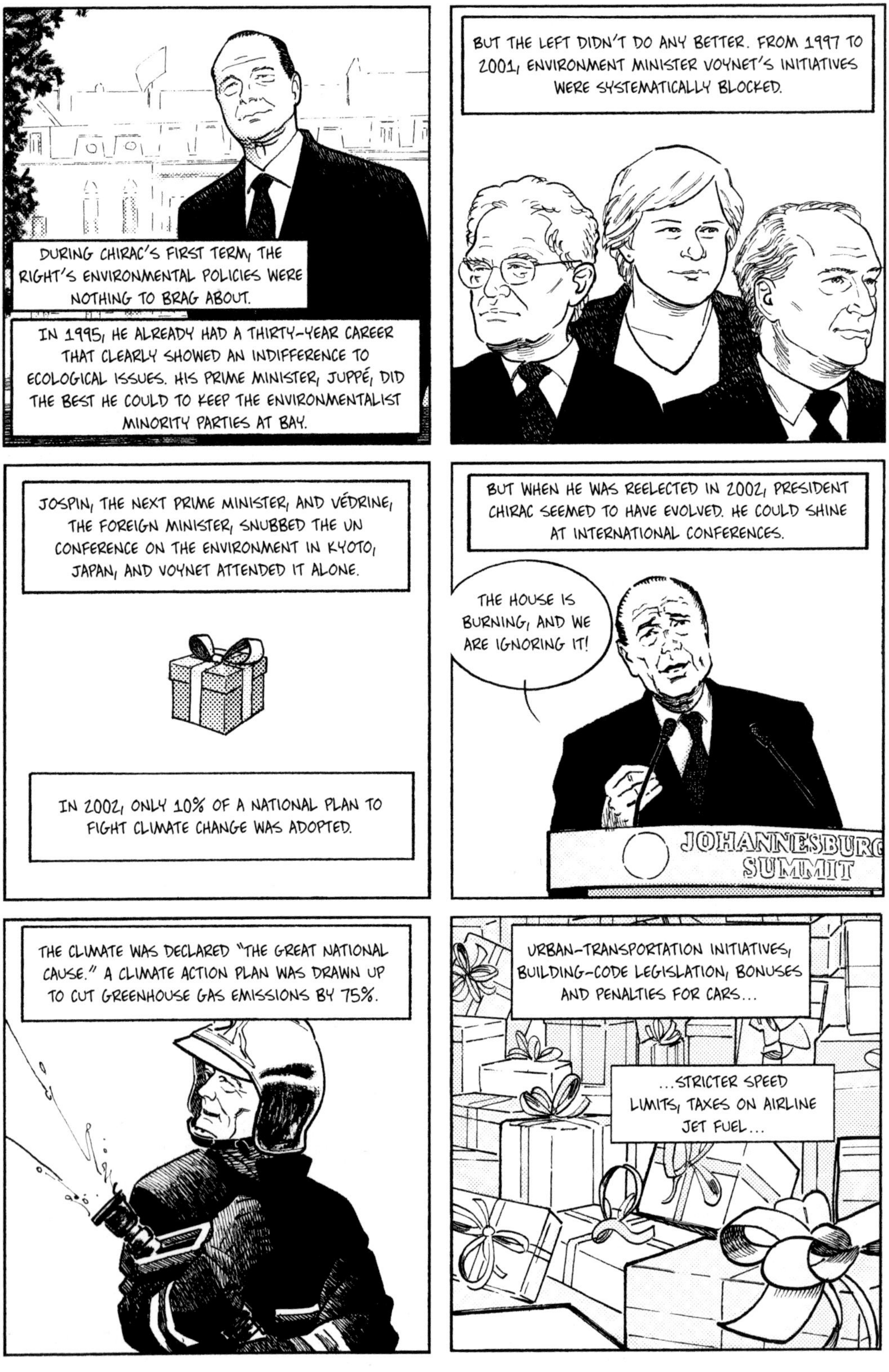
DURING CHIRAC'S FIRST TERM, THE RIGHT'S ENVIRONMENTAL POLICIES WERE NOTHING TO BRAG ABOUT.
IN 1995, HE ALREADY HAD A THIRTY-YEAR CAREER THAT CLEARLY SHOWED AN INDIFFERENCE TO ECOLOGICAL ISSUES. HIS PRIME MINISTER, JUPPÉ, DID THE BEST HE COULD TO KEEP THE ENVIRONMENTALIST MINORITY PARTIES AT BAY.
BUT THE LEFT DIDN'T DO ANY BETTER. FROM 1997 TO 2001, ENVIRONMENT MINISTER VOYNET'S INITIATIVES WERE SYSTEMATICALLY BLOCKED.
JOSPIN, THE NEXT PRIME MINISTER, AND VÉDRINE, THE FOREIGN MINISTER, SNUBBED THE UN CONFERENCE ON THE ENVIRONMENT IN KYOTO, JAPAN, AND VOYNET ATTENDED IT ALONE.
IN 2002, ONLY 10% OF A NATIONAL PLAN TO FIGHT CLIMATE CHANGE WAS ADOPTED.
BUT WHEN HE WAS REELECTED IN 2002, PRESIDENT CHIRAC SEEMED TO HAVE EVOLVED. HE COULD SHINE AT INTERNATIONAL CONFERENCES.
THE HOUSE IS BURNING, AND WE ARE IGNORING IT!
JOHANNESBURG SUMMIT
THE CLIMATE WAS DECLARED "THE GREAT NATIONAL CAUSE." A CLIMATE ACTION PLAN WAS DRAWN UP TO CUT GREENHOUSE GAS EMISSIONS BY 75%.
URBAN-TRANSPORTATION INITIATIVES, BUILDING-CODE LEGISLATION, BONUSES AND PENALTIES FOR CARS...
...STRICTER SPEED LIMITS, TAXES ON AIRLINE JET FUEL...

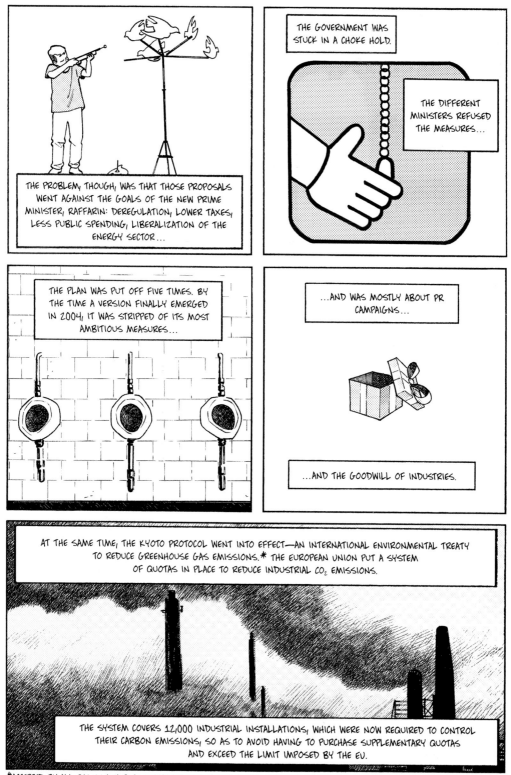

THE PROBLEM, THOUGH, WAS THAT THOSE PROPOSALS WENT AGAINST THE GOALS OF THE NEW PRIME MINISTER, RAFFARIN: DEREGULATION, LOWER TAXES, LESS PUBLIC SPENDING, LIBERALIZATION OF THE ENERGY SECTOR...

THE GOVERNMENT WAS STUCK IN A CHOKE HOLD.

THE DIFFERENT MINISTERS REFUSED THE MEASURES...

THE PLAN WAS PUT OFF FIVE TIMES. BY THE TIME A VERSION FINALLY EMERGED IN 2004, IT WAS STRIPPED OF ITS MOST AMBITIOUS MEASURES...

...AND WAS MOSTLY ABOUT PR CAMPAIGNS...

...AND THE GOODWILL OF INDUSTRIES.

AT THE SAME TIME, THE KYOTO PROTOCOL WENT INTO EFFECT—AN INTERNATIONAL ENVIRONMENTAL TREATY TO REDUCE GREENHOUSE GAS EMISSIONS.* THE EUROPEAN UNION PUT A SYSTEM OF QUOTAS IN PLACE TO REDUCE INDUSTRIAL CO_2 EMISSIONS.

THE SYSTEM COVERS 12,000 INDUSTRIAL INSTALLATIONS, WHICH WERE NOW REQUIRED TO CONTROL THEIR CARBON EMISSIONS, SO AS TO AVOID HAVING TO PURCHASE SUPPLEMENTARY QUOTAS AND EXCEED THE LIMIT IMPOSED BY THE EU.

*ACCEPTED BY ALL THE UN MEMBERS EXCEPT ANDORRA, SOUTH SUDAN...AND THE UNITED STATES. CANADA LATER WITHDREW.

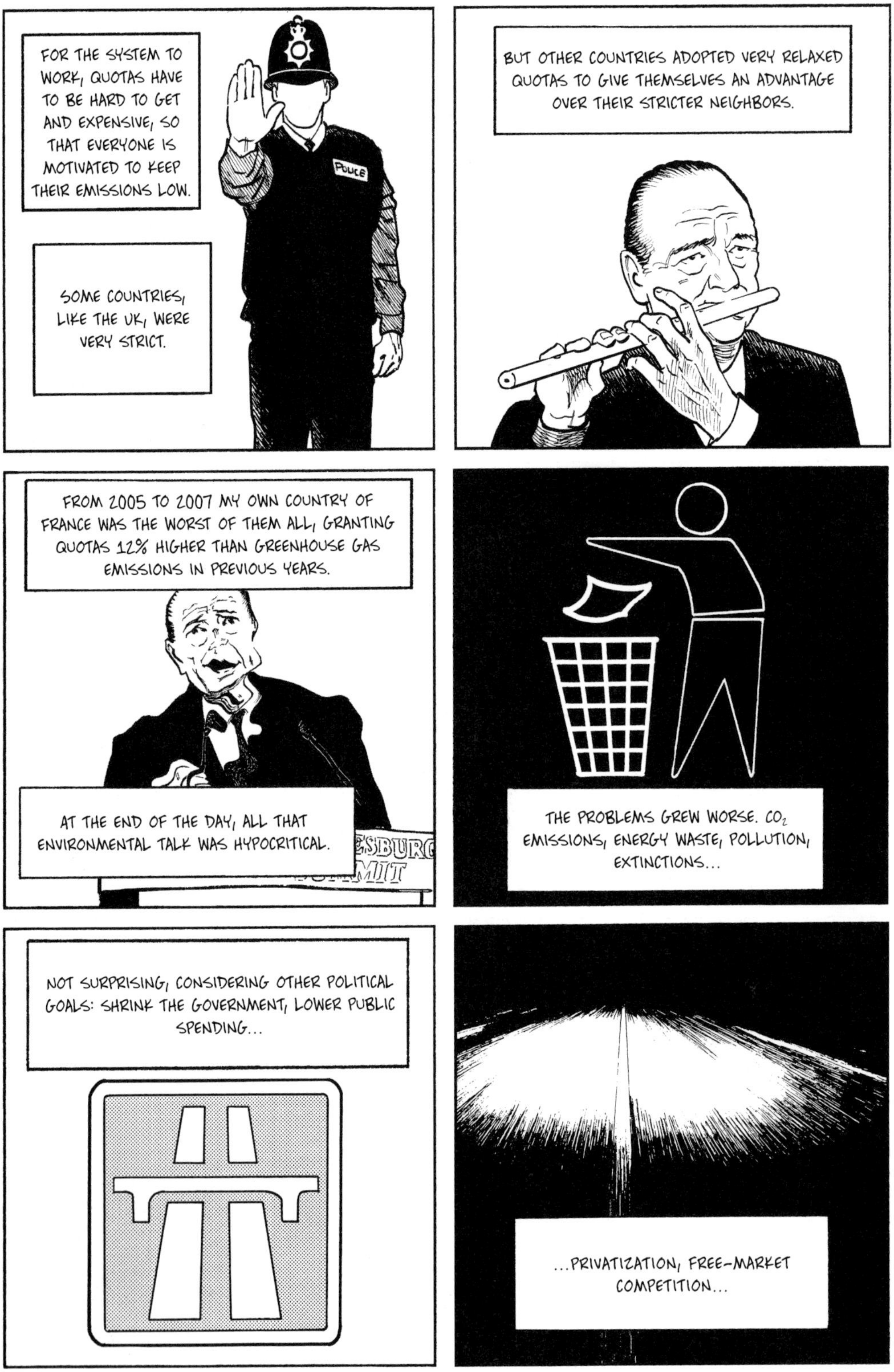
FOR THE SYSTEM TO WORK, QUOTAS HAVE TO BE HARD TO GET AND EXPENSIVE, SO THAT EVERYONE IS MOTIVATED TO KEEP THEIR EMISSIONS LOW.
SOME COUNTRIES, LIKE THE UK, WERE VERY STRICT.
POLICE
BUT OTHER COUNTRIES ADOPTED VERY RELAXED QUOTAS TO GIVE THEMSELVES AN ADVANTAGE OVER THEIR STRICTER NEIGHBORS.
FROM 2005 TO 2007 MY OWN COUNTRY OF FRANCE WAS THE WORST OF THEM ALL, GRANTING QUOTAS 12% HIGHER THAN GREENHOUSE GAS EMISSIONS IN PREVIOUS YEARS.
AT THE END OF THE DAY, ALL THAT ENVIRONMENTAL TALK WAS HYPOCRITICAL.
THE PROBLEMS GREW WORSE. CO_2 EMISSIONS, ENERGY WASTE, POLLUTION, EXTINCTIONS...
NOT SURPRISING, CONSIDERING OTHER POLITICAL GOALS: SHRINK THE GOVERNMENT, LOWER PUBLIC SPENDING...
...PRIVATIZATION, FREE-MARKET COMPETITION...

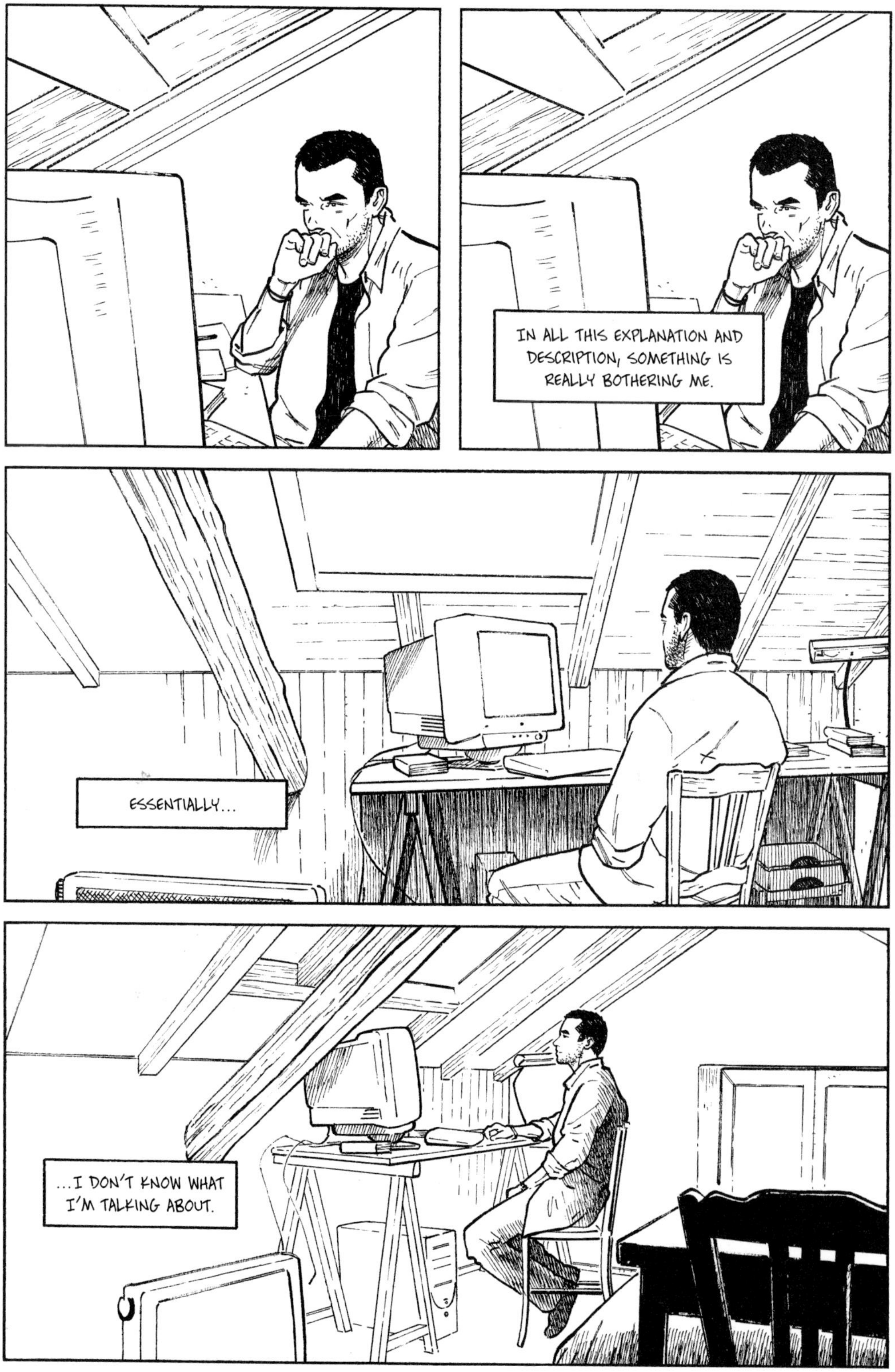
IN ALL THIS EXPLANATION AND DESCRIPTION, SOMETHING IS REALLY BOTHERING ME.
ESSENTIALLY...
...I DON'T KNOW WHAT I'M TALKING ABOUT.

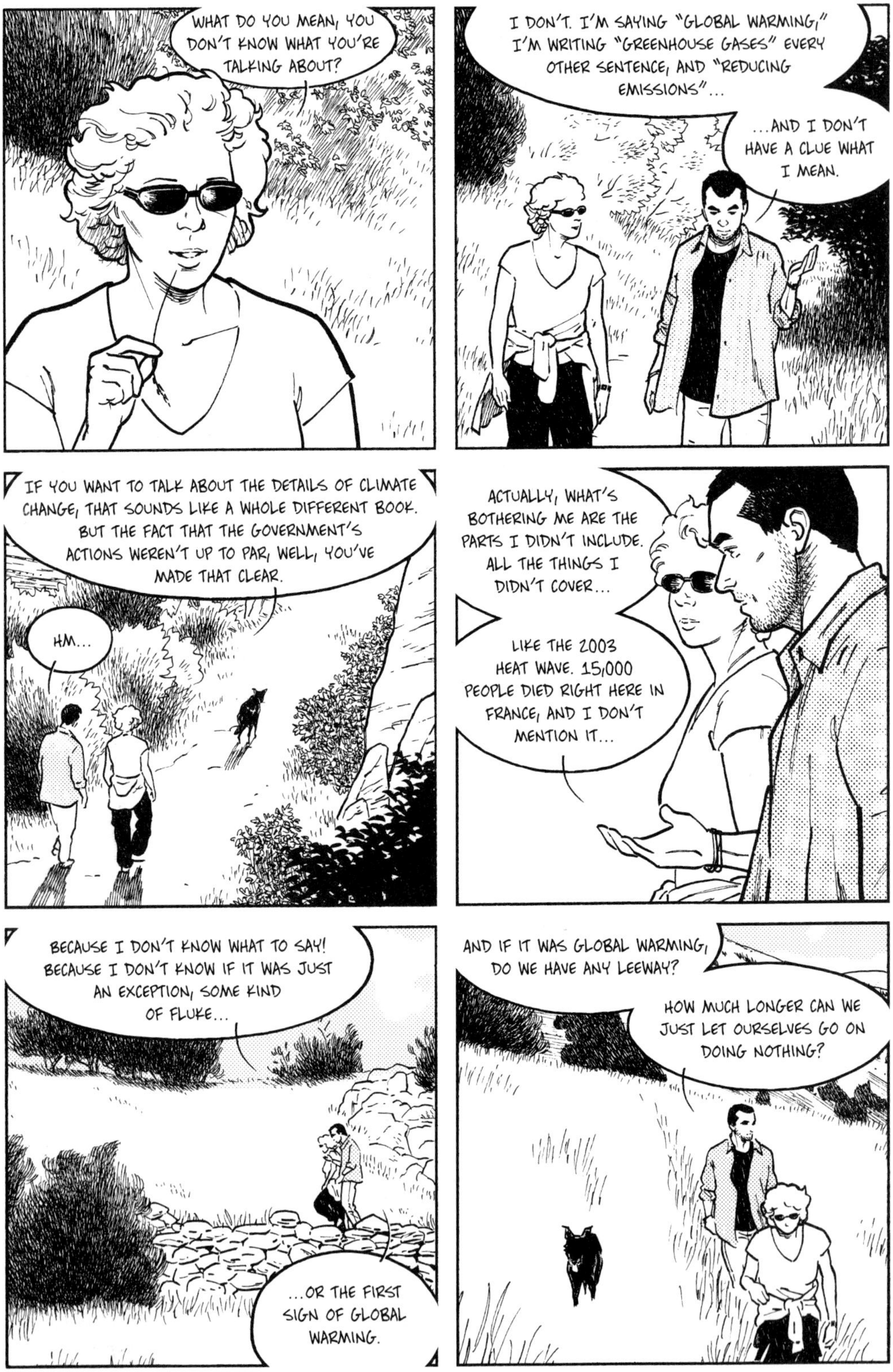
WHAT DO YOU MEAN, YOU DON'T KNOW WHAT YOU'RE TALKING ABOUT?
I DON'T. I'M SAYING "GLOBAL WARMING," I'M WRITING "GREENHOUSE GASES" EVERY OTHER SENTENCE, AND "REDUCING EMISSIONS"...
...AND I DON'T HAVE A CLUE WHAT I MEAN.
IF YOU WANT TO TALK ABOUT THE DETAILS OF CLIMATE CHANGE, THAT SOUNDS LIKE A WHOLE DIFFERENT BOOK. BUT THE FACT THAT THE GOVERNMENT'S ACTIONS WEREN'T UP TO PAR, WELL, YOU'VE MADE THAT CLEAR.
HM...
ACTUALLY, WHAT'S BOTHERING ME ARE THE PARTS I DIDN'T INCLUDE. ALL THE THINGS I DIDN'T COVER...
LIKE THE 2003 HEAT WAVE. 15,000 PEOPLE DIED RIGHT HERE IN FRANCE, AND I DON'T MENTION IT...
BECAUSE I DON'T KNOW WHAT TO SAY! BECAUSE I DON'T KNOW IF IT WAS JUST AN EXCEPTION, SOME KIND OF FLUKE...
...OR THE FIRST SIGN OF GLOBAL WARMING.
AND IF IT WAS GLOBAL WARMING, DO WE HAVE ANY LEEWAY?
HOW MUCH LONGER CAN WE JUST LET OURSELVES GO ON DOING NOTHING?

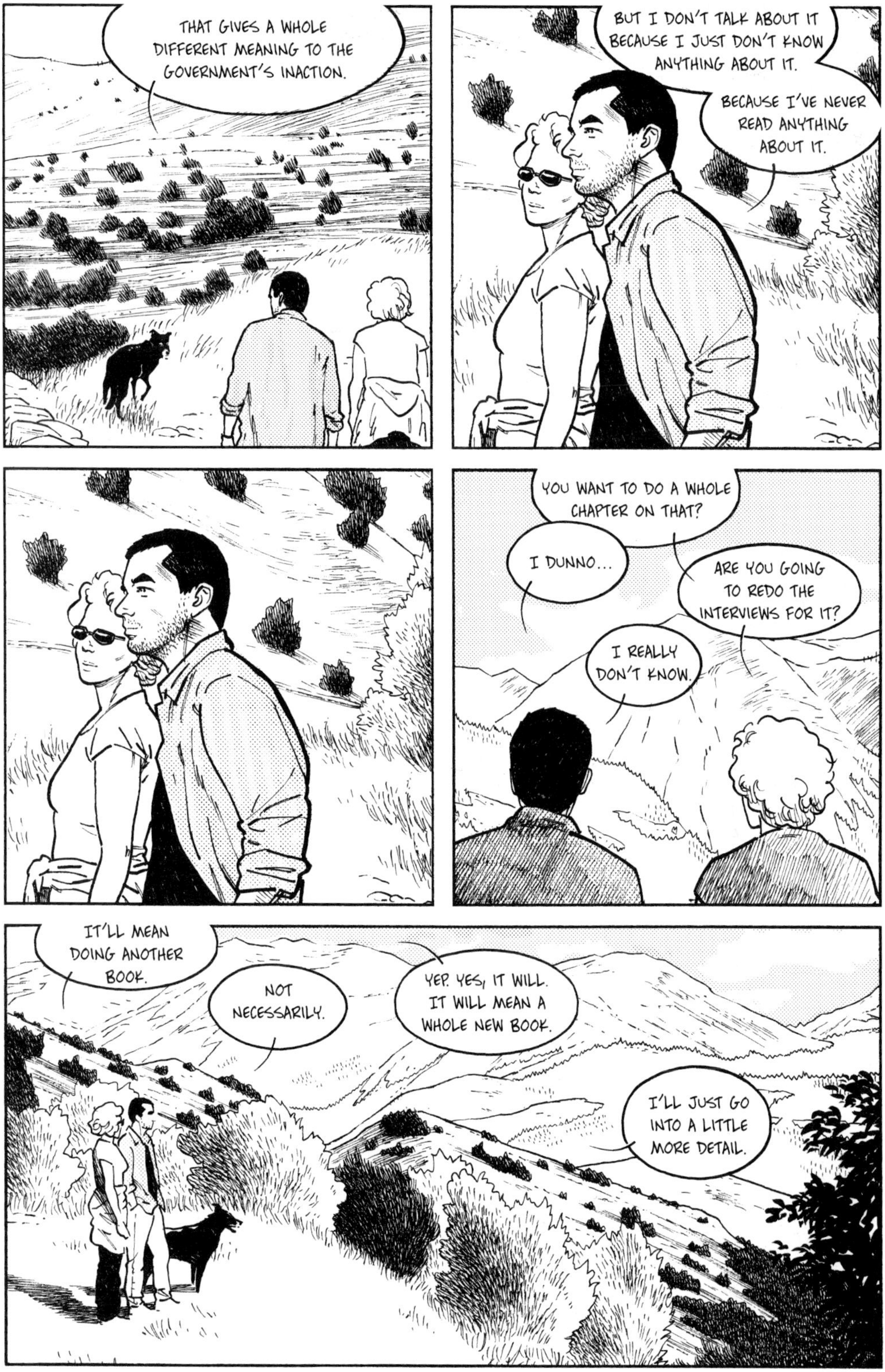
THAT GIVES A WHOLE DIFFERENT MEANING TO THE GOVERNMENT'S INACTION.
BUT I DON'T TALK ABOUT IT BECAUSE I JUST DON'T KNOW ANYTHING ABOUT IT.
BECAUSE I'VE NEVER READ ANYTHING ABOUT IT.
YOU WANT TO DO A WHOLE CHAPTER ON THAT?
I DUNNO...
ARE YOU GOING TO REDO THE INTERVIEWS FOR IT?
I REALLY DON'T KNOW.
IT'LL MEAN DOING ANOTHER BOOK.
NOT NECESSARILY.
YEP. YES, IT WILL. IT WILL MEAN A WHOLE NEW BOOK.
I'LL JUST GO INTO A LITTLE MORE DETAIL.

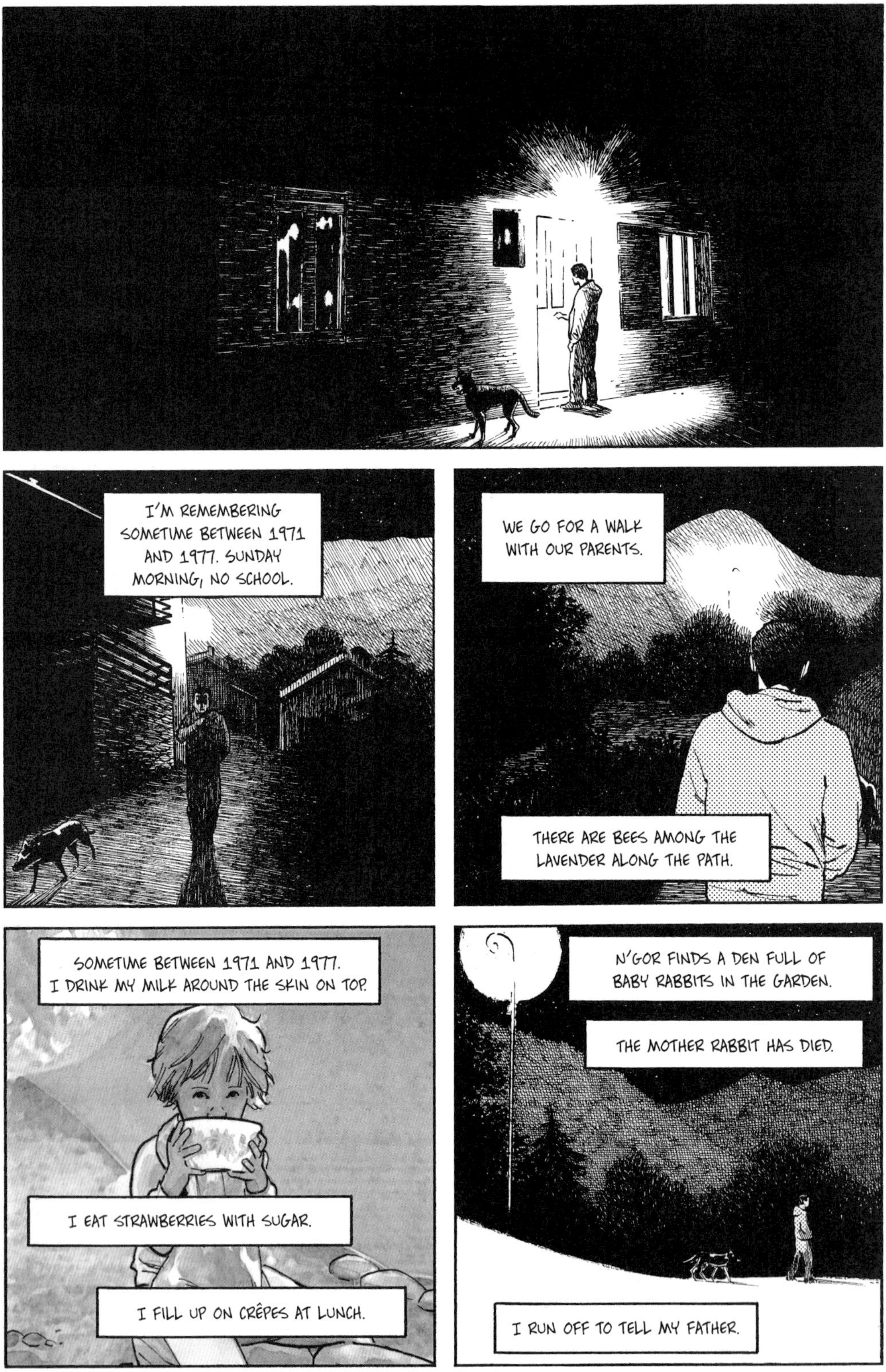
I'M REMEMBERING SOMETIME BETWEEN 1971 AND 1977. SUNDAY MORNING, NO SCHOOL.
WE GO FOR A WALK WITH OUR PARENTS.
THERE ARE BEES AMONG THE LAVENDER ALONG THE PATH.
SOMETIME BETWEEN 1971 AND 1977. I DRINK MY MILK AROUND THE SKIN ON TOP.
I EAT STRAWBERRIES WITH SUGAR.
I FILL UP ON CRÊPES AT LUNCH.
N'GOR FINDS A DEN FULL OF BABY RABBITS IN THE GARDEN.
THE MOTHER RABBIT HAS DIED.
I RUN OFF TO TELL MY FATHER.

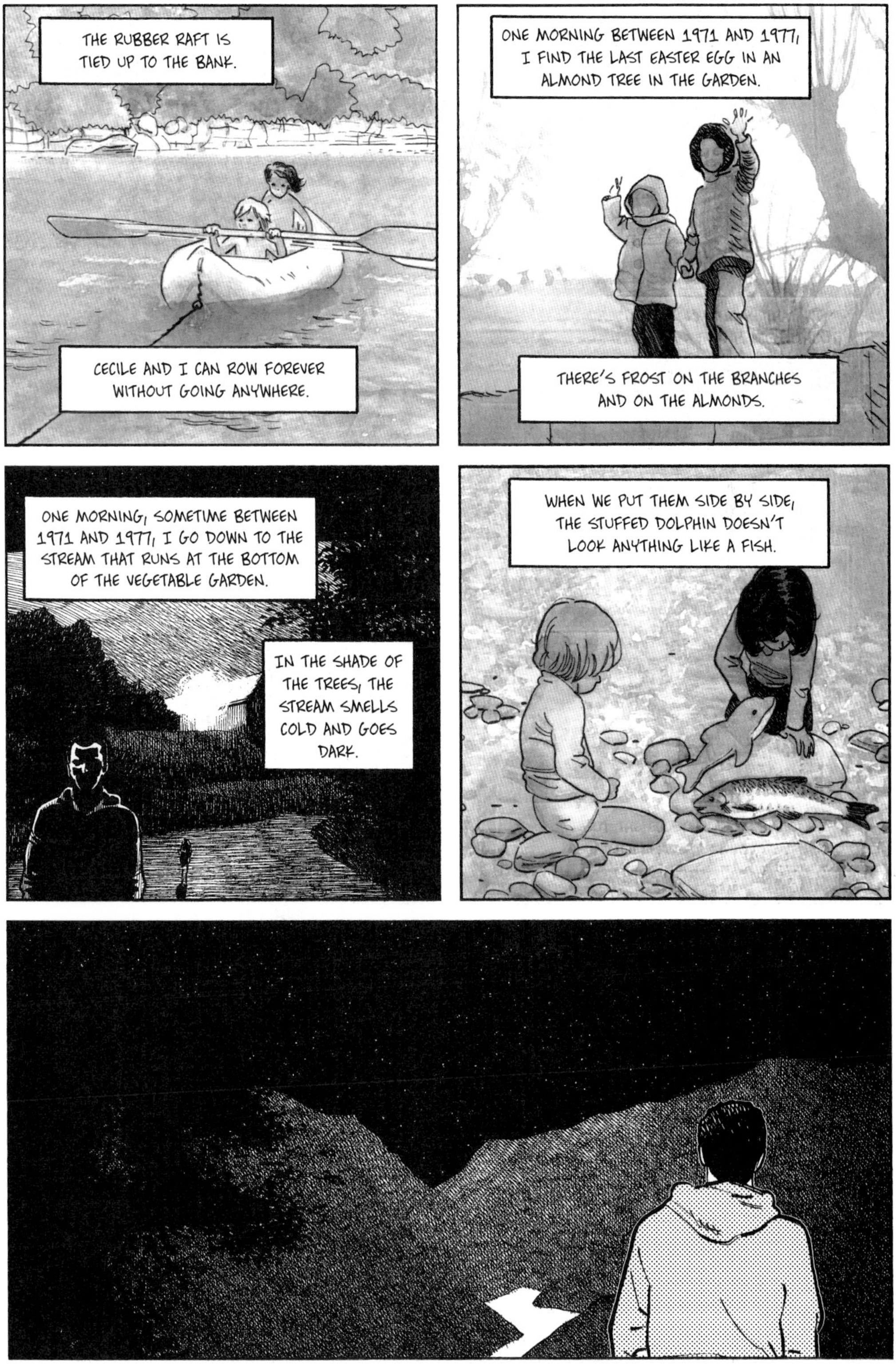
THE RUBBER RAFT IS TIED UP TO THE BANK.
CECILE AND I CAN ROW FOREVER WITHOUT GOING ANYWHERE.
ONE MORNING BETWEEN 1971 AND 1977, I FIND THE LAST EASTER EGG IN AN ALMOND TREE IN THE GARDEN.
THERE'S FROST ON THE BRANCHES AND ON THE ALMONDS.
ONE MORNING, SOMETIME BETWEEN 1971 AND 1977, I GO DOWN TO THE STREAM THAT RUNS AT THE BOTTOM OF THE VEGETABLE GARDEN.
IN THE SHADE OF THE TREES, THE STREAM SMELLS COLD AND GOES DARK.
WHEN WE PUT THEM SIDE BY SIDE, THE STUFFED DOLPHIN DOESN'T LOOK ANYTHING LIKE A FISH.

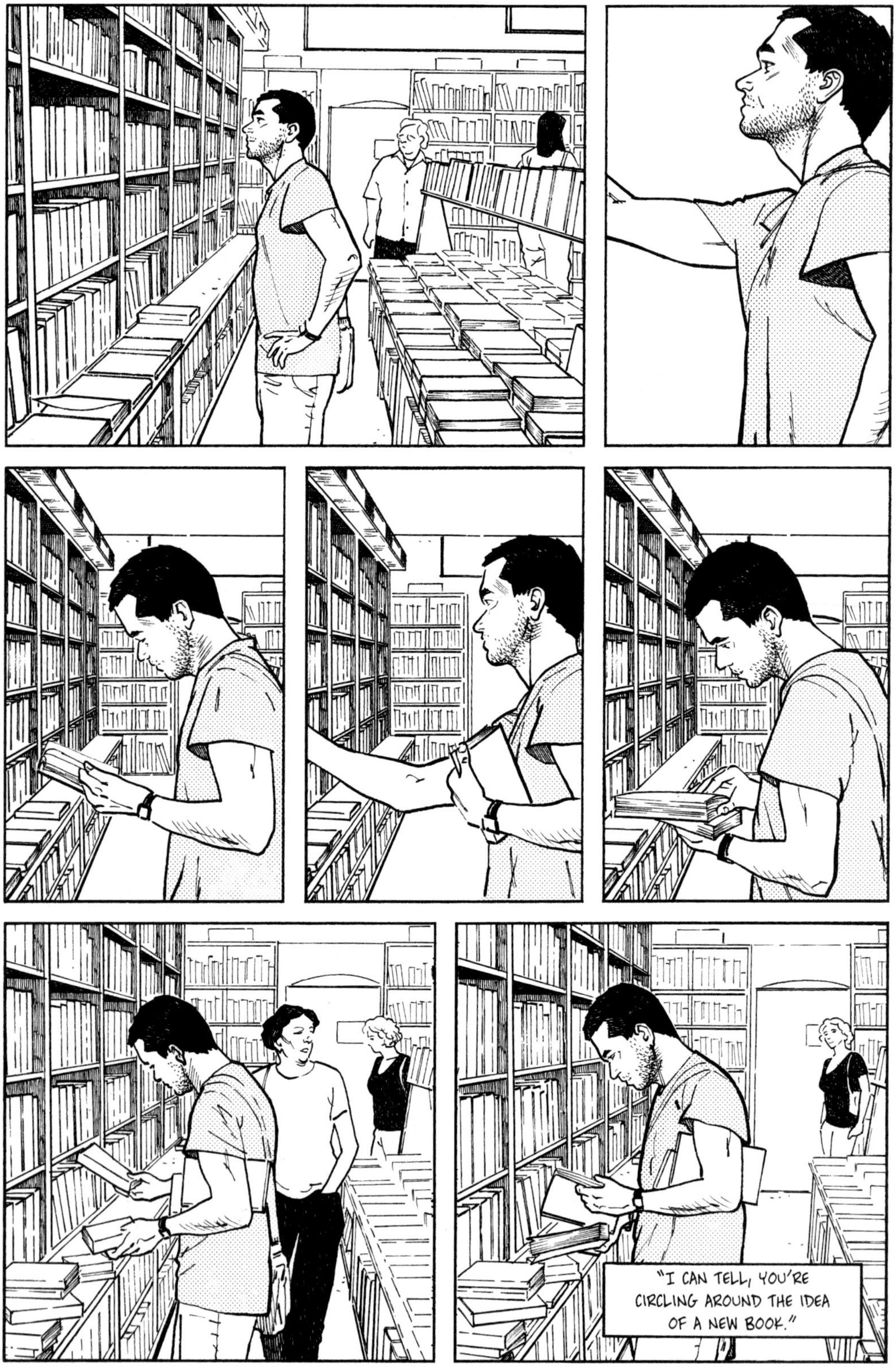
"I CAN TELL, YOU'RE CIRCLING AROUND THE IDEA OF A NEW BOOK."

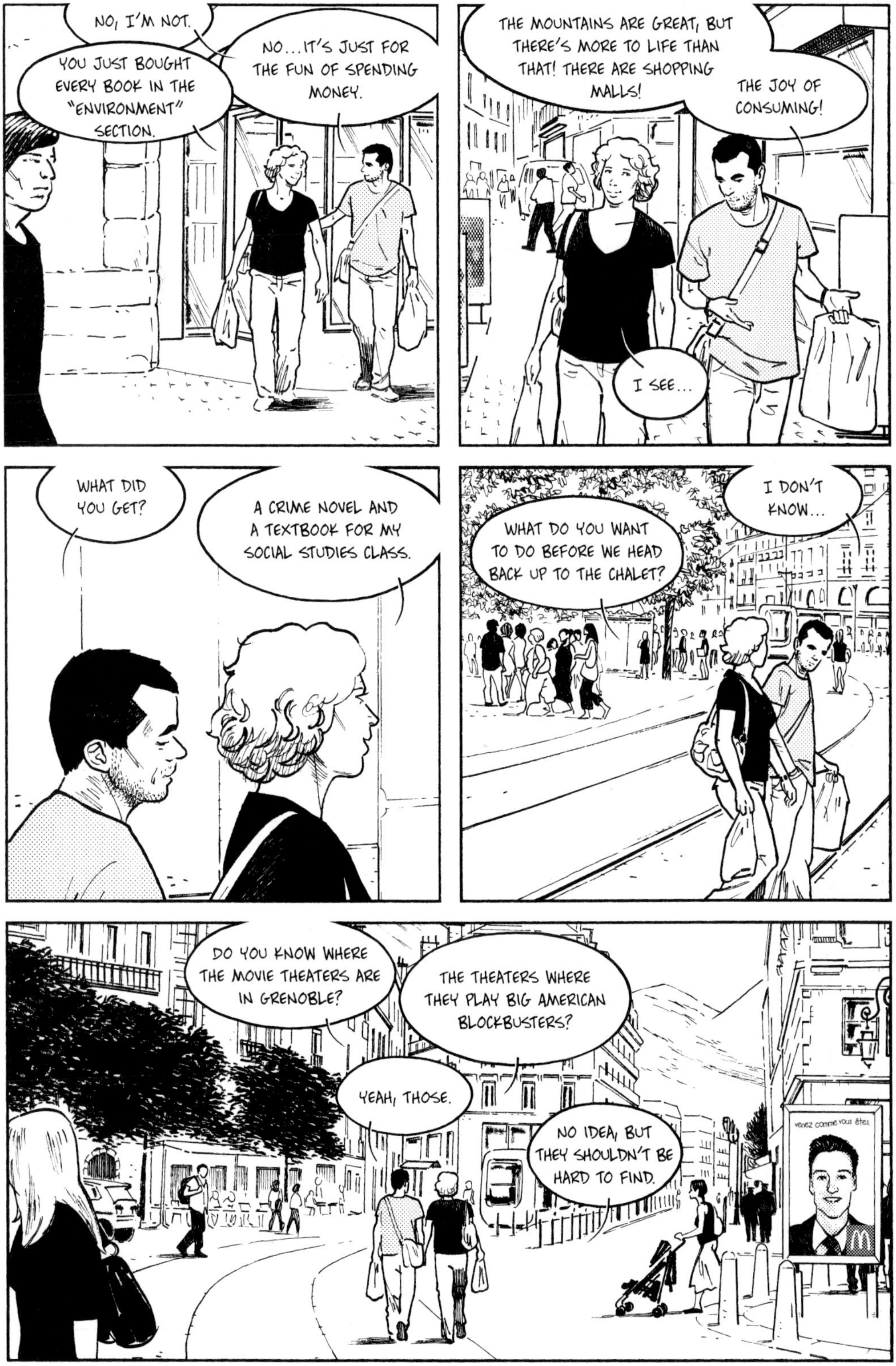
NO, I'M NOT.
YOU JUST BOUGHT EVERY BOOK IN THE "ENVIRONMENT" SECTION.
NO... IT'S JUST FOR THE FUN OF SPENDING MONEY.
THE MOUNTAINS ARE GREAT, BUT THERE'S MORE TO LIFE THAN THAT! THERE ARE SHOPPING MALLS!
THE JOY OF CONSUMING!
I SEE...
WHAT DID YOU GET?
A CRIME NOVEL AND A TEXTBOOK FOR MY SOCIAL STUDIES CLASS.
WHAT DO YOU WANT TO DO BEFORE WE HEAD BACK UP TO THE CHALET?
I DON'T KNOW...
DO YOU KNOW WHERE THE MOVIE THEATERS ARE IN GRENOBLE?
THE THEATERS WHERE THEY PLAY BIG AMERICAN BLOCKBUSTERS?
YEAH, THOSE.
NO IDEA, BUT THEY SHOULDN'T BE HARD TO FIND.
venez comme vous êtes

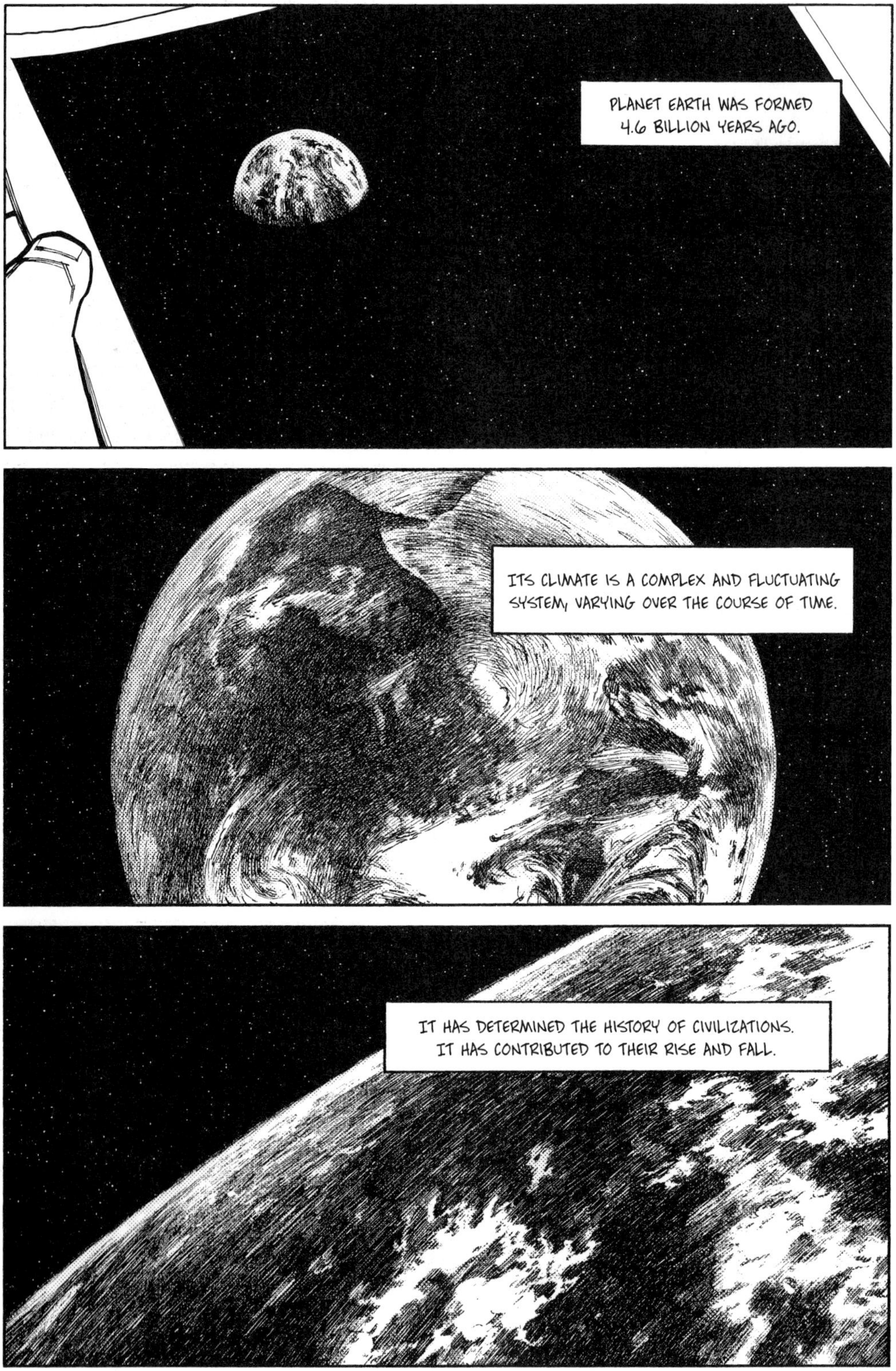
PLANET EARTH WAS FORMED 4.6 BILLION YEARS AGO.
ITS CLIMATE IS A COMPLEX AND FLUCTUATING SYSTEM, VARYING OVER THE COURSE OF TIME.
IT HAS DETERMINED THE HISTORY OF CIVILIZATIONS. IT HAS CONTRIBUTED TO THEIR RISE AND FALL.

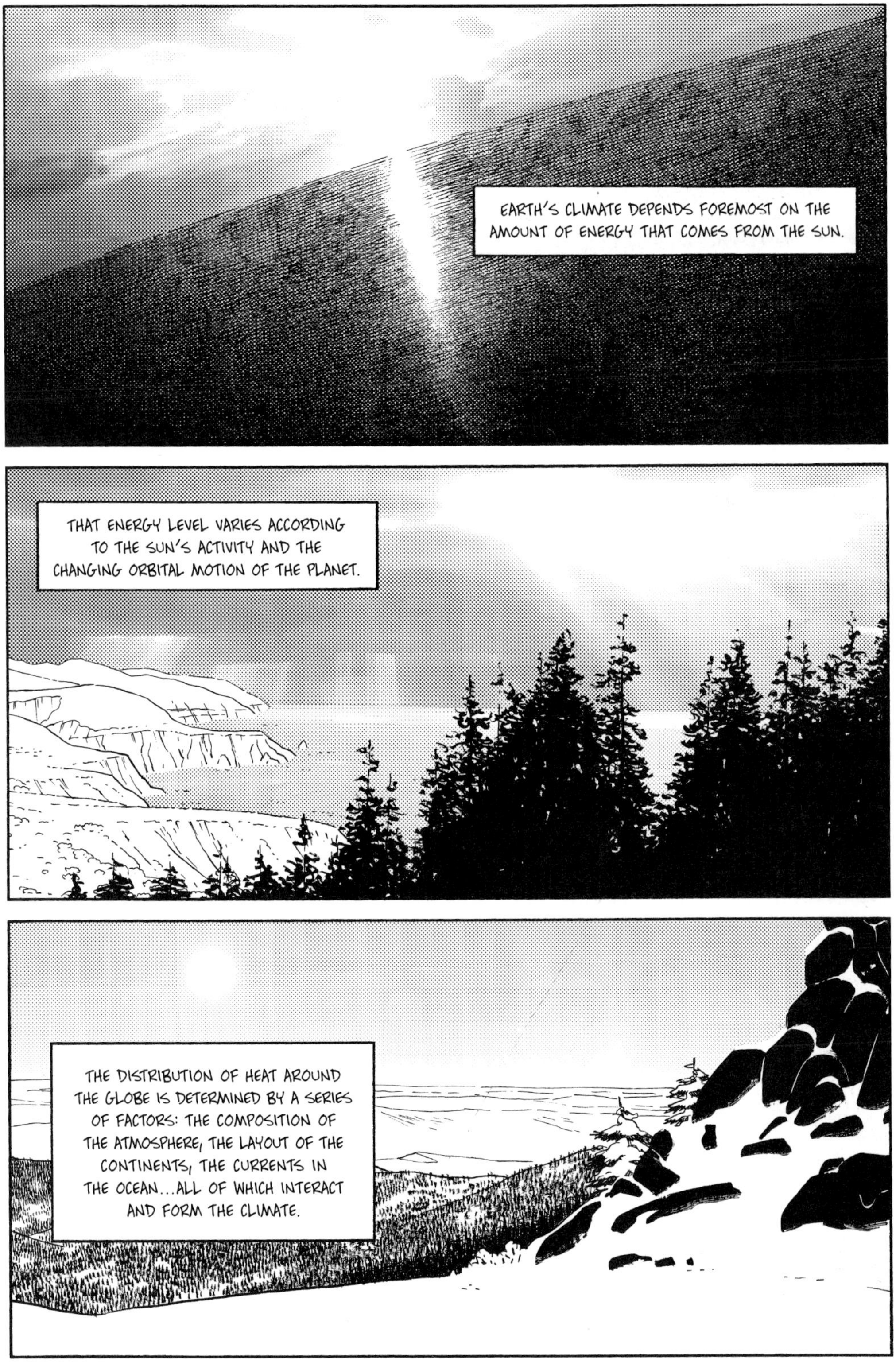
EARTH'S CLIMATE DEPENDS FOREMOST ON THE AMOUNT OF ENERGY THAT COMES FROM THE SUN.
THAT ENERGY LEVEL VARIES ACCORDING TO THE SUN'S ACTIVITY AND THE CHANGING ORBITAL MOTION OF THE PLANET.
THE DISTRIBUTION OF HEAT AROUND THE GLOBE IS DETERMINED BY A SERIES OF FACTORS: THE COMPOSITION OF THE ATMOSPHERE, THE LAYOUT OF THE CONTINENTS, THE CURRENTS IN THE OCEAN...ALL OF WHICH INTERACT AND FORM THE CLIMATE.

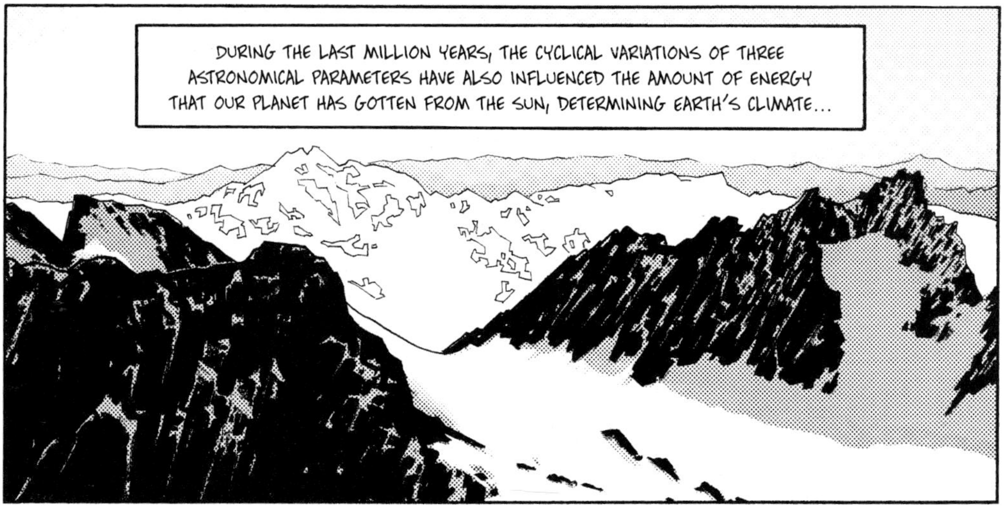
DURING THE LAST MILLION YEARS, THE CYCLICAL VARIATIONS OF THREE ASTRONOMICAL PARAMETERS HAVE ALSO INFLUENCED THE AMOUNT OF ENERGY THAT OUR PLANET HAS GOTTEN FROM THE SUN, DETERMINING EARTH'S CLIMATE...

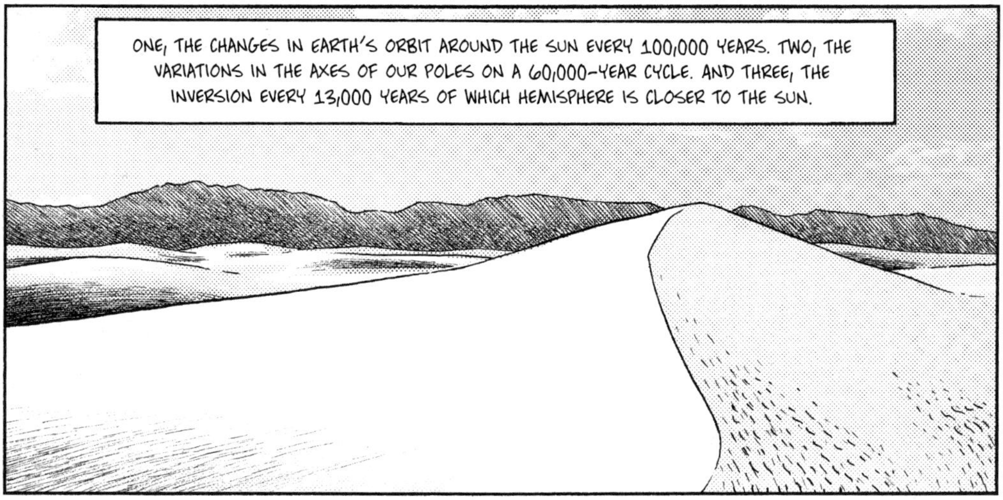
ONE, THE CHANGES IN EARTH'S ORBIT AROUND THE SUN EVERY 100,000 YEARS. TWO, THE VARIATIONS IN THE AXES OF OUR POLES ON A 60,000-YEAR CYCLE. AND THREE, THE INVERSION EVERY 13,000 YEARS OF WHICH HEMISPHERE IS CLOSER TO THE SUN.

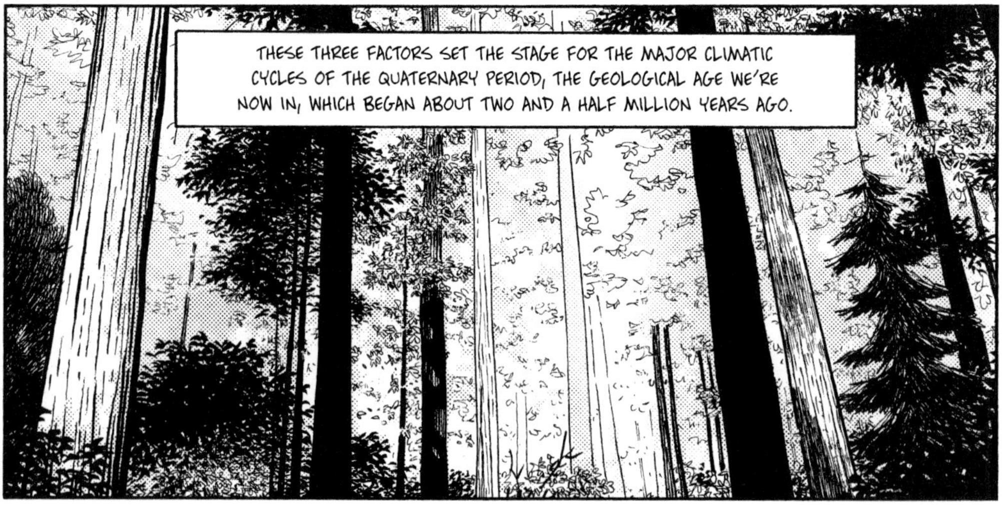
THESE THREE FACTORS SET THE STAGE FOR THE MAJOR CLIMATIC CYCLES OF THE QUATERNARY PERIOD, THE GEOLOGICAL AGE WE'RE NOW IN, WHICH BEGAN ABOUT TWO AND A HALF MILLION YEARS AGO.

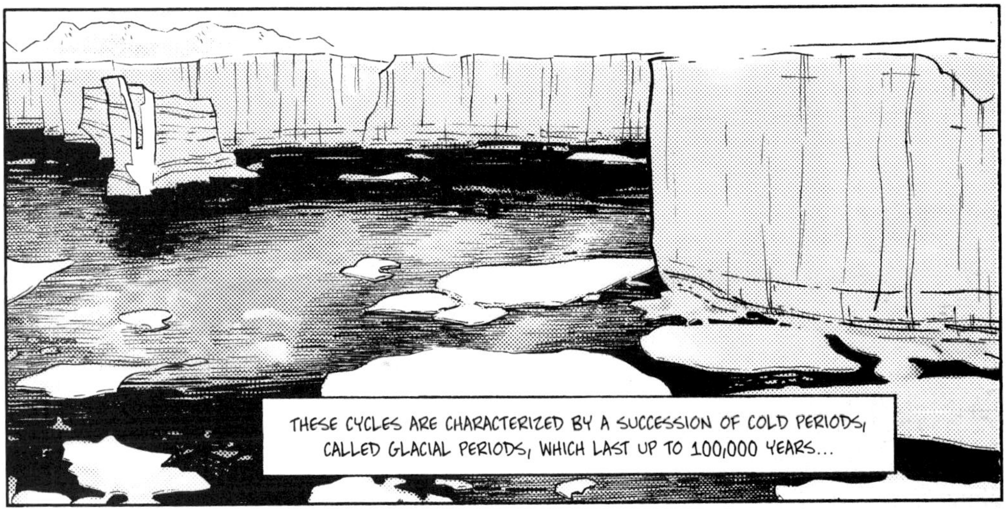
THESE CYCLES ARE CHARACTERIZED BY A SUCCESSION OF COLD PERIODS, CALLED GLACIAL PERIODS, WHICH LAST UP TO 100,000 YEARS...

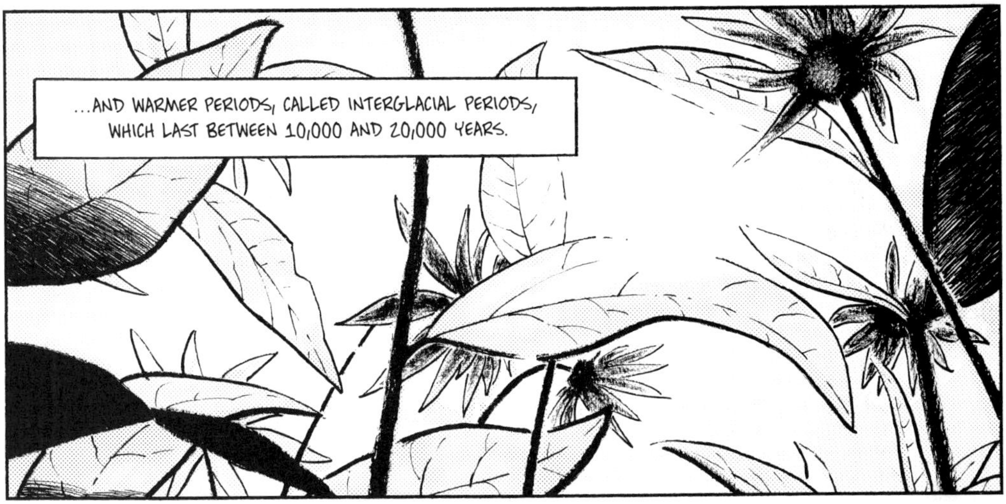
...AND WARMER PERIODS, CALLED INTERGLACIAL PERIODS, WHICH LAST BETWEEN 10,000 AND 20,000 YEARS.

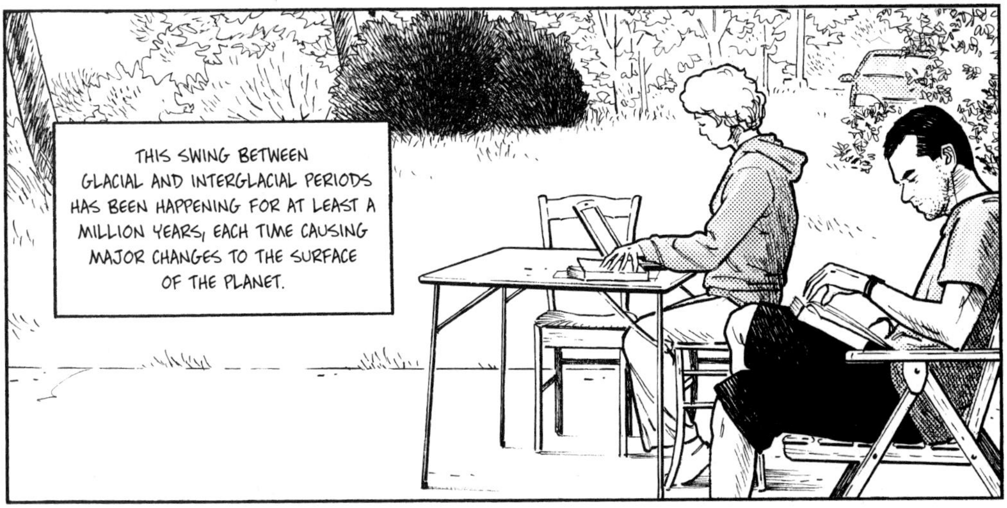
THIS SWING BETWEEN GLACIAL AND INTERGLACIAL PERIODS HAS BEEN HAPPENING FOR AT LEAST A MILLION YEARS, EACH TIME CAUSING MAJOR CHANGES TO THE SURFACE OF THE PLANET.

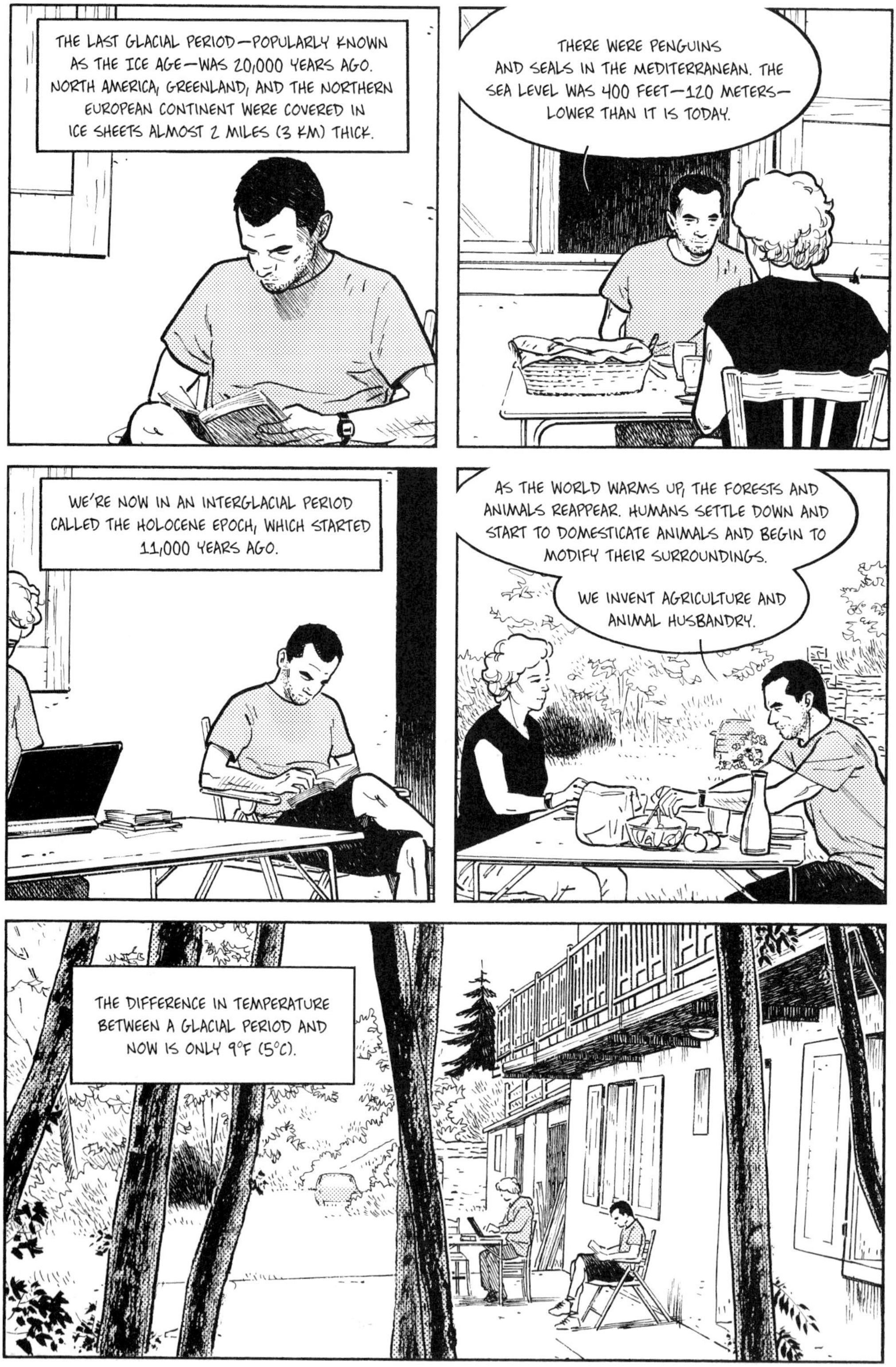
THE LAST GLACIAL PERIOD—POPULARLY KNOWN AS THE ICE AGE—WAS 20,000 YEARS AGO. NORTH AMERICA, GREENLAND, AND THE NORTHERN EUROPEAN CONTINENT WERE COVERED IN ICE SHEETS ALMOST 2 MILES (3 KM) THICK.
THERE WERE PENGUINS AND SEALS IN THE MEDITERRANEAN. THE SEA LEVEL WAS 400 FEET—120 METERS—LOWER THAN IT IS TODAY.
WE'RE NOW IN AN INTERGLACIAL PERIOD CALLED THE HOLOCENE EPOCH, WHICH STARTED 11,000 YEARS AGO.
AS THE WORLD WARMS UP, THE FORESTS AND ANIMALS REAPPEAR. HUMANS SETTLE DOWN AND START TO DOMESTICATE ANIMALS AND BEGIN TO MODIFY THEIR SURROUNDINGS.
WE INVENT AGRICULTURE AND ANIMAL HUSBANDRY.
THE DIFFERENCE IN TEMPERATURE BETWEEN A GLACIAL PERIOD AND NOW IS ONLY 9°F (5°C).

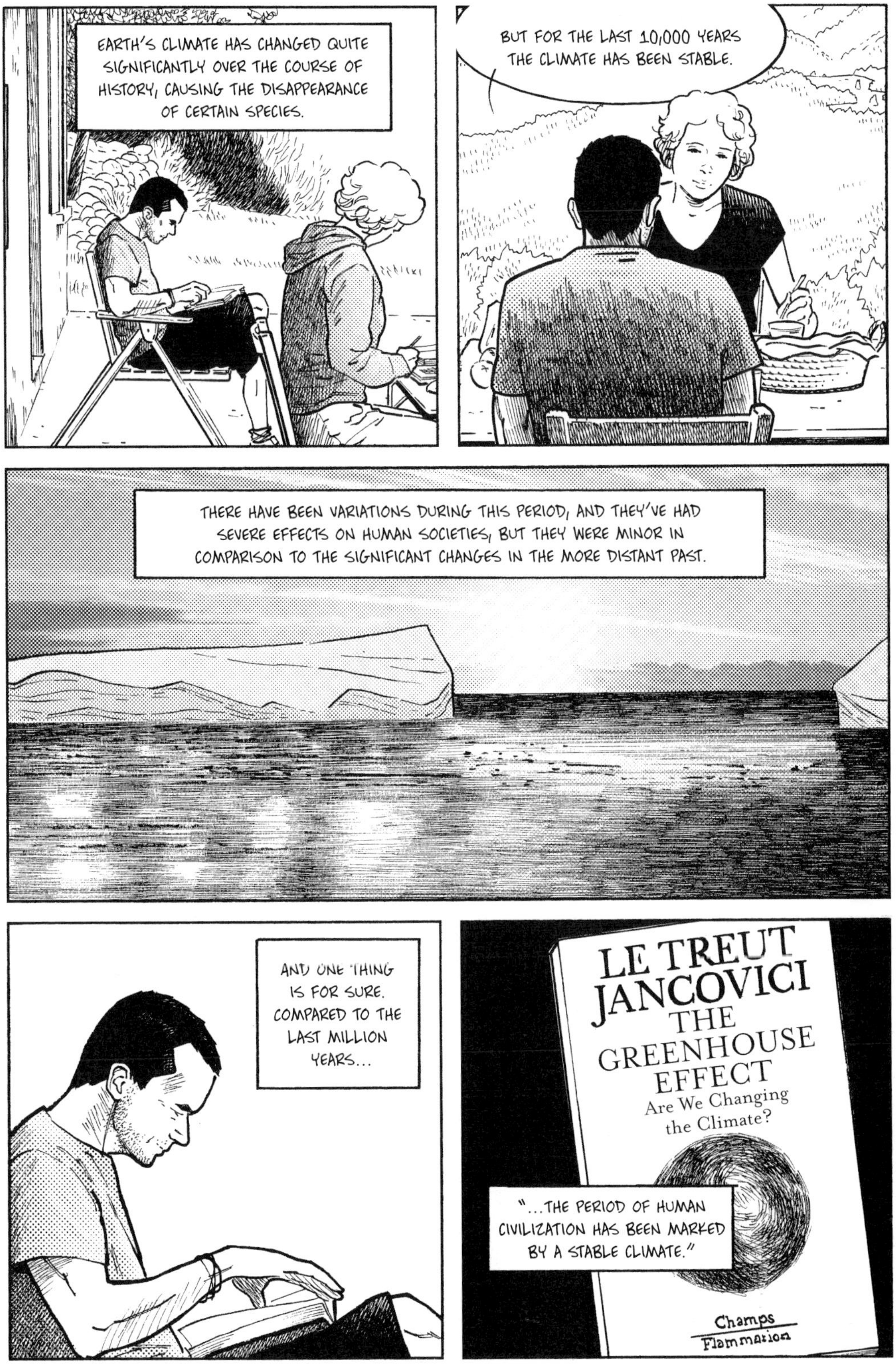
EARTH'S CLIMATE HAS CHANGED QUITE SIGNIFICANTLY OVER THE COURSE OF HISTORY, CAUSING THE DISAPPEARANCE OF CERTAIN SPECIES.
BUT FOR THE LAST 10,000 YEARS THE CLIMATE HAS BEEN STABLE.
THERE HAVE BEEN VARIATIONS DURING THIS PERIOD, AND THEY'VE HAD SEVERE EFFECTS ON HUMAN SOCIETIES, BUT THEY WERE MINOR IN COMPARISON TO THE SIGNIFICANT CHANGES IN THE MORE DISTANT PAST.
AND ONE THING IS FOR SURE. COMPARED TO THE LAST MILLION YEARS...
LE TREUT JANCOVICI
THE GREENHOUSE EFFECT
Are We Changing the Climate?
Champs
Flammarion
"...THE PERIOD OF HUMAN CIVILIZATION HAS BEEN MARKED BY A STABLE CLIMATE."

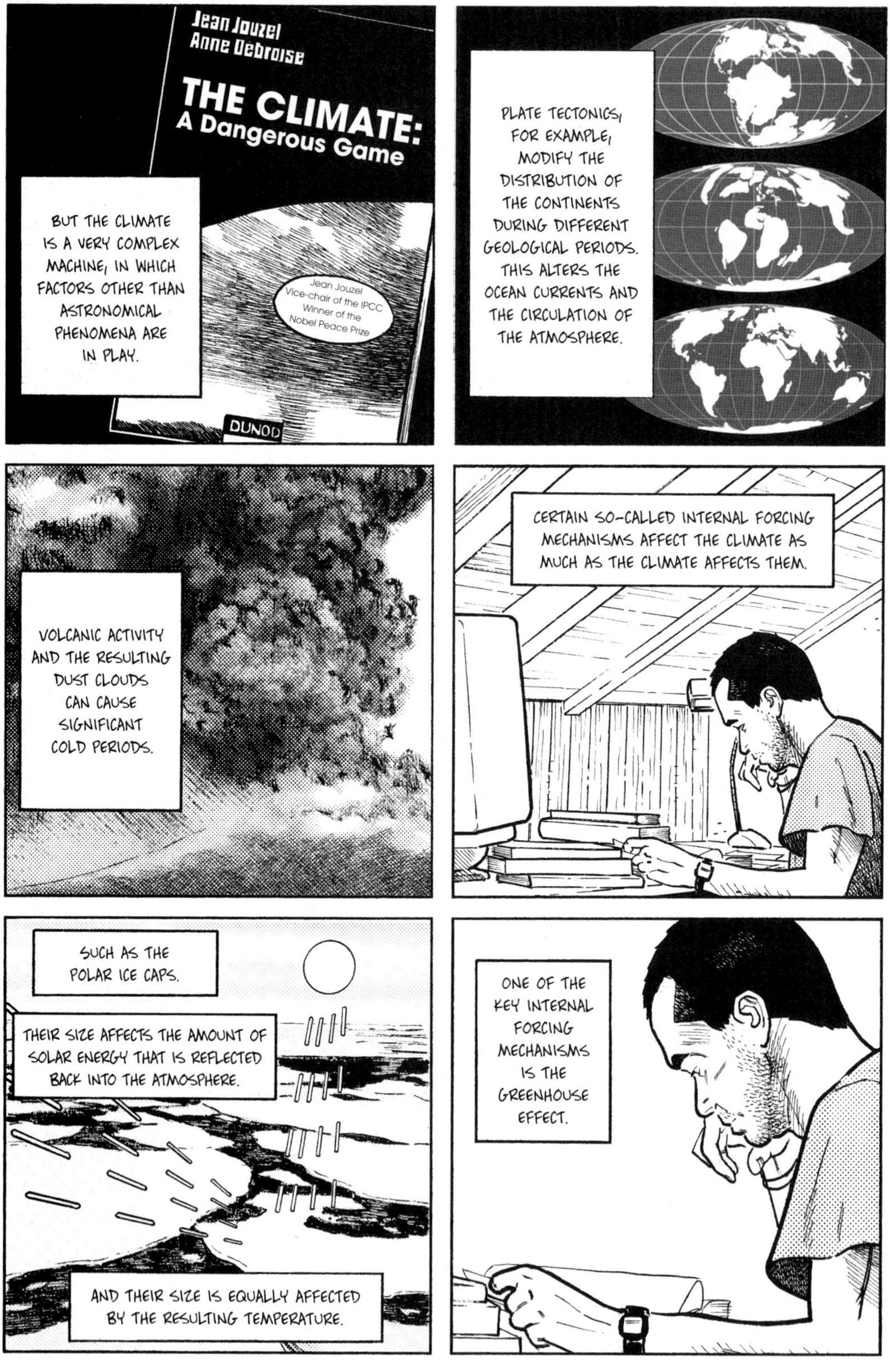
Jean Jouzel
Anne Debroise
THE CLIMATE: A Dangerous Game
Jean Jouzel Vice-chair of the IPCC Winner of the Nobel Peace Prize
DUNOD
BUT THE CLIMATE IS A VERY COMPLEX MACHINE, IN WHICH FACTORS OTHER THAN ASTRONOMICAL PHENOMENA ARE IN PLAY.
PLATE TECTONICS, FOR EXAMPLE, MODIFY THE DISTRIBUTION OF THE CONTINENTS DURING DIFFERENT GEOLOGICAL PERIODS. THIS ALTERS THE OCEAN CURRENTS AND THE CIRCULATION OF THE ATMOSPHERE.
VOLCANIC ACTIVITY AND THE RESULTING DUST CLOUDS CAN CAUSE SIGNIFICANT COLD PERIODS.
CERTAIN SO-CALLED INTERNAL FORCING MECHANISMS AFFECT THE CLIMATE AS MUCH AS THE CLIMATE AFFECTS THEM.
SUCH AS THE POLAR ICE CAPS.
THEIR SIZE AFFECTS THE AMOUNT OF SOLAR ENERGY THAT IS REFLECTED BACK INTO THE ATMOSPHERE.
AND THEIR SIZE IS EQUALLY AFFECTED BY THE RESULTING TEMPERATURE.
ONE OF THE KEY INTERNAL FORCING MECHANISMS IS THE GREENHOUSE EFFECT.

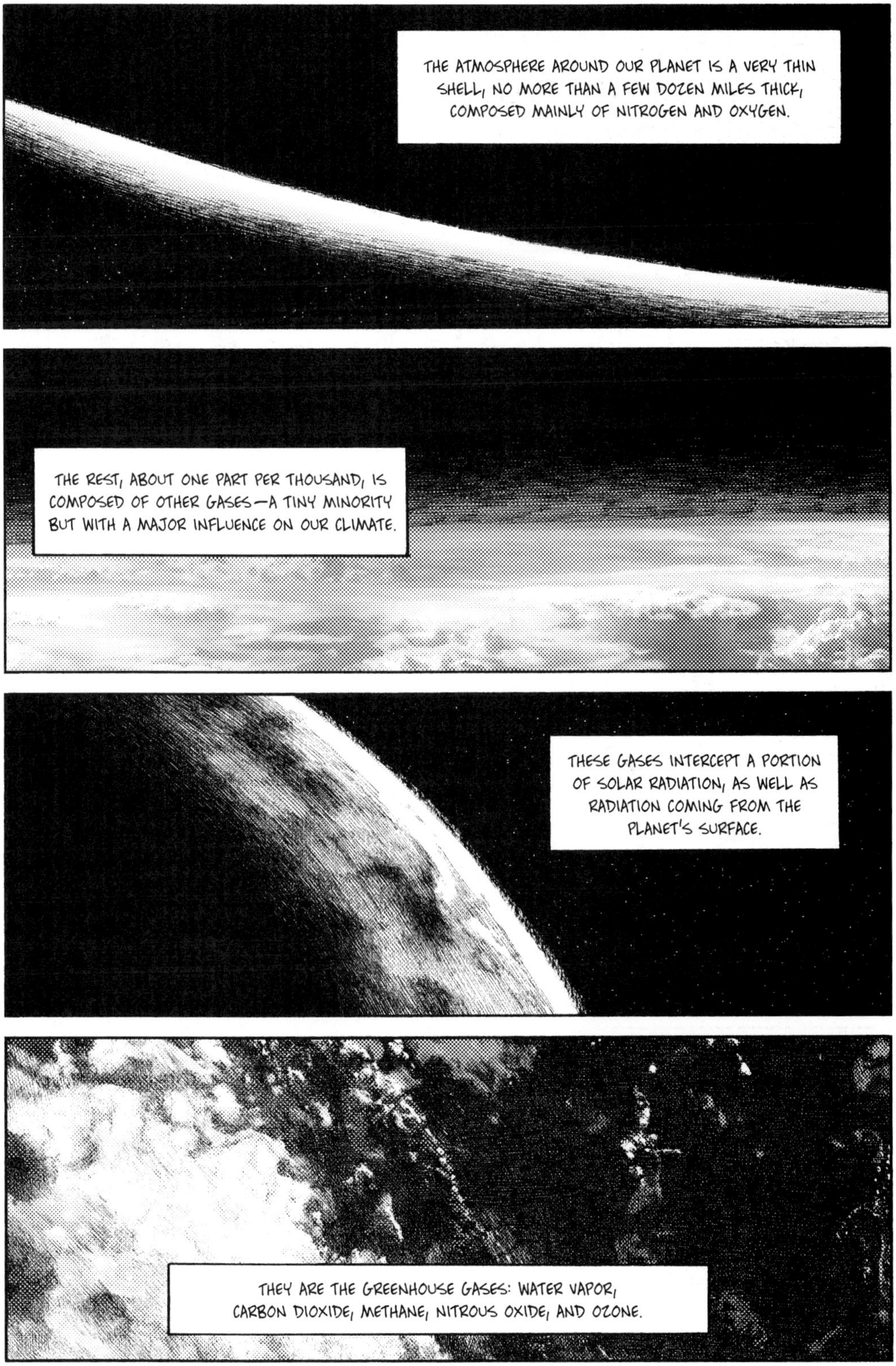
THE ATMOSPHERE AROUND OUR PLANET IS A VERY THIN SHELL, NO MORE THAN A FEW DOZEN MILES THICK, COMPOSED MAINLY OF NITROGEN AND OXYGEN.
THE REST, ABOUT ONE PART PER THOUSAND, IS COMPOSED OF OTHER GASES—A TINY MINORITY BUT WITH A MAJOR INFLUENCE ON OUR CLIMATE.
THESE GASES INTERCEPT A PORTION OF SOLAR RADIATION, AS WELL AS RADIATION COMING FROM THE PLANET'S SURFACE.
THEY ARE THE GREENHOUSE GASES: WATER VAPOR, CARBON DIOXIDE, METHANE, NITROUS OXIDE, AND OZONE.

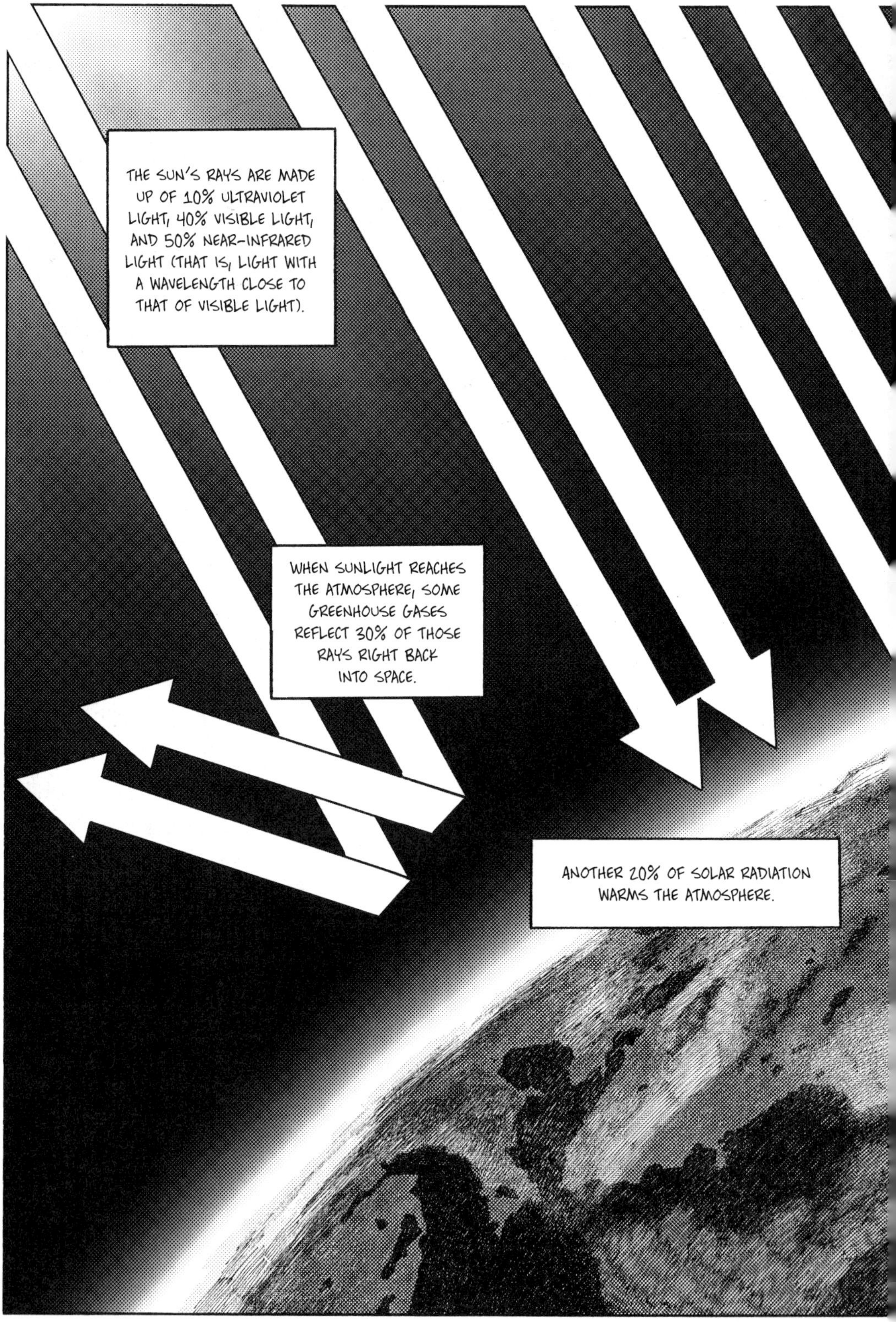
THE SUN'S RAYS ARE MADE UP OF 10% ULTRAVIOLET LIGHT, 40% VISIBLE LIGHT, AND 50% NEAR-INFRARED LIGHT (THAT IS, LIGHT WITH A WAVELENGTH CLOSE TO THAT OF VISIBLE LIGHT).
WHEN SUNLIGHT REACHES THE ATMOSPHERE, SOME GREENHOUSE GASES REFLECT 30% OF THOSE RAYS RIGHT BACK INTO SPACE.
ANOTHER 20% OF SOLAR RADIATION WARMS THE ATMOSPHERE.

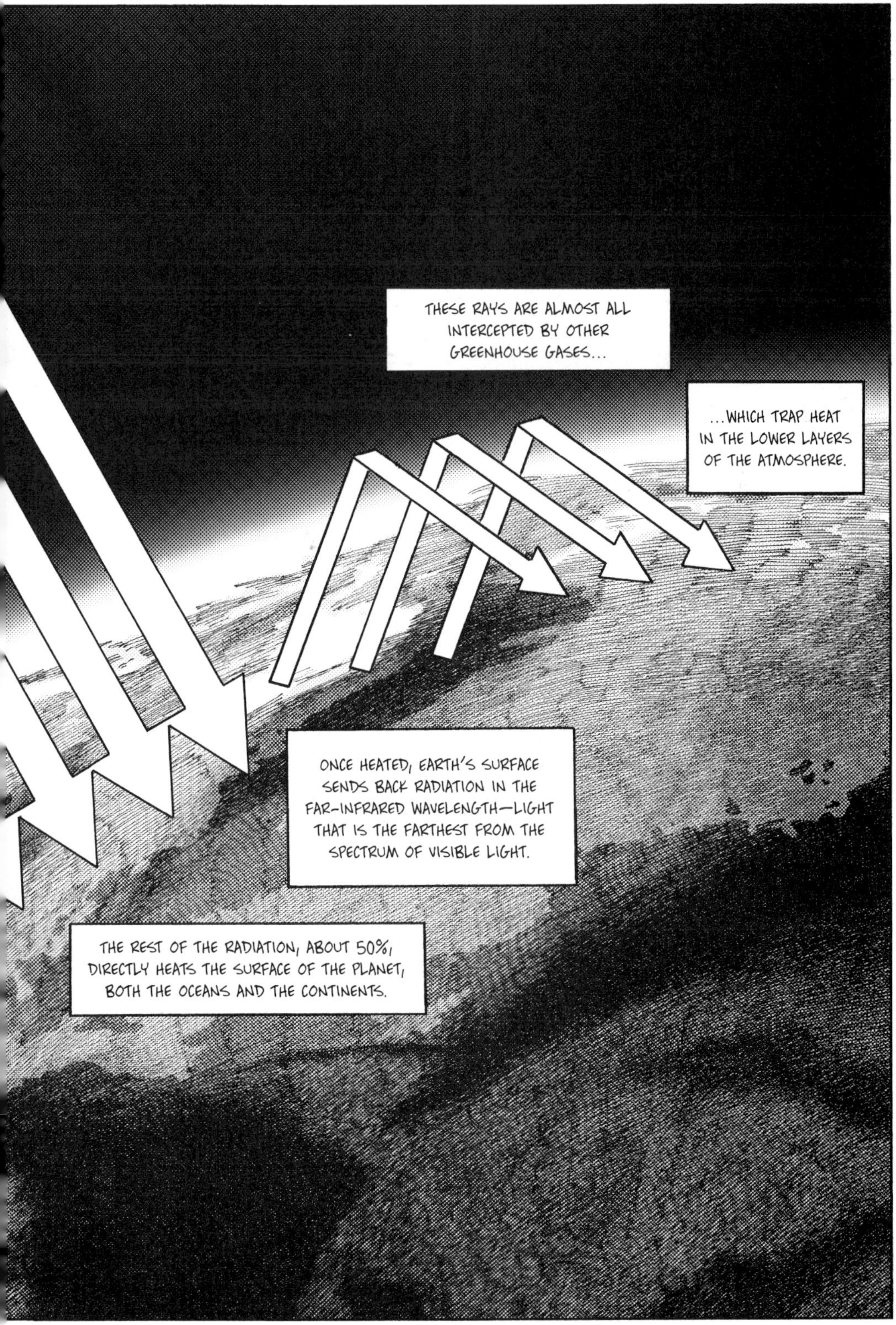
THESE RAYS ARE ALMOST ALL INTERCEPTED BY OTHER GREENHOUSE GASES...
...WHICH TRAP HEAT IN THE LOWER LAYERS OF THE ATMOSPHERE.
ONCE HEATED, EARTH'S SURFACE SENDS BACK RADIATION IN THE FAR-INFRARED WAVELENGTH—LIGHT THAT IS THE FARTHEST FROM THE SPECTRUM OF VISIBLE LIGHT.
THE REST OF THE RADIATION, ABOUT 50%, DIRECTLY HEATS THE SURFACE OF THE PLANET, BOTH THE OCEANS AND THE CONTINENTS.

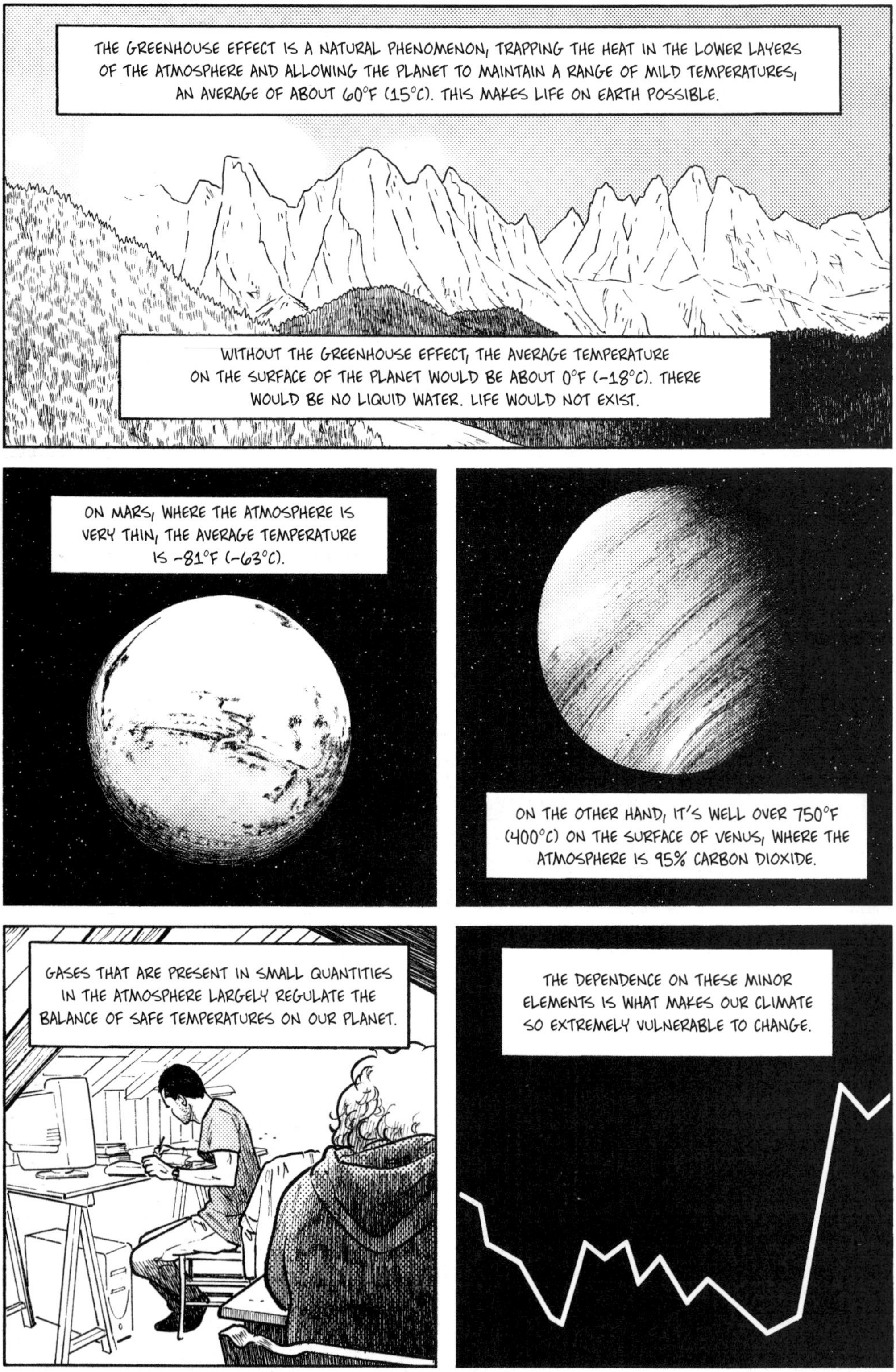
THE GREENHOUSE EFFECT IS A NATURAL PHENOMENON, TRAPPING THE HEAT IN THE LOWER LAYERS OF THE ATMOSPHERE AND ALLOWING THE PLANET TO MAINTAIN A RANGE OF MILD TEMPERATURES, AN AVERAGE OF ABOUT 60°F (15°C). THIS MAKES LIFE ON EARTH POSSIBLE.
WITHOUT THE GREENHOUSE EFFECT, THE AVERAGE TEMPERATURE ON THE SURFACE OF THE PLANET WOULD BE ABOUT 0°F (-18°C). THERE WOULD BE NO LIQUID WATER. LIFE WOULD NOT EXIST.
ON MARS, WHERE THE ATMOSPHERE IS VERY THIN, THE AVERAGE TEMPERATURE IS -81°F (-63°C).
ON THE OTHER HAND, IT'S WELL OVER 750°F (400°C) ON THE SURFACE OF VENUS, WHERE THE ATMOSPHERE IS 95% CARBON DIOXIDE.
GASES THAT ARE PRESENT IN SMALL QUANTITIES IN THE ATMOSPHERE LARGELY REGULATE THE BALANCE OF SAFE TEMPERATURES ON OUR PLANET.
THE DEPENDENCE ON THESE MINOR ELEMENTS IS WHAT MAKES OUR CLIMATE SO EXTREMELY VULNERABLE TO CHANGE.

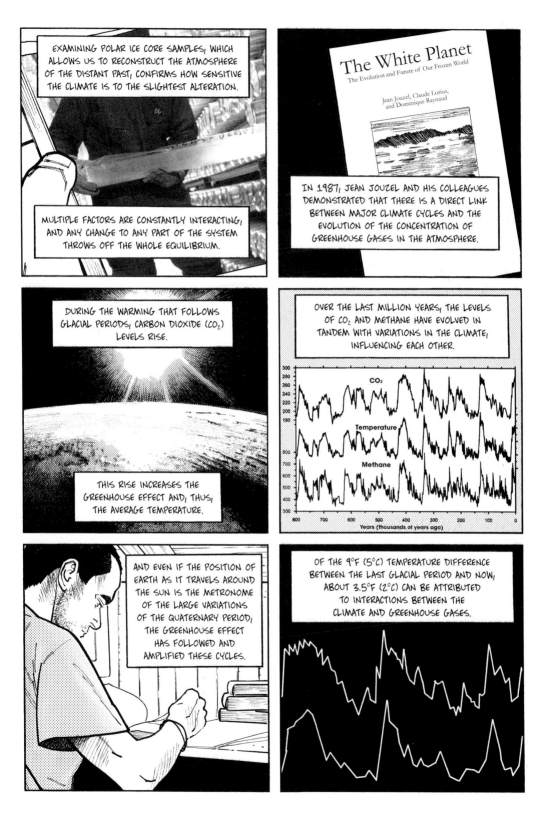

EXAMINING POLAR ICE CORE SAMPLES, WHICH ALLOWS US TO RECONSTRUCT THE ATMOSPHERE OF THE DISTANT PAST, CONFIRMS HOW SENSITIVE THE CLIMATE IS TO THE SLIGHTEST ALTERATION.

MULTIPLE FACTORS ARE CONSTANTLY INTERACTING, AND ANY CHANGE TO ANY PART OF THE SYSTEM THROWS OFF THE WHOLE EQUILIBRIUM.

The White Planet
The Evolution and Future of Our Frozen World

Jean Jouzel, Claude Lorius, and Dominique Raynaud

IN 1987, JEAN JOUZEL AND HIS COLLEAGUES DEMONSTRATED THAT THERE IS A DIRECT LINK BETWEEN MAJOR CLIMATE CYCLES AND THE EVOLUTION OF THE CONCENTRATION OF GREENHOUSE GASES IN THE ATMOSPHERE.

DURING THE WARMING THAT FOLLOWS GLACIAL PERIODS, CARBON DIOXIDE (CO_2) LEVELS RISE.

THIS RISE INCREASES THE GREENHOUSE EFFECT AND, THUS, THE AVERAGE TEMPERATURE.

OVER THE LAST MILLION YEARS, THE LEVELS OF CO_2 AND METHANE HAVE EVOLVED IN TANDEM WITH VARIATIONS IN THE CLIMATE, INFLUENCING EACH OTHER.

CO₂

Temperature

Methane

Years (thousands of years ago)

AND EVEN IF THE POSITION OF EARTH AS IT TRAVELS AROUND THE SUN IS THE METRONOME OF THE LARGE VARIATIONS OF THE QUATERNARY PERIOD, THE GREENHOUSE EFFECT HAS FOLLOWED AND AMPLIFIED THESE CYCLES.

OF THE 9°F (5°C) TEMPERATURE DIFFERENCE BETWEEN THE LAST GLACIAL PERIOD AND NOW, ABOUT 3.5°F (2°C) CAN BE ATTRIBUTED TO INTERACTIONS BETWEEN THE CLIMATE AND GREENHOUSE GASES.

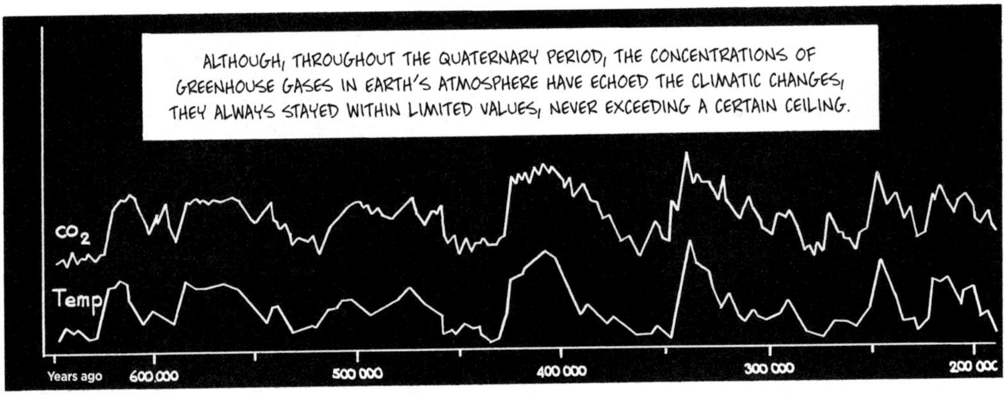
ALTHOUGH, THROUGHOUT THE QUATERNARY PERIOD, THE CONCENTRATIONS OF GREENHOUSE GASES IN EARTH'S ATMOSPHERE HAVE ECHOED THE CLIMATIC CHANGES, THEY ALWAYS STAYED WITHIN LIMITED VALUES, NEVER EXCEEDING A CERTAIN CEILING.
CO_2
Temp
Years ago
600 000
500 000
400 000
300 000
200 000

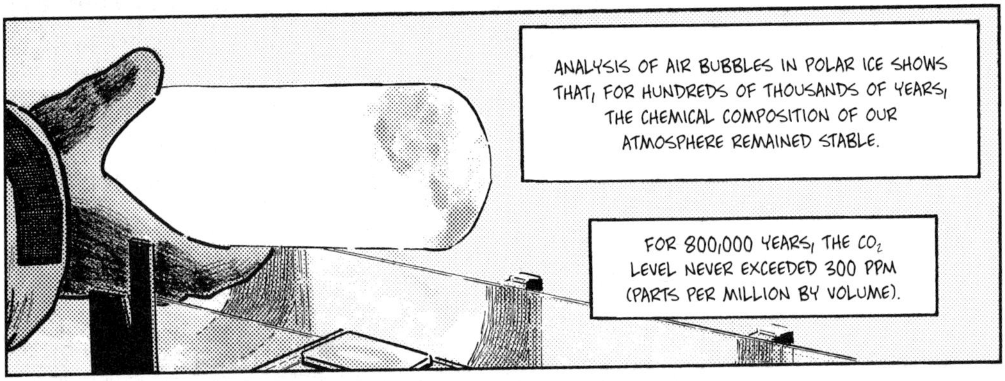
ANALYSIS OF AIR BUBBLES IN POLAR ICE SHOWS THAT, FOR HUNDREDS OF THOUSANDS OF YEARS, THE CHEMICAL COMPOSITION OF OUR ATMOSPHERE REMAINED STABLE.
FOR 800,000 YEARS, THE CO_2 LEVEL NEVER EXCEEDED 300 PPM (PARTS PER MILLION BY VOLUME).

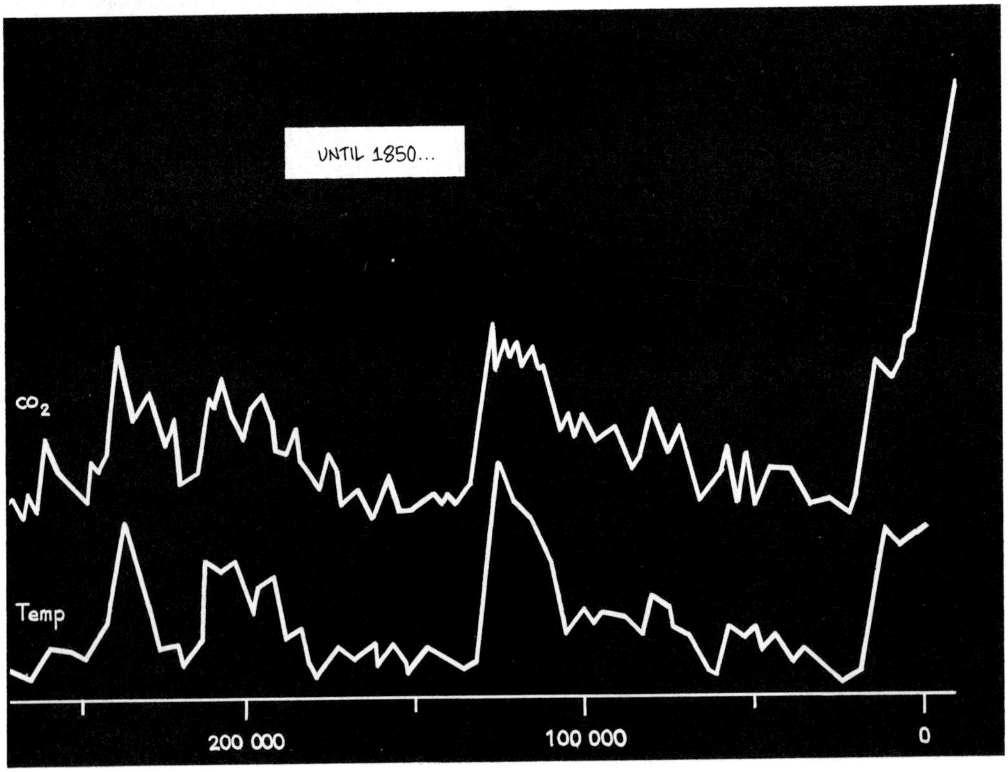
UNTIL 1850...
CO_2
Temp
200 000
100 000
0

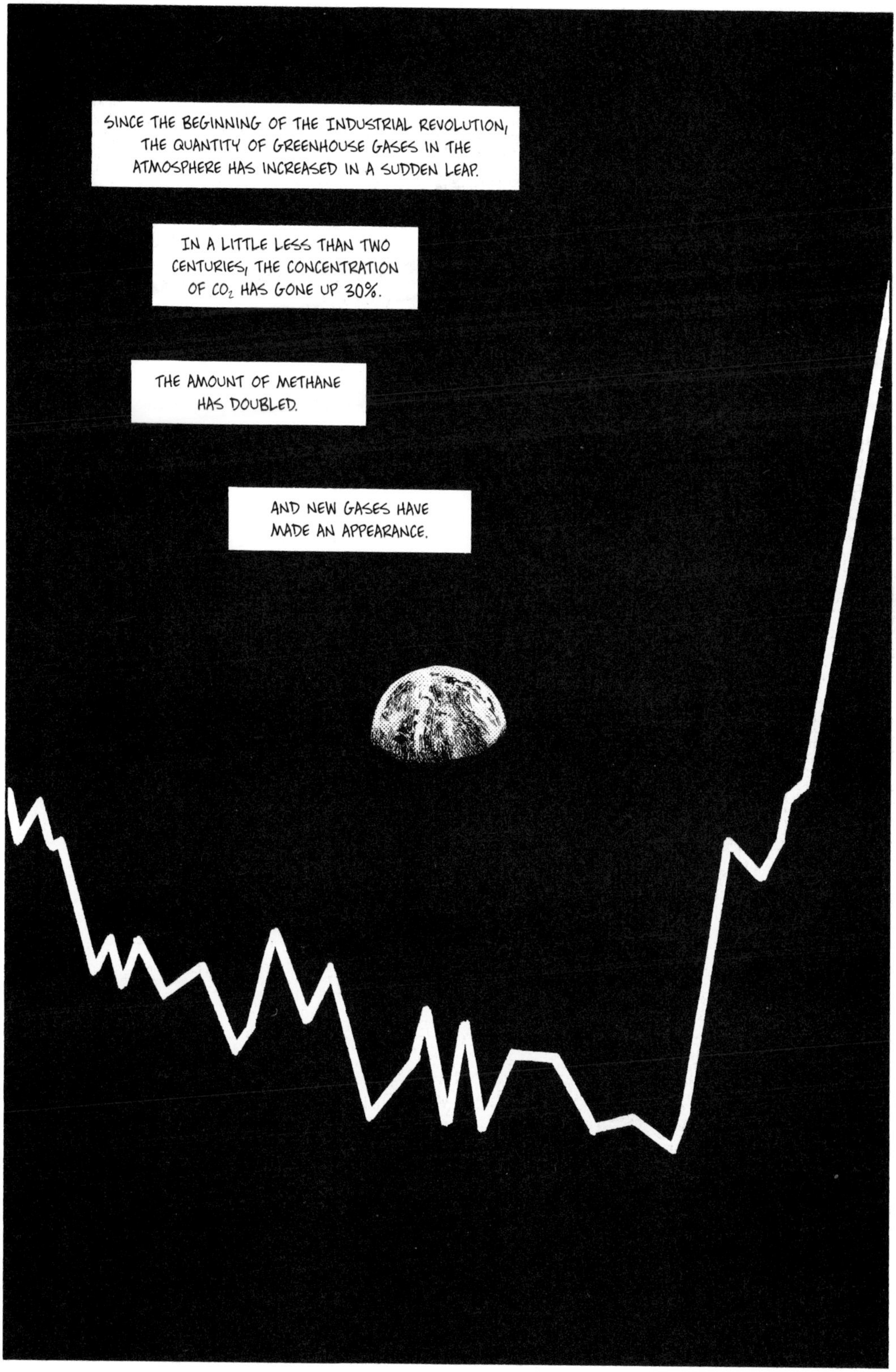
SINCE THE BEGINNING OF THE INDUSTRIAL REVOLUTION, THE QUANTITY OF GREENHOUSE GASES IN THE ATMOSPHERE HAS INCREASED IN A SUDDEN LEAP.
IN A LITTLE LESS THAN TWO CENTURIES, THE CONCENTRATION OF CO_2 HAS GONE UP 30%.
THE AMOUNT OF METHANE HAS DOUBLED.
AND NEW GASES HAVE MADE AN APPEARANCE.

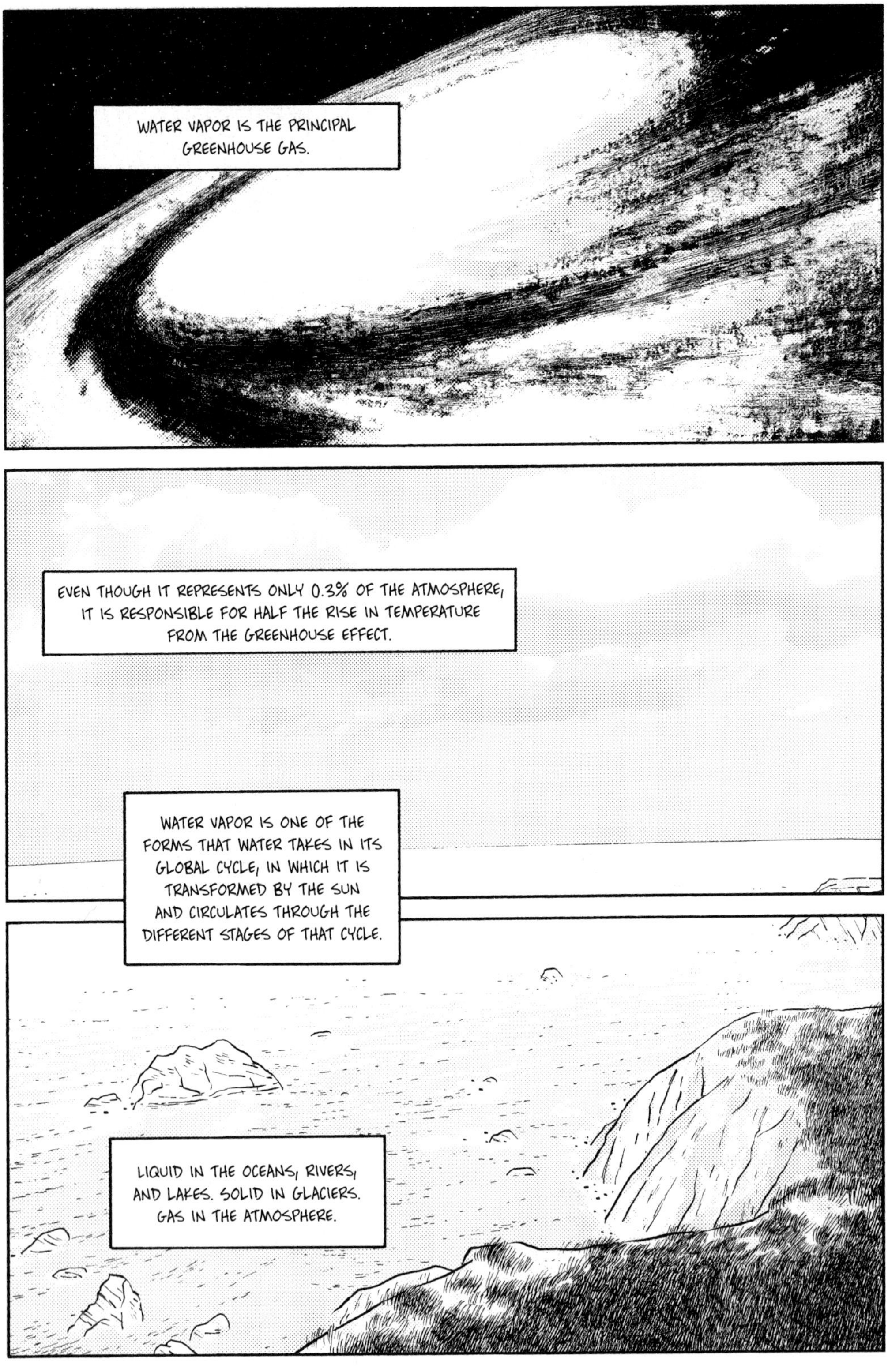
WATER VAPOR IS THE PRINCIPAL GREENHOUSE GAS.
EVEN THOUGH IT REPRESENTS ONLY 0.3% OF THE ATMOSPHERE, IT IS RESPONSIBLE FOR HALF THE RISE IN TEMPERATURE FROM THE GREENHOUSE EFFECT.
WATER VAPOR IS ONE OF THE FORMS THAT WATER TAKES IN ITS GLOBAL CYCLE, IN WHICH IT IS TRANSFORMED BY THE SUN AND CIRCULATES THROUGH THE DIFFERENT STAGES OF THAT CYCLE.
LIQUID IN THE OCEANS, RIVERS, AND LAKES. SOLID IN GLACIERS. GAS IN THE ATMOSPHERE.

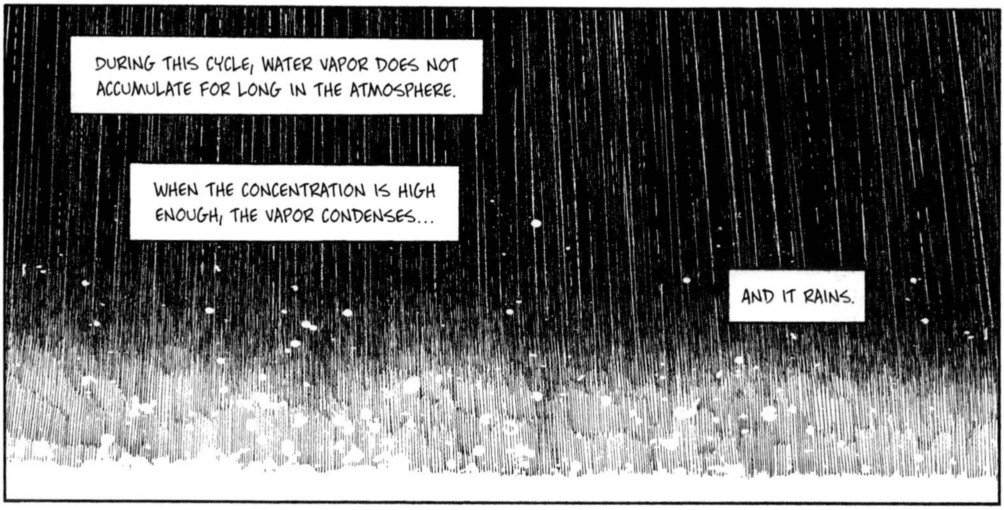
DURING THIS CYCLE, WATER VAPOR DOES NOT ACCUMULATE FOR LONG IN THE ATMOSPHERE.
WHEN THE CONCENTRATION IS HIGH ENOUGH, THE VAPOR CONDENSES...
AND IT RAINS.

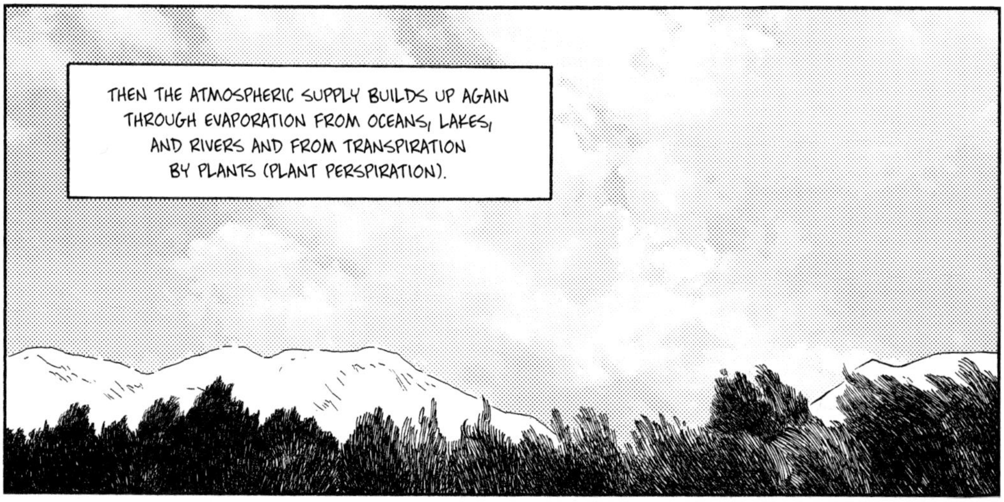
THEN THE ATMOSPHERIC SUPPLY BUILDS UP AGAIN THROUGH EVAPORATION FROM OCEANS, LAKES, AND RIVERS AND FROM TRANSPIRATION BY PLANTS (PLANT PERSPIRATION).

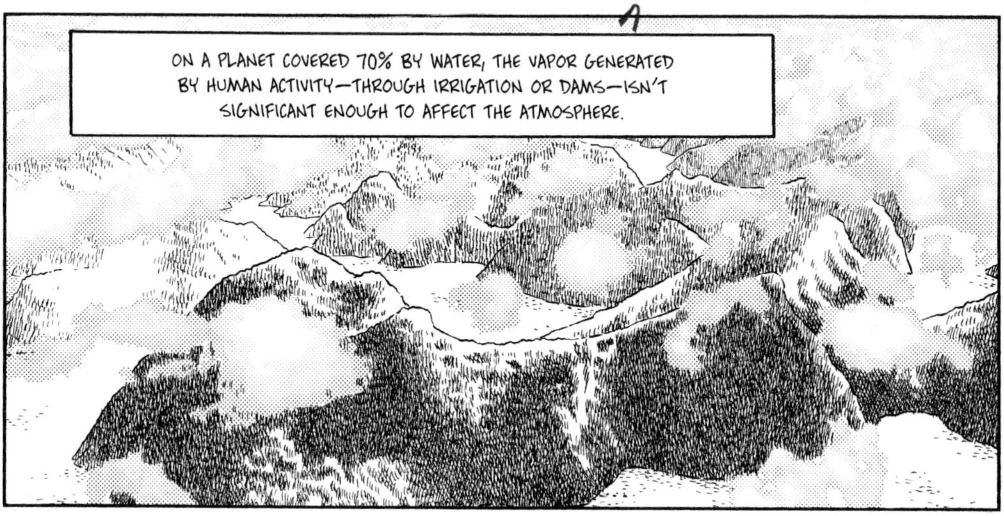
ON A PLANET COVERED 70% BY WATER, THE VAPOR GENERATED BY HUMAN ACTIVITY—THROUGH IRRIGATION OR DAMS—ISN'T SIGNIFICANT ENOUGH TO AFFECT THE ATMOSPHERE.

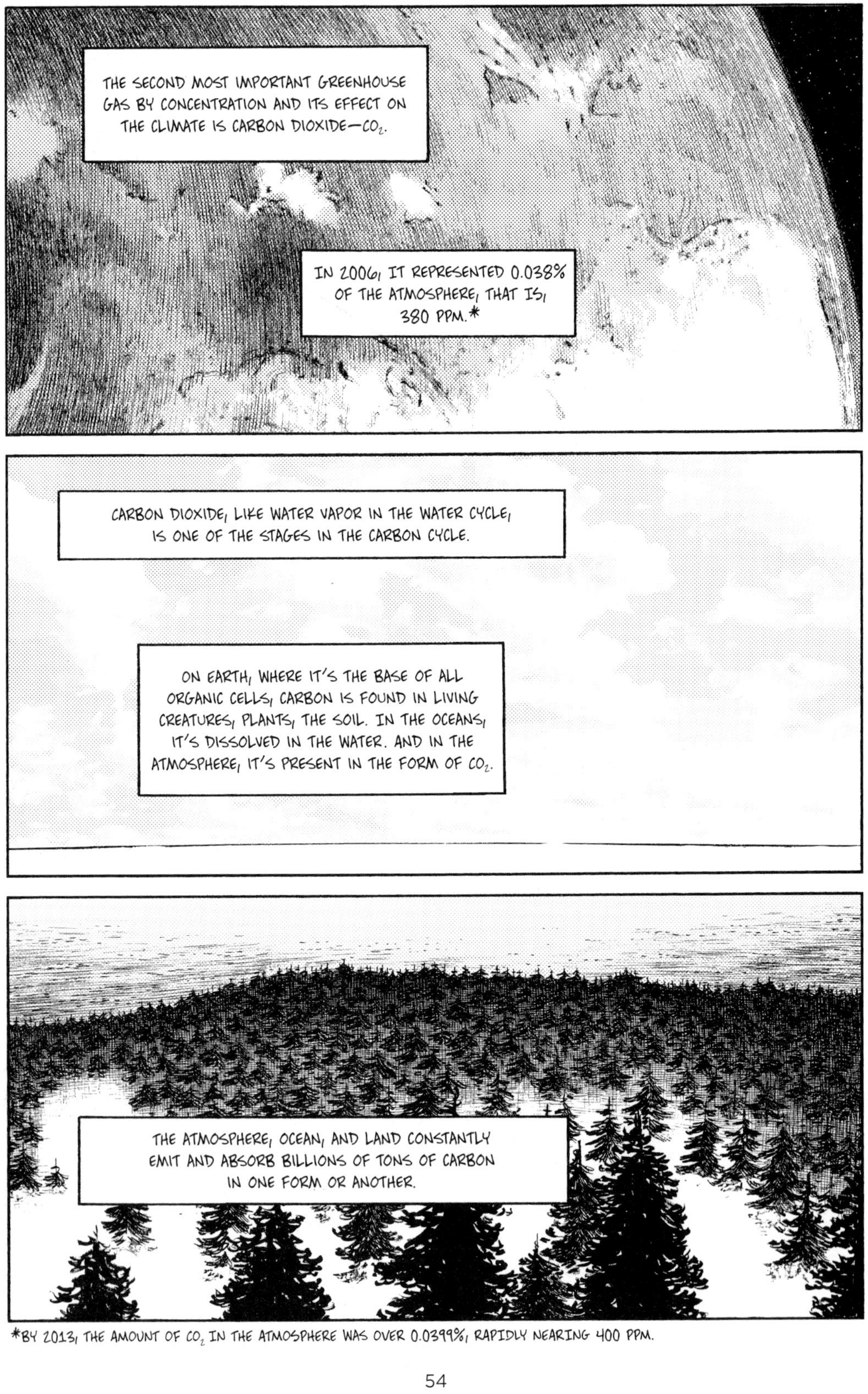
THE SECOND MOST IMPORTANT GREENHOUSE GAS BY CONCENTRATION AND ITS EFFECT ON THE CLIMATE IS CARBON DIOXIDE—CO_2.
IN 2006, IT REPRESENTED 0.038% OF THE ATMOSPHERE, THAT IS, 380 PPM.*
CARBON DIOXIDE, LIKE WATER VAPOR IN THE WATER CYCLE, IS ONE OF THE STAGES IN THE CARBON CYCLE.
ON EARTH, WHERE IT'S THE BASE OF ALL ORGANIC CELLS, CARBON IS FOUND IN LIVING CREATURES, PLANTS, THE SOIL. IN THE OCEANS, IT'S DISSOLVED IN THE WATER. AND IN THE ATMOSPHERE, IT'S PRESENT IN THE FORM OF CO_2.
THE ATMOSPHERE, OCEAN, AND LAND CONSTANTLY EMIT AND ABSORB BILLIONS OF TONS OF CARBON IN ONE FORM OR ANOTHER.

*BY 2013, THE AMOUNT OF CO_2 IN THE ATMOSPHERE WAS OVER 0.0399%, RAPIDLY NEARING 400 PPM.

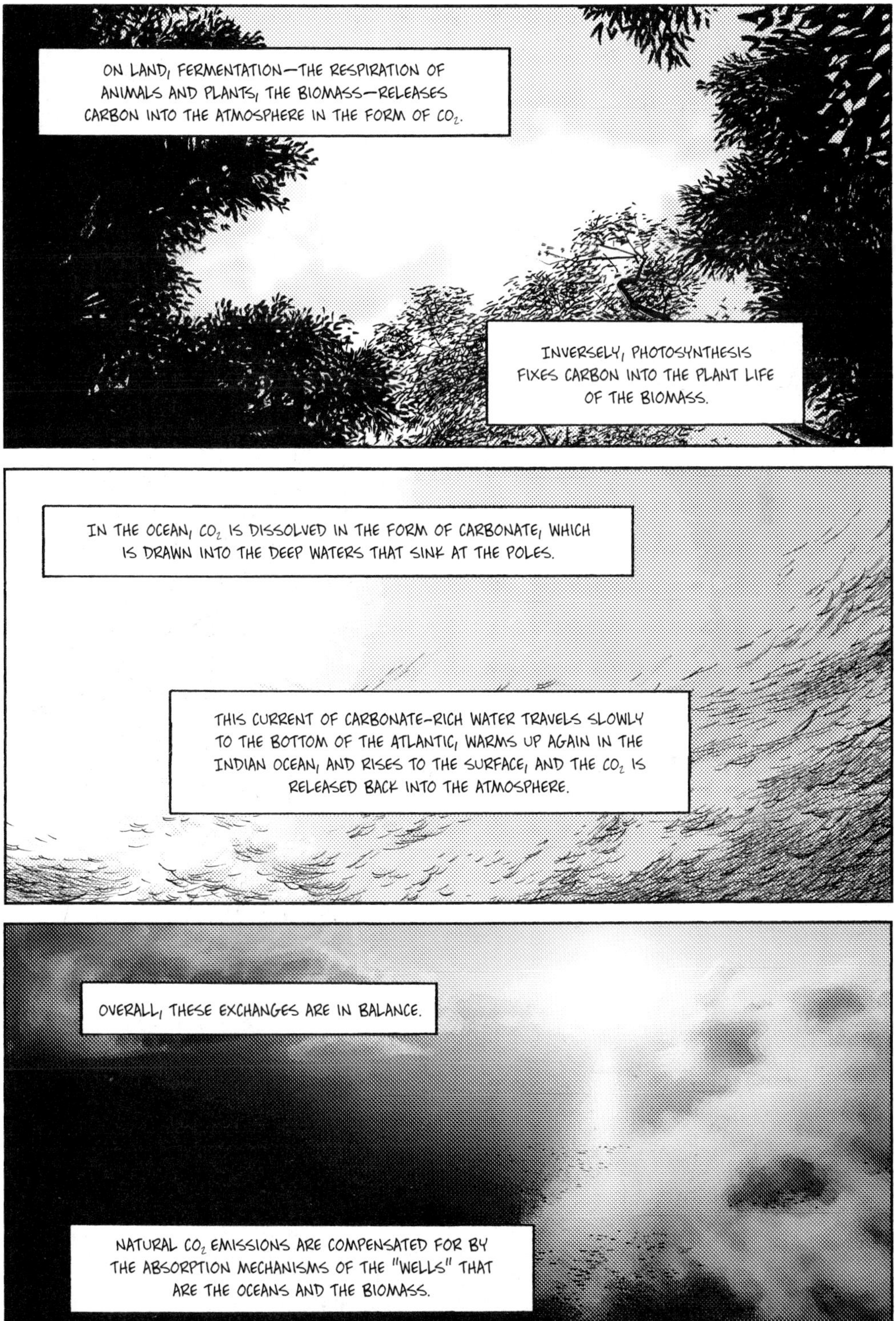
ON LAND, FERMENTATION—THE RESPIRATION OF ANIMALS AND PLANTS, THE BIOMASS—RELEASES CARBON INTO THE ATMOSPHERE IN THE FORM OF CO_2.
INVERSELY, PHOTOSYNTHESIS FIXES CARBON INTO THE PLANT LIFE OF THE BIOMASS.
IN THE OCEAN, CO_2 IS DISSOLVED IN THE FORM OF CARBONATE, WHICH IS DRAWN INTO THE DEEP WATERS THAT SINK AT THE POLES.
THIS CURRENT OF CARBONATE-RICH WATER TRAVELS SLOWLY TO THE BOTTOM OF THE ATLANTIC, WARMS UP AGAIN IN THE INDIAN OCEAN, AND RISES TO THE SURFACE, AND THE CO_2 IS RELEASED BACK INTO THE ATMOSPHERE.
OVERALL, THESE EXCHANGES ARE IN BALANCE.
NATURAL CO_2 EMISSIONS ARE COMPENSATED FOR BY THE ABSORPTION MECHANISMS OF THE "WELLS" THAT ARE THE OCEANS AND THE BIOMASS.

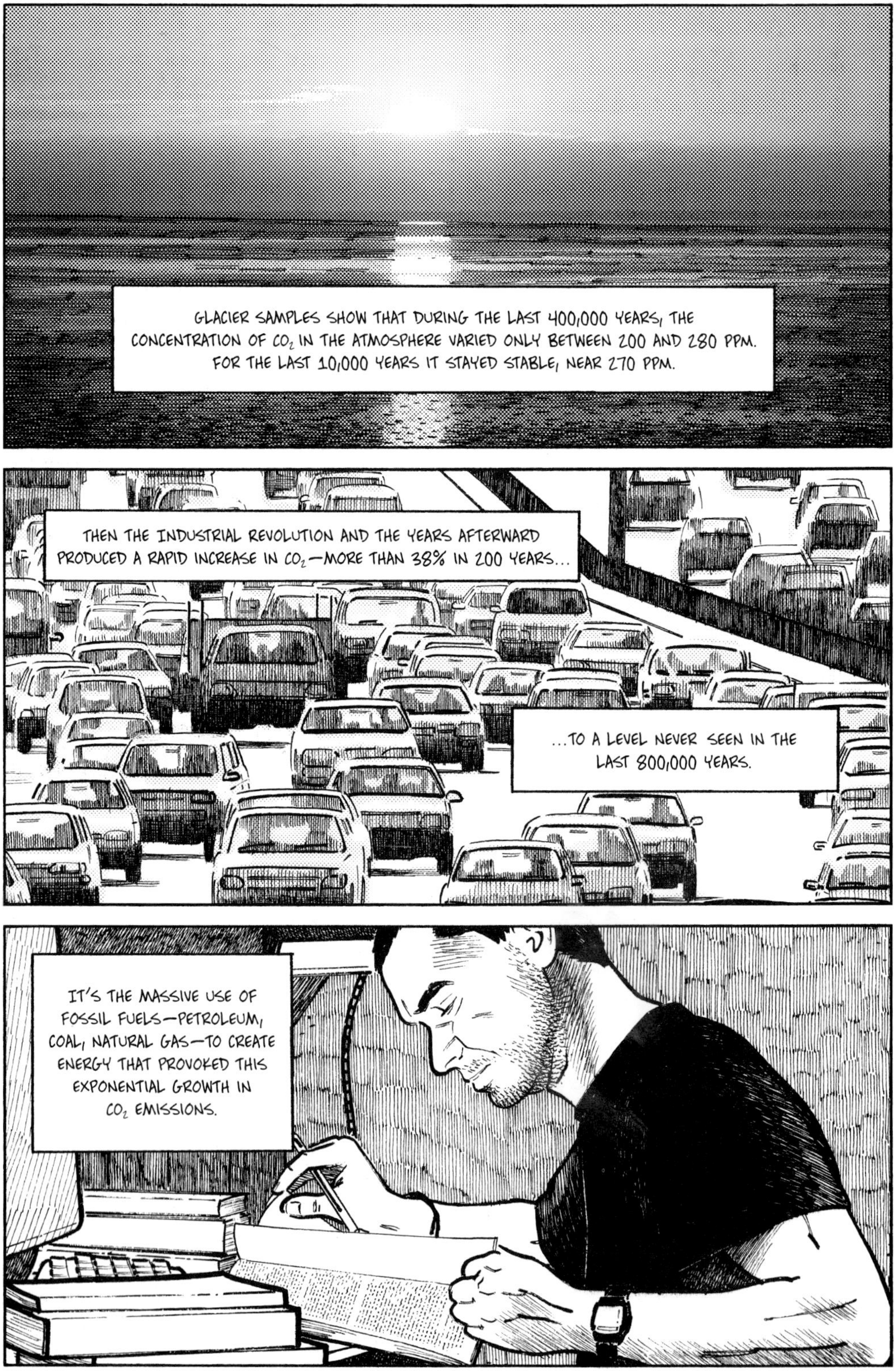
GLACIER SAMPLES SHOW THAT DURING THE LAST 400,000 YEARS, THE CONCENTRATION OF CO_2 IN THE ATMOSPHERE VARIED ONLY BETWEEN 200 AND 280 PPM. FOR THE LAST 10,000 YEARS IT STAYED STABLE, NEAR 270 PPM.
THEN THE INDUSTRIAL REVOLUTION AND THE YEARS AFTERWARD PRODUCED A RAPID INCREASE IN CO_2—MORE THAN 38% IN 200 YEARS...
...TO A LEVEL NEVER SEEN IN THE LAST 800,000 YEARS.
IT'S THE MASSIVE USE OF FOSSIL FUELS—PETROLEUM, COAL, NATURAL GAS—TO CREATE ENERGY THAT PROVOKED THIS EXPONENTIAL GROWTH IN CO_2 EMISSIONS.

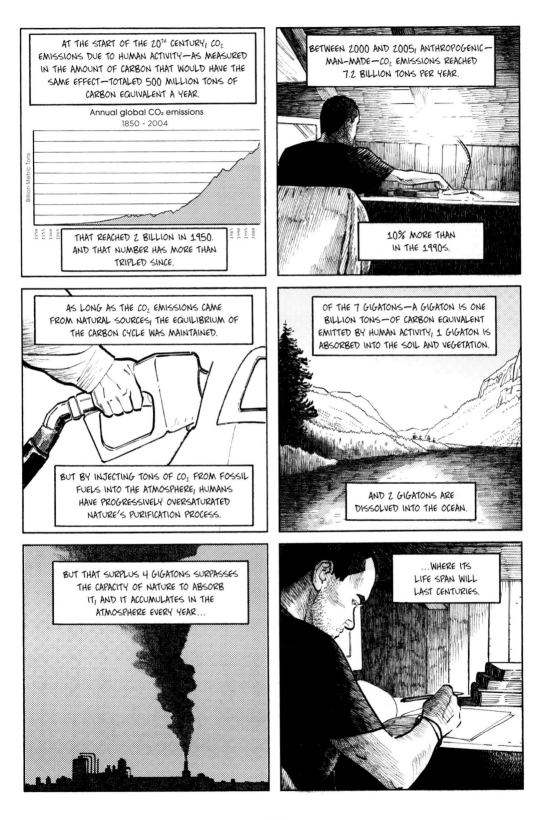

AT THE START OF THE 20TH CENTURY, CO_2 EMISSIONS DUE TO HUMAN ACTIVITY—AS MEASURED IN THE AMOUNT OF CARBON THAT WOULD HAVE THE SAME EFFECT—TOTALED 500 MILLION TONS OF CARBON EQUIVALENT A YEAR.

Annual global CO₂ emissions
1850 - 2004

Billion Metric Tons

THAT REACHED 2 BILLION IN 1950. AND THAT NUMBER HAS MORE THAN TRIPLED SINCE.

BETWEEN 2000 AND 2005, ANTHROPOGENIC—MAN-MADE—CO_2 EMISSIONS REACHED 7.2 BILLION TONS PER YEAR.

10% MORE THAN IN THE 1990S.

AS LONG AS THE CO_2 EMISSIONS CAME FROM NATURAL SOURCES, THE EQUILIBRIUM OF THE CARBON CYCLE WAS MAINTAINED.

BUT BY INJECTING TONS OF CO_2 FROM FOSSIL FUELS INTO THE ATMOSPHERE, HUMANS HAVE PROGRESSIVELY OVERSATURATED NATURE'S PURIFICATION PROCESS.

OF THE 7 GIGATONS—A GIGATON IS ONE BILLION TONS—OF CARBON EQUIVALENT EMITTED BY HUMAN ACTIVITY, 1 GIGATON IS ABSORBED INTO THE SOIL AND VEGETATION.

AND 2 GIGATONS ARE DISSOLVED INTO THE OCEAN.

BUT THAT SURPLUS 4 GIGATONS SURPASSES THE CAPACITY OF NATURE TO ABSORB IT, AND IT ACCUMULATES IN THE ATMOSPHERE EVERY YEAR...

...WHERE ITS LIFE SPAN WILL LAST CENTURIES.

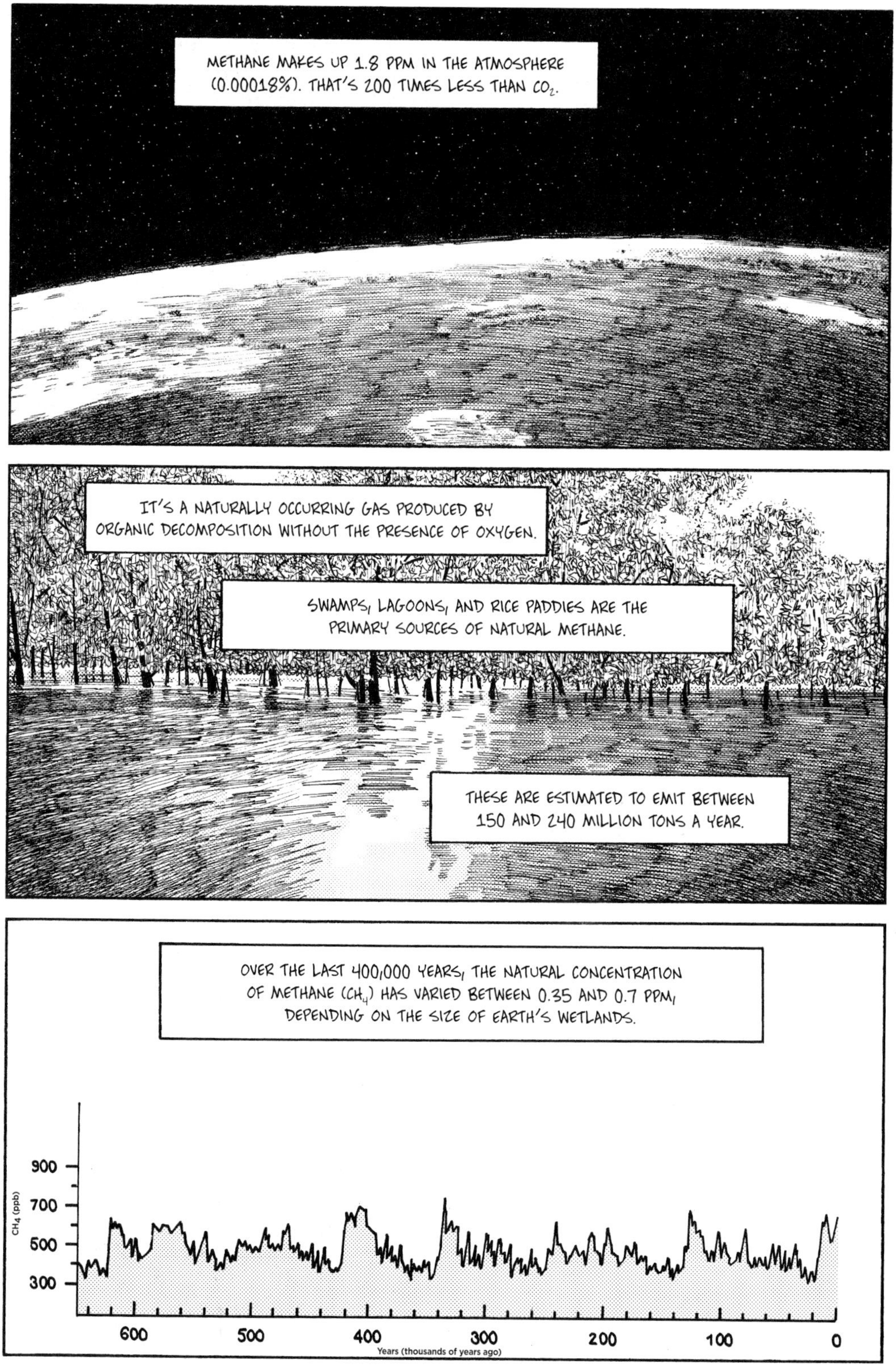
METHANE MAKES UP 1.8 PPM IN THE ATMOSPHERE (0.00018%). THAT'S 200 TIMES LESS THAN CO_2.
IT'S A NATURALLY OCCURRING GAS PRODUCED BY ORGANIC DECOMPOSITION WITHOUT THE PRESENCE OF OXYGEN.
SWAMPS, LAGOONS, AND RICE PADDIES ARE THE PRIMARY SOURCES OF NATURAL METHANE.
THESE ARE ESTIMATED TO EMIT BETWEEN 150 AND 240 MILLION TONS A YEAR.
OVER THE LAST 400,000 YEARS, THE NATURAL CONCENTRATION OF METHANE (CH_4) HAS VARIED BETWEEN 0.35 AND 0.7 PPM, DEPENDING ON THE SIZE OF EARTH'S WETLANDS.
CH_4 (ppb)
900
700
500
300
600
500
400
300
200
100
0
Years (thousands of years ago)

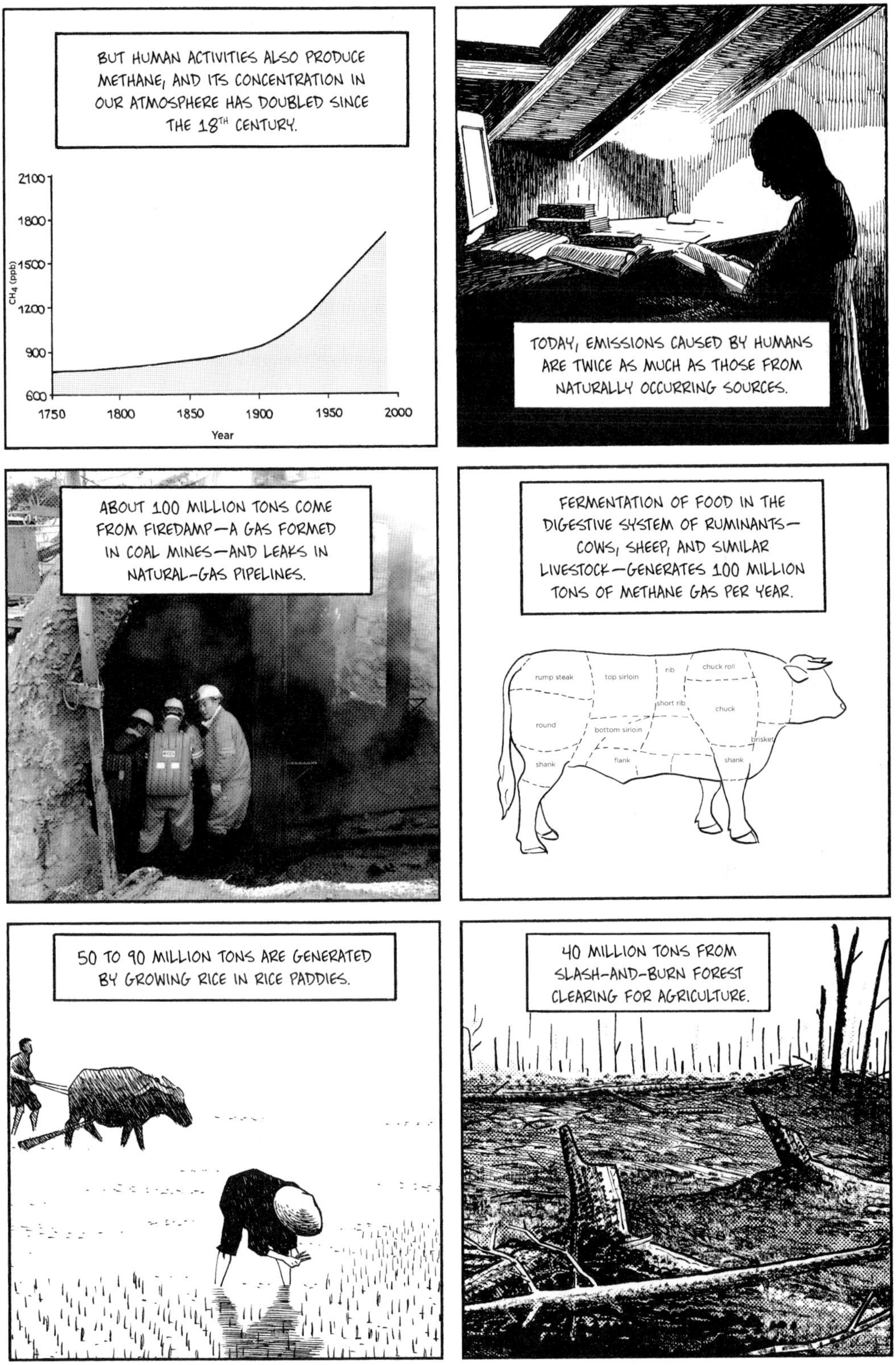
BUT HUMAN ACTIVITIES ALSO PRODUCE METHANE, AND ITS CONCENTRATION IN OUR ATMOSPHERE HAS DOUBLED SINCE THE 18TH CENTURY.
2100
1800
1500
1200
900
600
CH_4 (ppb)
1750
1800
1850
1900
1950
2000
Year
TODAY, EMISSIONS CAUSED BY HUMANS ARE TWICE AS MUCH AS THOSE FROM NATURALLY OCCURRING SOURCES.
ABOUT 100 MILLION TONS COME FROM FIREDAMP—A GAS FORMED IN COAL MINES—AND LEAKS IN NATURAL-GAS PIPELINES.
FERMENTATION OF FOOD IN THE DIGESTIVE SYSTEM OF RUMINANTS—COWS, SHEEP, AND SIMILAR LIVESTOCK—GENERATES 100 MILLION TONS OF METHANE GAS PER YEAR.
rump steak
top sirloin
rib
chuck roll
short rib
chuck
round
bottom sirloin
brisket
shank
flank
shank
50 TO 90 MILLION TONS ARE GENERATED BY GROWING RICE IN RICE PADDIES.
40 MILLION TONS FROM SLASH-AND-BURN FOREST CLEARING FOR AGRICULTURE.

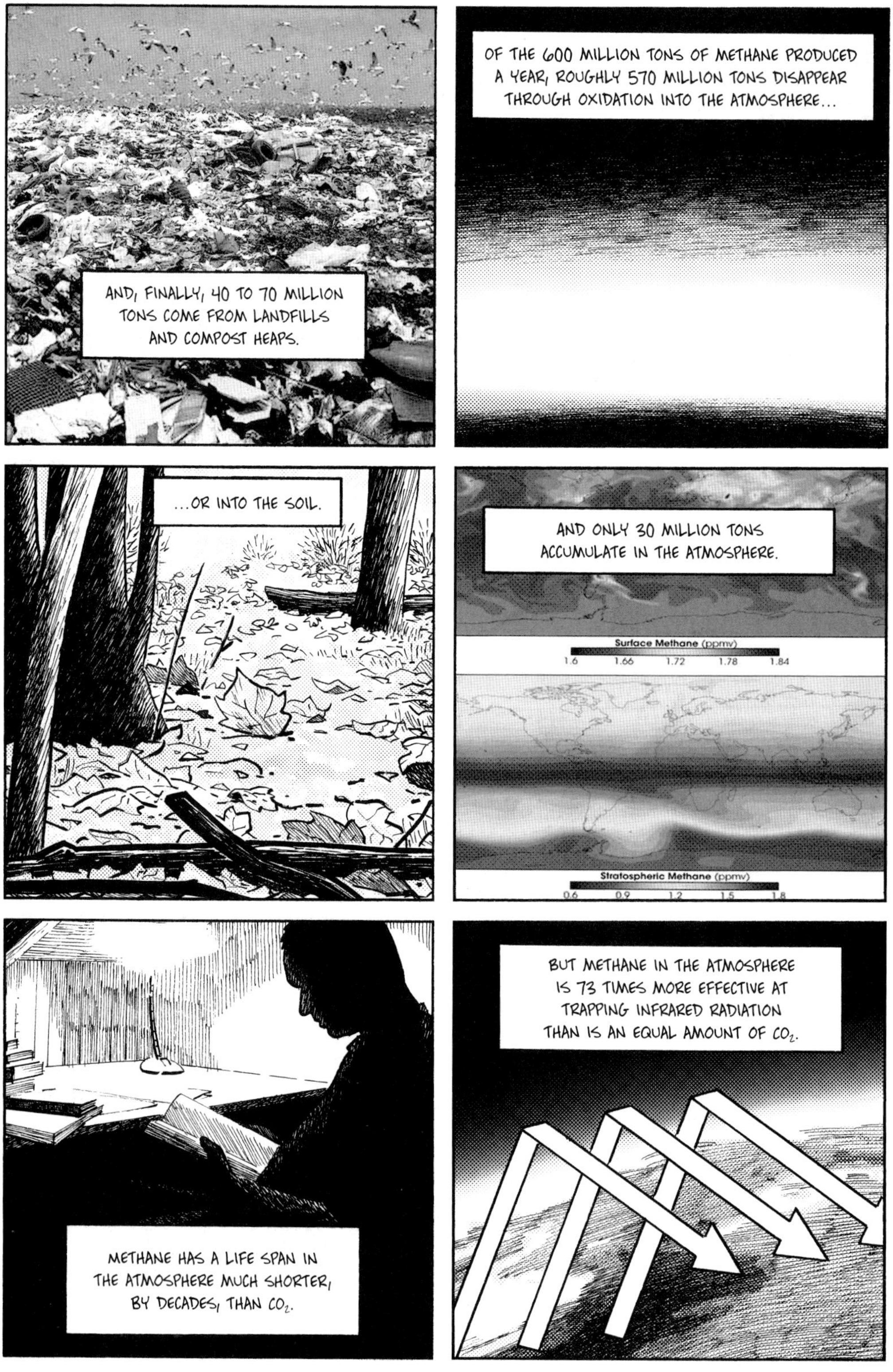
AND, FINALLY, 40 TO 70 MILLION TONS COME FROM LANDFILLS AND COMPOST HEAPS.
OF THE 600 MILLION TONS OF METHANE PRODUCED A YEAR, ROUGHLY 570 MILLION TONS DISAPPEAR THROUGH OXIDATION INTO THE ATMOSPHERE...
...OR INTO THE SOIL.
AND ONLY 30 MILLION TONS ACCUMULATE IN THE ATMOSPHERE.
Surface Methane (ppmv)
1.6 1.66 1.72 1.78 1.84
Stratospheric Methane (ppmv)
0.6 0.9 1.2 1.5 1.8
METHANE HAS A LIFE SPAN IN THE ATMOSPHERE MUCH SHORTER, BY DECADES, THAN CO_2.
BUT METHANE IN THE ATMOSPHERE IS 73 TIMES MORE EFFECTIVE AT TRAPPING INFRARED RADIATION THAN IS AN EQUAL AMOUNT OF CO_2.

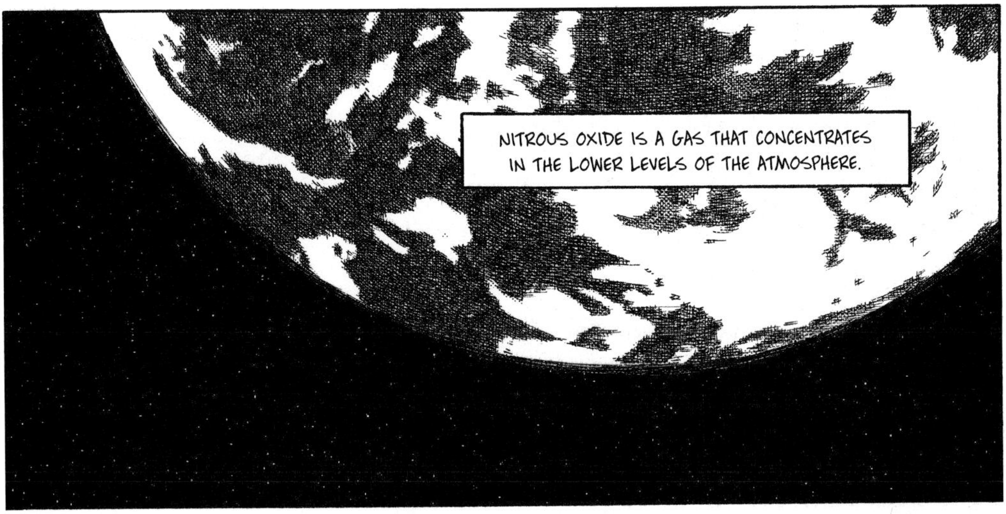
NITROUS OXIDE IS A GAS THAT CONCENTRATES IN THE LOWER LEVELS OF THE ATMOSPHERE.

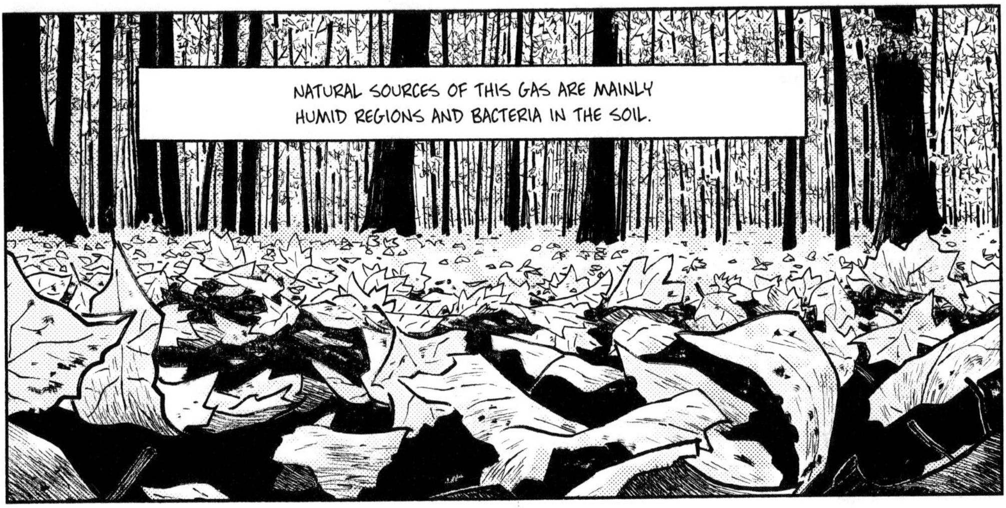
NATURAL SOURCES OF THIS GAS ARE MAINLY HUMID REGIONS AND BACTERIA IN THE SOIL.

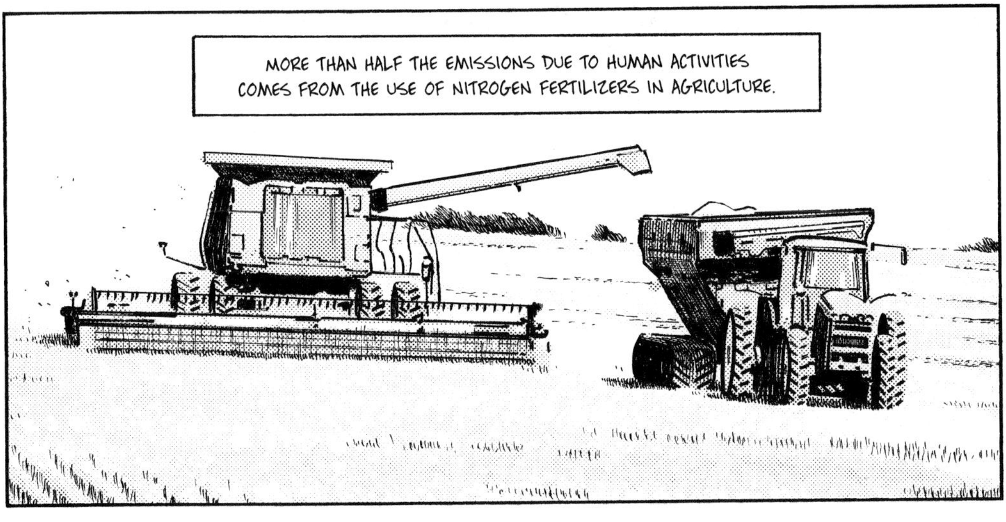
MORE THAN HALF THE EMISSIONS DUE TO HUMAN ACTIVITIES COMES FROM THE USE OF NITROGEN FERTILIZERS IN AGRICULTURE.

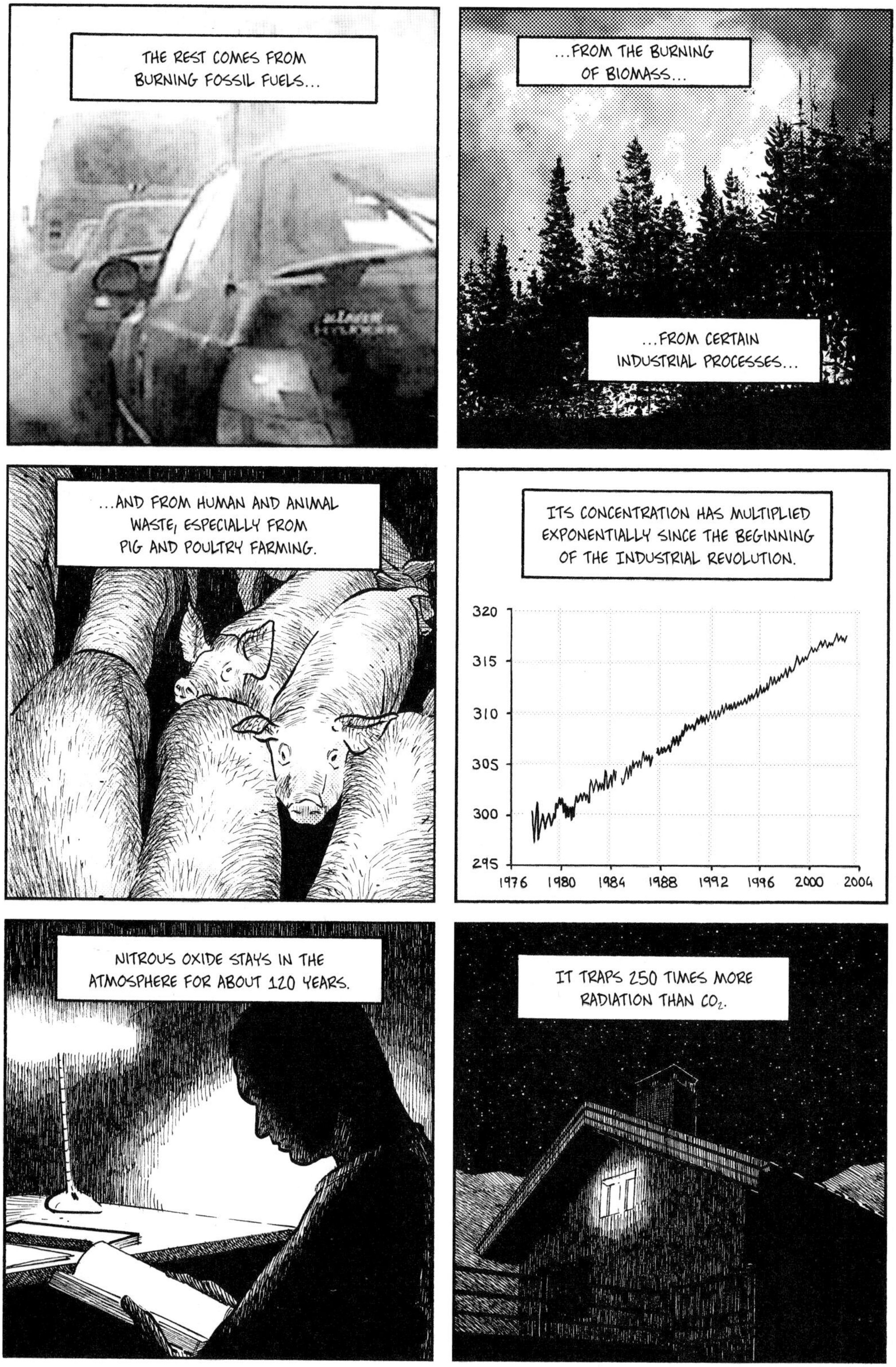
THE REST COMES FROM BURNING FOSSIL FUELS...
...FROM THE BURNING OF BIOMASS...
...FROM CERTAIN INDUSTRIAL PROCESSES...
...AND FROM HUMAN AND ANIMAL WASTE, ESPECIALLY FROM PIG AND POULTRY FARMING.
ITS CONCENTRATION HAS MULTIPLIED EXPONENTIALLY SINCE THE BEGINNING OF THE INDUSTRIAL REVOLUTION.
320
315
310
305
300
295
1976
1980
1984
1988
1992
1996
2000
2004
NITROUS OXIDE STAYS IN THE ATMOSPHERE FOR ABOUT 120 YEARS.
IT TRAPS 250 TIMES MORE RADIATION THAN CO_2.

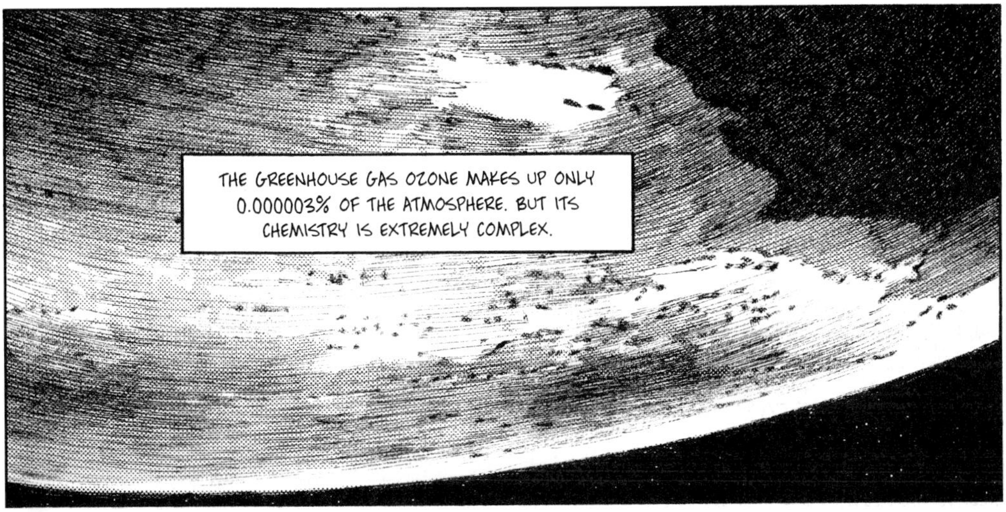
THE GREENHOUSE GAS OZONE MAKES UP ONLY 0.000003% OF THE ATMOSPHERE. BUT ITS CHEMISTRY IS EXTREMELY COMPLEX.

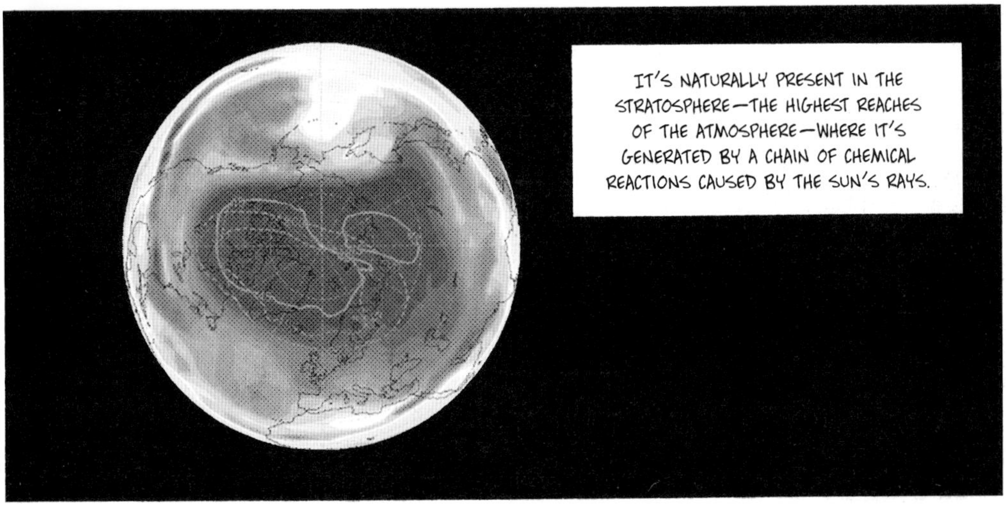
IT'S NATURALLY PRESENT IN THE STRATOSPHERE—THE HIGHEST REACHES OF THE ATMOSPHERE—WHERE IT'S GENERATED BY A CHAIN OF CHEMICAL REACTIONS CAUSED BY THE SUN'S RAYS.

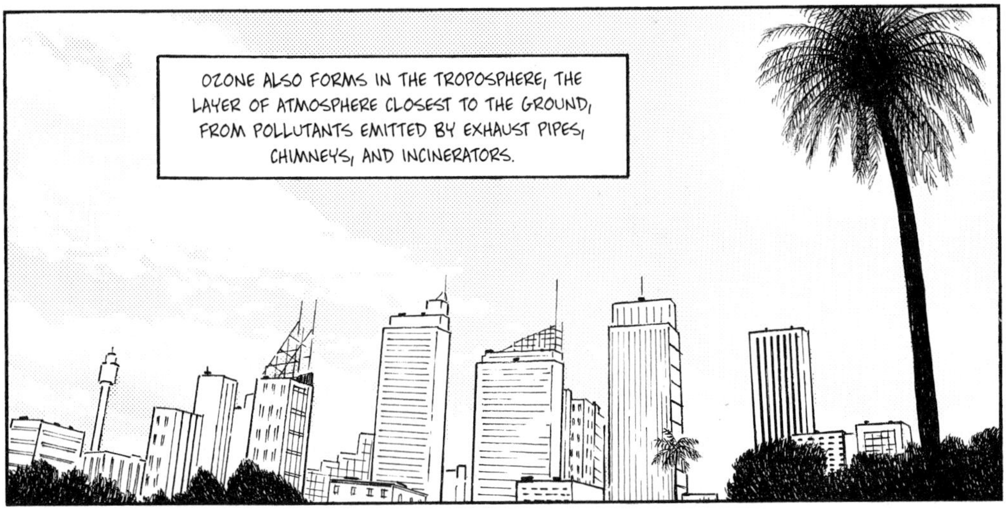
OZONE ALSO FORMS IN THE TROPOSPHERE, THE LAYER OF ATMOSPHERE CLOSEST TO THE GROUND, FROM POLLUTANTS EMITTED BY EXHAUST PIPES, CHIMNEYS, AND INCINERATORS.

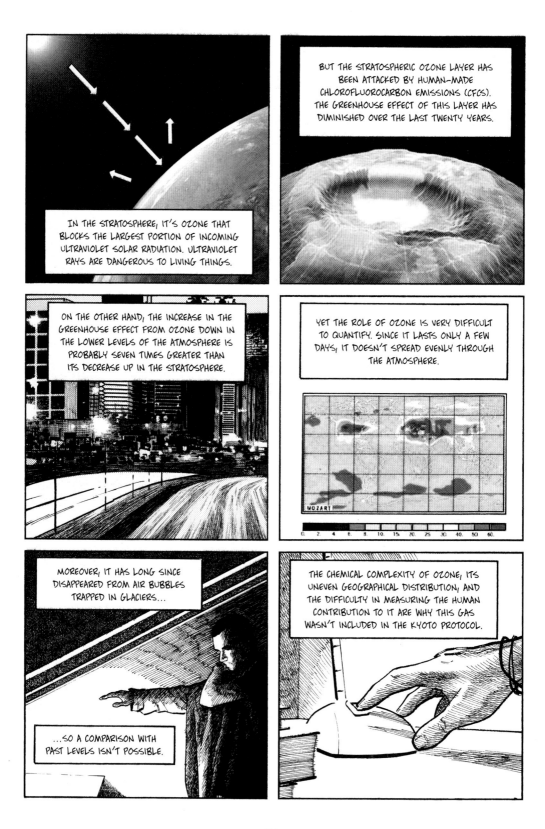

IN THE STRATOSPHERE, IT'S OZONE THAT BLOCKS THE LARGEST PORTION OF INCOMING ULTRAVIOLET SOLAR RADIATION. ULTRAVIOLET RAYS ARE DANGEROUS TO LIVING THINGS.

BUT THE STRATOSPHERIC OZONE LAYER HAS BEEN ATTACKED BY HUMAN-MADE CHLOROFLUOROCARBON EMISSIONS (CFCS). THE GREENHOUSE EFFECT OF THIS LAYER HAS DIMINISHED OVER THE LAST TWENTY YEARS.

ON THE OTHER HAND, THE INCREASE IN THE GREENHOUSE EFFECT FROM OZONE DOWN IN THE LOWER LEVELS OF THE ATMOSPHERE IS PROBABLY SEVEN TIMES GREATER THAN ITS DECREASE UP IN THE STRATOSPHERE.

YET THE ROLE OF OZONE IS VERY DIFFICULT TO QUANTIFY. SINCE IT LASTS ONLY A FEW DAYS, IT DOESN'T SPREAD EVENLY THROUGH THE ATMOSPHERE.

MOREOVER, IT HAS LONG SINCE DISAPPEARED FROM AIR BUBBLES TRAPPED IN GLACIERS...

...SO A COMPARISON WITH PAST LEVELS ISN'T POSSIBLE.

THE CHEMICAL COMPLEXITY OF OZONE, ITS UNEVEN GEOGRAPHICAL DISTRIBUTION, AND THE DIFFICULTY IN MEASURING THE HUMAN CONTRIBUTION TO IT ARE WHY THIS GAS WASN'T INCLUDED IN THE KYOTO PROTOCOL.

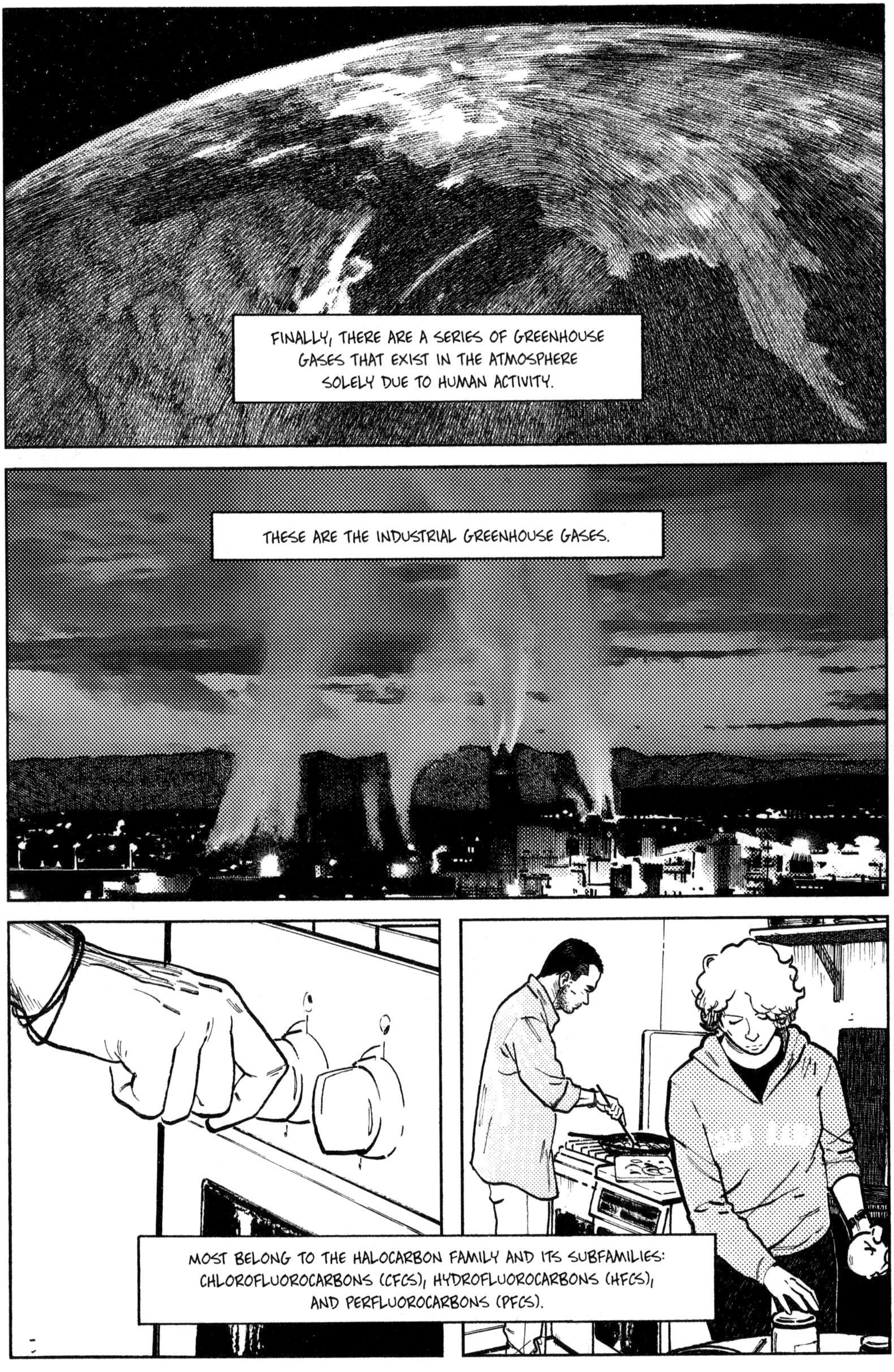
FINALLY, THERE ARE A SERIES OF GREENHOUSE GASES THAT EXIST IN THE ATMOSPHERE SOLELY DUE TO HUMAN ACTIVITY.
THESE ARE THE INDUSTRIAL GREENHOUSE GASES.
MOST BELONG TO THE HALOCARBON FAMILY AND ITS SUBFAMILIES: CHLOROFLUOROCARBONS (CFCS), HYDROFLUOROCARBONS (HFCS), AND PERFLUOROCARBONS (PFCS).

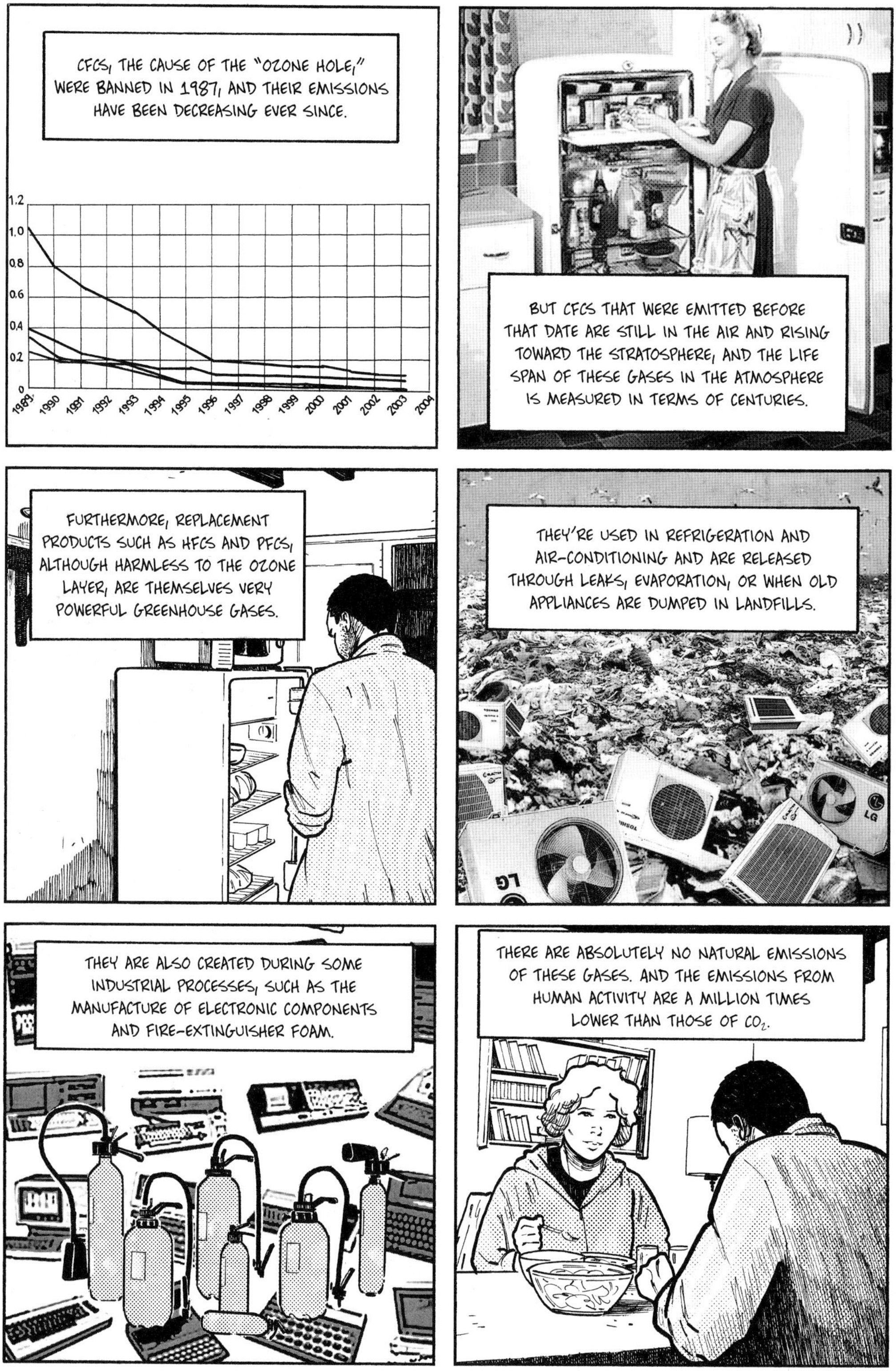
CFCS, THE CAUSE OF THE "OZONE HOLE," WERE BANNED IN 1987, AND THEIR EMISSIONS HAVE BEEN DECREASING EVER SINCE.
1.2
1.0
0.8
0.6
0.4
0.2
0
1989 1990 1991 1992 1993 1994 1995 1996 1997 1998 1999 2000 2001 2002 2003 2004
BUT CFCS THAT WERE EMITTED BEFORE THAT DATE ARE STILL IN THE AIR AND RISING TOWARD THE STRATOSPHERE, AND THE LIFE SPAN OF THESE GASES IN THE ATMOSPHERE IS MEASURED IN TERMS OF CENTURIES.
FURTHERMORE, REPLACEMENT PRODUCTS SUCH AS HFCS AND PFCS, ALTHOUGH HARMLESS TO THE OZONE LAYER, ARE THEMSELVES VERY POWERFUL GREENHOUSE GASES.
THEY'RE USED IN REFRIGERATION AND AIR-CONDITIONING AND ARE RELEASED THROUGH LEAKS, EVAPORATION, OR WHEN OLD APPLIANCES ARE DUMPED IN LANDFILLS.
THEY ARE ALSO CREATED DURING SOME INDUSTRIAL PROCESSES, SUCH AS THE MANUFACTURE OF ELECTRONIC COMPONENTS AND FIRE-EXTINGUISHER FOAM.
THERE ARE ABSOLUTELY NO NATURAL EMISSIONS OF THESE GASES. AND THE EMISSIONS FROM HUMAN ACTIVITY ARE A MILLION TIMES LOWER THAN THOSE OF CO_2.

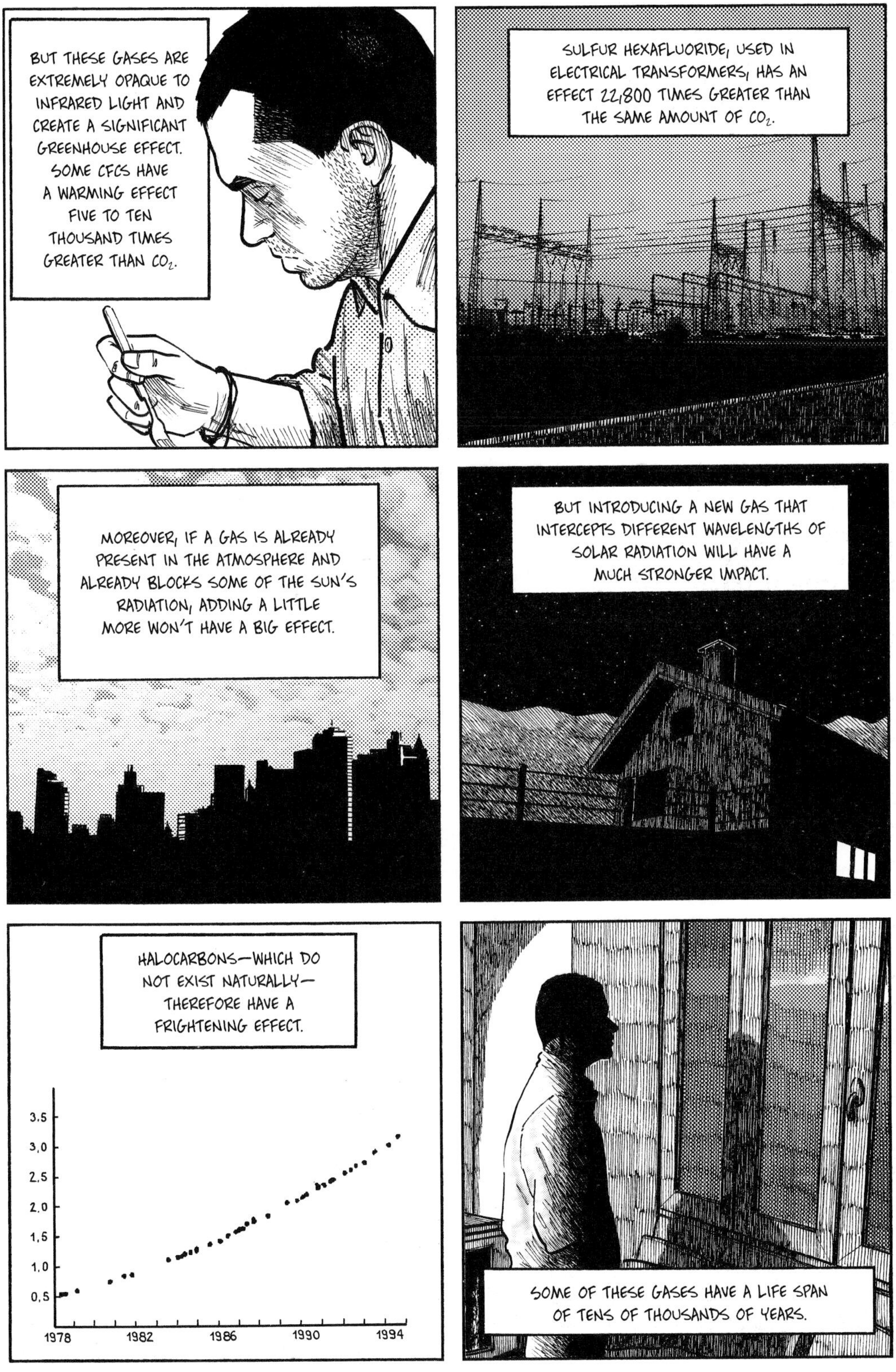
BUT THESE GASES ARE EXTREMELY OPAQUE TO INFRARED LIGHT AND CREATE A SIGNIFICANT GREENHOUSE EFFECT. SOME CFCS HAVE A WARMING EFFECT FIVE TO TEN THOUSAND TIMES GREATER THAN CO_2.
SULFUR HEXAFLUORIDE, USED IN ELECTRICAL TRANSFORMERS, HAS AN EFFECT 22,800 TIMES GREATER THAN THE SAME AMOUNT OF CO_2.
MOREOVER, IF A GAS IS ALREADY PRESENT IN THE ATMOSPHERE AND ALREADY BLOCKS SOME OF THE SUN'S RADIATION, ADDING A LITTLE MORE WON'T HAVE A BIG EFFECT.
BUT INTRODUCING A NEW GAS THAT INTERCEPTS DIFFERENT WAVELENGTHS OF SOLAR RADIATION WILL HAVE A MUCH STRONGER IMPACT.
HALOCARBONS—WHICH DO NOT EXIST NATURALLY—THEREFORE HAVE A FRIGHTENING EFFECT.
3,5
3,0
2,5
2,0
1,5
1,0
0,5
1978
1982
1986
1990
1994
SOME OF THESE GASES HAVE A LIFE SPAN OF TENS OF THOUSANDS OF YEARS.

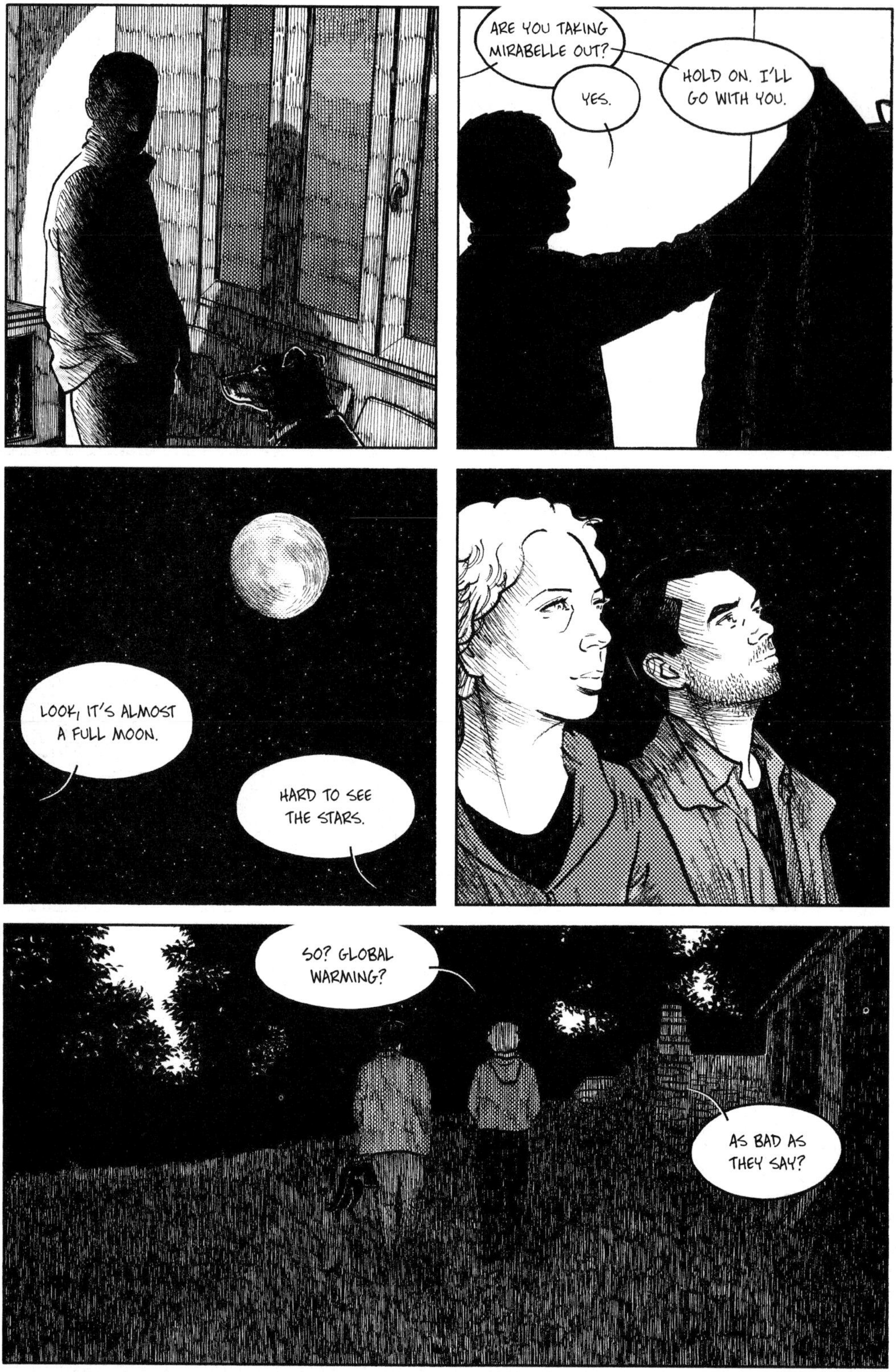
ARE YOU TAKING MIRABELLE OUT?
YES.
HOLD ON. I'LL GO WITH YOU.
LOOK, IT'S ALMOST A FULL MOON.
HARD TO SEE THE STARS.
SO? GLOBAL WARMING?
AS BAD AS THEY SAY?

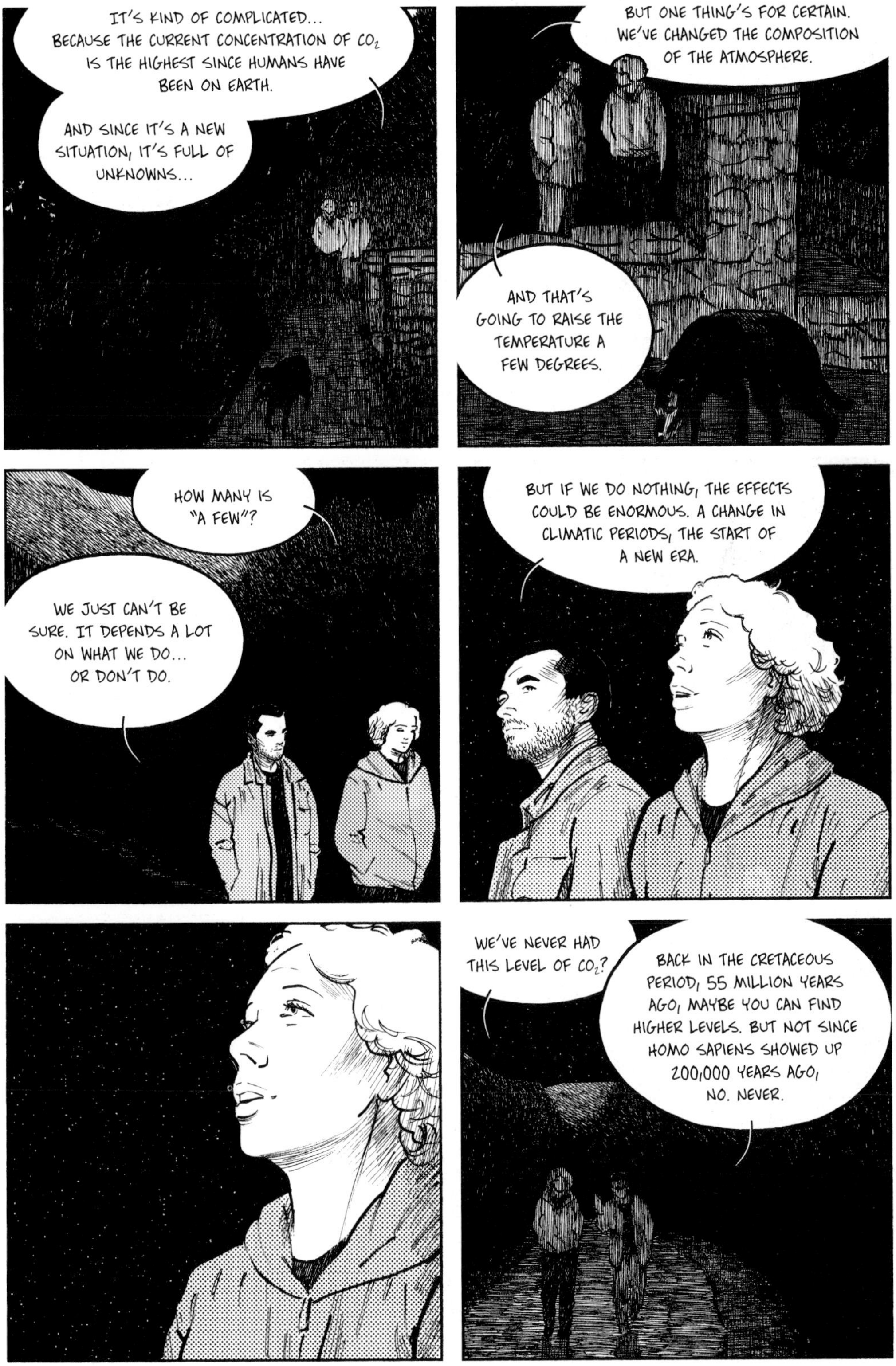
IT'S KIND OF COMPLICATED... BECAUSE THE CURRENT CONCENTRATION OF CO_2 IS THE HIGHEST SINCE HUMANS HAVE BEEN ON EARTH.
AND SINCE IT'S A NEW SITUATION, IT'S FULL OF UNKNOWNS...
BUT ONE THING'S FOR CERTAIN. WE'VE CHANGED THE COMPOSITION OF THE ATMOSPHERE.
AND THAT'S GOING TO RAISE THE TEMPERATURE A FEW DEGREES.
HOW MANY IS "A FEW"?
WE JUST CAN'T BE SURE. IT DEPENDS A LOT ON WHAT WE DO... OR DON'T DO.
BUT IF WE DO NOTHING, THE EFFECTS COULD BE ENORMOUS. A CHANGE IN CLIMATIC PERIODS, THE START OF A NEW ERA.
WE'VE NEVER HAD THIS LEVEL OF CO_2?
BACK IN THE CRETACEOUS PERIOD, 55 MILLION YEARS AGO, MAYBE YOU CAN FIND HIGHER LEVELS. BUT NOT SINCE HOMO SAPIENS SHOWED UP 200,000 YEARS AGO, NO. NEVER.

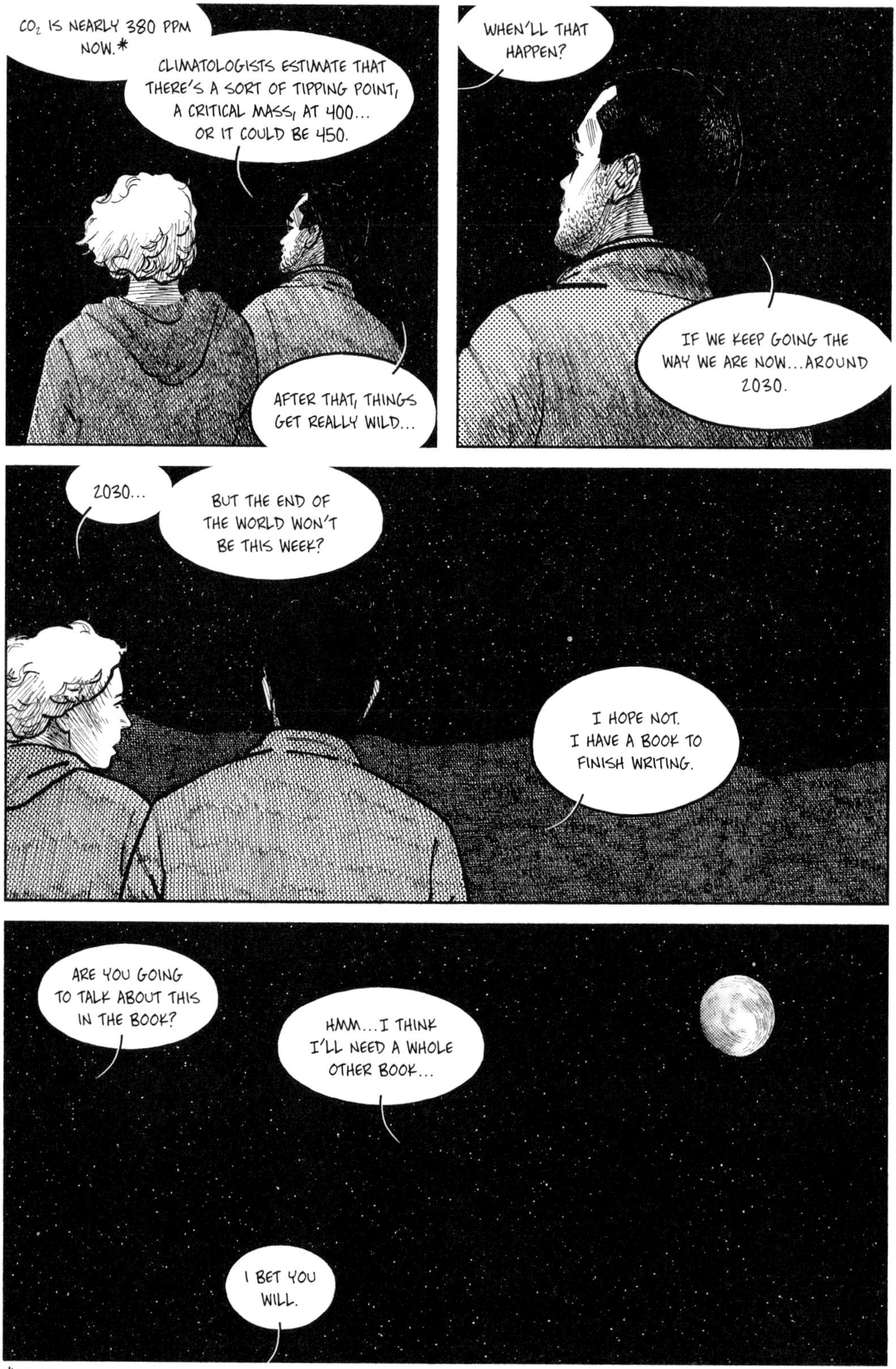
CO2 IS NEARLY 380 PPM NOW.*
CLIMATOLOGISTS ESTIMATE THAT THERE'S A SORT OF TIPPING POINT, A CRITICAL MASS, AT 400... OR IT COULD BE 450.
AFTER THAT, THINGS GET REALLY WILD...
WHEN'LL THAT HAPPEN?
IF WE KEEP GOING THE WAY WE ARE NOW... AROUND 2030.
2030...
BUT THE END OF THE WORLD WON'T BE THIS WEEK?
I HOPE NOT. I HAVE A BOOK TO FINISH WRITING.
ARE YOU GOING TO TALK ABOUT THIS IN THE BOOK?
HMM... I THINK I'LL NEED A WHOLE OTHER BOOK...
I BET YOU WILL.

*IN 2006.

THERE ARE MANY WAYS TO START...

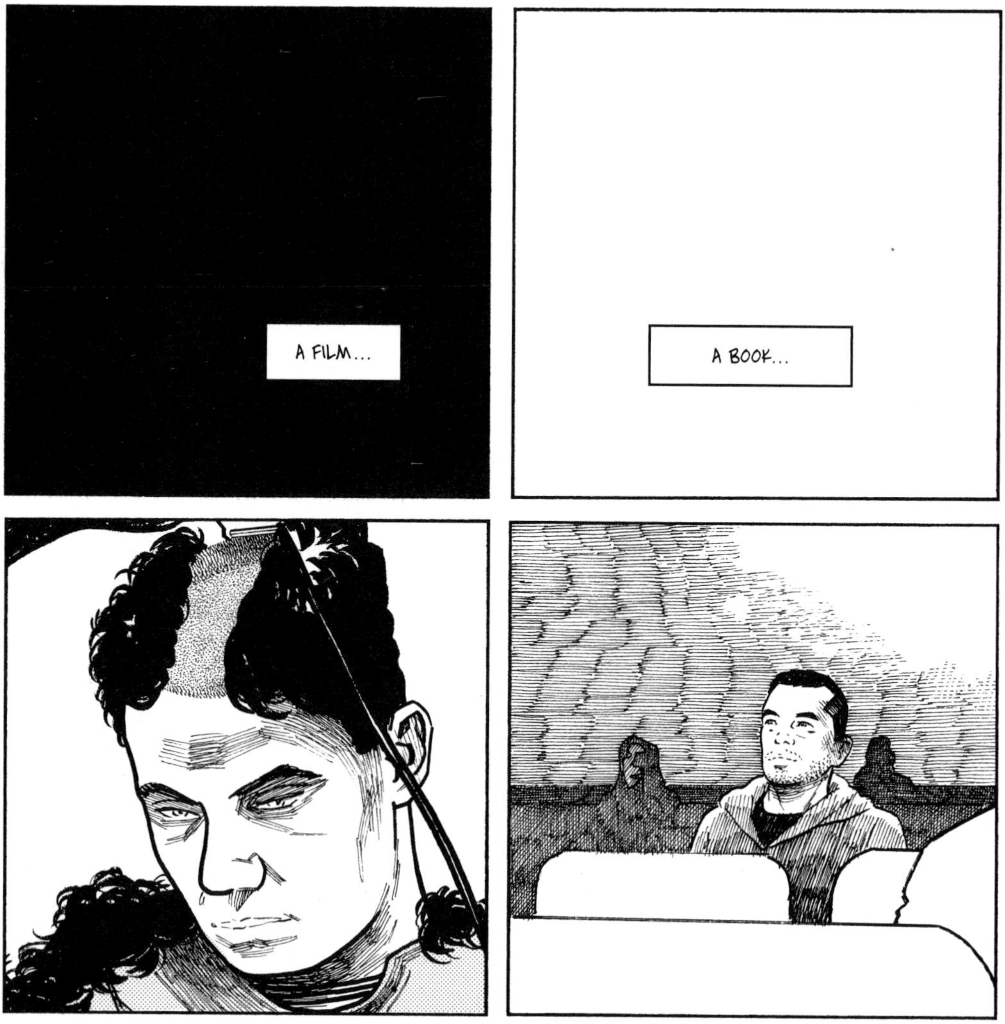
A FILM...
A BOOK...

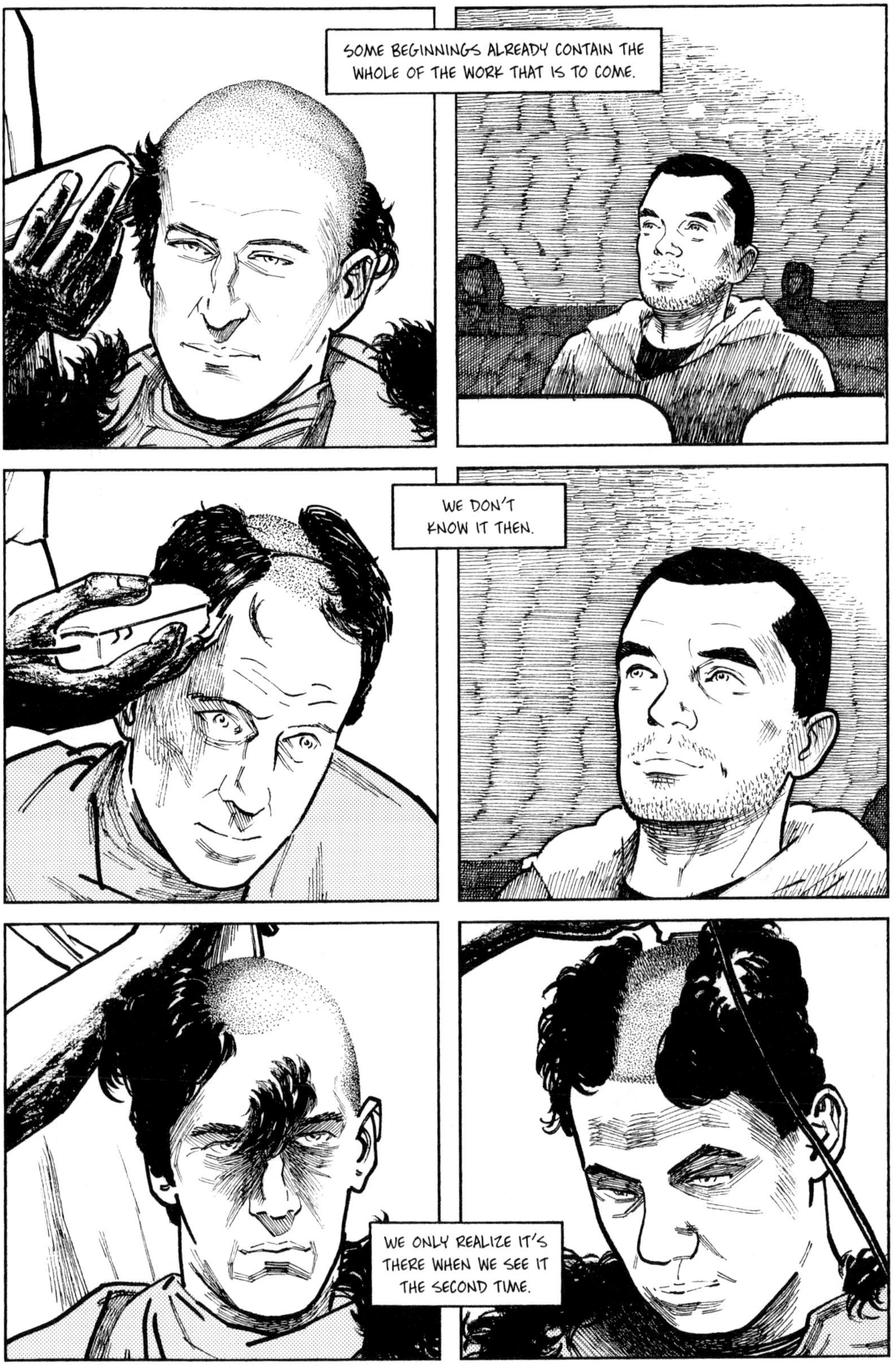
SOME BEGINNINGS ALREADY CONTAIN THE WHOLE OF THE WORK THAT IS TO COME.
WE DON'T KNOW IT THEN.
WE ONLY REALIZE IT'S THERE WHEN WE SEE IT THE SECOND TIME.

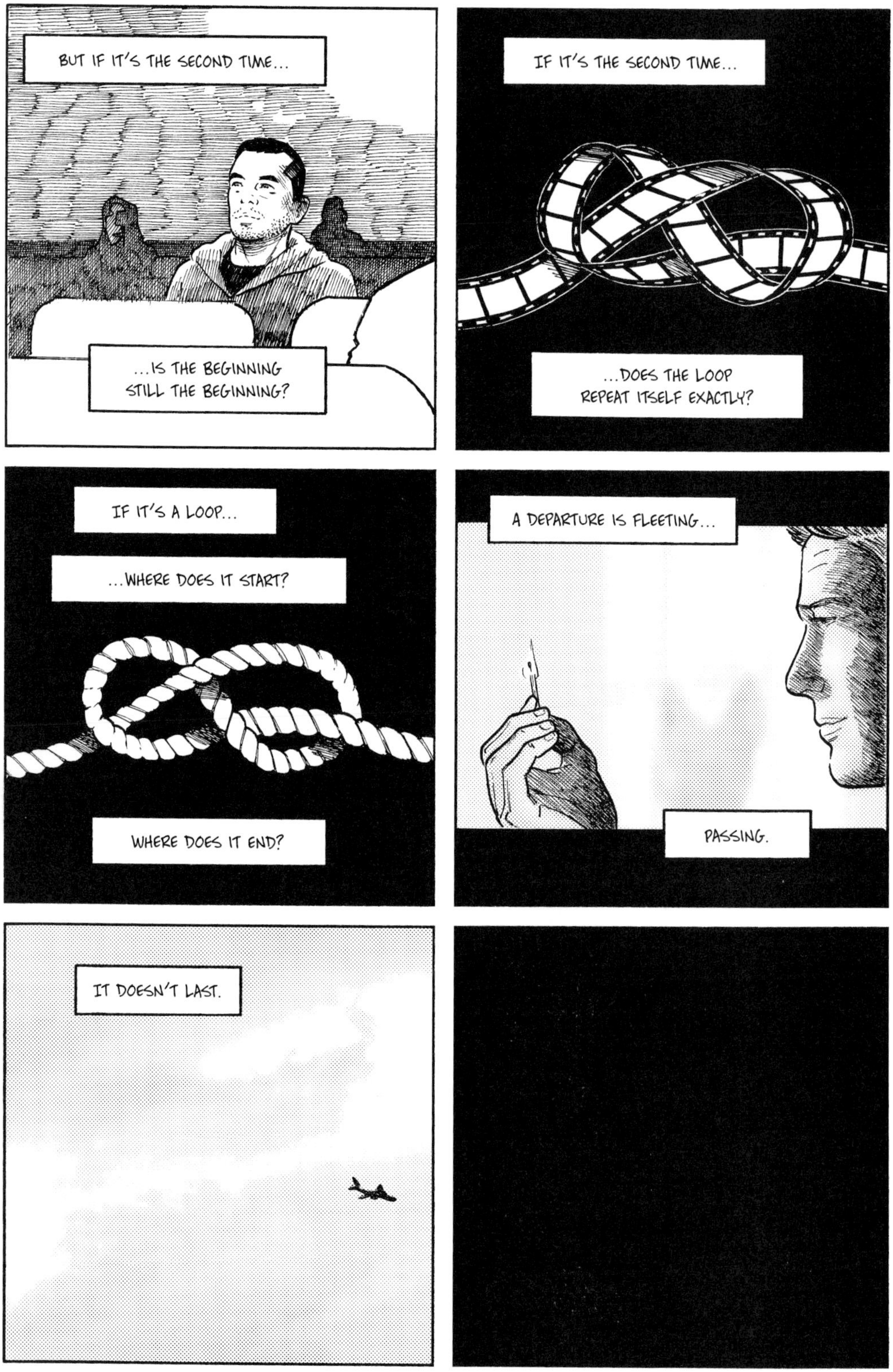
BUT IF IT'S THE SECOND TIME...
...IS THE BEGINNING STILL THE BEGINNING?
IF IT'S THE SECOND TIME...
...DOES THE LOOP REPEAT ITSELF EXACTLY?
IF IT'S A LOOP...
...WHERE DOES IT START?
WHERE DOES IT END?
A DEPARTURE IS FLEETING...
PASSING.
IT DOESN'T LAST.

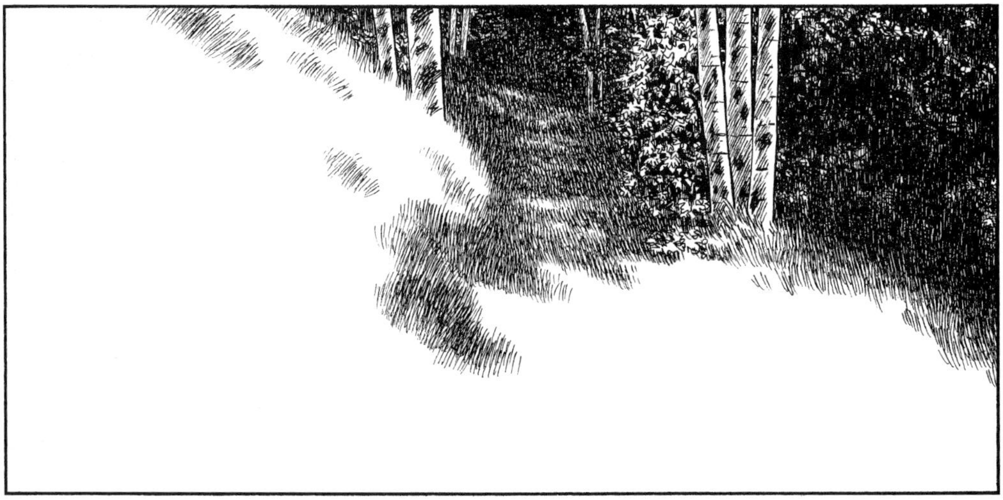

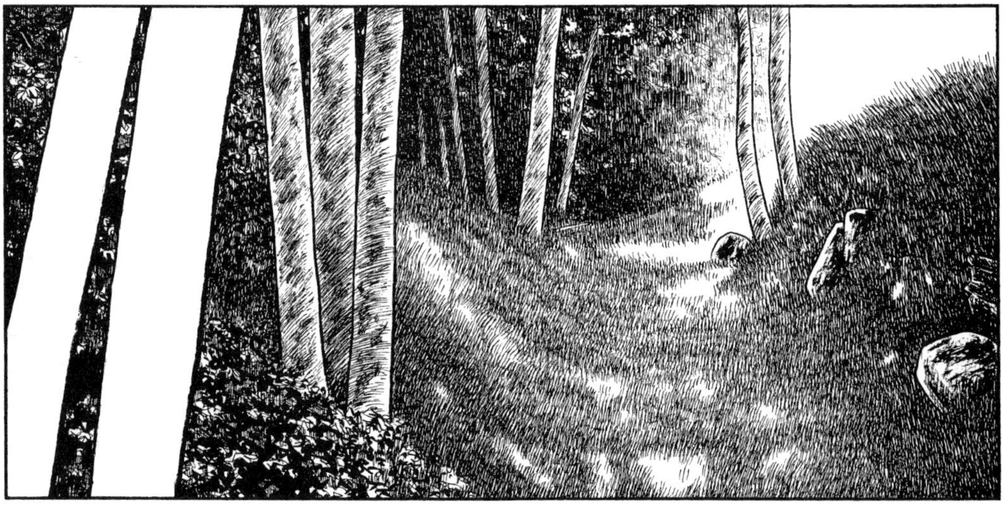

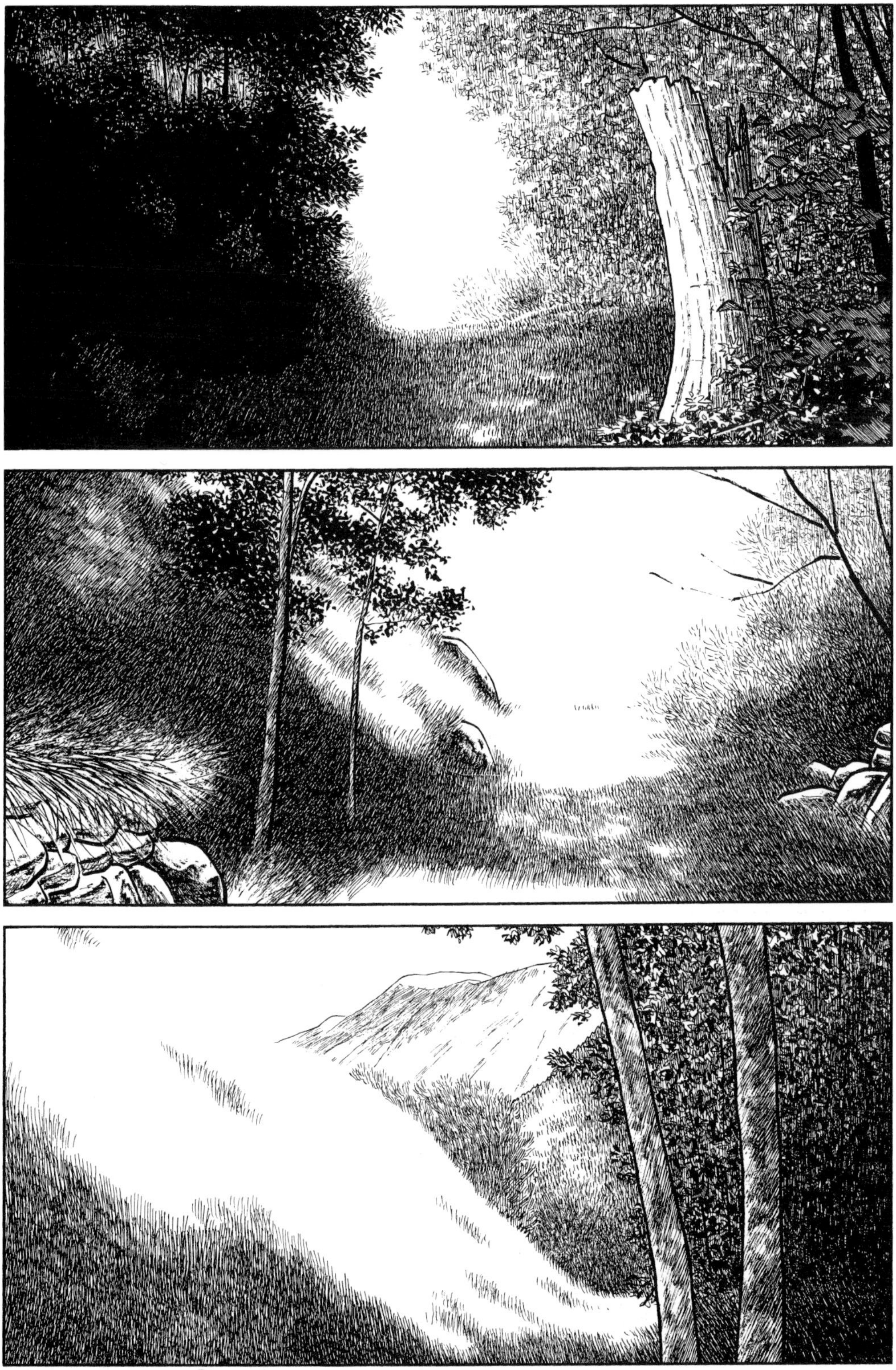

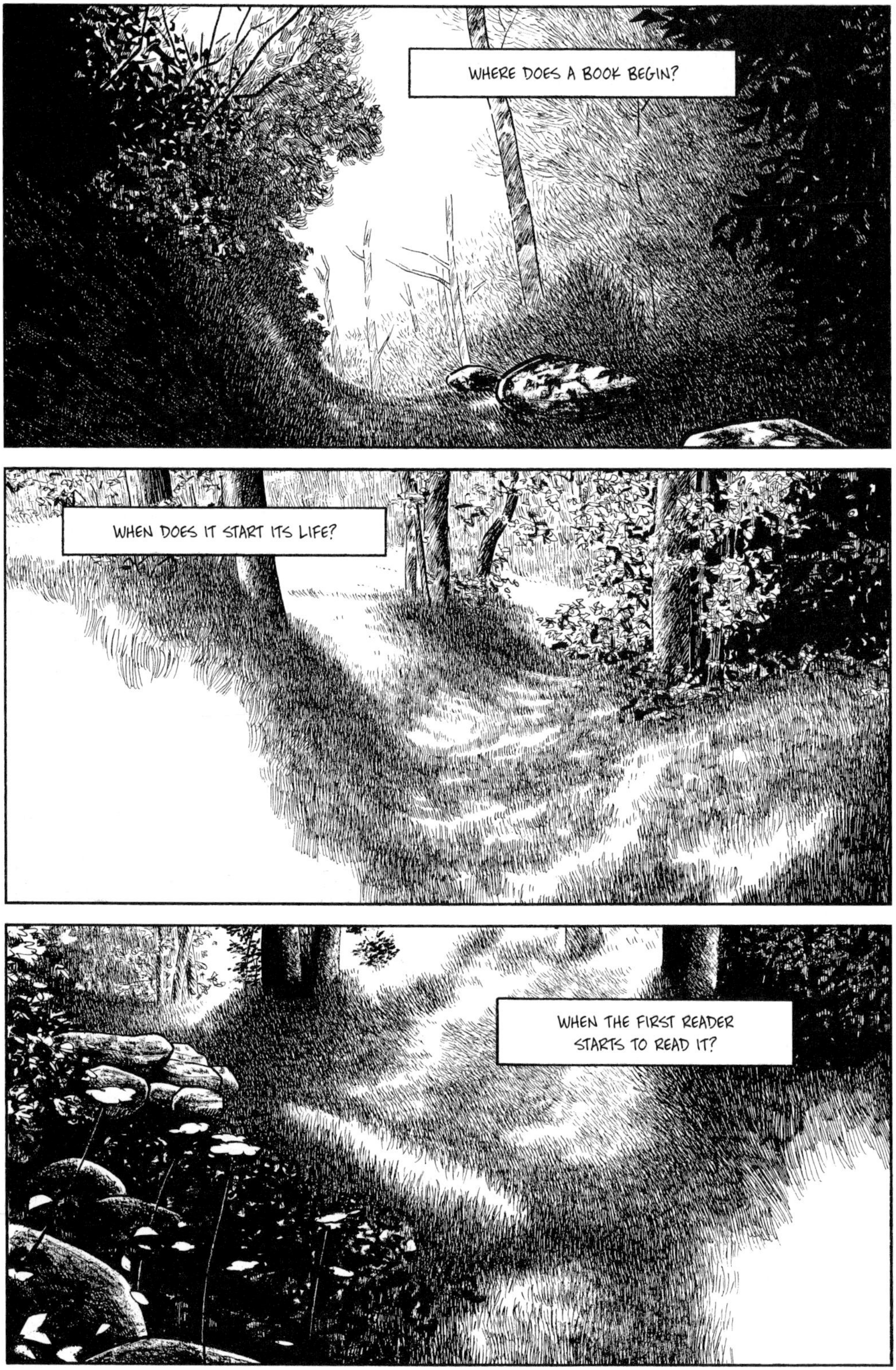
WHERE DOES A BOOK BEGIN?
WHEN DOES IT START ITS LIFE?
WHEN THE FIRST READER STARTS TO READ IT?

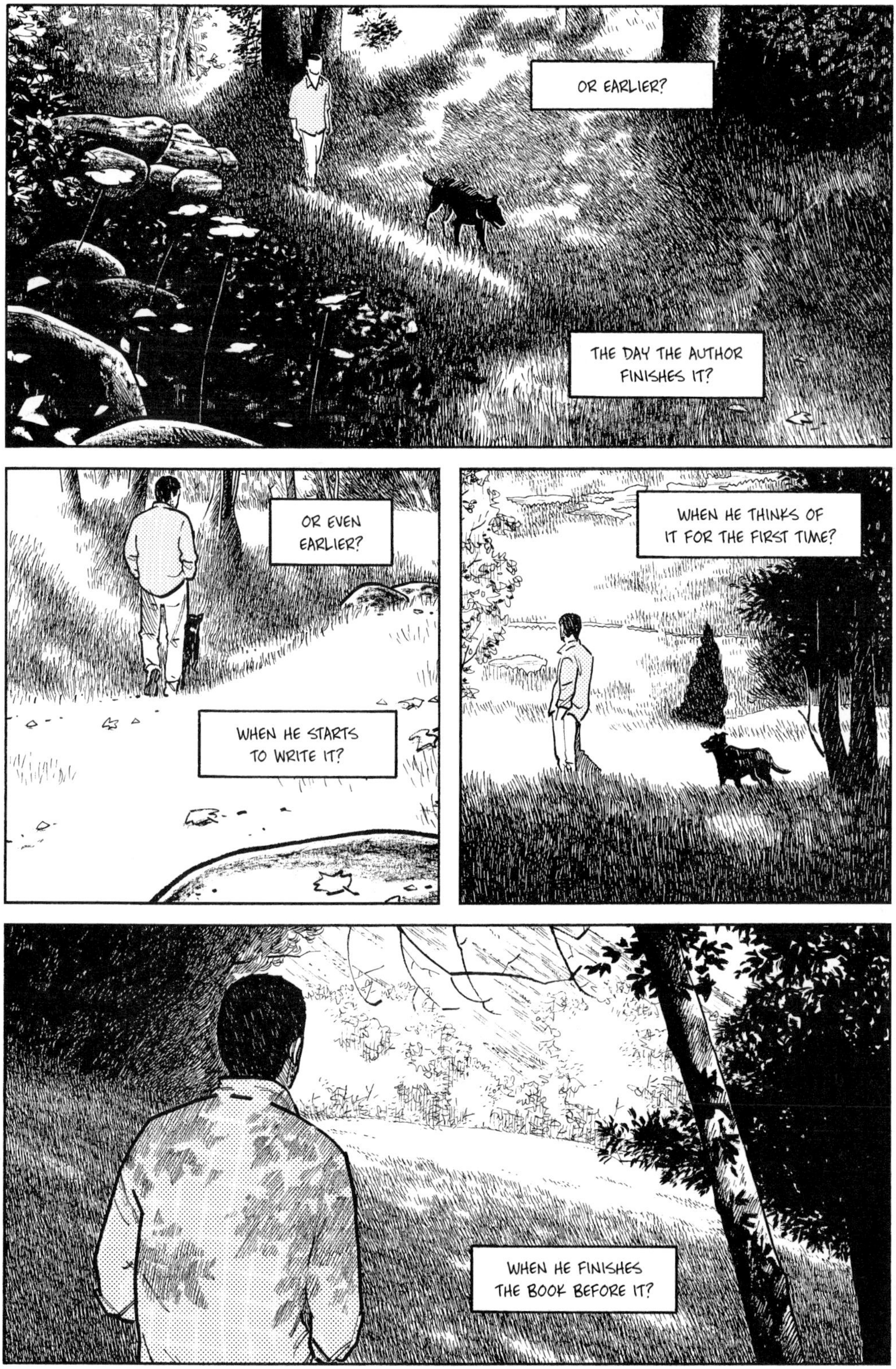
OR EARLIER?
THE DAY THE AUTHOR FINISHES IT?
OR EVEN EARLIER?
WHEN HE STARTS TO WRITE IT?
WHEN HE THINKS OF IT FOR THE FIRST TIME?
WHEN HE FINISHES THE BOOK BEFORE IT?

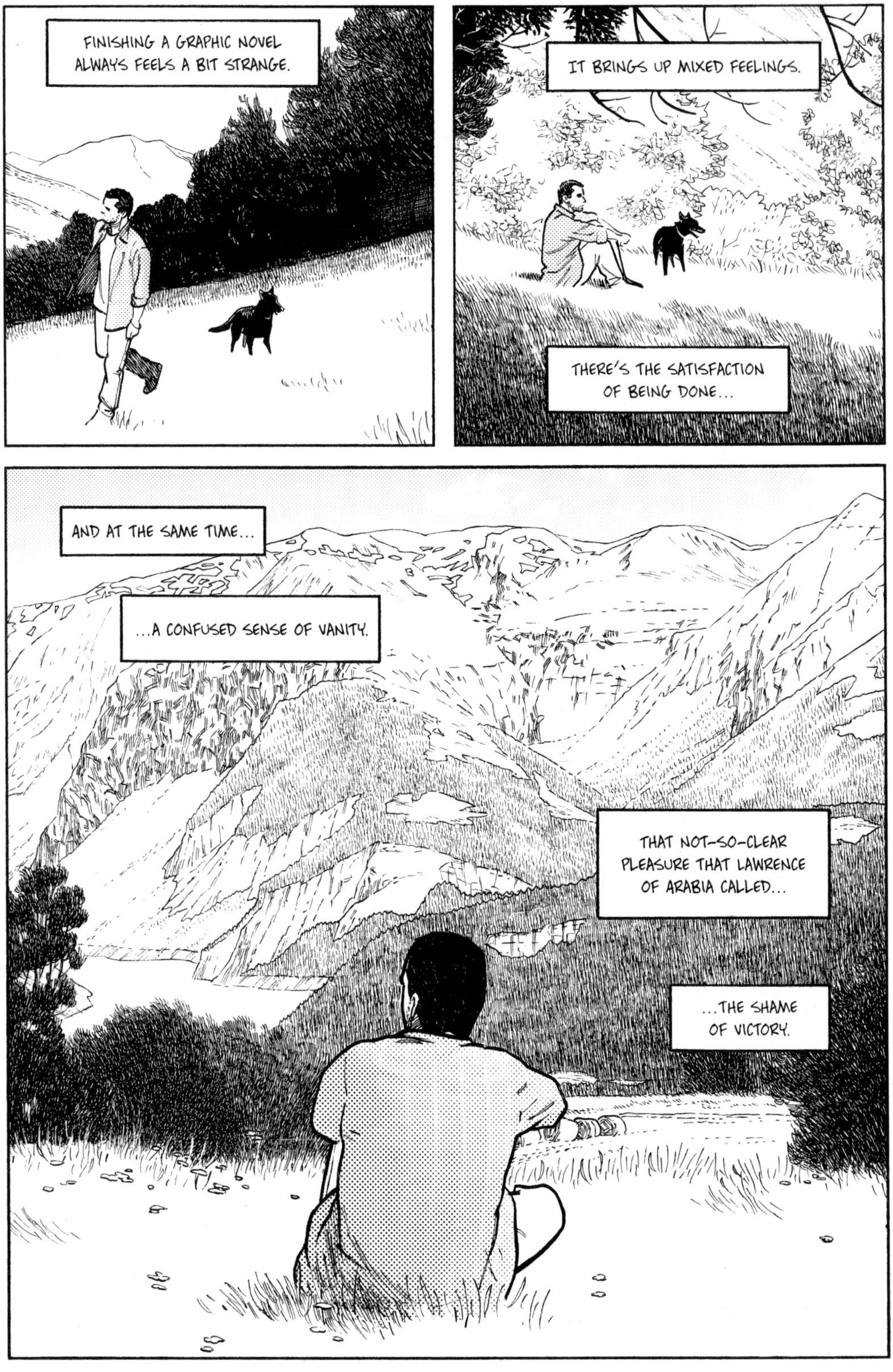
FINISHING A GRAPHIC NOVEL ALWAYS FEELS A BIT STRANGE.
IT BRINGS UP MIXED FEELINGS.
THERE'S THE SATISFACTION OF BEING DONE...
AND AT THE SAME TIME...
...A CONFUSED SENSE OF VANITY.
THAT NOT-SO-CLEAR PLEASURE THAT LAWRENCE OF ARABIA CALLED...
...THE SHAME OF VICTORY.

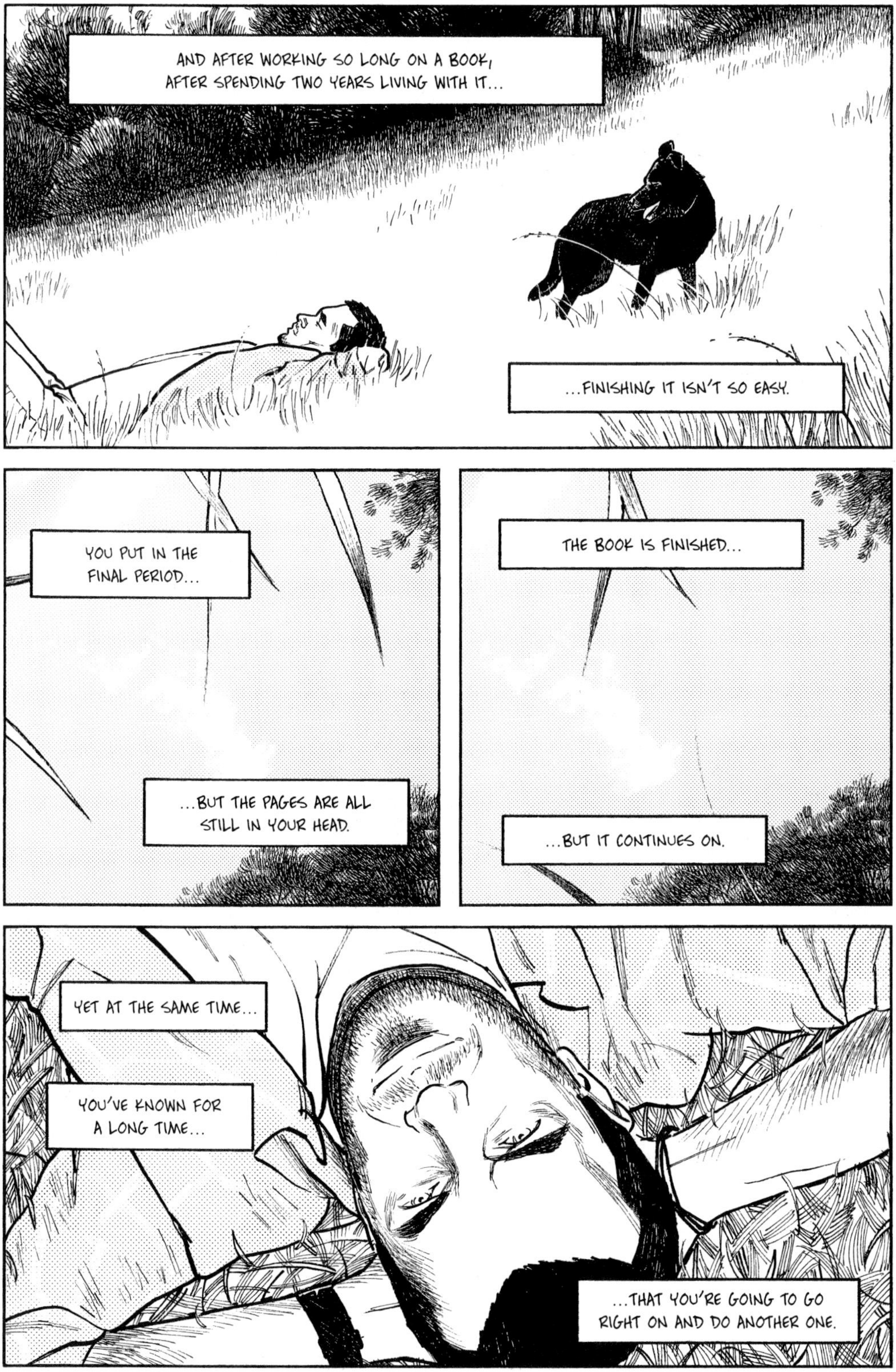
AND AFTER WORKING SO LONG ON A BOOK, AFTER SPENDING TWO YEARS LIVING WITH IT...
...FINISHING IT ISN'T SO EASY.
YOU PUT IN THE FINAL PERIOD...
...BUT THE PAGES ARE ALL STILL IN YOUR HEAD.
THE BOOK IS FINISHED...
...BUT IT CONTINUES ON.
YET AT THE SAME TIME...
YOU'VE KNOWN FOR A LONG TIME...
...THAT YOU'RE GOING TO GO RIGHT ON AND DO ANOTHER ONE.

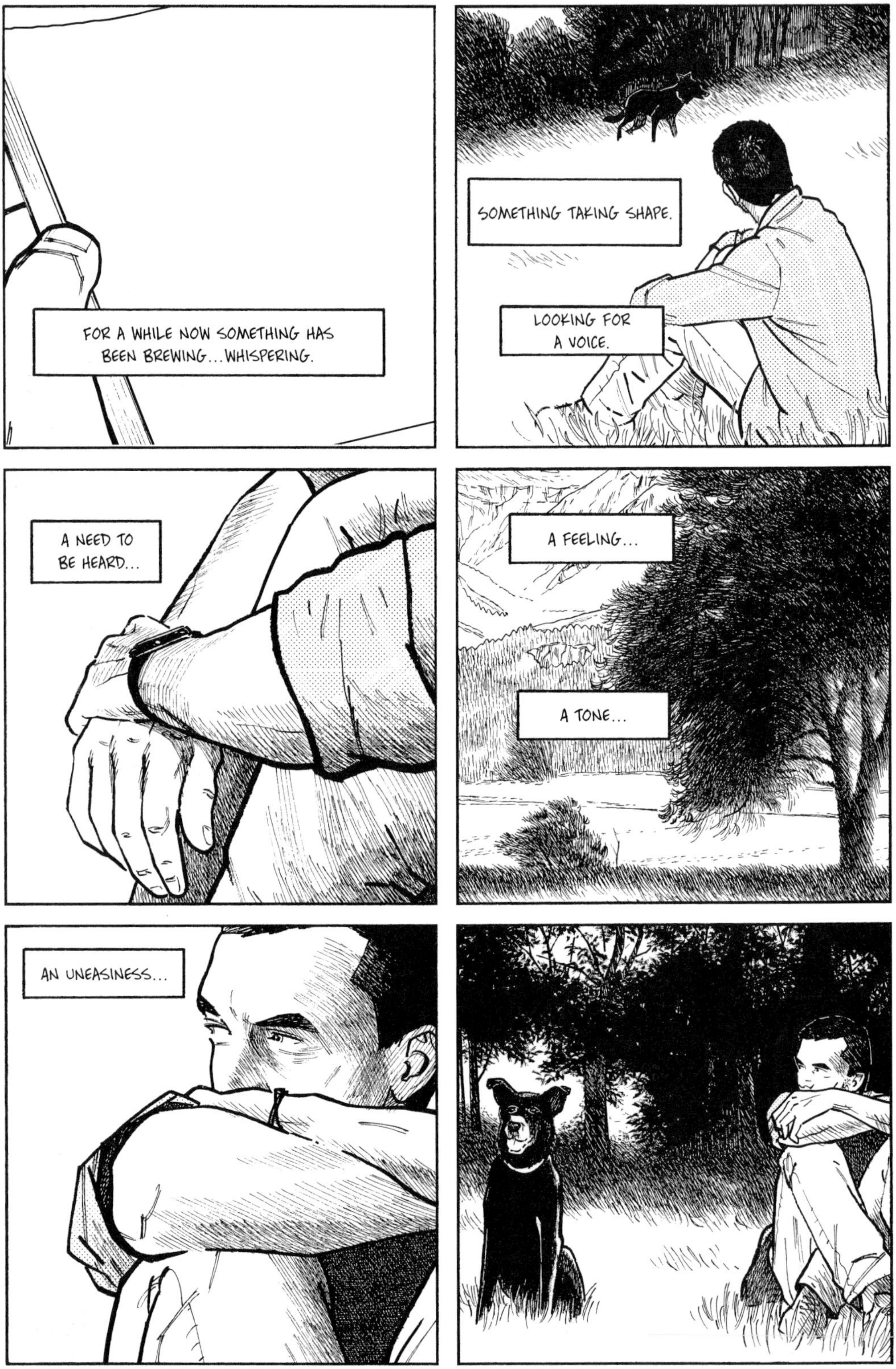
FOR A WHILE NOW SOMETHING HAS BEEN BREWING... WHISPERING.
SOMETHING TAKING SHAPE.
LOOKING FOR A VOICE.
A NEED TO BE HEARD...
A FEELING...
A TONE...
AN UNEASINESS...

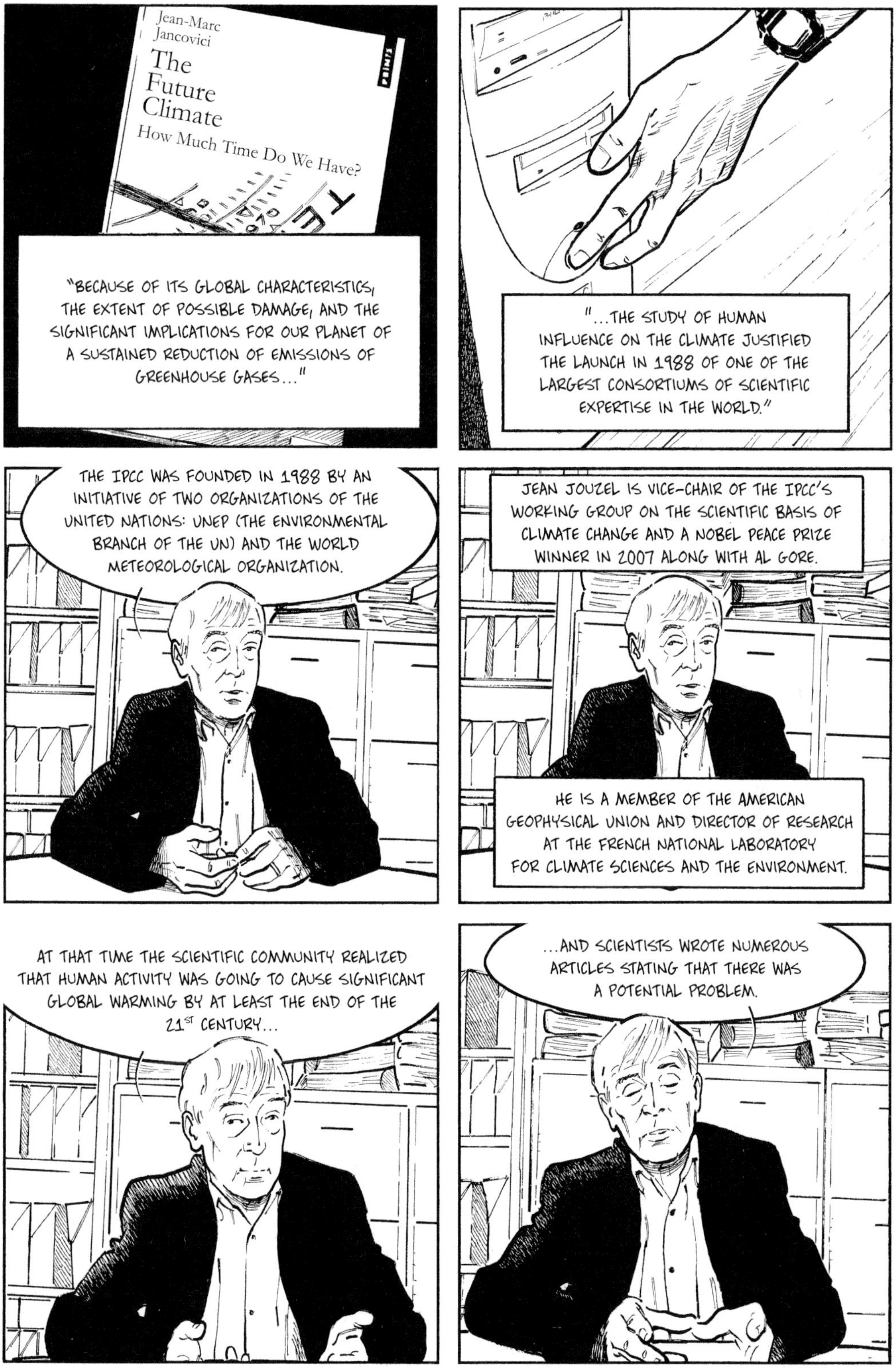
Jean-Marc Jancovici
The Future Climate
How Much Time Do We Have?
"BECAUSE OF ITS GLOBAL CHARACTERISTICS, THE EXTENT OF POSSIBLE DAMAGE, AND THE SIGNIFICANT IMPLICATIONS FOR OUR PLANET OF A SUSTAINED REDUCTION OF EMISSIONS OF GREENHOUSE GASES..."
"...THE STUDY OF HUMAN INFLUENCE ON THE CLIMATE JUSTIFIED THE LAUNCH IN 1988 OF ONE OF THE LARGEST CONSORTIUMS OF SCIENTIFIC EXPERTISE IN THE WORLD."
THE IPCC WAS FOUNDED IN 1988 BY AN INITIATIVE OF TWO ORGANIZATIONS OF THE UNITED NATIONS: UNEP (THE ENVIRONMENTAL BRANCH OF THE UN) AND THE WORLD METEOROLOGICAL ORGANIZATION.
JEAN JOUZEL IS VICE-CHAIR OF THE IPCC'S WORKING GROUP ON THE SCIENTIFIC BASIS OF CLIMATE CHANGE AND A NOBEL PEACE PRIZE WINNER IN 2007 ALONG WITH AL GORE.
HE IS A MEMBER OF THE AMERICAN GEOPHYSICAL UNION AND DIRECTOR OF RESEARCH AT THE FRENCH NATIONAL LABORATORY FOR CLIMATE SCIENCES AND THE ENVIRONMENT.
AT THAT TIME THE SCIENTIFIC COMMUNITY REALIZED THAT HUMAN ACTIVITY WAS GOING TO CAUSE SIGNIFICANT GLOBAL WARMING BY AT LEAST THE END OF THE 21ST CENTURY...
...AND SCIENTISTS WROTE NUMEROUS ARTICLES STATING THAT THERE WAS A POTENTIAL PROBLEM.

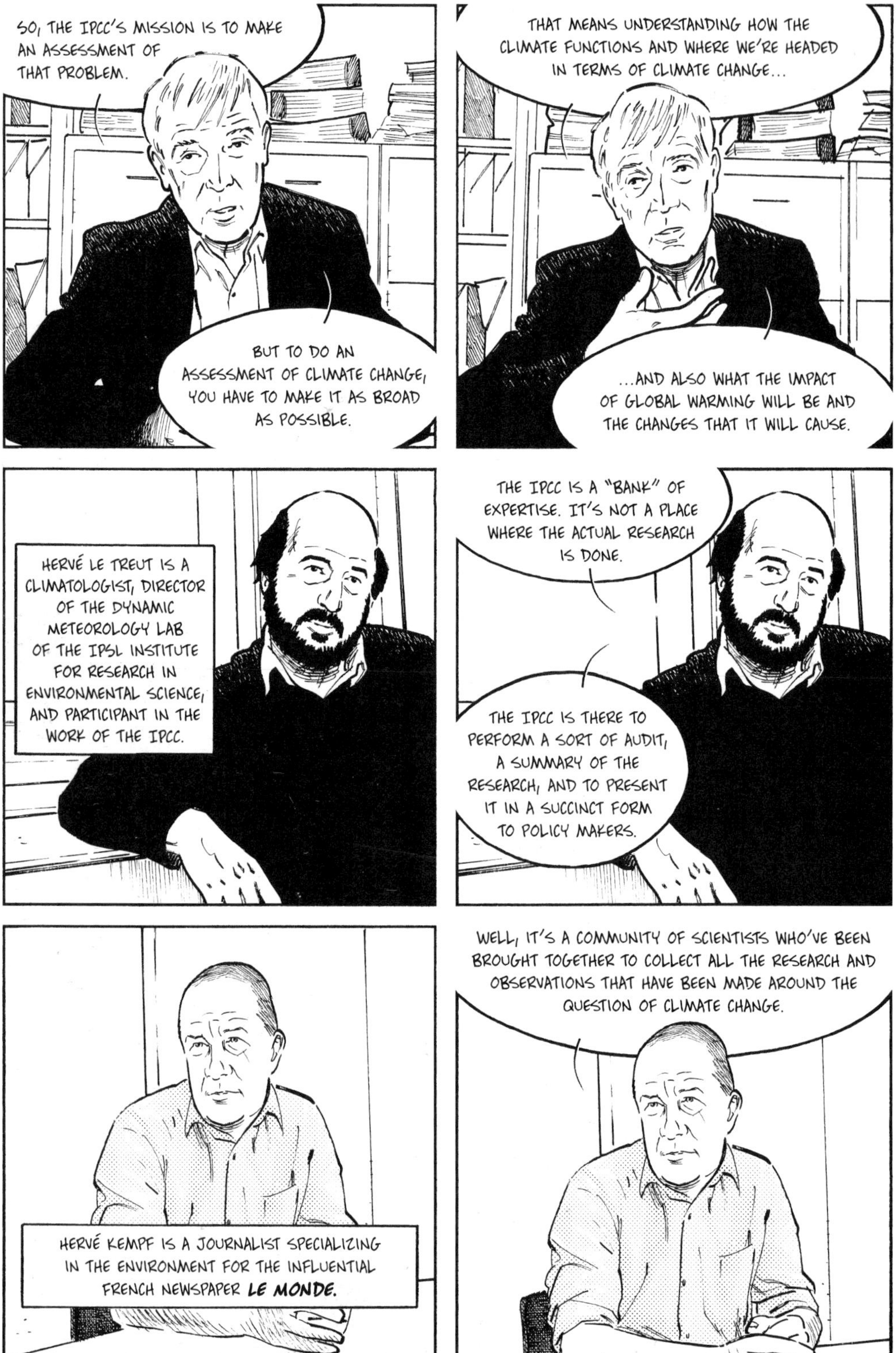
SO, THE IPCC'S MISSION IS TO MAKE AN ASSESSMENT OF THAT PROBLEM.
BUT TO DO AN ASSESSMENT OF CLIMATE CHANGE, YOU HAVE TO MAKE IT AS BROAD AS POSSIBLE.
THAT MEANS UNDERSTANDING HOW THE CLIMATE FUNCTIONS AND WHERE WE'RE HEADED IN TERMS OF CLIMATE CHANGE...
...AND ALSO WHAT THE IMPACT OF GLOBAL WARMING WILL BE AND THE CHANGES THAT IT WILL CAUSE.
HERVÉ LE TREUT IS A CLIMATOLOGIST, DIRECTOR OF THE DYNAMIC METEOROLOGY LAB OF THE IPSL INSTITUTE FOR RESEARCH IN ENVIRONMENTAL SCIENCE, AND PARTICIPANT IN THE WORK OF THE IPCC.
THE IPCC IS A "BANK" OF EXPERTISE. IT'S NOT A PLACE WHERE THE ACTUAL RESEARCH IS DONE.
THE IPCC IS THERE TO PERFORM A SORT OF AUDIT, A SUMMARY OF THE RESEARCH, AND TO PRESENT IT IN A SUCCINCT FORM TO POLICY MAKERS.
HERVÉ KEMPF IS A JOURNALIST SPECIALIZING IN THE ENVIRONMENT FOR THE INFLUENTIAL FRENCH NEWSPAPER LE MONDE.
WELL, IT'S A COMMUNITY OF SCIENTISTS WHO'VE BEEN BROUGHT TOGETHER TO COLLECT ALL THE RESEARCH AND OBSERVATIONS THAT HAVE BEEN MADE AROUND THE QUESTION OF CLIMATE CHANGE.

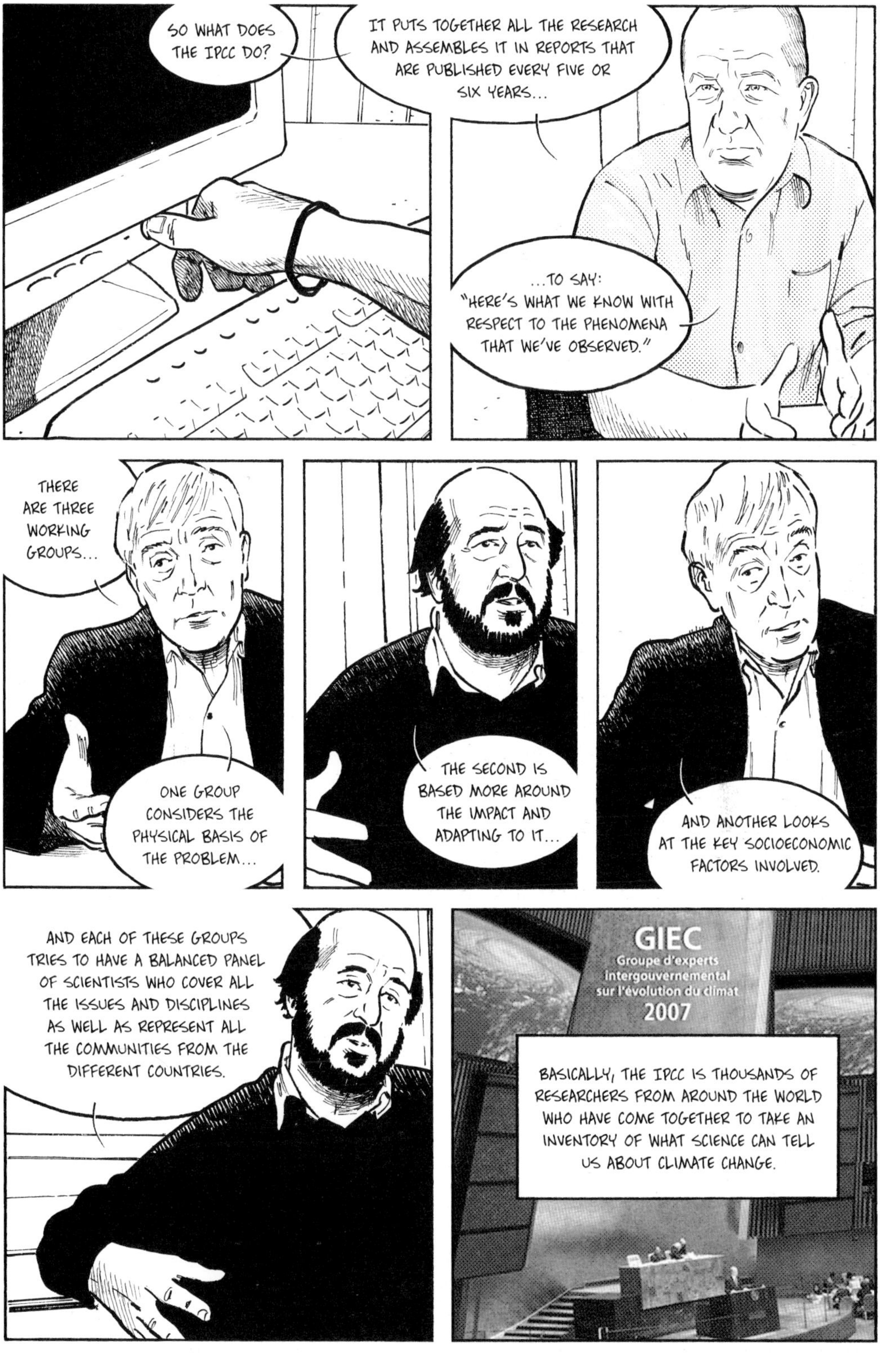
SO WHAT DOES THE IPCC DO?
IT PUTS TOGETHER ALL THE RESEARCH AND ASSEMBLES IT IN REPORTS THAT ARE PUBLISHED EVERY FIVE OR SIX YEARS...
...TO SAY: "HERE'S WHAT WE KNOW WITH RESPECT TO THE PHENOMENA THAT WE'VE OBSERVED."
THERE ARE THREE WORKING GROUPS...
ONE GROUP CONSIDERS THE PHYSICAL BASIS OF THE PROBLEM...
THE SECOND IS BASED MORE AROUND THE IMPACT AND ADAPTING TO IT...
AND ANOTHER LOOKS AT THE KEY SOCIOECONOMIC FACTORS INVOLVED.
AND EACH OF THESE GROUPS TRIES TO HAVE A BALANCED PANEL OF SCIENTISTS WHO COVER ALL THE ISSUES AND DISCIPLINES AS WELL AS REPRESENT ALL THE COMMUNITIES FROM THE DIFFERENT COUNTRIES.
GIEC
Groupe d'experts intergouvernemental sur l'évolution du climat
2007
BASICALLY, THE IPCC IS THOUSANDS OF RESEARCHERS FROM AROUND THE WORLD WHO HAVE COME TOGETHER TO TAKE AN INVENTORY OF WHAT SCIENCE CAN TELL US ABOUT CLIMATE CHANGE.

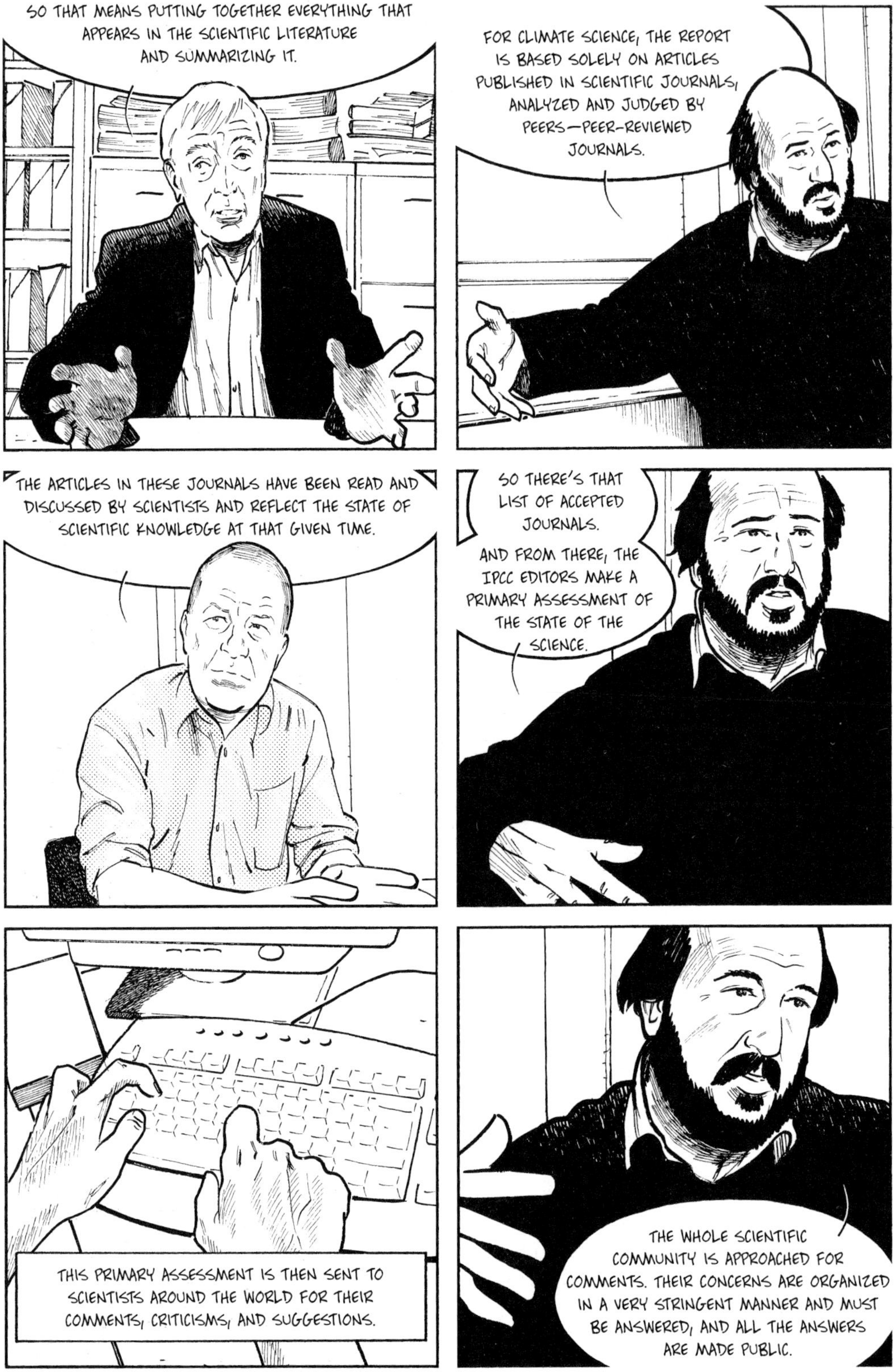
SO THAT MEANS PUTTING TOGETHER EVERYTHING THAT APPEARS IN THE SCIENTIFIC LITERATURE AND SUMMARIZING IT.
FOR CLIMATE SCIENCE, THE REPORT IS BASED SOLELY ON ARTICLES PUBLISHED IN SCIENTIFIC JOURNALS, ANALYZED AND JUDGED BY PEERS—PEER-REVIEWED JOURNALS.
THE ARTICLES IN THESE JOURNALS HAVE BEEN READ AND DISCUSSED BY SCIENTISTS AND REFLECT THE STATE OF SCIENTIFIC KNOWLEDGE AT THAT GIVEN TIME.
SO THERE'S THAT LIST OF ACCEPTED JOURNALS.
AND FROM THERE, THE IPCC EDITORS MAKE A PRIMARY ASSESSMENT OF THE STATE OF THE SCIENCE.
THIS PRIMARY ASSESSMENT IS THEN SENT TO SCIENTISTS AROUND THE WORLD FOR THEIR COMMENTS, CRITICISMS, AND SUGGESTIONS.
THE WHOLE SCIENTIFIC COMMUNITY IS APPROACHED FOR COMMENTS. THEIR CONCERNS ARE ORGANIZED IN A VERY STRINGENT MANNER AND MUST BE ANSWERED, AND ALL THE ANSWERS ARE MADE PUBLIC.

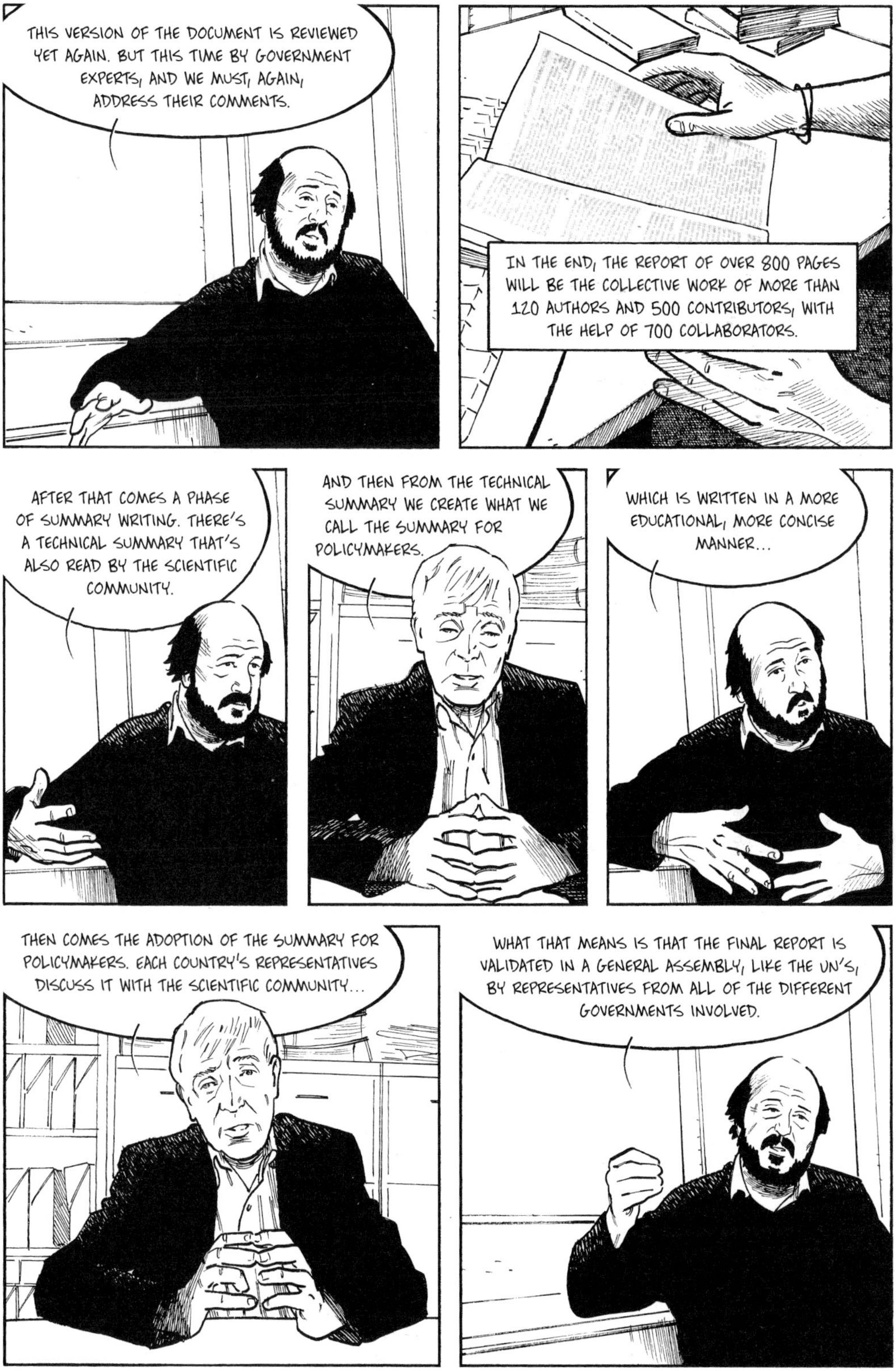
THIS VERSION OF THE DOCUMENT IS REVIEWED YET AGAIN. BUT THIS TIME BY GOVERNMENT EXPERTS, AND WE MUST, AGAIN, ADDRESS THEIR COMMENTS.
IN THE END, THE REPORT OF OVER 800 PAGES WILL BE THE COLLECTIVE WORK OF MORE THAN 120 AUTHORS AND 500 CONTRIBUTORS, WITH THE HELP OF 700 COLLABORATORS.
AFTER THAT COMES A PHASE OF SUMMARY WRITING. THERE'S A TECHNICAL SUMMARY THAT'S ALSO READ BY THE SCIENTIFIC COMMUNITY.
AND THEN FROM THE TECHNICAL SUMMARY WE CREATE WHAT WE CALL THE SUMMARY FOR POLICYMAKERS.
WHICH IS WRITTEN IN A MORE EDUCATIONAL, MORE CONCISE MANNER...
THEN COMES THE ADOPTION OF THE SUMMARY FOR POLICYMAKERS. EACH COUNTRY'S REPRESENTATIVES DISCUSS IT WITH THE SCIENTIFIC COMMUNITY...
WHAT THAT MEANS IS THAT THE FINAL REPORT IS VALIDATED IN A GENERAL ASSEMBLY, LIKE THE UN'S, BY REPRESENTATIVES FROM ALL OF THE DIFFERENT GOVERNMENTS INVOLVED.

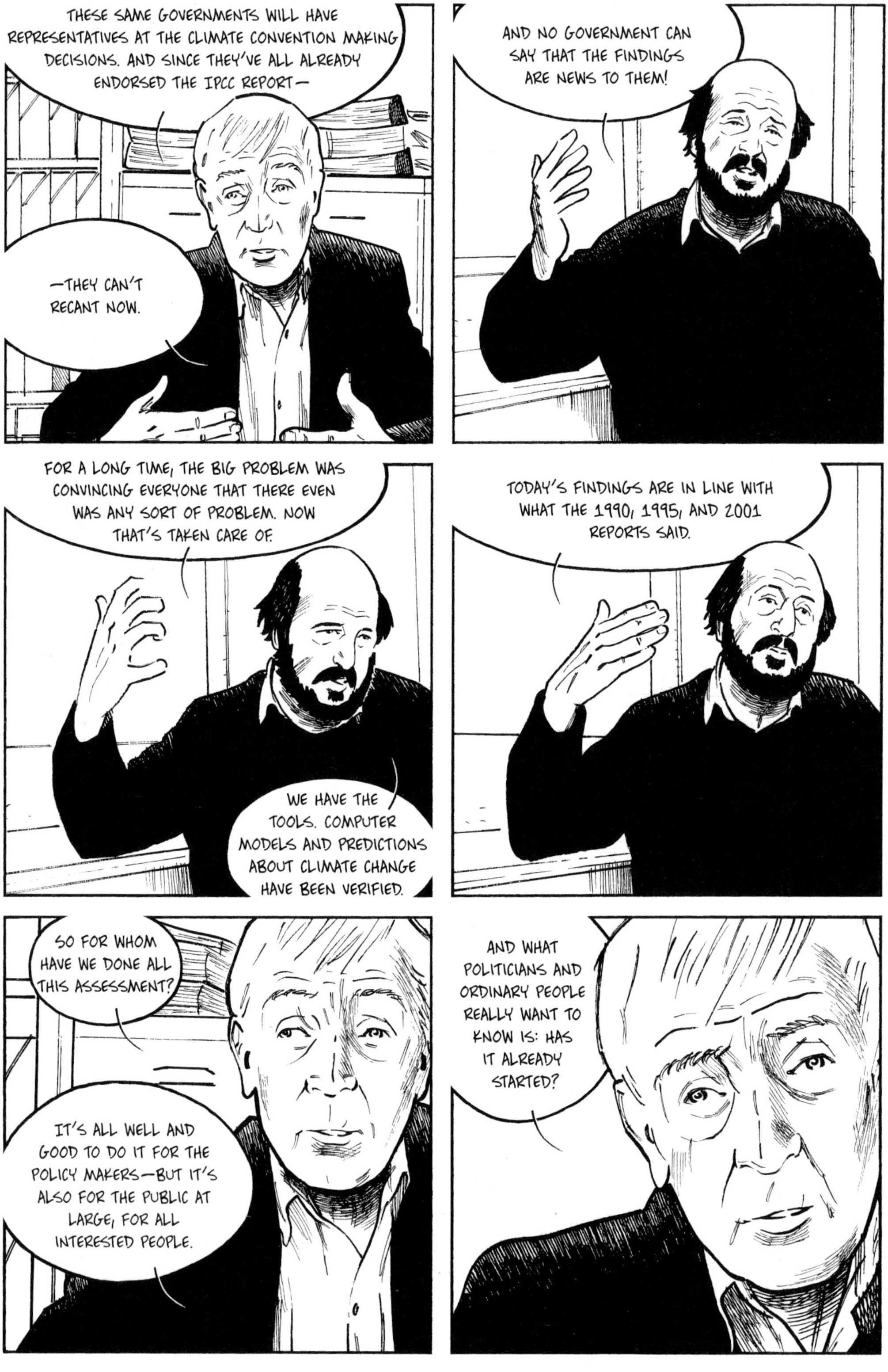
THESE SAME GOVERNMENTS WILL HAVE REPRESENTATIVES AT THE CLIMATE CONVENTION MAKING DECISIONS. AND SINCE THEY'VE ALL ALREADY ENDORSED THE IPCC REPORT—
—THEY CAN'T RECANT NOW.
AND NO GOVERNMENT CAN SAY THAT THE FINDINGS ARE NEWS TO THEM!
FOR A LONG TIME, THE BIG PROBLEM WAS CONVINCING EVERYONE THAT THERE EVEN WAS ANY SORT OF PROBLEM. NOW THAT'S TAKEN CARE OF.
WE HAVE THE TOOLS. COMPUTER MODELS AND PREDICTIONS ABOUT CLIMATE CHANGE HAVE BEEN VERIFIED.
TODAY'S FINDINGS ARE IN LINE WITH WHAT THE 1990, 1995, AND 2001 REPORTS SAID.
SO FOR WHOM HAVE WE DONE ALL THIS ASSESSMENT?
IT'S ALL WELL AND GOOD TO DO IT FOR THE POLICY MAKERS—BUT IT'S ALSO FOR THE PUBLIC AT LARGE, FOR ALL INTERESTED PEOPLE.
AND WHAT POLITICIANS AND ORDINARY PEOPLE REALLY WANT TO KNOW IS: HAS IT ALREADY STARTED?

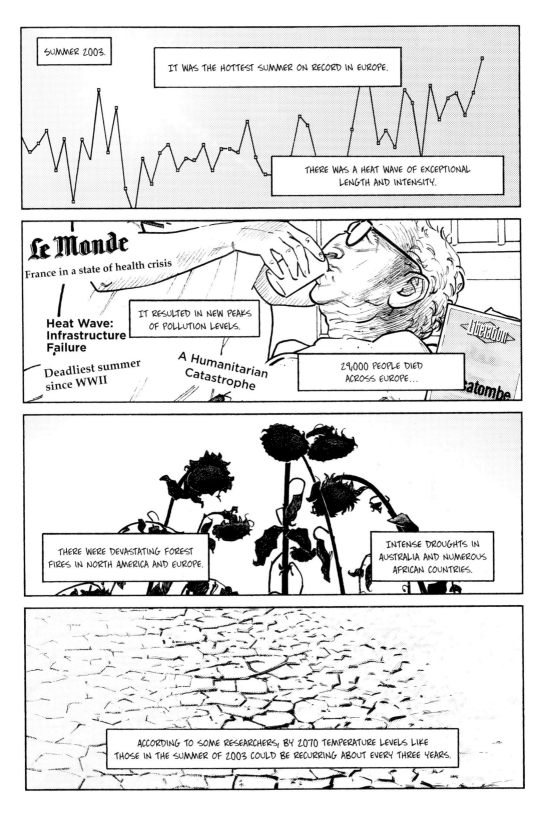

SUMMER 2003.

IT WAS THE HOTTEST SUMMER ON RECORD IN EUROPE.

THERE WAS A HEAT WAVE OF EXCEPTIONAL LENGTH AND INTENSITY.

Le Monde

France in a state of health crisis

Heat Wave: Infrastructure Failure

Deadliest summer since WWII

A Humanitarian Catastrophe

IT RESULTED IN NEW PEAKS OF POLLUTION LEVELS.

29,000 PEOPLE DIED ACROSS EUROPE...

THERE WERE DEVASTATING FOREST FIRES IN NORTH AMERICA AND EUROPE.

INTENSE DROUGHTS IN AUSTRALIA AND NUMEROUS AFRICAN COUNTRIES.

ACCORDING TO SOME RESEARCHERS, BY 2070 TEMPERATURE LEVELS LIKE THOSE IN THE SUMMER OF 2003 COULD BE RECURRING ABOUT EVERY THREE YEARS.

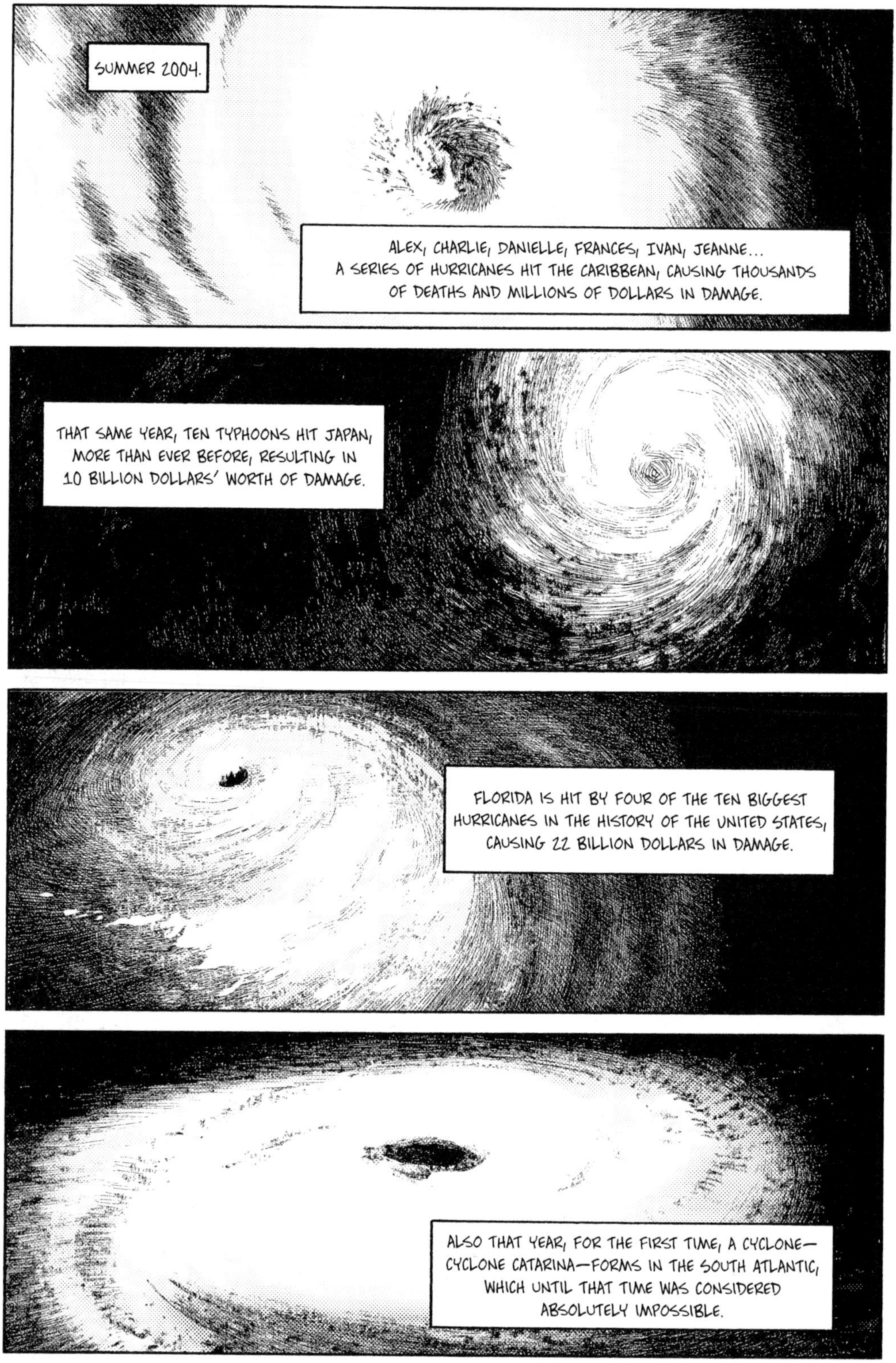
SUMMER 2004.
ALEX, CHARLIE, DANIELLE, FRANCES, IVAN, JEANNE... A SERIES OF HURRICANES HIT THE CARIBBEAN, CAUSING THOUSANDS OF DEATHS AND MILLIONS OF DOLLARS IN DAMAGE.
THAT SAME YEAR, TEN TYPHOONS HIT JAPAN, MORE THAN EVER BEFORE, RESULTING IN 10 BILLION DOLLARS' WORTH OF DAMAGE.
FLORIDA IS HIT BY FOUR OF THE TEN BIGGEST HURRICANES IN THE HISTORY OF THE UNITED STATES, CAUSING 22 BILLION DOLLARS IN DAMAGE.
ALSO THAT YEAR, FOR THE FIRST TIME, A CYCLONE—CYCLONE CATARINA—FORMS IN THE SOUTH ATLANTIC, WHICH UNTIL THAT TIME WAS CONSIDERED ABSOLUTELY IMPOSSIBLE.

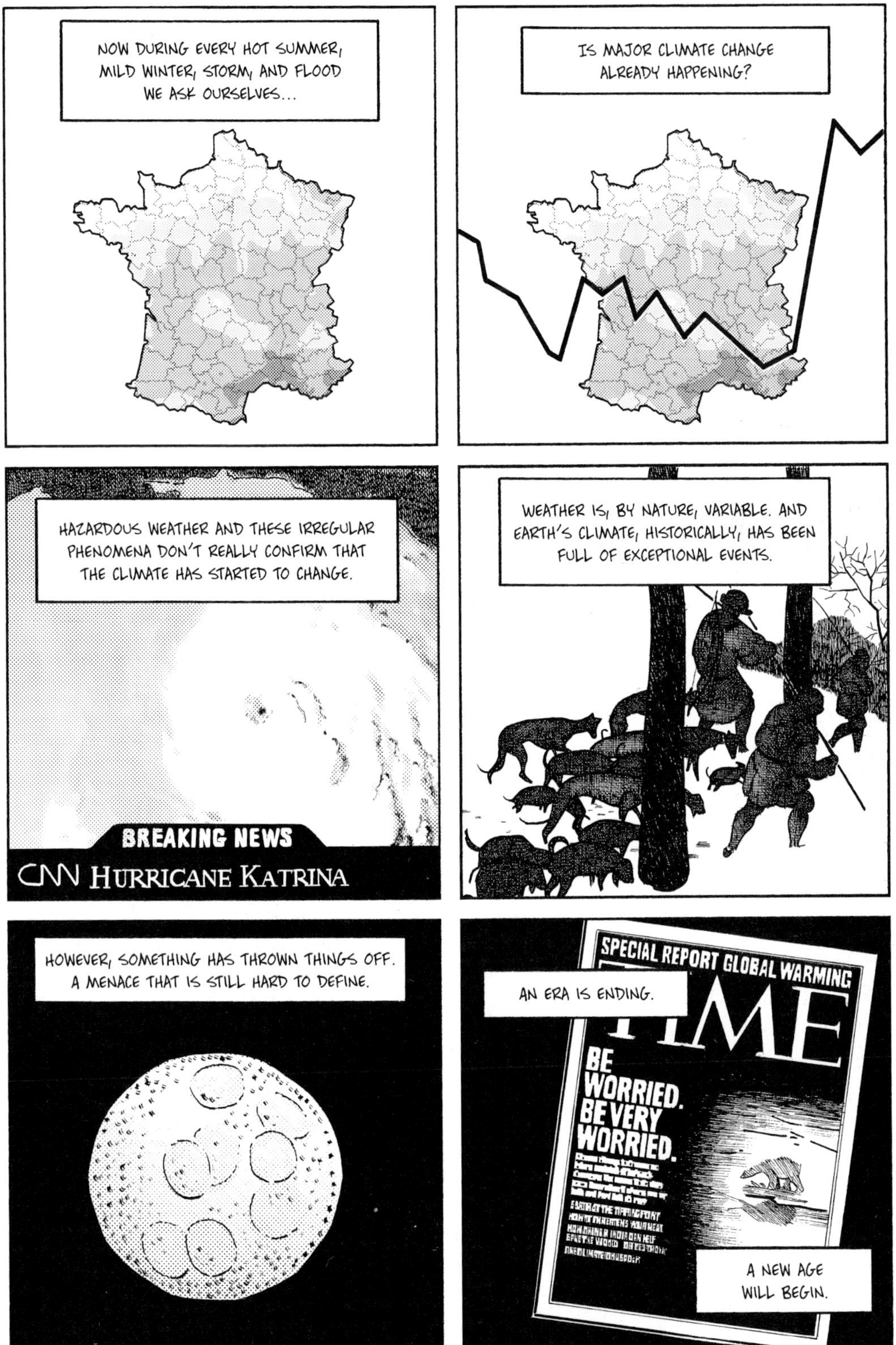
NOW DURING EVERY HOT SUMMER, MILD WINTER, STORM, AND FLOOD WE ASK OURSELVES...
IS MAJOR CLIMATE CHANGE ALREADY HAPPENING?
HAZARDOUS WEATHER AND THESE IRREGULAR PHENOMENA DON'T REALLY CONFIRM THAT THE CLIMATE HAS STARTED TO CHANGE.
BREAKING NEWS
CNN HURRICANE KATRINA
WEATHER IS, BY NATURE, VARIABLE. AND EARTH'S CLIMATE, HISTORICALLY, HAS BEEN FULL OF EXCEPTIONAL EVENTS.
HOWEVER, SOMETHING HAS THROWN THINGS OFF. A MENACE THAT IS STILL HARD TO DEFINE.
SPECIAL REPORT GLOBAL WARMING
AN ERA IS ENDING.
TIME
BE WORRIED. BE VERY WORRIED.
A NEW AGE WILL BEGIN.

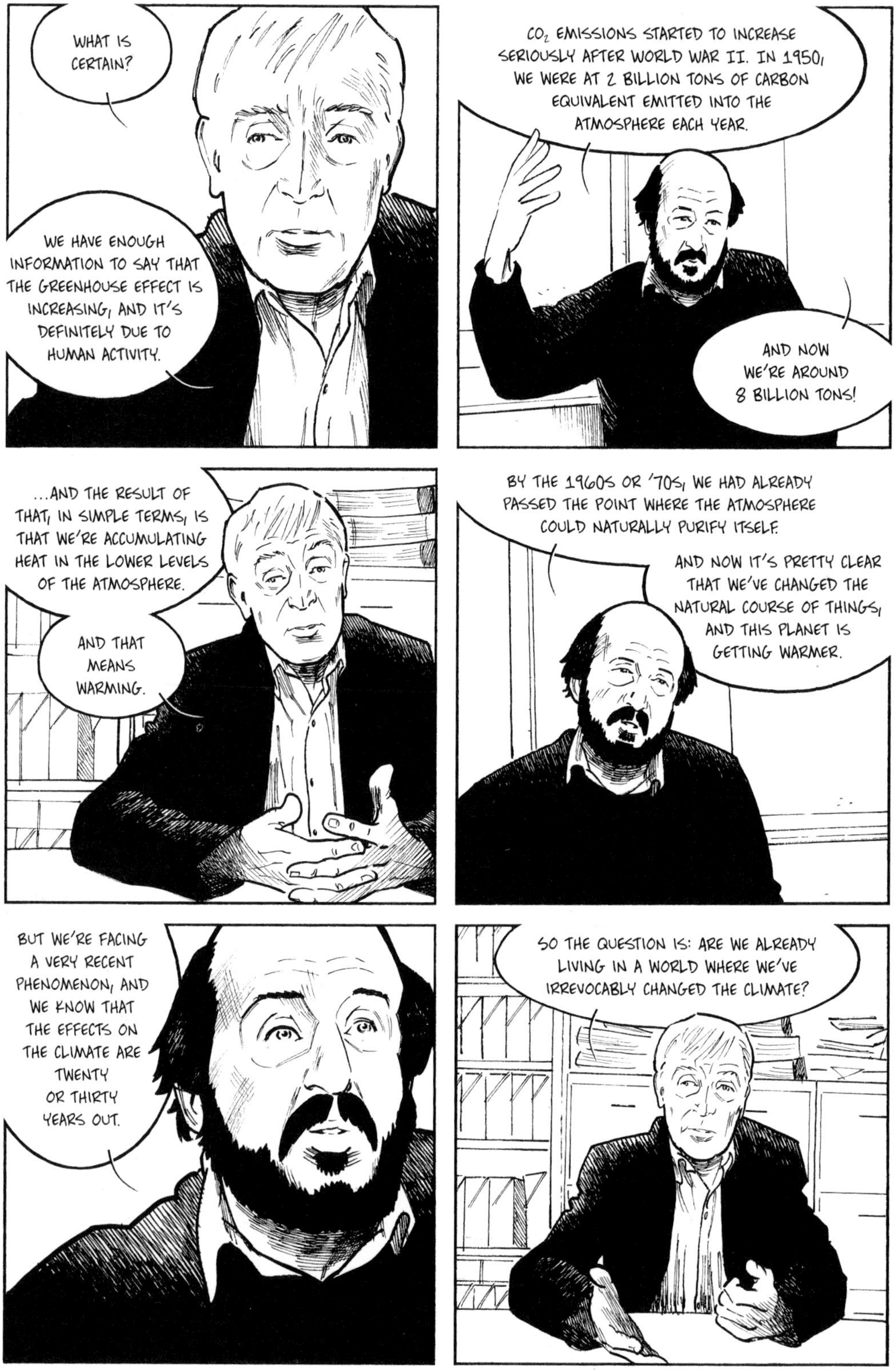
WHAT IS CERTAIN?
WE HAVE ENOUGH INFORMATION TO SAY THAT THE GREENHOUSE EFFECT IS INCREASING, AND IT'S DEFINITELY DUE TO HUMAN ACTIVITY.
CO_2 EMISSIONS STARTED TO INCREASE SERIOUSLY AFTER WORLD WAR II. IN 1950, WE WERE AT 2 BILLION TONS OF CARBON EQUIVALENT EMITTED INTO THE ATMOSPHERE EACH YEAR.
AND NOW WE'RE AROUND 8 BILLION TONS!
...AND THE RESULT OF THAT, IN SIMPLE TERMS, IS THAT WE'RE ACCUMULATING HEAT IN THE LOWER LEVELS OF THE ATMOSPHERE.
AND THAT MEANS WARMING.
BY THE 1960S OR '70S, WE HAD ALREADY PASSED THE POINT WHERE THE ATMOSPHERE COULD NATURALLY PURIFY ITSELF.
AND NOW IT'S PRETTY CLEAR THAT WE'VE CHANGED THE NATURAL COURSE OF THINGS, AND THIS PLANET IS GETTING WARMER.
BUT WE'RE FACING A VERY RECENT PHENOMENON, AND WE KNOW THAT THE EFFECTS ON THE CLIMATE ARE TWENTY OR THIRTY YEARS OUT.
SO THE QUESTION IS: ARE WE ALREADY LIVING IN A WORLD WHERE WE'VE IRREVOCABLY CHANGED THE CLIMATE?

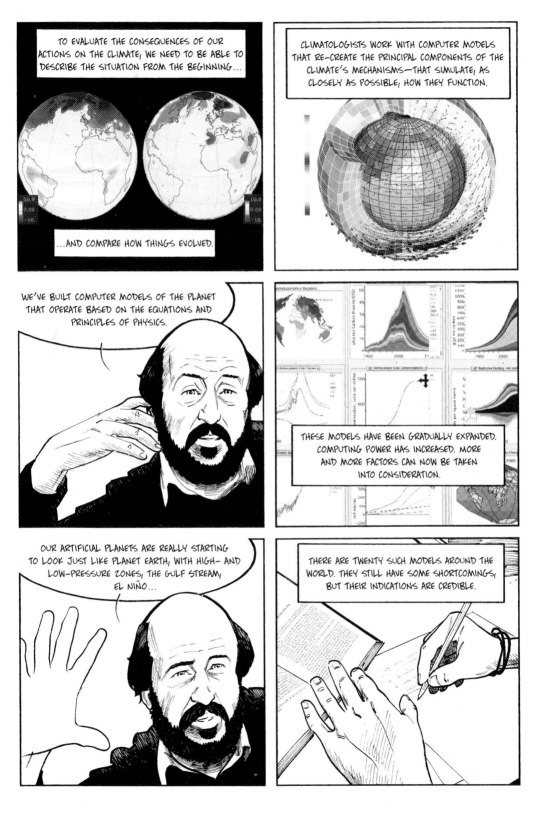

TO EVALUATE THE CONSEQUENCES OF OUR ACTIONS ON THE CLIMATE, WE NEED TO BE ABLE TO DESCRIBE THE SITUATION FROM THE BEGINNING...

...AND COMPARE HOW THINGS EVOLVED.

CLIMATOLOGISTS WORK WITH COMPUTER MODELS THAT RE-CREATE THE PRINCIPAL COMPONENTS OF THE CLIMATE'S MECHANISMS—THAT SIMULATE, AS CLOSELY AS POSSIBLE, HOW THEY FUNCTION.

WE'VE BUILT COMPUTER MODELS OF THE PLANET THAT OPERATE BASED ON THE EQUATIONS AND PRINCIPLES OF PHYSICS.

THESE MODELS HAVE BEEN GRADUALLY EXPANDED. COMPUTING POWER HAS INCREASED. MORE AND MORE FACTORS CAN NOW BE TAKEN INTO CONSIDERATION.

OUR ARTIFICIAL PLANETS ARE REALLY STARTING TO LOOK JUST LIKE PLANET EARTH, WITH HIGH- AND LOW-PRESSURE ZONES, THE GULF STREAM, EL NIÑO...

THERE ARE TWENTY SUCH MODELS AROUND THE WORLD. THEY STILL HAVE SOME SHORTCOMINGS, BUT THEIR INDICATIONS ARE CREDIBLE.

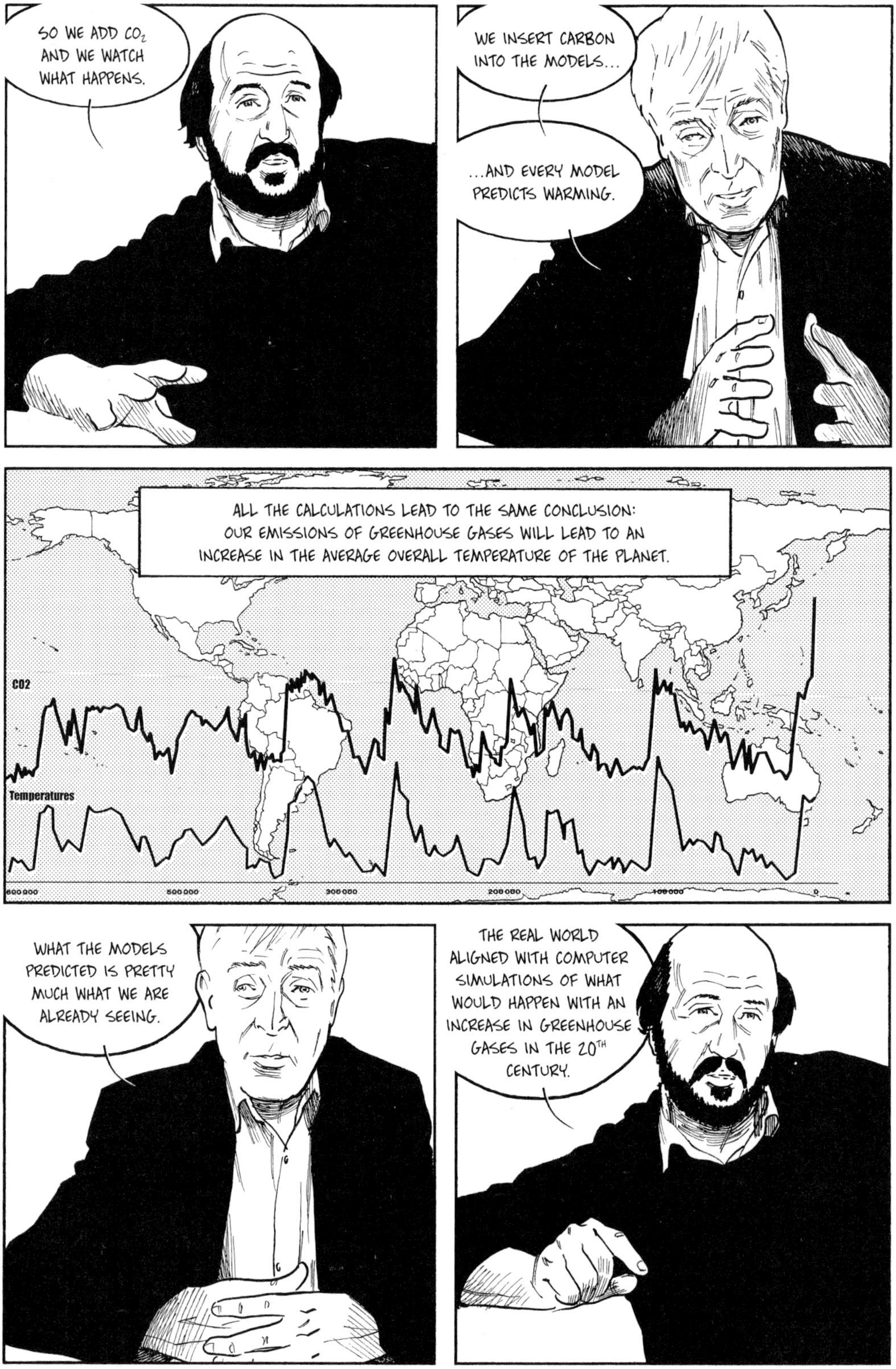
SO WE ADD CO_2 AND WE WATCH WHAT HAPPENS.
WE INSERT CARBON INTO THE MODELS...
...AND EVERY MODEL PREDICTS WARMING.
ALL THE CALCULATIONS LEAD TO THE SAME CONCLUSION: OUR EMISSIONS OF GREENHOUSE GASES WILL LEAD TO AN INCREASE IN THE AVERAGE OVERALL TEMPERATURE OF THE PLANET.
CO2
Temperatures
600 000
500 000
300 000
200 000
100 000
0
WHAT THE MODELS PREDICTED IS PRETTY MUCH WHAT WE ARE ALREADY SEEING.
THE REAL WORLD ALIGNED WITH COMPUTER SIMULATIONS OF WHAT WOULD HAPPEN WITH AN INCREASE IN GREENHOUSE GASES IN THE 20TH CENTURY.

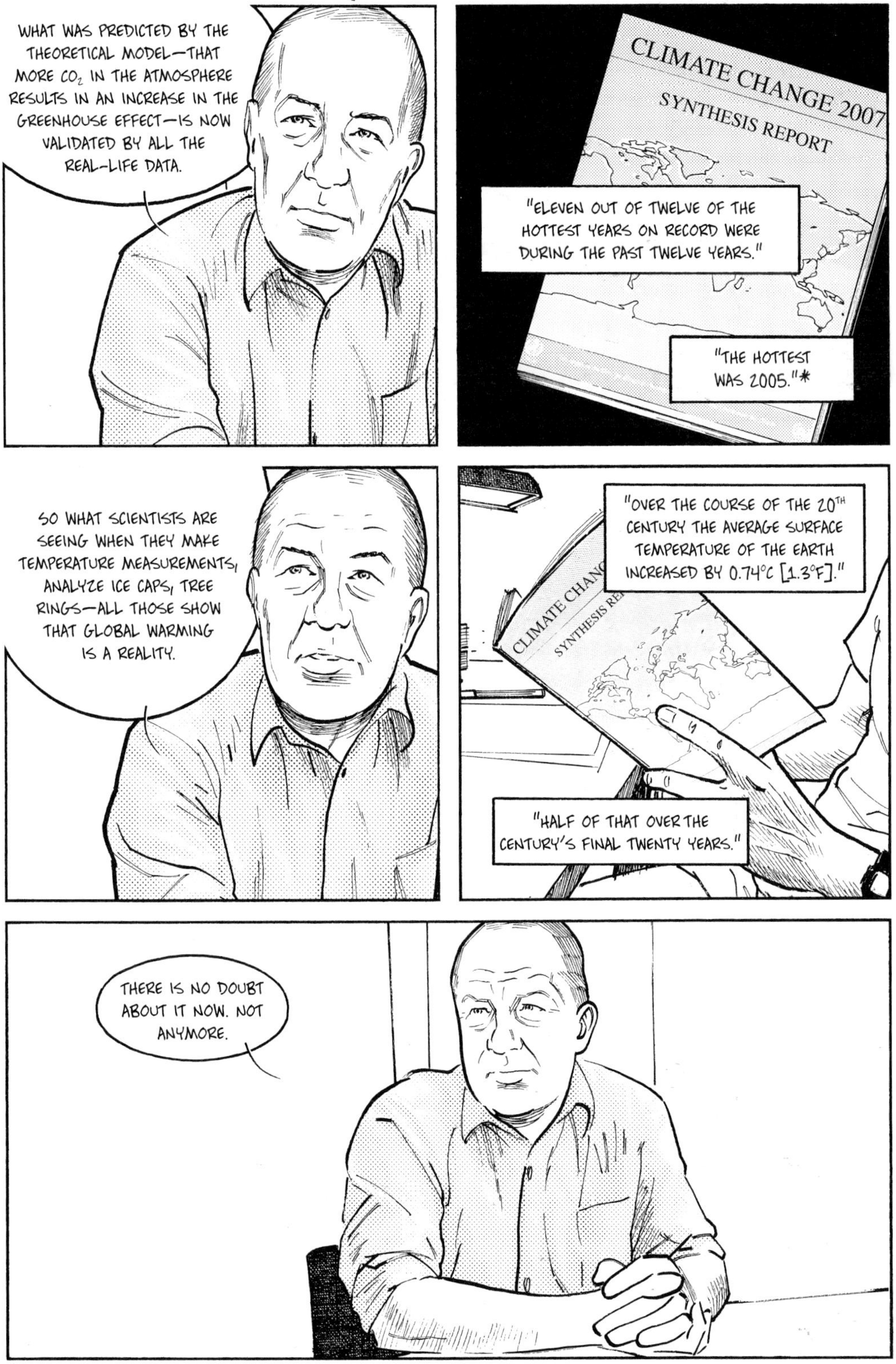
WHAT WAS PREDICTED BY THE THEORETICAL MODEL—THAT MORE CO_2 IN THE ATMOSPHERE RESULTS IN AN INCREASE IN THE GREENHOUSE EFFECT—IS NOW VALIDATED BY ALL THE REAL-LIFE DATA.
CLIMATE CHANGE 2007
SYNTHESIS REPORT
"ELEVEN OUT OF TWELVE OF THE HOTTEST YEARS ON RECORD WERE DURING THE PAST TWELVE YEARS."
"THE HOTTEST WAS 2005."*
SO WHAT SCIENTISTS ARE SEEING WHEN THEY MAKE TEMPERATURE MEASUREMENTS, ANALYZE ICE CAPS, TREE RINGS—ALL THOSE SHOW THAT GLOBAL WARMING IS A REALITY.
"OVER THE COURSE OF THE 20TH CENTURY THE AVERAGE SURFACE TEMPERATURE OF THE EARTH INCREASED BY 0.74°C [1.3°F]."
"HALF OF THAT OVER THE CENTURY'S FINAL TWENTY YEARS."
THERE IS NO DOUBT ABOUT IT NOW. NOT ANYMORE.

*AT THE TIME OF THE 2007 REPORT. BY 2013, THE LAST SIXTEEN YEARS HAD BEEN THE HOTTEST EVER RECORDED. THE HOTTEST WAS 2010.

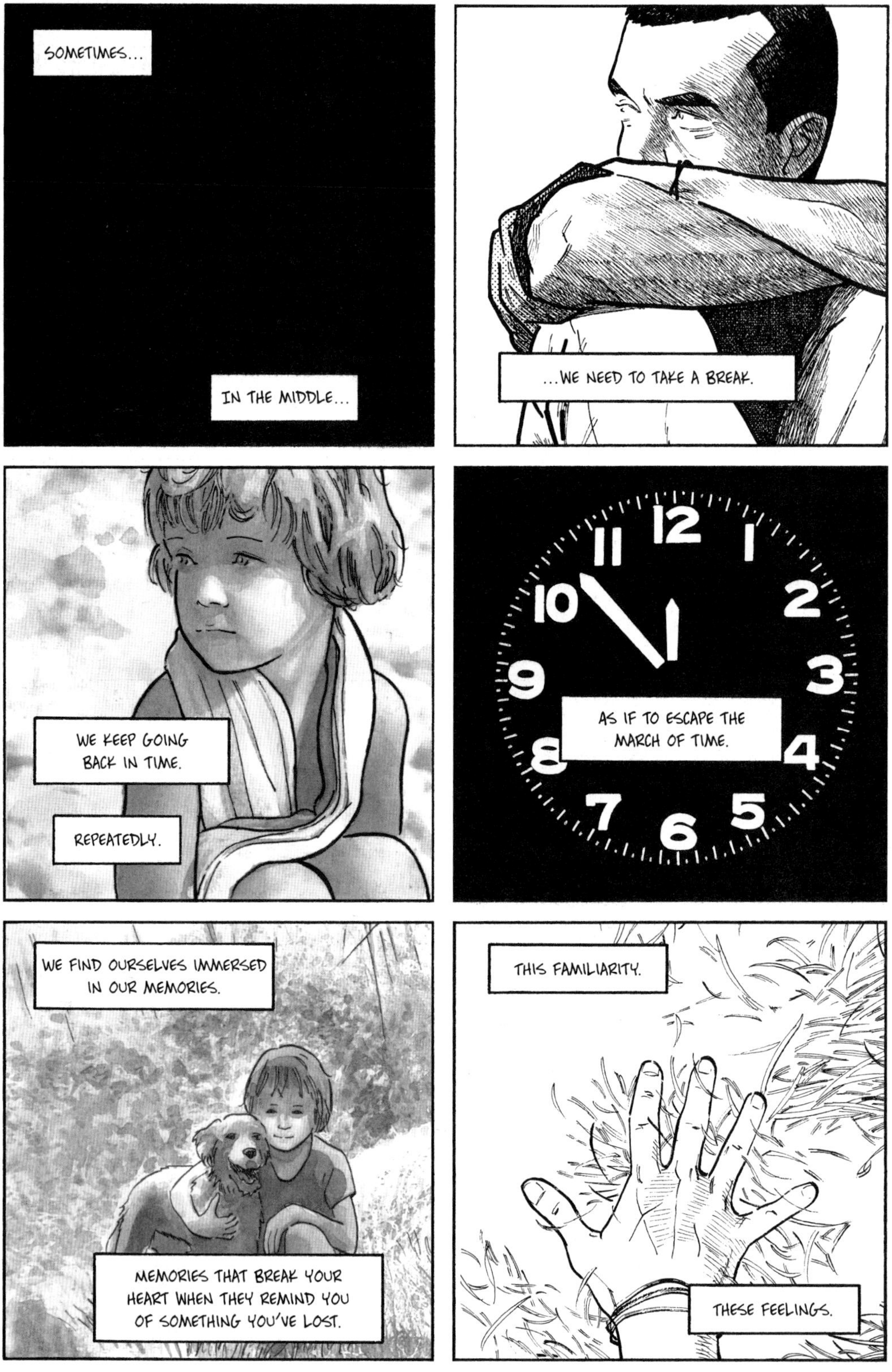
SOMETIMES...
IN THE MIDDLE...
...WE NEED TO TAKE A BREAK.
WE KEEP GOING BACK IN TIME.
REPEATEDLY.
12 1 2 3 4 5 6 7 8 9 10 11
AS IF TO ESCAPE THE MARCH OF TIME.
WE FIND OURSELVES IMMERSED IN OUR MEMORIES.
MEMORIES THAT BREAK YOUR HEART WHEN THEY REMIND YOU OF SOMETHING YOU'VE LOST.
THIS FAMILIARITY.
THESE FEELINGS.

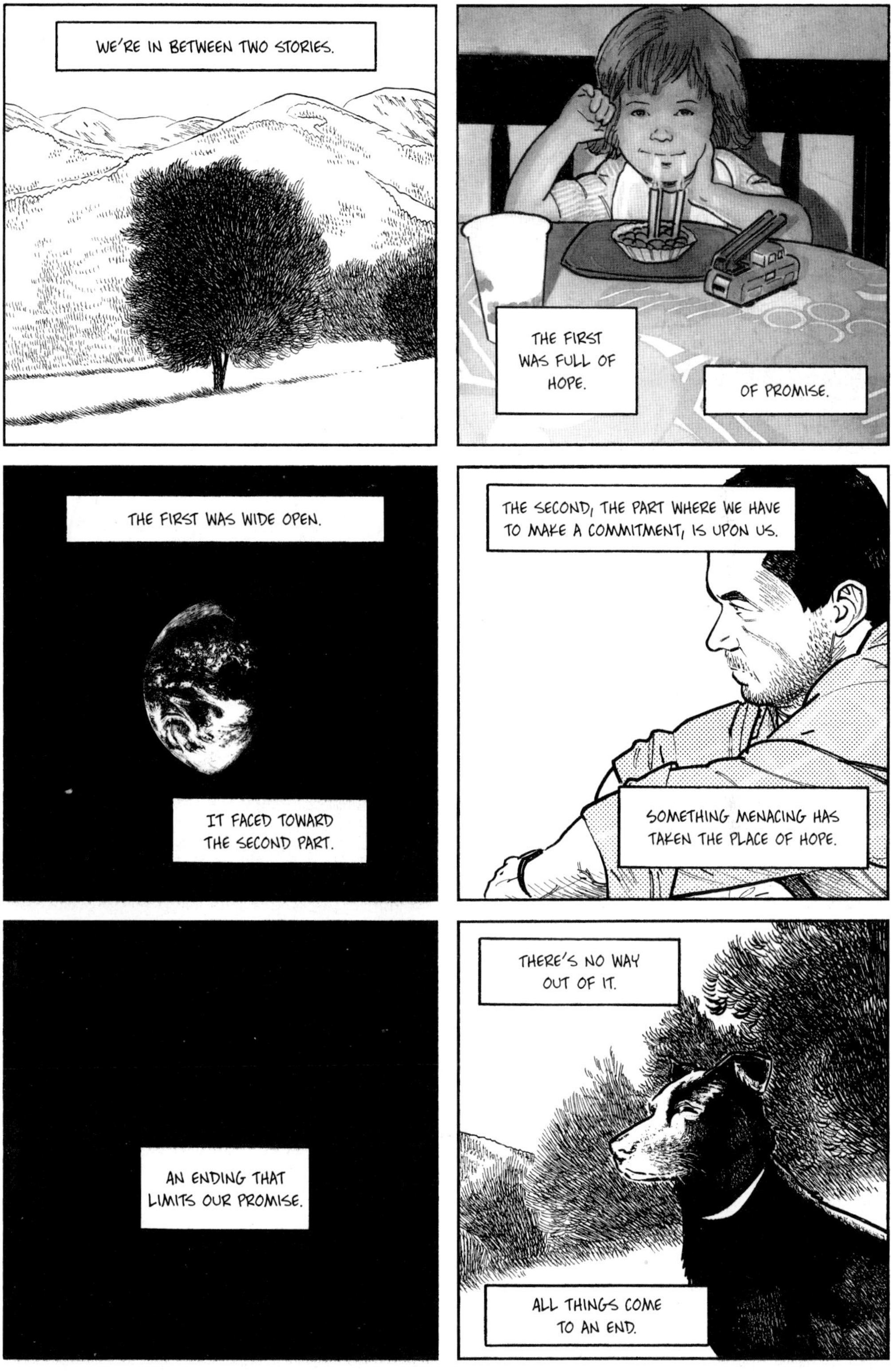
WE'RE IN BETWEEN TWO STORIES.
THE FIRST WAS FULL OF HOPE.
OF PROMISE.
THE FIRST WAS WIDE OPEN.
IT FACED TOWARD THE SECOND PART.
THE SECOND, THE PART WHERE WE HAVE TO MAKE A COMMITMENT, IS UPON US.
SOMETHING MENACING HAS TAKEN THE PLACE OF HOPE.
AN ENDING THAT LIMITS OUR PROMISE.
THERE'S NO WAY OUT OF IT.
ALL THINGS COME TO AN END.

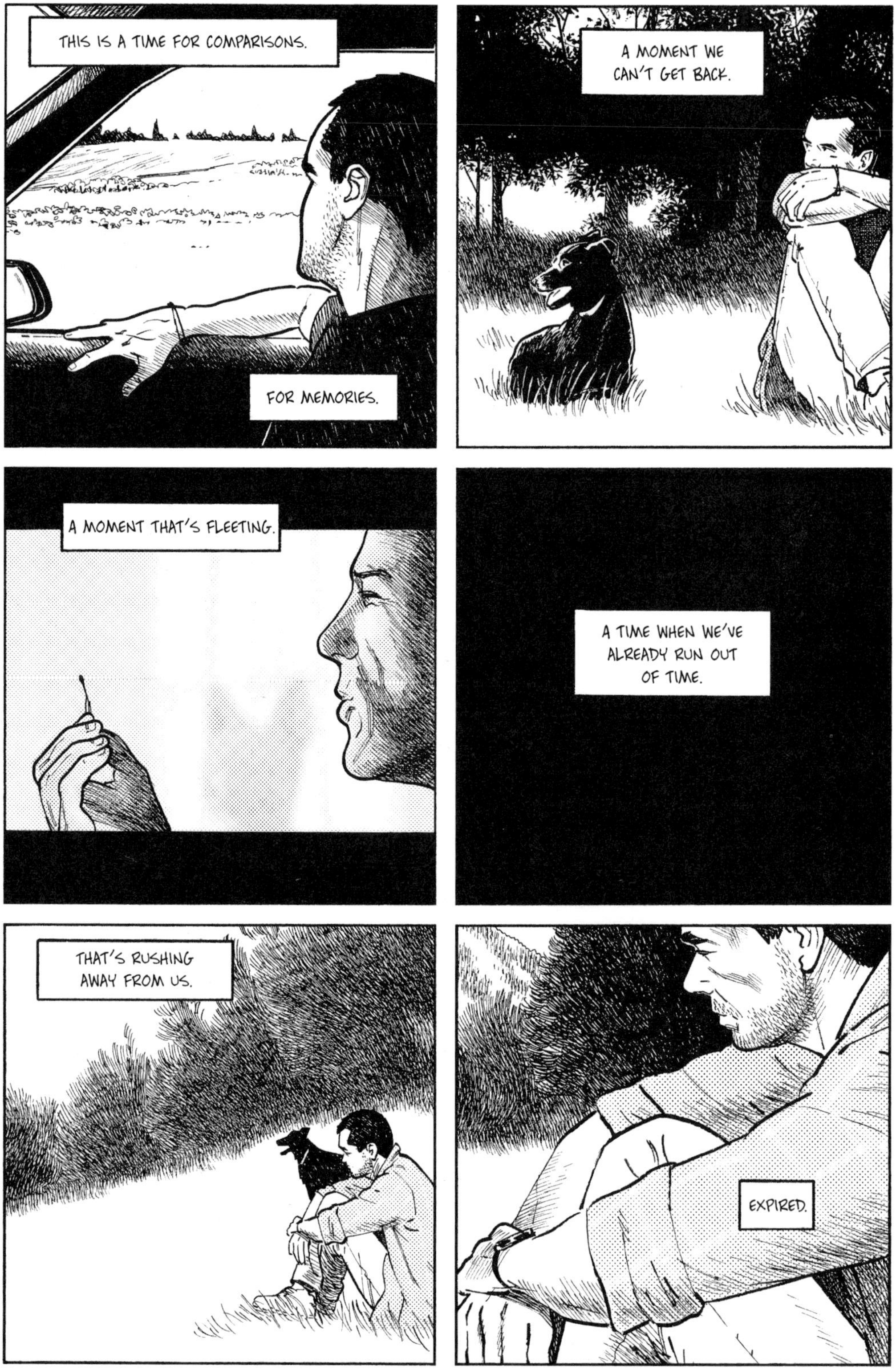
THIS IS A TIME FOR COMPARISONS.
FOR MEMORIES.
A MOMENT WE CAN'T GET BACK.
A MOMENT THAT'S FLEETING.
A TIME WHEN WE'VE ALREADY RUN OUT OF TIME.
THAT'S RUSHING AWAY FROM US.
EXPIRED.

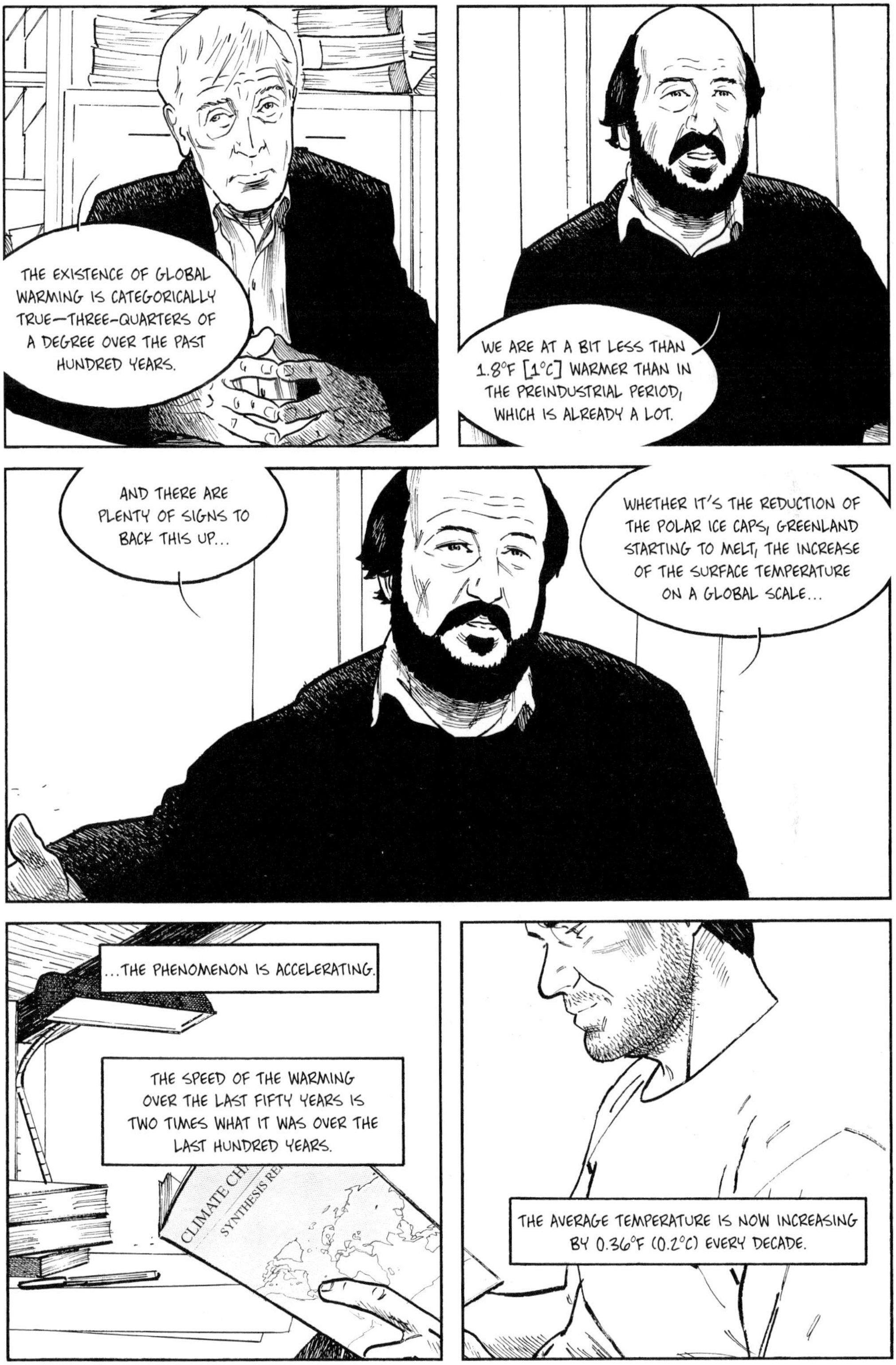
THE EXISTENCE OF GLOBAL WARMING IS CATEGORICALLY TRUE—THREE-QUARTERS OF A DEGREE OVER THE PAST HUNDRED YEARS.
WE ARE AT A BIT LESS THAN 1.8°F [1°C] WARMER THAN IN THE PREINDUSTRIAL PERIOD, WHICH IS ALREADY A LOT.
AND THERE ARE PLENTY OF SIGNS TO BACK THIS UP...
WHETHER IT'S THE REDUCTION OF THE POLAR ICE CAPS, GREENLAND STARTING TO MELT, THE INCREASE OF THE SURFACE TEMPERATURE ON A GLOBAL SCALE...
...THE PHENOMENON IS ACCELERATING.
THE SPEED OF THE WARMING OVER THE LAST FIFTY YEARS IS TWO TIMES WHAT IT WAS OVER THE LAST HUNDRED YEARS.
CLIMATE CH
SYNTHESIS RE
THE AVERAGE TEMPERATURE IS NOW INCREASING BY 0.36°F (0.2°C) EVERY DECADE.

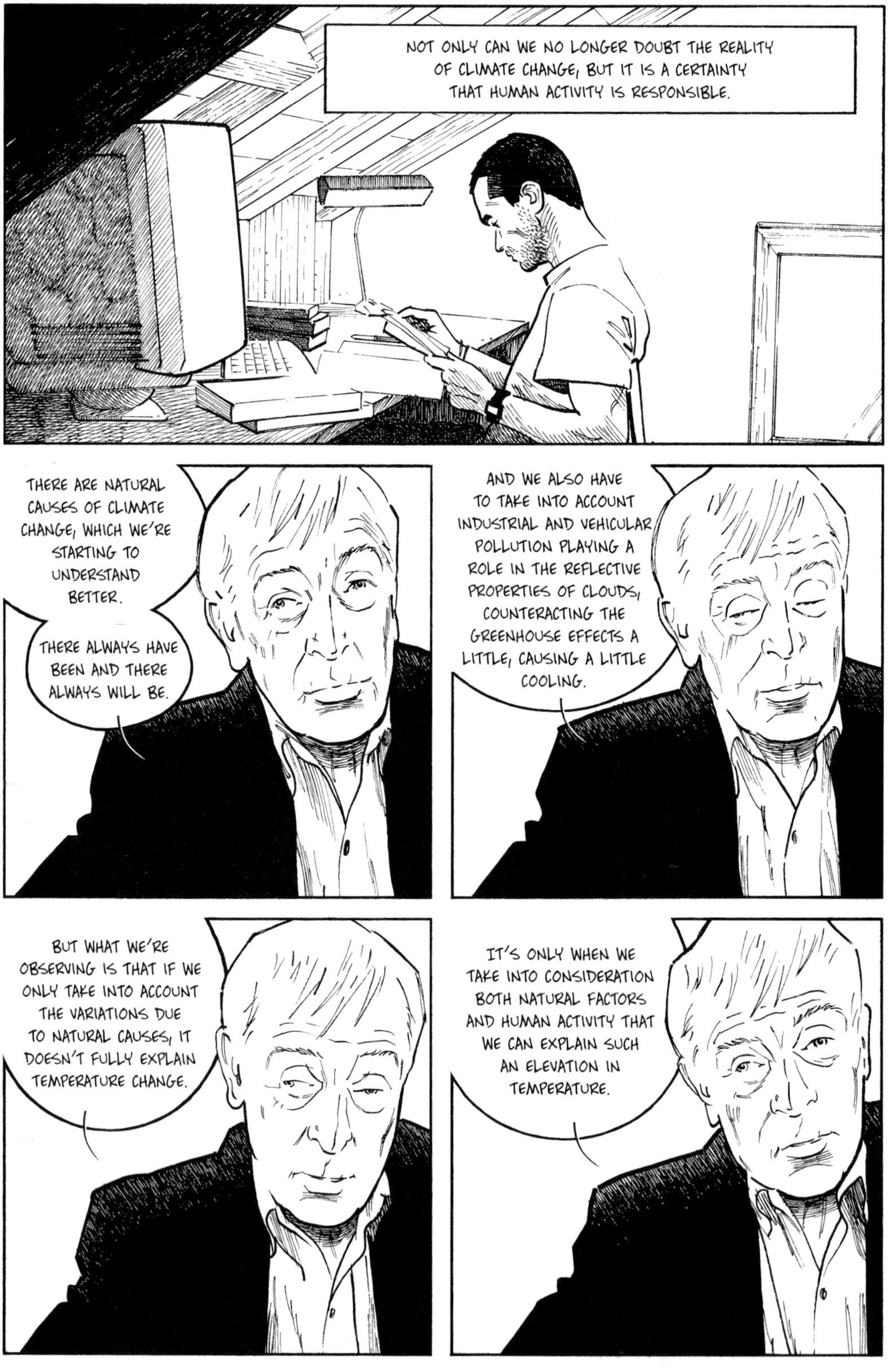
NOT ONLY CAN WE NO LONGER DOUBT THE REALITY OF CLIMATE CHANGE, BUT IT IS A CERTAINTY THAT HUMAN ACTIVITY IS RESPONSIBLE.
THERE ARE NATURAL CAUSES OF CLIMATE CHANGE, WHICH WE'RE STARTING TO UNDERSTAND BETTER.
THERE ALWAYS HAVE BEEN AND THERE ALWAYS WILL BE.
AND WE ALSO HAVE TO TAKE INTO ACCOUNT INDUSTRIAL AND VEHICULAR POLLUTION PLAYING A ROLE IN THE REFLECTIVE PROPERTIES OF CLOUDS, COUNTERACTING THE GREENHOUSE EFFECTS A LITTLE, CAUSING A LITTLE COOLING.
BUT WHAT WE'RE OBSERVING IS THAT IF WE ONLY TAKE INTO ACCOUNT THE VARIATIONS DUE TO NATURAL CAUSES, IT DOESN'T FULLY EXPLAIN TEMPERATURE CHANGE.
IT'S ONLY WHEN WE TAKE INTO CONSIDERATION BOTH NATURAL FACTORS AND HUMAN ACTIVITY THAT WE CAN EXPLAIN SUCH AN ELEVATION IN TEMPERATURE.

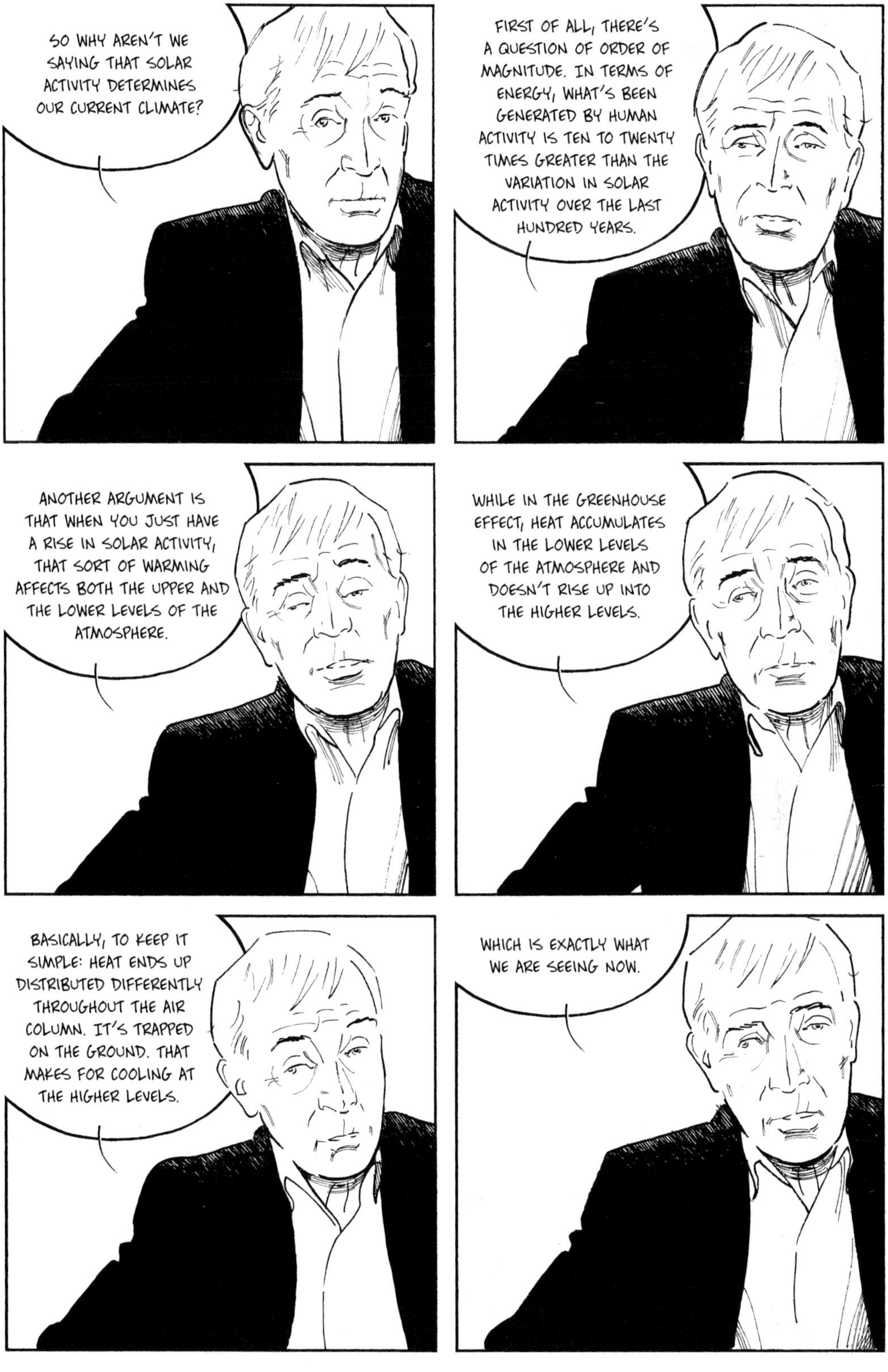
SO WHY AREN'T WE SAYING THAT SOLAR ACTIVITY DETERMINES OUR CURRENT CLIMATE?
FIRST OF ALL, THERE'S A QUESTION OF ORDER OF MAGNITUDE. IN TERMS OF ENERGY, WHAT'S BEEN GENERATED BY HUMAN ACTIVITY IS TEN TO TWENTY TIMES GREATER THAN THE VARIATION IN SOLAR ACTIVITY OVER THE LAST HUNDRED YEARS.
ANOTHER ARGUMENT IS THAT WHEN YOU JUST HAVE A RISE IN SOLAR ACTIVITY, THAT SORT OF WARMING AFFECTS BOTH THE UPPER AND THE LOWER LEVELS OF THE ATMOSPHERE.
WHILE IN THE GREENHOUSE EFFECT, HEAT ACCUMULATES IN THE LOWER LEVELS OF THE ATMOSPHERE AND DOESN'T RISE UP INTO THE HIGHER LEVELS.
BASICALLY, TO KEEP IT SIMPLE: HEAT ENDS UP DISTRIBUTED DIFFERENTLY THROUGHOUT THE AIR COLUMN. IT'S TRAPPED ON THE GROUND. THAT MAKES FOR COOLING AT THE HIGHER LEVELS.
WHICH IS EXACTLY WHAT WE ARE SEEING NOW.

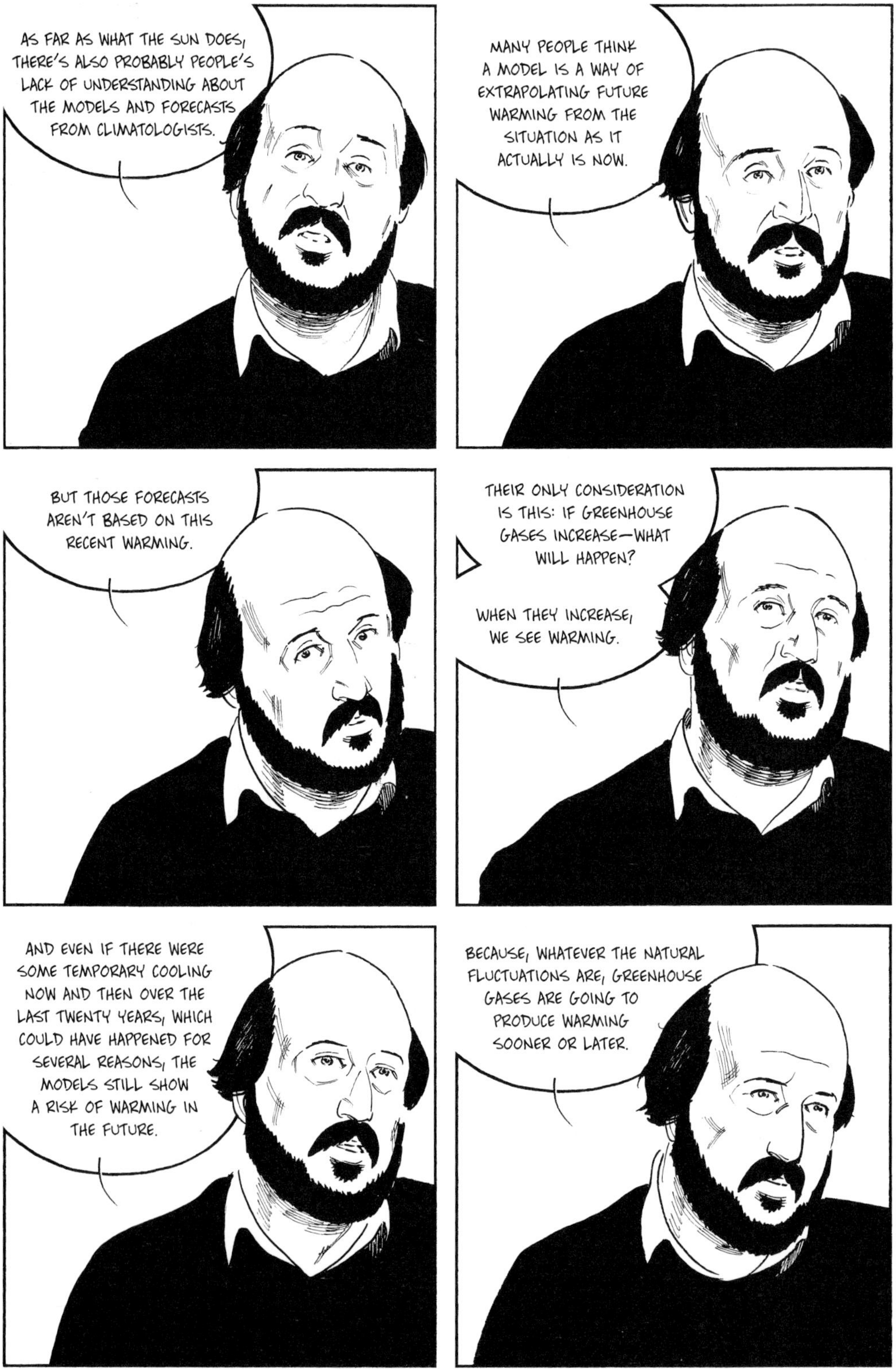
AS FAR AS WHAT THE SUN DOES, THERE'S ALSO PROBABLY PEOPLE'S LACK OF UNDERSTANDING ABOUT THE MODELS AND FORECASTS FROM CLIMATOLOGISTS.
MANY PEOPLE THINK A MODEL IS A WAY OF EXTRAPOLATING FUTURE WARMING FROM THE SITUATION AS IT ACTUALLY IS NOW.
BUT THOSE FORECASTS AREN'T BASED ON THIS RECENT WARMING.
THEIR ONLY CONSIDERATION IS THIS: IF GREENHOUSE GASES INCREASE—WHAT WILL HAPPEN?
WHEN THEY INCREASE, WE SEE WARMING.
AND EVEN IF THERE WERE SOME TEMPORARY COOLING NOW AND THEN OVER THE LAST TWENTY YEARS, WHICH COULD HAVE HAPPENED FOR SEVERAL REASONS, THE MODELS STILL SHOW A RISK OF WARMING IN THE FUTURE.
BECAUSE, WHATEVER THE NATURAL FLUCTUATIONS ARE, GREENHOUSE GASES ARE GOING TO PRODUCE WARMING SOONER OR LATER.

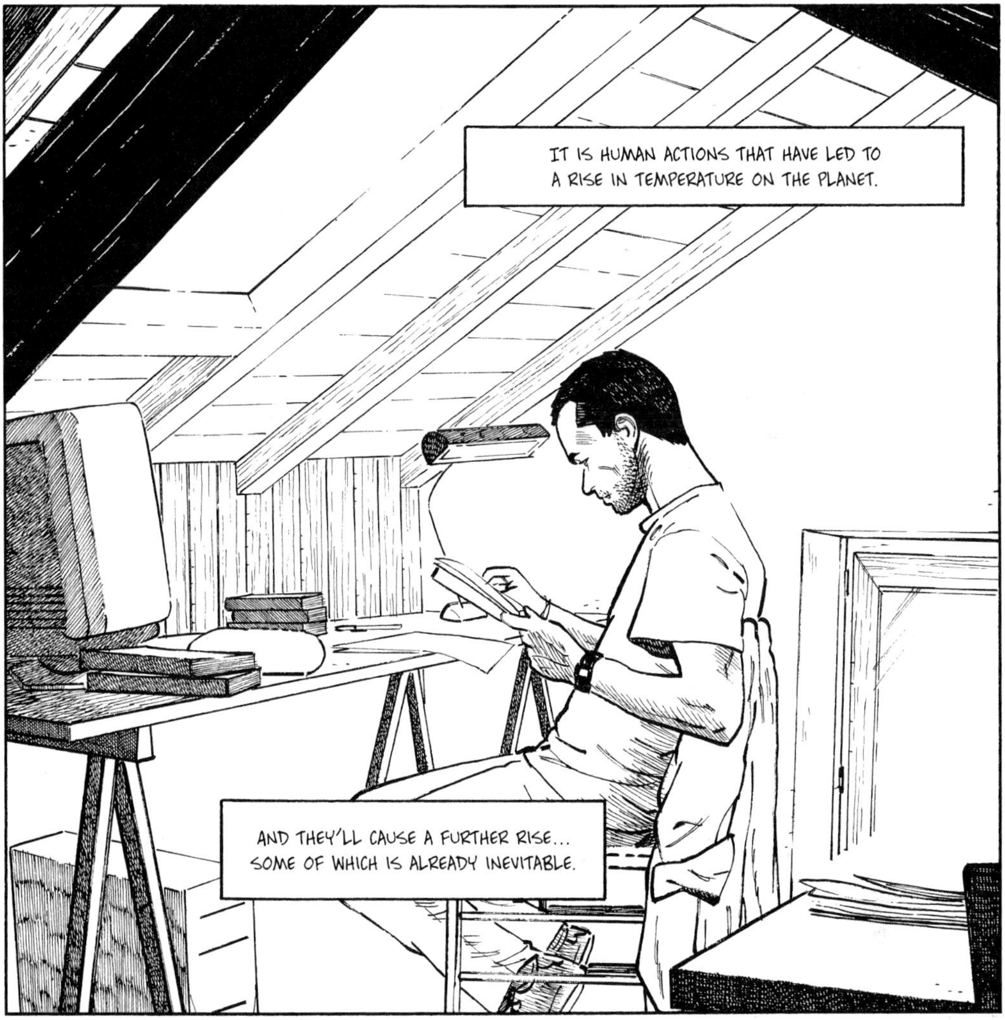
IT IS HUMAN ACTIONS THAT HAVE LED TO A RISE IN TEMPERATURE ON THE PLANET.
AND THEY'LL CAUSE A FURTHER RISE... SOME OF WHICH IS ALREADY INEVITABLE.

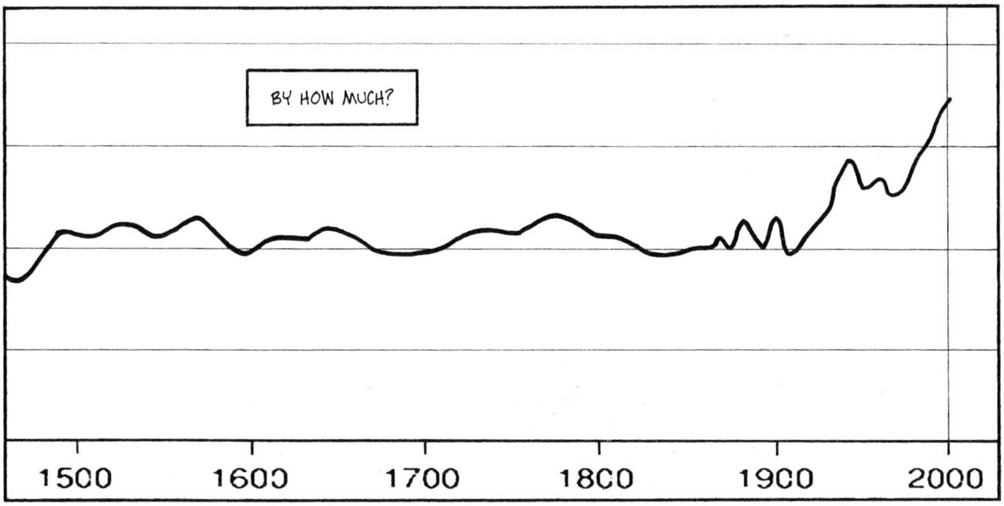
BY HOW MUCH?
1500
1600
1700
1800
1900
2000

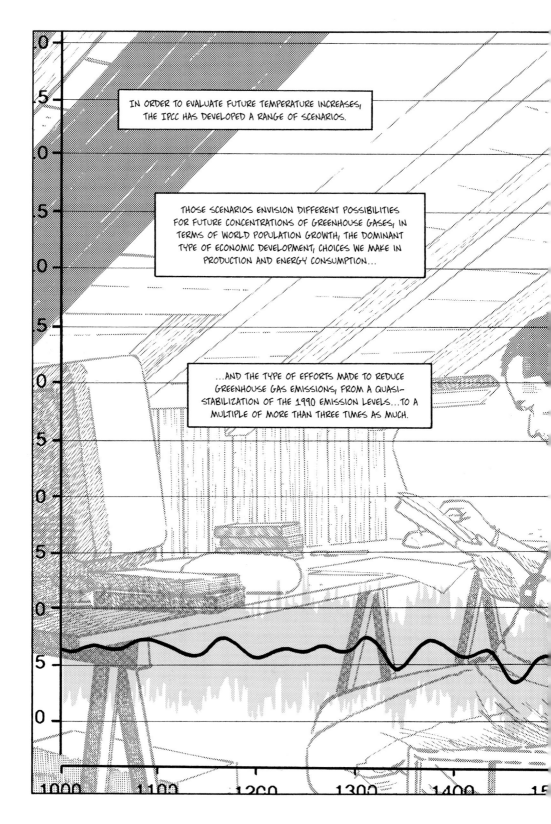

IN ORDER TO EVALUATE FUTURE TEMPERATURE INCREASES, THE IPCC HAS DEVELOPED A RANGE OF SCENARIOS.

THOSE SCENARIOS ENVISION DIFFERENT POSSIBILITIES FOR FUTURE CONCENTRATIONS OF GREENHOUSE GASES, IN TERMS OF WORLD POPULATION GROWTH, THE DOMINANT TYPE OF ECONOMIC DEVELOPMENT, CHOICES WE MAKE IN PRODUCTION AND ENERGY CONSUMPTION...

...AND THE TYPE OF EFFORTS MADE TO REDUCE GREENHOUSE GAS EMISSIONS, FROM A QUASI-STABILIZATION OF THE 1990 EMISSION LEVELS...TO A MULTIPLE OF MORE THAN THREE TIMES AS MUCH.

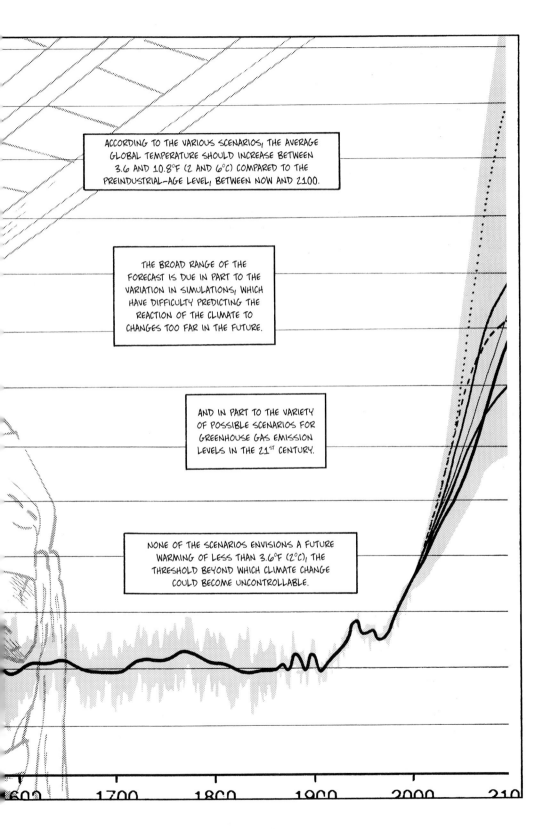

ACCORDING TO THE VARIOUS SCENARIOS, THE AVERAGE
GLOBAL TEMPERATURE SHOULD INCREASE BETWEEN
3.6 AND 10.8°F (2 AND 6°C) COMPARED TO THE
PREINDUSTRIAL-AGE LEVEL, BETWEEN NOW AND 2100.

THE BROAD RANGE OF THE
FORECAST IS DUE IN PART TO THE
VARIATION IN SIMULATIONS, WHICH
HAVE DIFFICULTY PREDICTING THE
REACTION OF THE CLIMATE TO
CHANGES TOO FAR IN THE FUTURE.

AND IN PART TO THE VARIETY
OF POSSIBLE SCENARIOS FOR
GREENHOUSE GAS EMISSION
LEVELS IN THE 21ST CENTURY.

NONE OF THE SCENARIOS ENVISIONS A FUTURE
WARMING OF LESS THAN 3.6°F (2°C), THE
THRESHOLD BEYOND WHICH CLIMATE CHANGE
COULD BECOME UNCONTROLLABLE.

1600   1700   1800   1900   2000   2100

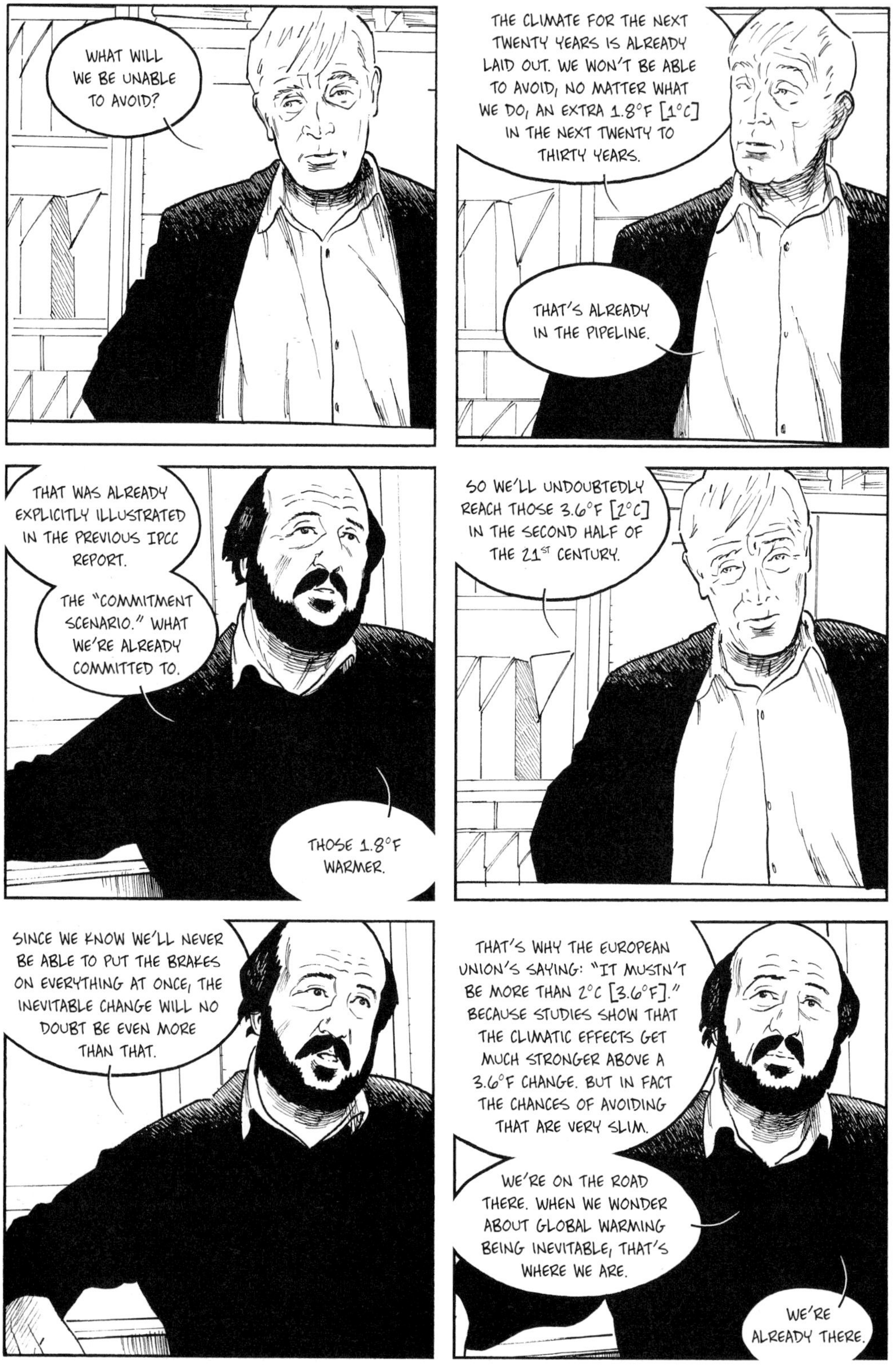
WHAT WILL WE BE UNABLE TO AVOID?
THE CLIMATE FOR THE NEXT TWENTY YEARS IS ALREADY LAID OUT. WE WON'T BE ABLE TO AVOID, NO MATTER WHAT WE DO, AN EXTRA 1.8°F [1°C] IN THE NEXT TWENTY TO THIRTY YEARS.
THAT'S ALREADY IN THE PIPELINE.
THAT WAS ALREADY EXPLICITLY ILLUSTRATED IN THE PREVIOUS IPCC REPORT.
THE "COMMITMENT SCENARIO." WHAT WE'RE ALREADY COMMITTED TO.
THOSE 1.8°F WARMER.
SO WE'LL UNDOUBTEDLY REACH THOSE 3.6°F [2°C] IN THE SECOND HALF OF THE 21ST CENTURY.
SINCE WE KNOW WE'LL NEVER BE ABLE TO PUT THE BRAKES ON EVERYTHING AT ONCE, THE INEVITABLE CHANGE WILL NO DOUBT BE EVEN MORE THAN THAT.
THAT'S WHY THE EUROPEAN UNION'S SAYING: "IT MUSTN'T BE MORE THAN 2°C [3.6°F]." BECAUSE STUDIES SHOW THAT THE CLIMATIC EFFECTS GET MUCH STRONGER ABOVE A 3.6°F CHANGE. BUT IN FACT THE CHANCES OF AVOIDING THAT ARE VERY SLIM.
WE'RE ON THE ROAD THERE. WHEN WE WONDER ABOUT GLOBAL WARMING BEING INEVITABLE, THAT'S WHERE WE ARE.
WE'RE ALREADY THERE.

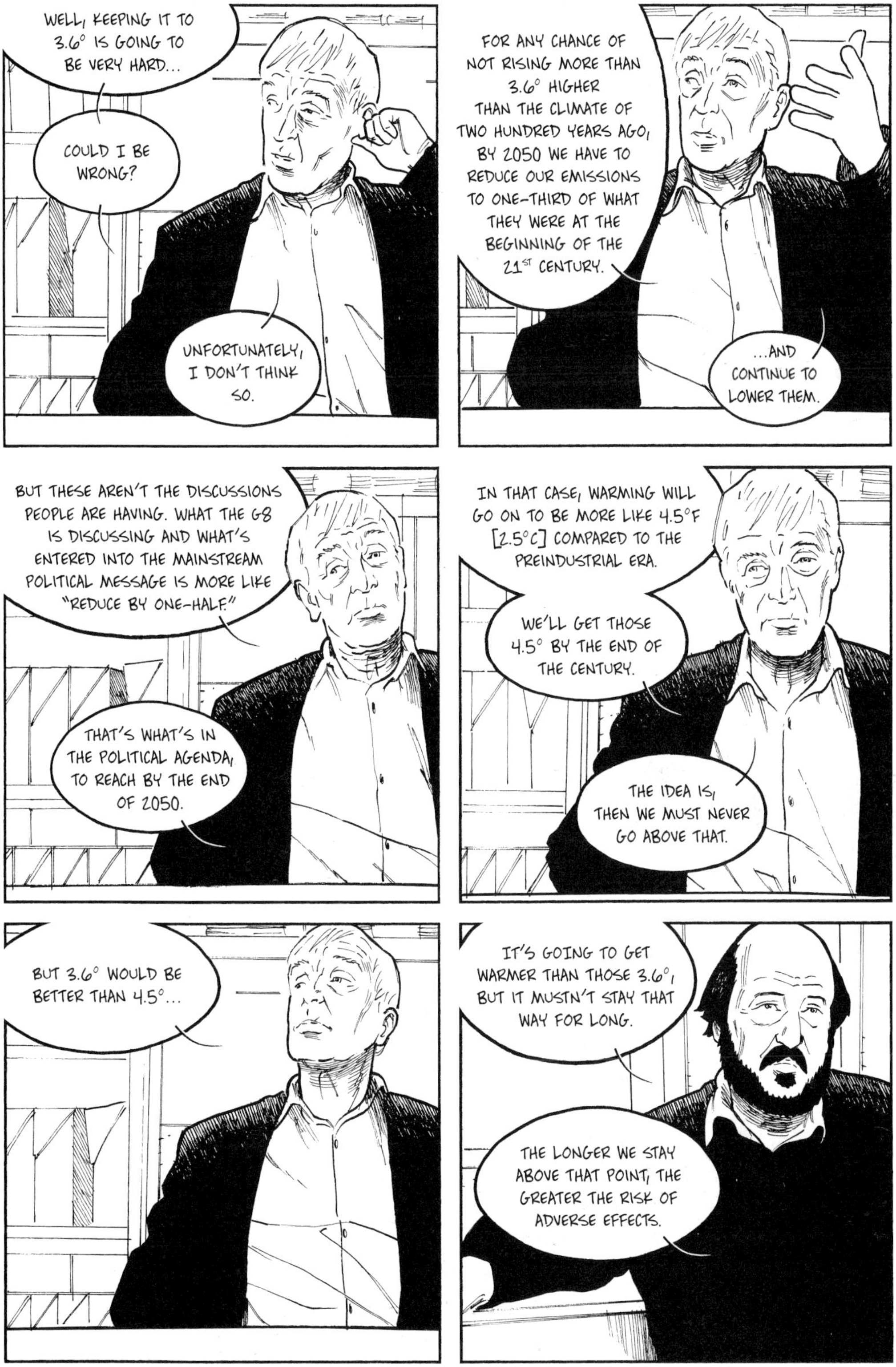
WELL, KEEPING IT TO 3.6° IS GOING TO BE VERY HARD...
COULD I BE WRONG?
UNFORTUNATELY, I DON'T THINK SO.
FOR ANY CHANCE OF NOT RISING MORE THAN 3.6° HIGHER THAN THE CLIMATE OF TWO HUNDRED YEARS AGO, BY 2050 WE HAVE TO REDUCE OUR EMISSIONS TO ONE-THIRD OF WHAT THEY WERE AT THE BEGINNING OF THE 21ST CENTURY.
...AND CONTINUE TO LOWER THEM.
BUT THESE AREN'T THE DISCUSSIONS PEOPLE ARE HAVING. WHAT THE G8 IS DISCUSSING AND WHAT'S ENTERED INTO THE MAINSTREAM POLITICAL MESSAGE IS MORE LIKE "REDUCE BY ONE-HALF."
THAT'S WHAT'S IN THE POLITICAL AGENDA, TO REACH BY THE END OF 2050.
IN THAT CASE, WARMING WILL GO ON TO BE MORE LIKE 4.5°F [2.5°C] COMPARED TO THE PREINDUSTRIAL ERA.
WE'LL GET THOSE 4.5° BY THE END OF THE CENTURY.
THE IDEA IS, THEN WE MUST NEVER GO ABOVE THAT.
BUT 3.6° WOULD BE BETTER THAN 4.5°...
IT'S GOING TO GET WARMER THAN THOSE 3.6°, BUT IT MUSTN'T STAY THAT WAY FOR LONG.
THE LONGER WE STAY ABOVE THAT POINT, THE GREATER THE RISK OF ADVERSE EFFECTS.

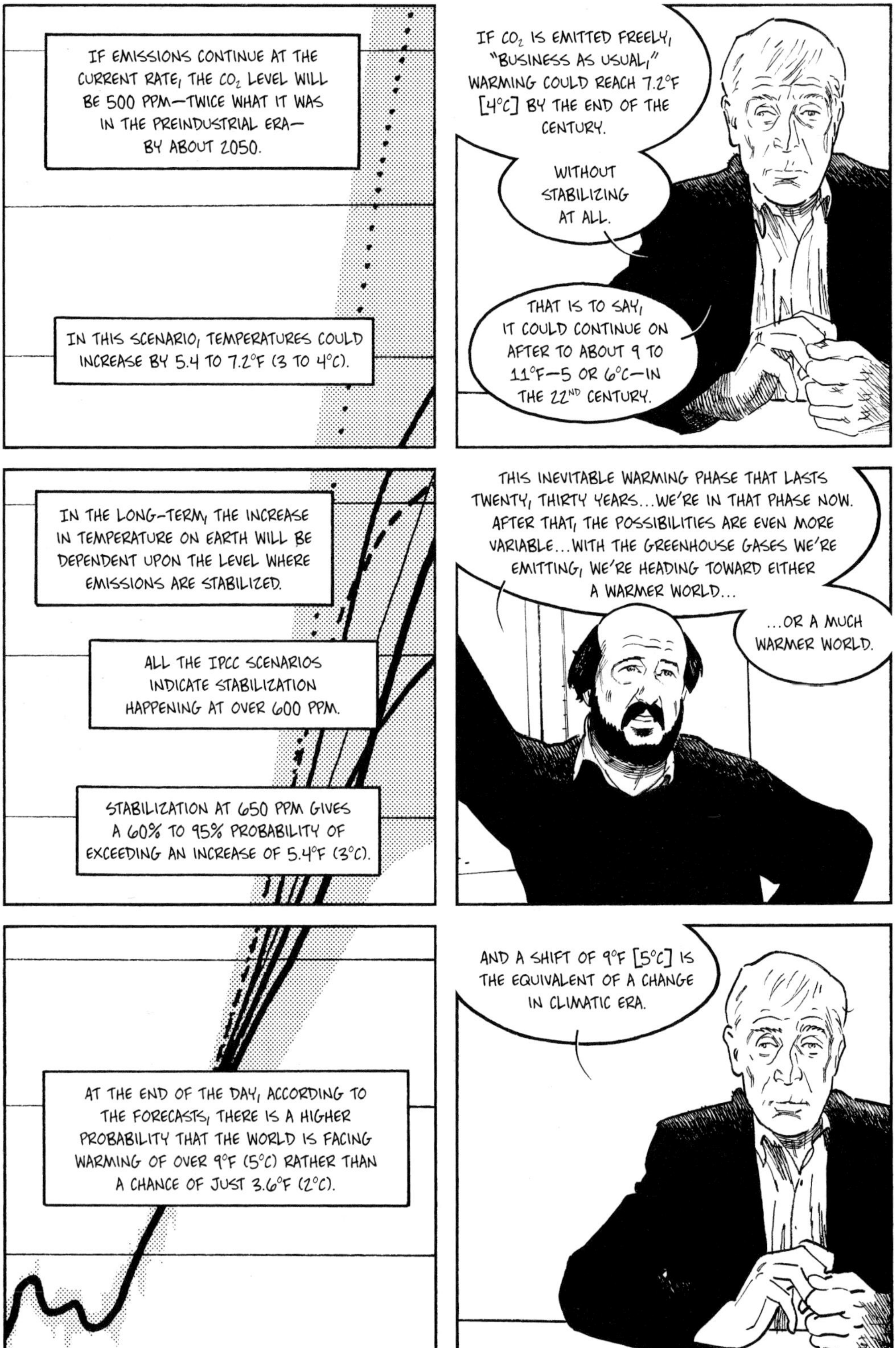

IF EMISSIONS CONTINUE AT THE CURRENT RATE, THE CO_2 LEVEL WILL BE 500 PPM—TWICE WHAT IT WAS IN THE PREINDUSTRIAL ERA—BY ABOUT 2050.
IN THIS SCENARIO, TEMPERATURES COULD INCREASE BY 5.4 TO 7.2°F (3 TO 4°C).
IF CO_2 IS EMITTED FREELY, "BUSINESS AS USUAL," WARMING COULD REACH 7.2°F [4°C] BY THE END OF THE CENTURY.
WITHOUT STABILIZING AT ALL.
THAT IS TO SAY, IT COULD CONTINUE ON AFTER TO ABOUT 9 TO 11°F—5 OR 6°C—IN THE 22ND CENTURY.
IN THE LONG-TERM, THE INCREASE IN TEMPERATURE ON EARTH WILL BE DEPENDENT UPON THE LEVEL WHERE EMISSIONS ARE STABILIZED.
ALL THE IPCC SCENARIOS INDICATE STABILIZATION HAPPENING AT OVER 600 PPM.
STABILIZATION AT 650 PPM GIVES A 60% TO 95% PROBABILITY OF EXCEEDING AN INCREASE OF 5.4°F (3°C).
THIS INEVITABLE WARMING PHASE THAT LASTS TWENTY, THIRTY YEARS...WE'RE IN THAT PHASE NOW. AFTER THAT, THE POSSIBILITIES ARE EVEN MORE VARIABLE...WITH THE GREENHOUSE GASES WE'RE EMITTING, WE'RE HEADING TOWARD EITHER A WARMER WORLD...
...OR A MUCH WARMER WORLD.
AT THE END OF THE DAY, ACCORDING TO THE FORECASTS, THERE IS A HIGHER PROBABILITY THAT THE WORLD IS FACING WARMING OF OVER 9°F (5°C) RATHER THAN A CHANCE OF JUST 3.6°F (2°C).
AND A SHIFT OF 9°F [5°C] IS THE EQUIVALENT OF A CHANGE IN CLIMATIC ERA.

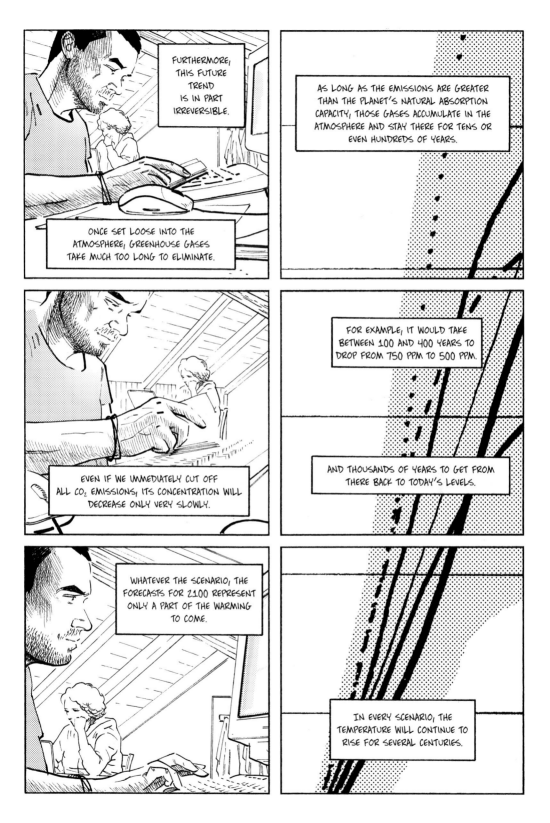

FURTHERMORE, THIS FUTURE TREND IS IN PART IRREVERSIBLE.

ONCE SET LOOSE INTO THE ATMOSPHERE, GREENHOUSE GASES TAKE MUCH TOO LONG TO ELIMINATE.

AS LONG AS THE EMISSIONS ARE GREATER THAN THE PLANET'S NATURAL ABSORPTION CAPACITY, THOSE GASES ACCUMULATE IN THE ATMOSPHERE AND STAY THERE FOR TENS OR EVEN HUNDREDS OF YEARS.

EVEN IF WE IMMEDIATELY CUT OFF ALL CO_2 EMISSIONS, ITS CONCENTRATION WILL DECREASE ONLY VERY SLOWLY.

FOR EXAMPLE, IT WOULD TAKE BETWEEN 100 AND 400 YEARS TO DROP FROM 750 PPM TO 500 PPM.

AND THOUSANDS OF YEARS TO GET FROM THERE BACK TO TODAY'S LEVELS.

WHATEVER THE SCENARIO, THE FORECASTS FOR 2100 REPRESENT ONLY A PART OF THE WARMING TO COME.

IN EVERY SCENARIO, THE TEMPERATURE WILL CONTINUE TO RISE FOR SEVERAL CENTURIES.

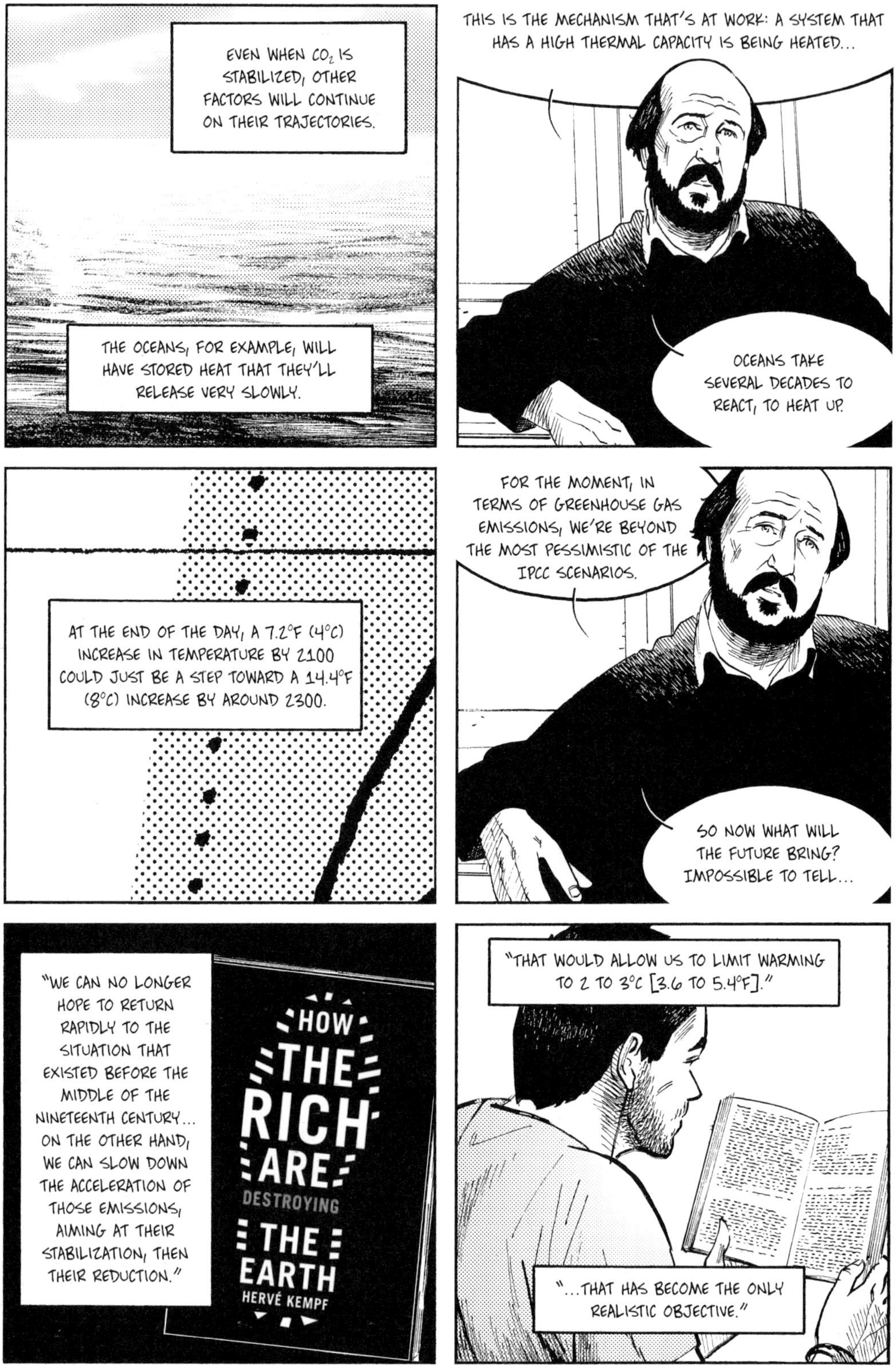
EVEN WHEN CO_2 IS STABILIZED, OTHER FACTORS WILL CONTINUE ON THEIR TRAJECTORIES.
THE OCEANS, FOR EXAMPLE, WILL HAVE STORED HEAT THAT THEY'LL RELEASE VERY SLOWLY.
THIS IS THE MECHANISM THAT'S AT WORK: A SYSTEM THAT HAS A HIGH THERMAL CAPACITY IS BEING HEATED...
OCEANS TAKE SEVERAL DECADES TO REACT, TO HEAT UP.
AT THE END OF THE DAY, A 7.2°F (4°C) INCREASE IN TEMPERATURE BY 2100 COULD JUST BE A STEP TOWARD A 14.4°F (8°C) INCREASE BY AROUND 2300.
FOR THE MOMENT, IN TERMS OF GREENHOUSE GAS EMISSIONS, WE'RE BEYOND THE MOST PESSIMISTIC OF THE IPCC SCENARIOS.
SO NOW WHAT WILL THE FUTURE BRING? IMPOSSIBLE TO TELL...
"WE CAN NO LONGER HOPE TO RETURN RAPIDLY TO THE SITUATION THAT EXISTED BEFORE THE MIDDLE OF THE NINETEENTH CENTURY... ON THE OTHER HAND, WE CAN SLOW DOWN THE ACCELERATION OF THOSE EMISSIONS, AIMING AT THEIR STABILIZATION, THEN THEIR REDUCTION."
HOW THE RICH ARE DESTROYING THE EARTH
HERVÉ KEMPF
"THAT WOULD ALLOW US TO LIMIT WARMING TO 2 TO 3°C [3.6 TO 5.4°F]."
"...THAT HAS BECOME THE ONLY REALISTIC OBJECTIVE."

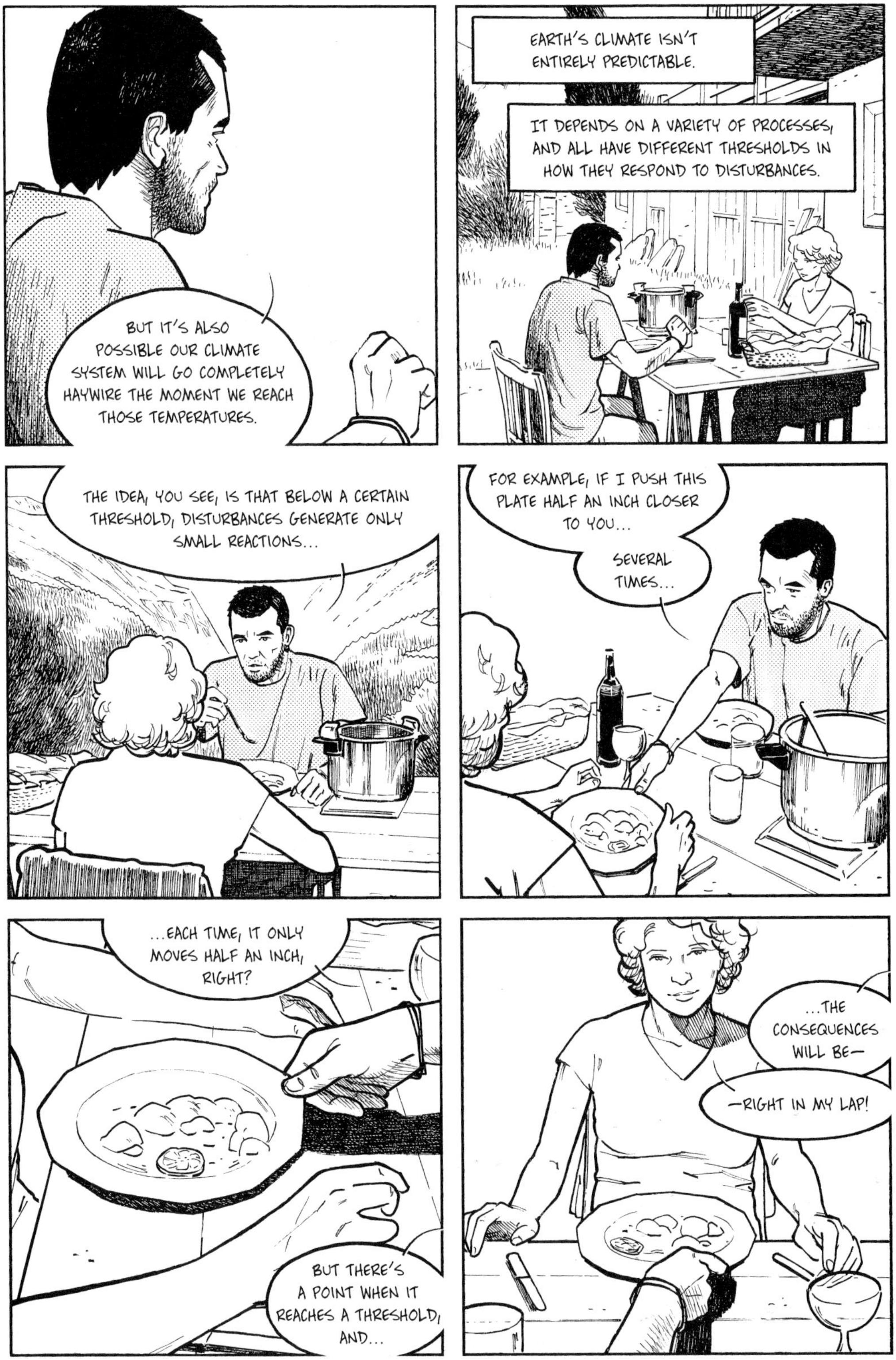
BUT IT'S ALSO POSSIBLE OUR CLIMATE SYSTEM WILL GO COMPLETELY HAYWIRE THE MOMENT WE REACH THOSE TEMPERATURES.
EARTH'S CLIMATE ISN'T ENTIRELY PREDICTABLE.
IT DEPENDS ON A VARIETY OF PROCESSES, AND ALL HAVE DIFFERENT THRESHOLDS IN HOW THEY RESPOND TO DISTURBANCES.
THE IDEA, YOU SEE, IS THAT BELOW A CERTAIN THRESHOLD, DISTURBANCES GENERATE ONLY SMALL REACTIONS...
FOR EXAMPLE, IF I PUSH THIS PLATE HALF AN INCH CLOSER TO YOU...
SEVERAL TIMES...
...EACH TIME, IT ONLY MOVES HALF AN INCH, RIGHT?
BUT THERE'S A POINT WHEN IT REACHES A THRESHOLD, AND...
...THE CONSEQUENCES WILL BE—
—RIGHT IN MY LAP!

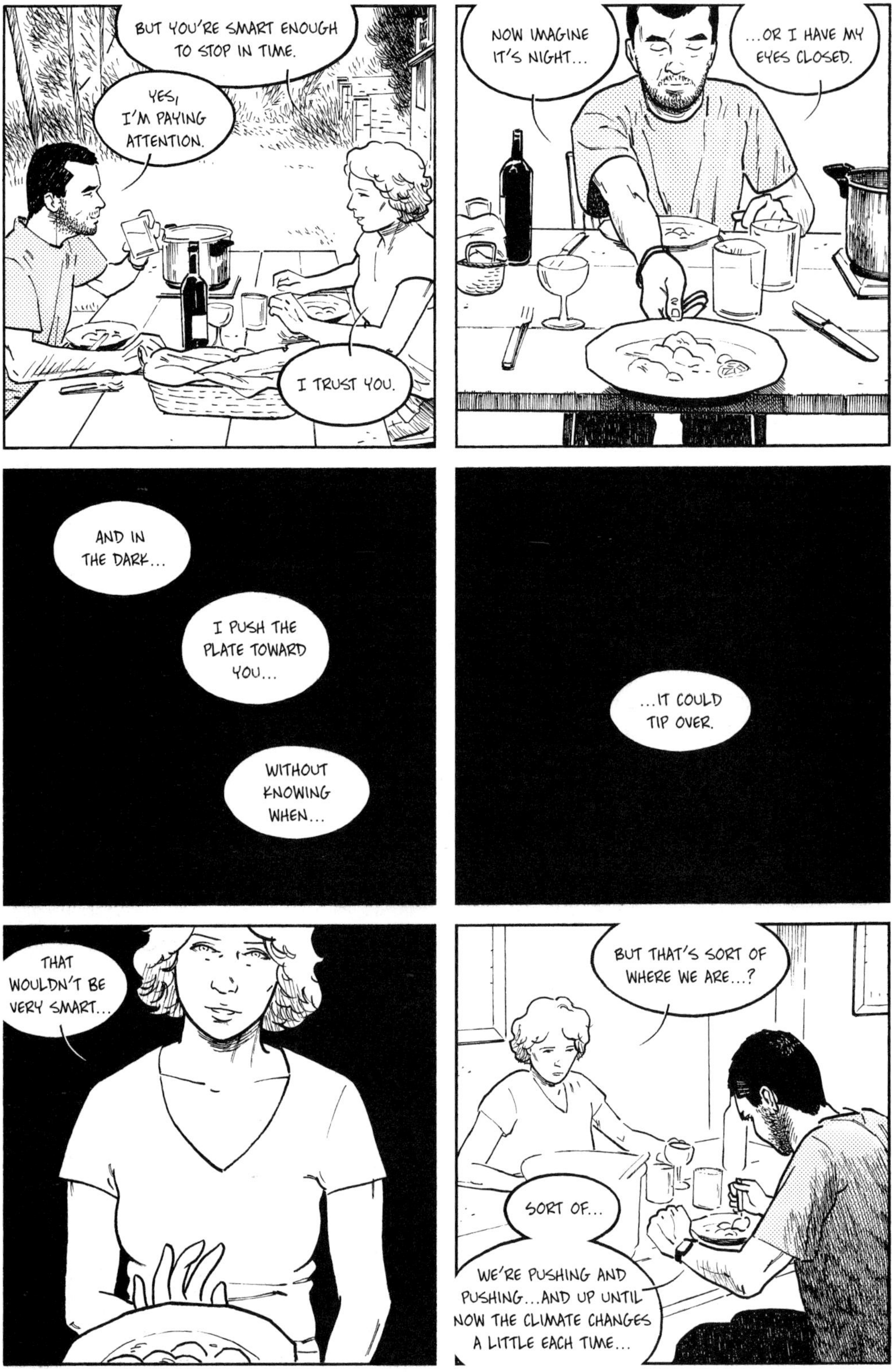
BUT YOU'RE SMART ENOUGH TO STOP IN TIME.
YES, I'M PAYING ATTENTION.
I TRUST YOU.
NOW IMAGINE IT'S NIGHT...
...OR I HAVE MY EYES CLOSED.
AND IN THE DARK...
I PUSH THE PLATE TOWARD YOU...
WITHOUT KNOWING WHEN...
...IT COULD TIP OVER.
THAT WOULDN'T BE VERY SMART...
BUT THAT'S SORT OF WHERE WE ARE...?
SORT OF...
WE'RE PUSHING AND PUSHING...AND UP UNTIL NOW THE CLIMATE CHANGES A LITTLE EACH TIME...

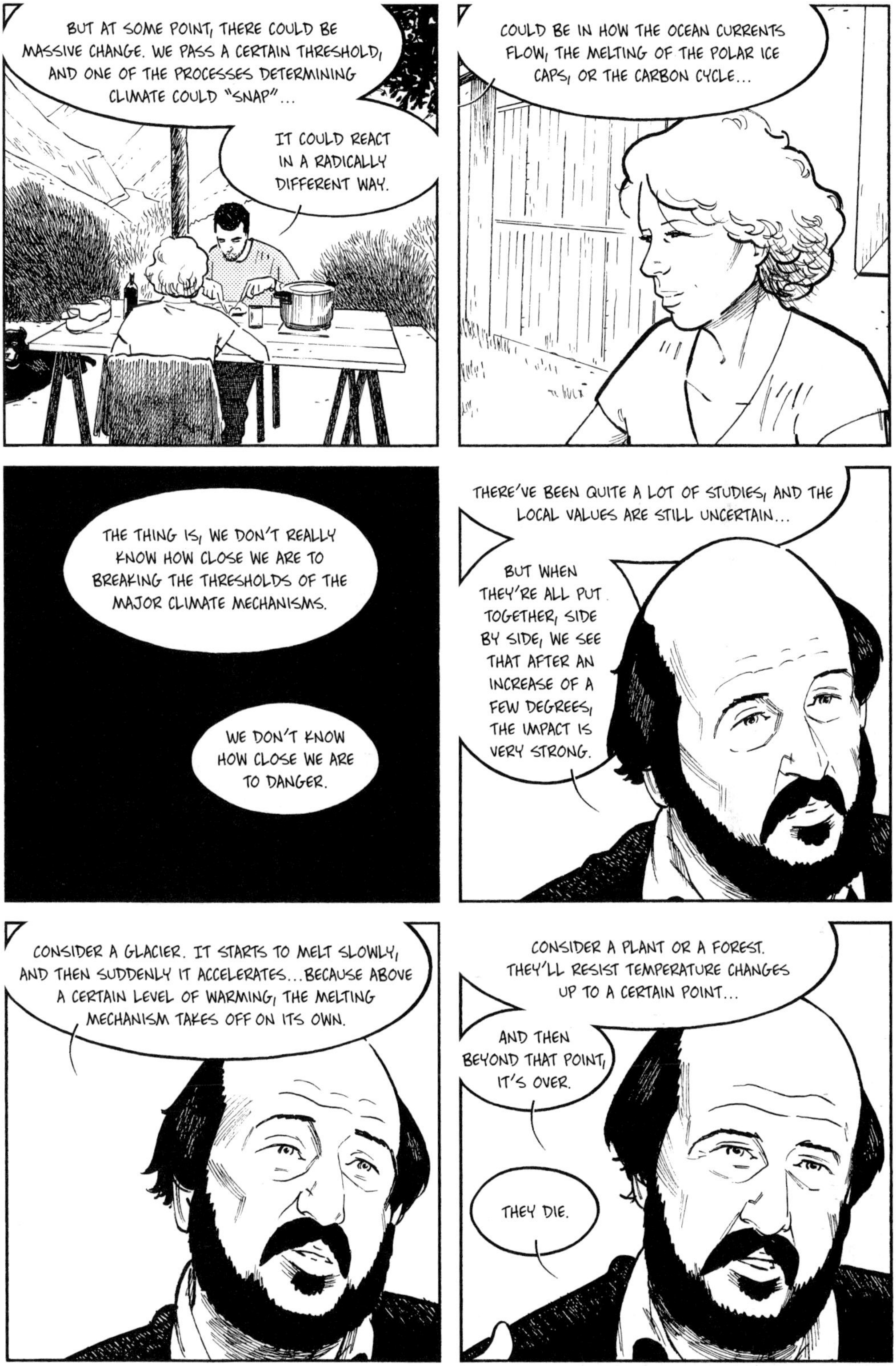
BUT AT SOME POINT, THERE COULD BE MASSIVE CHANGE. WE PASS A CERTAIN THRESHOLD, AND ONE OF THE PROCESSES DETERMINING CLIMATE COULD "SNAP"...
IT COULD REACT IN A RADICALLY DIFFERENT WAY.
COULD BE IN HOW THE OCEAN CURRENTS FLOW, THE MELTING OF THE POLAR ICE CAPS, OR THE CARBON CYCLE...
THE THING IS, WE DON'T REALLY KNOW HOW CLOSE WE ARE TO BREAKING THE THRESHOLDS OF THE MAJOR CLIMATE MECHANISMS.
WE DON'T KNOW HOW CLOSE WE ARE TO DANGER.
THERE'VE BEEN QUITE A LOT OF STUDIES, AND THE LOCAL VALUES ARE STILL UNCERTAIN...
BUT WHEN THEY'RE ALL PUT TOGETHER, SIDE BY SIDE, WE SEE THAT AFTER AN INCREASE OF A FEW DEGREES, THE IMPACT IS VERY STRONG.
CONSIDER A GLACIER. IT STARTS TO MELT SLOWLY, AND THEN SUDDENLY IT ACCELERATES... BECAUSE ABOVE A CERTAIN LEVEL OF WARMING, THE MELTING MECHANISM TAKES OFF ON ITS OWN.
CONSIDER A PLANT OR A FOREST. THEY'LL RESIST TEMPERATURE CHANGES UP TO A CERTAIN POINT...
AND THEN BEYOND THAT POINT, IT'S OVER.
THEY DIE.

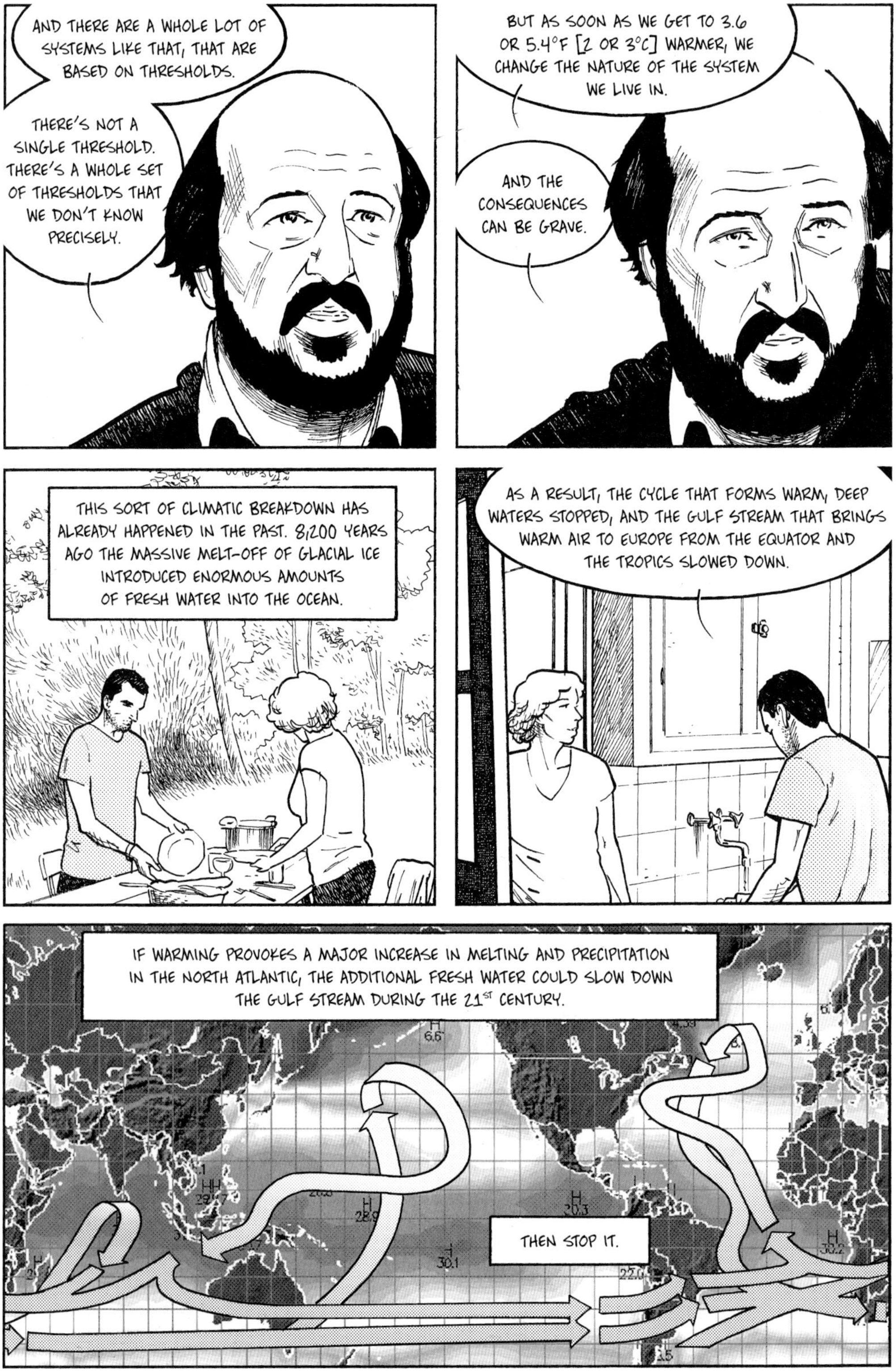

AND THERE ARE A WHOLE LOT OF SYSTEMS LIKE THAT, THAT ARE BASED ON THRESHOLDS.
THERE'S NOT A SINGLE THRESHOLD. THERE'S A WHOLE SET OF THRESHOLDS THAT WE DON'T KNOW PRECISELY.
BUT AS SOON AS WE GET TO 3.6 OR 5.4°F [2 OR 3°C] WARMER, WE CHANGE THE NATURE OF THE SYSTEM WE LIVE IN.
AND THE CONSEQUENCES CAN BE GRAVE.
THIS SORT OF CLIMATIC BREAKDOWN HAS ALREADY HAPPENED IN THE PAST. 8,200 YEARS AGO THE MASSIVE MELT-OFF OF GLACIAL ICE INTRODUCED ENORMOUS AMOUNTS OF FRESH WATER INTO THE OCEAN.
AS A RESULT, THE CYCLE THAT FORMS WARM, DEEP WATERS STOPPED, AND THE GULF STREAM THAT BRINGS WARM AIR TO EUROPE FROM THE EQUATOR AND THE TROPICS SLOWED DOWN.
IF WARMING PROVOKES A MAJOR INCREASE IN MELTING AND PRECIPITATION IN THE NORTH ATLANTIC, THE ADDITIONAL FRESH WATER COULD SLOW DOWN THE GULF STREAM DURING THE 21ST CENTURY.
THEN STOP IT.

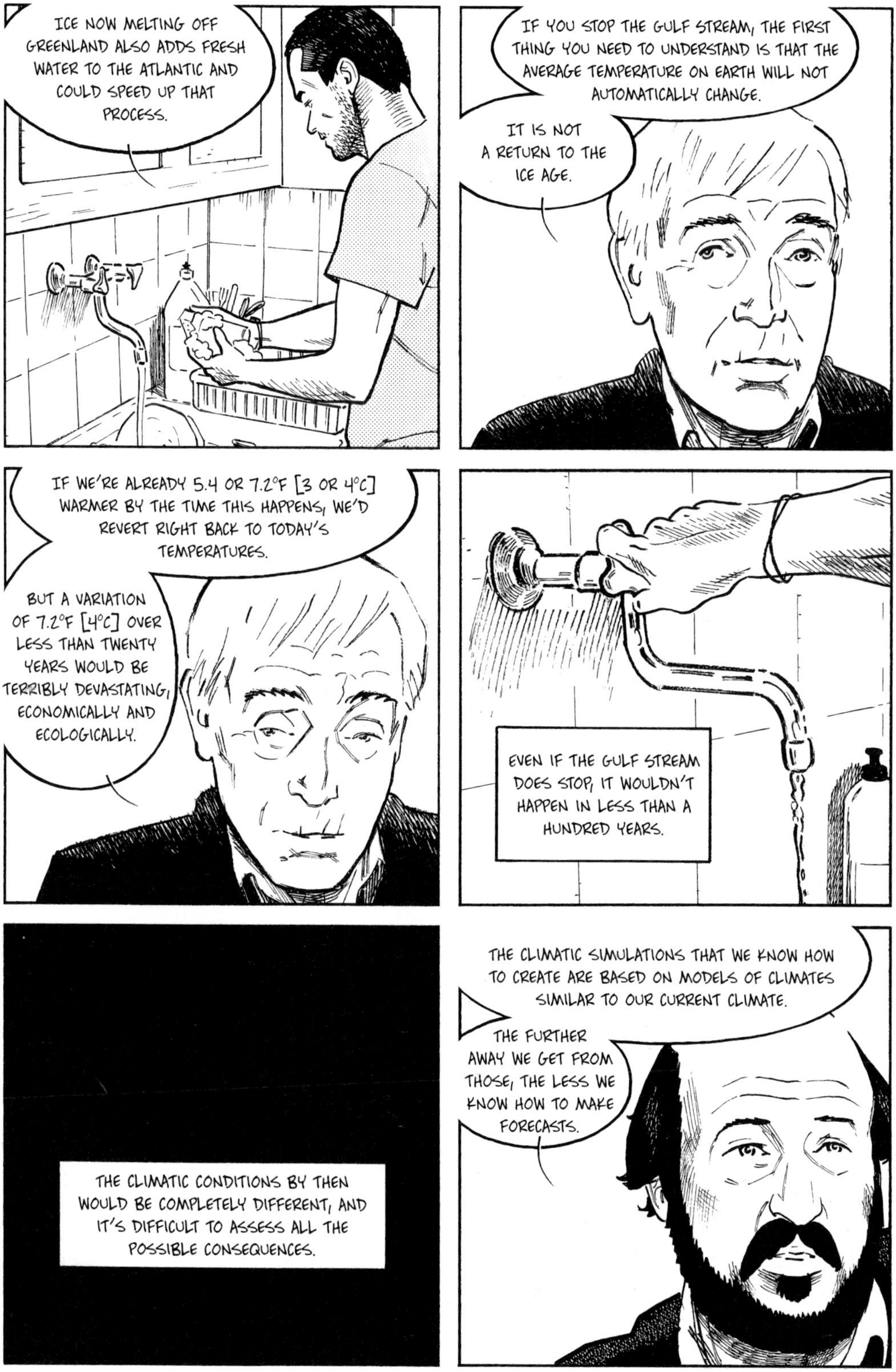
ICE NOW MELTING OFF GREENLAND ALSO ADDS FRESH WATER TO THE ATLANTIC AND COULD SPEED UP THAT PROCESS.
IF YOU STOP THE GULF STREAM, THE FIRST THING YOU NEED TO UNDERSTAND IS THAT THE AVERAGE TEMPERATURE ON EARTH WILL NOT AUTOMATICALLY CHANGE.
IT IS NOT A RETURN TO THE ICE AGE.
IF WE'RE ALREADY 5.4 OR 7.2°F [3 OR 4°C] WARMER BY THE TIME THIS HAPPENS, WE'D REVERT RIGHT BACK TO TODAY'S TEMPERATURES.
BUT A VARIATION OF 7.2°F [4°C] OVER LESS THAN TWENTY YEARS WOULD BE TERRIBLY DEVASTATING, ECONOMICALLY AND ECOLOGICALLY.
EVEN IF THE GULF STREAM DOES STOP, IT WOULDN'T HAPPEN IN LESS THAN A HUNDRED YEARS.
THE CLIMATIC CONDITIONS BY THEN WOULD BE COMPLETELY DIFFERENT, AND IT'S DIFFICULT TO ASSESS ALL THE POSSIBLE CONSEQUENCES.
THE CLIMATIC SIMULATIONS THAT WE KNOW HOW TO CREATE ARE BASED ON MODELS OF CLIMATES SIMILAR TO OUR CURRENT CLIMATE.
THE FURTHER AWAY WE GET FROM THOSE, THE LESS WE KNOW HOW TO MAKE FORECASTS.

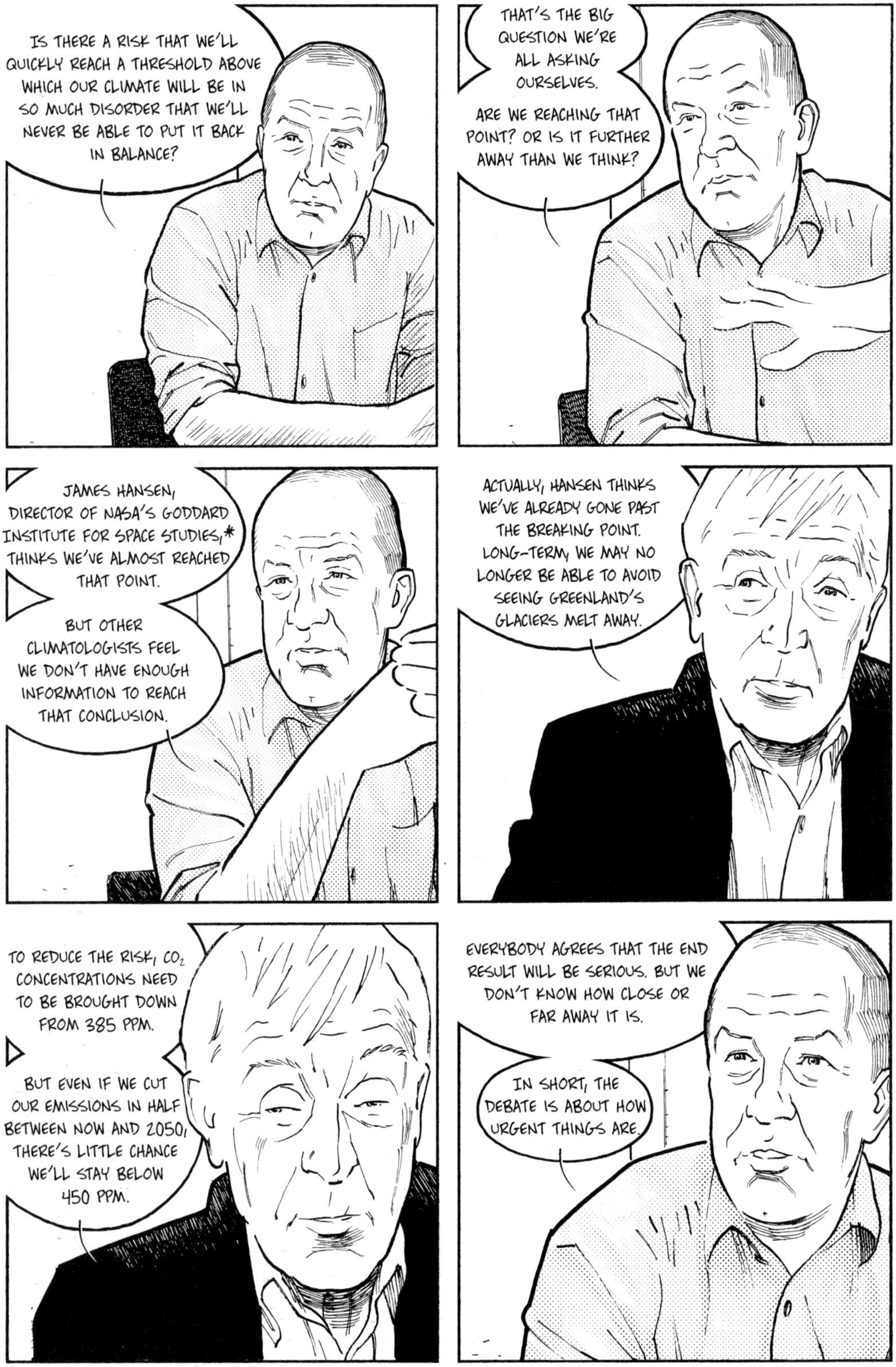
IS THERE A RISK THAT WE'LL QUICKLY REACH A THRESHOLD ABOVE WHICH OUR CLIMATE WILL BE IN SO MUCH DISORDER THAT WE'LL NEVER BE ABLE TO PUT IT BACK IN BALANCE?
THAT'S THE BIG QUESTION WE'RE ALL ASKING OURSELVES.
ARE WE REACHING THAT POINT? OR IS IT FURTHER AWAY THAN WE THINK?
JAMES HANSEN, DIRECTOR OF NASA'S GODDARD INSTITUTE FOR SPACE STUDIES,* THINKS WE'VE ALMOST REACHED THAT POINT.
BUT OTHER CLIMATOLOGISTS FEEL WE DON'T HAVE ENOUGH INFORMATION TO REACH THAT CONCLUSION.
ACTUALLY, HANSEN THINKS WE'VE ALREADY GONE PAST THE BREAKING POINT. LONG-TERM, WE MAY NO LONGER BE ABLE TO AVOID SEEING GREENLAND'S GLACIERS MELT AWAY.
TO REDUCE THE RISK, CO_2 CONCENTRATIONS NEED TO BE BROUGHT DOWN FROM 385 PPM.
BUT EVEN IF WE CUT OUR EMISSIONS IN HALF BETWEEN NOW AND 2050, THERE'S LITTLE CHANCE WE'LL STAY BELOW 450 PPM.
EVERYBODY AGREES THAT THE END RESULT WILL BE SERIOUS. BUT WE DON'T KNOW HOW CLOSE OR FAR AWAY IT IS.
IN SHORT, THE DEBATE IS ABOUT HOW URGENT THINGS ARE.

*FROM 1981 TO 2013.

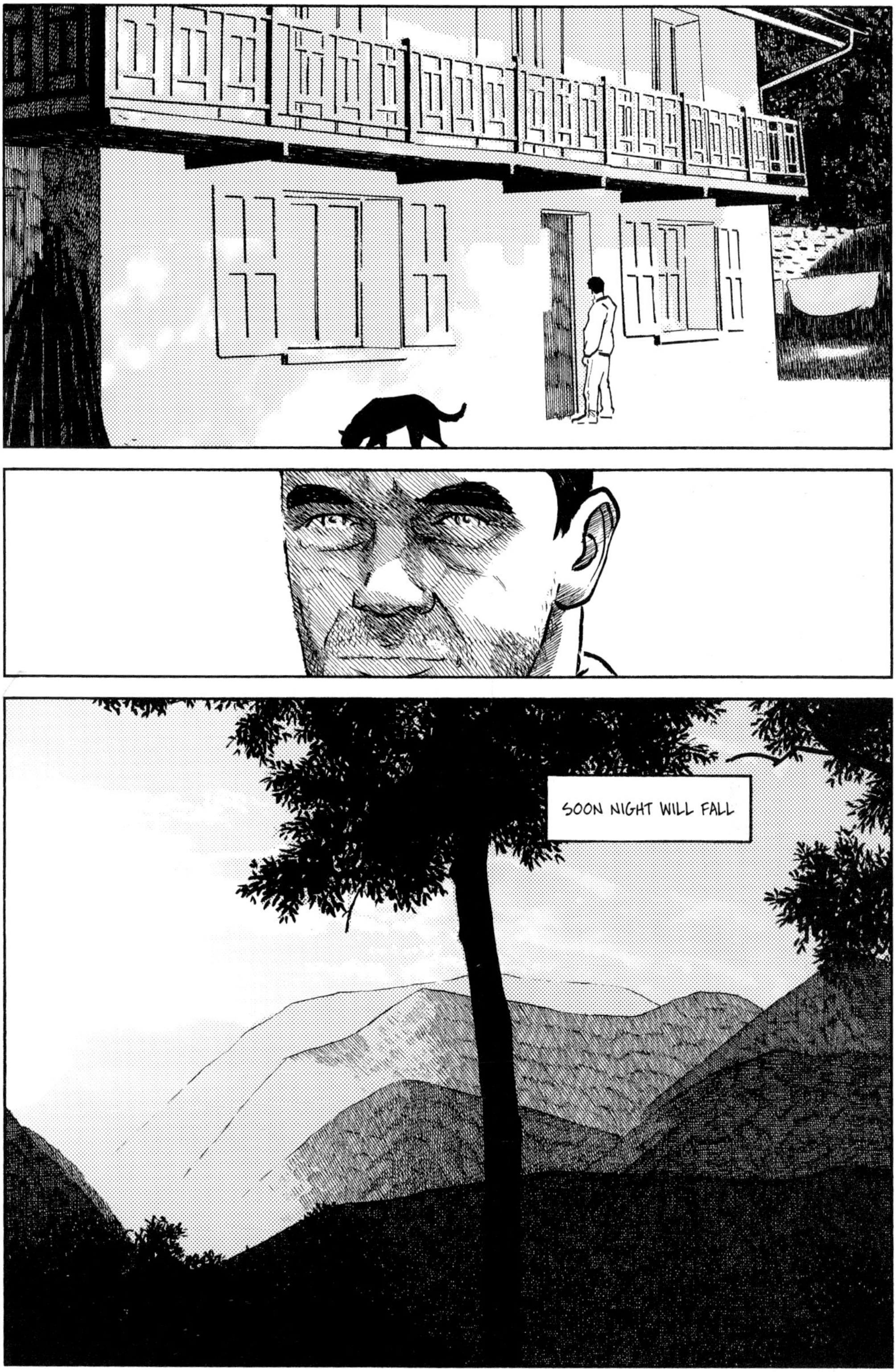
SOON NIGHT WILL FALL

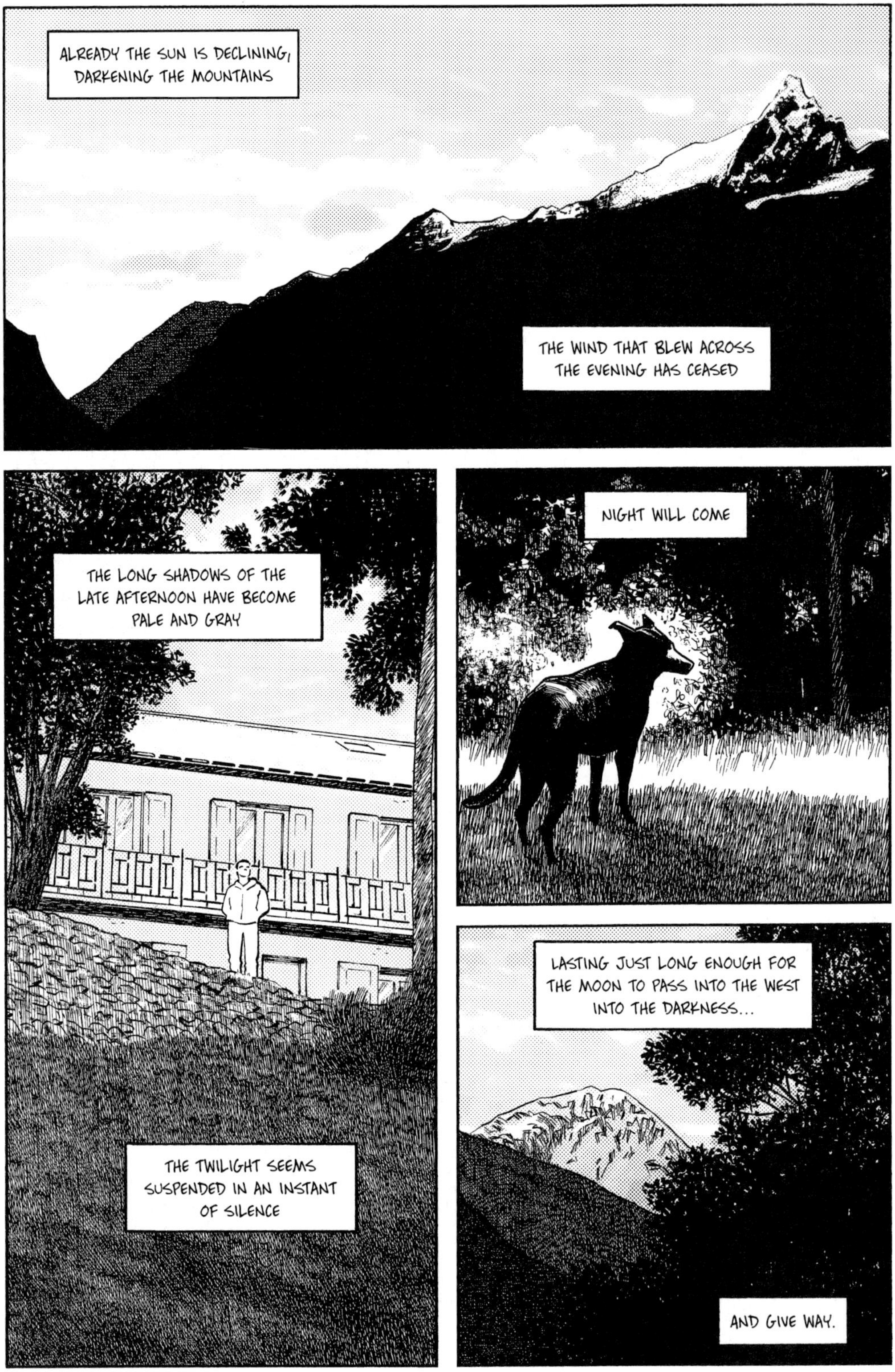
ALREADY THE SUN IS DECLINING, DARKENING THE MOUNTAINS
THE WIND THAT BLEW ACROSS THE EVENING HAS CEASED
THE LONG SHADOWS OF THE LATE AFTERNOON HAVE BECOME PALE AND GRAY
THE TWILIGHT SEEMS SUSPENDED IN AN INSTANT OF SILENCE
NIGHT WILL COME
LASTING JUST LONG ENOUGH FOR THE MOON TO PASS INTO THE WEST INTO THE DARKNESS...
AND GIVE WAY.

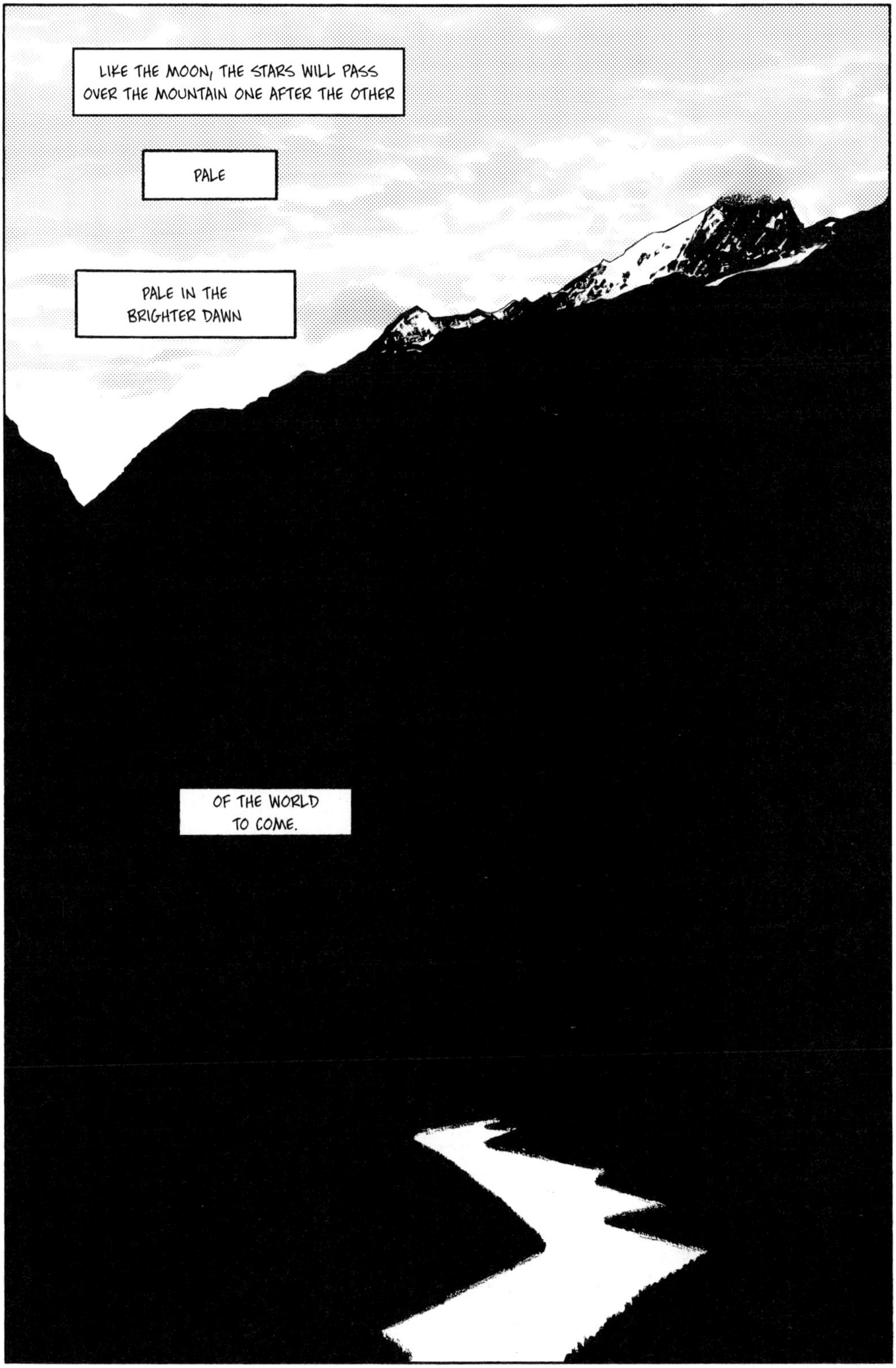
LIKE THE MOON, THE STARS WILL PASS OVER THE MOUNTAIN ONE AFTER THE OTHER
PALE
PALE IN THE BRIGHTER DAWN
OF THE WORLD TO COME.

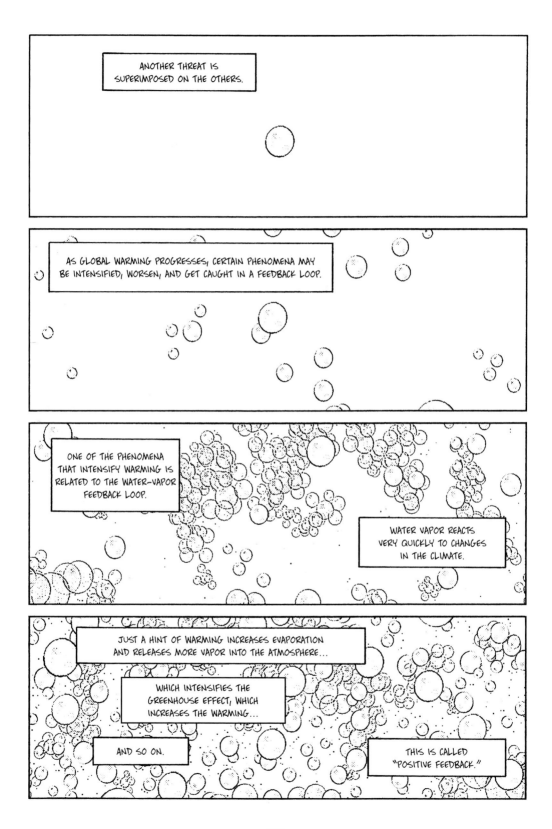

ANOTHER THREAT IS SUPERIMPOSED ON THE OTHERS.

AS GLOBAL WARMING PROGRESSES, CERTAIN PHENOMENA MAY BE INTENSIFIED, WORSEN, AND GET CAUGHT IN A FEEDBACK LOOP.

ONE OF THE PHENOMENA THAT INTENSIFY WARMING IS RELATED TO THE WATER-VAPOR FEEDBACK LOOP.

WATER VAPOR REACTS VERY QUICKLY TO CHANGES IN THE CLIMATE.

JUST A HINT OF WARMING INCREASES EVAPORATION AND RELEASES MORE VAPOR INTO THE ATMOSPHERE...

WHICH INTENSIFIES THE GREENHOUSE EFFECT, WHICH INCREASES THE WARMING...

AND SO ON.

THIS IS CALLED "POSITIVE FEEDBACK."

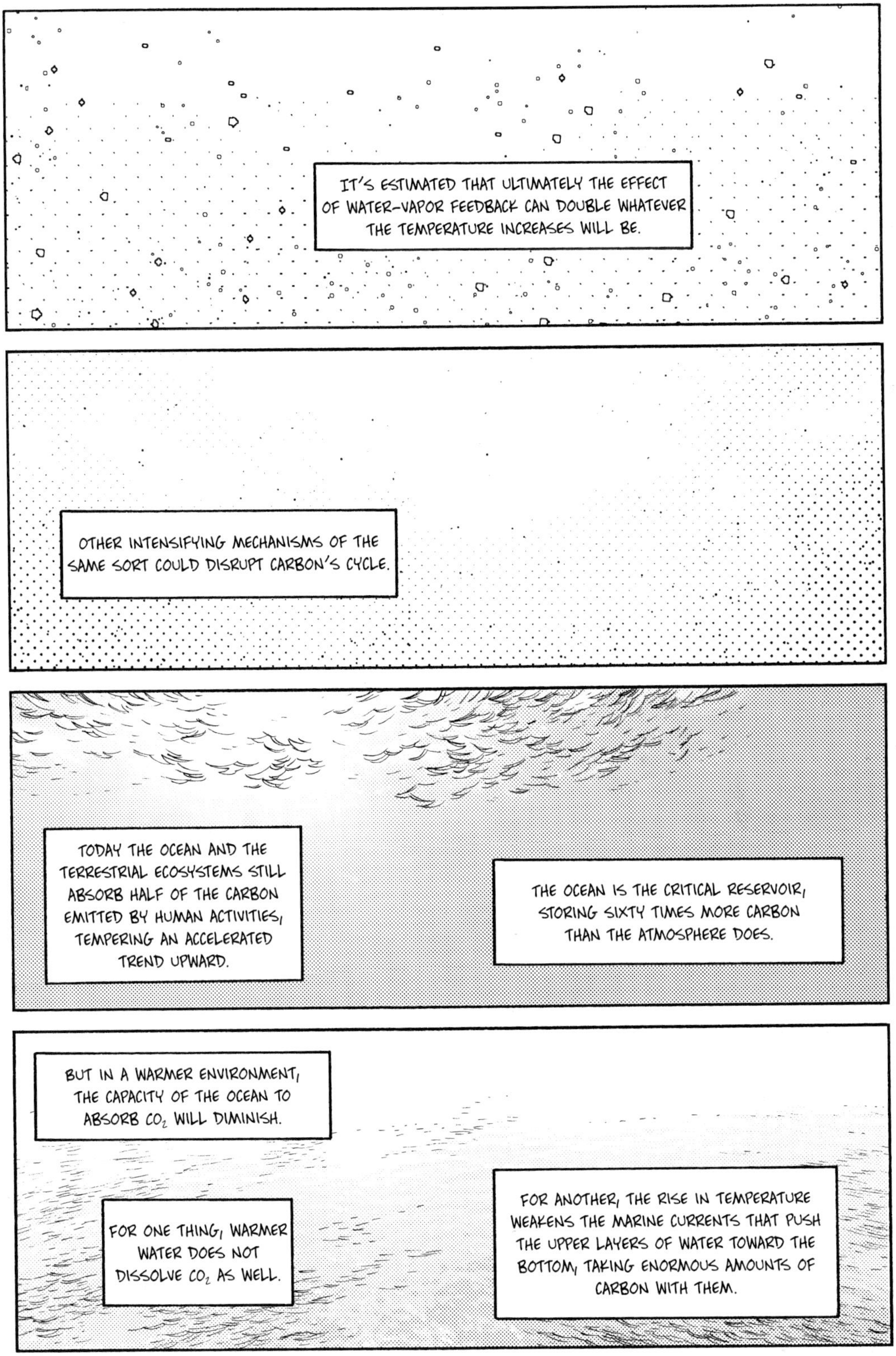
IT'S ESTIMATED THAT ULTIMATELY THE EFFECT OF WATER-VAPOR FEEDBACK CAN DOUBLE WHATEVER THE TEMPERATURE INCREASES WILL BE.
OTHER INTENSIFYING MECHANISMS OF THE SAME SORT COULD DISRUPT CARBON'S CYCLE.
TODAY THE OCEAN AND THE TERRESTRIAL ECOSYSTEMS STILL ABSORB HALF OF THE CARBON EMITTED BY HUMAN ACTIVITIES, TEMPERING AN ACCELERATED TREND UPWARD.
THE OCEAN IS THE CRITICAL RESERVOIR, STORING SIXTY TIMES MORE CARBON THAN THE ATMOSPHERE DOES.
BUT IN A WARMER ENVIRONMENT, THE CAPACITY OF THE OCEAN TO ABSORB CO_2 WILL DIMINISH.
FOR ONE THING, WARMER WATER DOES NOT DISSOLVE CO_2 AS WELL.
FOR ANOTHER, THE RISE IN TEMPERATURE WEAKENS THE MARINE CURRENTS THAT PUSH THE UPPER LAYERS OF WATER TOWARD THE BOTTOM, TAKING ENORMOUS AMOUNTS OF CARBON WITH THEM.

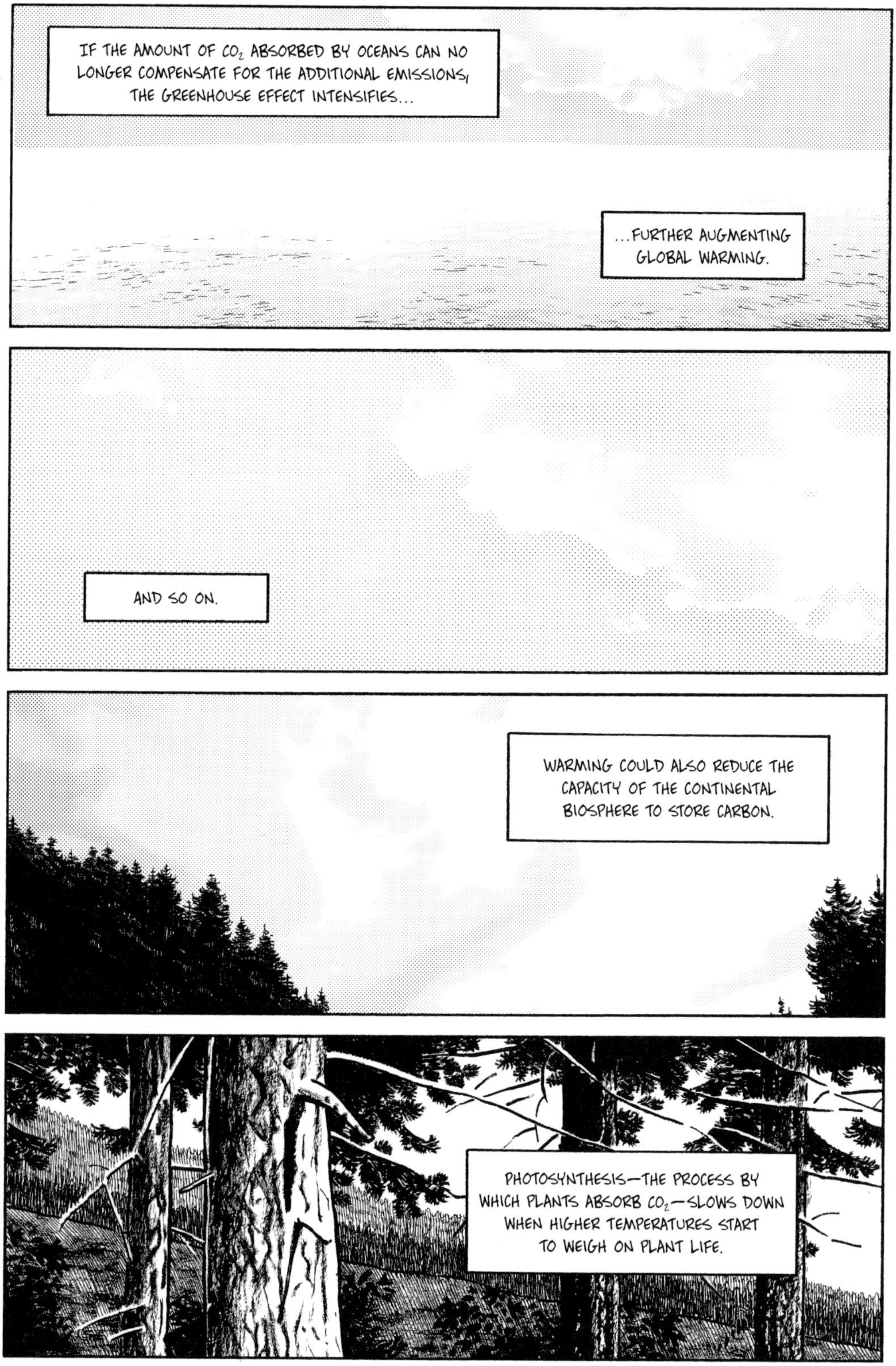
IF THE AMOUNT OF CO_2 ABSORBED BY OCEANS CAN NO LONGER COMPENSATE FOR THE ADDITIONAL EMISSIONS, THE GREENHOUSE EFFECT INTENSIFIES...
...FURTHER AUGMENTING GLOBAL WARMING.
AND SO ON.
WARMING COULD ALSO REDUCE THE CAPACITY OF THE CONTINENTAL BIOSPHERE TO STORE CARBON.
PHOTOSYNTHESIS—THE PROCESS BY WHICH PLANTS ABSORB CO_2—SLOWS DOWN WHEN HIGHER TEMPERATURES START TO WEIGH ON PLANT LIFE.

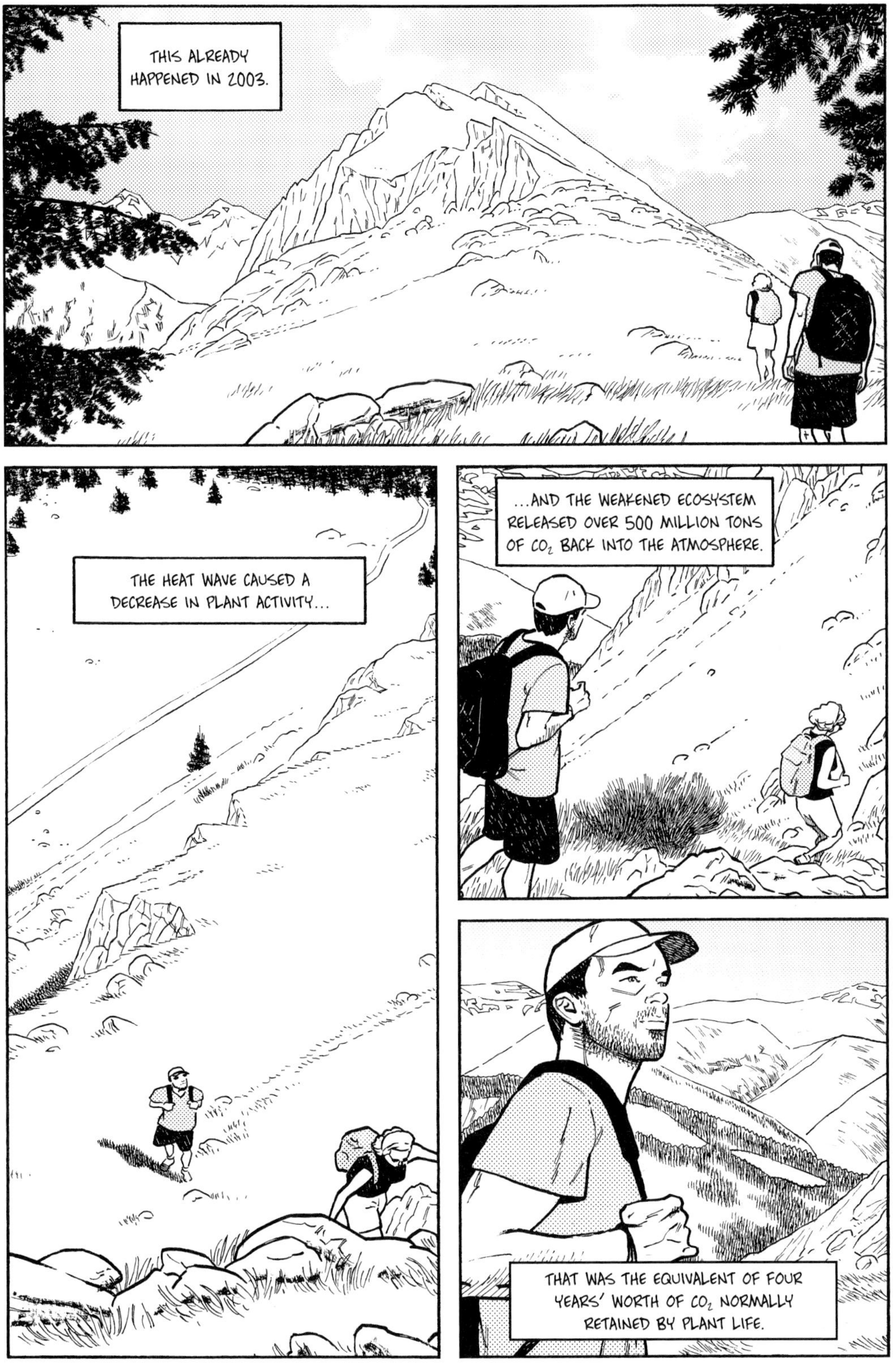
THIS ALREADY HAPPENED IN 2003.
THE HEAT WAVE CAUSED A DECREASE IN PLANT ACTIVITY...
...AND THE WEAKENED ECOSYSTEM RELEASED OVER 500 MILLION TONS OF CO2 BACK INTO THE ATMOSPHERE.
THAT WAS THE EQUIVALENT OF FOUR YEARS' WORTH OF CO2 NORMALLY RETAINED BY PLANT LIFE.

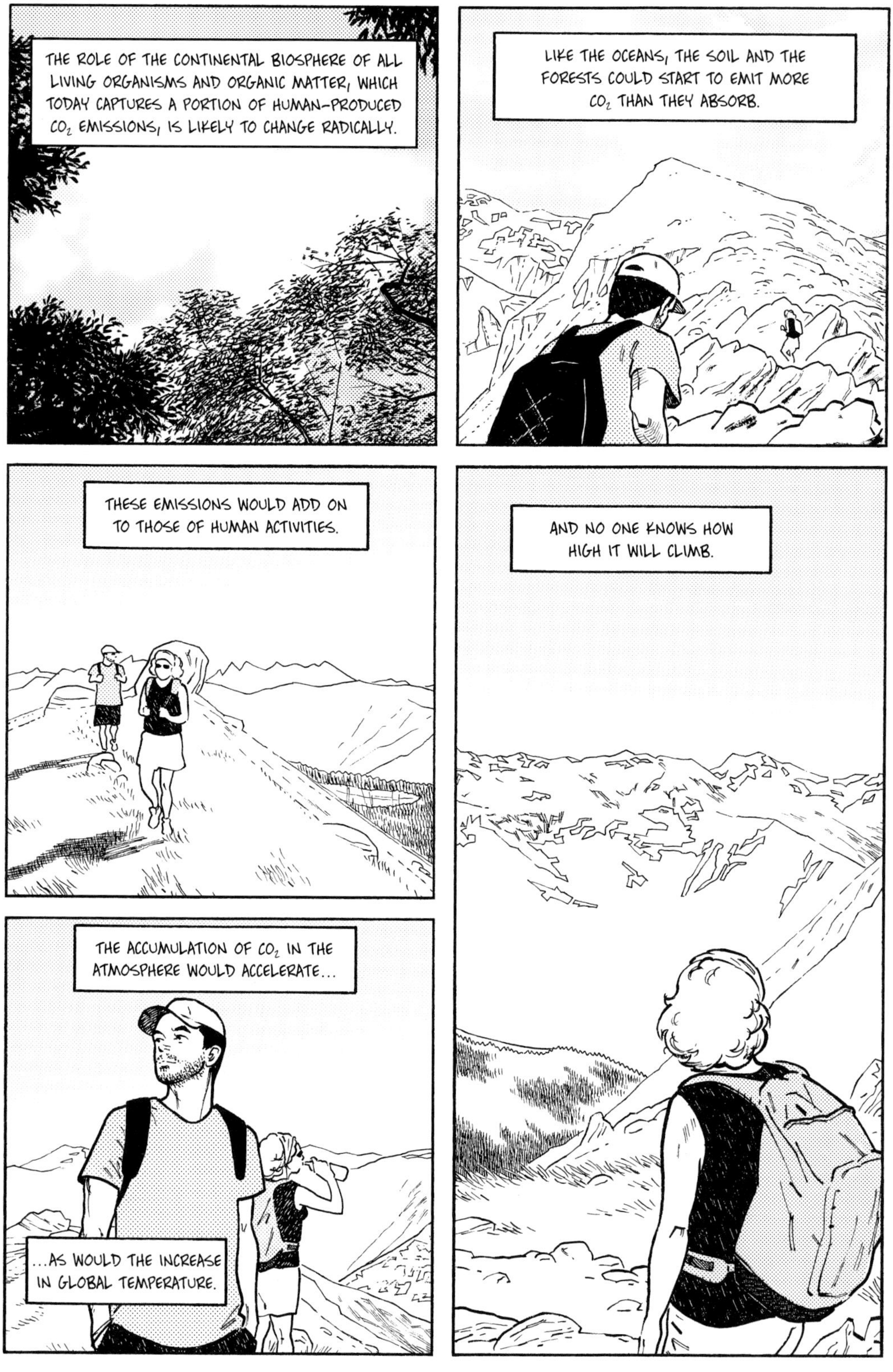
THE ROLE OF THE CONTINENTAL BIOSPHERE OF ALL LIVING ORGANISMS AND ORGANIC MATTER, WHICH TODAY CAPTURES A PORTION OF HUMAN-PRODUCED CO_2 EMISSIONS, IS LIKELY TO CHANGE RADICALLY.
LIKE THE OCEANS, THE SOIL AND THE FORESTS COULD START TO EMIT MORE CO_2 THAN THEY ABSORB.
THESE EMISSIONS WOULD ADD ON TO THOSE OF HUMAN ACTIVITIES.
AND NO ONE KNOWS HOW HIGH IT WILL CLIMB.
THE ACCUMULATION OF CO_2 IN THE ATMOSPHERE WOULD ACCELERATE...
...AS WOULD THE INCREASE IN GLOBAL TEMPERATURE.

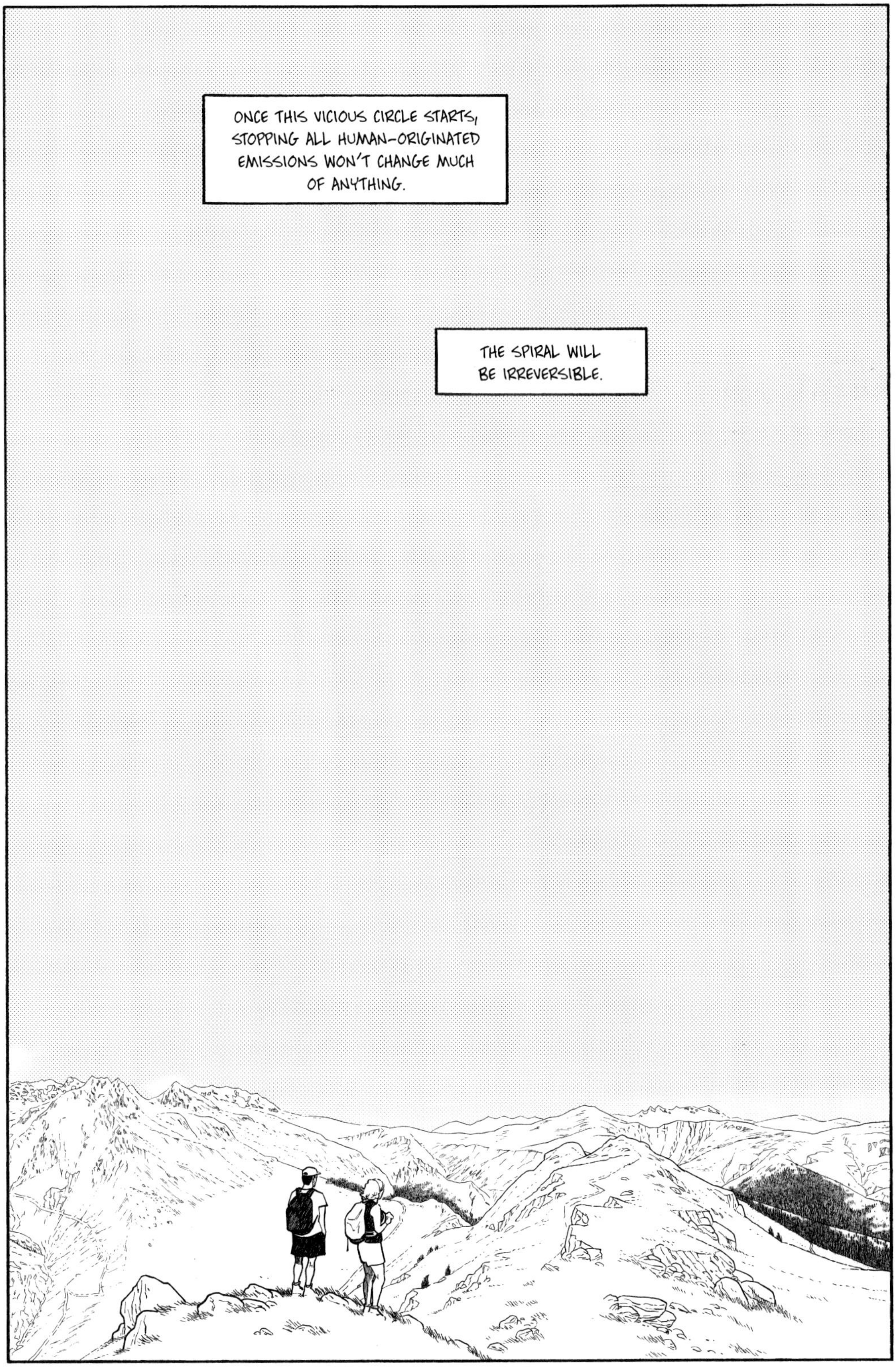
ONCE THIS VICIOUS CIRCLE STARTS, STOPPING ALL HUMAN-ORIGINATED EMISSIONS WON'T CHANGE MUCH OF ANYTHING.
THE SPIRAL WILL BE IRREVERSIBLE.

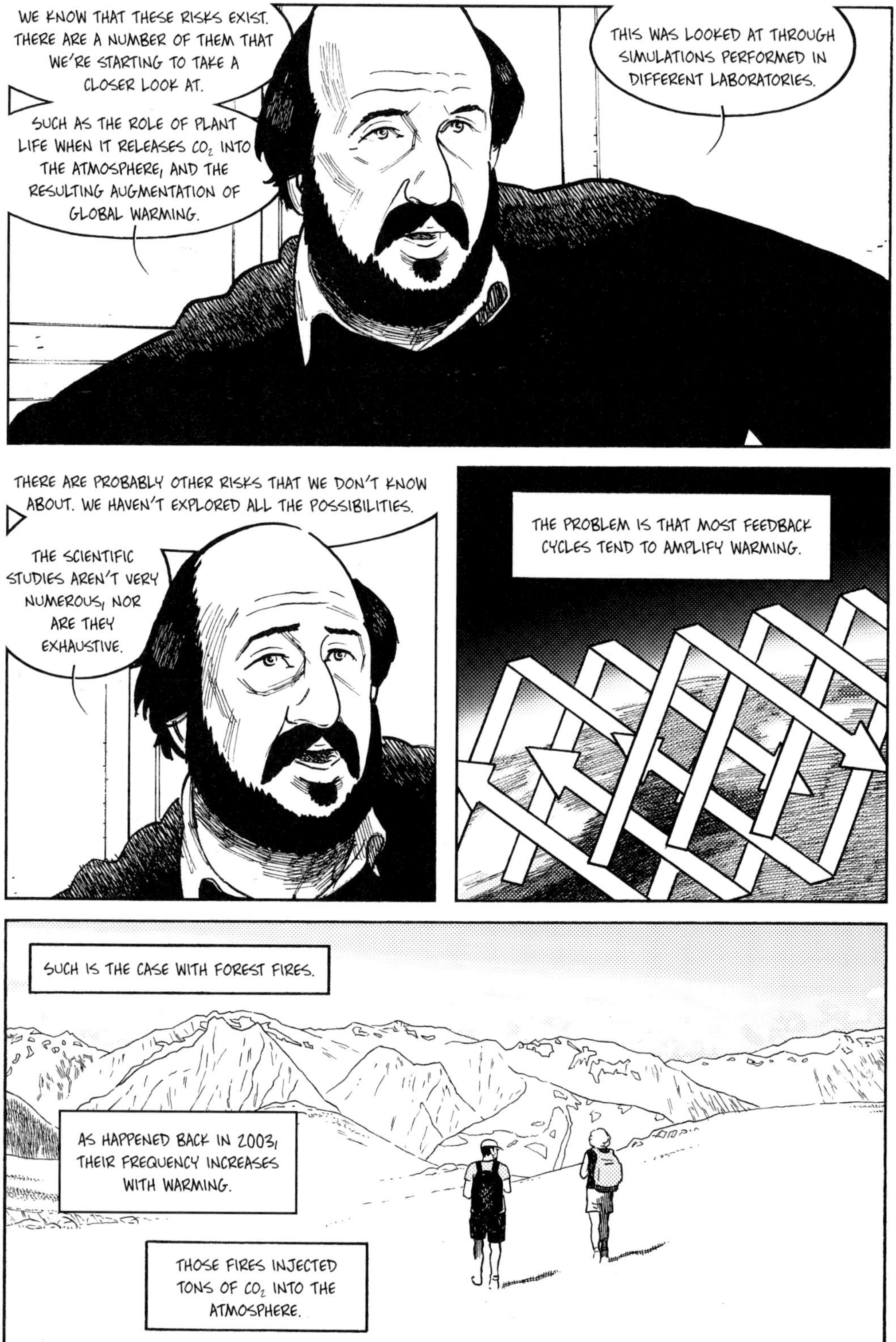
WE KNOW THAT THESE RISKS EXIST. THERE ARE A NUMBER OF THEM THAT WE'RE STARTING TO TAKE A CLOSER LOOK AT.
SUCH AS THE ROLE OF PLANT LIFE WHEN IT RELEASES CO_2 INTO THE ATMOSPHERE, AND THE RESULTING AUGMENTATION OF GLOBAL WARMING.
THIS WAS LOOKED AT THROUGH SIMULATIONS PERFORMED IN DIFFERENT LABORATORIES.
THERE ARE PROBABLY OTHER RISKS THAT WE DON'T KNOW ABOUT. WE HAVEN'T EXPLORED ALL THE POSSIBILITIES.
THE SCIENTIFIC STUDIES AREN'T VERY NUMEROUS, NOR ARE THEY EXHAUSTIVE.
THE PROBLEM IS THAT MOST FEEDBACK CYCLES TEND TO AMPLIFY WARMING.
SUCH IS THE CASE WITH FOREST FIRES.
AS HAPPENED BACK IN 2003, THEIR FREQUENCY INCREASES WITH WARMING.
THOSE FIRES INJECTED TONS OF CO_2 INTO THE ATMOSPHERE.

IT'S THE SAME THING WITH THE MELTING ICE CAPS.

POLAR ICE REFLECTS A GREAT DEAL OF THE SUN'S RAYS.

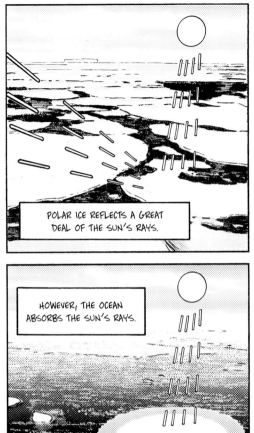

HOWEVER, THE OCEAN ABSORBS THE SUN'S RAYS.

SO WHEN THE ICE MELTS, THE REFLECTIVE SURFACE AREA OF THE PLANET IS REPLACED BY THE SAME AMOUNT OF ABSORBENT AREA...

AND THAT TENDS TO ACCELERATE WARMING.

ICE DISAPPEARING BECOMES A SELF-PERPETUATING PROCESS. THE SEA ICE MELTS, SO THE FOLLOWING YEAR IT'S A LOT LOWER AND MELTS MORE EASILY...

AND SO ON.

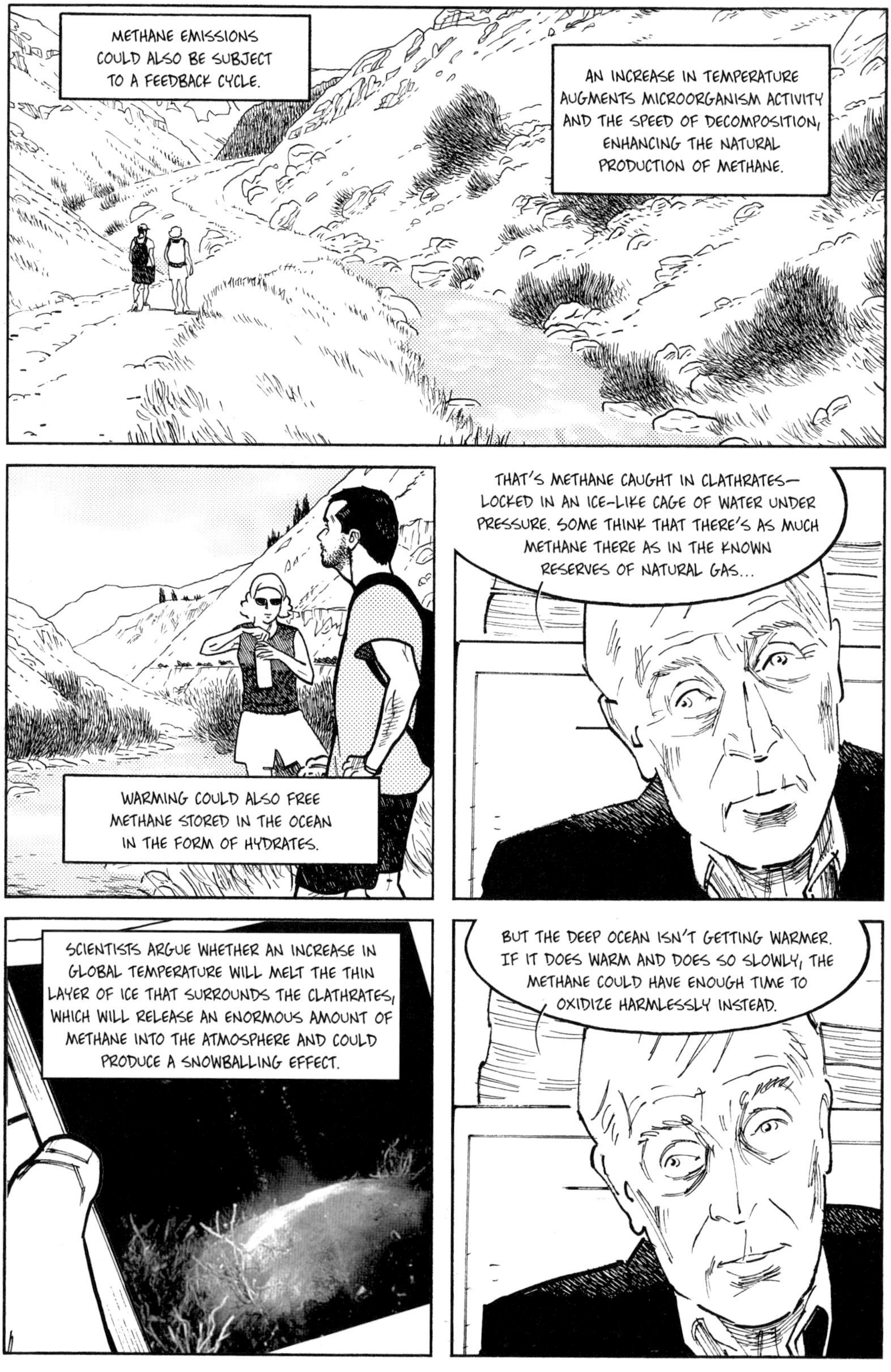
METHANE EMISSIONS COULD ALSO BE SUBJECT TO A FEEDBACK CYCLE.
AN INCREASE IN TEMPERATURE AUGMENTS MICROORGANISM ACTIVITY AND THE SPEED OF DECOMPOSITION, ENHANCING THE NATURAL PRODUCTION OF METHANE.
WARMING COULD ALSO FREE METHANE STORED IN THE OCEAN IN THE FORM OF HYDRATES.
THAT'S METHANE CAUGHT IN CLATHRATES—LOCKED IN AN ICE-LIKE CAGE OF WATER UNDER PRESSURE. SOME THINK THAT THERE'S AS MUCH METHANE THERE AS IN THE KNOWN RESERVES OF NATURAL GAS...
SCIENTISTS ARGUE WHETHER AN INCREASE IN GLOBAL TEMPERATURE WILL MELT THE THIN LAYER OF ICE THAT SURROUNDS THE CLATHRATES, WHICH WILL RELEASE AN ENORMOUS AMOUNT OF METHANE INTO THE ATMOSPHERE AND COULD PRODUCE A SNOWBALLING EFFECT.
BUT THE DEEP OCEAN ISN'T GETTING WARMER. IF IT DOES WARM AND DOES SO SLOWLY, THE METHANE COULD HAVE ENOUGH TIME TO OXIDIZE HARMLESSLY INSTEAD.

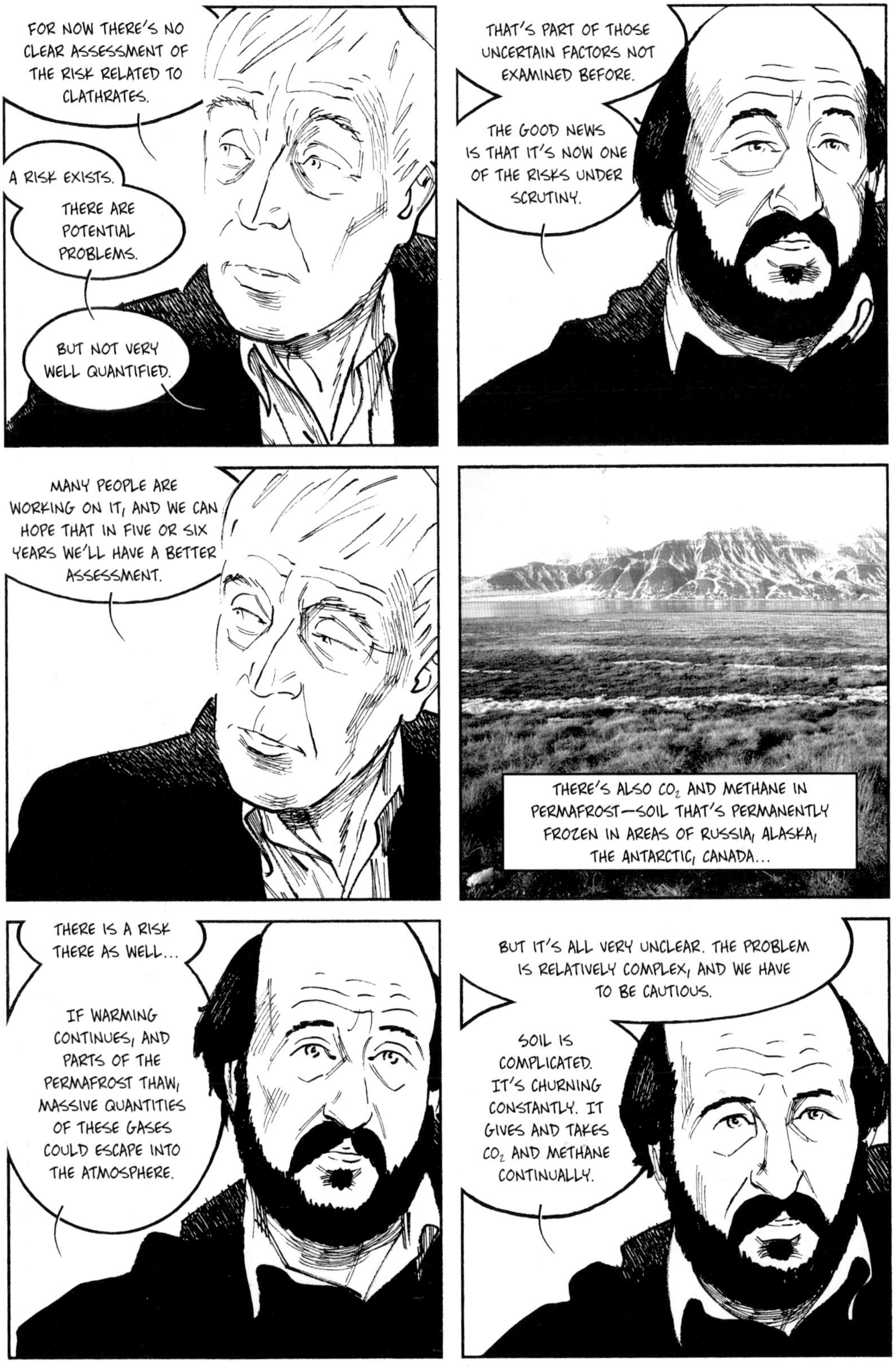
FOR NOW THERE'S NO CLEAR ASSESSMENT OF THE RISK RELATED TO CLATHRATES.
A RISK EXISTS.
THERE ARE POTENTIAL PROBLEMS.
BUT NOT VERY WELL QUANTIFIED.
THAT'S PART OF THOSE UNCERTAIN FACTORS NOT EXAMINED BEFORE.
THE GOOD NEWS IS THAT IT'S NOW ONE OF THE RISKS UNDER SCRUTINY.
MANY PEOPLE ARE WORKING ON IT, AND WE CAN HOPE THAT IN FIVE OR SIX YEARS WE'LL HAVE A BETTER ASSESSMENT.
THERE'S ALSO CO_2 AND METHANE IN PERMAFROST—SOIL THAT'S PERMANENTLY FROZEN IN AREAS OF RUSSIA, ALASKA, THE ANTARCTIC, CANADA...
THERE IS A RISK THERE AS WELL...
IF WARMING CONTINUES, AND PARTS OF THE PERMAFROST THAW, MASSIVE QUANTITIES OF THESE GASES COULD ESCAPE INTO THE ATMOSPHERE.
BUT IT'S ALL VERY UNCLEAR. THE PROBLEM IS RELATIVELY COMPLEX, AND WE HAVE TO BE CAUTIOUS.
SOIL IS COMPLICATED. IT'S CHURNING CONSTANTLY. IT GIVES AND TAKES CO_2 AND METHANE CONTINUALLY.

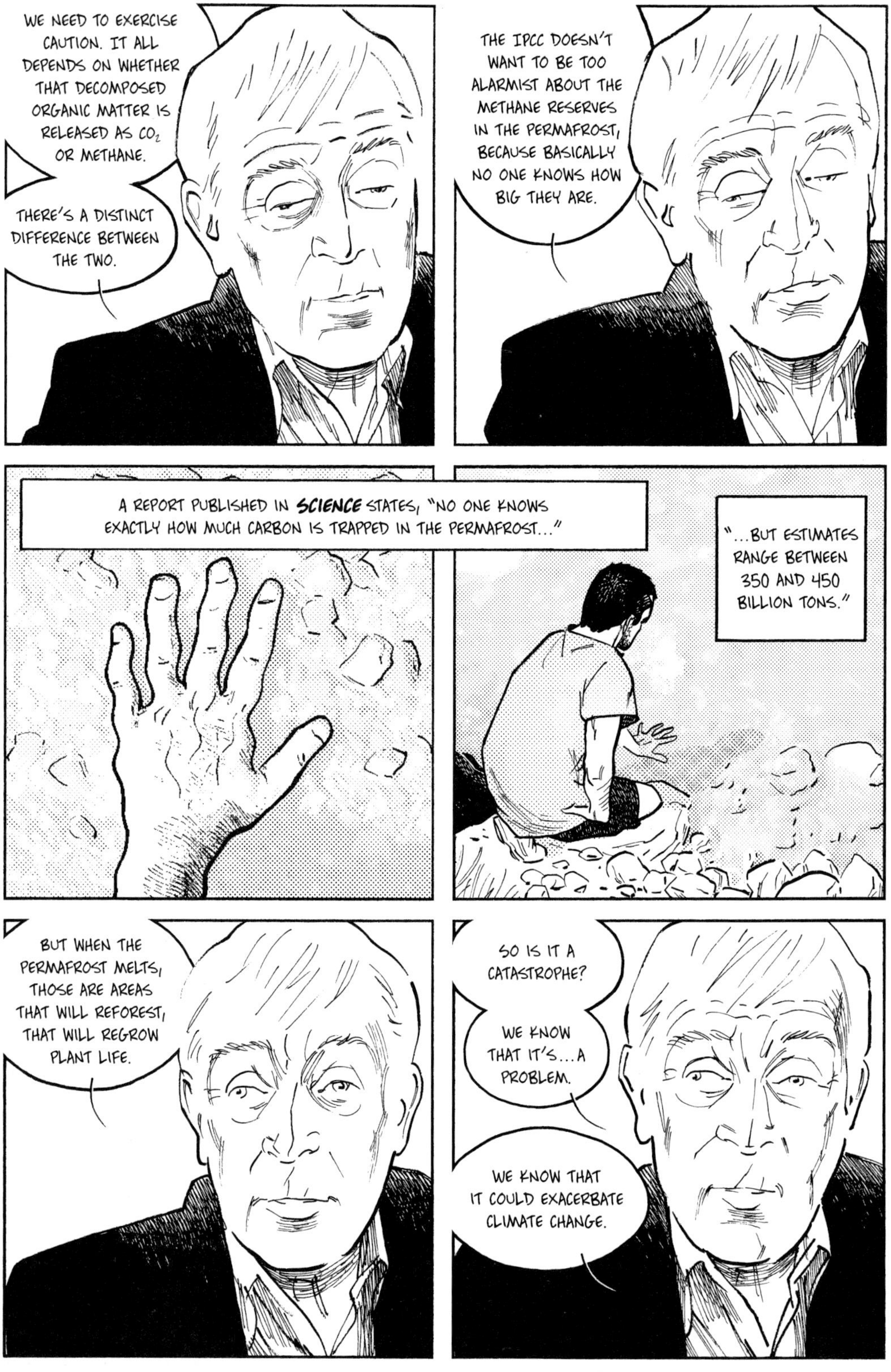
WE NEED TO EXERCISE CAUTION. IT ALL DEPENDS ON WHETHER THAT DECOMPOSED ORGANIC MATTER IS RELEASED AS CO_2 OR METHANE.
THERE'S A DISTINCT DIFFERENCE BETWEEN THE TWO.
THE IPCC DOESN'T WANT TO BE TOO ALARMIST ABOUT THE METHANE RESERVES IN THE PERMAFROST, BECAUSE BASICALLY NO ONE KNOWS HOW BIG THEY ARE.
A REPORT PUBLISHED IN SCIENCE STATES, "NO ONE KNOWS EXACTLY HOW MUCH CARBON IS TRAPPED IN THE PERMAFROST..."
"...BUT ESTIMATES RANGE BETWEEN 350 AND 450 BILLION TONS."
BUT WHEN THE PERMAFROST MELTS, THOSE ARE AREAS THAT WILL REFOREST, THAT WILL REGROW PLANT LIFE.
SO IS IT A CATASTROPHE?
WE KNOW THAT IT'S... A PROBLEM.
WE KNOW THAT IT COULD EXACERBATE CLIMATE CHANGE.

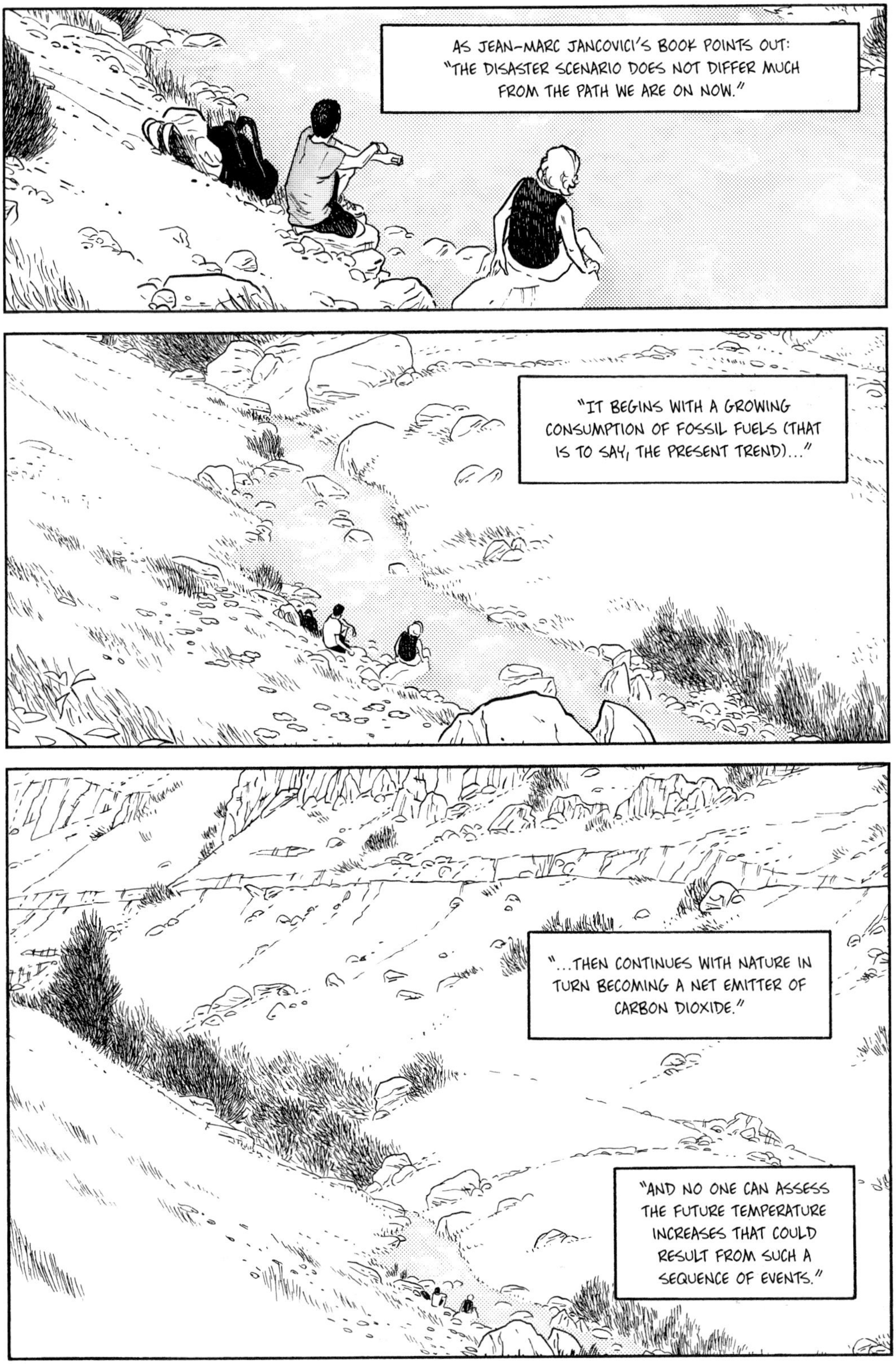
AS JEAN-MARC JANCOVICI'S BOOK POINTS OUT: "THE DISASTER SCENARIO DOES NOT DIFFER MUCH FROM THE PATH WE ARE ON NOW."
"IT BEGINS WITH A GROWING CONSUMPTION OF FOSSIL FUELS (THAT IS TO SAY, THE PRESENT TREND)..."
"...THEN CONTINUES WITH NATURE IN TURN BECOMING A NET EMITTER OF CARBON DIOXIDE."
"AND NO ONE CAN ASSESS THE FUTURE TEMPERATURE INCREASES THAT COULD RESULT FROM SUCH A SEQUENCE OF EVENTS."

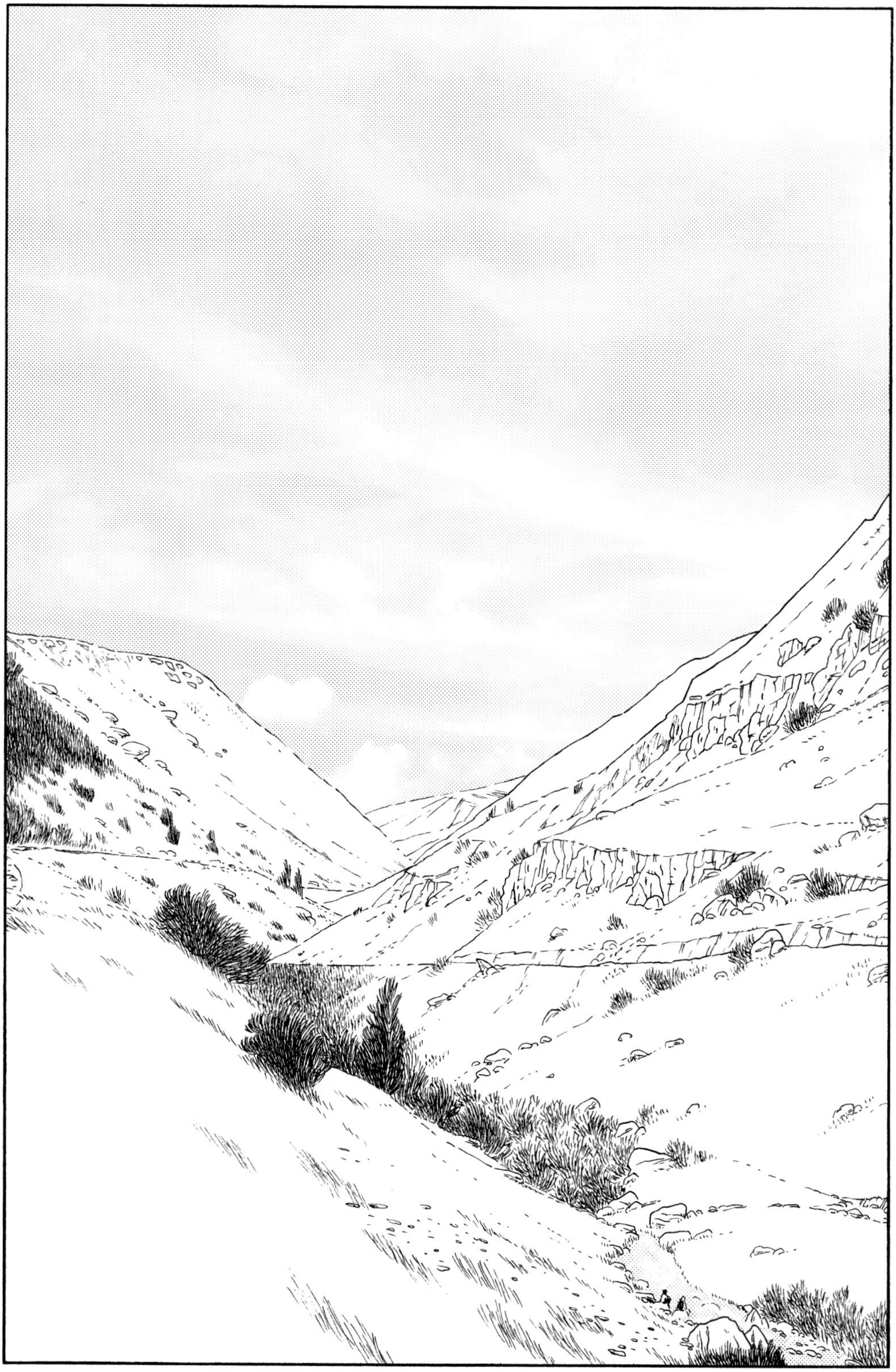

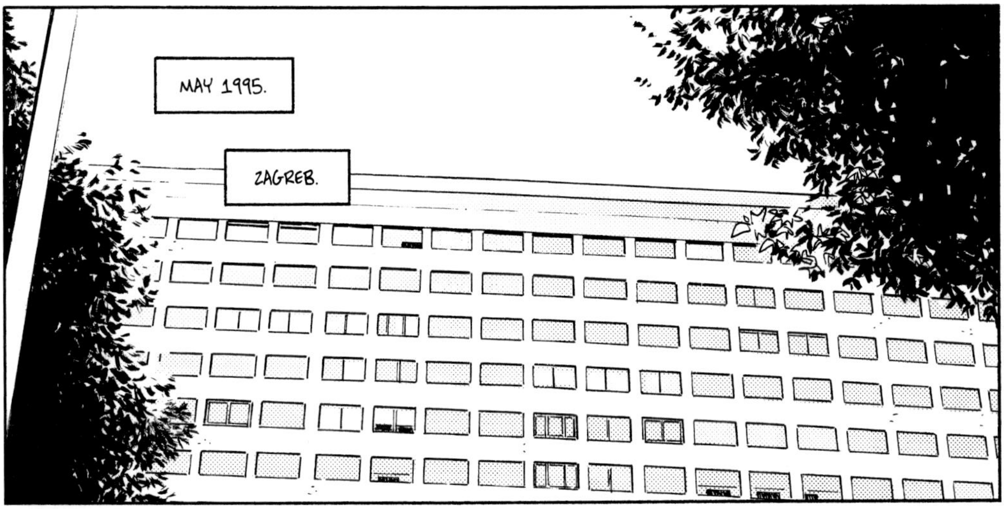
MAY 1995.
ZAGREB.

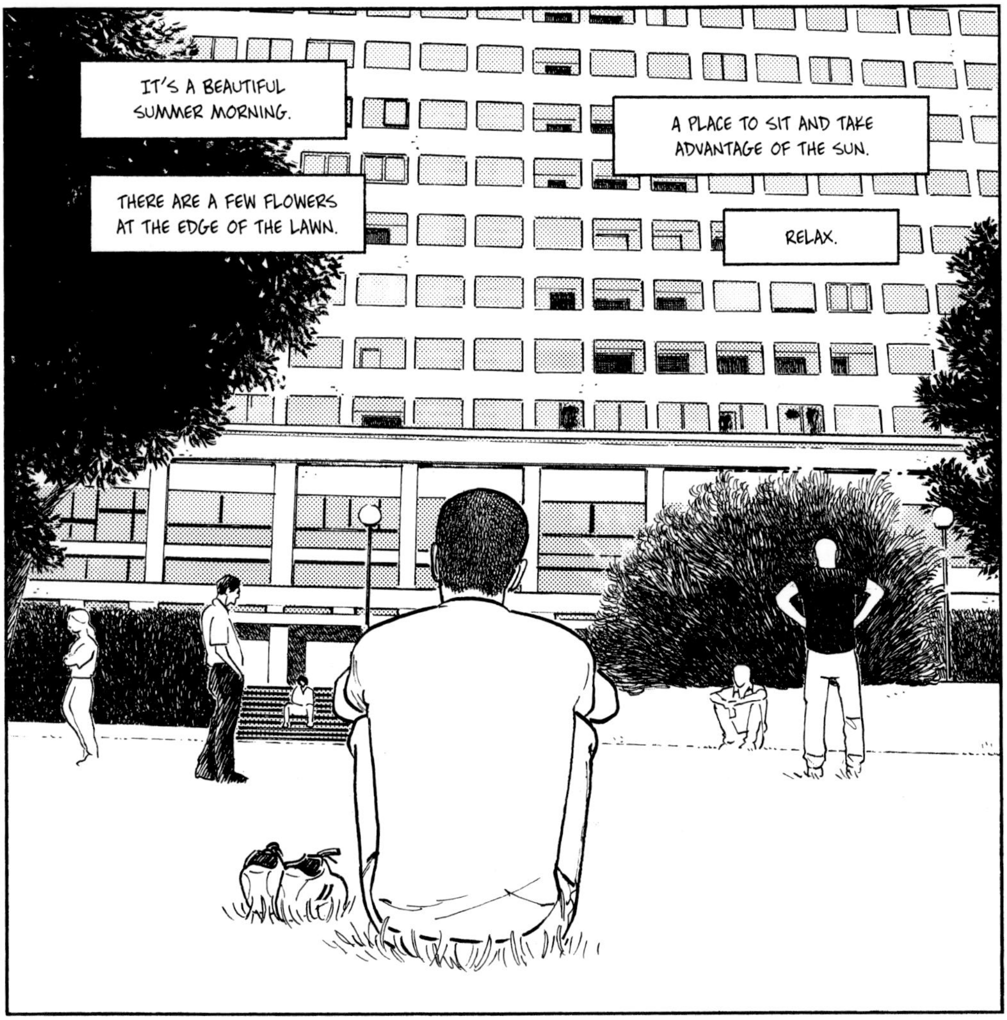
IT'S A BEAUTIFUL SUMMER MORNING.
THERE ARE A FEW FLOWERS AT THE EDGE OF THE LAWN.
A PLACE TO SIT AND TAKE ADVANTAGE OF THE SUN.
RELAX.

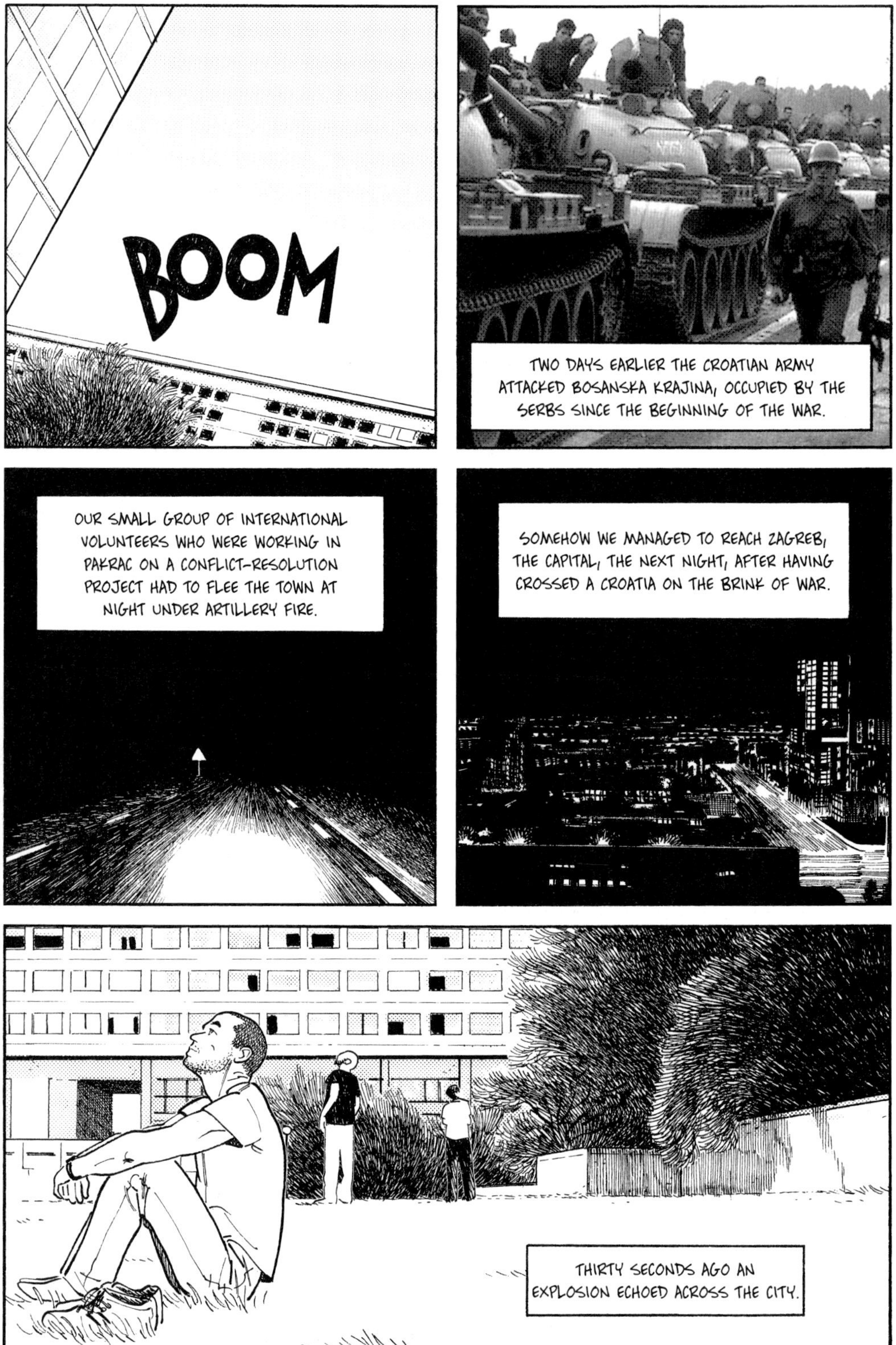
BOOM
TWO DAYS EARLIER THE CROATIAN ARMY ATTACKED BOSANSKA KRAJINA, OCCUPIED BY THE SERBS SINCE THE BEGINNING OF THE WAR.
OUR SMALL GROUP OF INTERNATIONAL VOLUNTEERS WHO WERE WORKING IN PAKRAC ON A CONFLICT-RESOLUTION PROJECT HAD TO FLEE THE TOWN AT NIGHT UNDER ARTILLERY FIRE.
SOMEHOW WE MANAGED TO REACH ZAGREB, THE CAPITAL, THE NEXT NIGHT, AFTER HAVING CROSSED A CROATIA ON THE BRINK OF WAR.
THIRTY SECONDS AGO AN EXPLOSION ECHOED ACROSS THE CITY.

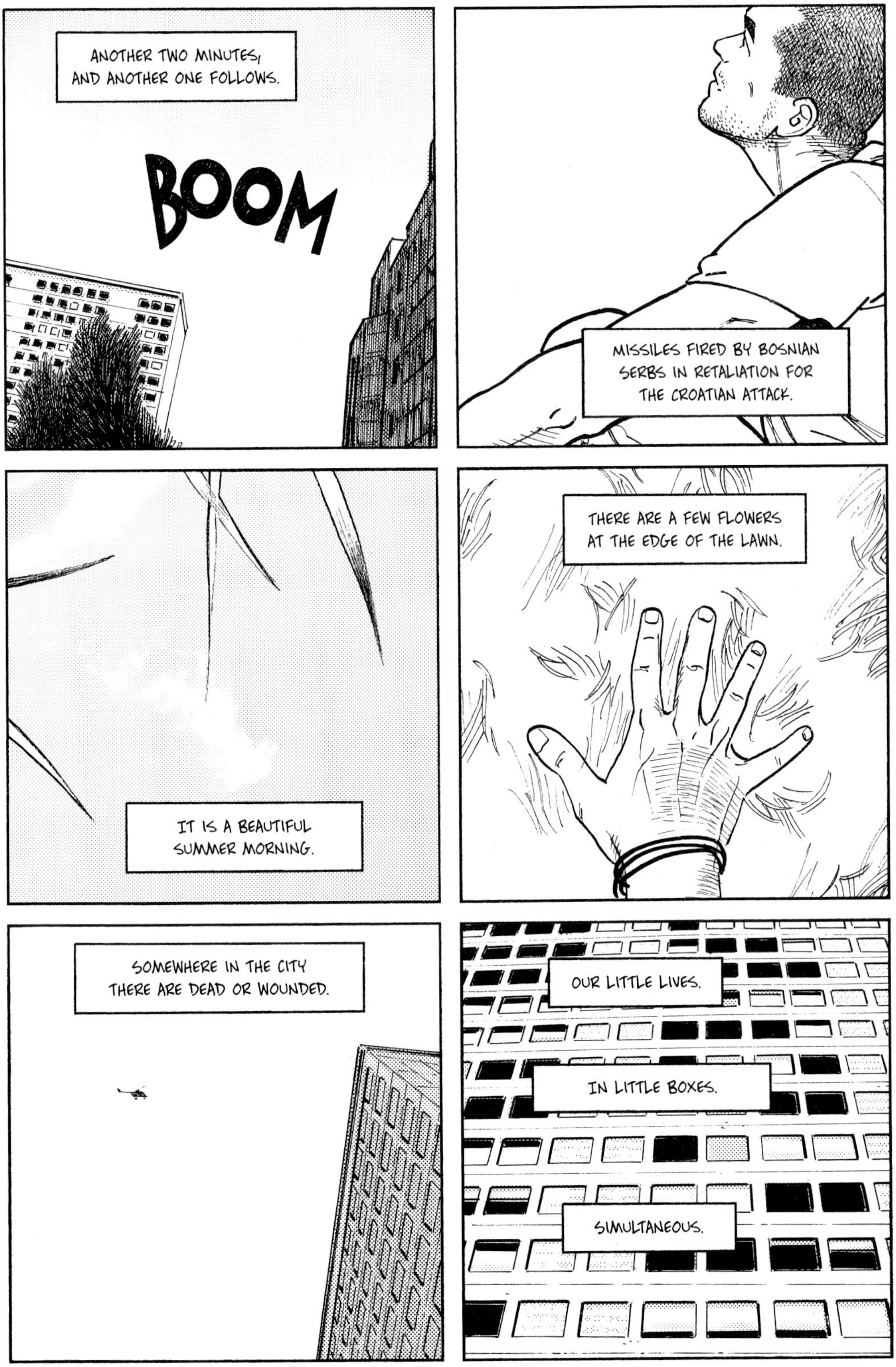
ANOTHER TWO MINUTES, AND ANOTHER ONE FOLLOWS.
BOOM
MISSILES FIRED BY BOSNIAN SERBS IN RETALIATION FOR THE CROATIAN ATTACK.
IT IS A BEAUTIFUL SUMMER MORNING.
THERE ARE A FEW FLOWERS AT THE EDGE OF THE LAWN.
SOMEWHERE IN THE CITY THERE ARE DEAD OR WOUNDED.
OUR LITTLE LIVES.
IN LITTLE BOXES.
SIMULTANEOUS.

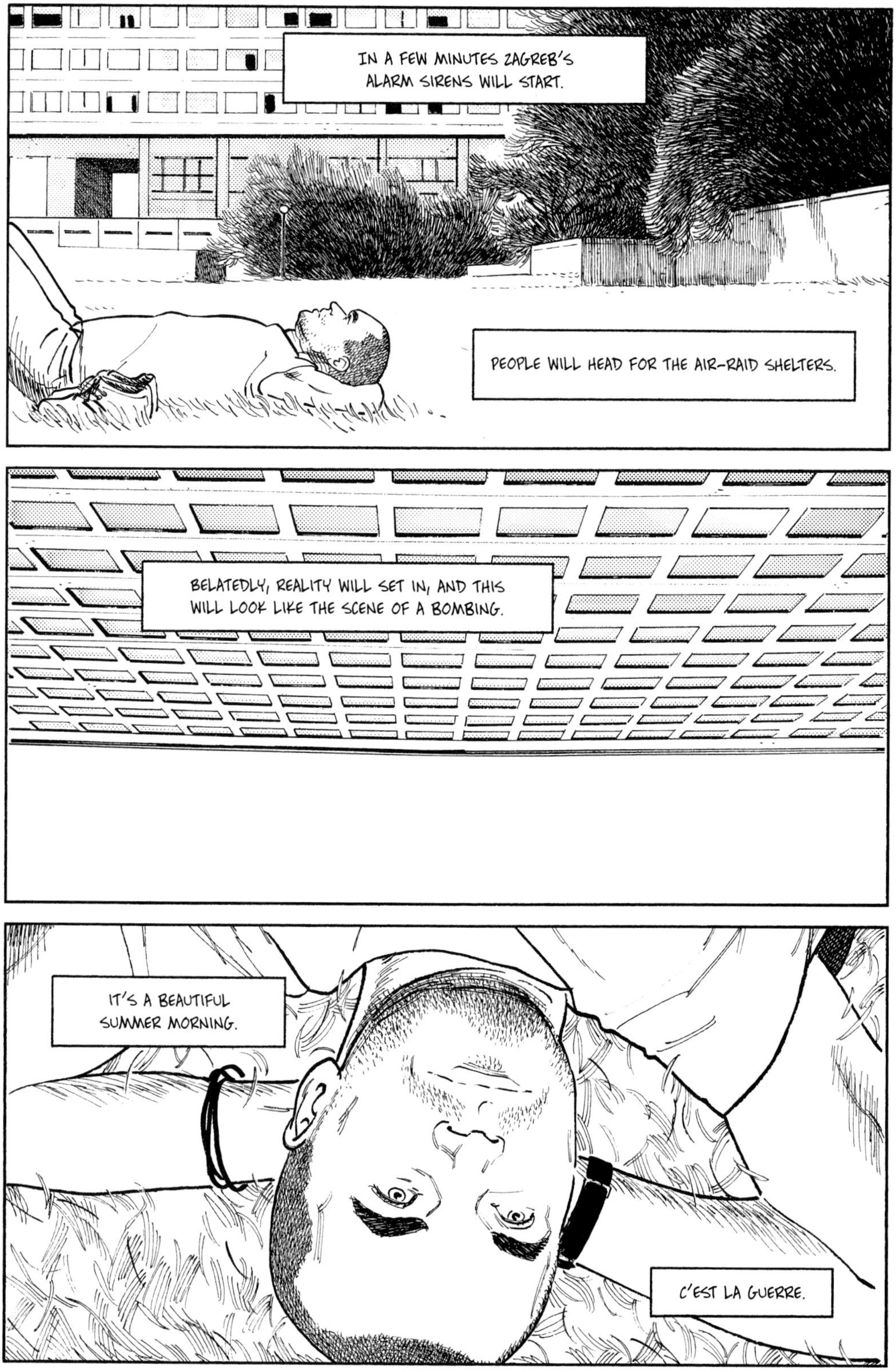
IN A FEW MINUTES ZAGREB'S ALARM SIRENS WILL START.
PEOPLE WILL HEAD FOR THE AIR-RAID SHELTERS.
BELATEDLY, REALITY WILL SET IN, AND THIS WILL LOOK LIKE THE SCENE OF A BOMBING.
IT'S A BEAUTIFUL SUMMER MORNING.
C'EST LA GUERRE.

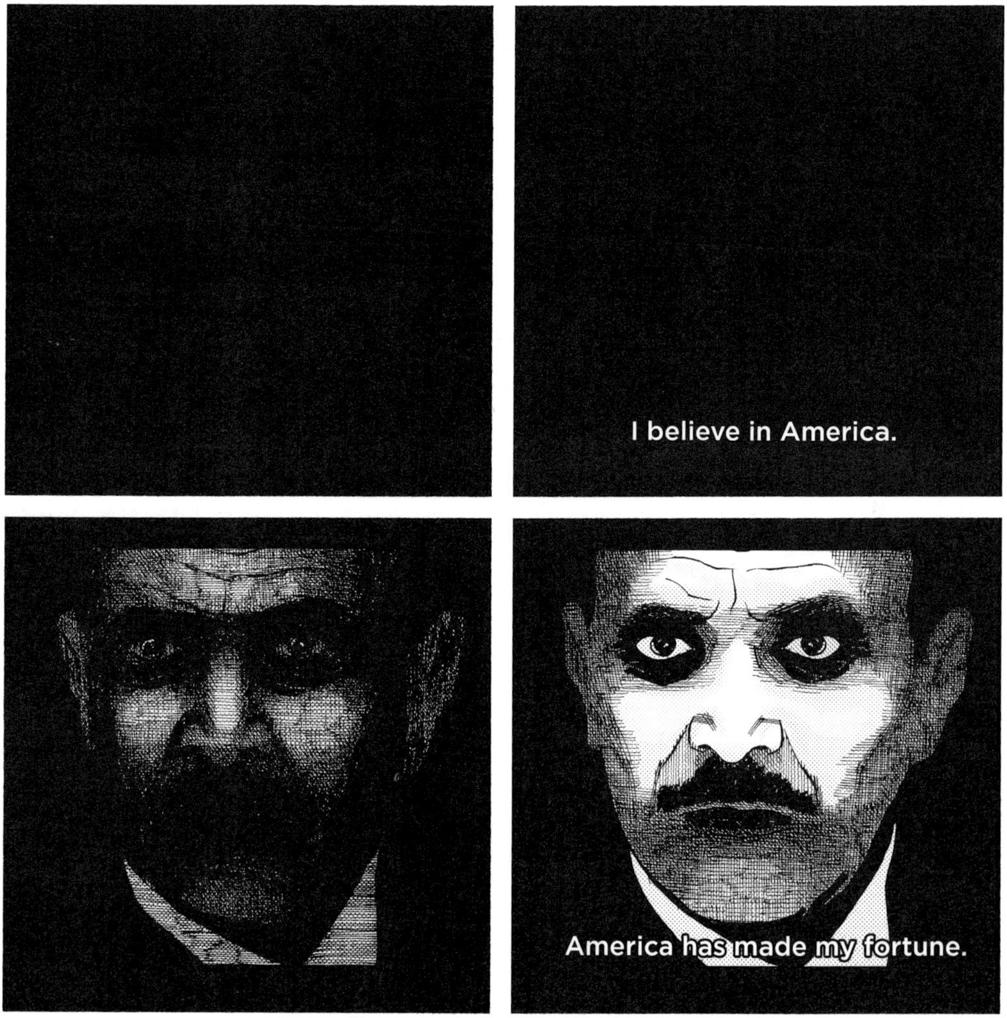
I believe in America.
America has made my fortune.

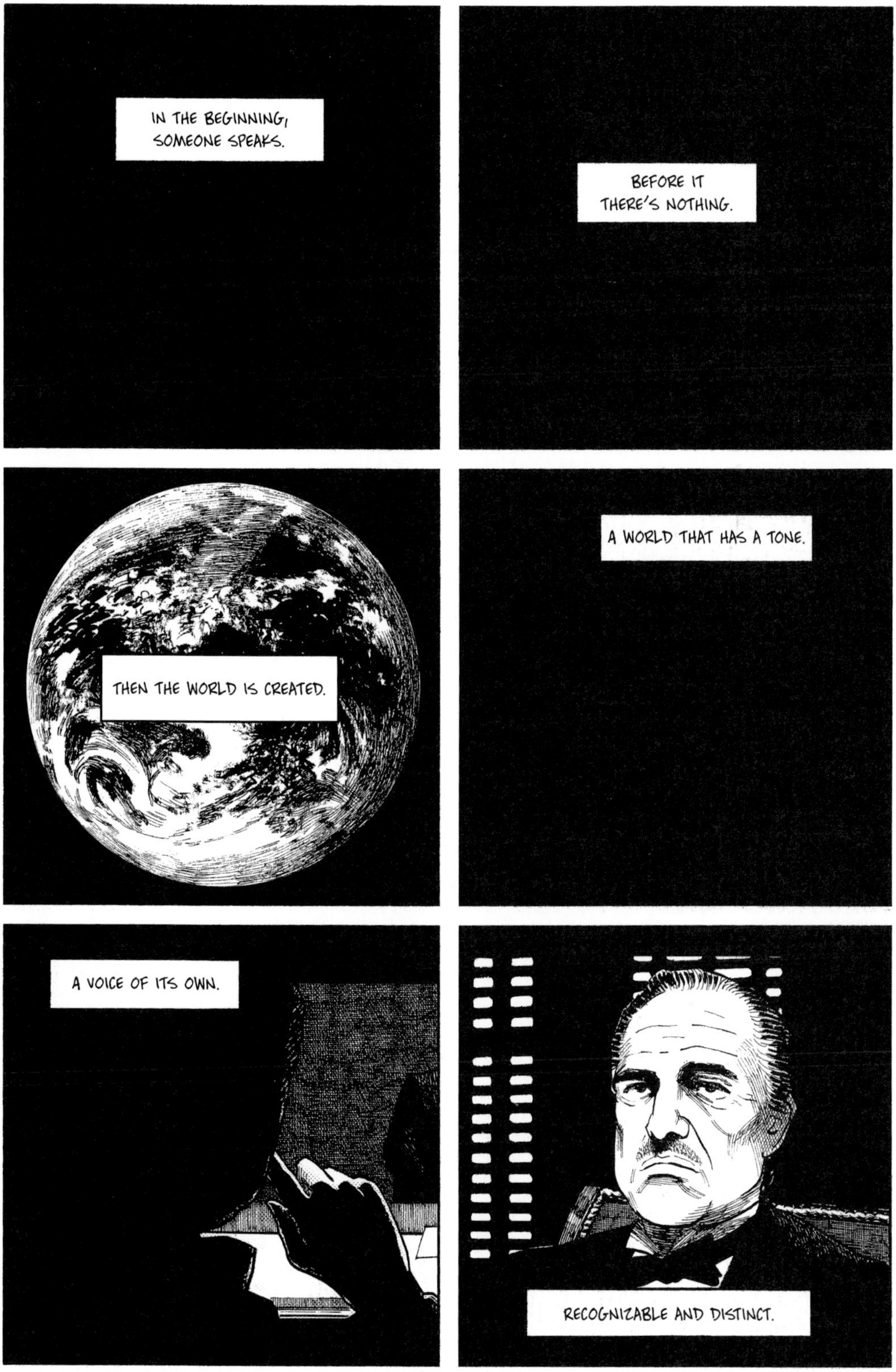
IN THE BEGINNING, SOMEONE SPEAKS.
BEFORE IT THERE'S NOTHING.
THEN THE WORLD IS CREATED.
A WORLD THAT HAS A TONE.
A VOICE OF ITS OWN.
RECOGNIZABLE AND DISTINCT.

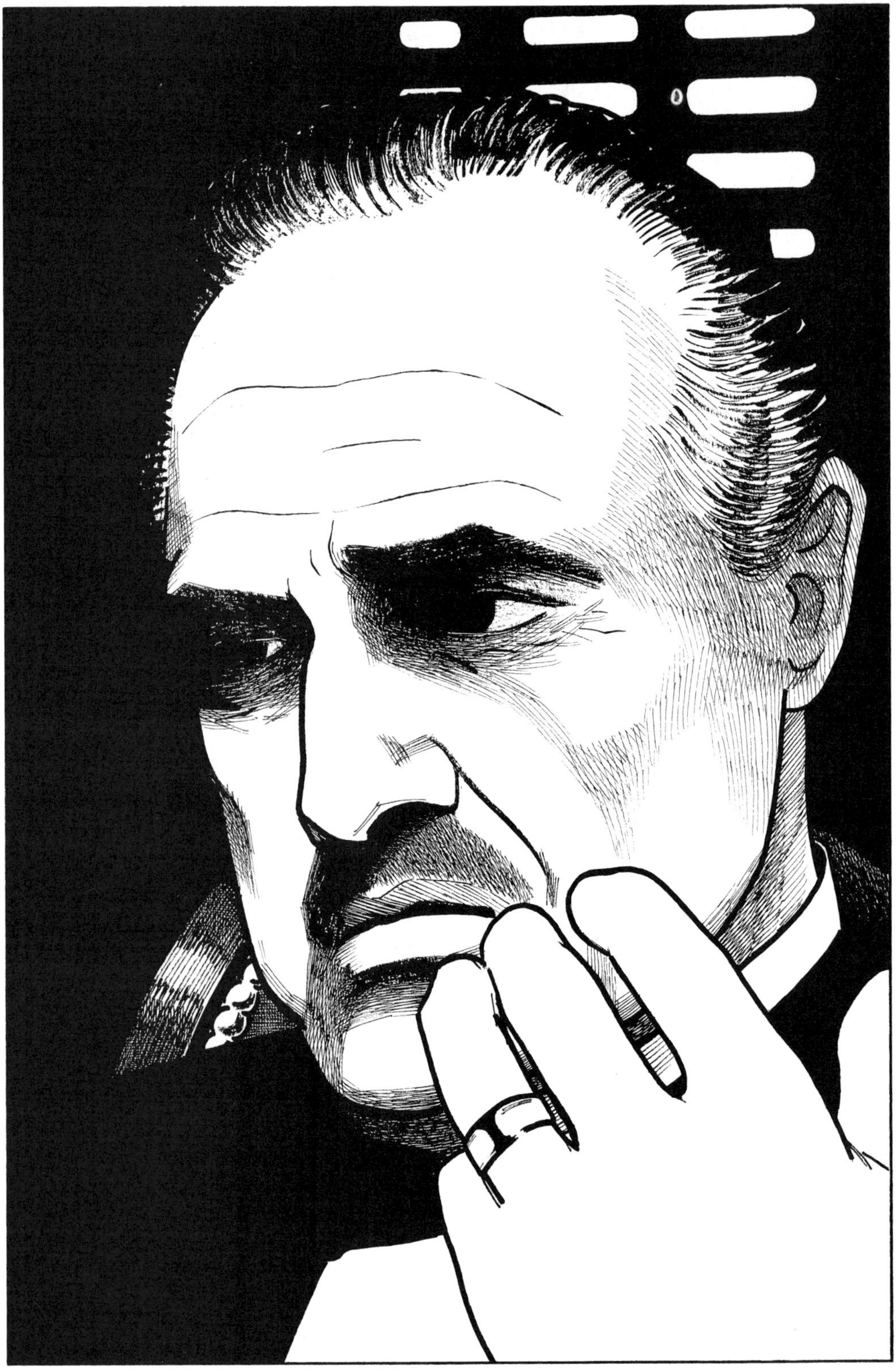

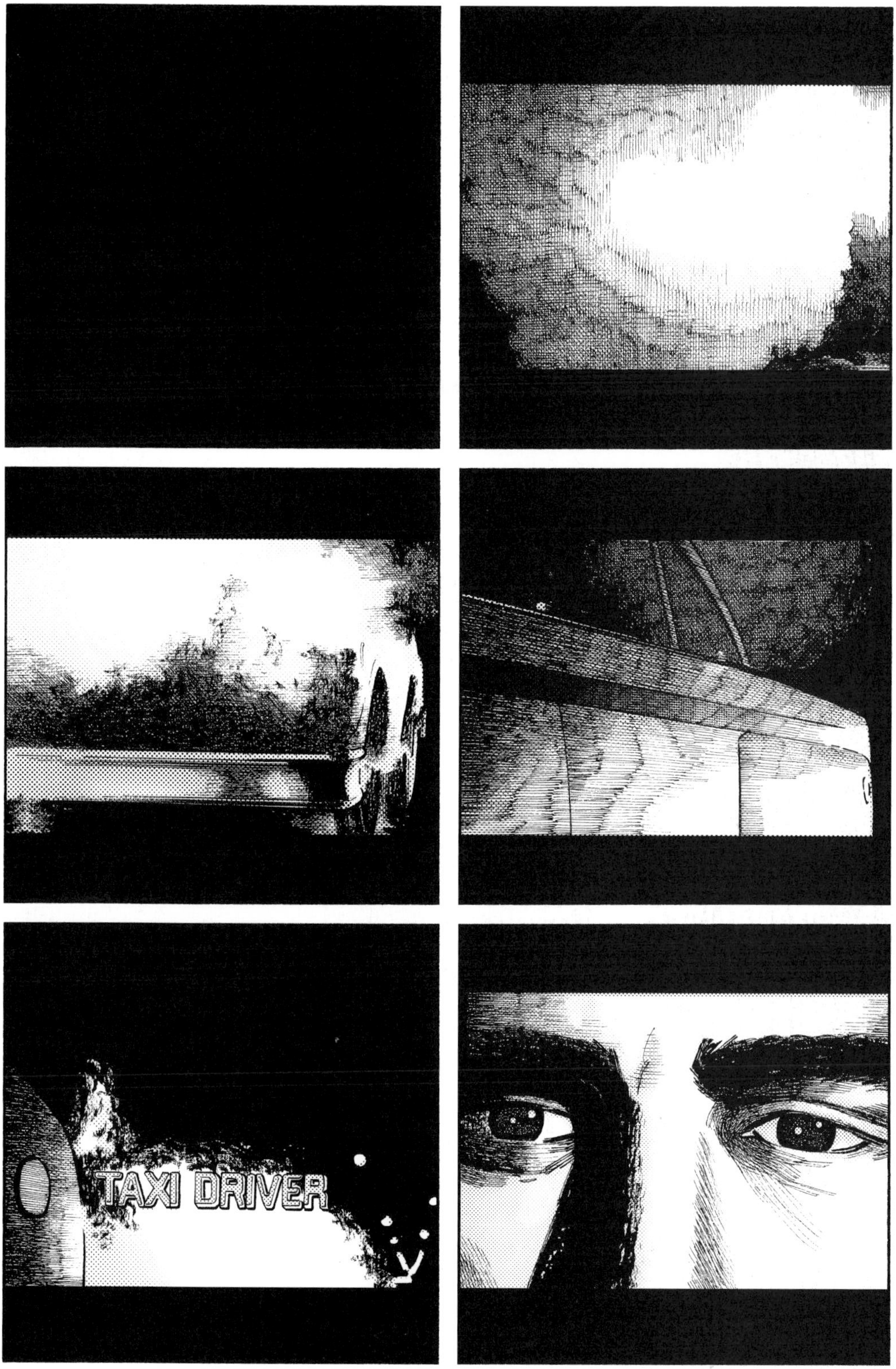
TAXI DRIVER

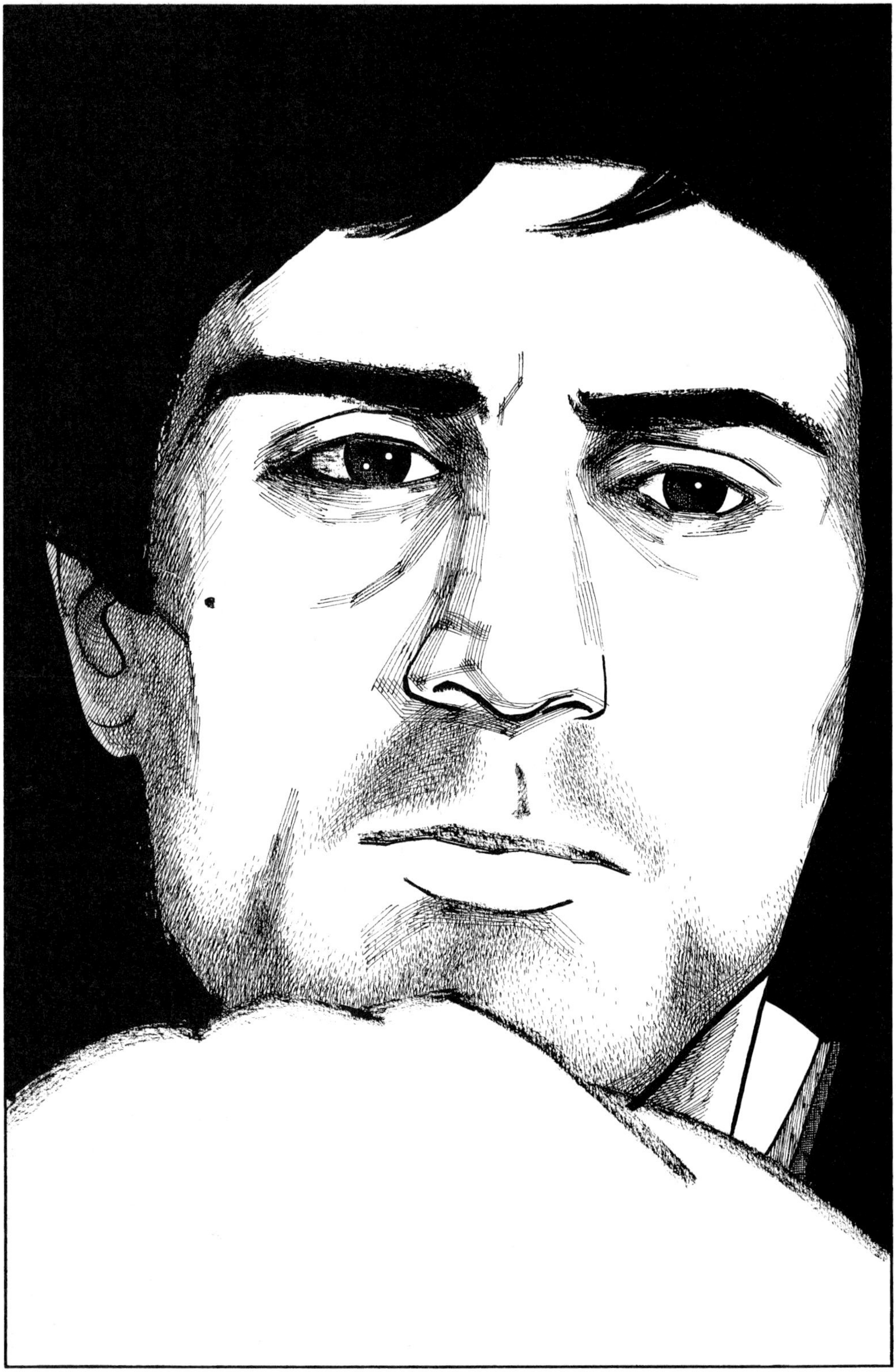

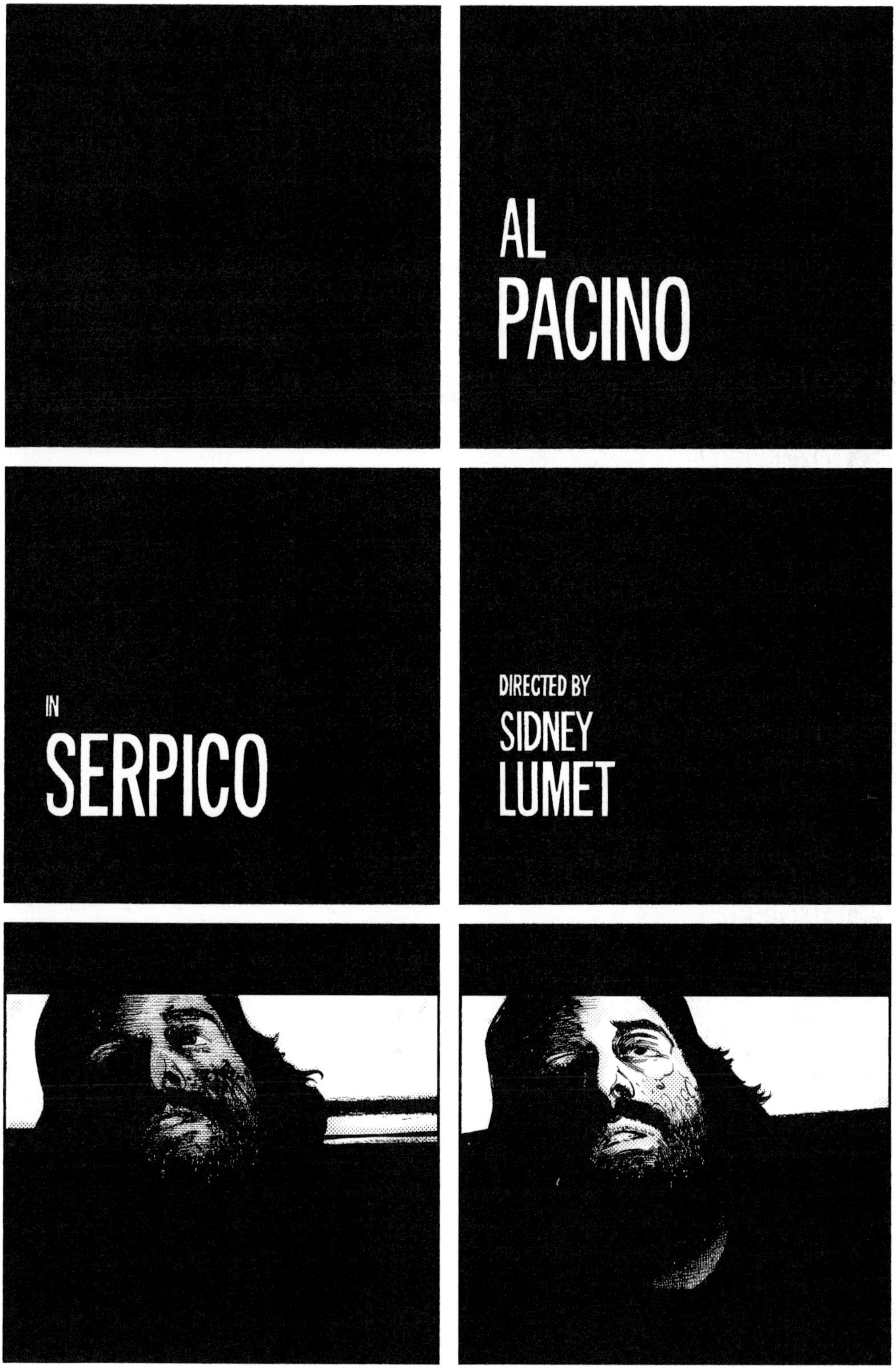
AL
PACINO
IN
SERPICO
DIRECTED BY
SIDNEY
LUMET

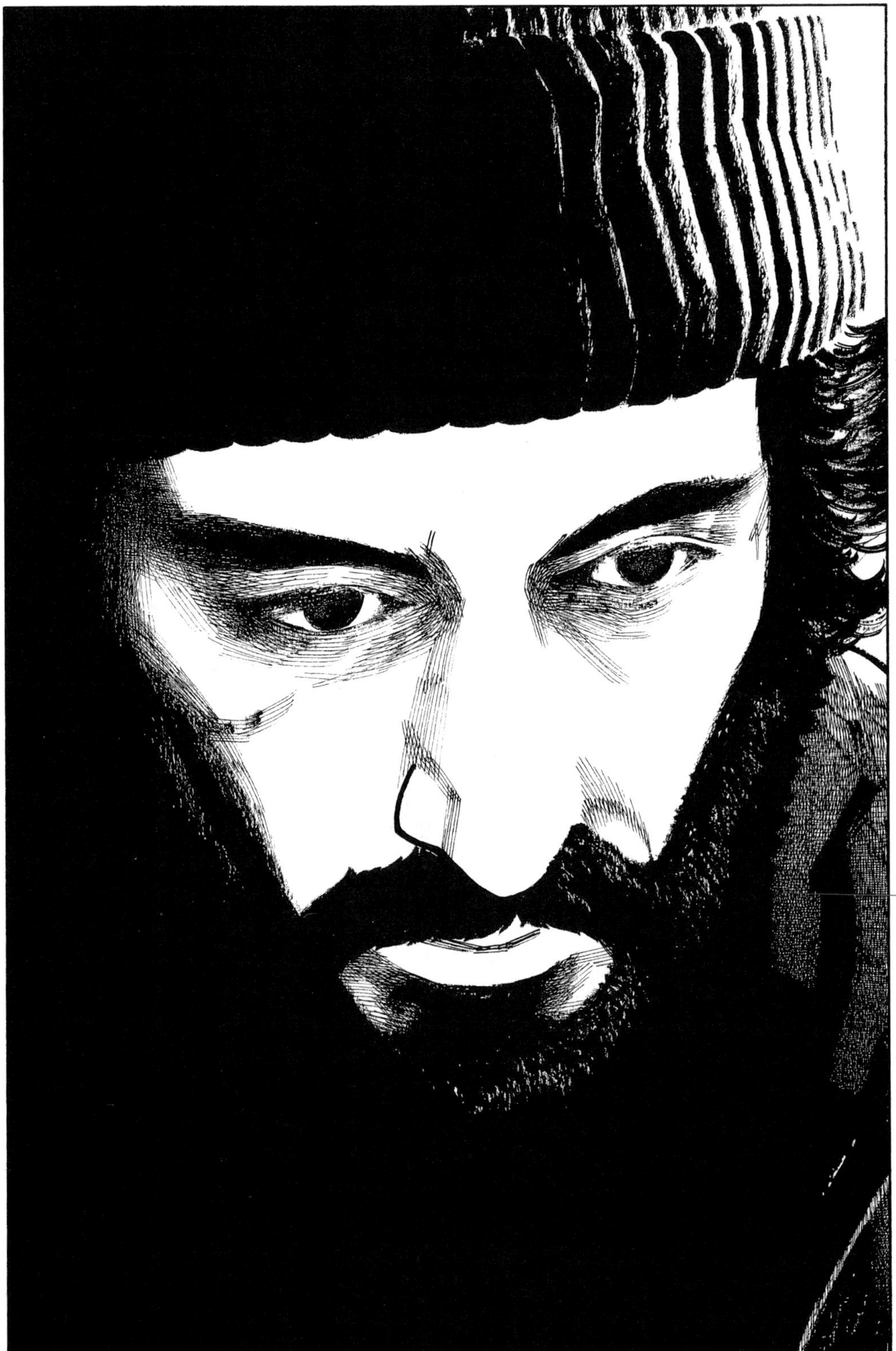

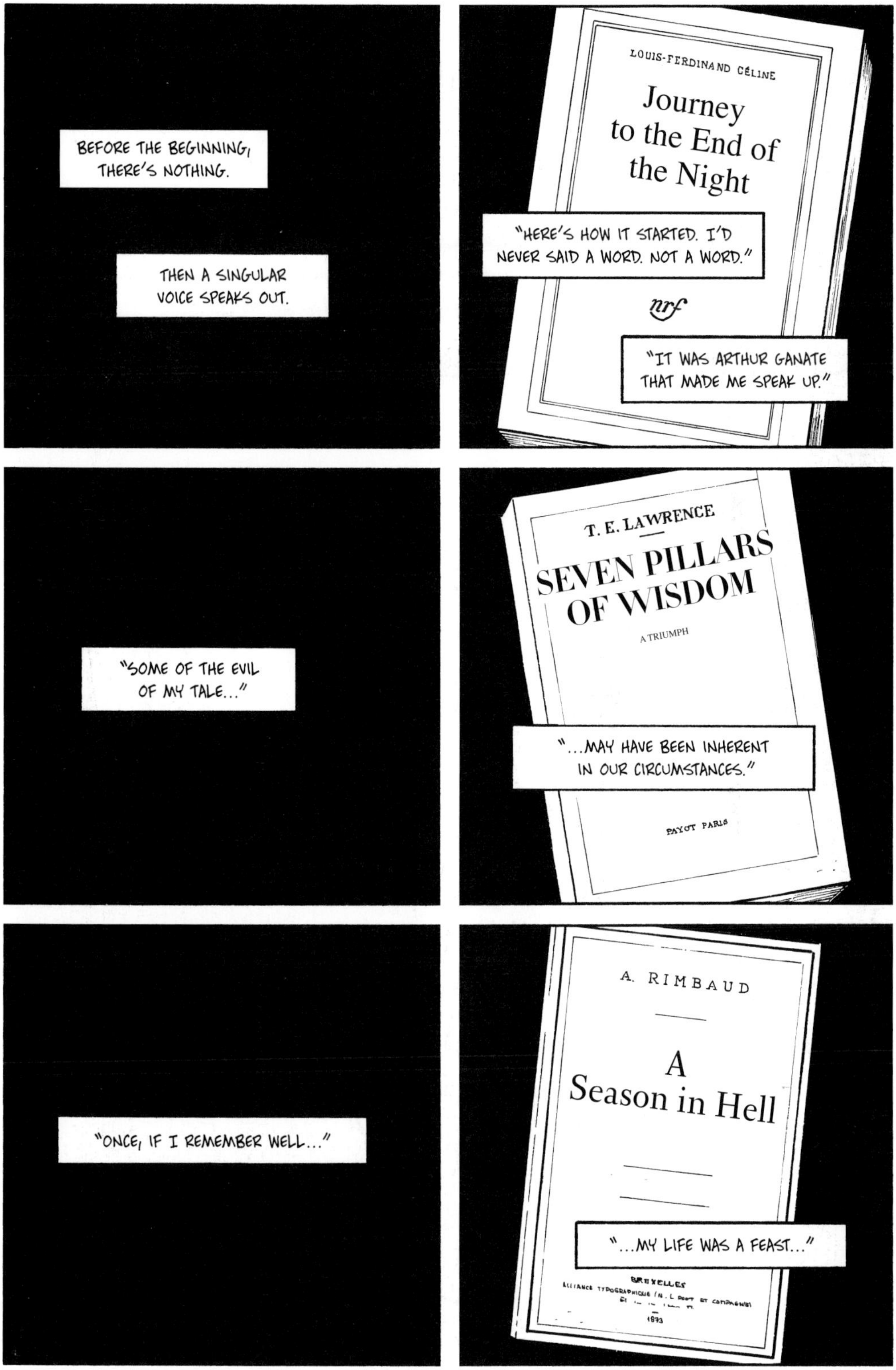
BEFORE THE BEGINNING, THERE'S NOTHING.
THEN A SINGULAR VOICE SPEAKS OUT.
LOUIS-FERDINAND CÉLINE
Journey to the End of the Night
"HERE'S HOW IT STARTED. I'D NEVER SAID A WORD. NOT A WORD."
nrf
"IT WAS ARTHUR GANATE THAT MADE ME SPEAK UP."
"SOME OF THE EVIL OF MY TALE..."
T. E. LAWRENCE
SEVEN PILLARS OF WISDOM
A TRIUMPH
"...MAY HAVE BEEN INHERENT IN OUR CIRCUMSTANCES."
PAYOT PARIS
"ONCE, IF I REMEMBER WELL..."
A. RIMBAUD
A Season in Hell
"...MY LIFE WAS A FEAST..."
BRUXELLES
1873

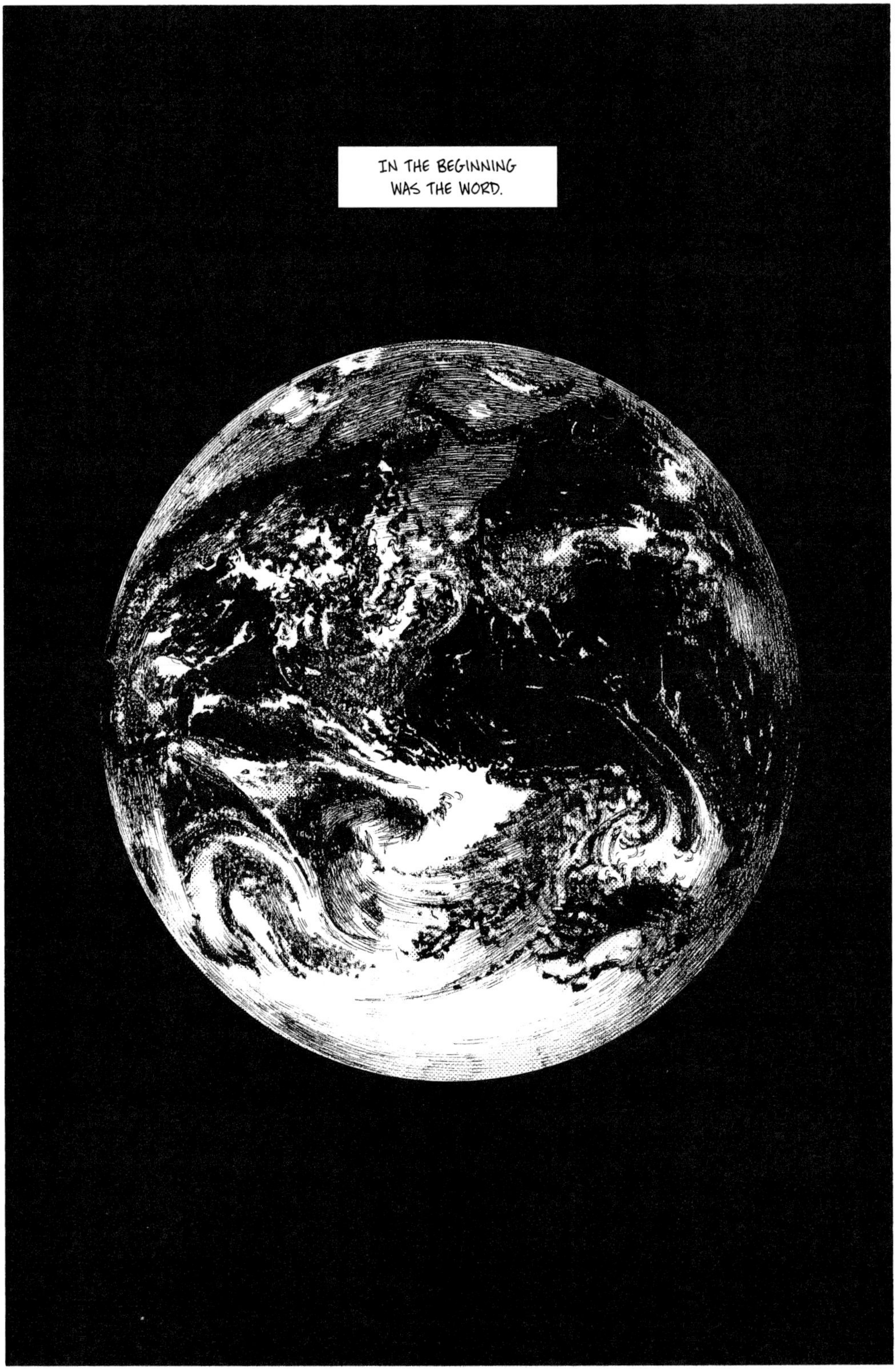
IN THE BEGINNING
WAS THE WORD.

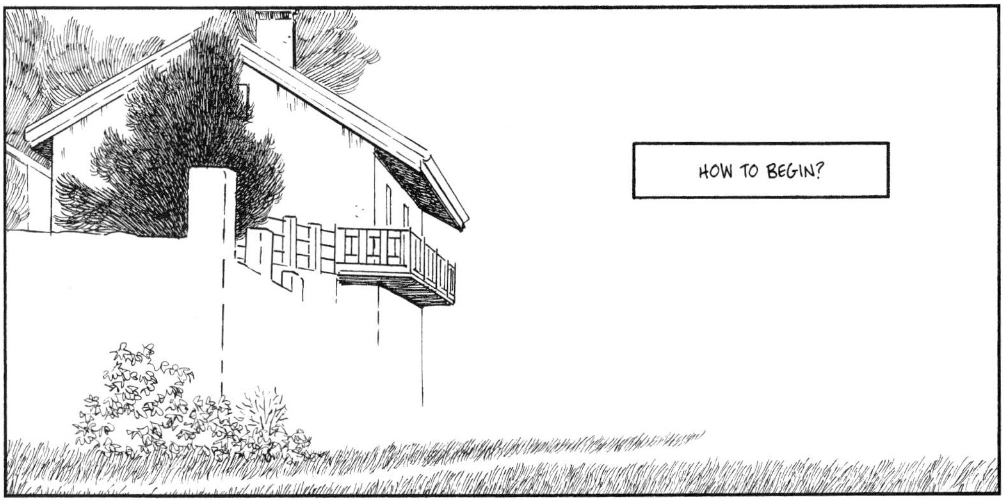
HOW TO BEGIN?

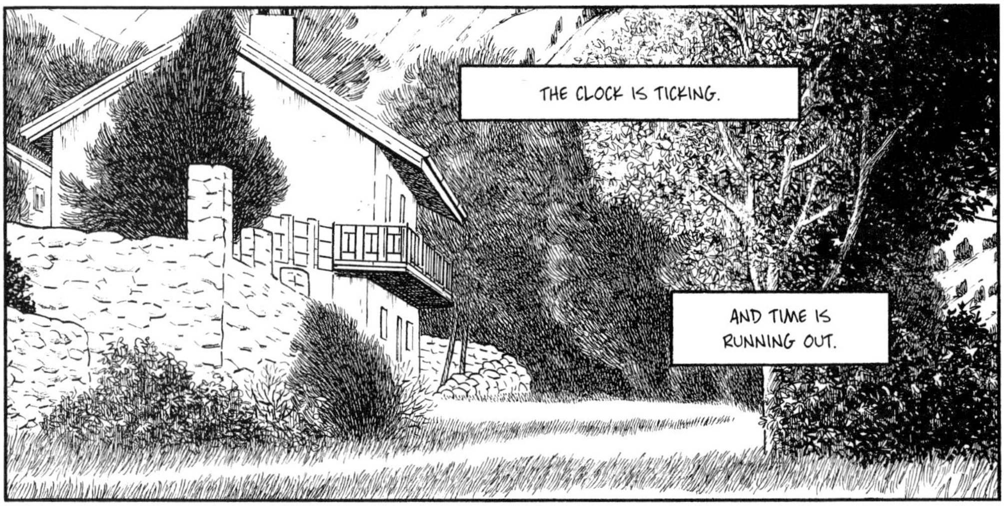
THE CLOCK IS TICKING.
AND TIME IS RUNNING OUT.

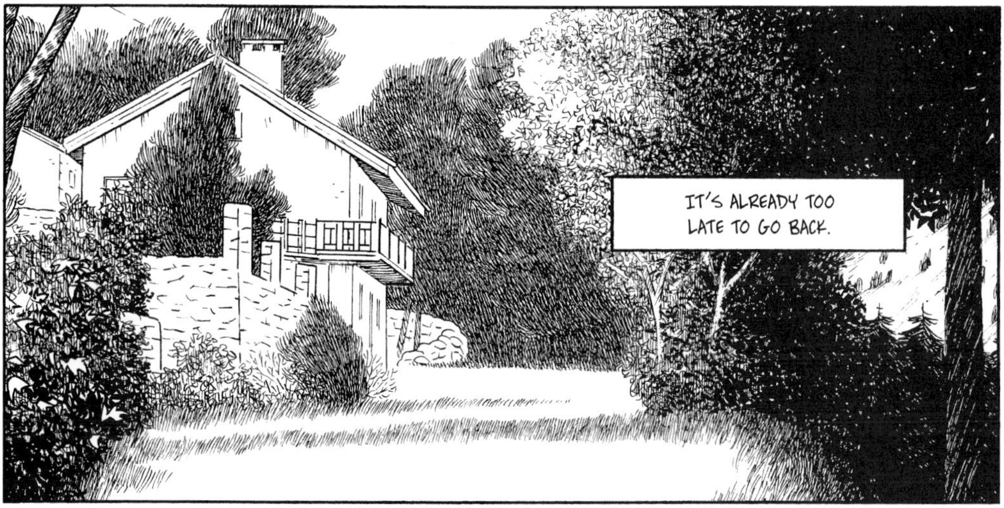
IT'S ALREADY TOO LATE TO GO BACK.

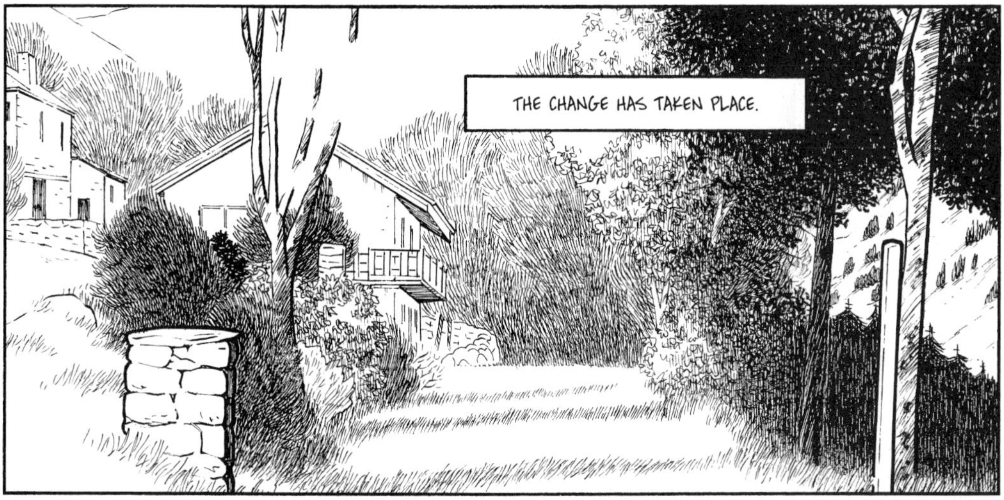
THE CHANGE HAS TAKEN PLACE.

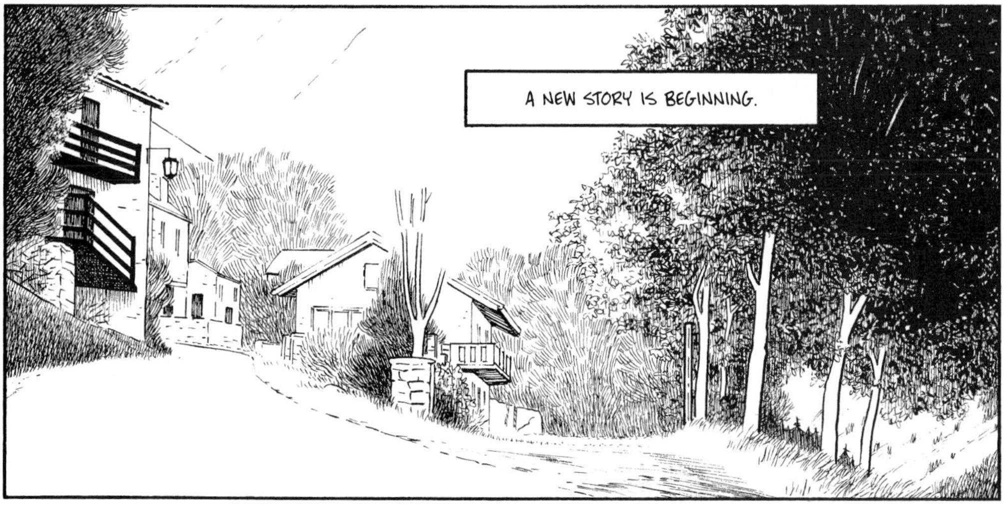
A NEW STORY IS BEGINNING.

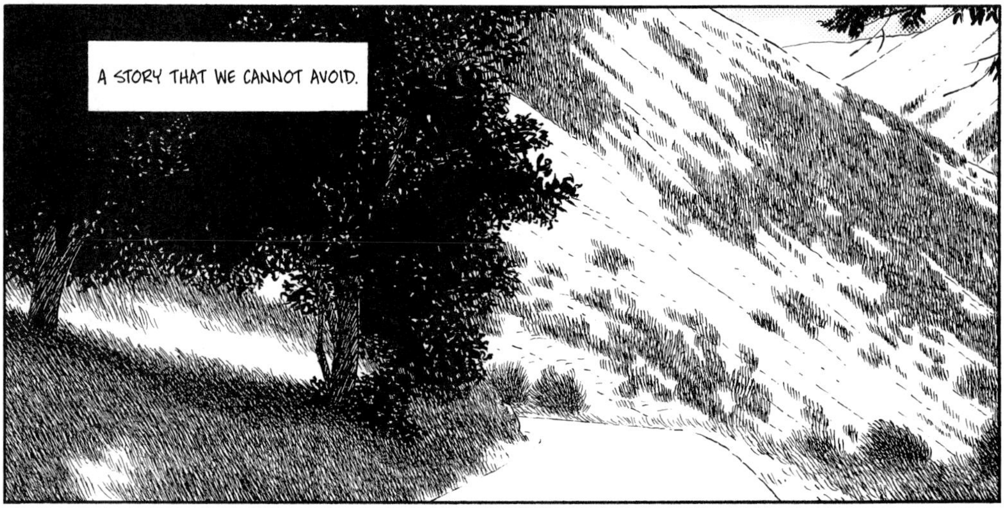
A STORY THAT WE CANNOT AVOID.

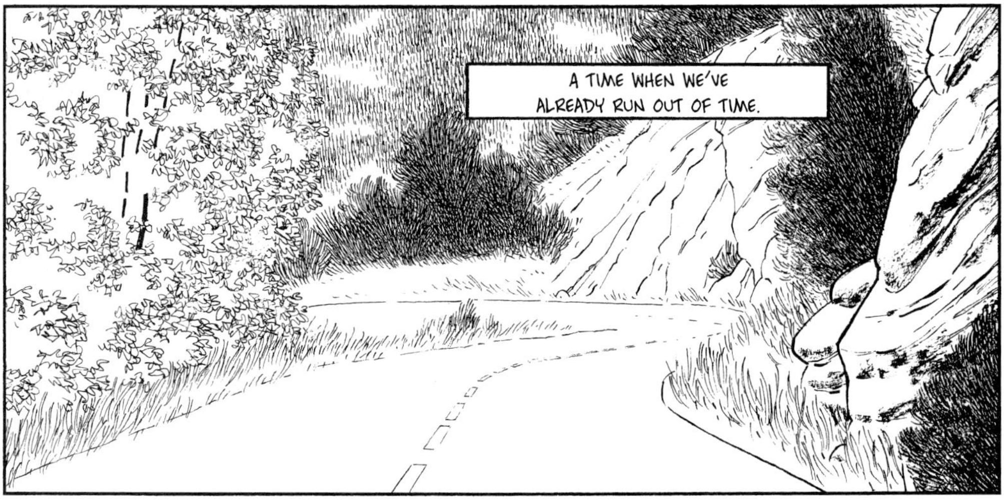
A TIME WHEN WE'VE ALREADY RUN OUT OF TIME.

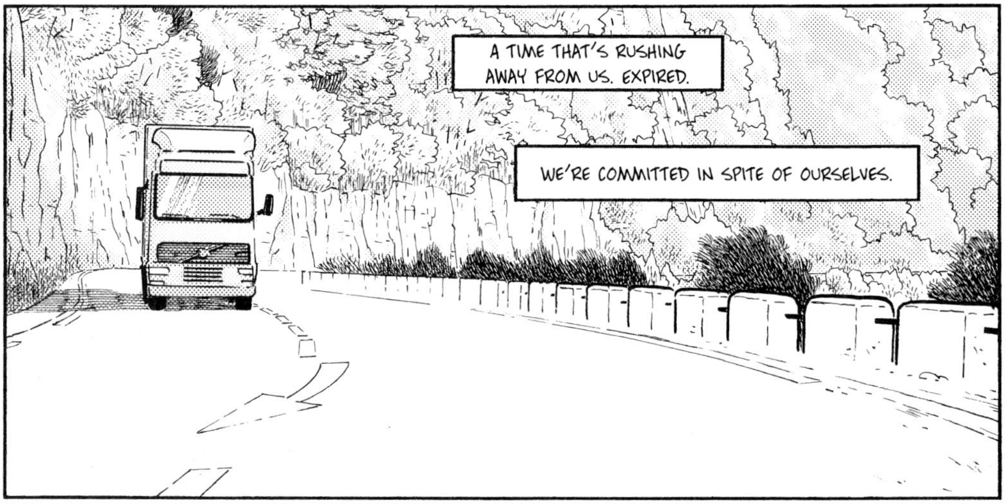
A TIME THAT'S RUSHING AWAY FROM US. EXPIRED.
WE'RE COMMITTED IN SPITE OF OURSELVES.

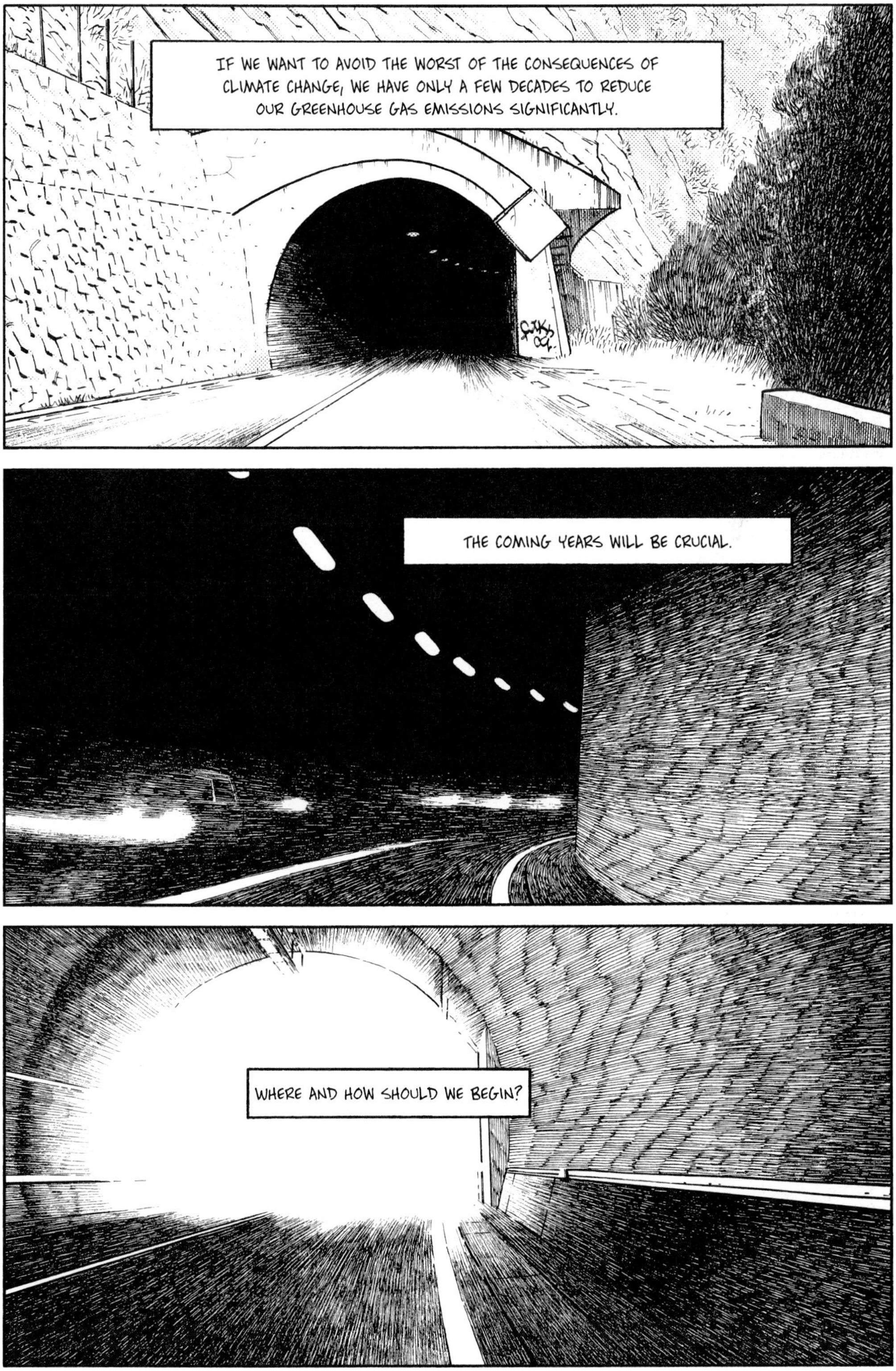
IF WE WANT TO AVOID THE WORST OF THE CONSEQUENCES OF CLIMATE CHANGE, WE HAVE ONLY A FEW DECADES TO REDUCE OUR GREENHOUSE GAS EMISSIONS SIGNIFICANTLY.
THE COMING YEARS WILL BE CRUCIAL.
WHERE AND HOW SHOULD WE BEGIN?

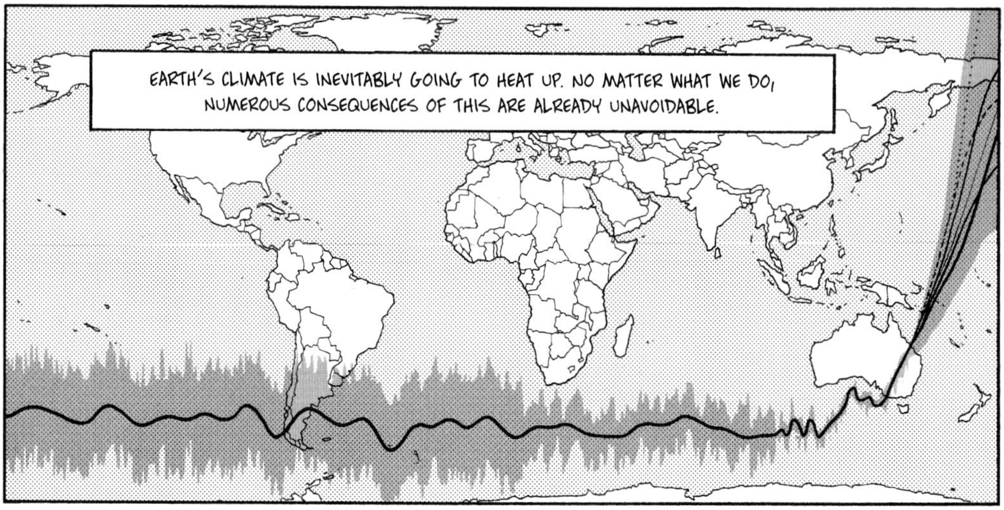
EARTH'S CLIMATE IS INEVITABLY GOING TO HEAT UP. NO MATTER WHAT WE DO, NUMEROUS CONSEQUENCES OF THIS ARE ALREADY UNAVOIDABLE.

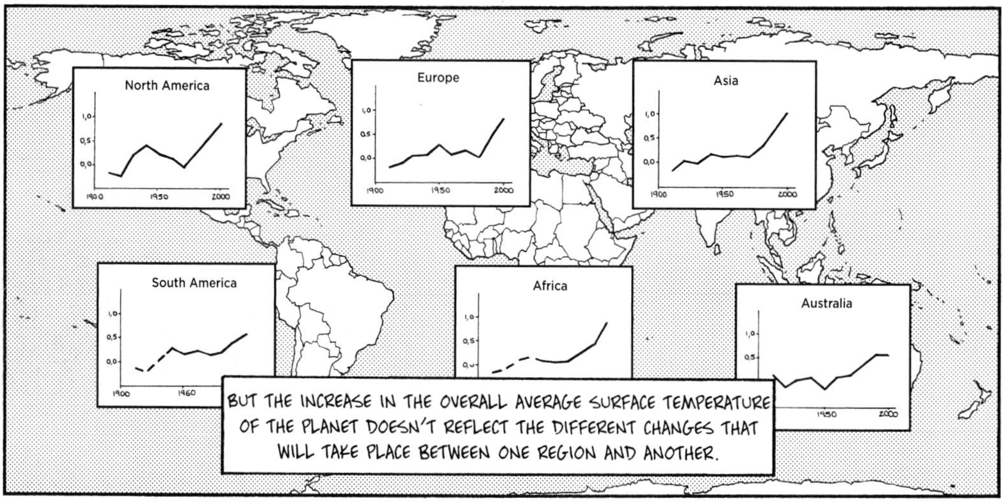
North America
1,0
0,5
0,0
1900
1950
2000
Europe
1,0
0,5
0,0
1900
1950
2000
Asia
1,0
0,5
0,0
1900
1950
2000
South America
1,0
0,5
0,0
1900
1960
Africa
1,0
0,5
Australia
1,0
0,5
1950
2000
BUT THE INCREASE IN THE OVERALL AVERAGE SURFACE TEMPERATURE OF THE PLANET DOESN'T REFLECT THE DIFFERENT CHANGES THAT WILL TAKE PLACE BETWEEN ONE REGION AND ANOTHER.

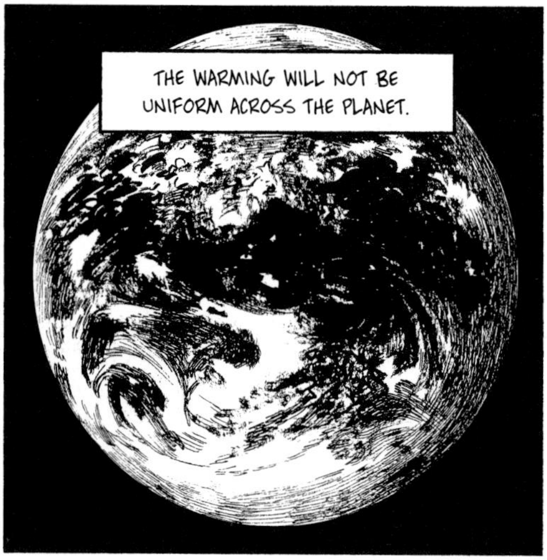
THE WARMING WILL NOT BE UNIFORM ACROSS THE PLANET.

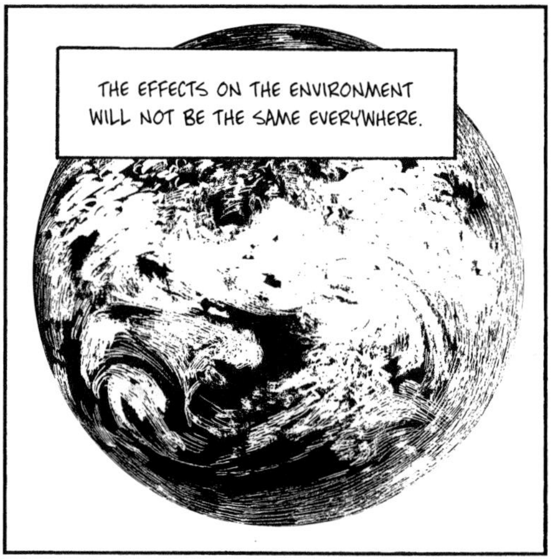
THE EFFECTS ON THE ENVIRONMENT WILL NOT BE THE SAME EVERYWHERE.

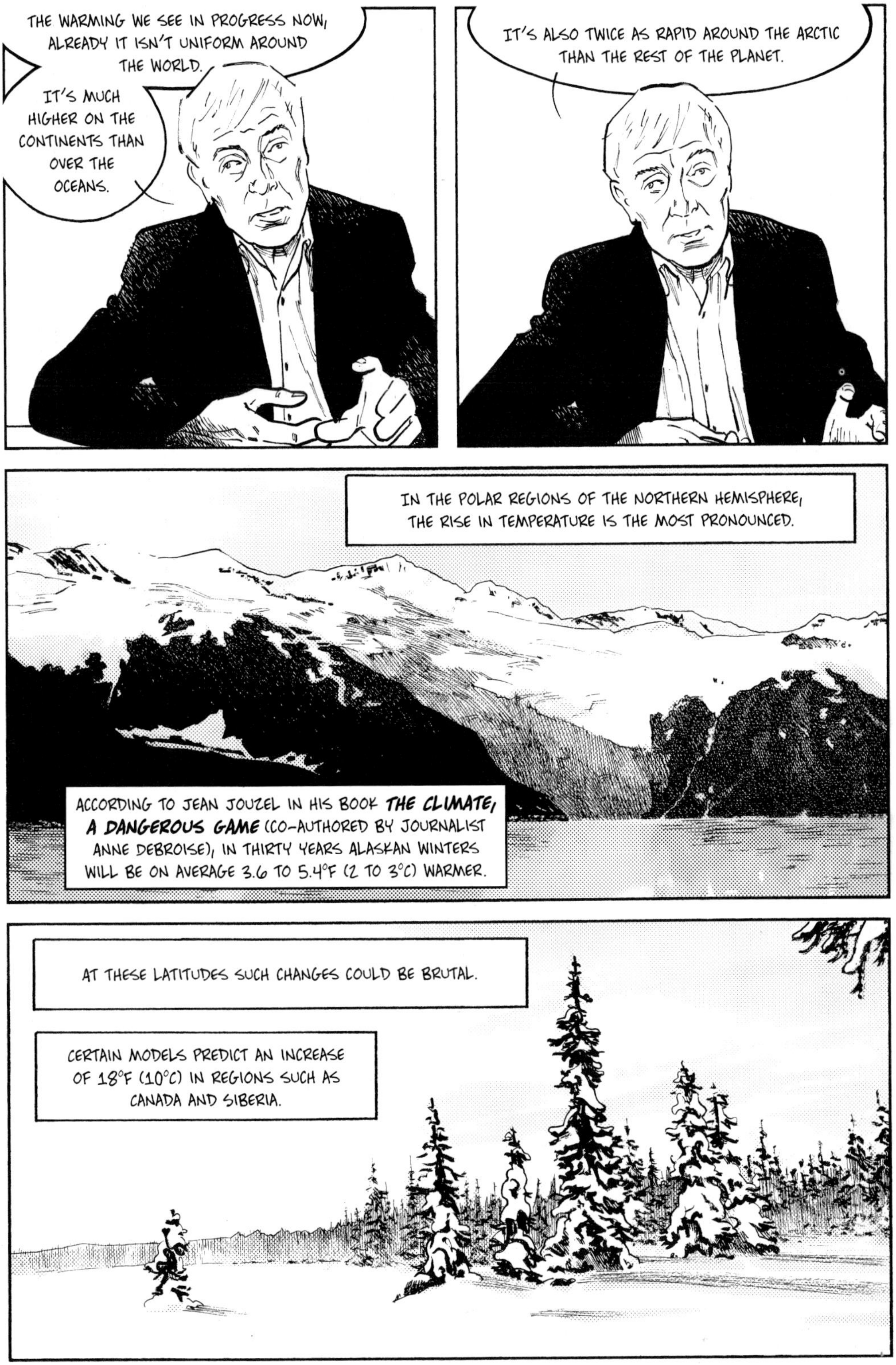
THE WARMING WE SEE IN PROGRESS NOW, ALREADY IT ISN'T UNIFORM AROUND THE WORLD.
IT'S MUCH HIGHER ON THE CONTINENTS THAN OVER THE OCEANS.
IT'S ALSO TWICE AS RAPID AROUND THE ARCTIC THAN THE REST OF THE PLANET.
IN THE POLAR REGIONS OF THE NORTHERN HEMISPHERE, THE RISE IN TEMPERATURE IS THE MOST PRONOUNCED.
ACCORDING TO JEAN JOUZEL IN HIS BOOK **THE CLIMATE, A DANGEROUS GAME** (CO-AUTHORED BY JOURNALIST ANNE DEBROISE), IN THIRTY YEARS ALASKAN WINTERS WILL BE ON AVERAGE 3.6 TO 5.4°F (2 TO 3°C) WARMER.
AT THESE LATITUDES SUCH CHANGES COULD BE BRUTAL.
CERTAIN MODELS PREDICT AN INCREASE OF 18°F (10°C) IN REGIONS SUCH AS CANADA AND SIBERIA.

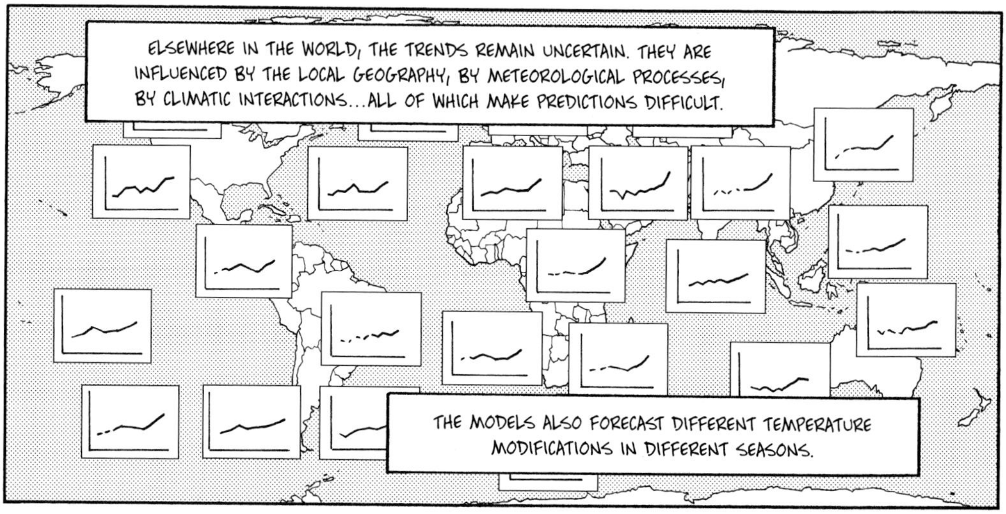
ELSEWHERE IN THE WORLD, THE TRENDS REMAIN UNCERTAIN. THEY ARE INFLUENCED BY THE LOCAL GEOGRAPHY, BY METEOROLOGICAL PROCESSES, BY CLIMATIC INTERACTIONS...ALL OF WHICH MAKE PREDICTIONS DIFFICULT.
THE MODELS ALSO FORECAST DIFFERENT TEMPERATURE MODIFICATIONS IN DIFFERENT SEASONS.

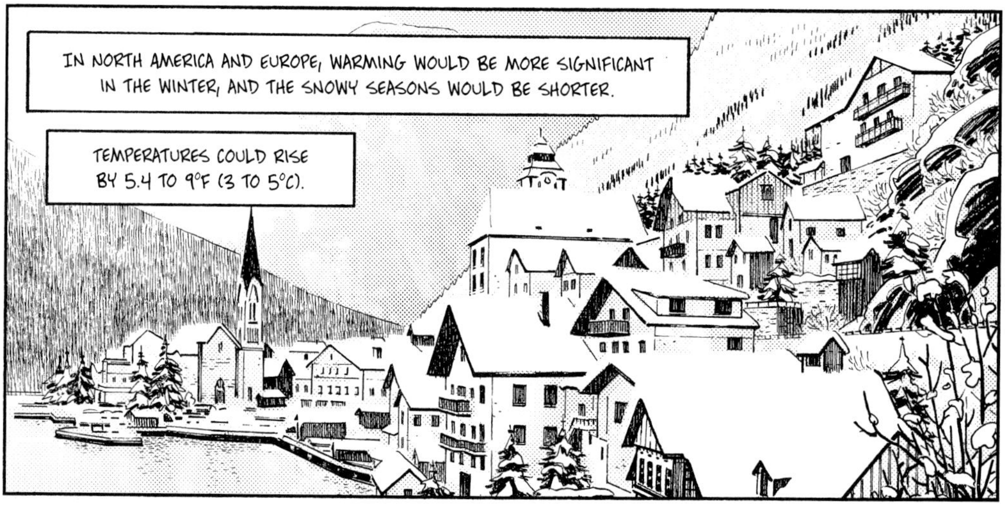
IN NORTH AMERICA AND EUROPE, WARMING WOULD BE MORE SIGNIFICANT IN THE WINTER, AND THE SNOWY SEASONS WOULD BE SHORTER.
TEMPERATURES COULD RISE BY 5.4 TO 9°F (3 TO 5°C).

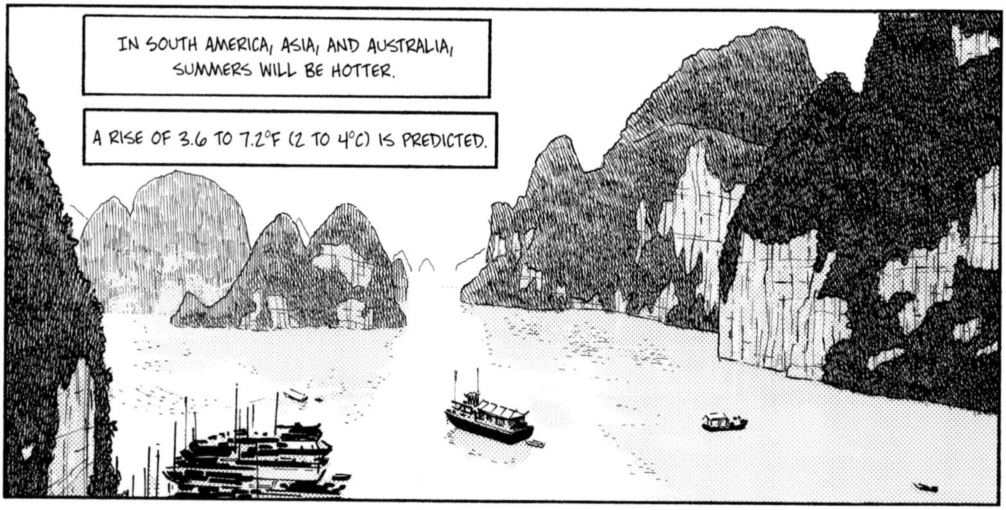
IN SOUTH AMERICA, ASIA, AND AUSTRALIA, SUMMERS WILL BE HOTTER.
A RISE OF 3.6 TO 7.2°F (2 TO 4°C) IS PREDICTED.

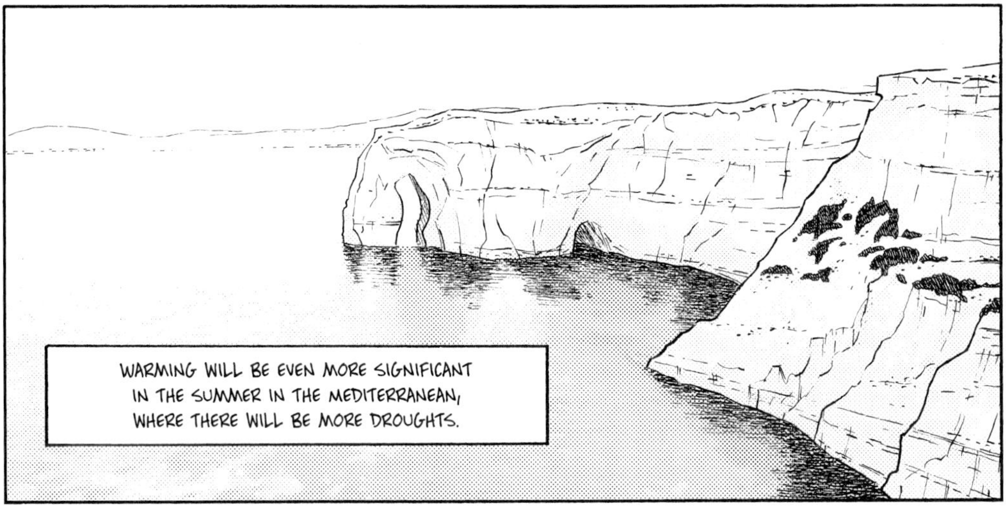
WARMING WILL BE EVEN MORE SIGNIFICANT IN THE SUMMER IN THE MEDITERRANEAN, WHERE THERE WILL BE MORE DROUGHTS.

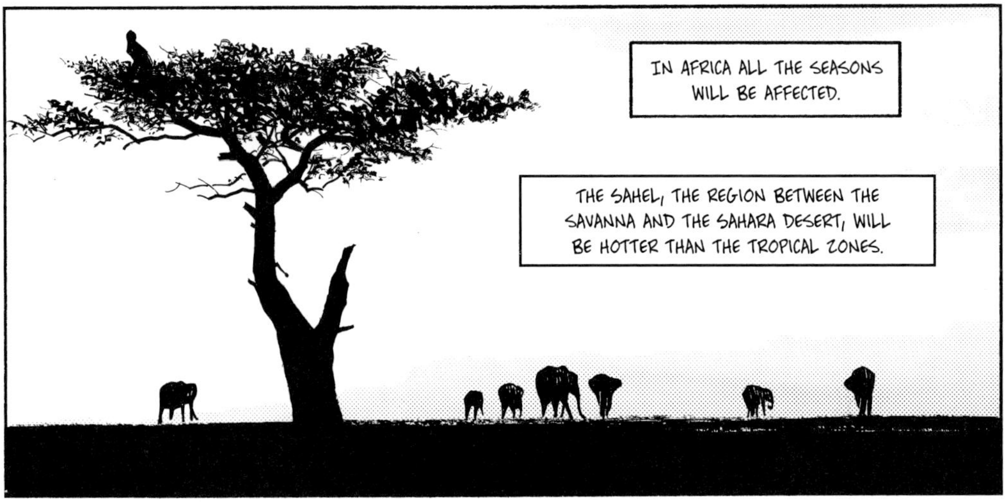
IN AFRICA ALL THE SEASONS WILL BE AFFECTED.
THE SAHEL, THE REGION BETWEEN THE SAVANNA AND THE SAHARA DESERT, WILL BE HOTTER THAN THE TROPICAL ZONES.

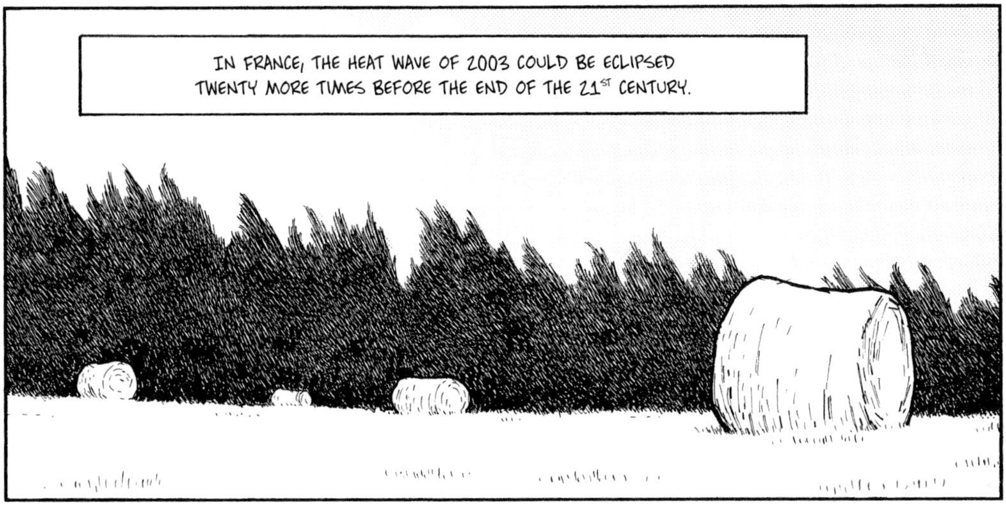
IN FRANCE, THE HEAT WAVE OF 2003 COULD BE ECLIPSED TWENTY MORE TIMES BEFORE THE END OF THE 21ST CENTURY.

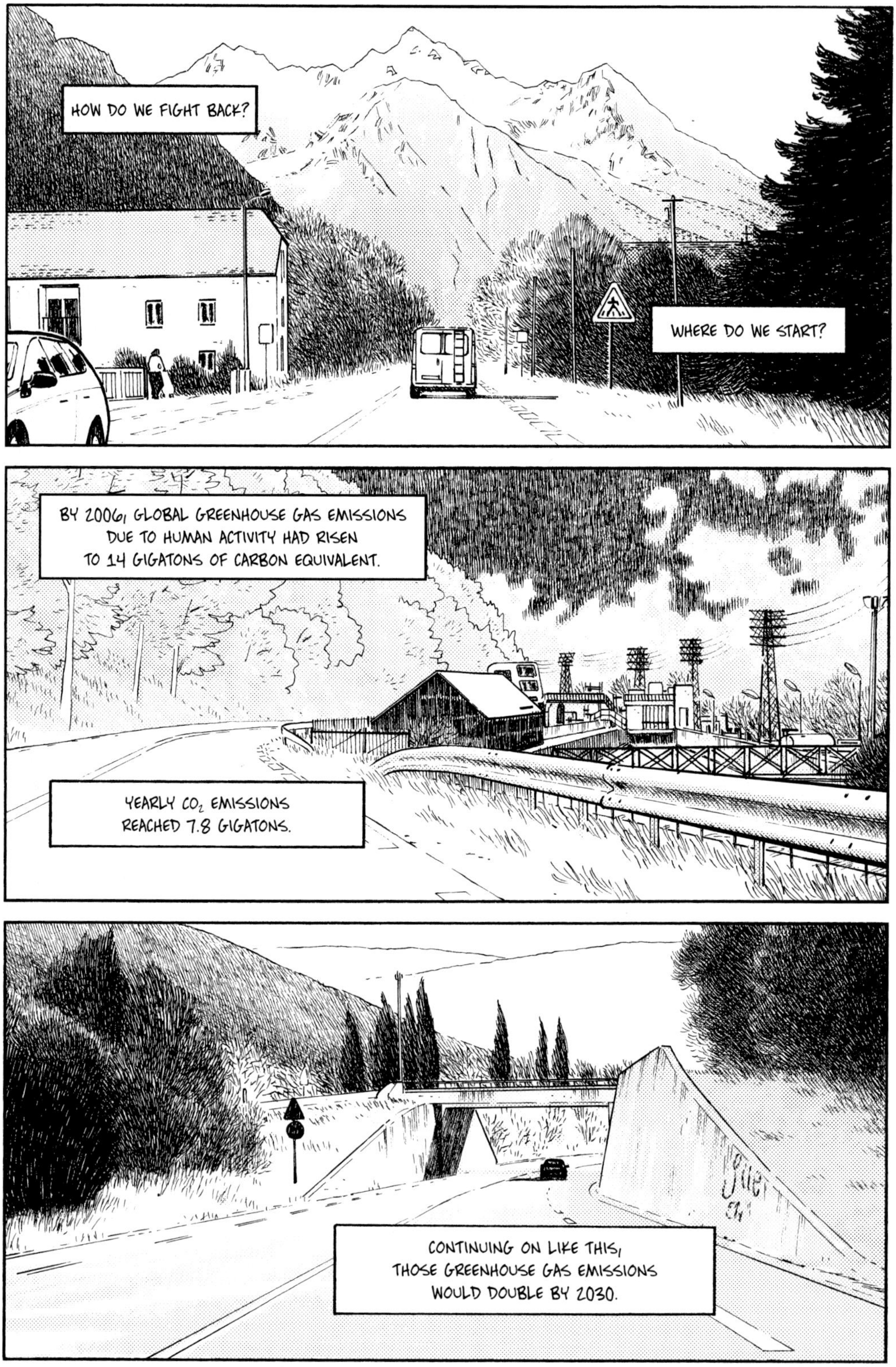
HOW DO WE FIGHT BACK?
WHERE DO WE START?
BY 2006, GLOBAL GREENHOUSE GAS EMISSIONS DUE TO HUMAN ACTIVITY HAD RISEN TO 14 GIGATONS OF CARBON EQUIVALENT.
YEARLY CO_2 EMISSIONS REACHED 7.8 GIGATONS.
CONTINUING ON LIKE THIS, THOSE GREENHOUSE GAS EMISSIONS WOULD DOUBLE BY 2030.

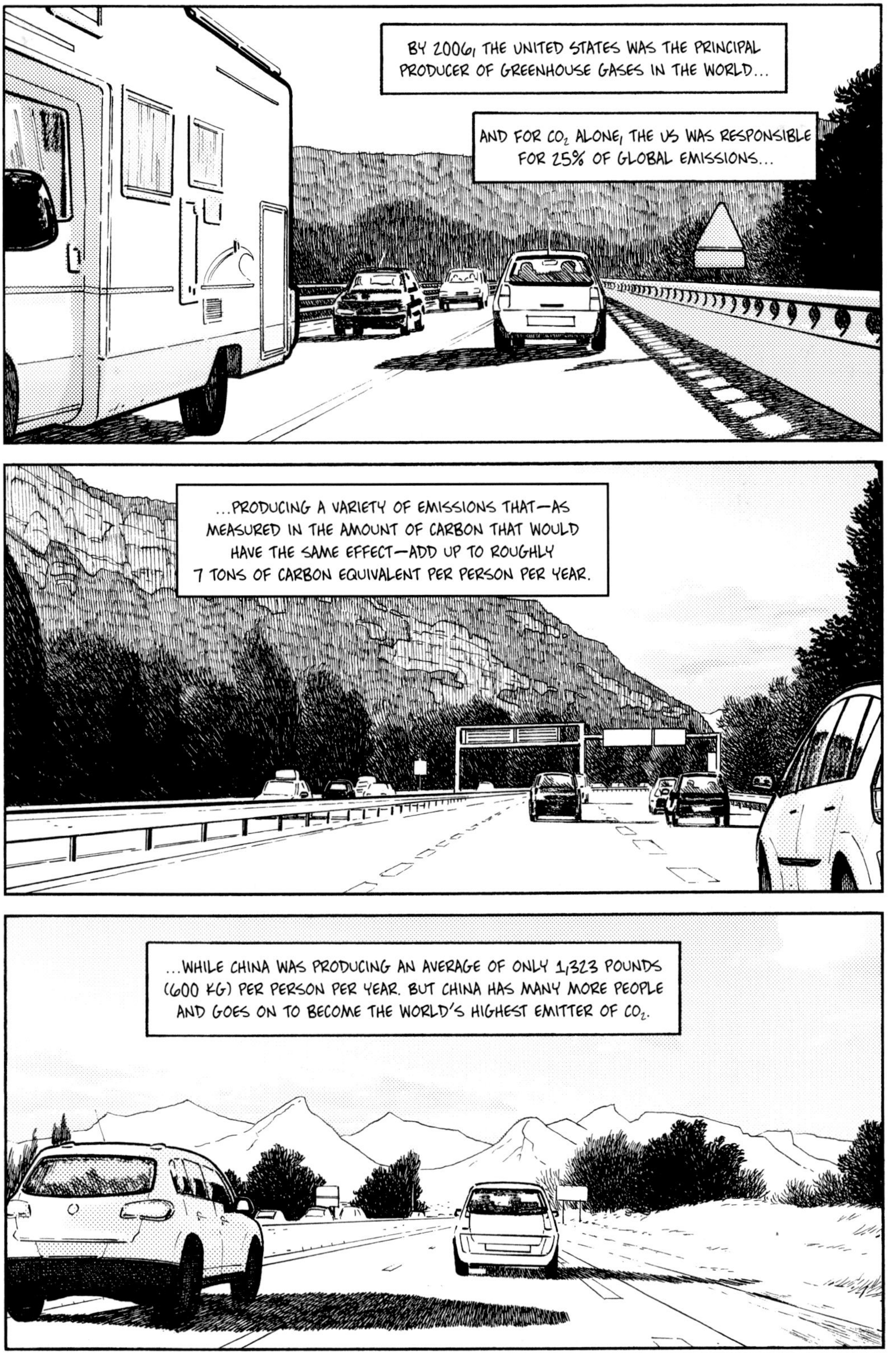
BY 2006, THE UNITED STATES WAS THE PRINCIPAL PRODUCER OF GREENHOUSE GASES IN THE WORLD...
AND FOR CO_2 ALONE, THE US WAS RESPONSIBLE FOR 25% OF GLOBAL EMISSIONS...
...PRODUCING A VARIETY OF EMISSIONS THAT—AS MEASURED IN THE AMOUNT OF CARBON THAT WOULD HAVE THE SAME EFFECT—ADD UP TO ROUGHLY 7 TONS OF CARBON EQUIVALENT PER PERSON PER YEAR.
...WHILE CHINA WAS PRODUCING AN AVERAGE OF ONLY 1,323 POUNDS (600 KG) PER PERSON PER YEAR. BUT CHINA HAS MANY MORE PEOPLE AND GOES ON TO BECOME THE WORLD'S HIGHEST EMITTER OF CO_2.

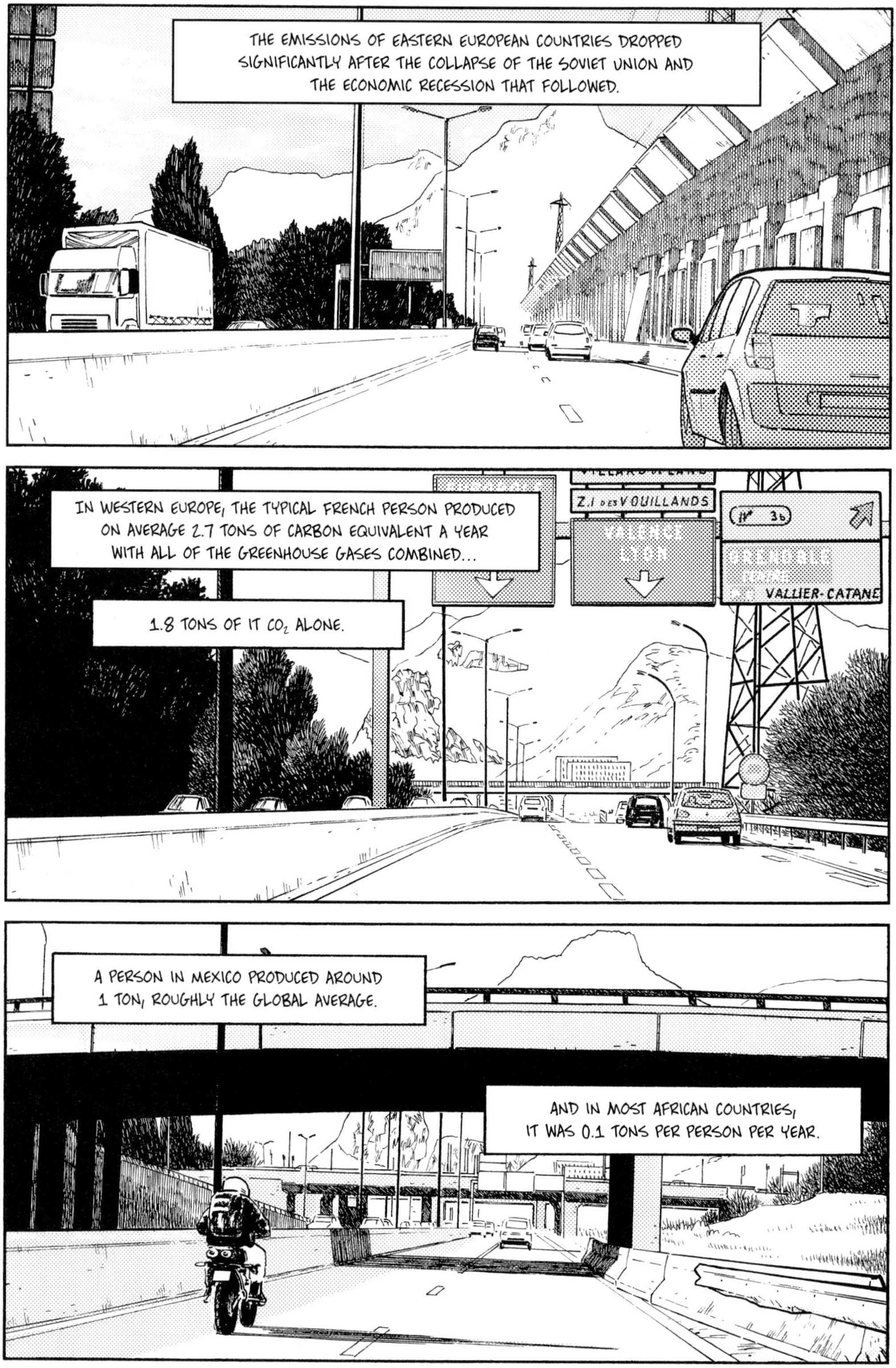
THE EMISSIONS OF EASTERN EUROPEAN COUNTRIES DROPPED SIGNIFICANTLY AFTER THE COLLAPSE OF THE SOVIET UNION AND THE ECONOMIC RECESSION THAT FOLLOWED.
IN WESTERN EUROPE, THE TYPICAL FRENCH PERSON PRODUCED ON AVERAGE 2.7 TONS OF CARBON EQUIVALENT A YEAR WITH ALL OF THE GREENHOUSE GASES COMBINED...
Z.I des VOUILLANDS
VALLIER-CATANE
1.8 TONS OF IT CO_2 ALONE.
A PERSON IN MEXICO PRODUCED AROUND 1 TON, ROUGHLY THE GLOBAL AVERAGE.
AND IN MOST AFRICAN COUNTRIES, IT WAS 0.1 TONS PER PERSON PER YEAR.

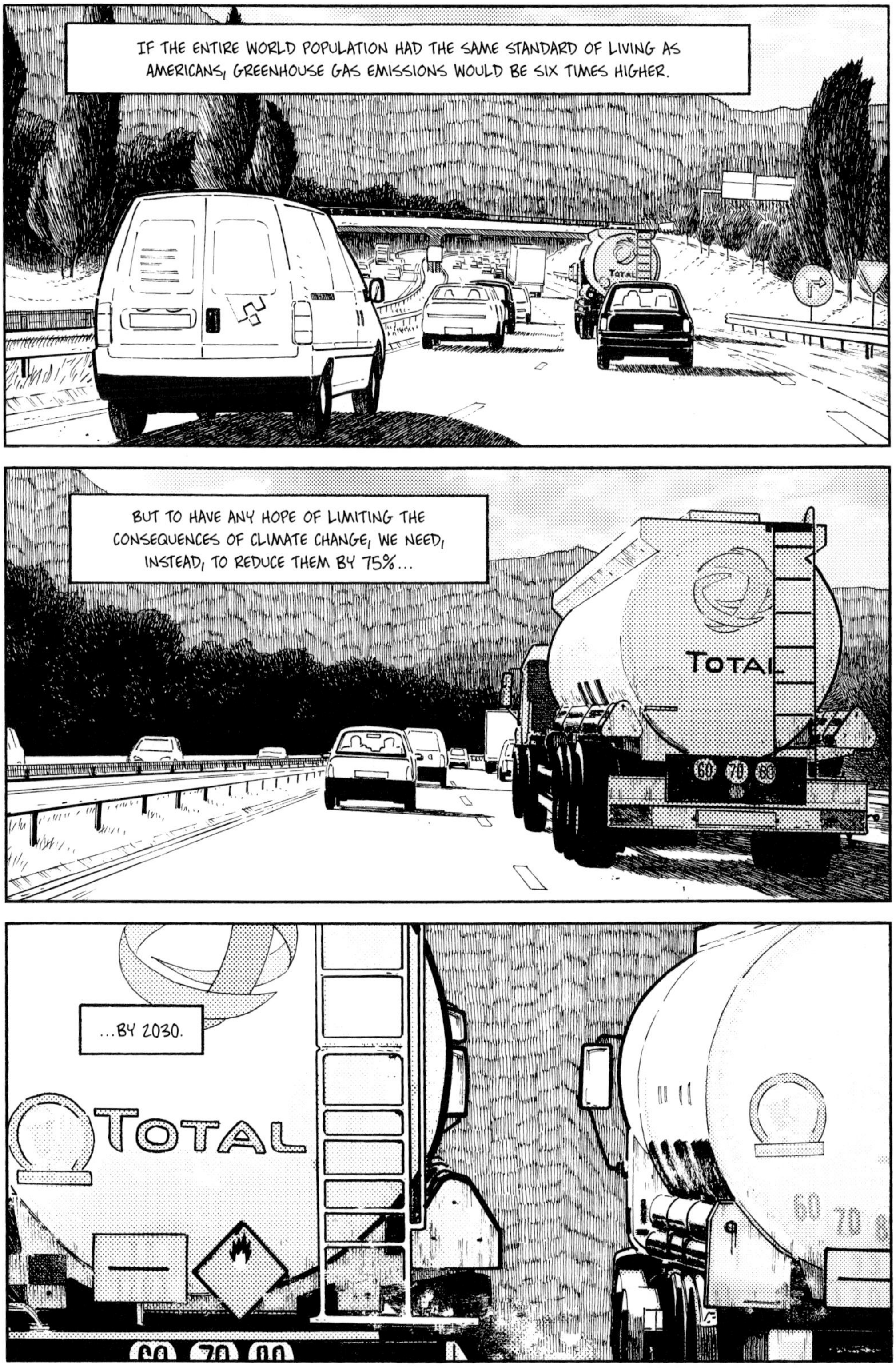
IF THE ENTIRE WORLD POPULATION HAD THE SAME STANDARD OF LIVING AS AMERICANS, GREENHOUSE GAS EMISSIONS WOULD BE SIX TIMES HIGHER.
BUT TO HAVE ANY HOPE OF LIMITING THE CONSEQUENCES OF CLIMATE CHANGE, WE NEED, INSTEAD, TO REDUCE THEM BY 75%...
TOTAL
...BY 2030.
TOTAL

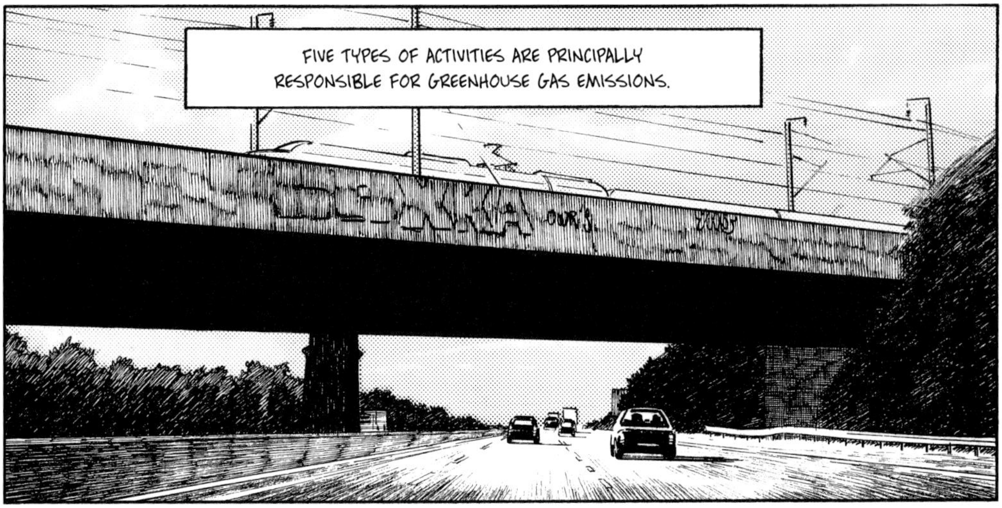
FIVE TYPES OF ACTIVITIES ARE PRINCIPALLY RESPONSIBLE FOR GREENHOUSE GAS EMISSIONS.

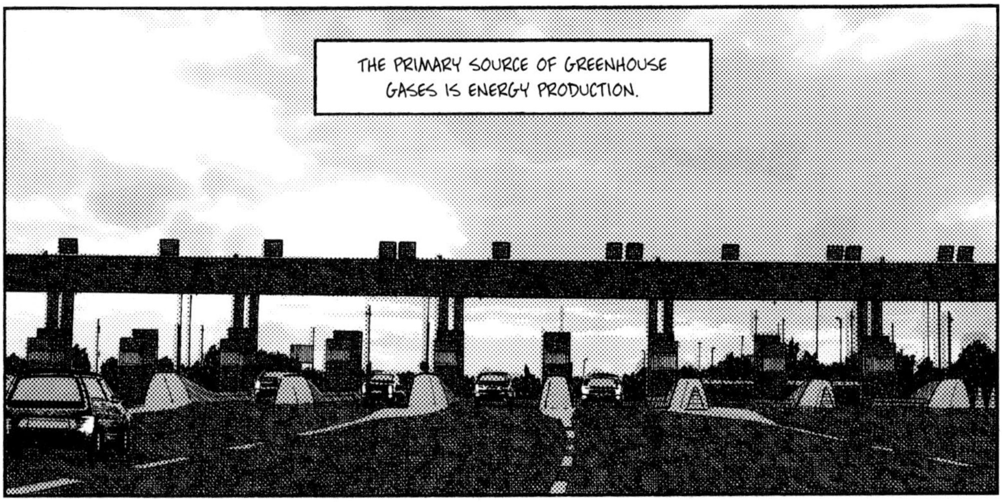
THE PRIMARY SOURCE OF GREENHOUSE GASES IS ENERGY PRODUCTION.

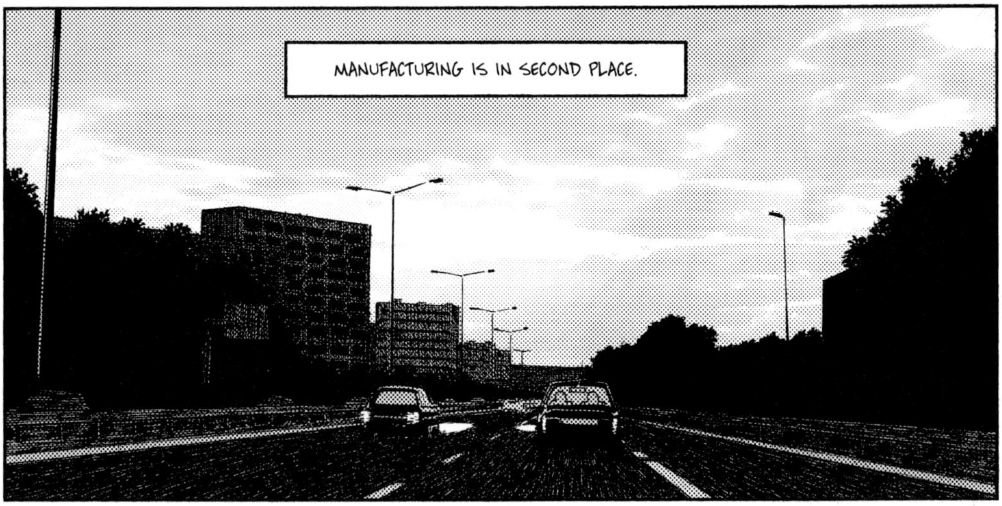
MANUFACTURING IS IN SECOND PLACE.

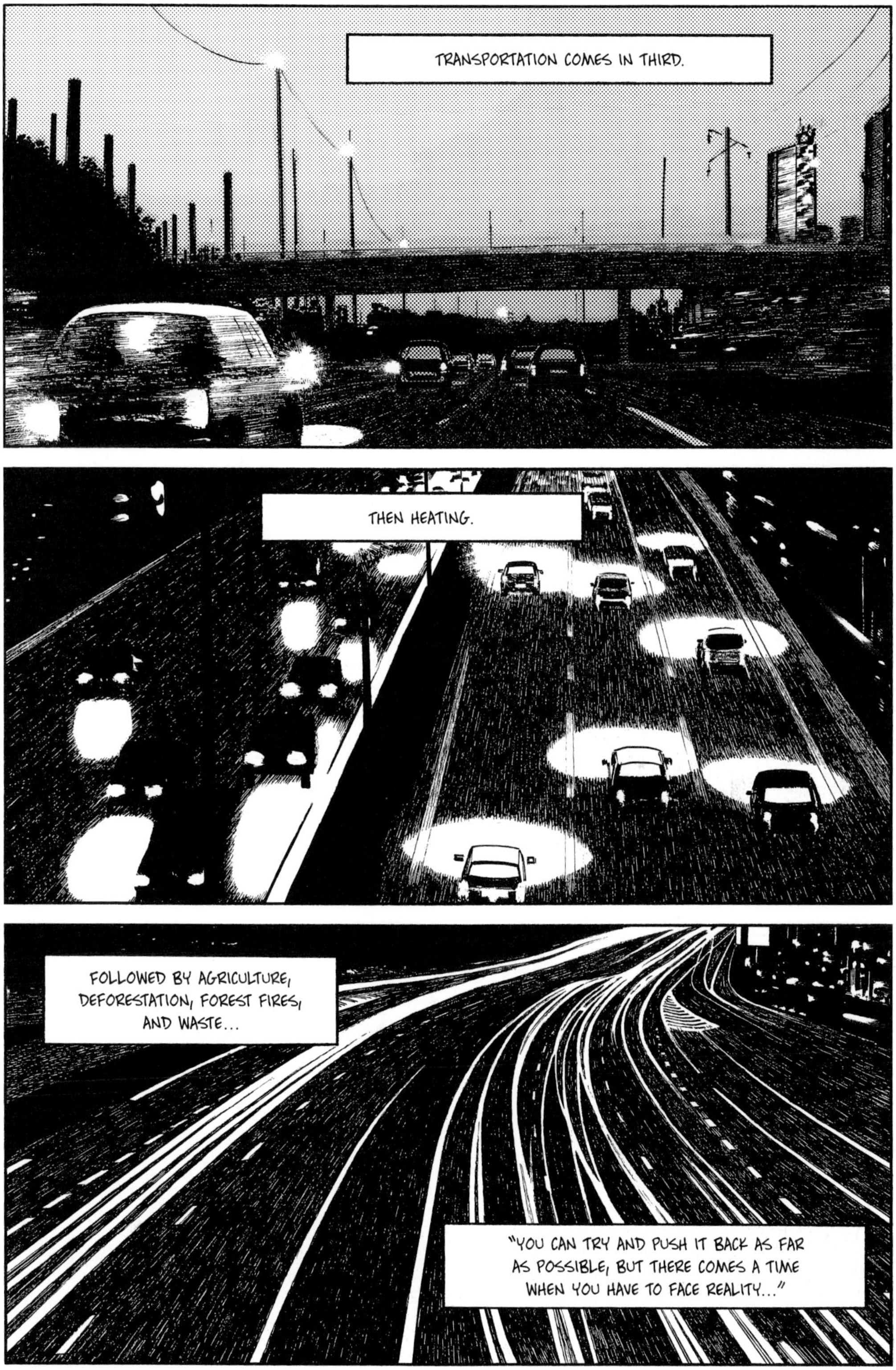
TRANSPORTATION COMES IN THIRD.
THEN HEATING.
FOLLOWED BY AGRICULTURE, DEFORESTATION, FOREST FIRES, AND WASTE...
"YOU CAN TRY AND PUSH IT BACK AS FAR AS POSSIBLE, BUT THERE COMES A TIME WHEN YOU HAVE TO FACE REALITY..."

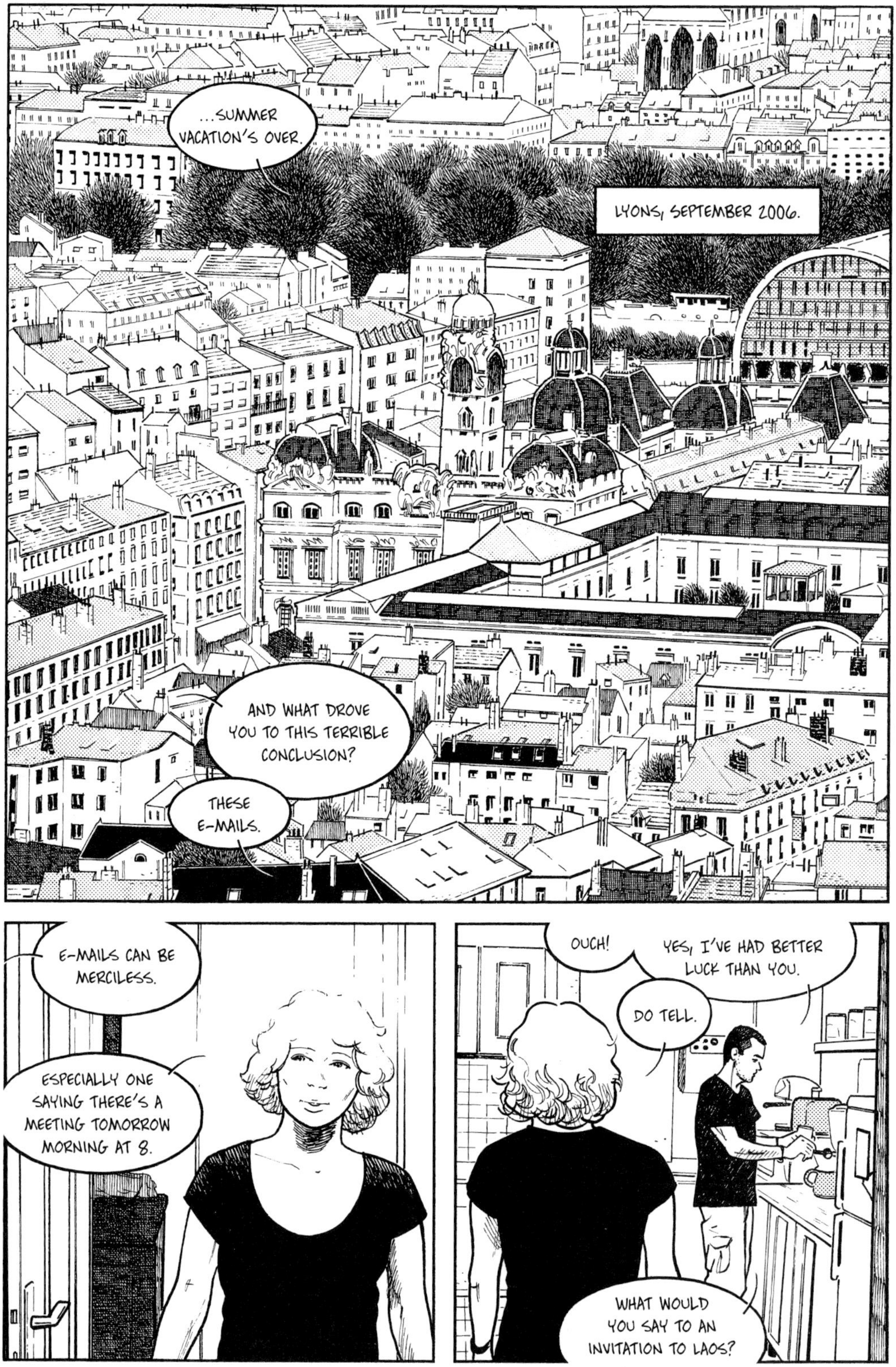
...SUMMER VACATION'S OVER.
LYONS, SEPTEMBER 2006.
AND WHAT DROVE YOU TO THIS TERRIBLE CONCLUSION?
THESE E-MAILS.
E-MAILS CAN BE MERCILESS.
ESPECIALLY ONE SAYING THERE'S A MEETING TOMORROW MORNING AT 8.
OUCH!
YES, I'VE HAD BETTER LUCK THAN YOU.
DO TELL.
WHAT WOULD YOU SAY TO AN INVITATION TO LAOS?

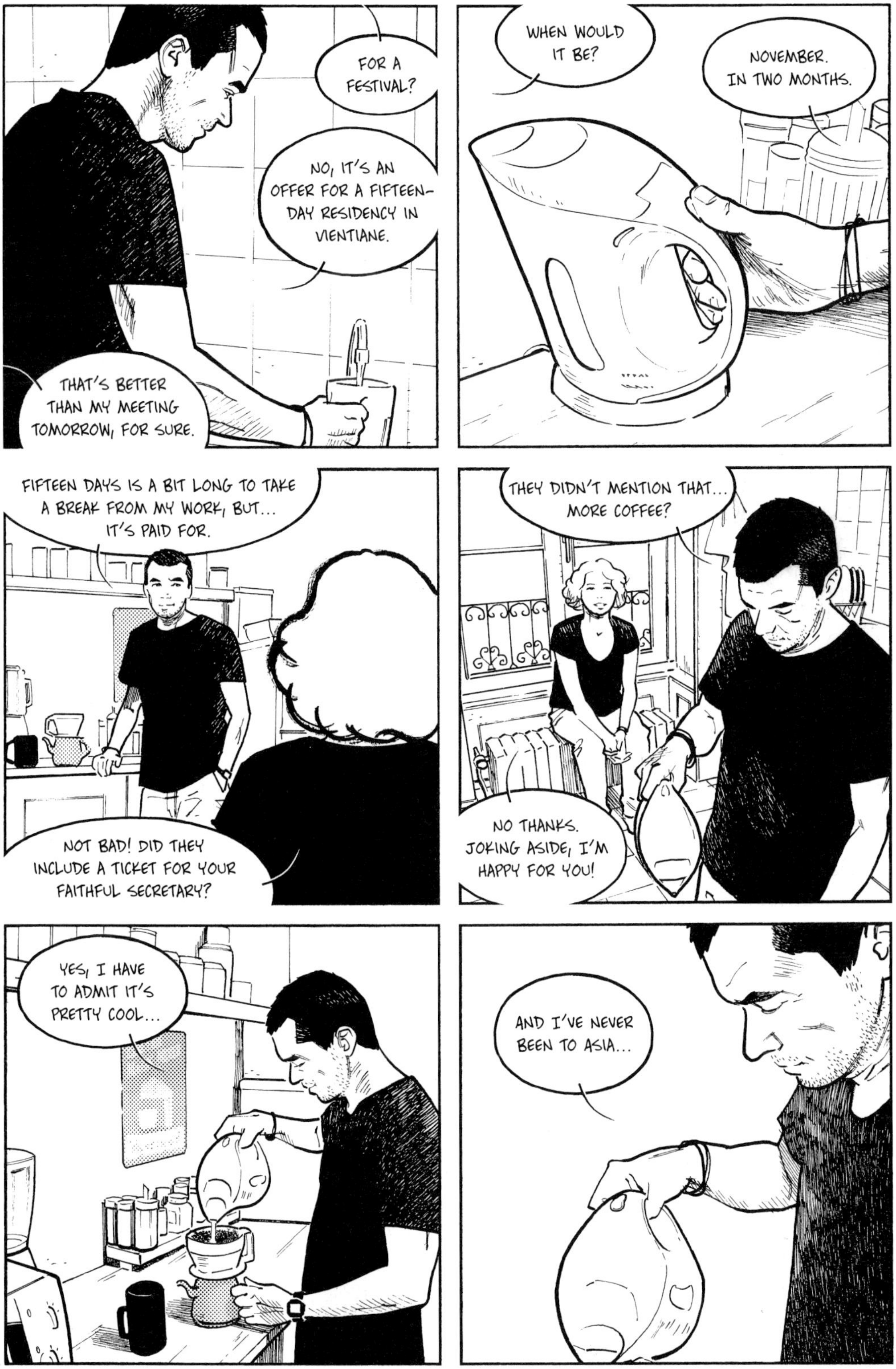

FOR A FESTIVAL?
NO, IT'S AN OFFER FOR A FIFTEEN-DAY RESIDENCY IN VIENTIANE.
THAT'S BETTER THAN MY MEETING TOMORROW, FOR SURE.
WHEN WOULD IT BE?
NOVEMBER. IN TWO MONTHS.
FIFTEEN DAYS IS A BIT LONG TO TAKE A BREAK FROM MY WORK, BUT... IT'S PAID FOR.
NOT BAD! DID THEY INCLUDE A TICKET FOR YOUR FAITHFUL SECRETARY?
THEY DIDN'T MENTION THAT... MORE COFFEE?
NO THANKS. JOKING ASIDE, I'M HAPPY FOR YOU!
YES, I HAVE TO ADMIT IT'S PRETTY COOL...
AND I'VE NEVER BEEN TO ASIA...

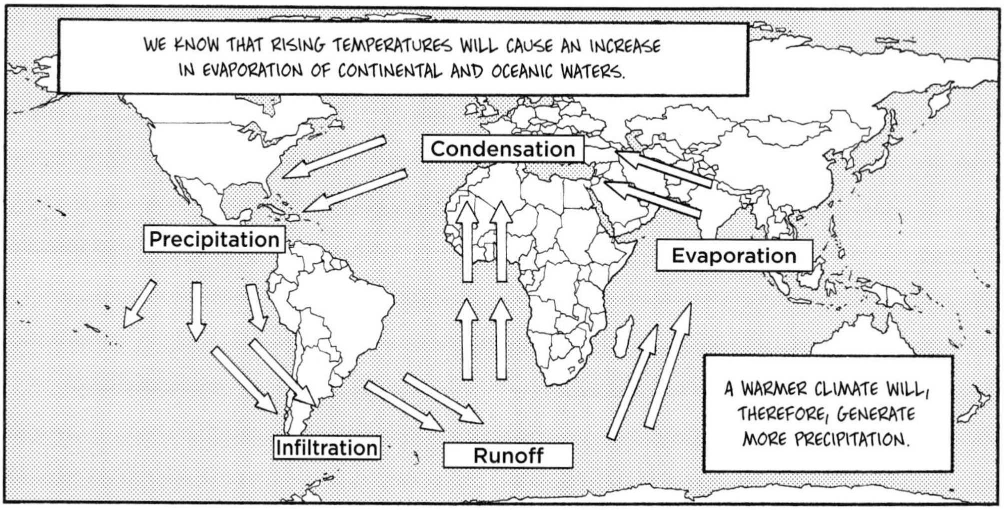

WE KNOW THAT RISING TEMPERATURES WILL CAUSE AN INCREASE IN EVAPORATION OF CONTINENTAL AND OCEANIC WATERS.
Condensation
Precipitation
Evaporation
Infiltration
Runoff
A WARMER CLIMATE WILL, THEREFORE, GENERATE MORE PRECIPITATION.

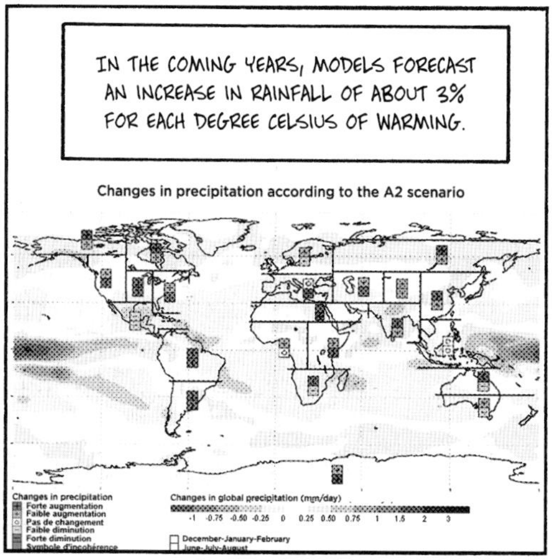

IN THE COMING YEARS, MODELS FORECAST AN INCREASE IN RAINFALL OF ABOUT 3% FOR EACH DEGREE CELSIUS OF WARMING.
Changes in precipitation according to the A2 scenario
Changes in precipitation
Forte augmentation
Faible augmentation
Pas de changement
Faible diminution
Forte diminution
Changes in global precipitation (mm/day)
-1 -0.75 -0.50 -0.25 0 0.25 0.50 0.75 1 1.5 2 3
December-January-February

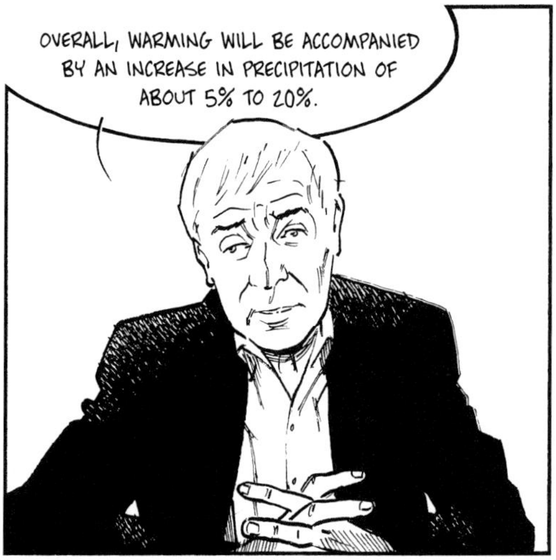

OVERALL, WARMING WILL BE ACCOMPANIED BY AN INCREASE IN PRECIPITATION OF ABOUT 5% TO 20%.

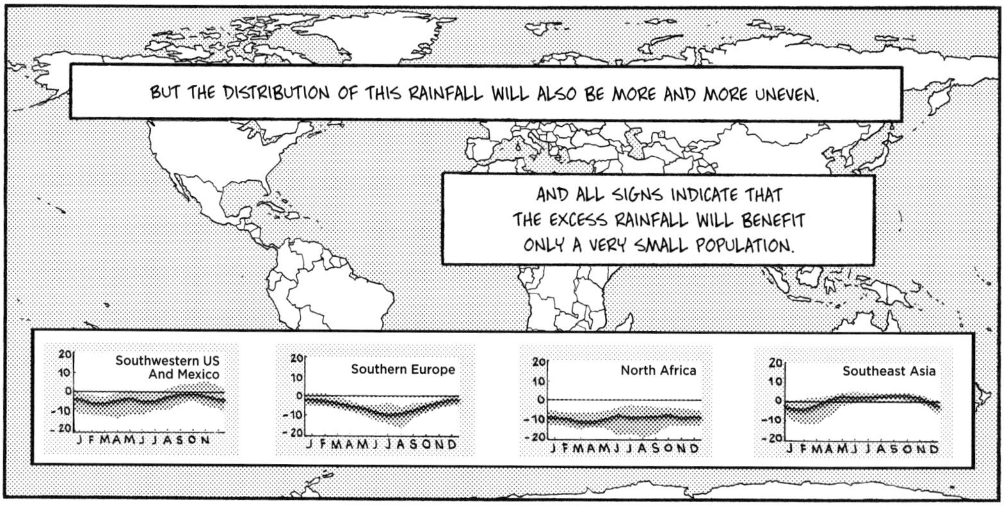

BUT THE DISTRIBUTION OF THIS RAINFALL WILL ALSO BE MORE AND MORE UNEVEN.
AND ALL SIGNS INDICATE THAT THE EXCESS RAINFALL WILL BENEFIT ONLY A VERY SMALL POPULATION.
Southwestern US And Mexico
20 10 0 -10 -20
J F M A M J J A S O N
Southern Europe
20 10 0 -10 -20
J F M A M J J A S O N D
North Africa
20 10 0 -10 -20
J F M A M J J A S O N D
Southeast Asia
20 10 0 -10 -20
J F M A M J J A S O N D

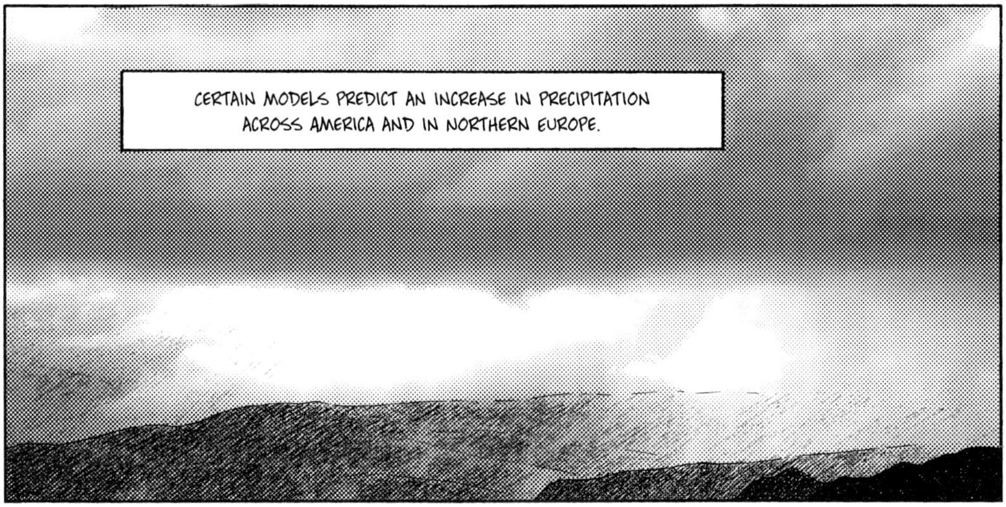
CERTAIN MODELS PREDICT AN INCREASE IN PRECIPITATION ACROSS AMERICA AND IN NORTHERN EUROPE.

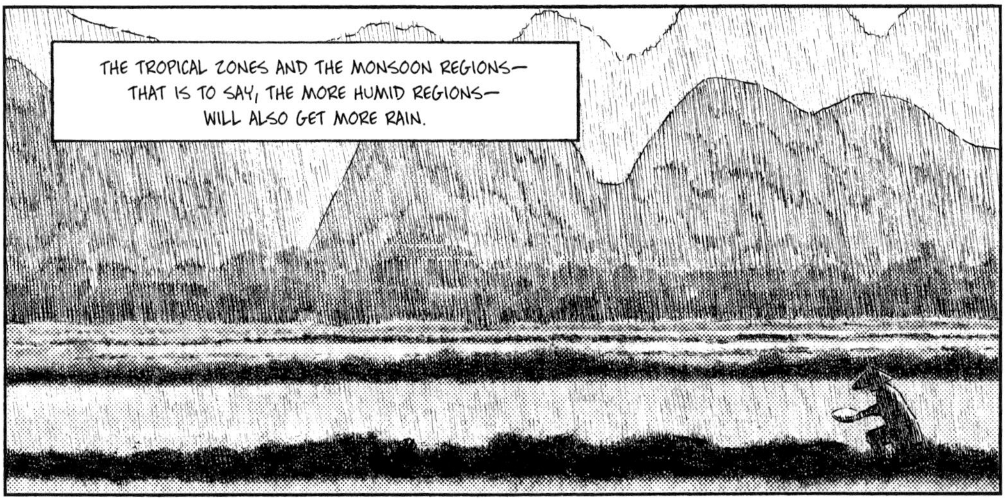
THE TROPICAL ZONES AND THE MONSOON REGIONS— THAT IS TO SAY, THE MORE HUMID REGIONS— WILL ALSO GET MORE RAIN.

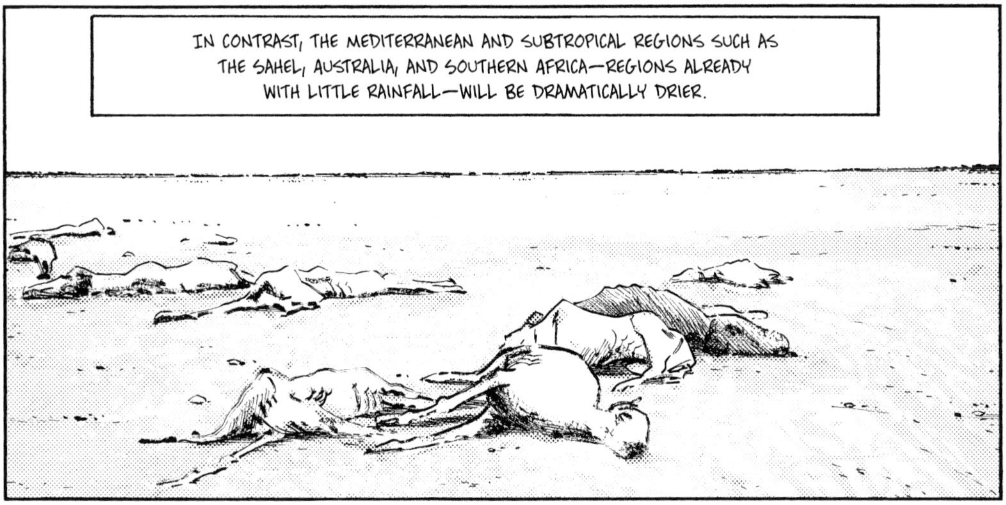
IN CONTRAST, THE MEDITERRANEAN AND SUBTROPICAL REGIONS SUCH AS THE SAHEL, AUSTRALIA, AND SOUTHERN AFRICA—REGIONS ALREADY WITH LITTLE RAINFALL—WILL BE DRAMATICALLY DRIER.

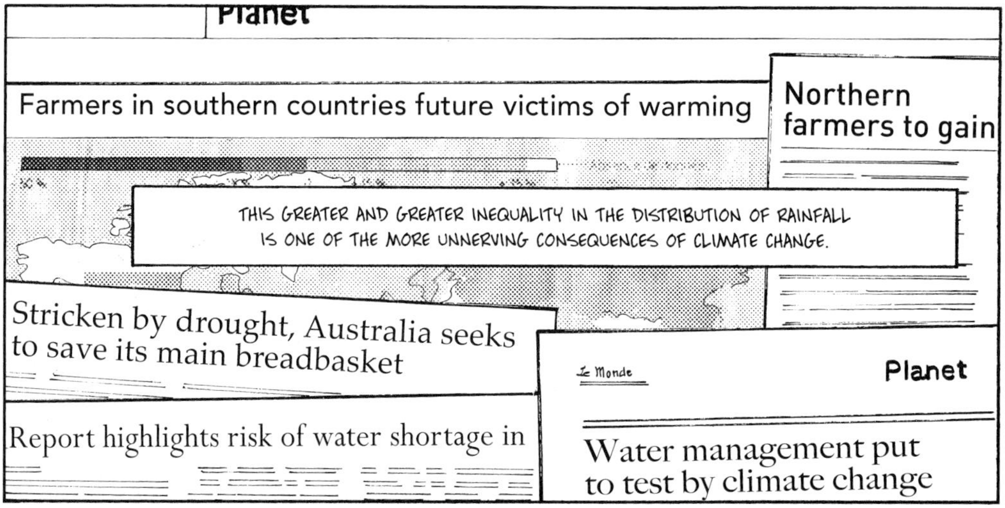
Planet
Farmers in southern countries future victims of warming
Northern farmers to gain
THIS GREATER AND GREATER INEQUALITY IN THE DISTRIBUTION OF RAINFALL IS ONE OF THE MORE UNNERVING CONSEQUENCES OF CLIMATE CHANGE.
Stricken by drought, Australia seeks to save its main breadbasket
Le Monde
Planet
Report highlights risk of water shortage in
Water management put to test by climate change

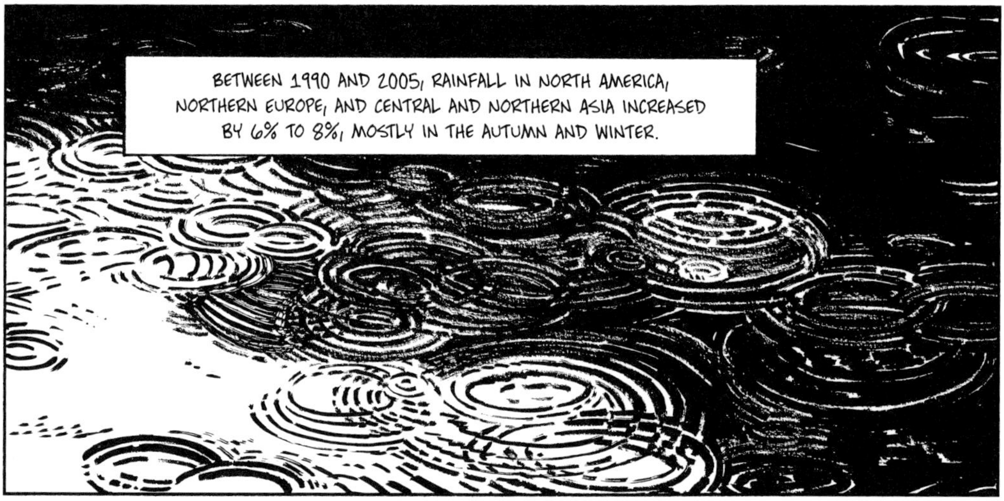
BETWEEN 1990 AND 2005, RAINFALL IN NORTH AMERICA, NORTHERN EUROPE, AND CENTRAL AND NORTHERN ASIA INCREASED BY 6% TO 8%, MOSTLY IN THE AUTUMN AND WINTER.

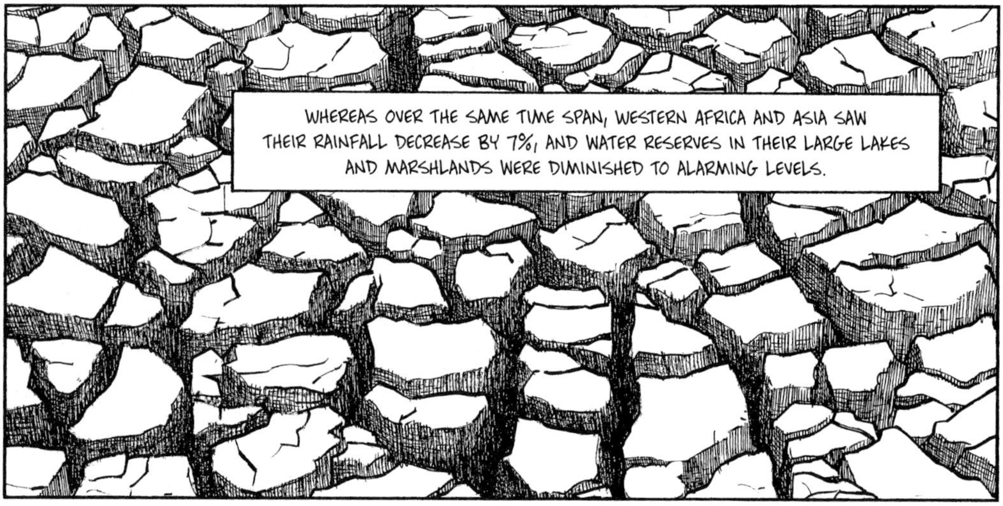
WHEREAS OVER THE SAME TIME SPAN, WESTERN AFRICA AND ASIA SAW THEIR RAINFALL DECREASE BY 7%, AND WATER RESERVES IN THEIR LARGE LAKES AND MARSHLANDS WERE DIMINISHED TO ALARMING LEVELS.

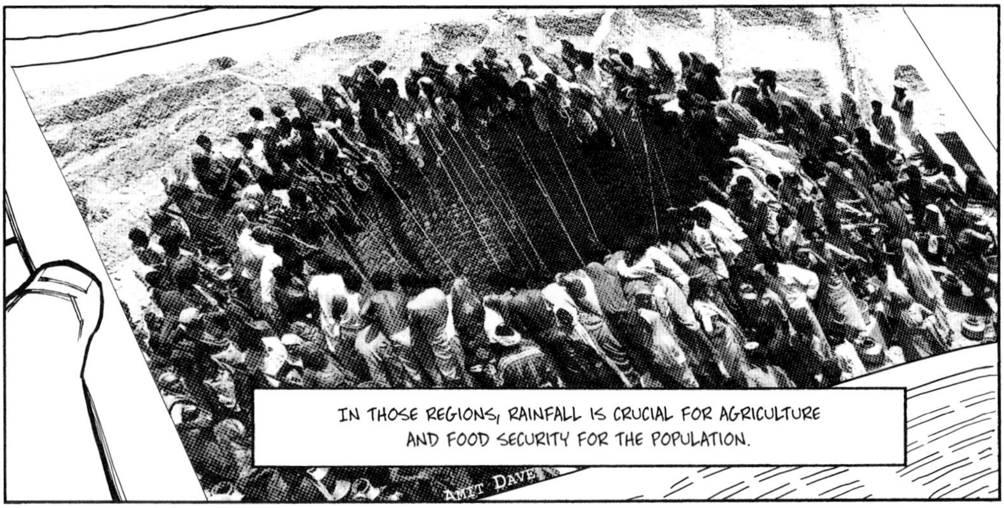
IN THOSE REGIONS, RAINFALL IS CRUCIAL FOR AGRICULTURE AND FOOD SECURITY FOR THE POPULATION.
AMIT DAVE

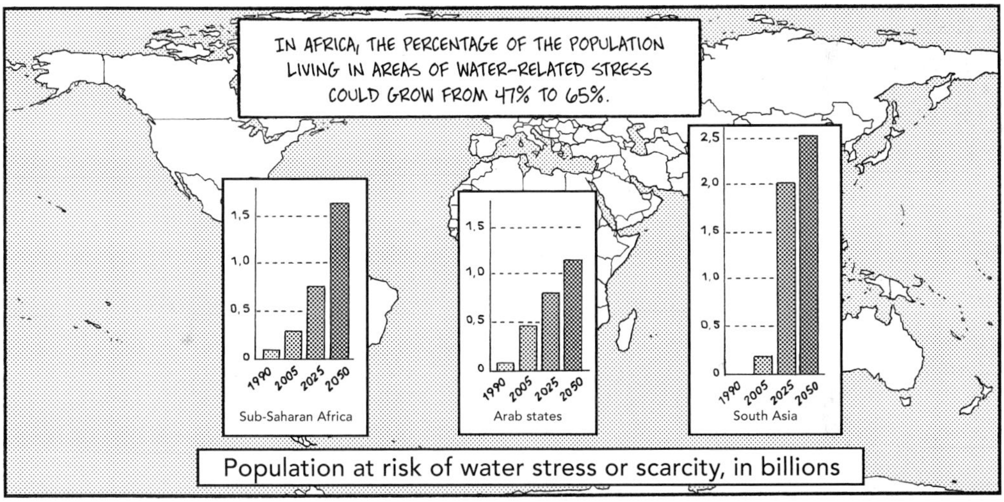
IN AFRICA, THE PERCENTAGE OF THE POPULATION LIVING IN AREAS OF WATER-RELATED STRESS COULD GROW FROM 47% TO 65%.
1,5
1,0
0,5
0
1990
2005
2025
2050
Sub-Saharan Africa
1,5
1,0
0,5
0
1990
2005
2025
2050
Arab states
2,5
2,0
1,5
1,0
0,5
0
1990
2005
2025
2050
South Asia
Population at risk of water stress or scarcity, in billions

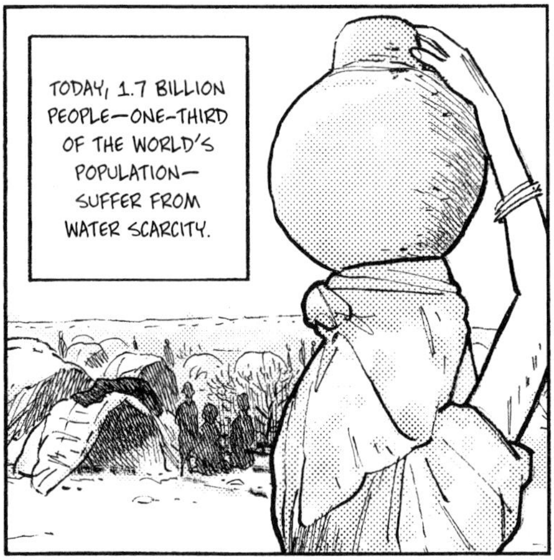
TODAY, 1.7 BILLION PEOPLE—ONE-THIRD OF THE WORLD'S POPULATION—SUFFER FROM WATER SCARCITY.

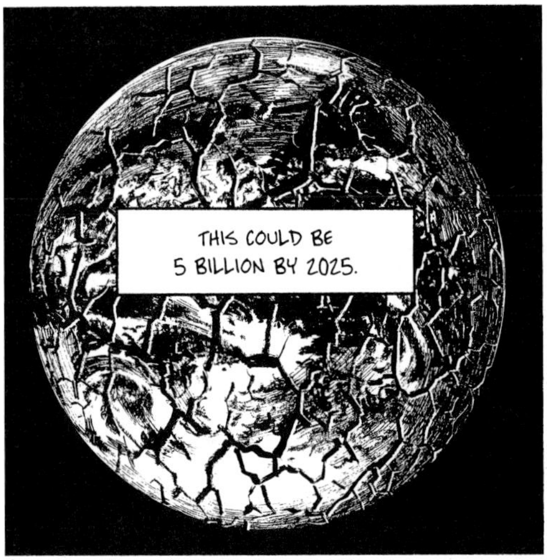
THIS COULD BE 5 BILLION BY 2025.

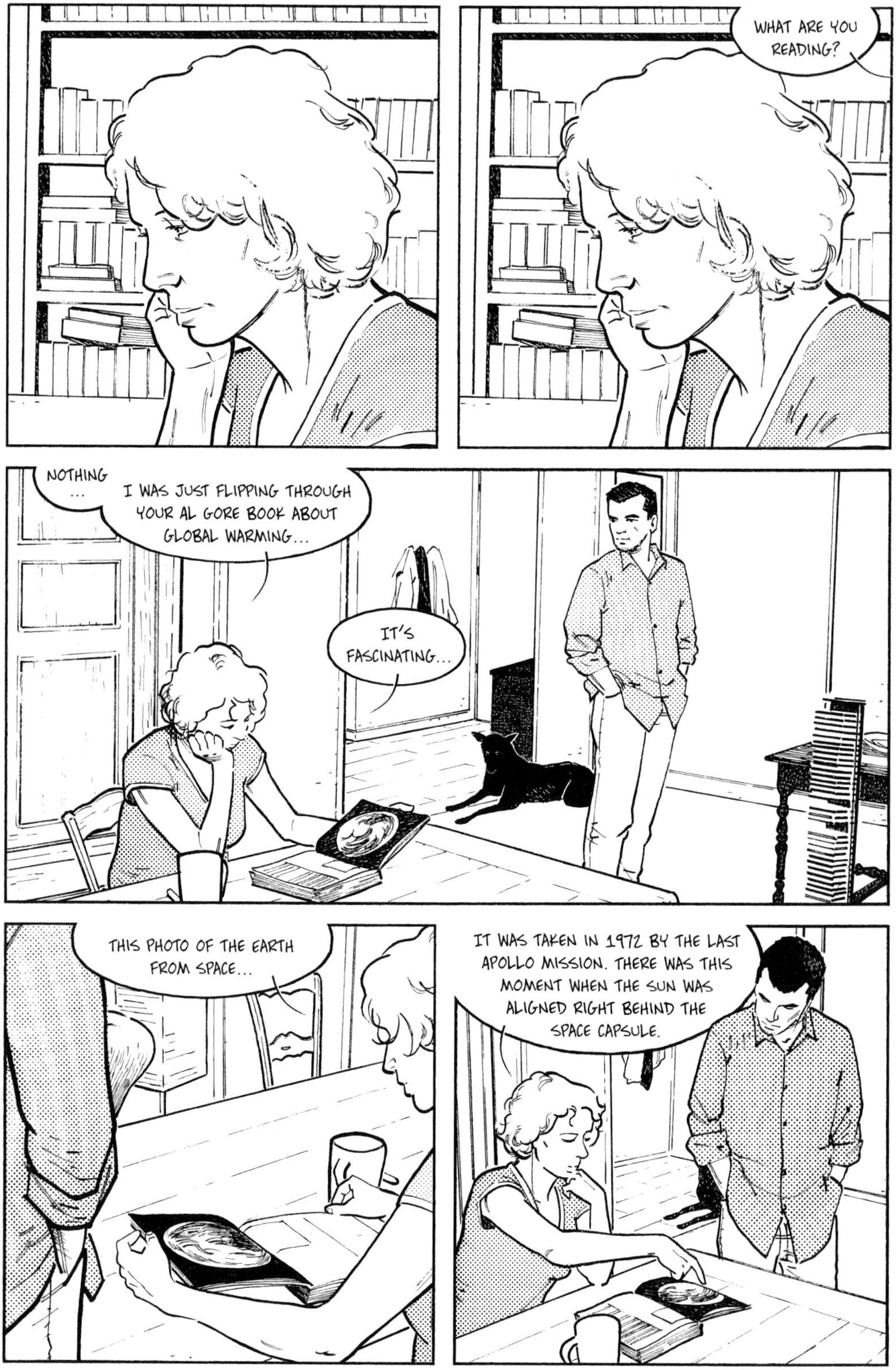
WHAT ARE YOU READING?
NOTHING ...
I WAS JUST FLIPPING THROUGH YOUR AL GORE BOOK ABOUT GLOBAL WARMING...
IT'S FASCINATING...
THIS PHOTO OF THE EARTH FROM SPACE...
IT WAS TAKEN IN 1972 BY THE LAST APOLLO MISSION. THERE WAS THIS MOMENT WHEN THE SUN WAS ALIGNED RIGHT BEHIND THE SPACE CAPSULE.

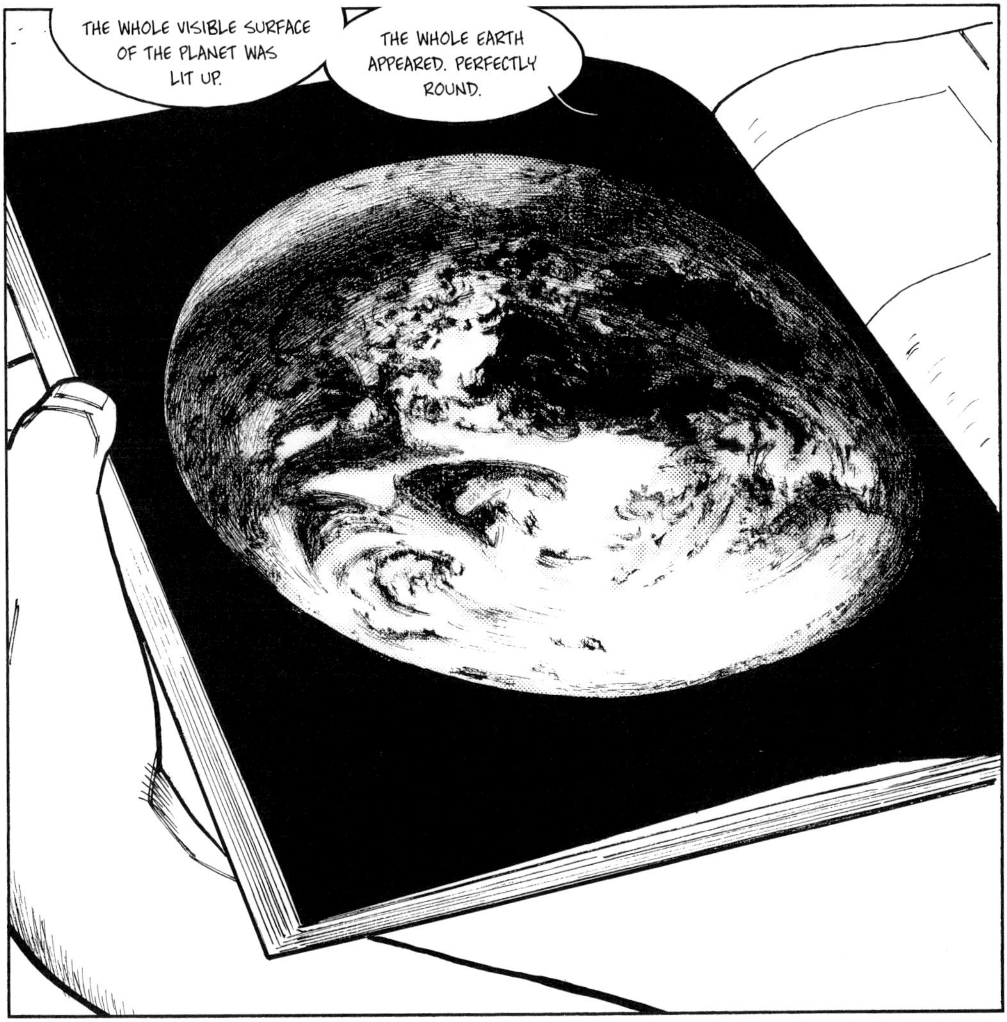
THE WHOLE VISIBLE SURFACE OF THE PLANET WAS LIT UP.
THE WHOLE EARTH APPEARED. PERFECTLY ROUND.

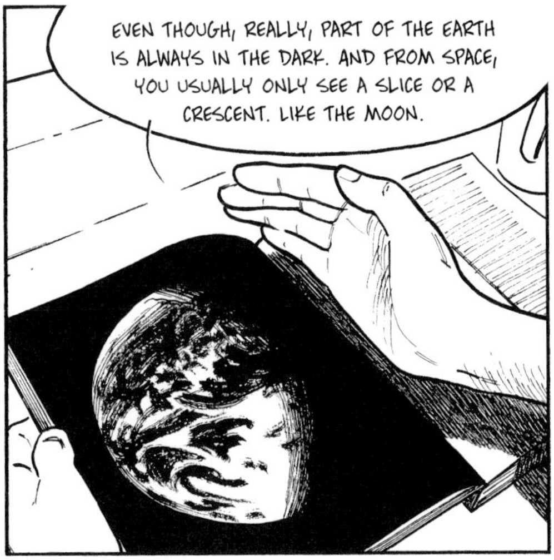
EVEN THOUGH, REALLY, PART OF THE EARTH IS ALWAYS IN THE DARK. AND FROM SPACE, YOU USUALLY ONLY SEE A SLICE OR A CRESCENT. LIKE THE MOON.

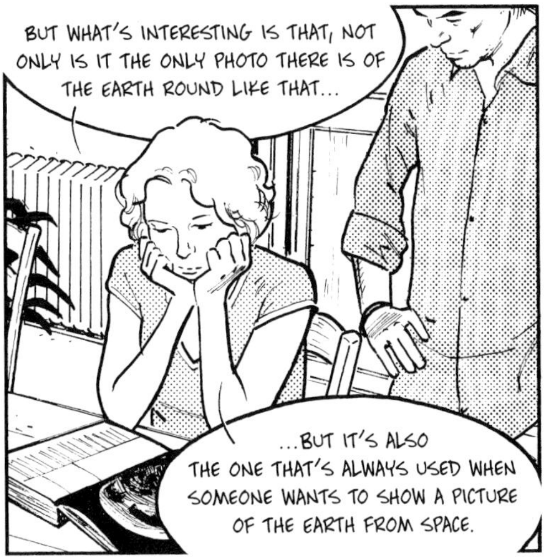
BUT WHAT'S INTERESTING IS THAT, NOT ONLY IS IT THE ONLY PHOTO THERE IS OF THE EARTH ROUND LIKE THAT...
...BUT IT'S ALSO THE ONE THAT'S ALWAYS USED WHEN SOMEONE WANTS TO SHOW A PICTURE OF THE EARTH FROM SPACE.

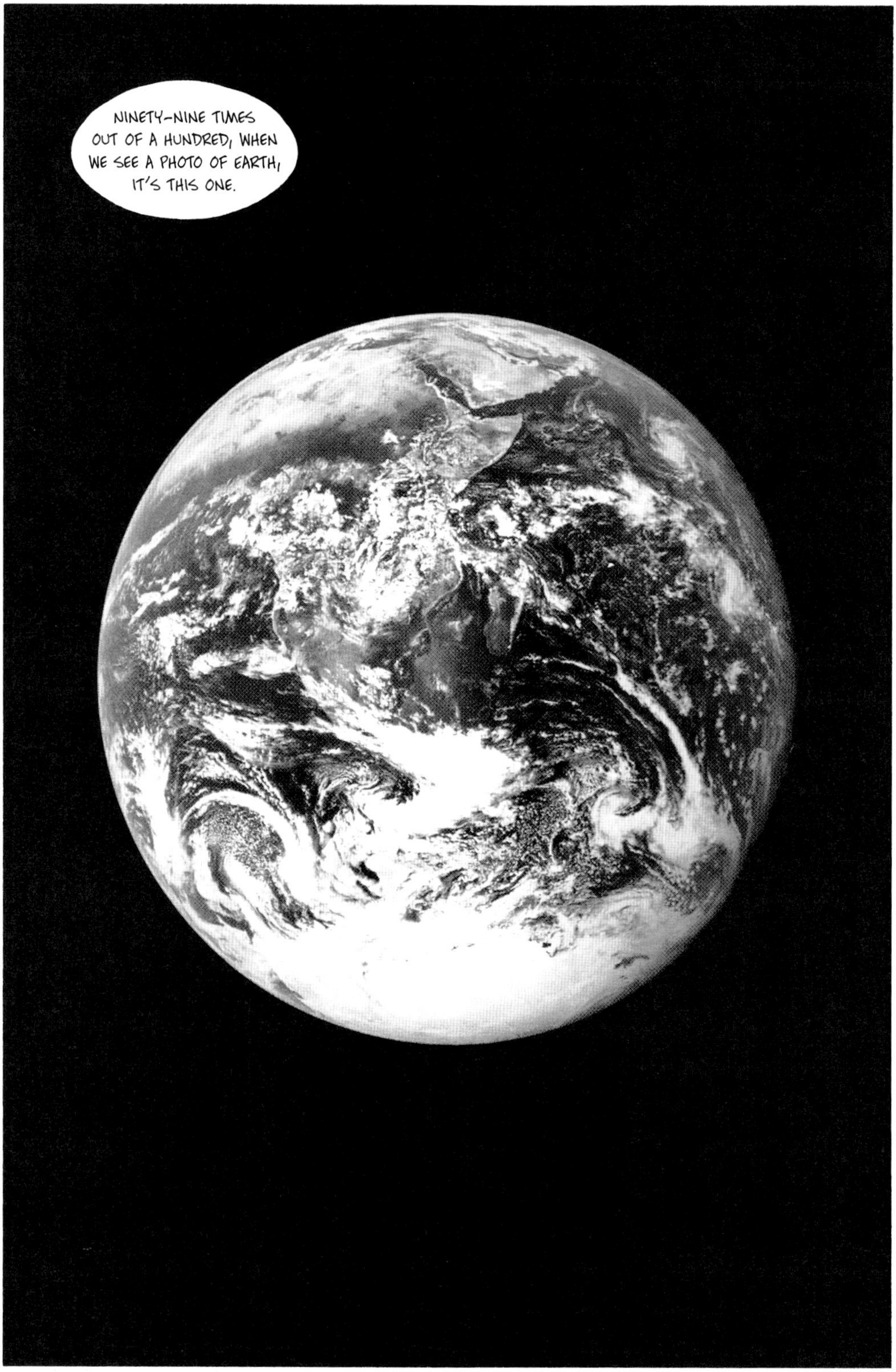
NINETY-NINE TIMES OUT OF A HUNDRED, WHEN WE SEE A PHOTO OF EARTH, IT'S THIS ONE.

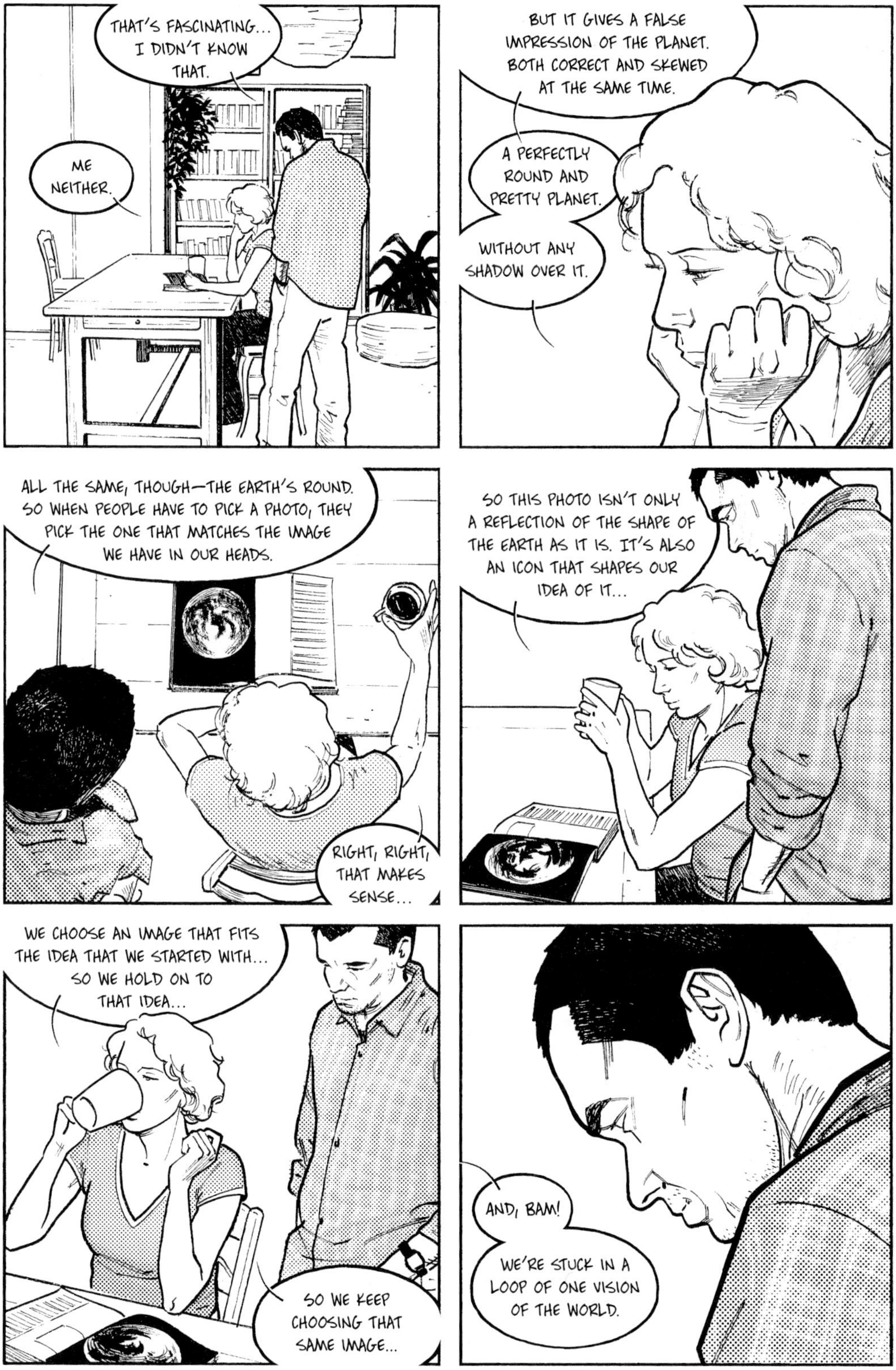
THAT'S FASCINATING... I DIDN'T KNOW THAT.
ME NEITHER.
BUT IT GIVES A FALSE IMPRESSION OF THE PLANET. BOTH CORRECT AND SKEWED AT THE SAME TIME.
A PERFECTLY ROUND AND PRETTY PLANET.
WITHOUT ANY SHADOW OVER IT.
ALL THE SAME, THOUGH—THE EARTH'S ROUND. SO WHEN PEOPLE HAVE TO PICK A PHOTO, THEY PICK THE ONE THAT MATCHES THE IMAGE WE HAVE IN OUR HEADS.
RIGHT, RIGHT, THAT MAKES SENSE...
SO THIS PHOTO ISN'T ONLY A REFLECTION OF THE SHAPE OF THE EARTH AS IT IS. IT'S ALSO AN ICON THAT SHAPES OUR IDEA OF IT...
WE CHOOSE AN IMAGE THAT FITS THE IDEA THAT WE STARTED WITH... SO WE HOLD ON TO THAT IDEA...
SO WE KEEP CHOOSING THAT SAME IMAGE...
AND, BAM!
WE'RE STUCK IN A LOOP OF ONE VISION OF THE WORLD.

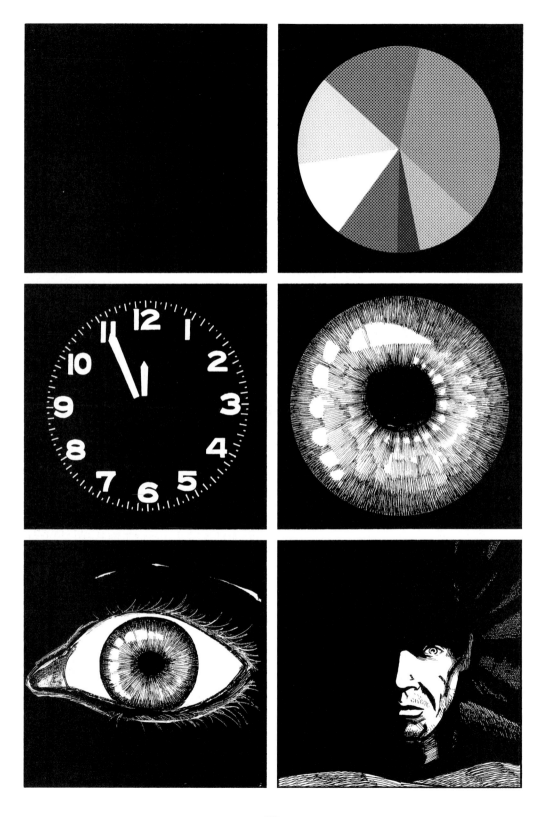

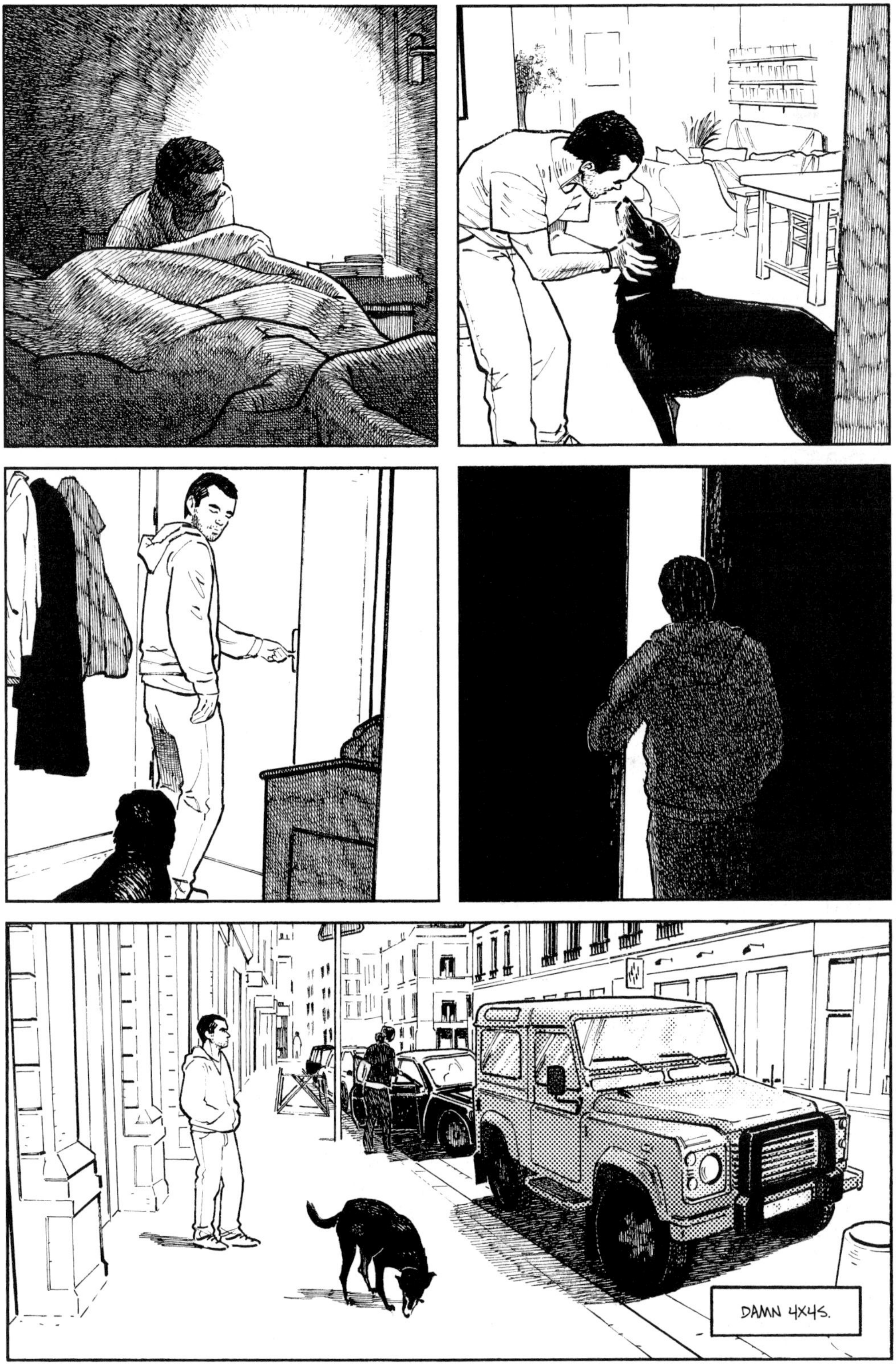
DAMN 4X4S.

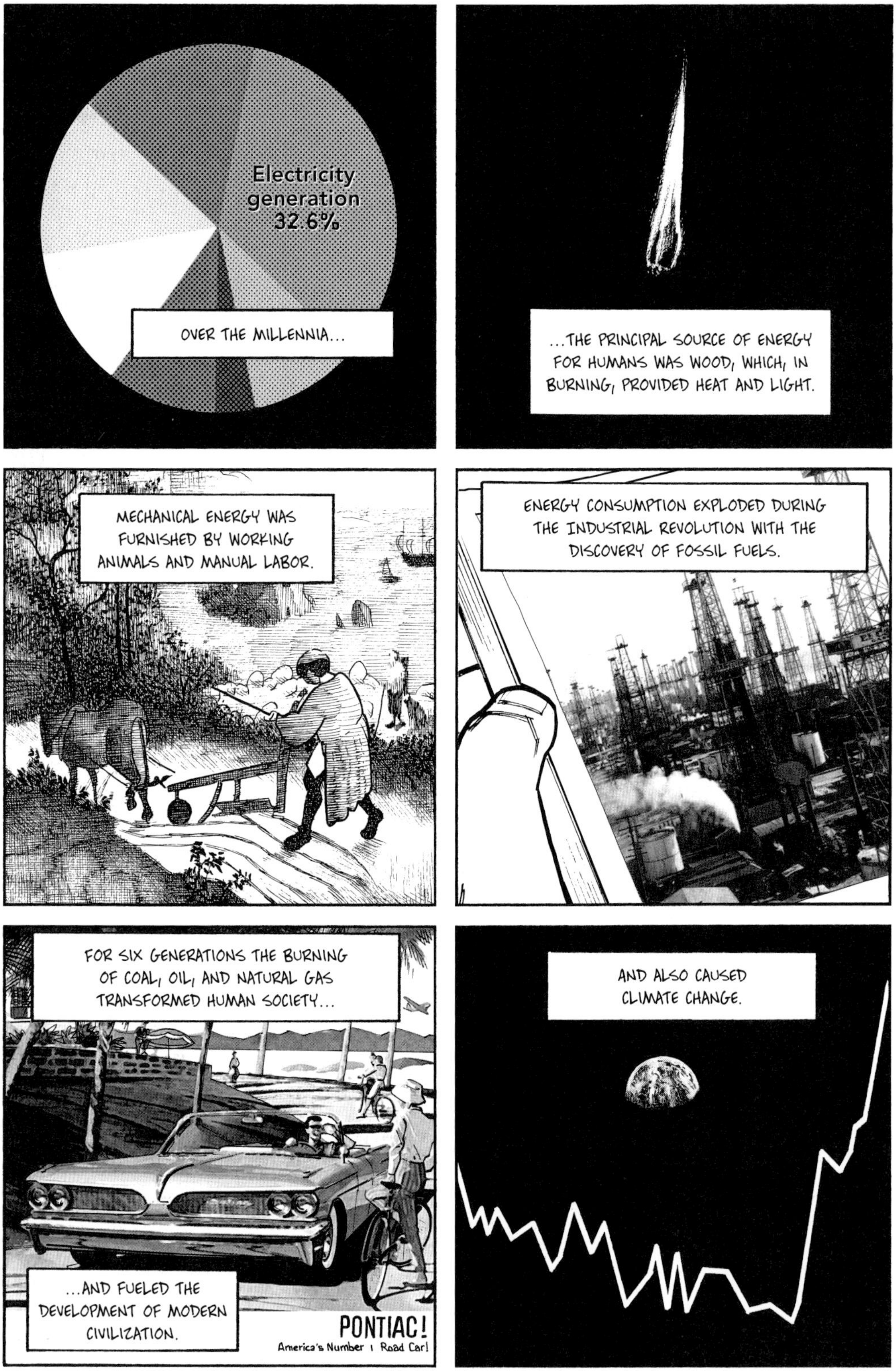
Electricity generation 32.6%
OVER THE MILLENNIA...
...THE PRINCIPAL SOURCE OF ENERGY FOR HUMANS WAS WOOD, WHICH, IN BURNING, PROVIDED HEAT AND LIGHT.
MECHANICAL ENERGY WAS FURNISHED BY WORKING ANIMALS AND MANUAL LABOR.
ENERGY CONSUMPTION EXPLODED DURING THE INDUSTRIAL REVOLUTION WITH THE DISCOVERY OF FOSSIL FUELS.
FOR SIX GENERATIONS THE BURNING OF COAL, OIL, AND NATURAL GAS TRANSFORMED HUMAN SOCIETY...
...AND FUELED THE DEVELOPMENT OF MODERN CIVILIZATION.
PONTIAC!
America's Number 1 Road Car!
AND ALSO CAUSED CLIMATE CHANGE.

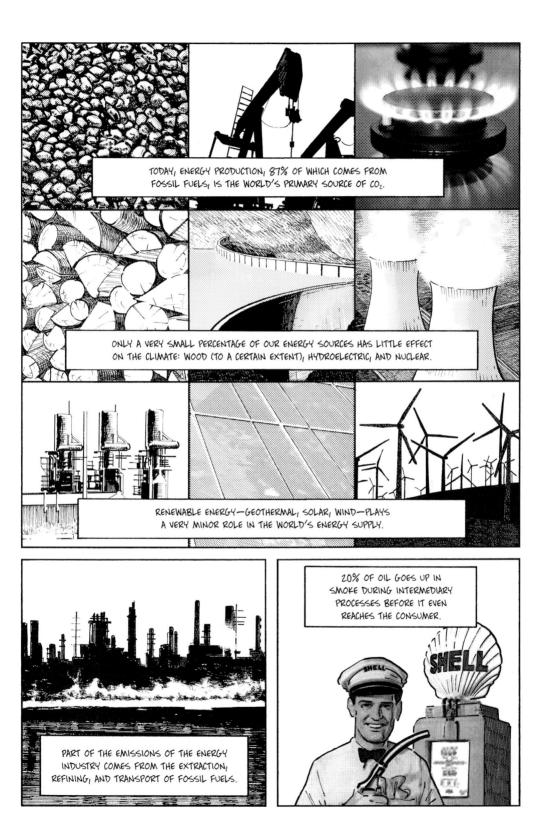

TODAY, ENERGY PRODUCTION, 87% OF WHICH COMES FROM FOSSIL FUELS, IS THE WORLD'S PRIMARY SOURCE OF CO_2.

ONLY A VERY SMALL PERCENTAGE OF OUR ENERGY SOURCES HAS LITTLE EFFECT ON THE CLIMATE: WOOD (TO A CERTAIN EXTENT), HYDROELECTRIC, AND NUCLEAR.

RENEWABLE ENERGY—GEOTHERMAL, SOLAR, WIND—PLAYS A VERY MINOR ROLE IN THE WORLD'S ENERGY SUPPLY.

20% OF OIL GOES UP IN SMOKE DURING INTERMEDIARY PROCESSES BEFORE IT EVEN REACHES THE CONSUMER.

PART OF THE EMISSIONS OF THE ENERGY INDUSTRY COMES FROM THE EXTRACTION, REFINING, AND TRANSPORT OF FOSSIL FUELS.

SHELL

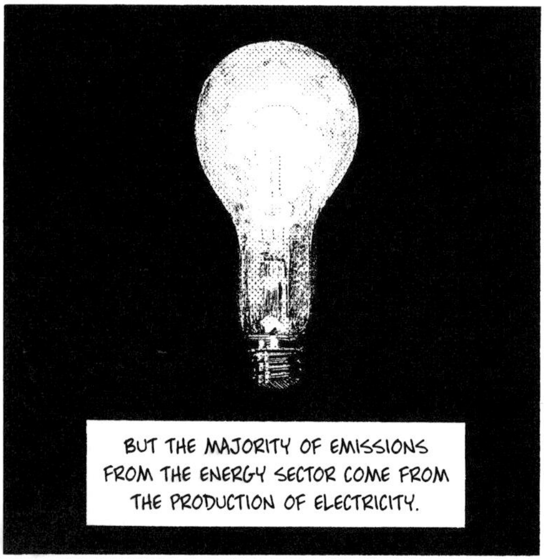
BUT THE MAJORITY OF EMISSIONS FROM THE ENERGY SECTOR COME FROM THE PRODUCTION OF ELECTRICITY.

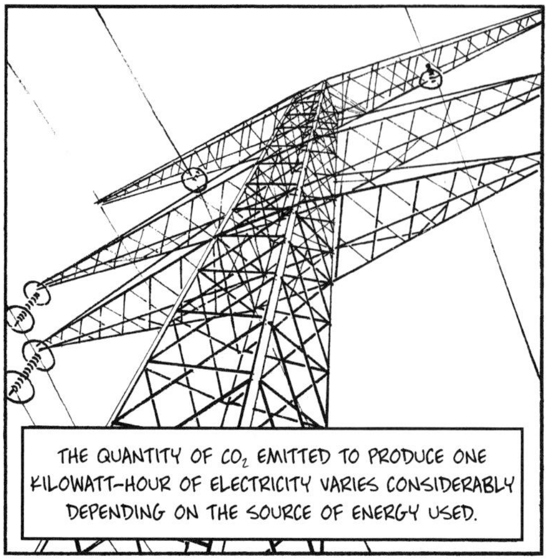
THE QUANTITY OF CO_2 EMITTED TO PRODUCE ONE KILOWATT-HOUR OF ELECTRICITY VARIES CONSIDERABLY DEPENDING ON THE SOURCE OF ENERGY USED.

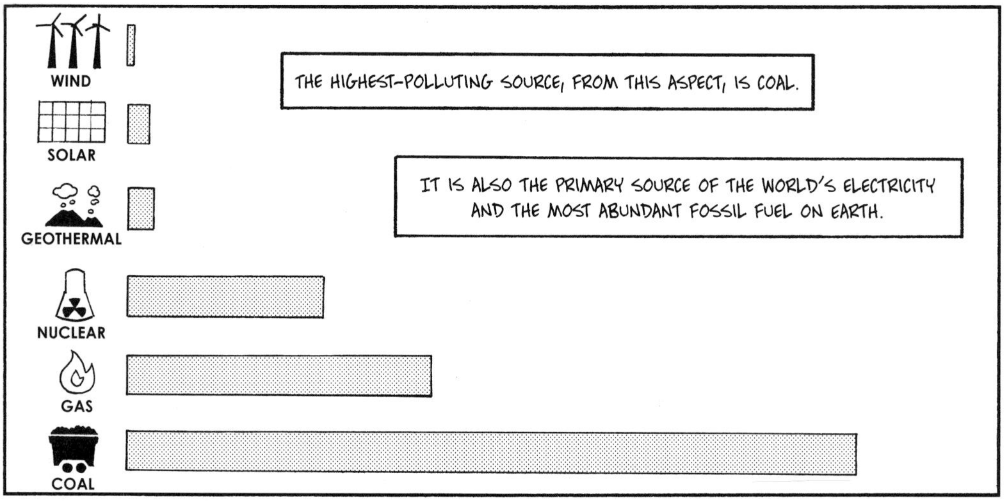
WIND
THE HIGHEST-POLLUTING SOURCE, FROM THIS ASPECT, IS COAL.
SOLAR
IT IS ALSO THE PRIMARY SOURCE OF THE WORLD'S ELECTRICITY AND THE MOST ABUNDANT FOSSIL FUEL ON EARTH.
GEOTHERMAL
NUCLEAR
GAS
COAL

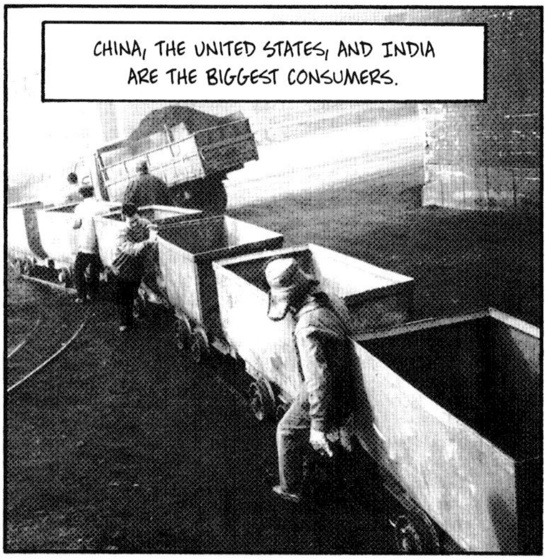
CHINA, THE UNITED STATES, AND INDIA ARE THE BIGGEST CONSUMERS.

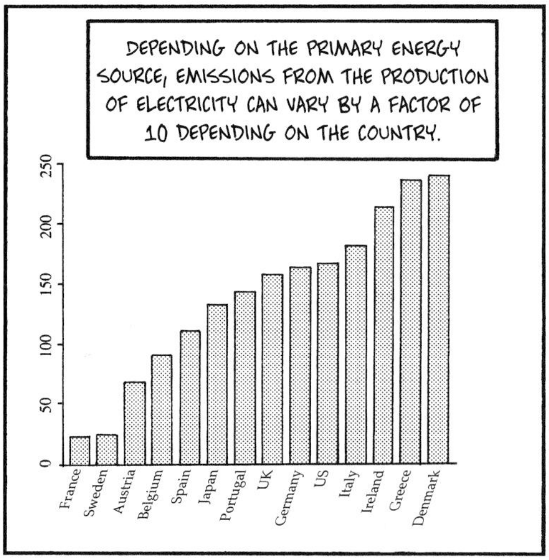
DEPENDING ON THE PRIMARY ENERGY SOURCE, EMISSIONS FROM THE PRODUCTION OF ELECTRICITY CAN VARY BY A FACTOR OF 10 DEPENDING ON THE COUNTRY.
250
200
150
100
50
0
France
Sweden
Austria
Belgium
Spain
Japan
Portugal
UK
Germany
US
Italy
Ireland
Greece
Denmark

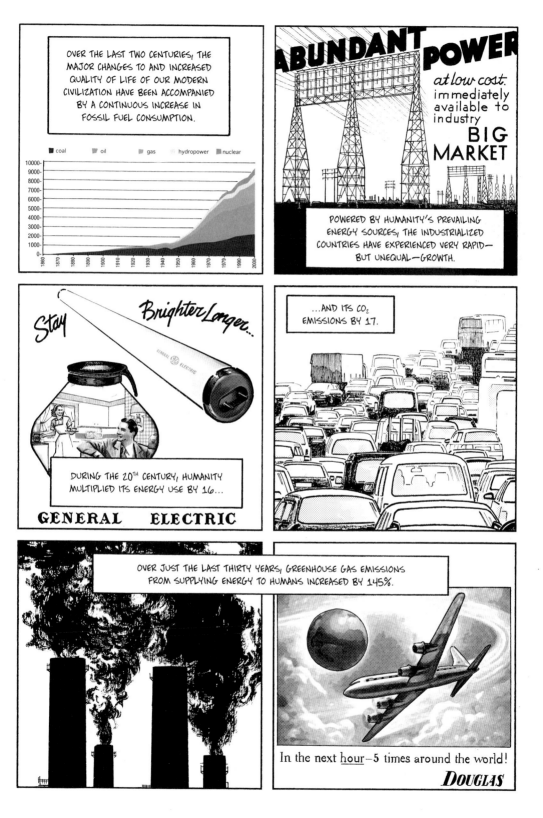

OVER THE LAST TWO CENTURIES, THE MAJOR CHANGES TO AND INCREASED QUALITY OF LIFE OF OUR MODERN CIVILIZATION HAVE BEEN ACCOMPANIED BY A CONTINUOUS INCREASE IN FOSSIL FUEL CONSUMPTION.

coal   oil   gas   hydropower   nuclear

ABUNDANT POWER

at low cost. immediately available to industry BIG MARKET

POWERED BY HUMANITY'S PREVAILING ENERGY SOURCES, THE INDUSTRIALIZED COUNTRIES HAVE EXPERIENCED VERY RAPID— BUT UNEQUAL—GROWTH.

Stay Brighter Longer...

DURING THE 20TH CENTURY, HUMANITY MULTIPLIED ITS ENERGY USE BY 16...

GENERAL ELECTRIC

...AND ITS CO_2 EMISSIONS BY 17.

OVER JUST THE LAST THIRTY YEARS, GREENHOUSE GAS EMISSIONS FROM SUPPLYING ENERGY TO HUMANS INCREASED BY 145%.

In the next hour—5 times around the world!

DOUGLAS

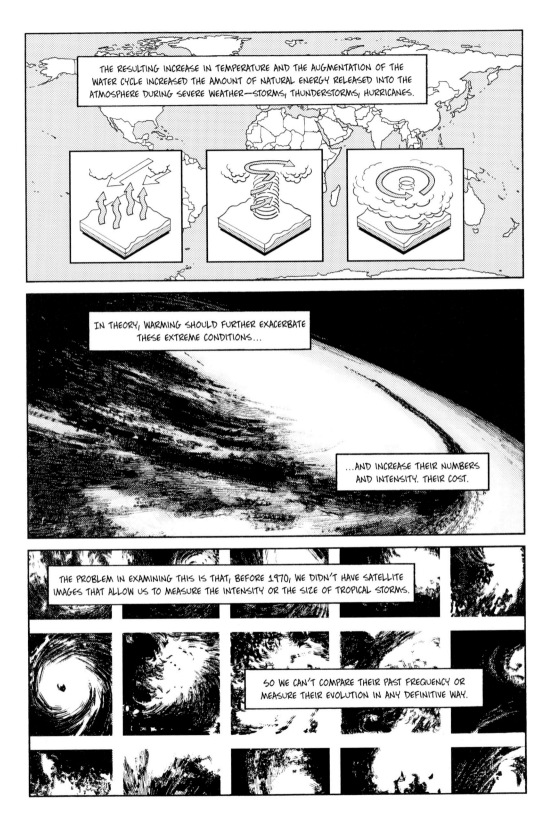

THE RESULTING INCREASE IN TEMPERATURE AND THE AUGMENTATION OF THE WATER CYCLE INCREASED THE AMOUNT OF NATURAL ENERGY RELEASED INTO THE ATMOSPHERE DURING SEVERE WEATHER—STORMS, THUNDERSTORMS, HURRICANES.

IN THEORY, WARMING SHOULD FURTHER EXACERBATE THESE EXTREME CONDITIONS...

...AND INCREASE THEIR NUMBERS AND INTENSITY. THEIR COST.

THE PROBLEM IN EXAMINING THIS IS THAT, BEFORE 1970, WE DIDN'T HAVE SATELLITE IMAGES THAT ALLOW US TO MEASURE THE INTENSITY OR THE SIZE OF TROPICAL STORMS.

SO WE CAN'T COMPARE THEIR PAST FREQUENCY OR MEASURE THEIR EVOLUTION IN ANY DEFINITIVE WAY.

ARLENE	8 - 11 June	DENNIS	4 - 11 July
BRET	28 - 29 June	EMILY	11 - 21 July
CINDY			4 - 18 August
FRANK		NA	23 - 30 August
GERT	23 - 25 July	MARIA	1 - 10 Septem
HARVEY	2 - 8 August	NATE	5 - 10 Septemb
JOSE	22 - 23 August	OPHELIA	6 - 17 Septemb
LEE	28 August - 1 September	PHILIPPE	17 - 23 Septem
TAMMY	5 - 6 Octob		Septem
ALPHA	22 - 24 Oct		ctober
GAMMA	13 - 19 November	VINCE	9 - 11 October
DELTA	23 - 28 November	WILMA	15 - 25 Octobe

IT IS, HOWEVER, A FACT THAT 2005 SURPASSED 2004, ALREADY A TURBULENT YEAR, AND BROKE ALL RECORDS FOR HURRICANES IN TERMS OF NUMBER AND SIZE.

IN THE NORTH ATLANTIC, THE ROMAN ALPHABET DIDN'T HAVE ENOUGH LETTERS FOR ALL THE STORMS, SO FOR THE FIRST TIME THE GREEK ALPHABET WAS ADDED.

ALSO FOR THE FIRST TIME SINCE 1942, A HURRICANE, VINCE, HIT THE IBERIAN PENINSULA.

YET ANOTHER FIRST: FOUR CATEGORY 5 HURRICANES—EMILY, RITA, WILMA, AND KATRINA—FORMED IN THE SAME SEASON.

179

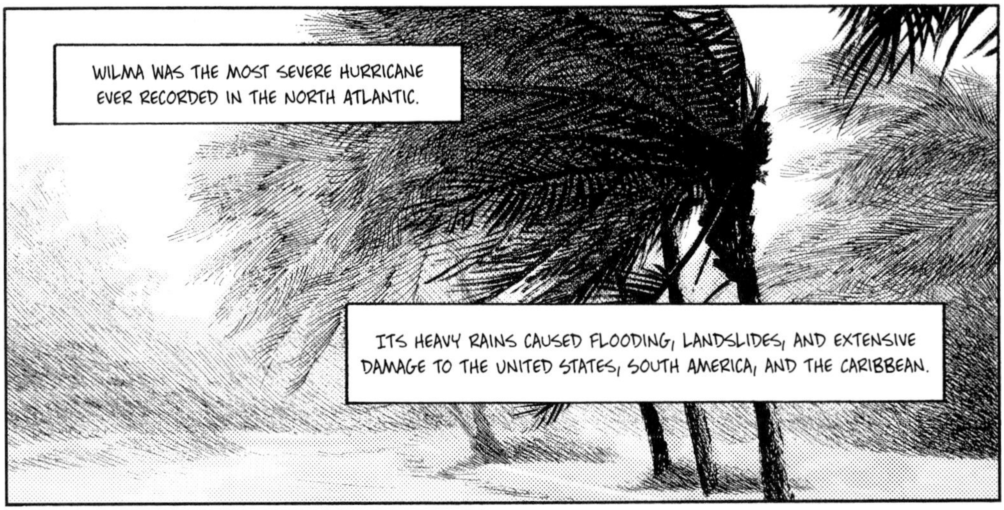
WILMA WAS THE MOST SEVERE HURRICANE EVER RECORDED IN THE NORTH ATLANTIC.
ITS HEAVY RAINS CAUSED FLOODING, LANDSLIDES, AND EXTENSIVE DAMAGE TO THE UNITED STATES, SOUTH AMERICA, AND THE CARIBBEAN.

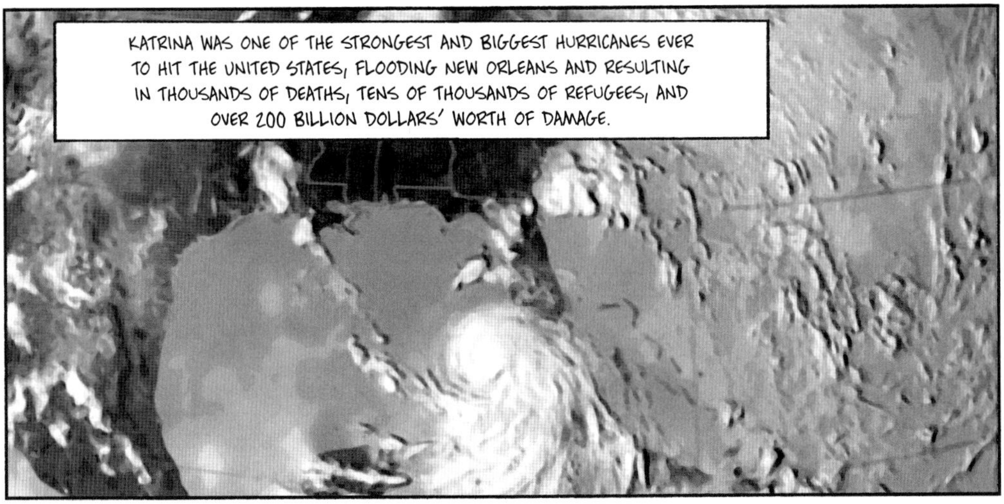
KATRINA WAS ONE OF THE STRONGEST AND BIGGEST HURRICANES EVER TO HIT THE UNITED STATES, FLOODING NEW ORLEANS AND RESULTING IN THOUSANDS OF DEATHS, TENS OF THOUSANDS OF REFUGEES, AND OVER 200 BILLION DOLLARS' WORTH OF DAMAGE.

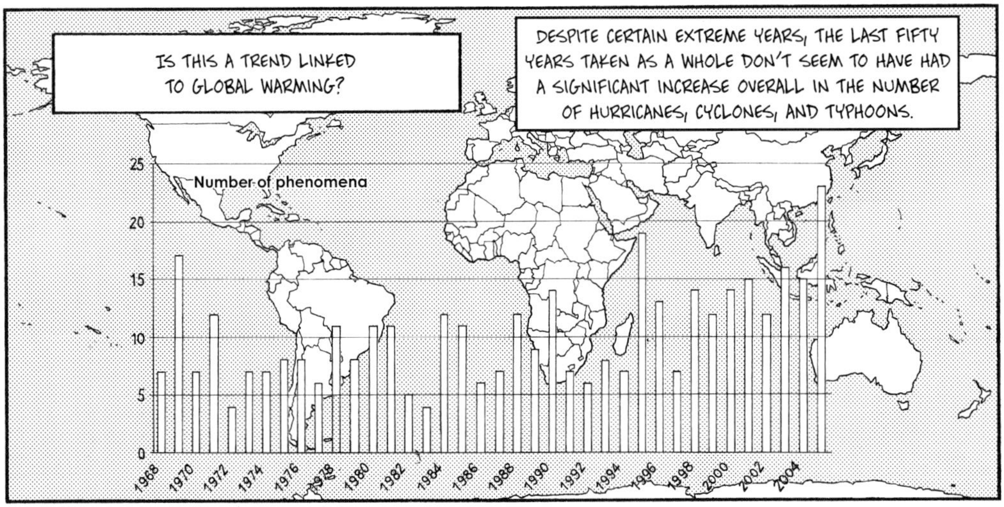
IS THIS A TREND LINKED TO GLOBAL WARMING?
DESPITE CERTAIN EXTREME YEARS, THE LAST FIFTY YEARS TAKEN AS A WHOLE DON'T SEEM TO HAVE HAD A SIGNIFICANT INCREASE OVERALL IN THE NUMBER OF HURRICANES, CYCLONES, AND TYPHOONS.
Number of phenomena
25
20
15
10
5
0
1968
1970
1972
1974
1976
1978
1980
1982
1984
1986
1988
1990
1992
1994
1996
1998
2000
2002
2004

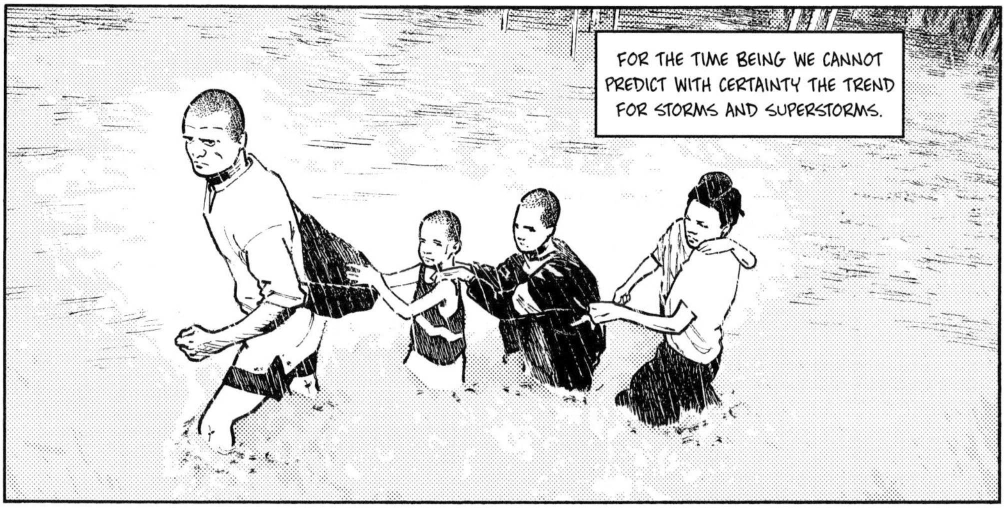
FOR THE TIME BEING WE CANNOT PREDICT WITH CERTAINTY THE TREND FOR STORMS AND SUPERSTORMS.

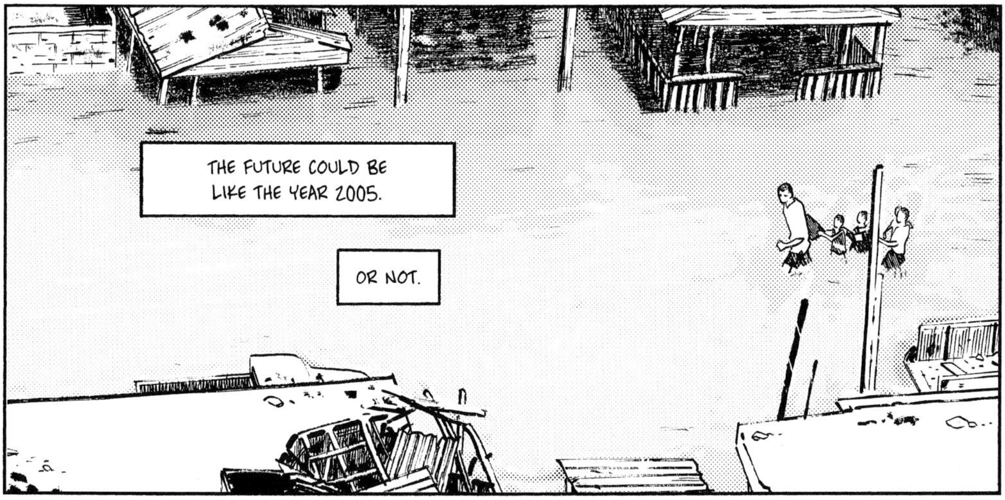
THE FUTURE COULD BE LIKE THE YEAR 2005.
OR NOT.

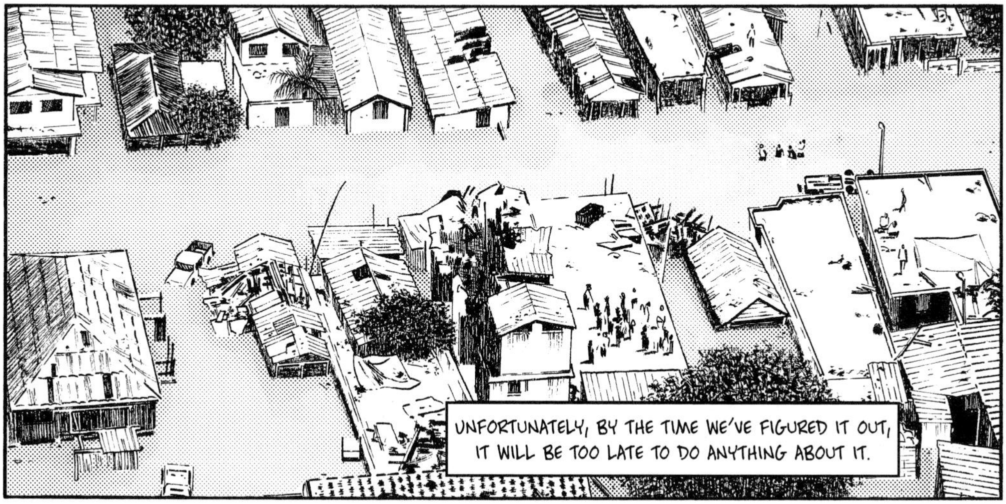
UNFORTUNATELY, BY THE TIME WE'VE FIGURED IT OUT, IT WILL BE TOO LATE TO DO ANYTHING ABOUT IT.

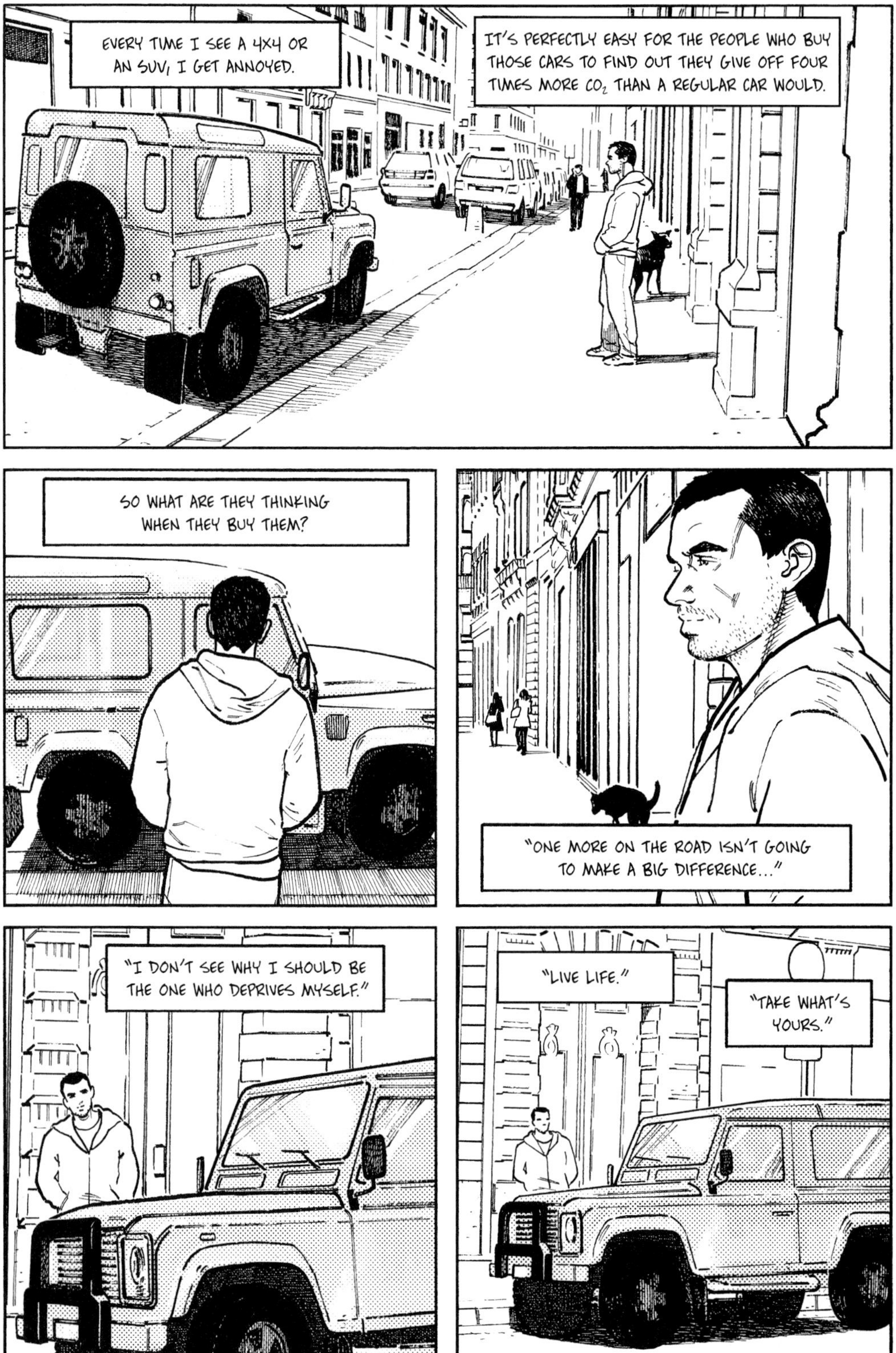
EVERY TIME I SEE A 4X4 OR AN SUV, I GET ANNOYED.
IT'S PERFECTLY EASY FOR THE PEOPLE WHO BUY THOSE CARS TO FIND OUT THEY GIVE OFF FOUR TIMES MORE CO_2 THAN A REGULAR CAR WOULD.
SO WHAT ARE THEY THINKING WHEN THEY BUY THEM?
"ONE MORE ON THE ROAD ISN'T GOING TO MAKE A BIG DIFFERENCE..."
"I DON'T SEE WHY I SHOULD BE THE ONE WHO DEPRIVES MYSELF."
"LIVE LIFE."
"TAKE WHAT'S YOURS."

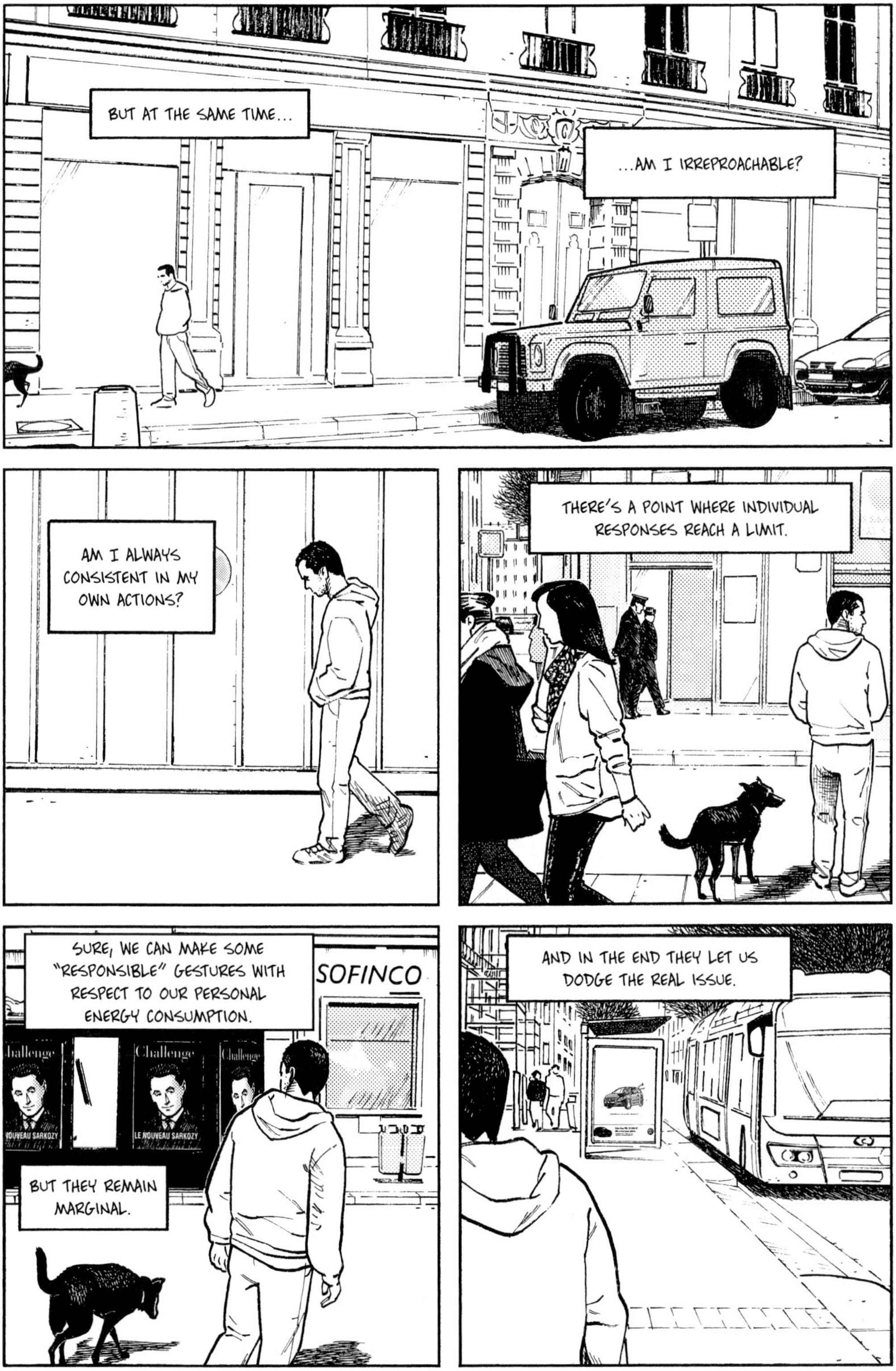
BUT AT THE SAME TIME...
...AM I IRREPROACHABLE?
AM I ALWAYS CONSISTENT IN MY OWN ACTIONS?
THERE'S A POINT WHERE INDIVIDUAL RESPONSES REACH A LIMIT.
SURE, WE CAN MAKE SOME "RESPONSIBLE" GESTURES WITH RESPECT TO OUR PERSONAL ENERGY CONSUMPTION.
SOFINCO
Challenge
Challenge
Challenge
NOUVEAU SARKOZY
LE NOUVEAU SARKOZY
BUT THEY REMAIN MARGINAL.
AND IN THE END THEY LET US DODGE THE REAL ISSUE.

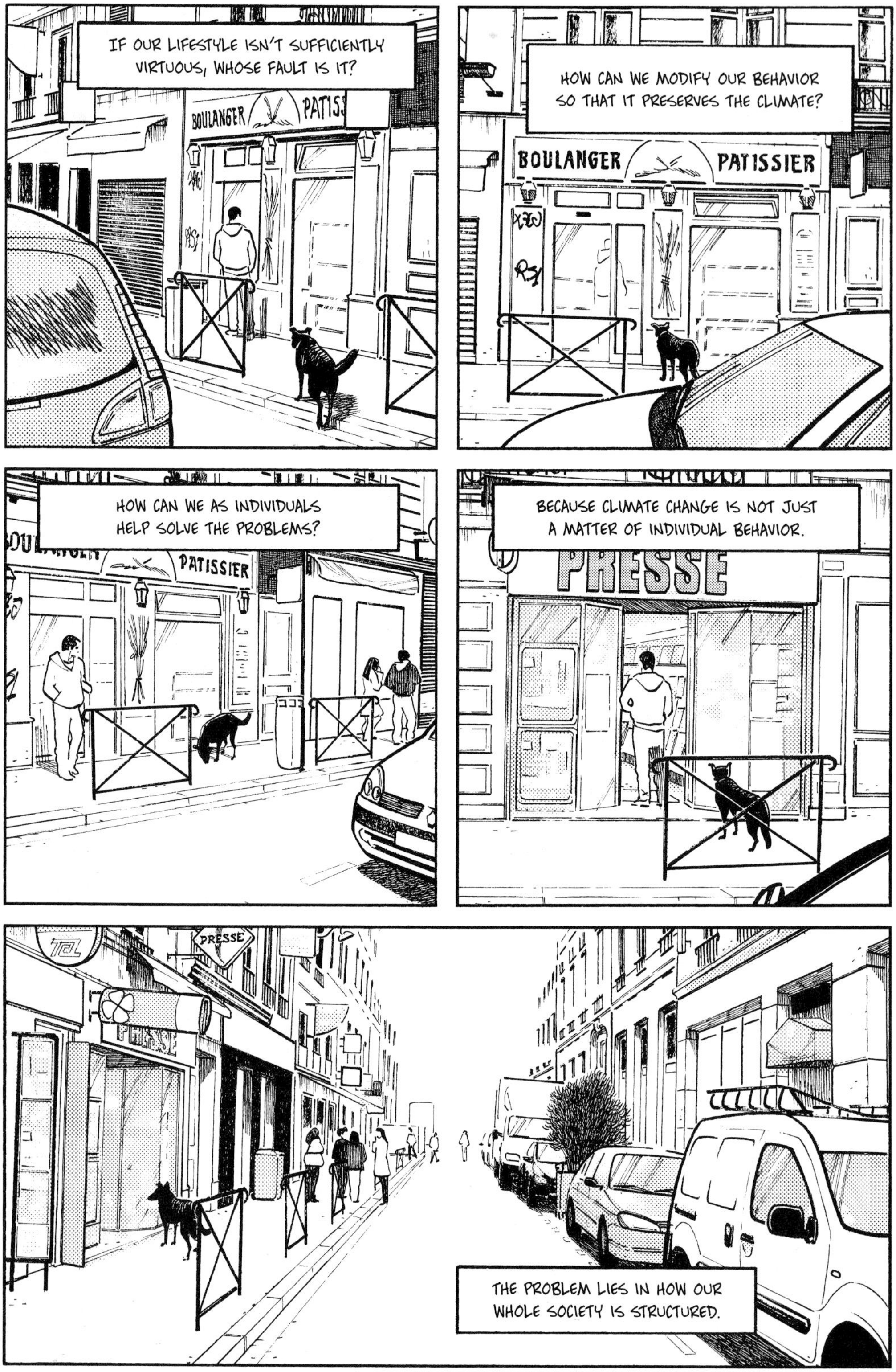
IF OUR LIFESTYLE ISN'T SUFFICIENTLY VIRTUOUS, WHOSE FAULT IS IT?
BOULANGER PATISS
HOW CAN WE MODIFY OUR BEHAVIOR SO THAT IT PRESERVES THE CLIMATE?
BOULANGER PATISSIER
HOW CAN WE AS INDIVIDUALS HELP SOLVE THE PROBLEMS?
PATISSIER
BECAUSE CLIMATE CHANGE IS NOT JUST A MATTER OF INDIVIDUAL BEHAVIOR.
PRESSE
PRESSE
PRESSE
THE PROBLEM LIES IN HOW OUR WHOLE SOCIETY IS STRUCTURED.

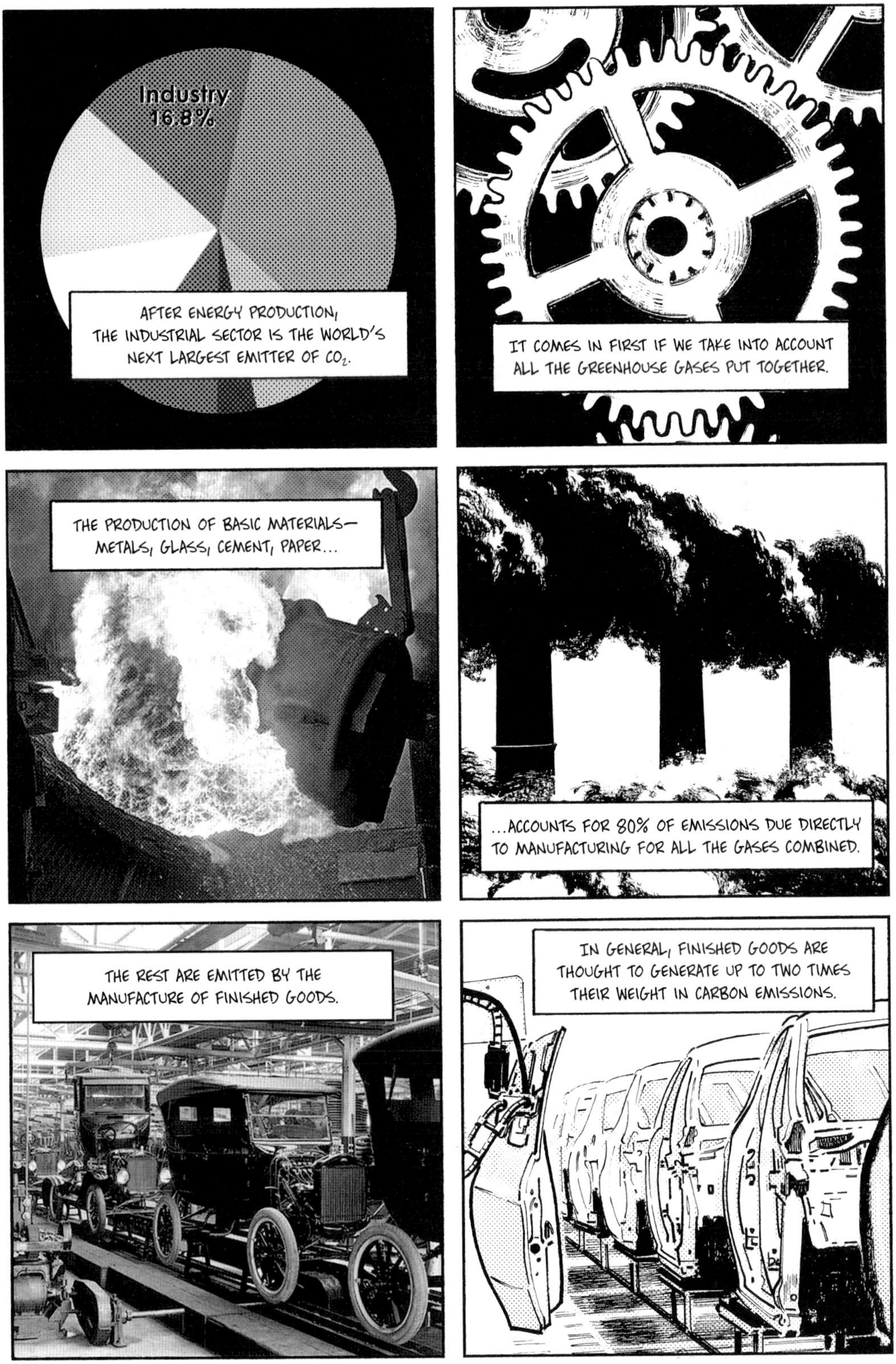
Industry
16.8%
AFTER ENERGY PRODUCTION, THE INDUSTRIAL SECTOR IS THE WORLD'S NEXT LARGEST EMITTER OF CO_2.
IT COMES IN FIRST IF WE TAKE INTO ACCOUNT ALL THE GREENHOUSE GASES PUT TOGETHER.
THE PRODUCTION OF BASIC MATERIALS—METALS, GLASS, CEMENT, PAPER...
...ACCOUNTS FOR 80% OF EMISSIONS DUE DIRECTLY TO MANUFACTURING FOR ALL THE GASES COMBINED.
THE REST ARE EMITTED BY THE MANUFACTURE OF FINISHED GOODS.
IN GENERAL, FINISHED GOODS ARE THOUGHT TO GENERATE UP TO TWO TIMES THEIR WEIGHT IN CARBON EMISSIONS.

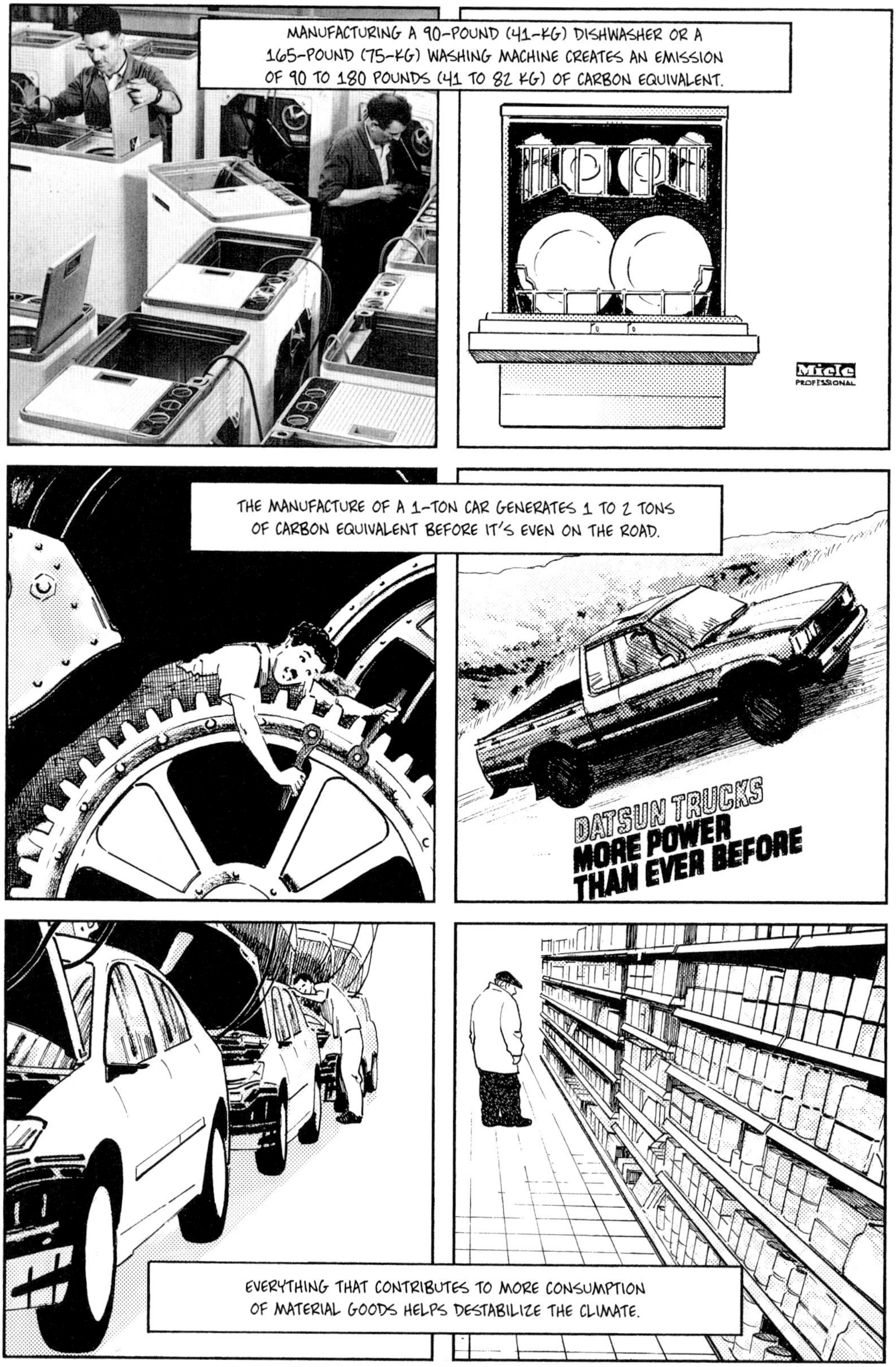
MANUFACTURING A 90-POUND (41-KG) DISHWASHER OR A 165-POUND (75-KG) WASHING MACHINE CREATES AN EMISSION OF 90 TO 180 POUNDS (41 TO 82 KG) OF CARBON EQUIVALENT.
Miele PROFESSIONAL
THE MANUFACTURE OF A 1-TON CAR GENERATES 1 TO 2 TONS OF CARBON EQUIVALENT BEFORE IT'S EVEN ON THE ROAD.
DATSUN TRUCKS
MORE POWER THAN EVER BEFORE
EVERYTHING THAT CONTRIBUTES TO MORE CONSUMPTION OF MATERIAL GOODS HELPS DESTABILIZE THE CLIMATE.

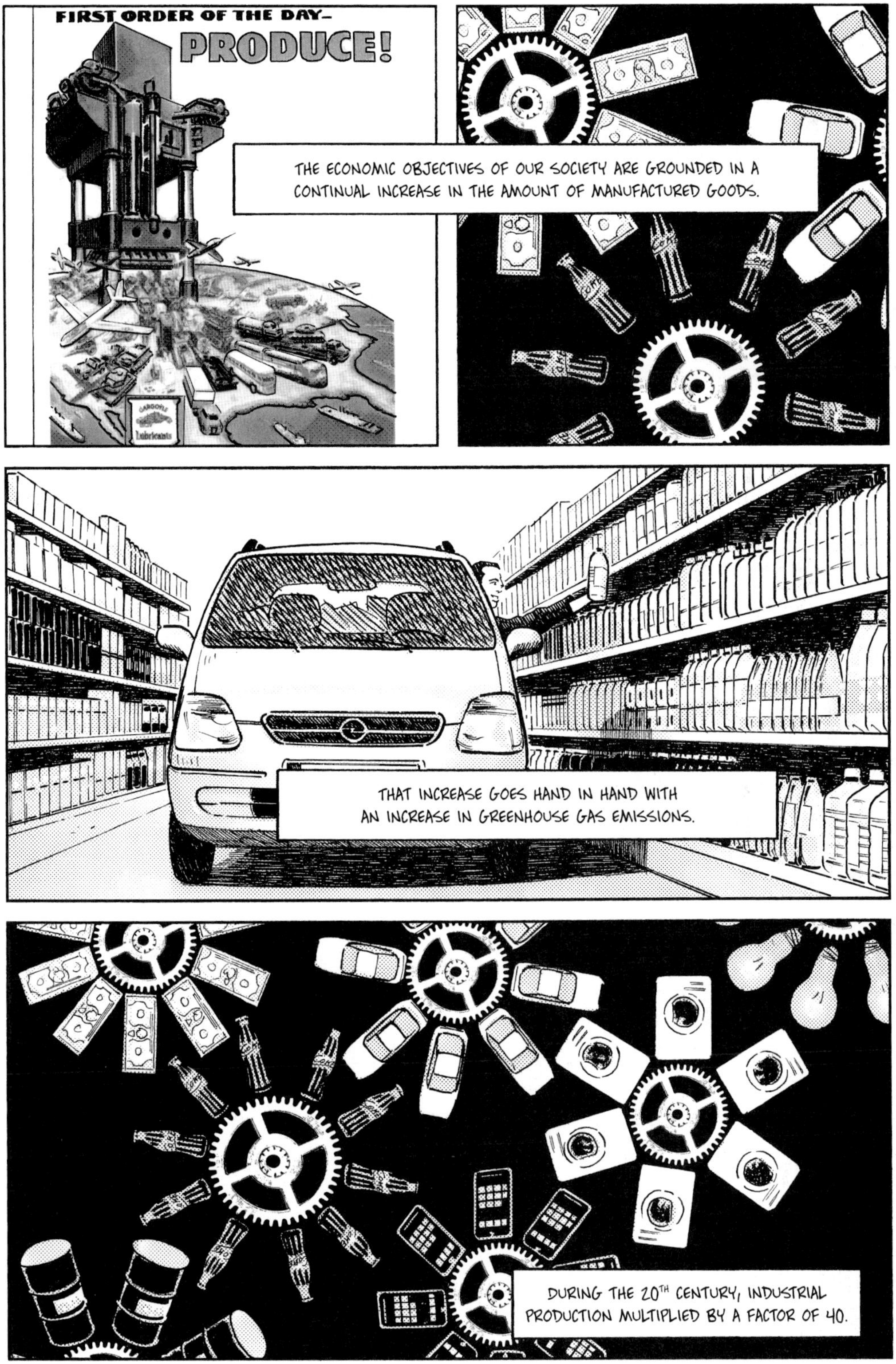
FIRST ORDER OF THE DAY–
PRODUCE!
THE ECONOMIC OBJECTIVES OF OUR SOCIETY ARE GROUNDED IN A CONTINUAL INCREASE IN THE AMOUNT OF MANUFACTURED GOODS.
THAT INCREASE GOES HAND IN HAND WITH AN INCREASE IN GREENHOUSE GAS EMISSIONS.
DURING THE 20TH CENTURY, INDUSTRIAL PRODUCTION MULTIPLIED BY A FACTOR OF 40.

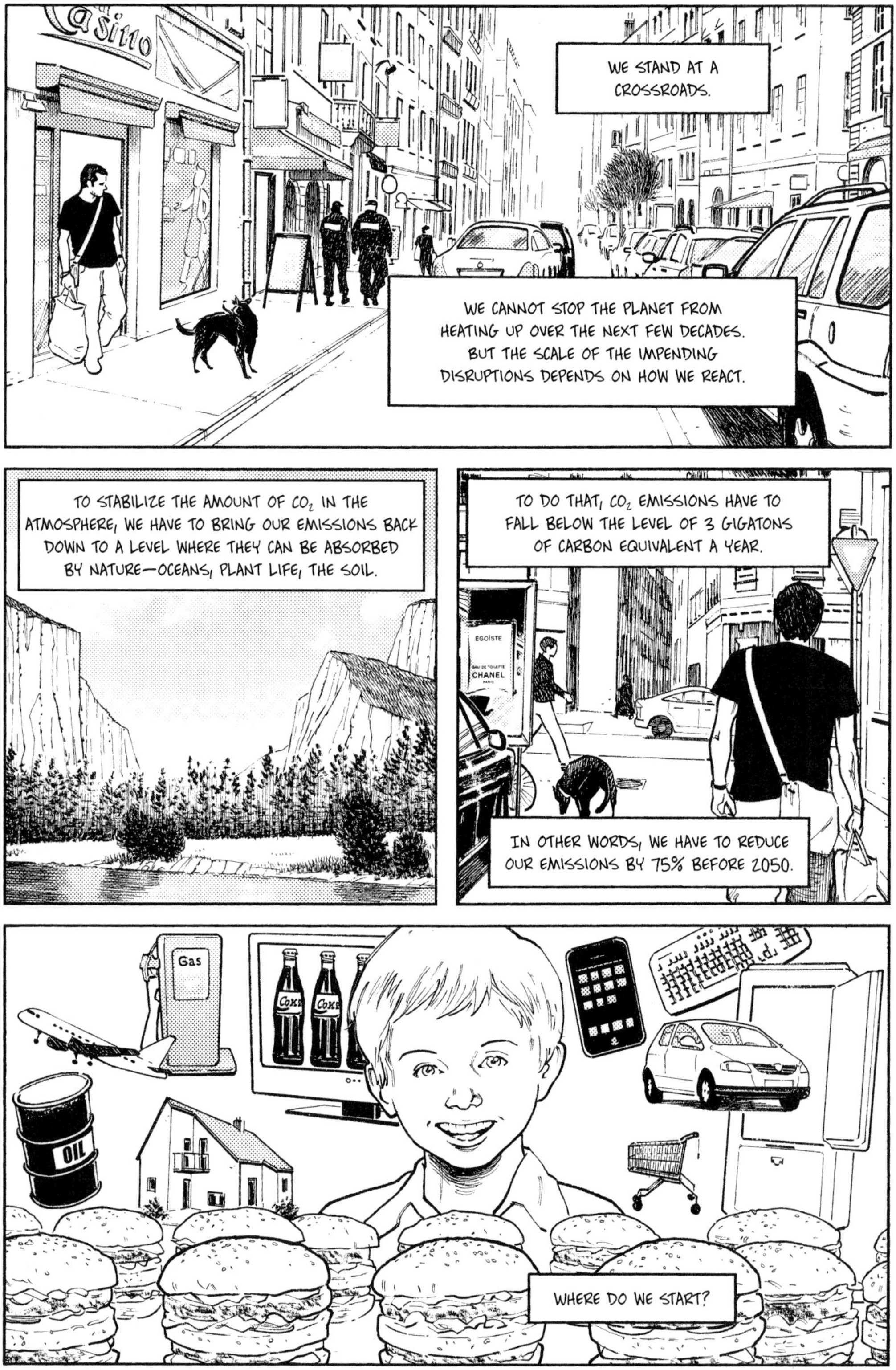
WE STAND AT A CROSSROADS.
WE CANNOT STOP THE PLANET FROM HEATING UP OVER THE NEXT FEW DECADES. BUT THE SCALE OF THE IMPENDING DISRUPTIONS DEPENDS ON HOW WE REACT.
TO STABILIZE THE AMOUNT OF CO_2 IN THE ATMOSPHERE, WE HAVE TO BRING OUR EMISSIONS BACK DOWN TO A LEVEL WHERE THEY CAN BE ABSORBED BY NATURE—OCEANS, PLANT LIFE, THE SOIL.
TO DO THAT, CO_2 EMISSIONS HAVE TO FALL BELOW THE LEVEL OF 3 GIGATONS OF CARBON EQUIVALENT A YEAR.
EGOÏSTE
CHANEL
PARIS
IN OTHER WORDS, WE HAVE TO REDUCE OUR EMISSIONS BY 75% BEFORE 2050.
Gas
Coke
Coke
OIL
WHERE DO WE START?

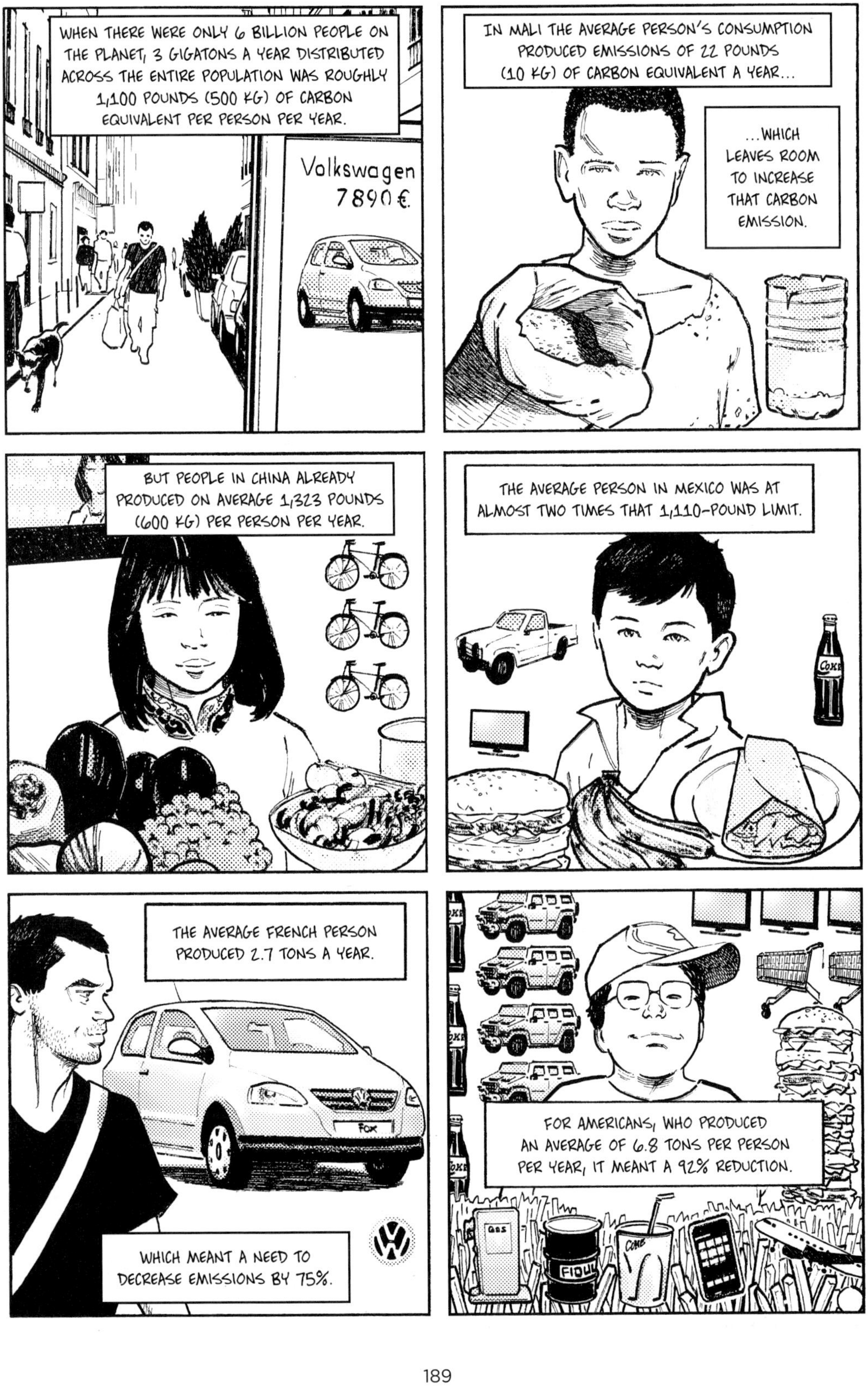
WHEN THERE WERE ONLY 6 BILLION PEOPLE ON THE PLANET, 3 GIGATONS A YEAR DISTRIBUTED ACROSS THE ENTIRE POPULATION WAS ROUGHLY 1,100 POUNDS (500 KG) OF CARBON EQUIVALENT PER PERSON PER YEAR.
Volkswagen
7890 €.
IN MALI THE AVERAGE PERSON'S CONSUMPTION PRODUCED EMISSIONS OF 22 POUNDS (10 KG) OF CARBON EQUIVALENT A YEAR...
...WHICH LEAVES ROOM TO INCREASE THAT CARBON EMISSION.
BUT PEOPLE IN CHINA ALREADY PRODUCED ON AVERAGE 1,323 POUNDS (600 KG) PER PERSON PER YEAR.
THE AVERAGE PERSON IN MEXICO WAS AT ALMOST TWO TIMES THAT 1,110-POUND LIMIT.
COKE
THE AVERAGE FRENCH PERSON PRODUCED 2.7 TONS A YEAR.
FOX
WHICH MEANT A NEED TO DECREASE EMISSIONS BY 75%.
FOR AMERICANS, WHO PRODUCED AN AVERAGE OF 6.8 TONS PER PERSON PER YEAR, IT MEANT A 92% REDUCTION.

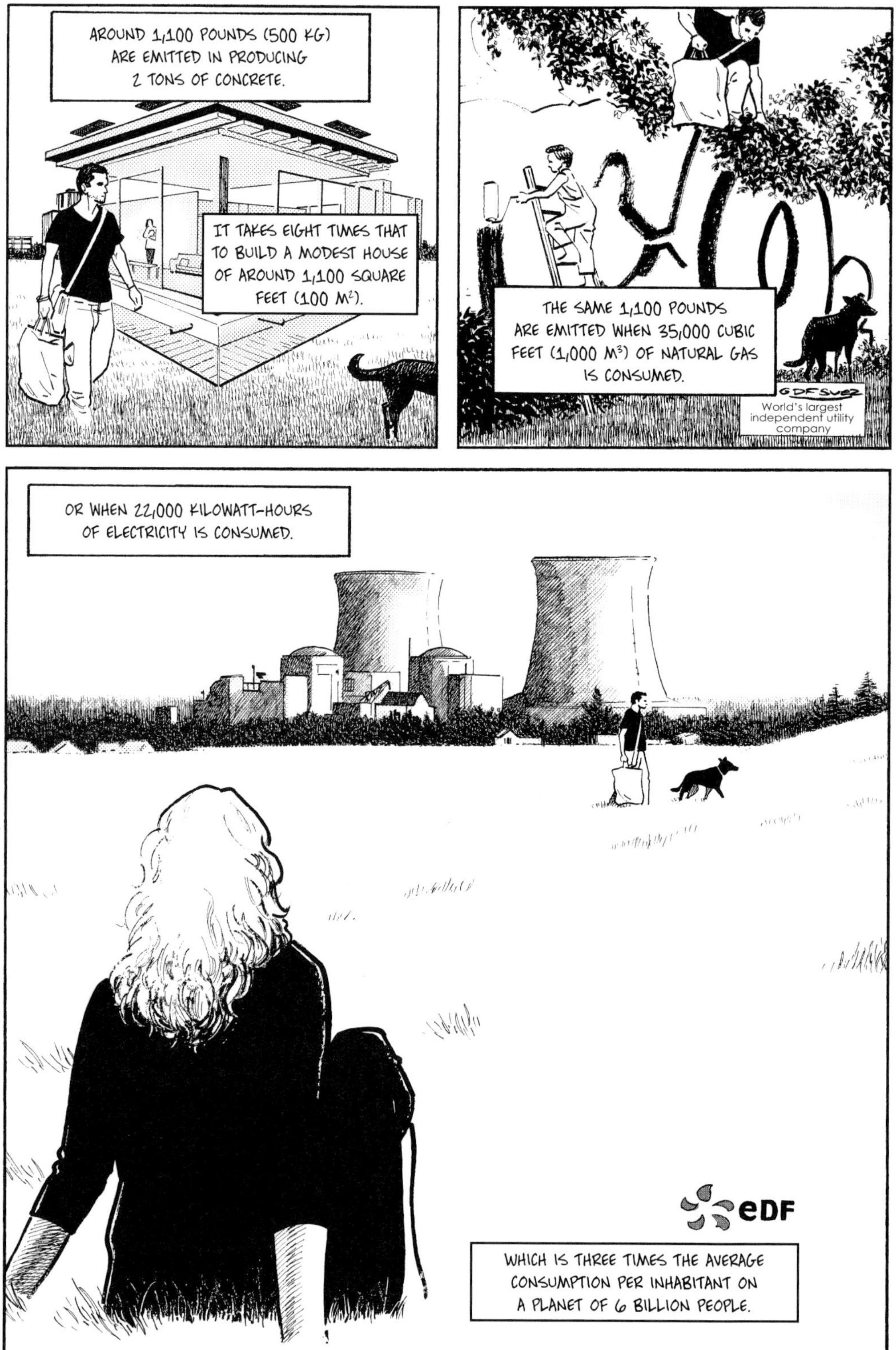
AROUND 1,100 POUNDS (500 KG) ARE EMITTED IN PRODUCING 2 TONS OF CONCRETE.
IT TAKES EIGHT TIMES THAT TO BUILD A MODEST HOUSE OF AROUND 1,100 SQUARE FEET (100 M²).
THE SAME 1,100 POUNDS ARE EMITTED WHEN 35,000 CUBIC FEET (1,000 M³) OF NATURAL GAS IS CONSUMED.
GDF SUEZ
World's largest independent utility company
OR WHEN 22,000 KILOWATT-HOURS OF ELECTRICITY IS CONSUMED.
eDF
WHICH IS THREE TIMES THE AVERAGE CONSUMPTION PER INHABITANT ON A PLANET OF 6 BILLION PEOPLE.

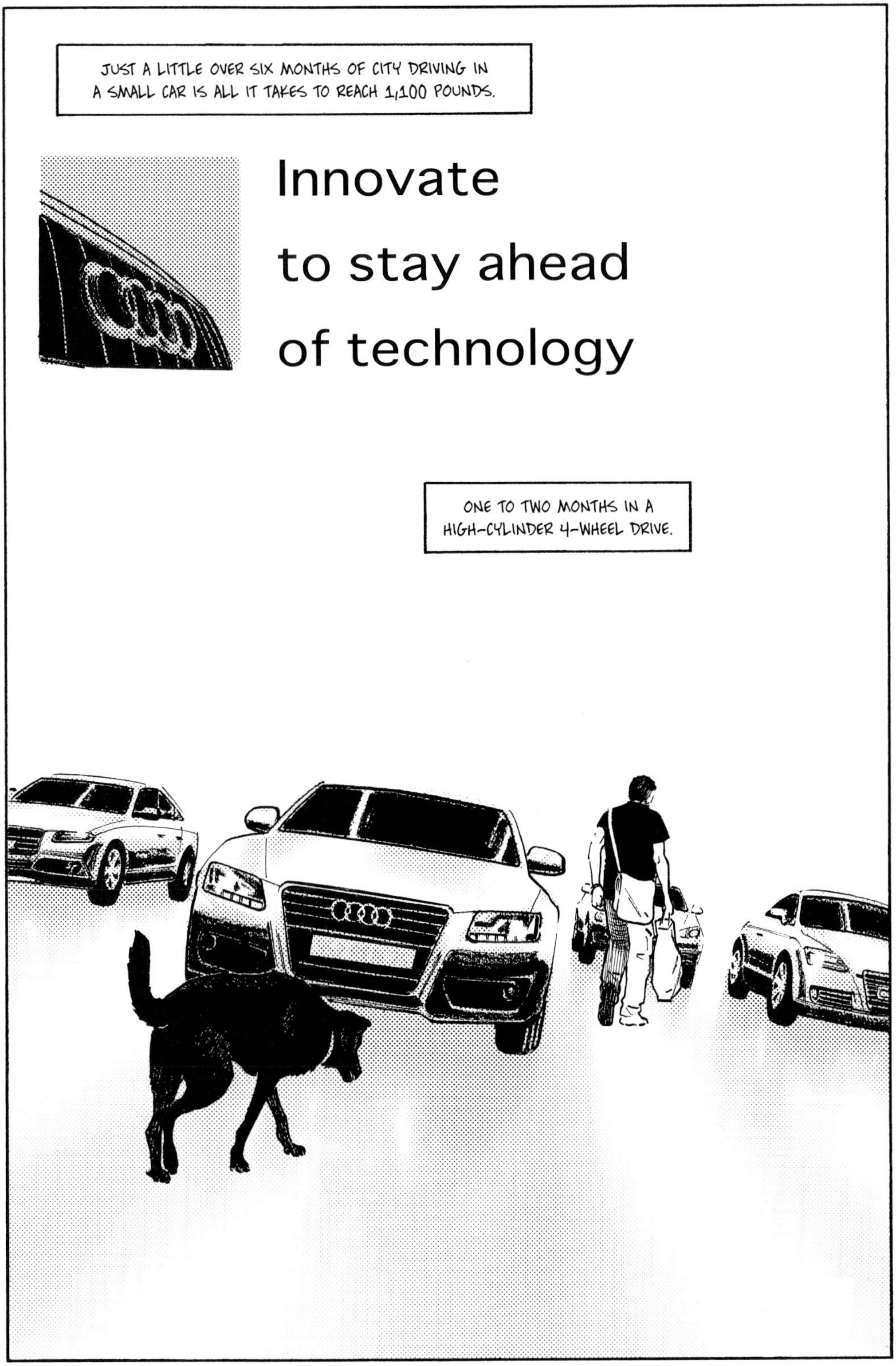
JUST A LITTLE OVER SIX MONTHS OF CITY DRIVING IN A SMALL CAR IS ALL IT TAKES TO REACH 1,100 POUNDS.
Innovate to stay ahead of technology
ONE TO TWO MONTHS IN A HIGH-CYLINDER 4-WHEEL DRIVE.

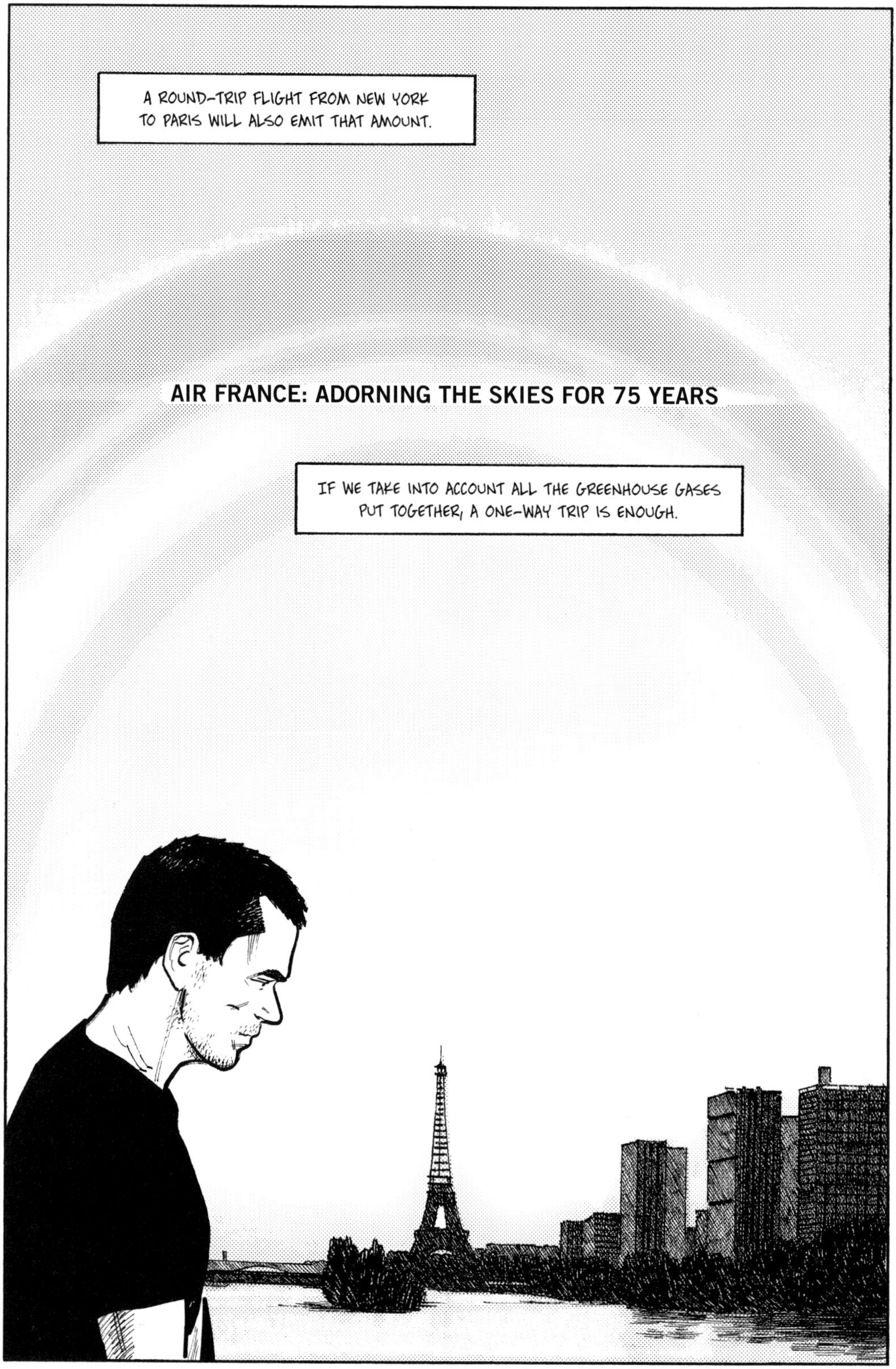
A ROUND-TRIP FLIGHT FROM NEW YORK TO PARIS WILL ALSO EMIT THAT AMOUNT.
AIR FRANCE: ADORNING THE SKIES FOR 75 YEARS
IF WE TAKE INTO ACCOUNT ALL THE GREENHOUSE GASES PUT TOGETHER, A ONE-WAY TRIP IS ENOUGH.

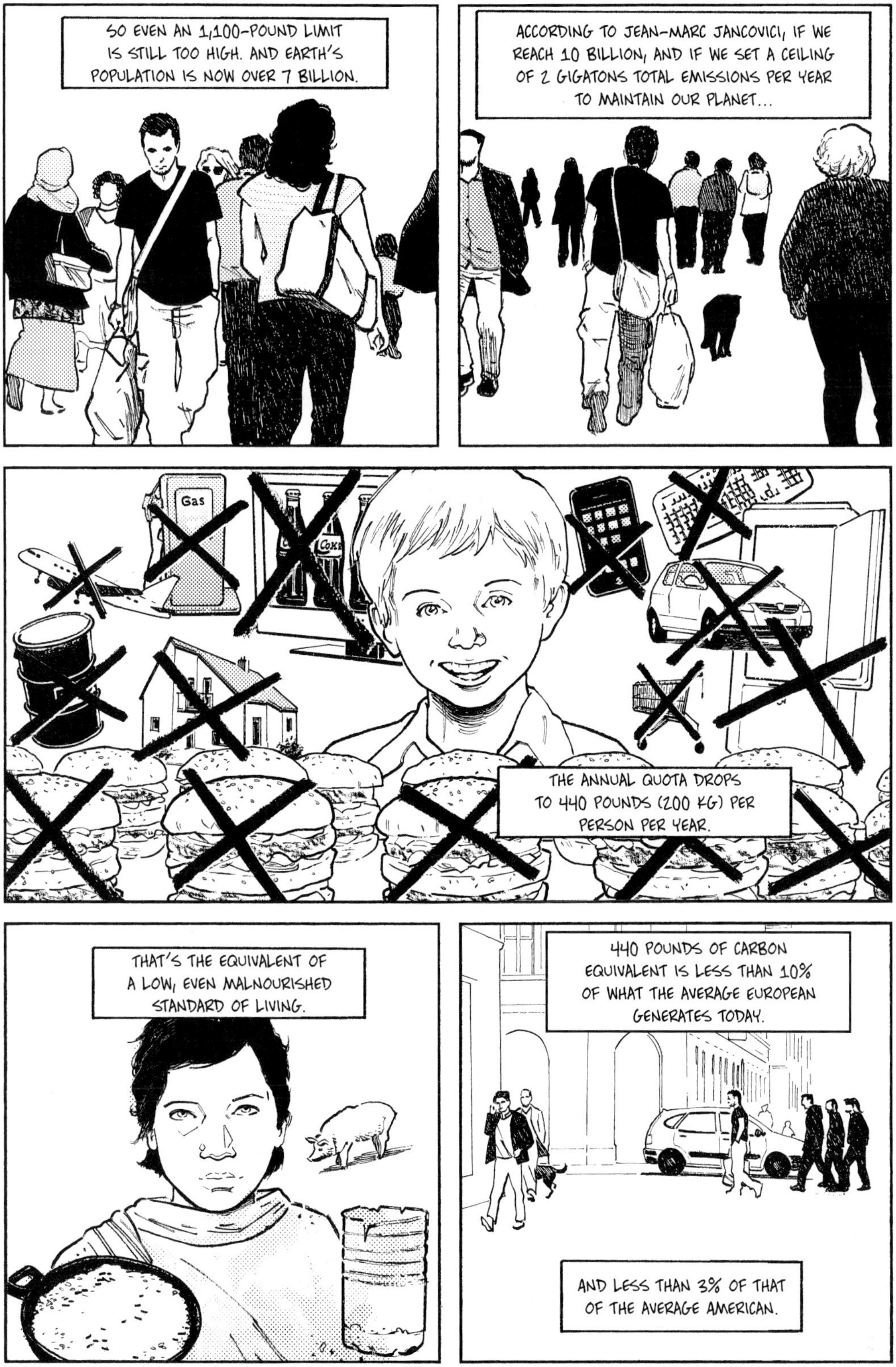
SO EVEN AN 1,100-POUND LIMIT IS STILL TOO HIGH. AND EARTH'S POPULATION IS NOW OVER 7 BILLION.
ACCORDING TO JEAN-MARC JANCOVICI, IF WE REACH 10 BILLION, AND IF WE SET A CEILING OF 2 GIGATONS TOTAL EMISSIONS PER YEAR TO MAINTAIN OUR PLANET...
Gas
THE ANNUAL QUOTA DROPS TO 440 POUNDS (200 KG) PER PERSON PER YEAR.
THAT'S THE EQUIVALENT OF A LOW, EVEN MALNOURISHED STANDARD OF LIVING.
440 POUNDS OF CARBON EQUIVALENT IS LESS THAN 10% OF WHAT THE AVERAGE EUROPEAN GENERATES TODAY.
AND LESS THAN 3% OF THAT OF THE AVERAGE AMERICAN.

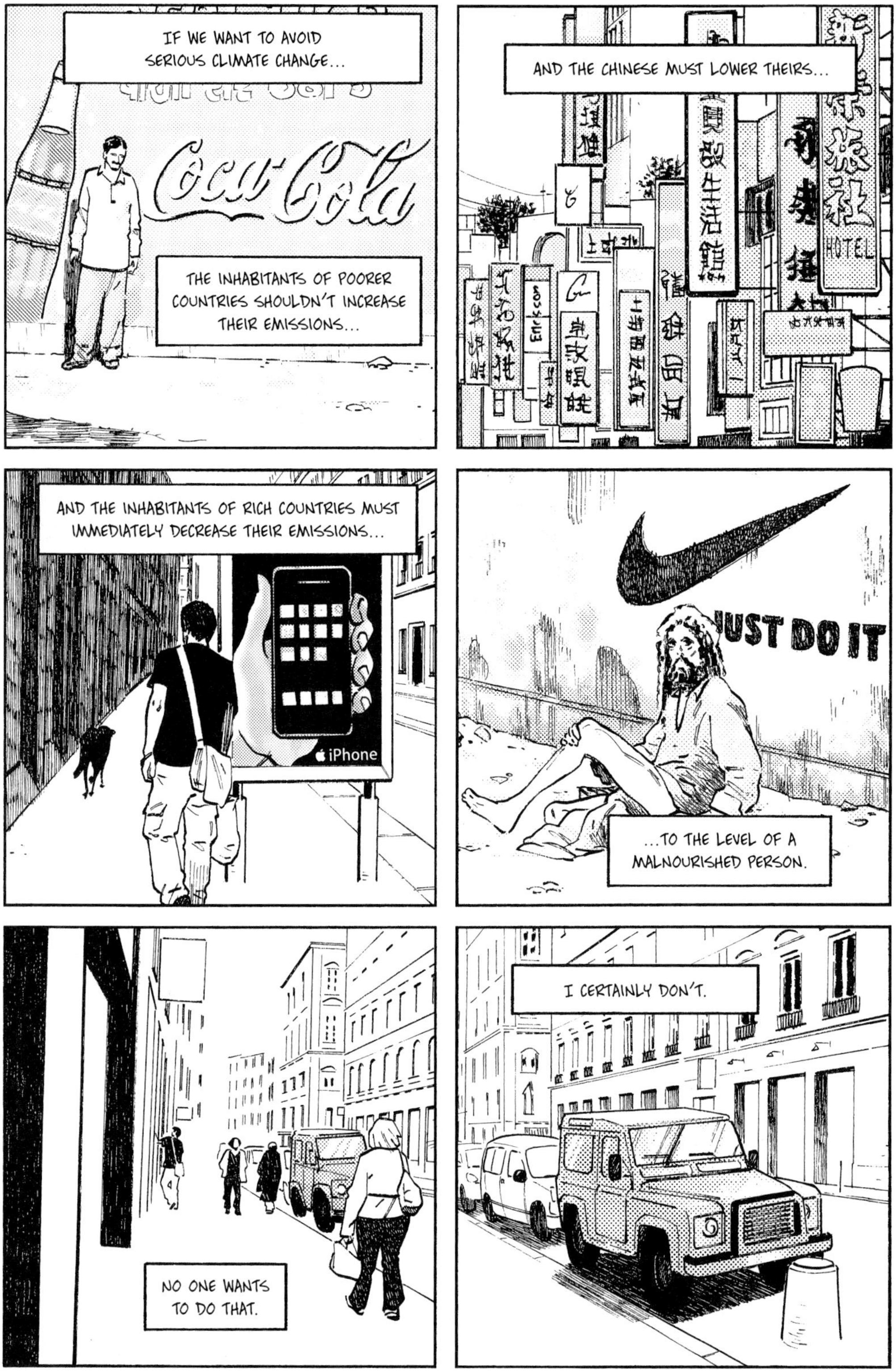
IF WE WANT TO AVOID SERIOUS CLIMATE CHANGE...
Coca-Cola
THE INHABITANTS OF POORER COUNTRIES SHOULDN'T INCREASE THEIR EMISSIONS...
AND THE CHINESE MUST LOWER THEIRS...
HOTEL
AND THE INHABITANTS OF RICH COUNTRIES MUST IMMEDIATELY DECREASE THEIR EMISSIONS...
iPhone
JUST DO IT
...TO THE LEVEL OF A MALNOURISHED PERSON.
NO ONE WANTS TO DO THAT.
I CERTAINLY DON'T.

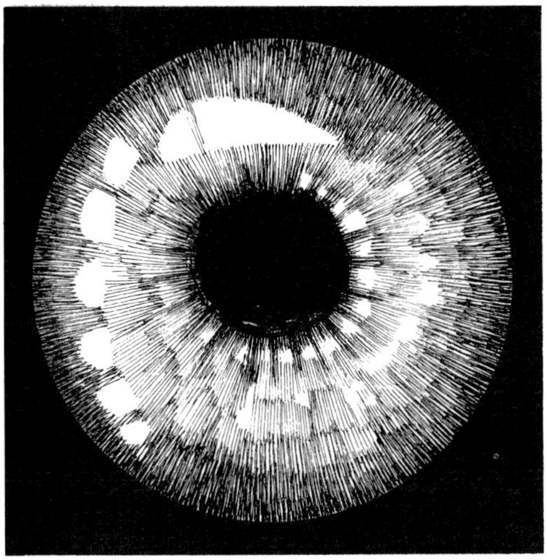

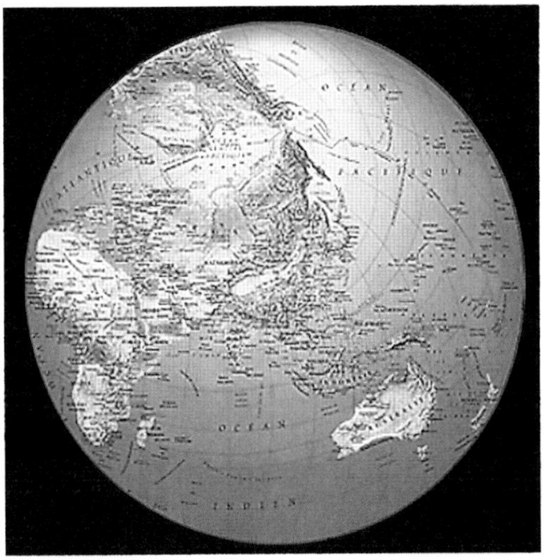

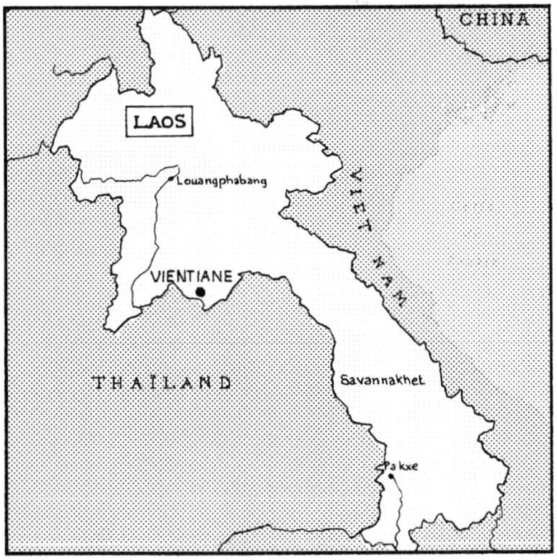

CHINA
LAOS
Louangphabang
VIET NAM
VIENTIANE
THAÏLAND
Savannakhet
Pakxe

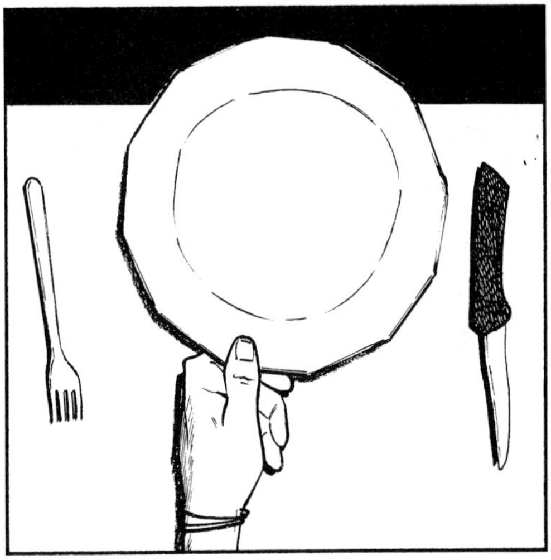

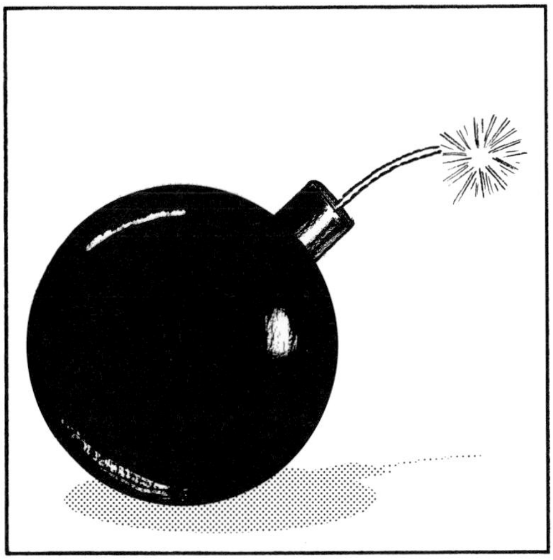

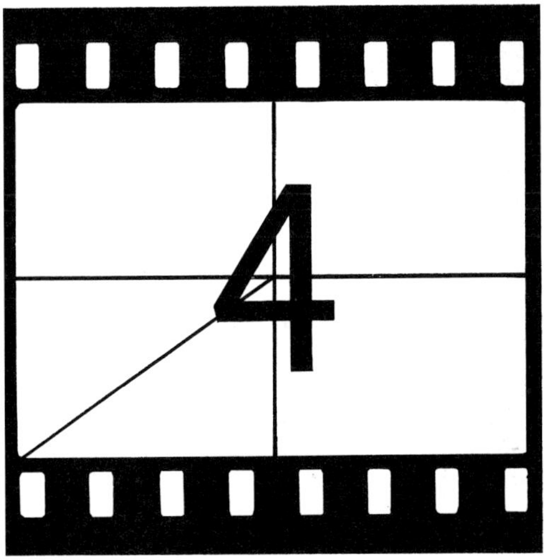

4

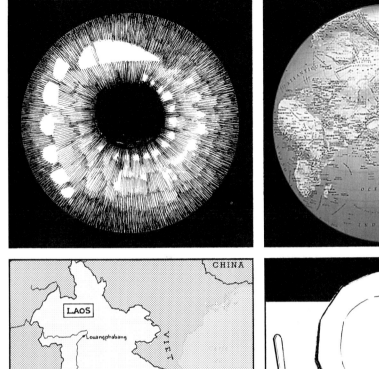

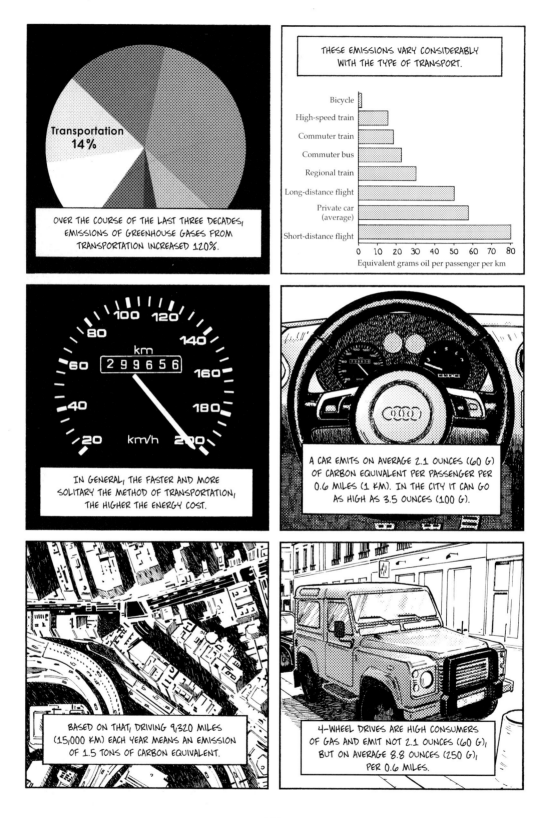

Transportation
14%

OVER THE COURSE OF THE LAST THREE DECADES, EMISSIONS OF GREENHOUSE GASES FROM TRANSPORTATION INCREASED 120%.

THESE EMISSIONS VARY CONSIDERABLY WITH THE TYPE OF TRANSPORT.

Bicycle
High-speed train
Commuter train
Commuter bus
Regional train
Long-distance flight
Private car (average)
Short-distance flight

0 10 20 30 40 50 60 70 80
Equivalent grams oil per passenger per km

km
2 9 9 6 5 6
km/h

IN GENERAL, THE FASTER AND MORE SOLITARY THE METHOD OF TRANSPORTATION, THE HIGHER THE ENERGY COST.

A CAR EMITS ON AVERAGE 2.1 OUNCES (60 G) OF CARBON EQUIVALENT PER PASSENGER PER 0.6 MILES (1 KM). IN THE CITY IT CAN GO AS HIGH AS 3.5 OUNCES (100 G).

BASED ON THAT, DRIVING 9,320 MILES (15,000 KM) EACH YEAR MEANS AN EMISSION OF 1.5 TONS OF CARBON EQUIVALENT.

4-WHEEL DRIVES ARE HIGH CONSUMERS OF GAS AND EMIT NOT 2.1 OUNCES (60 G), BUT ON AVERAGE 8.8 OUNCES (250 G), PER 0.6 MILES.

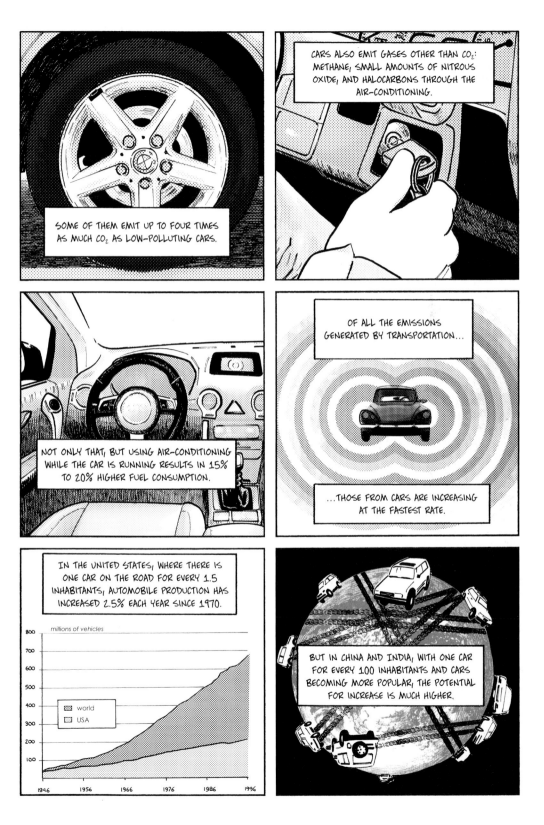

SOME OF THEM EMIT UP TO FOUR TIMES AS MUCH CO_2 AS LOW-POLLUTING CARS.

CARS ALSO EMIT GASES OTHER THAN CO_2: METHANE, SMALL AMOUNTS OF NITROUS OXIDE, AND HALOCARBONS THROUGH THE AIR-CONDITIONING.

NOT ONLY THAT, BUT USING AIR-CONDITIONING WHILE THE CAR IS RUNNING RESULTS IN 15% TO 20% HIGHER FUEL CONSUMPTION.

OF ALL THE EMISSIONS GENERATED BY TRANSPORTATION...

...THOSE FROM CARS ARE INCREASING AT THE FASTEST RATE.

IN THE UNITED STATES, WHERE THERE IS ONE CAR ON THE ROAD FOR EVERY 1.5 INHABITANTS, AUTOMOBILE PRODUCTION HAS INCREASED 2.5% EACH YEAR SINCE 1970.

millions of vehicles

800
700
600
500
400
300
200
100

world
USA

1946   1956   1966   1976   1986   1996

BUT IN CHINA AND INDIA, WITH ONE CAR FOR EVERY 100 INHABITANTS AND CARS BECOMING MORE POPULAR, THE POTENTIAL FOR INCREASE IS MUCH HIGHER.

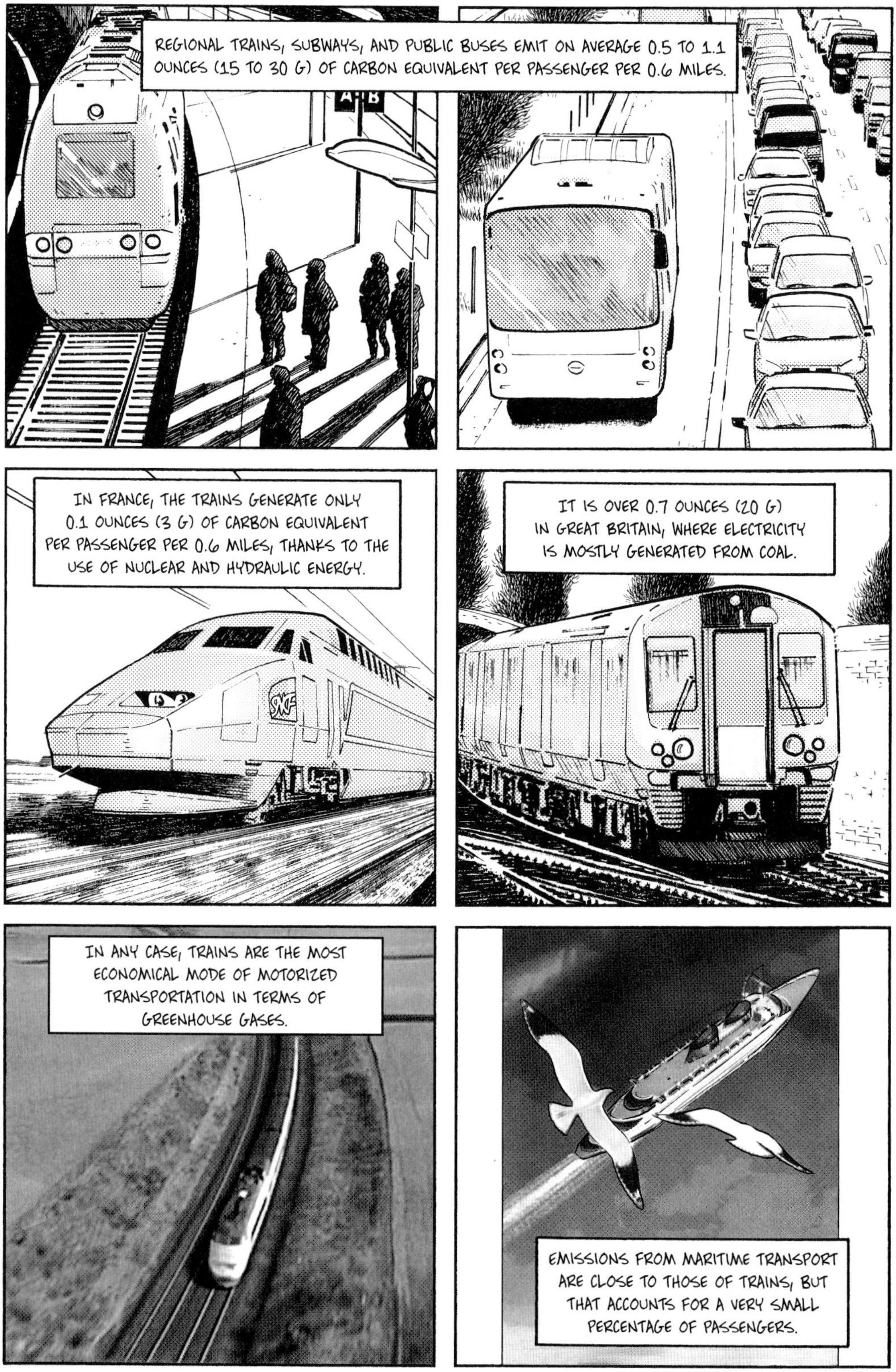
REGIONAL TRAINS, SUBWAYS, AND PUBLIC BUSES EMIT ON AVERAGE 0.5 TO 1.1 OUNCES (15 TO 30 G) OF CARBON EQUIVALENT PER PASSENGER PER 0.6 MILES.
A B
IN FRANCE, THE TRAINS GENERATE ONLY 0.1 OUNCES (3 G) OF CARBON EQUIVALENT PER PASSENGER PER 0.6 MILES, THANKS TO THE USE OF NUCLEAR AND HYDRAULIC ENERGY.
SNCF
IT IS OVER 0.7 OUNCES (20 G) IN GREAT BRITAIN, WHERE ELECTRICITY IS MOSTLY GENERATED FROM COAL.
IN ANY CASE, TRAINS ARE THE MOST ECONOMICAL MODE OF MOTORIZED TRANSPORTATION IN TERMS OF GREENHOUSE GASES.
EMISSIONS FROM MARITIME TRANSPORT ARE CLOSE TO THOSE OF TRAINS, BUT THAT ACCOUNTS FOR A VERY SMALL PERCENTAGE OF PASSENGERS.

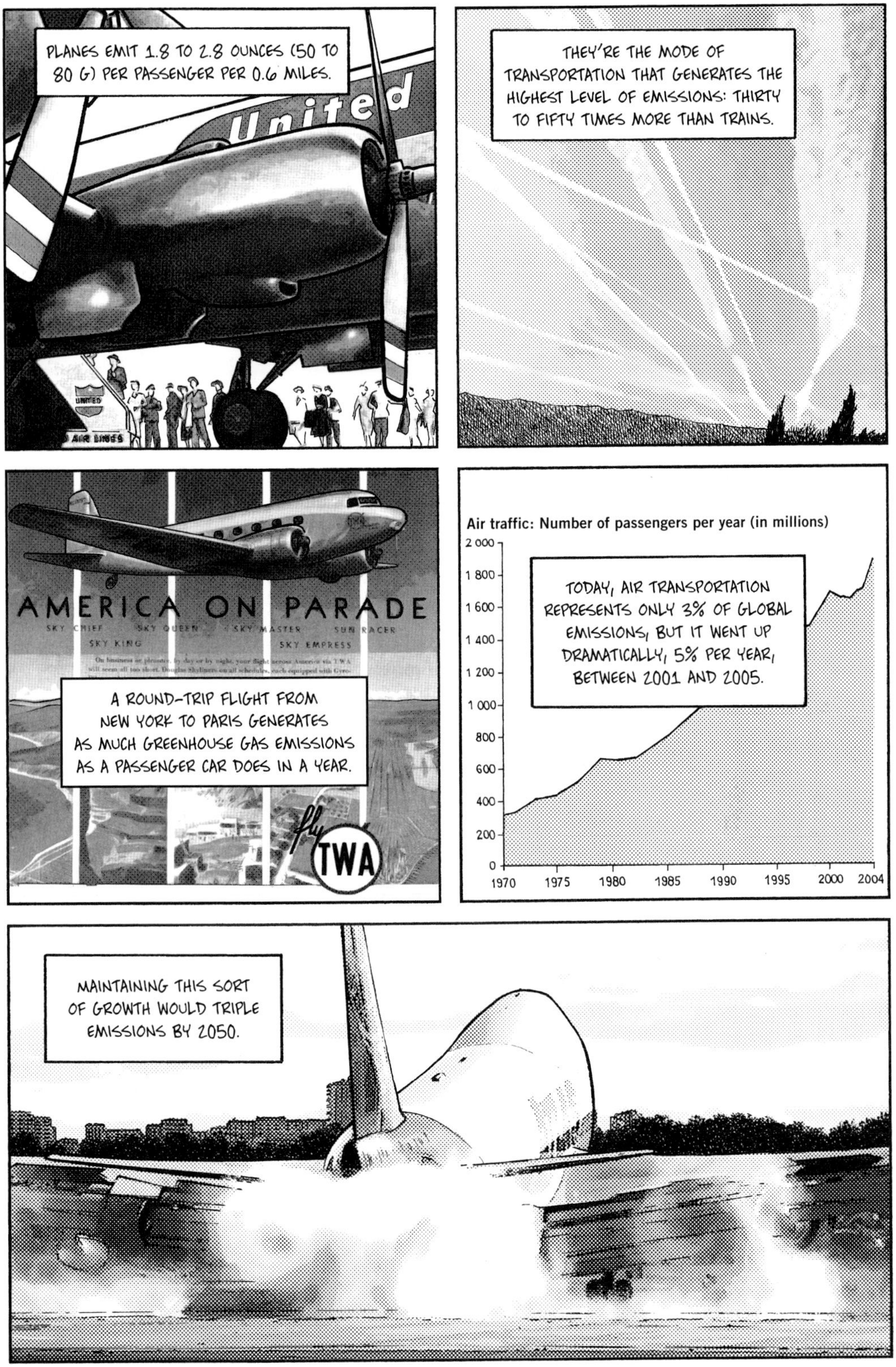
PLANES EMIT 1.8 TO 2.8 OUNCES (50 TO 80 G) PER PASSENGER PER 0.6 MILES.
United
THEY'RE THE MODE OF TRANSPORTATION THAT GENERATES THE HIGHEST LEVEL OF EMISSIONS: THIRTY TO FIFTY TIMES MORE THAN TRAINS.
AMERICA ON PARADE
SKY CHIEF SKY QUEEN SKY MASTER SUN RACER
SKY KING SKY EMPRESS
A ROUND-TRIP FLIGHT FROM NEW YORK TO PARIS GENERATES AS MUCH GREENHOUSE GAS EMISSIONS AS A PASSENGER CAR DOES IN A YEAR.
fly TWA
Air traffic: Number of passengers per year (in millions)
2 000
1 800
1 600
1 400
1 200
1 000
800
600
400
200
0
1970
1975
1980
1985
1990
1995
2000
2004
TODAY, AIR TRANSPORTATION REPRESENTS ONLY 3% OF GLOBAL EMISSIONS, BUT IT WENT UP DRAMATICALLY, 5% PER YEAR, BETWEEN 2001 AND 2005.
MAINTAINING THIS SORT OF GROWTH WOULD TRIPLE EMISSIONS BY 2050.

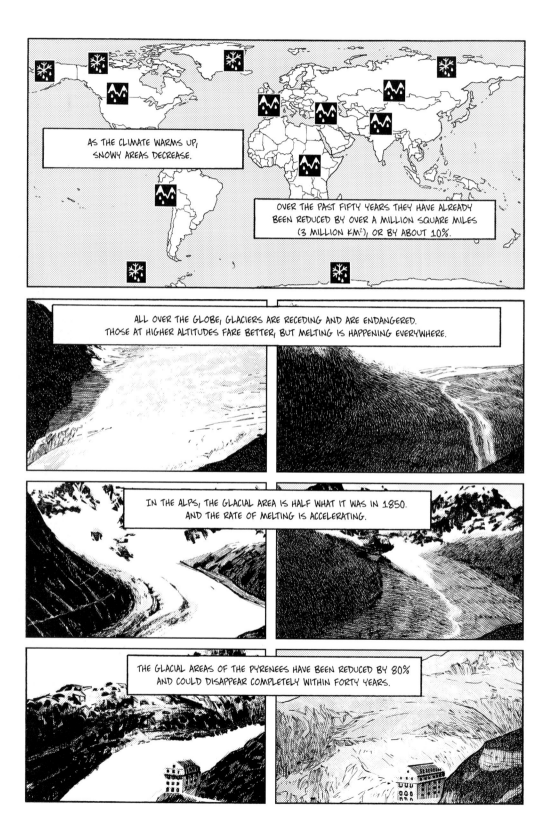

AS THE CLIMATE WARMS UP, SNOWY AREAS DECREASE.

OVER THE PAST FIFTY YEARS THEY HAVE ALREADY BEEN REDUCED BY OVER A MILLION SQUARE MILES (3 MILLION KM²), OR BY ABOUT 10%.

ALL OVER THE GLOBE, GLACIERS ARE RECEDING AND ARE ENDANGERED. THOSE AT HIGHER ALTITUDES FARE BETTER, BUT MELTING IS HAPPENING EVERYWHERE.

IN THE ALPS, THE GLACIAL AREA IS HALF WHAT IT WAS IN 1850. AND THE RATE OF MELTING IS ACCELERATING.

THE GLACIAL AREAS OF THE PYRENEES HAVE BEEN REDUCED BY 80% AND COULD DISAPPEAR COMPLETELY WITHIN FORTY YEARS.

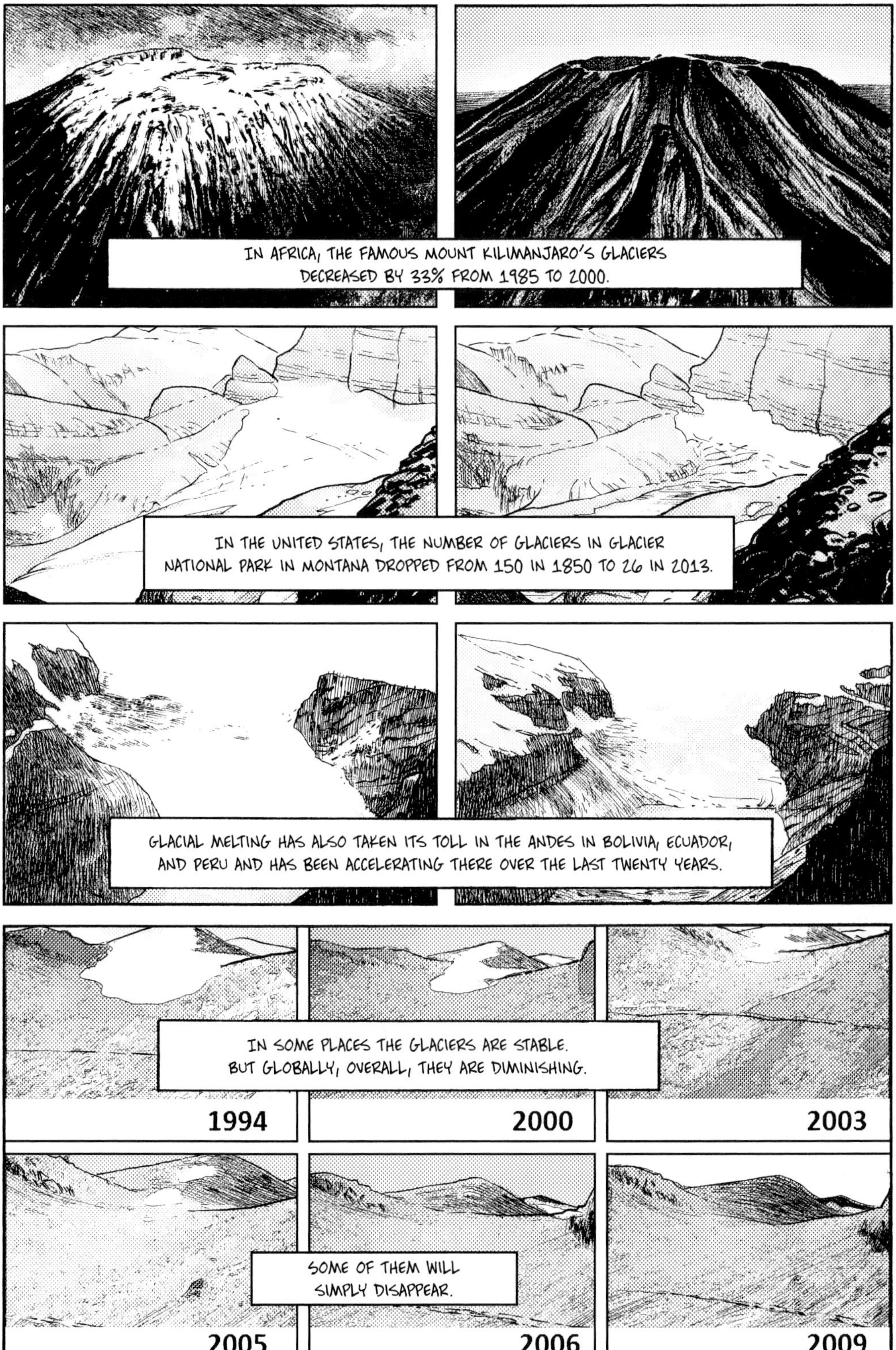
IN AFRICA, THE FAMOUS MOUNT KILIMANJARO'S GLACIERS DECREASED BY 33% FROM 1985 TO 2000.
IN THE UNITED STATES, THE NUMBER OF GLACIERS IN GLACIER NATIONAL PARK IN MONTANA DROPPED FROM 150 IN 1850 TO 26 IN 2013.
GLACIAL MELTING HAS ALSO TAKEN ITS TOLL IN THE ANDES IN BOLIVIA, ECUADOR, AND PERU AND HAS BEEN ACCELERATING THERE OVER THE LAST TWENTY YEARS.
IN SOME PLACES THE GLACIERS ARE STABLE. BUT GLOBALLY, OVERALL, THEY ARE DIMINISHING.
1994
2000
2003
SOME OF THEM WILL SIMPLY DISAPPEAR.
2005
2006
2009

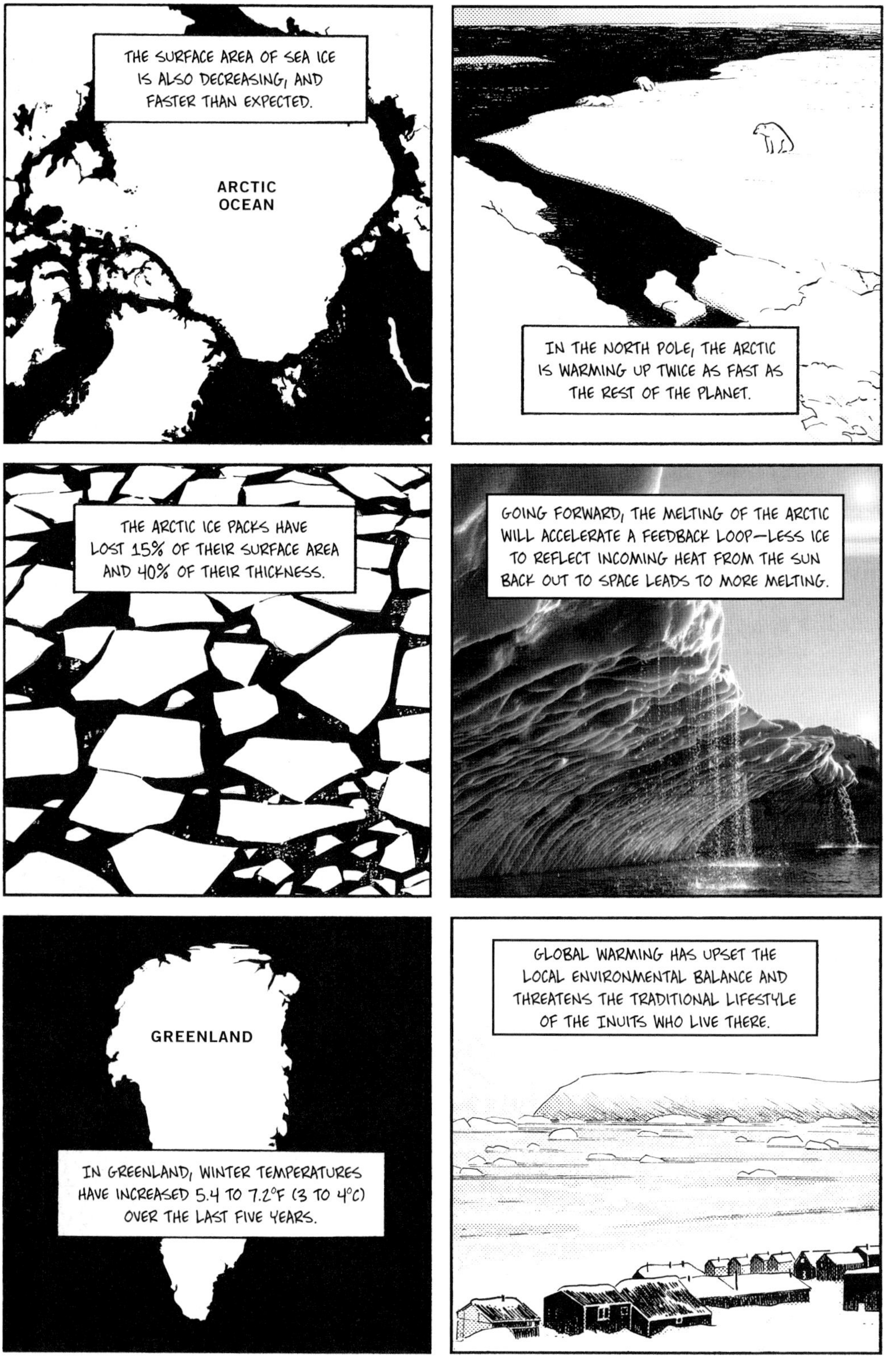
THE SURFACE AREA OF SEA ICE IS ALSO DECREASING, AND FASTER THAN EXPECTED.
ARCTIC OCEAN
IN THE NORTH POLE, THE ARCTIC IS WARMING UP TWICE AS FAST AS THE REST OF THE PLANET.
THE ARCTIC ICE PACKS HAVE LOST 15% OF THEIR SURFACE AREA AND 40% OF THEIR THICKNESS.
GOING FORWARD, THE MELTING OF THE ARCTIC WILL ACCELERATE A FEEDBACK LOOP—LESS ICE TO REFLECT INCOMING HEAT FROM THE SUN BACK OUT TO SPACE LEADS TO MORE MELTING.
GREENLAND
IN GREENLAND, WINTER TEMPERATURES HAVE INCREASED 5.4 TO 7.2°F (3 TO 4°C) OVER THE LAST FIVE YEARS.
GLOBAL WARMING HAS UPSET THE LOCAL ENVIRONMENTAL BALANCE AND THREATENS THE TRADITIONAL LIFESTYLE OF THE INUITS WHO LIVE THERE.

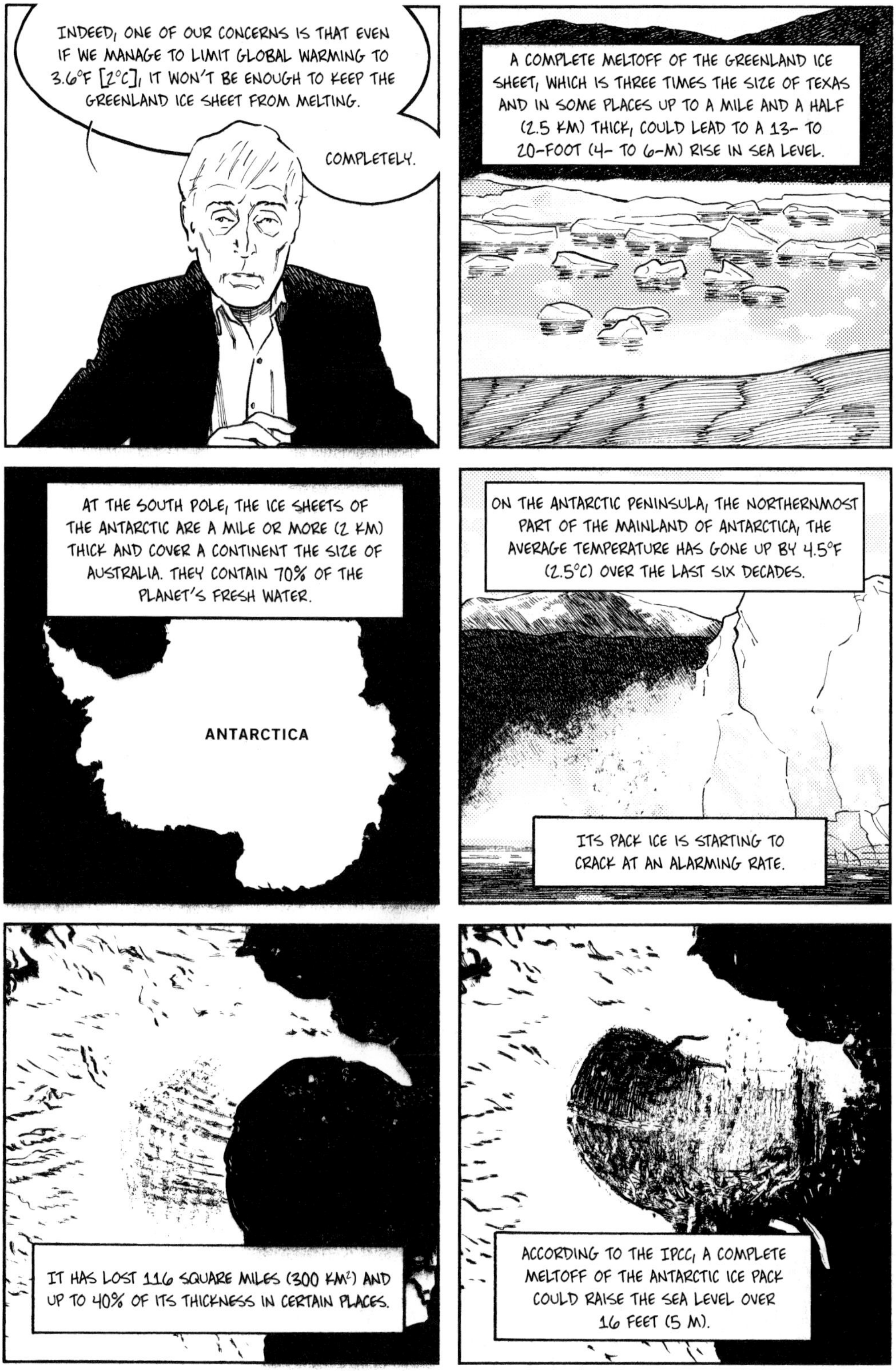
INDEED, ONE OF OUR CONCERNS IS THAT EVEN IF WE MANAGE TO LIMIT GLOBAL WARMING TO 3.6°F [2°C], IT WON'T BE ENOUGH TO KEEP THE GREENLAND ICE SHEET FROM MELTING.
COMPLETELY.
A COMPLETE MELTOFF OF THE GREENLAND ICE SHEET, WHICH IS THREE TIMES THE SIZE OF TEXAS AND IN SOME PLACES UP TO A MILE AND A HALF (2.5 KM) THICK, COULD LEAD TO A 13- TO 20-FOOT (4- TO 6-M) RISE IN SEA LEVEL.
AT THE SOUTH POLE, THE ICE SHEETS OF THE ANTARCTIC ARE A MILE OR MORE (2 KM) THICK AND COVER A CONTINENT THE SIZE OF AUSTRALIA. THEY CONTAIN 70% OF THE PLANET'S FRESH WATER.
ANTARCTICA
ON THE ANTARCTIC PENINSULA, THE NORTHERNMOST PART OF THE MAINLAND OF ANTARCTICA, THE AVERAGE TEMPERATURE HAS GONE UP BY 4.5°F (2.5°C) OVER THE LAST SIX DECADES.
ITS PACK ICE IS STARTING TO CRACK AT AN ALARMING RATE.
IT HAS LOST 116 SQUARE MILES (300 KM²) AND UP TO 40% OF ITS THICKNESS IN CERTAIN PLACES.
ACCORDING TO THE IPCC, A COMPLETE MELTOFF OF THE ANTARCTIC ICE PACK COULD RAISE THE SEA LEVEL OVER 16 FEET (5 M).

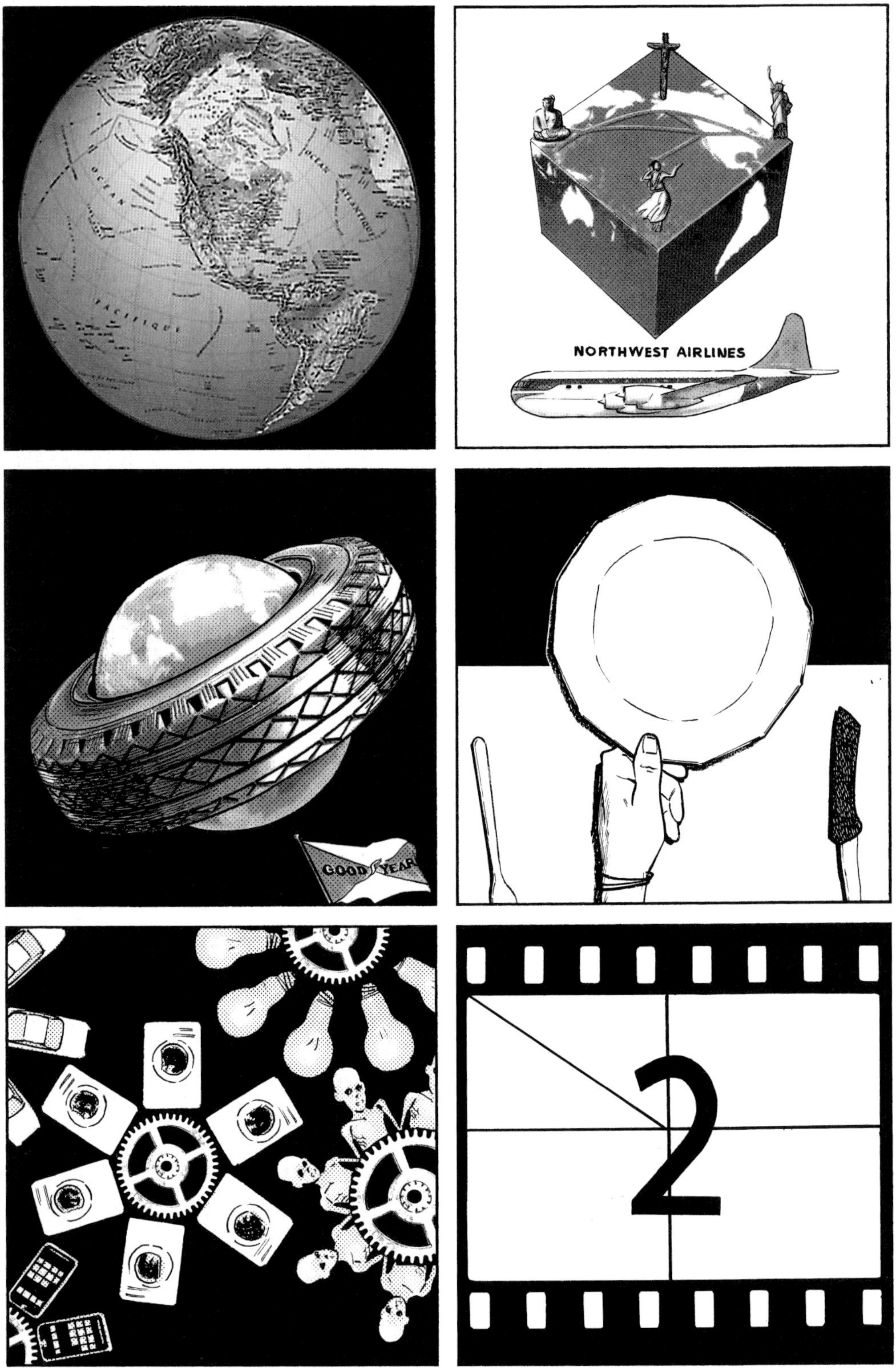
NORTHWEST AIRLINES
GOOD YEAR
2

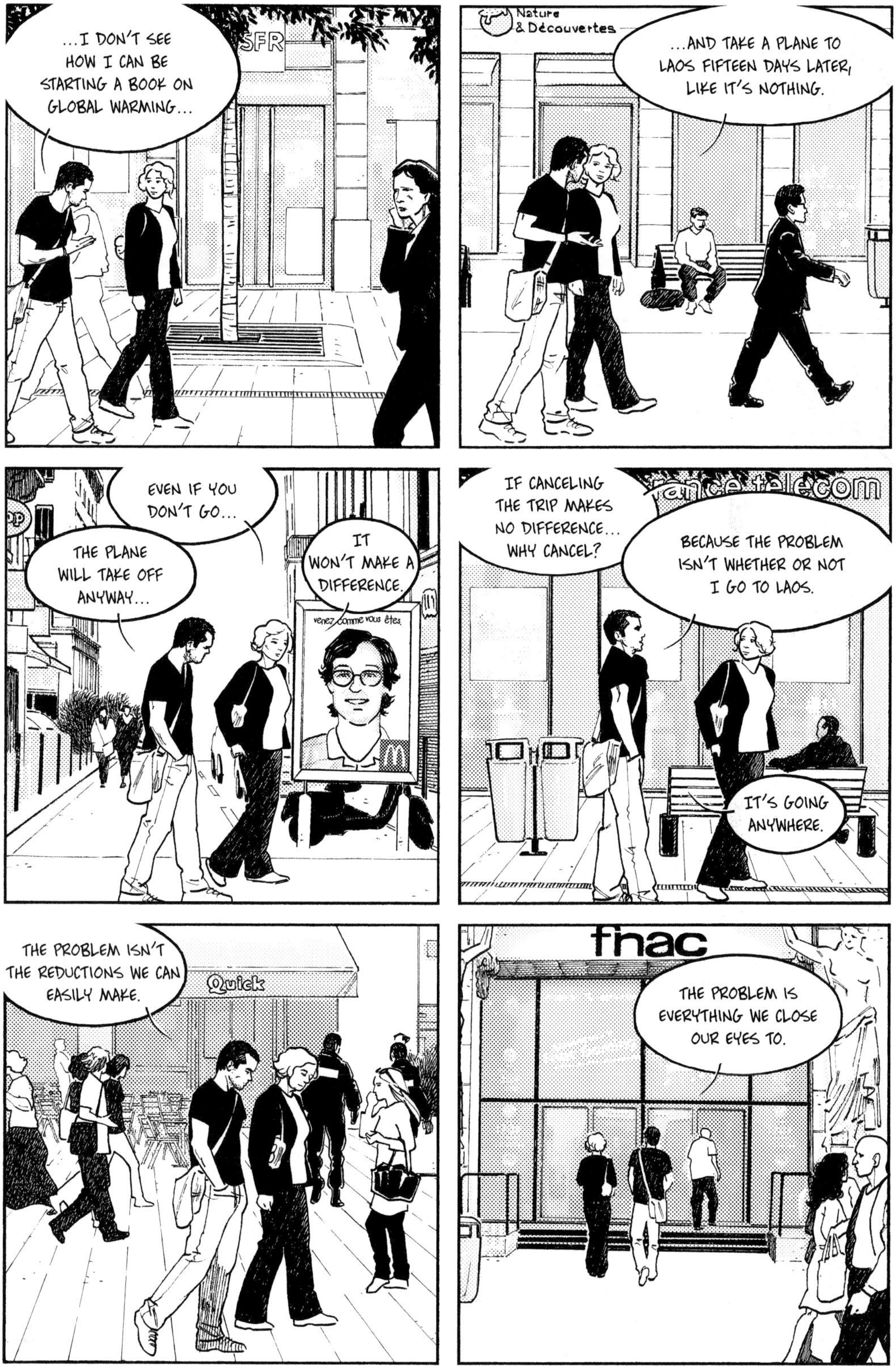
...I DON'T SEE HOW I CAN BE STARTING A BOOK ON GLOBAL WARMING...
SFR
Nature & Découvertes
...AND TAKE A PLANE TO LAOS FIFTEEN DAYS LATER, LIKE IT'S NOTHING.
EVEN IF YOU DON'T GO...
THE PLANE WILL TAKE OFF ANYWAY...
IT WON'T MAKE A DIFFERENCE.
venez comme vous êtes.
IF CANCELING THE TRIP MAKES NO DIFFERENCE... WHY CANCEL?
BECAUSE THE PROBLEM ISN'T WHETHER OR NOT I GO TO LAOS.
IT'S GOING ANYWHERE.
THE PROBLEM ISN'T THE REDUCTIONS WE CAN EASILY MAKE.
Quick
fnac
THE PROBLEM IS EVERYTHING WE CLOSE OUR EYES TO.

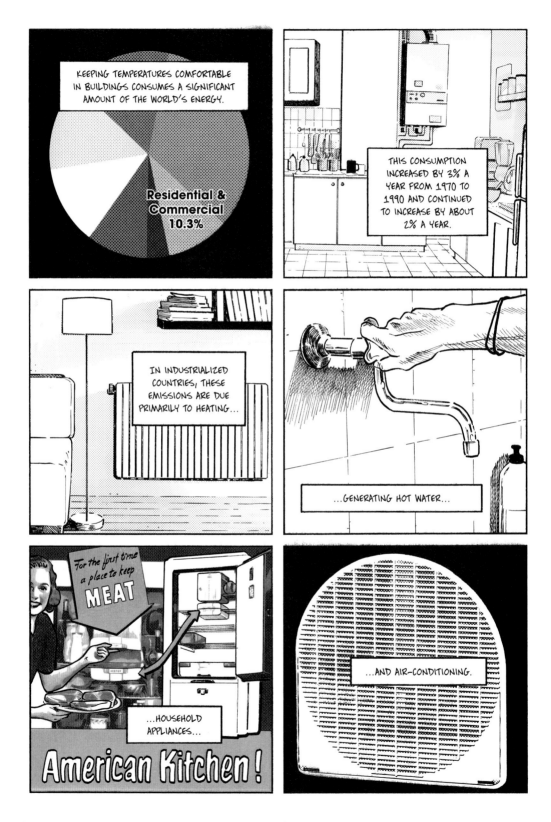

KEEPING TEMPERATURES COMFORTABLE IN BUILDINGS CONSUMES A SIGNIFICANT AMOUNT OF THE WORLD'S ENERGY.

Residential & Commercial 10.3%

THIS CONSUMPTION INCREASED BY 3% A YEAR FROM 1970 TO 1990 AND CONTINUED TO INCREASE BY ABOUT 2% A YEAR.

IN INDUSTRIALIZED COUNTRIES, THESE EMISSIONS ARE DUE PRIMARILY TO HEATING...

...GENERATING HOT WATER...

For the first time a place to keep MEAT

...HOUSEHOLD APPLIANCES...

American Kitchen!

...AND AIR-CONDITIONING.

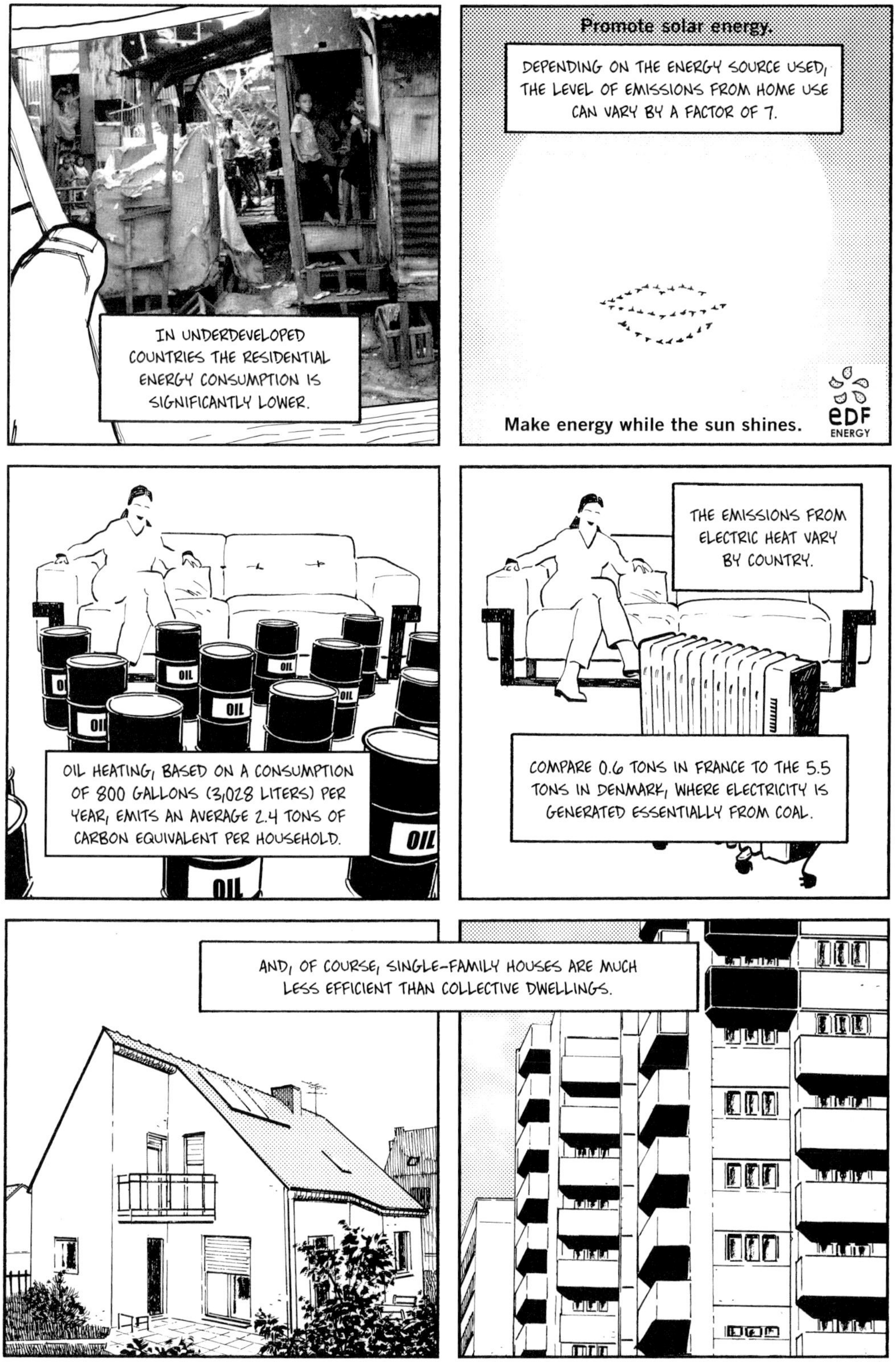

IN UNDERDEVELOPED COUNTRIES THE RESIDENTIAL ENERGY CONSUMPTION IS SIGNIFICANTLY LOWER.
Promote solar energy.
DEPENDING ON THE ENERGY SOURCE USED, THE LEVEL OF EMISSIONS FROM HOME USE CAN VARY BY A FACTOR OF 7.
Make energy while the sun shines.
EDF ENERGY
OIL
OIL HEATING, BASED ON A CONSUMPTION OF 800 GALLONS (3,028 LITERS) PER YEAR, EMITS AN AVERAGE 2.4 TONS OF CARBON EQUIVALENT PER HOUSEHOLD.
THE EMISSIONS FROM ELECTRIC HEAT VARY BY COUNTRY.
COMPARE 0.6 TONS IN FRANCE TO THE 5.5 TONS IN DENMARK, WHERE ELECTRICITY IS GENERATED ESSENTIALLY FROM COAL.
AND, OF COURSE, SINGLE-FAMILY HOUSES ARE MUCH LESS EFFICIENT THAN COLLECTIVE DWELLINGS.

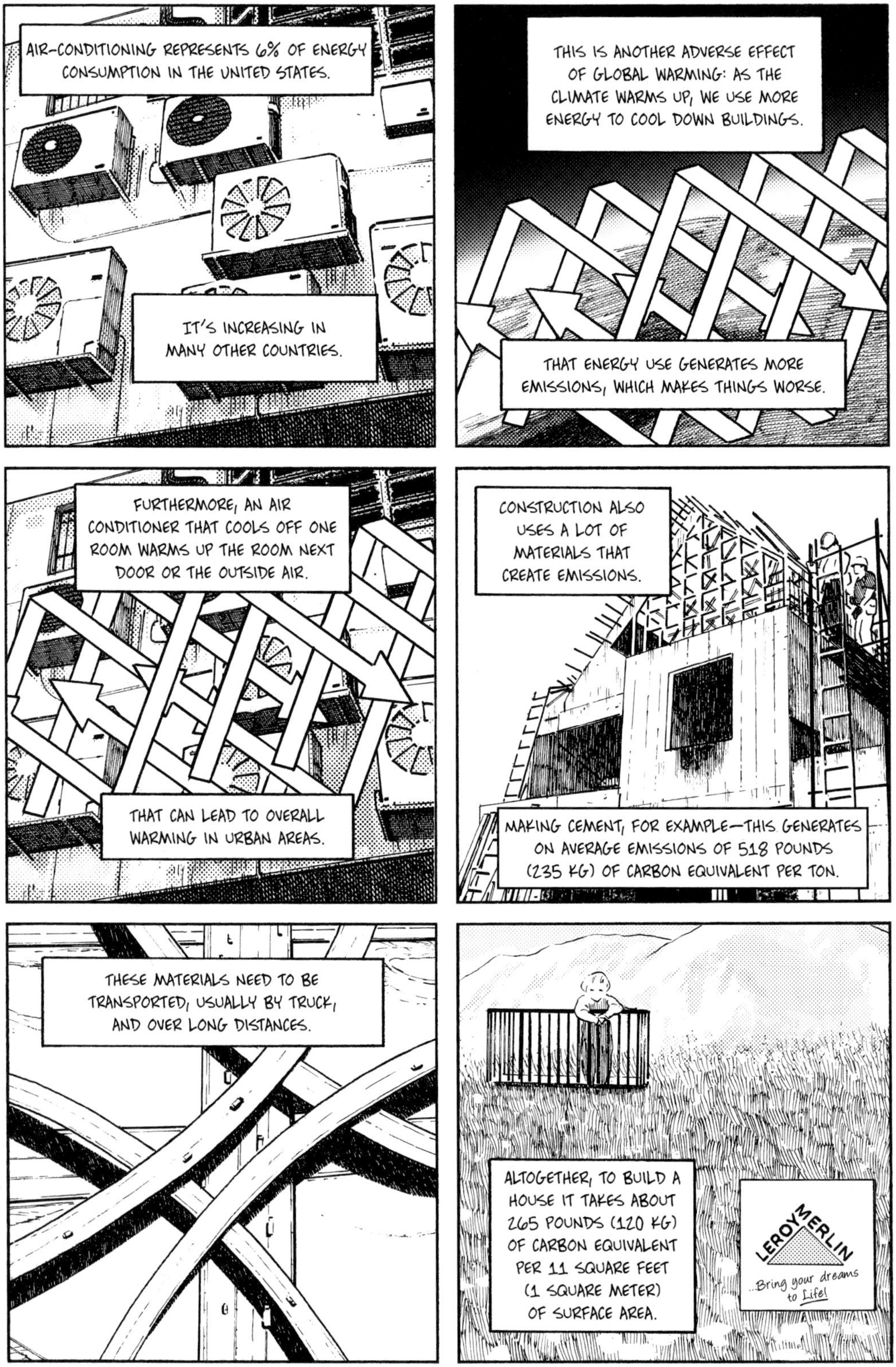
AIR-CONDITIONING REPRESENTS 6% OF ENERGY CONSUMPTION IN THE UNITED STATES.
IT'S INCREASING IN MANY OTHER COUNTRIES.
THIS IS ANOTHER ADVERSE EFFECT OF GLOBAL WARMING: AS THE CLIMATE WARMS UP, WE USE MORE ENERGY TO COOL DOWN BUILDINGS.
THAT ENERGY USE GENERATES MORE EMISSIONS, WHICH MAKES THINGS WORSE.
FURTHERMORE, AN AIR CONDITIONER THAT COOLS OFF ONE ROOM WARMS UP THE ROOM NEXT DOOR OR THE OUTSIDE AIR.
THAT CAN LEAD TO OVERALL WARMING IN URBAN AREAS.
CONSTRUCTION ALSO USES A LOT OF MATERIALS THAT CREATE EMISSIONS.
MAKING CEMENT, FOR EXAMPLE—THIS GENERATES ON AVERAGE EMISSIONS OF 518 POUNDS (235 KG) OF CARBON EQUIVALENT PER TON.
THESE MATERIALS NEED TO BE TRANSPORTED, USUALLY BY TRUCK, AND OVER LONG DISTANCES.
ALTOGETHER, TO BUILD A HOUSE IT TAKES ABOUT 265 POUNDS (120 KG) OF CARBON EQUIVALENT PER 11 SQUARE FEET (1 SQUARE METER) OF SURFACE AREA.
LEROY MERLIN
...Bring your dreams to Life!

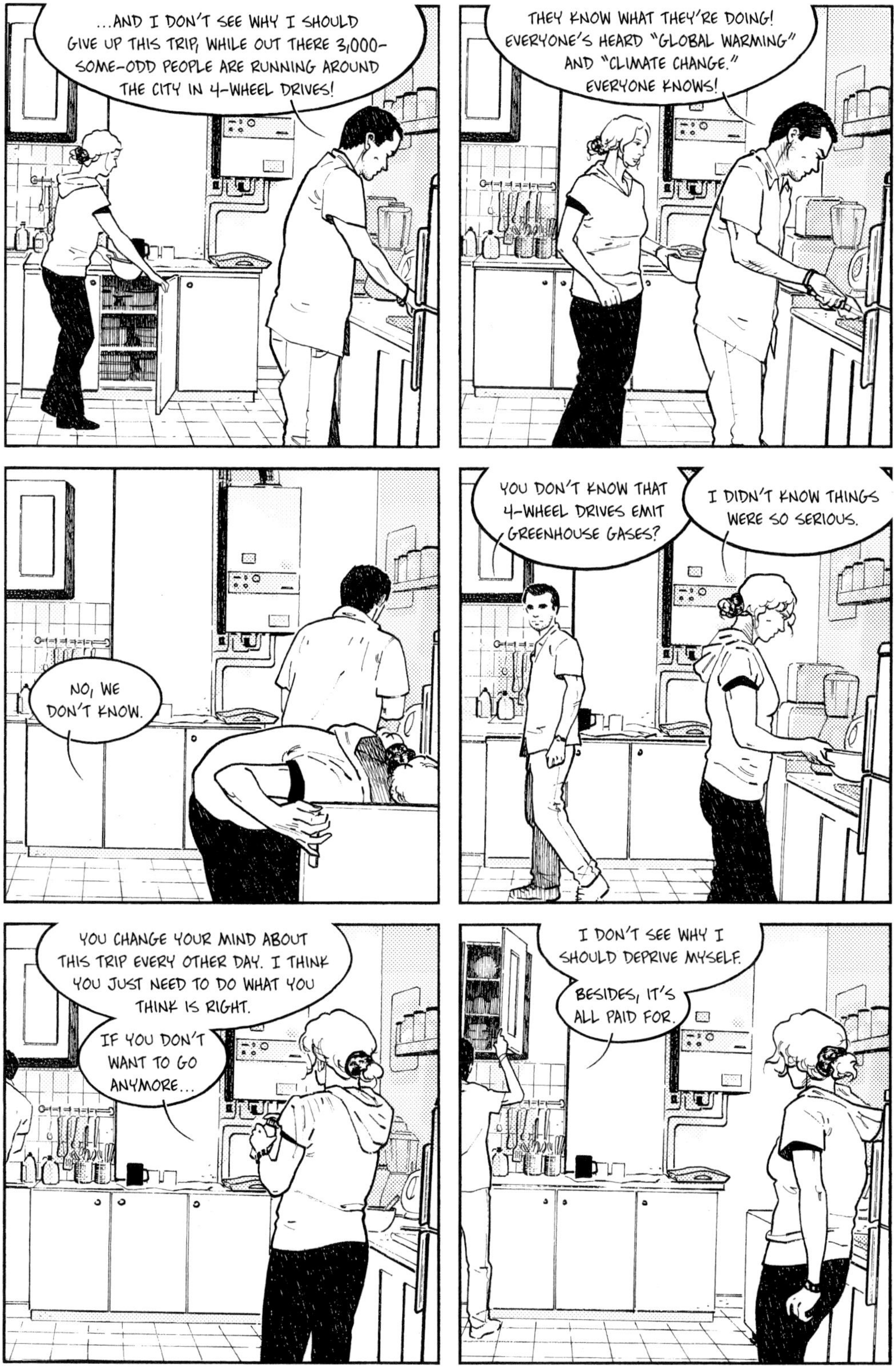
...AND I DON'T SEE WHY I SHOULD GIVE UP THIS TRIP, WHILE OUT THERE 3,000-SOME-ODD PEOPLE ARE RUNNING AROUND THE CITY IN 4-WHEEL DRIVES!
THEY KNOW WHAT THEY'RE DOING! EVERYONE'S HEARD "GLOBAL WARMING" AND "CLIMATE CHANGE." EVERYONE KNOWS!
NO, WE DON'T KNOW.
YOU DON'T KNOW THAT 4-WHEEL DRIVES EMIT GREENHOUSE GASES?
I DIDN'T KNOW THINGS WERE SO SERIOUS.
YOU CHANGE YOUR MIND ABOUT THIS TRIP EVERY OTHER DAY. I THINK YOU JUST NEED TO DO WHAT YOU THINK IS RIGHT.
IF YOU DON'T WANT TO GO ANYMORE...
I DON'T SEE WHY I SHOULD DEPRIVE MYSELF.
BESIDES, IT'S ALL PAID FOR.

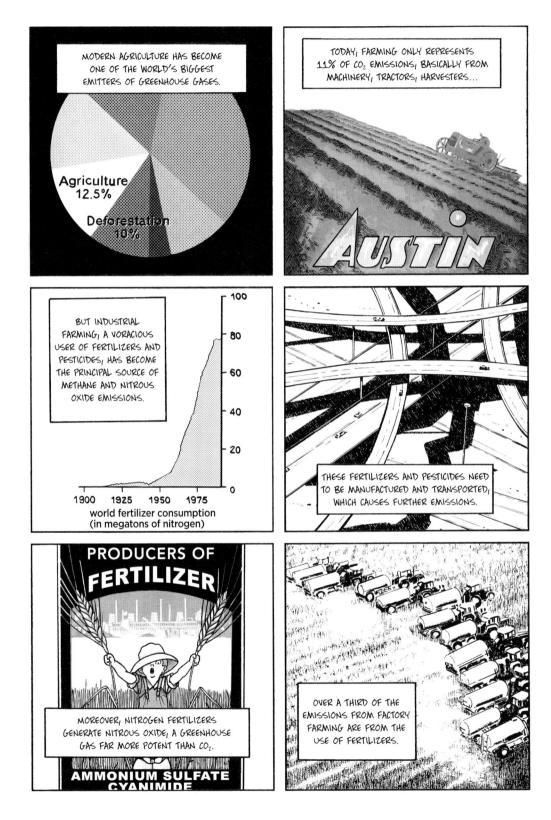

MODERN AGRICULTURE HAS BECOME ONE OF THE WORLD'S BIGGEST EMITTERS OF GREENHOUSE GASES.

Agriculture 12.5%

Deforestation 10%

TODAY, FARMING ONLY REPRESENTS 11% OF CO_2 EMISSIONS, BASICALLY FROM MACHINERY, TRACTORS, HARVESTERS...

AUSTIN

BUT INDUSTRIAL FARMING, A VORACIOUS USER OF FERTILIZERS AND PESTICIDES, HAS BECOME THE PRINCIPAL SOURCE OF METHANE AND NITROUS OXIDE EMISSIONS.

100
80
60
40
20
0

1900   1925   1950   1975

world fertilizer consumption
(in megatons of nitrogen)

THESE FERTILIZERS AND PESTICIDES NEED TO BE MANUFACTURED AND TRANSPORTED, WHICH CAUSES FURTHER EMISSIONS.

PRODUCERS OF
FERTILIZER

MOREOVER, NITROGEN FERTILIZERS GENERATE NITROUS OXIDE, A GREENHOUSE GAS FAR MORE POTENT THAN CO_2.

AMMONIUM SULFATE
CYANIMIDE

OVER A THIRD OF THE EMISSIONS FROM FACTORY FARMING ARE FROM THE USE OF FERTILIZERS.

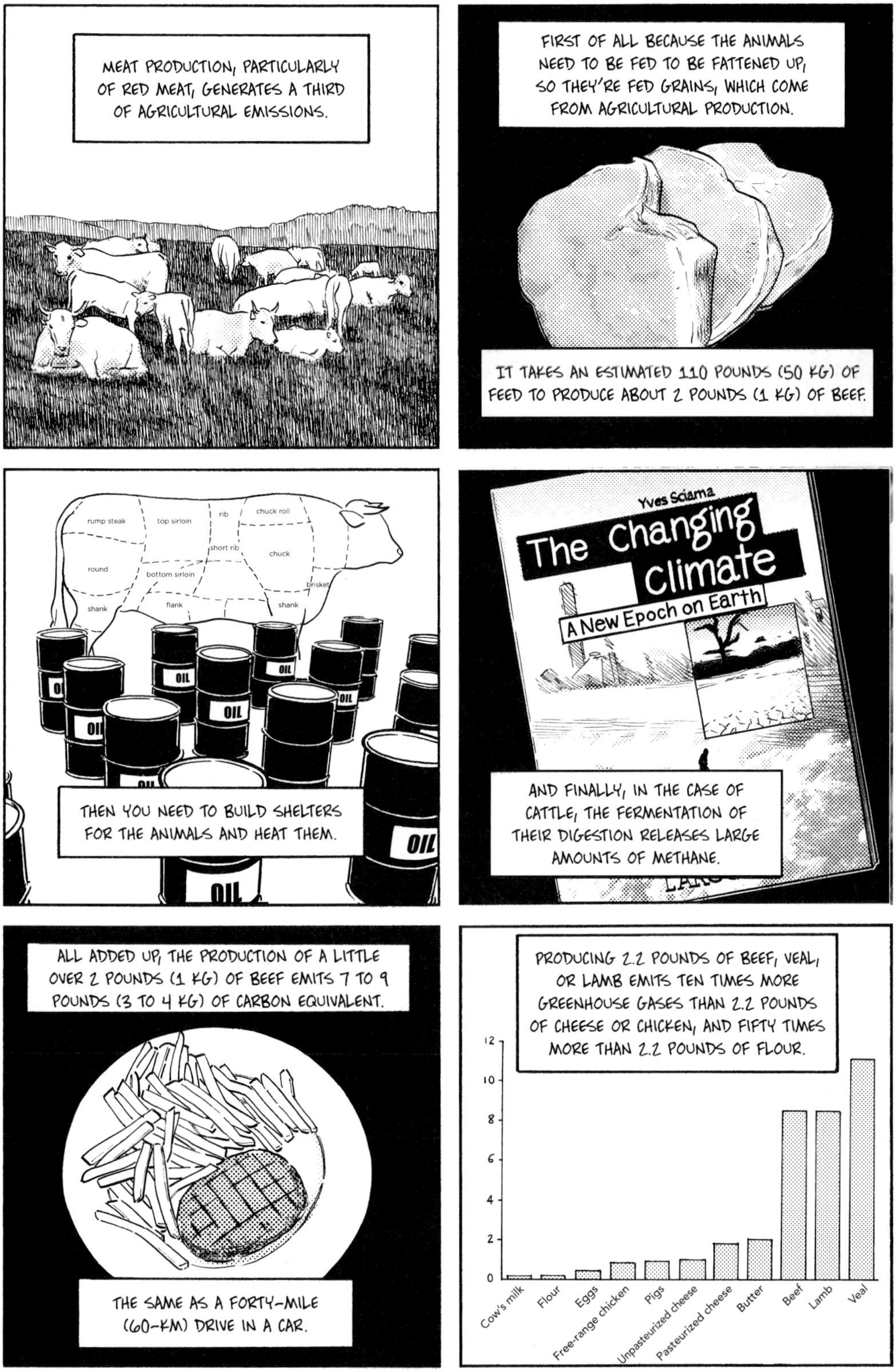
MEAT PRODUCTION, PARTICULARLY OF RED MEAT, GENERATES A THIRD OF AGRICULTURAL EMISSIONS.
FIRST OF ALL BECAUSE THE ANIMALS NEED TO BE FED TO BE FATTENED UP, SO THEY'RE FED GRAINS, WHICH COME FROM AGRICULTURAL PRODUCTION.
IT TAKES AN ESTIMATED 110 POUNDS (50 KG) OF FEED TO PRODUCE ABOUT 2 POUNDS (1 KG) OF BEEF.
rump steak
top sirloin
rib
chuck roll
short rib
chuck
round
bottom sirloin
brisket
shank
flank
shank
OIL
THEN YOU NEED TO BUILD SHELTERS FOR THE ANIMALS AND HEAT THEM.
Yves Sciama
The Changing Climate
A New Epoch on Earth
AND FINALLY, IN THE CASE OF CATTLE, THE FERMENTATION OF THEIR DIGESTION RELEASES LARGE AMOUNTS OF METHANE.
ALL ADDED UP, THE PRODUCTION OF A LITTLE OVER 2 POUNDS (1 KG) OF BEEF EMITS 7 TO 9 POUNDS (3 TO 4 KG) OF CARBON EQUIVALENT.
THE SAME AS A FORTY-MILE (60-KM) DRIVE IN A CAR.
PRODUCING 2.2 POUNDS OF BEEF, VEAL, OR LAMB EMITS TEN TIMES MORE GREENHOUSE GASES THAN 2.2 POUNDS OF CHEESE OR CHICKEN, AND FIFTY TIMES MORE THAN 2.2 POUNDS OF FLOUR.
12
10
8
6
4
2
0
Cow's milk
Flour
Eggs
Free-range chicken
Pigs
Unpasteurized cheese
Pasteurized cheese
Butter
Beef
Lamb
Veal

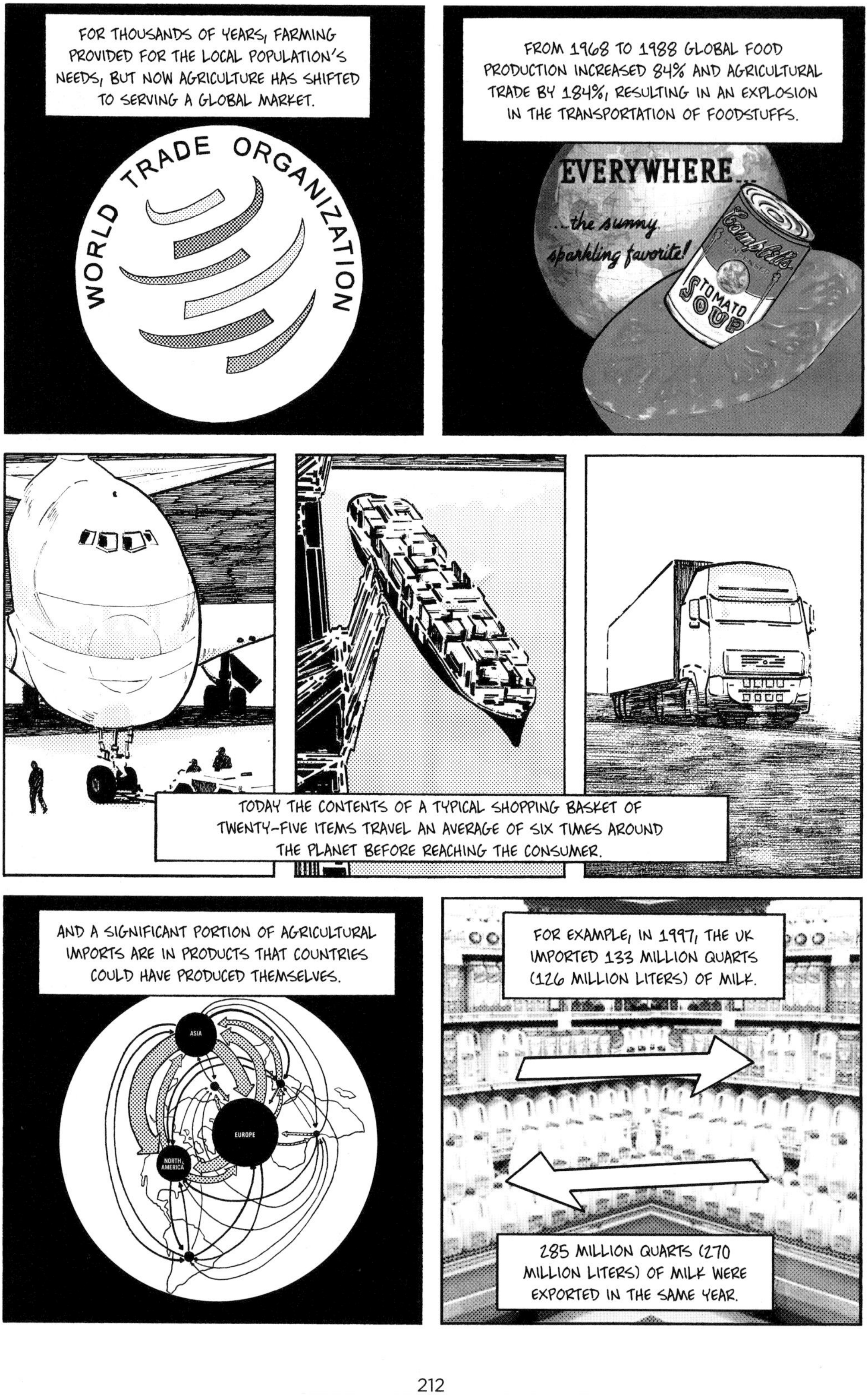

FOR THOUSANDS OF YEARS, FARMING PROVIDED FOR THE LOCAL POPULATION'S NEEDS, BUT NOW AGRICULTURE HAS SHIFTED TO SERVING A GLOBAL MARKET.
WORLD TRADE ORGANIZATION
FROM 1968 TO 1988 GLOBAL FOOD PRODUCTION INCREASED 84% AND AGRICULTURAL TRADE BY 184%, RESULTING IN AN EXPLOSION IN THE TRANSPORTATION OF FOODSTUFFS.
EVERYWHERE...
...the sunny sparkling favorite!
Campbell's
CONDENSED
TOMATO SOUP
TODAY THE CONTENTS OF A TYPICAL SHOPPING BASKET OF TWENTY-FIVE ITEMS TRAVEL AN AVERAGE OF SIX TIMES AROUND THE PLANET BEFORE REACHING THE CONSUMER.
AND A SIGNIFICANT PORTION OF AGRICULTURAL IMPORTS ARE IN PRODUCTS THAT COUNTRIES COULD HAVE PRODUCED THEMSELVES.
ASIA
EUROPE
NORTH AMERICA
FOR EXAMPLE, IN 1997, THE UK IMPORTED 133 MILLION QUARTS (126 MILLION LITERS) OF MILK.
285 MILLION QUARTS (270 MILLION LITERS) OF MILK WERE EXPORTED IN THE SAME YEAR.

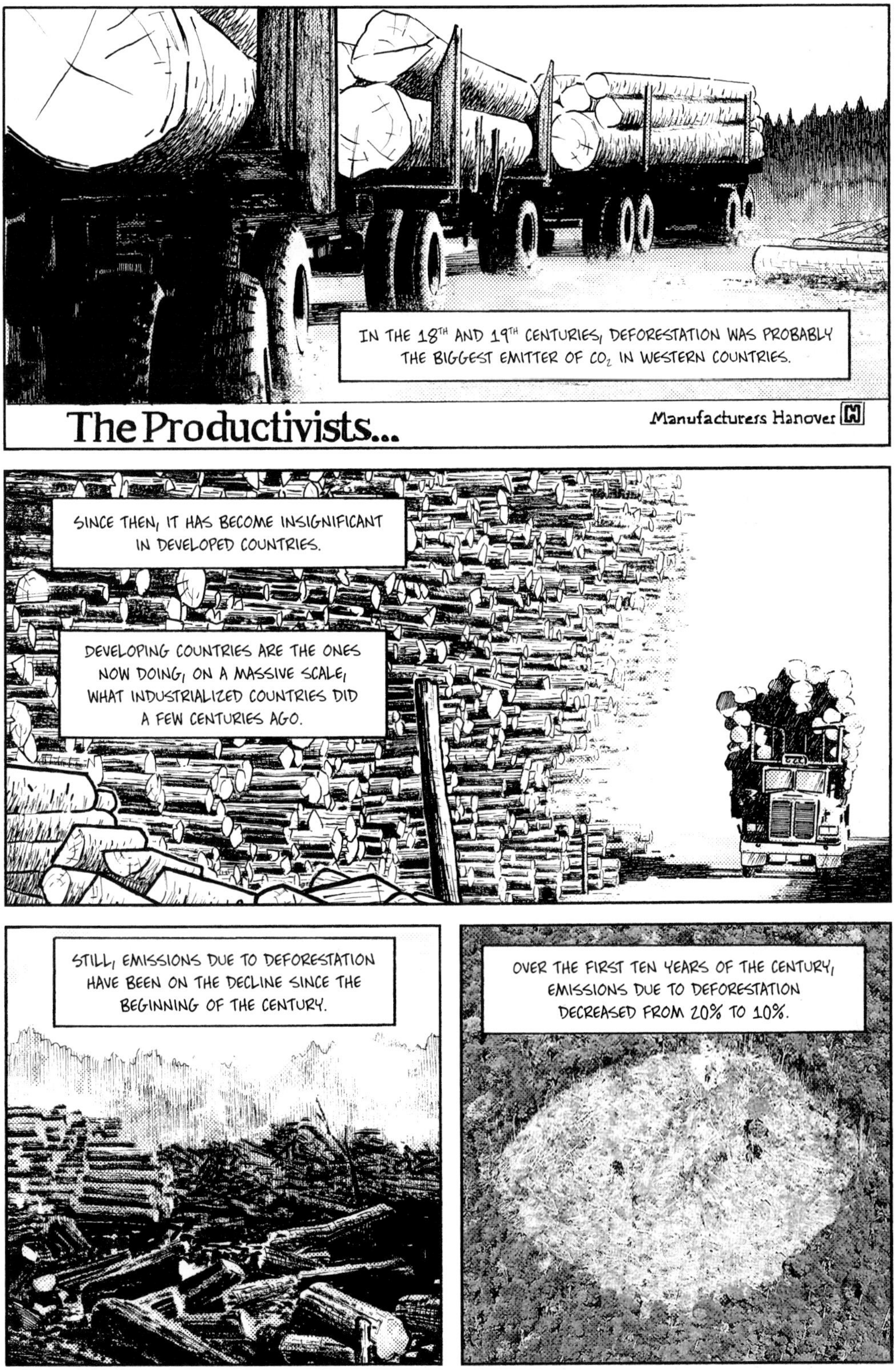
IN THE 18TH AND 19TH CENTURIES, DEFORESTATION WAS PROBABLY THE BIGGEST EMITTER OF CO_2 IN WESTERN COUNTRIES.
The Productivists...
Manufacturers Hanover
SINCE THEN, IT HAS BECOME INSIGNIFICANT IN DEVELOPED COUNTRIES.
DEVELOPING COUNTRIES ARE THE ONES NOW DOING, ON A MASSIVE SCALE, WHAT INDUSTRIALIZED COUNTRIES DID A FEW CENTURIES AGO.
STILL, EMISSIONS DUE TO DEFORESTATION HAVE BEEN ON THE DECLINE SINCE THE BEGINNING OF THE CENTURY.
OVER THE FIRST TEN YEARS OF THE CENTURY, EMISSIONS DUE TO DEFORESTATION DECREASED FROM 20% TO 10%.

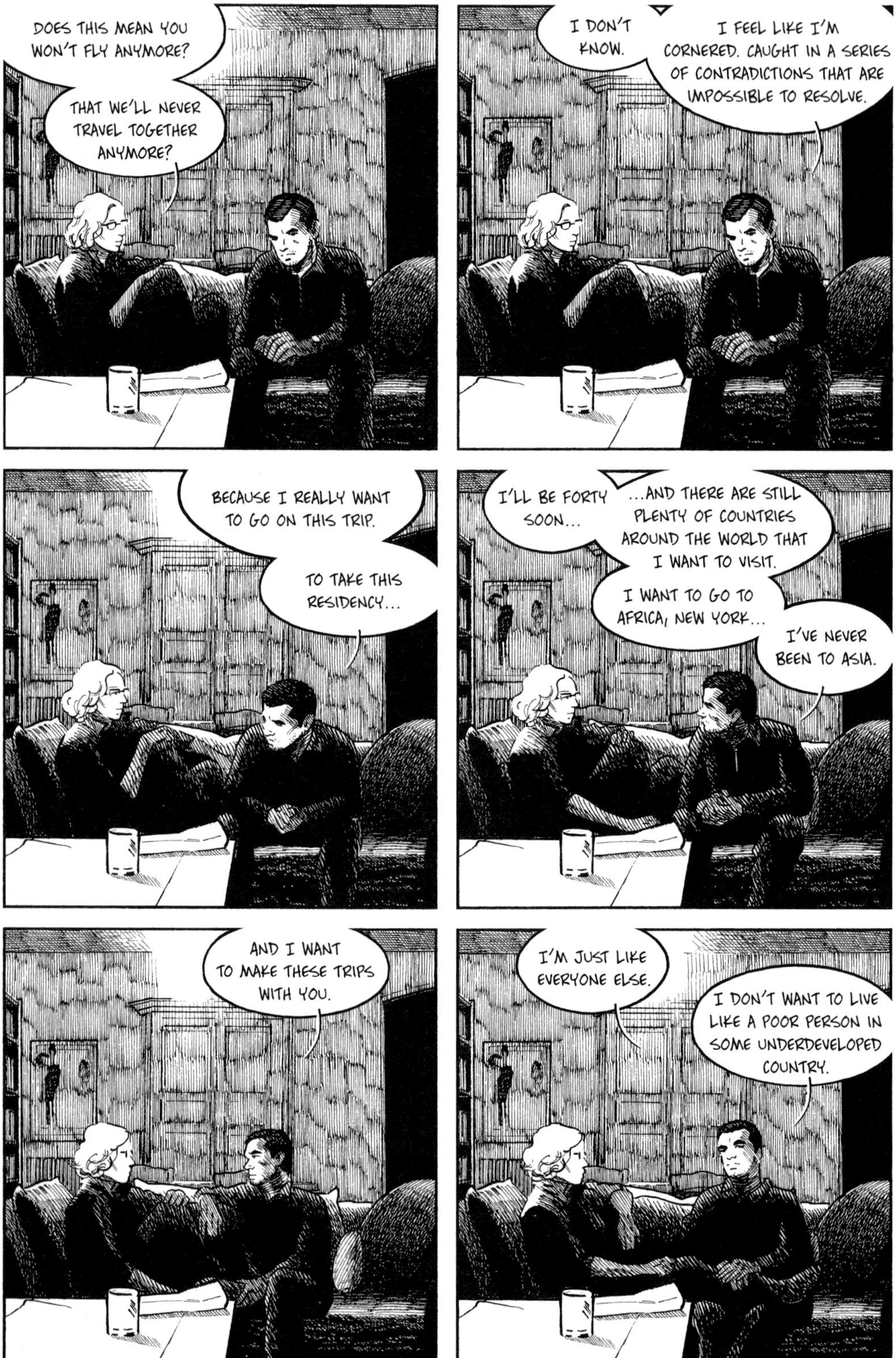
DOES THIS MEAN YOU WON'T FLY ANYMORE?
THAT WE'LL NEVER TRAVEL TOGETHER ANYMORE?
I DON'T KNOW.
I FEEL LIKE I'M CORNERED. CAUGHT IN A SERIES OF CONTRADICTIONS THAT ARE IMPOSSIBLE TO RESOLVE.
BECAUSE I REALLY WANT TO GO ON THIS TRIP.
TO TAKE THIS RESIDENCY...
I'LL BE FORTY SOON...
...AND THERE ARE STILL PLENTY OF COUNTRIES AROUND THE WORLD THAT I WANT TO VISIT.
I WANT TO GO TO AFRICA, NEW YORK...
I'VE NEVER BEEN TO ASIA.
AND I WANT TO MAKE THESE TRIPS WITH YOU.
I'M JUST LIKE EVERYONE ELSE.
I DON'T WANT TO LIVE LIKE A POOR PERSON IN SOME UNDERDEVELOPED COUNTRY.

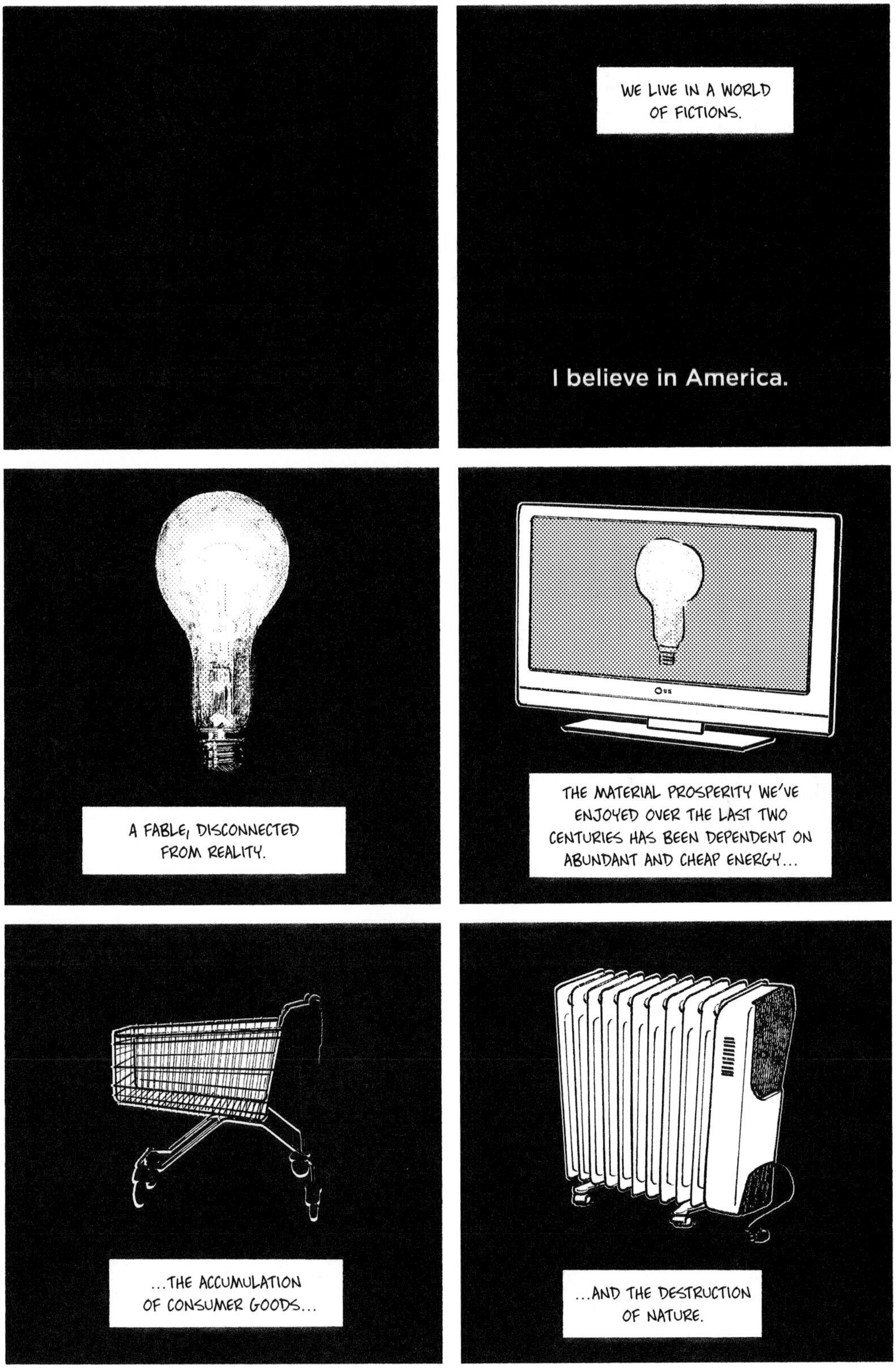
WE LIVE IN A WORLD OF FICTIONS.
I believe in America.
A FABLE, DISCONNECTED FROM REALITY.
THE MATERIAL PROSPERITY WE'VE ENJOYED OVER THE LAST TWO CENTURIES HAS BEEN DEPENDENT ON ABUNDANT AND CHEAP ENERGY...
...THE ACCUMULATION OF CONSUMER GOODS...
...AND THE DESTRUCTION OF NATURE.

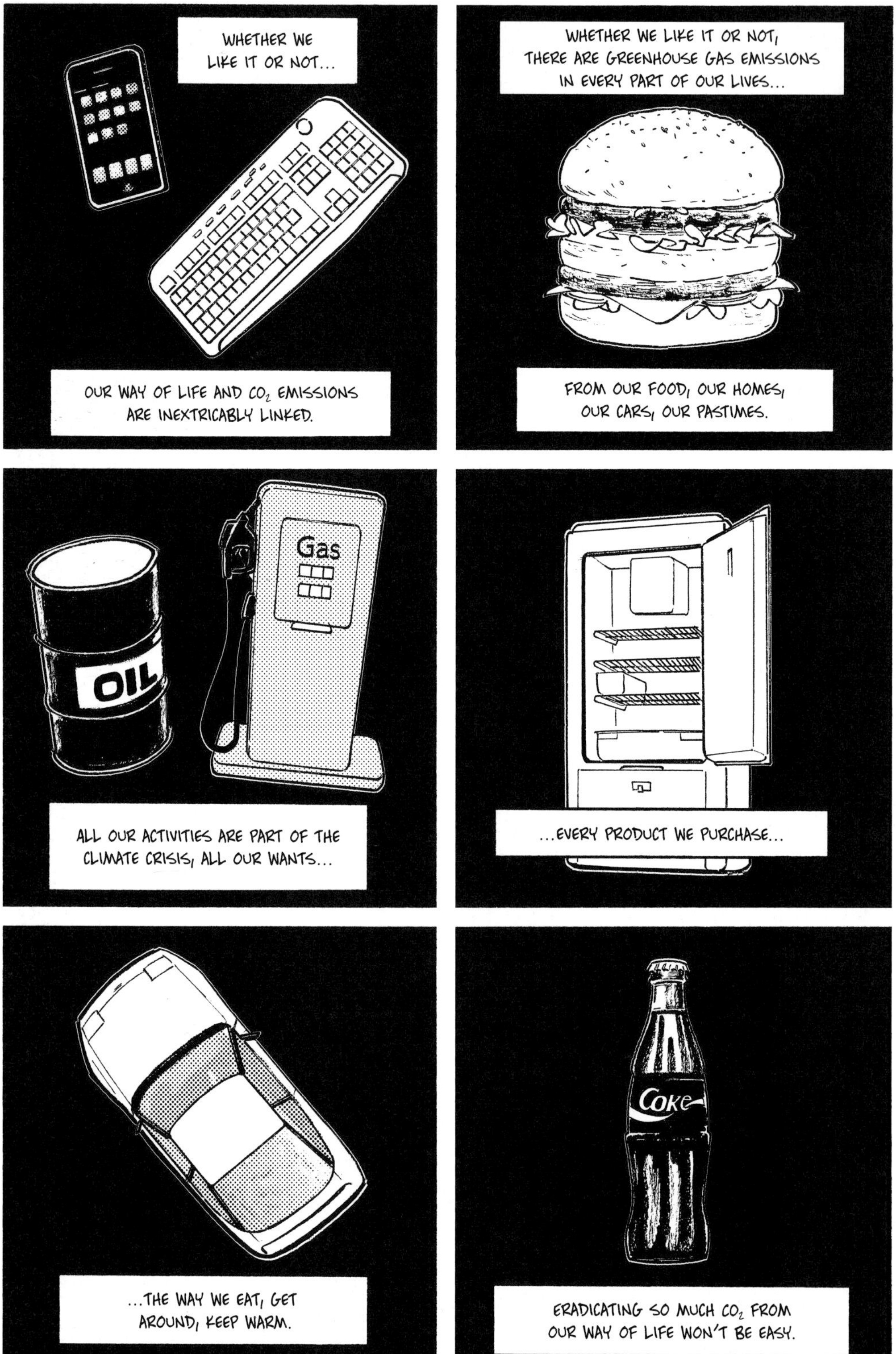
WHETHER WE LIKE IT OR NOT...
OUR WAY OF LIFE AND CO_2 EMISSIONS ARE INEXTRICABLY LINKED.
WHETHER WE LIKE IT OR NOT, THERE ARE GREENHOUSE GAS EMISSIONS IN EVERY PART OF OUR LIVES...
FROM OUR FOOD, OUR HOMES, OUR CARS, OUR PASTIMES.
OIL
Gas
ALL OUR ACTIVITIES ARE PART OF THE CLIMATE CRISIS, ALL OUR WANTS...
...EVERY PRODUCT WE PURCHASE...
...THE WAY WE EAT, GET AROUND, KEEP WARM.
Coke
ERADICATING SO MUCH CO_2 FROM OUR WAY OF LIFE WON'T BE EASY.

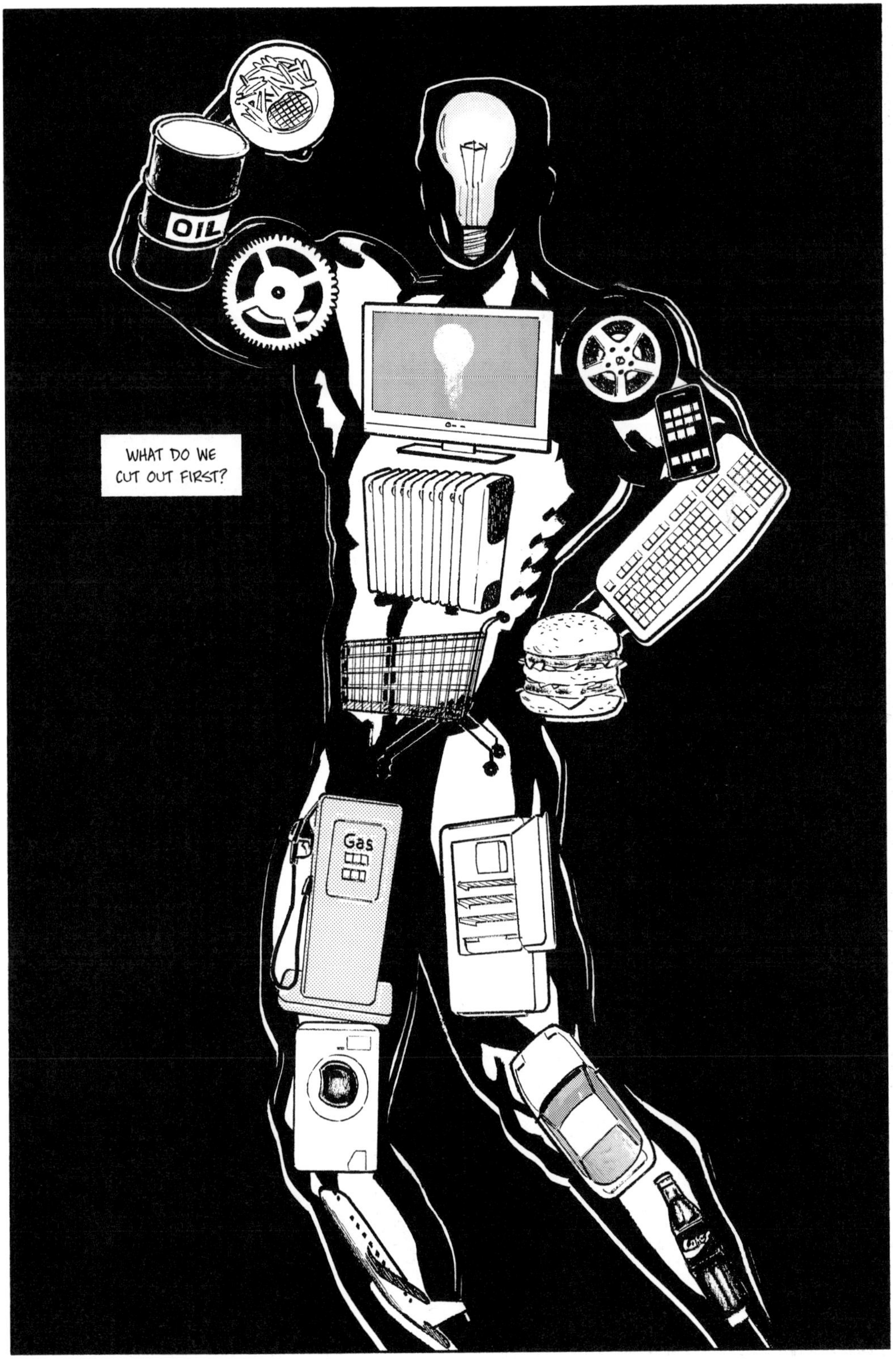
WHAT DO WE CUT OUT FIRST?
OIL
Gas

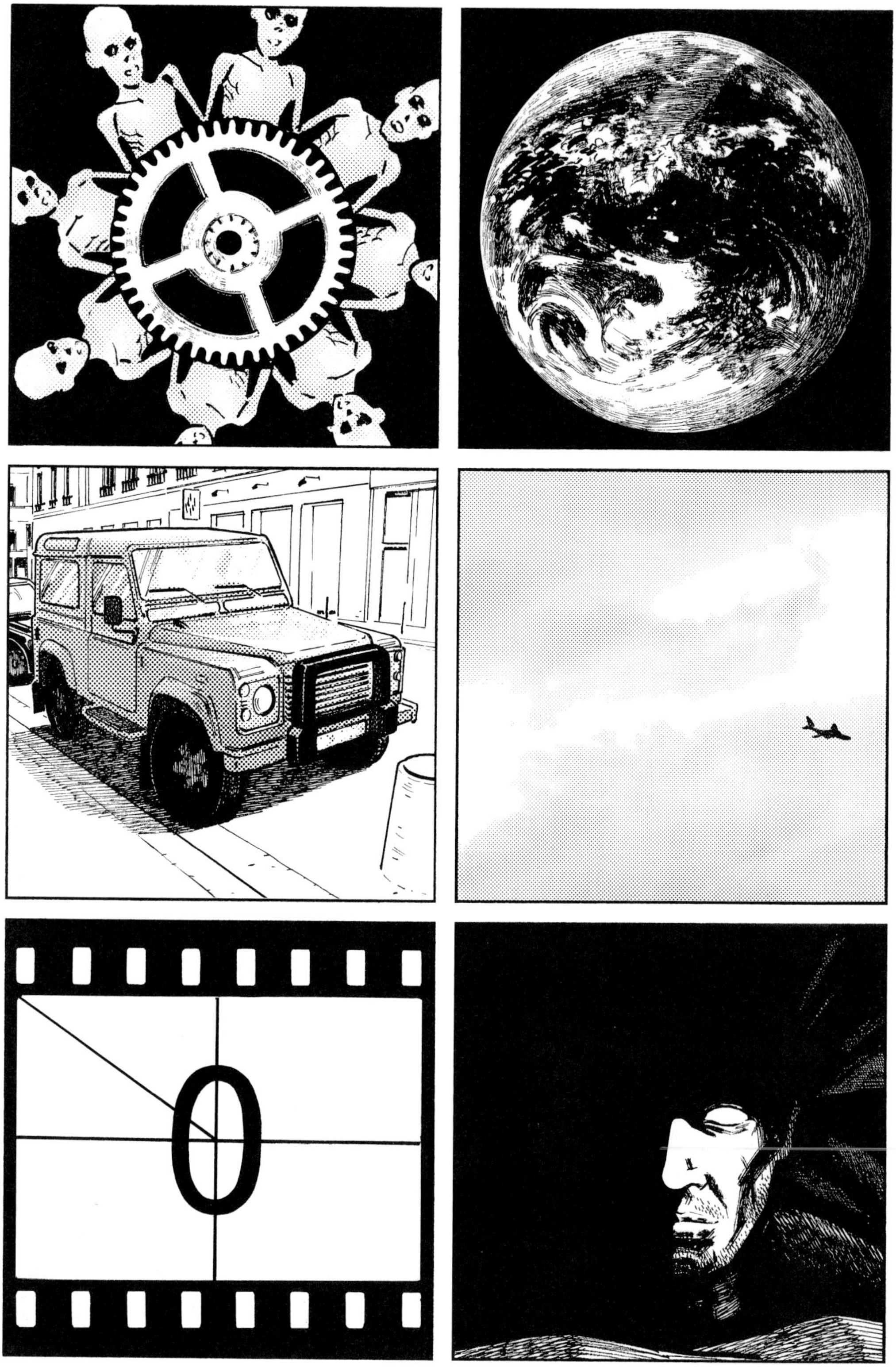
0

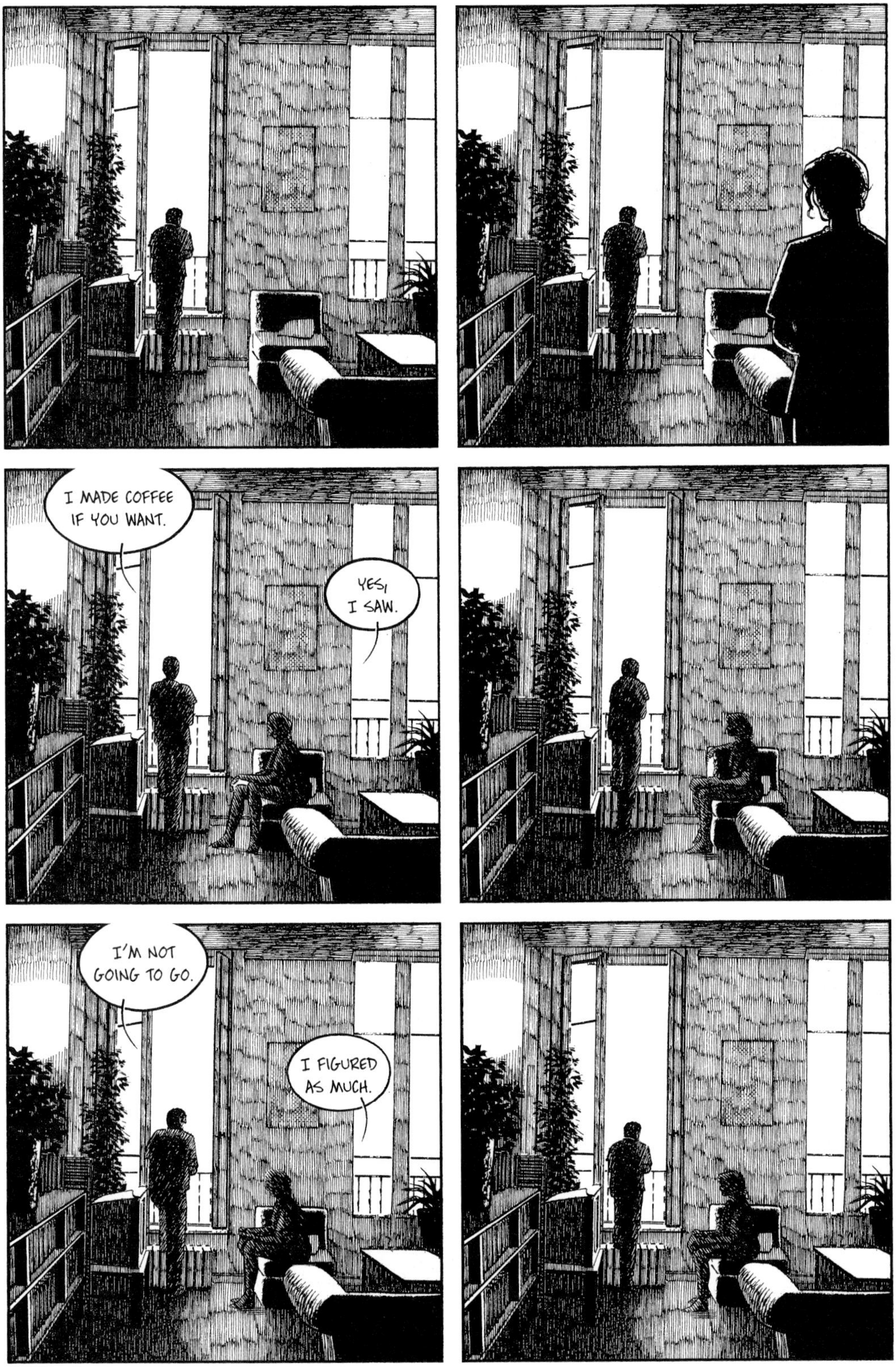
I MADE COFFEE IF YOU WANT.
YES, I SAW.
I'M NOT GOING TO GO.
I FIGURED AS MUCH.

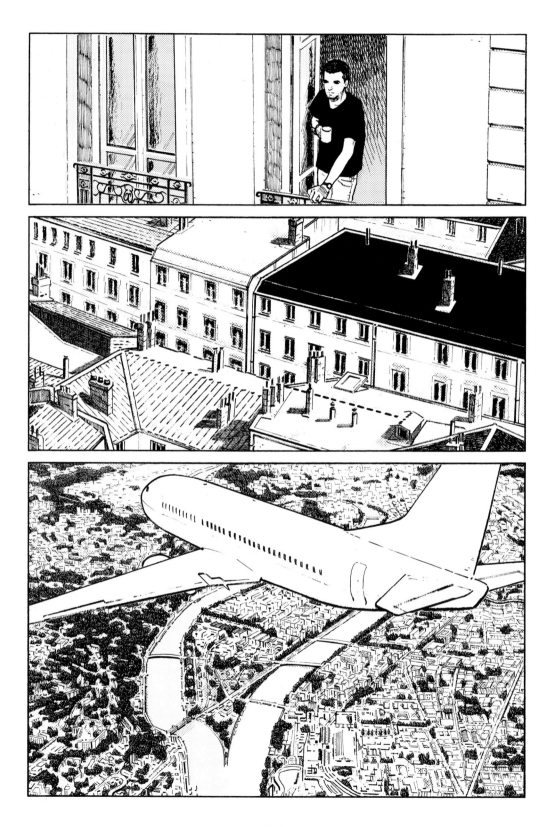

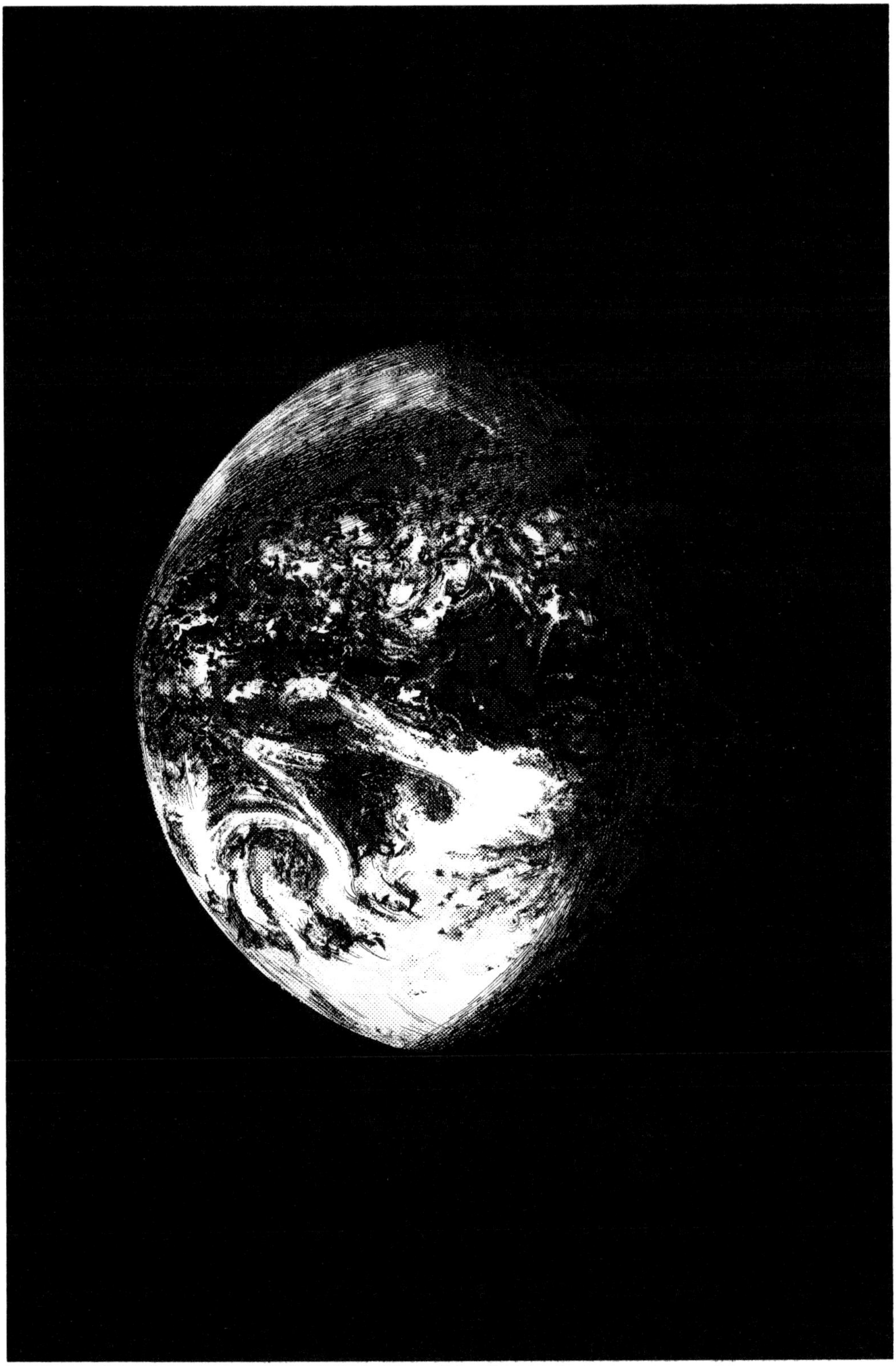

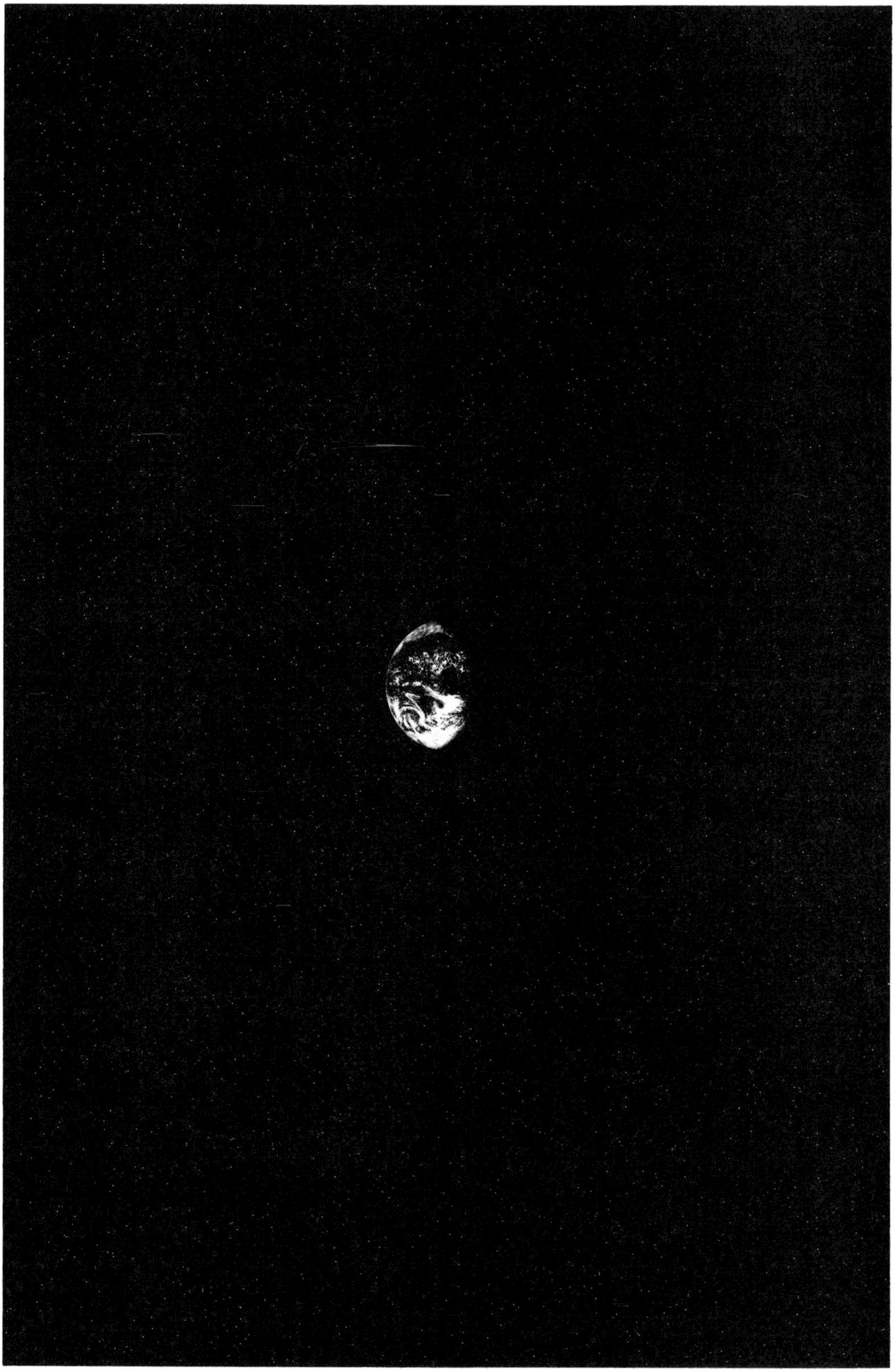

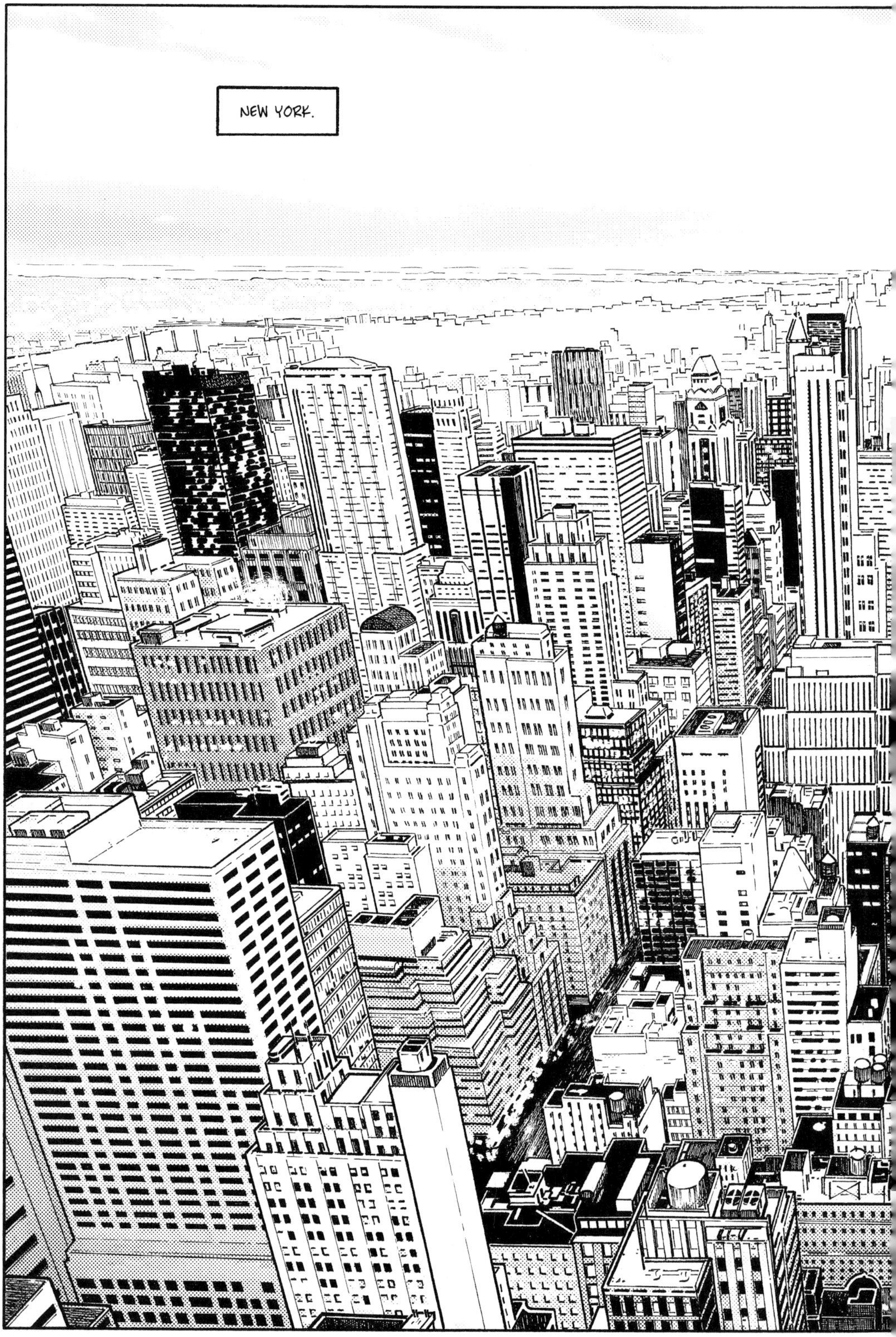
NEW YORK.

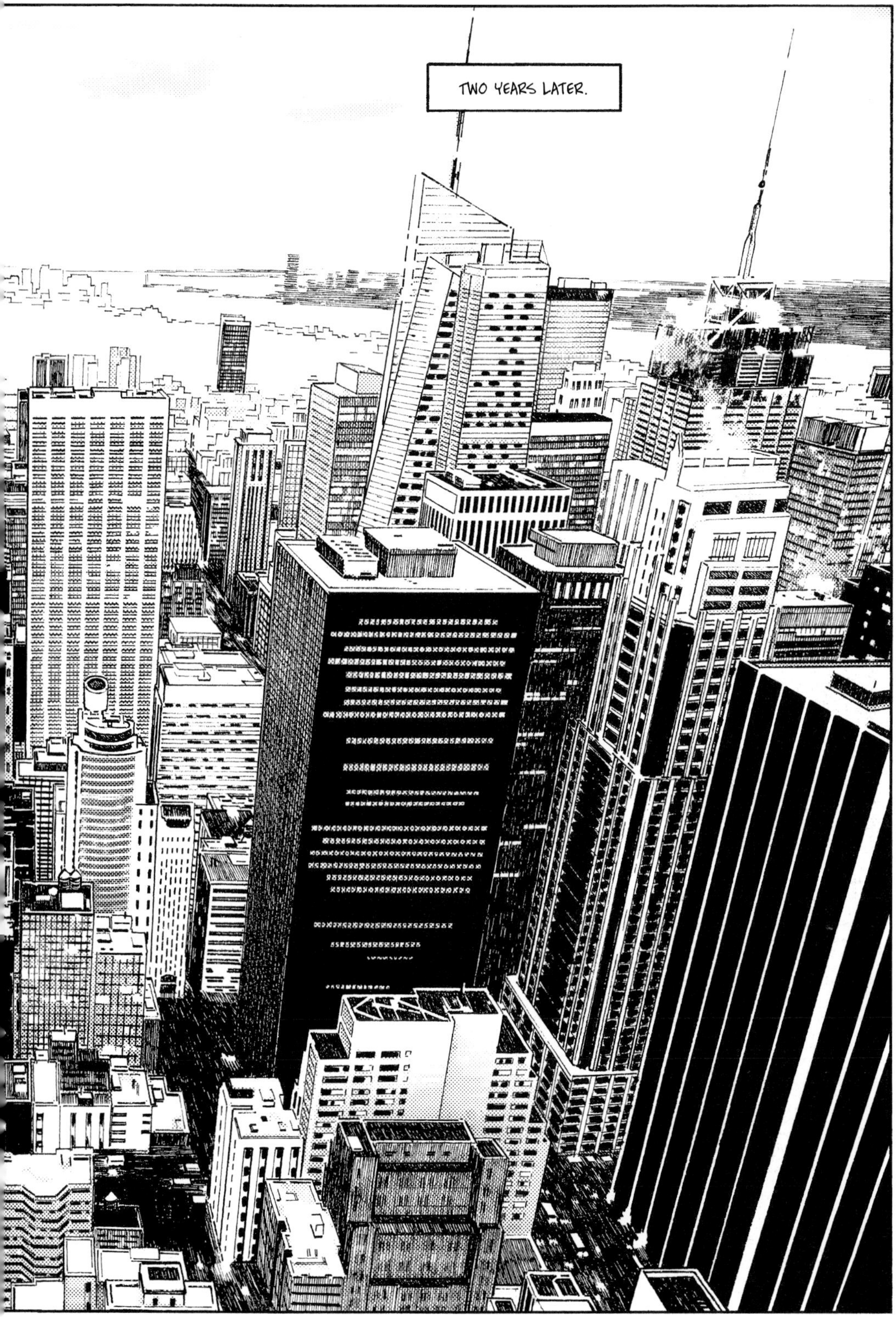
TWO YEARS LATER.

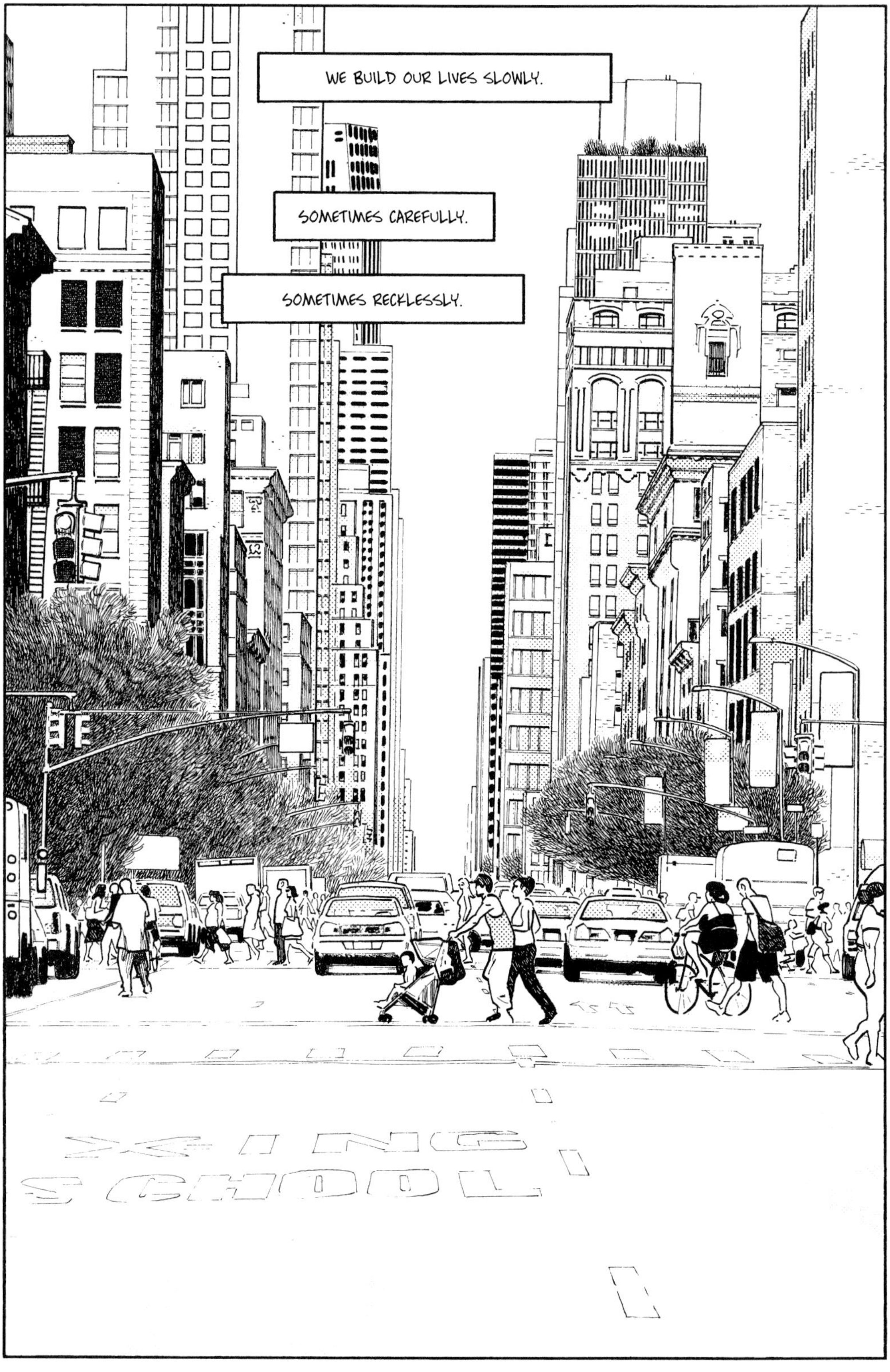
WE BUILD OUR LIVES SLOWLY.
SOMETIMES CAREFULLY.
SOMETIMES RECKLESSLY.
XING
SCHOOL

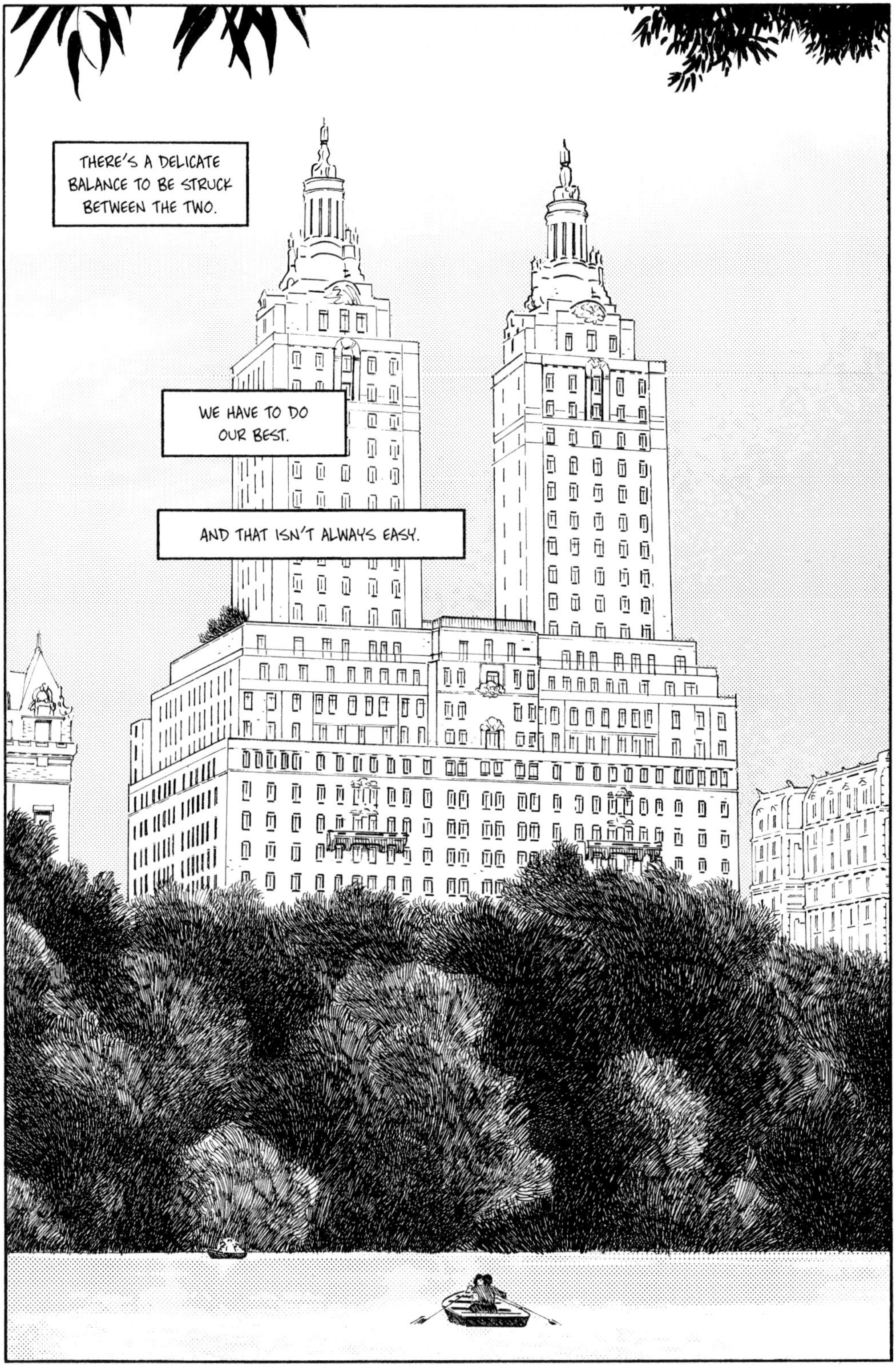
THERE'S A DELICATE BALANCE TO BE STRUCK BETWEEN THE TWO.
WE HAVE TO DO OUR BEST.
AND THAT ISN'T ALWAYS EASY.

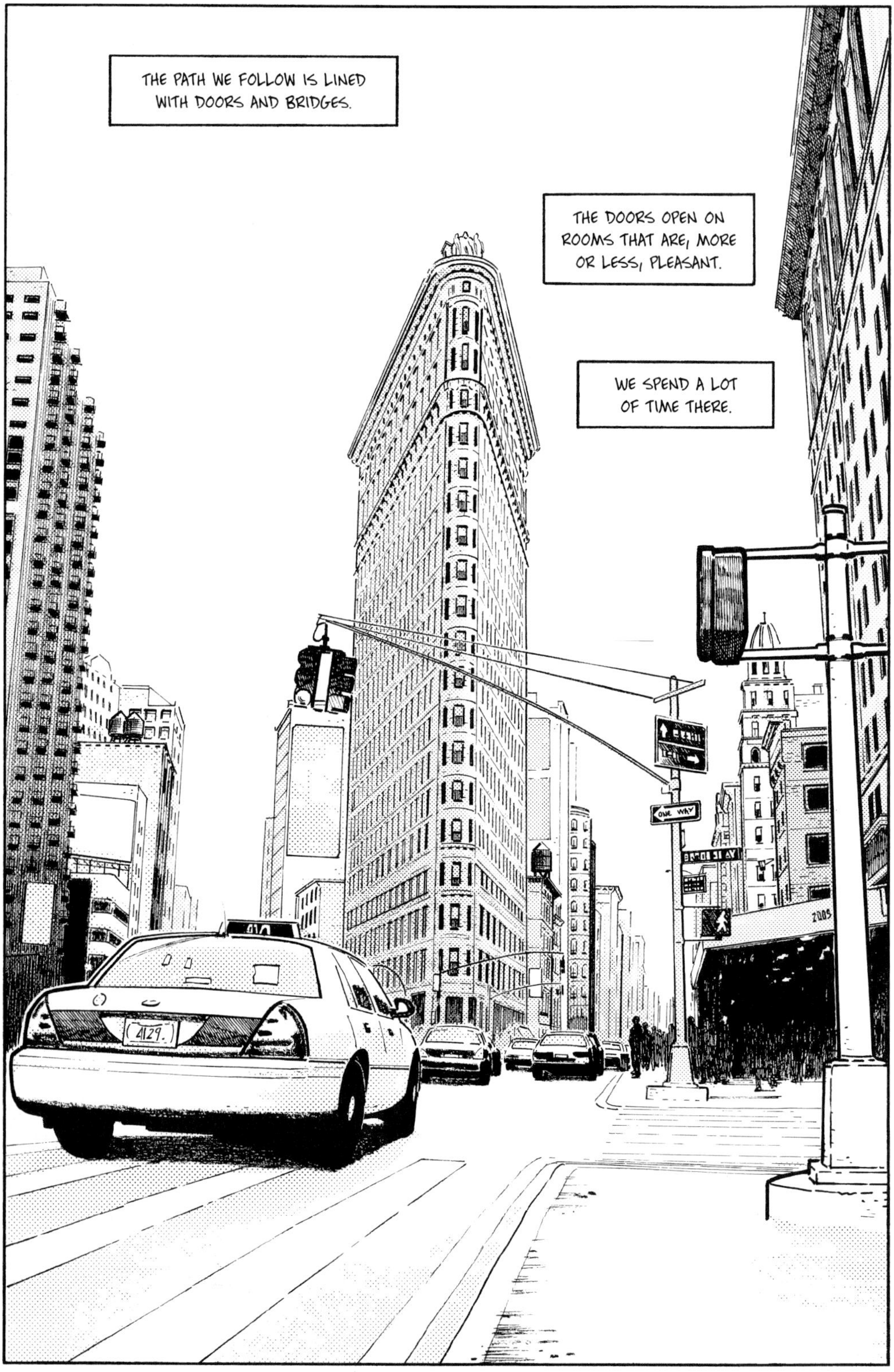
THE PATH WE FOLLOW IS LINED WITH DOORS AND BRIDGES.
THE DOORS OPEN ON ROOMS THAT ARE, MORE OR LESS, PLEASANT.
WE SPEND A LOT OF TIME THERE.
ONE WAY

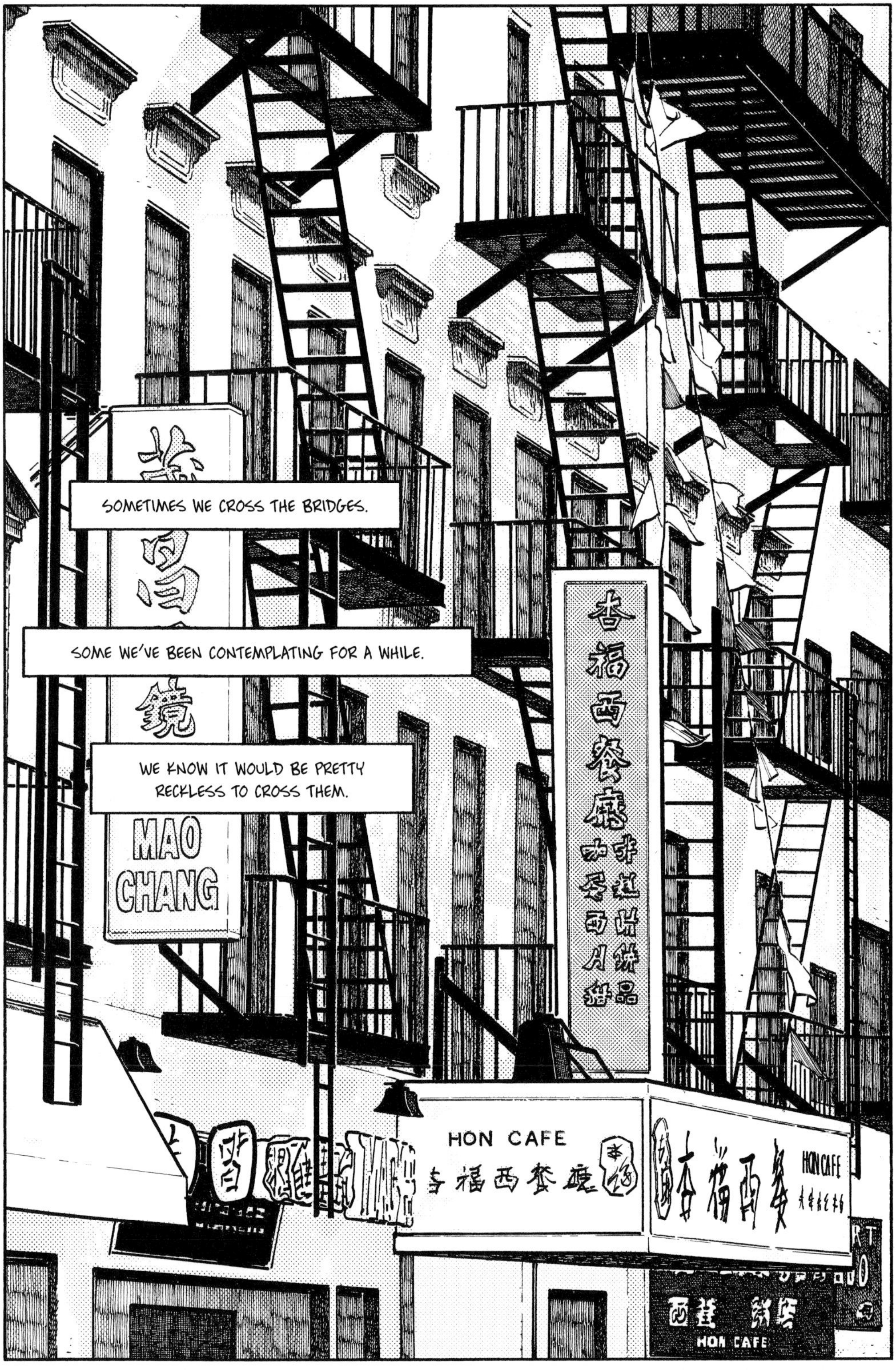
SOMETIMES WE CROSS THE BRIDGES.
SOME WE'VE BEEN CONTEMPLATING FOR A WHILE.
WE KNOW IT WOULD BE PRETTY RECKLESS TO CROSS THEM.
MAO CHANG
HON CAFE
HON CAFE
HON CAFE

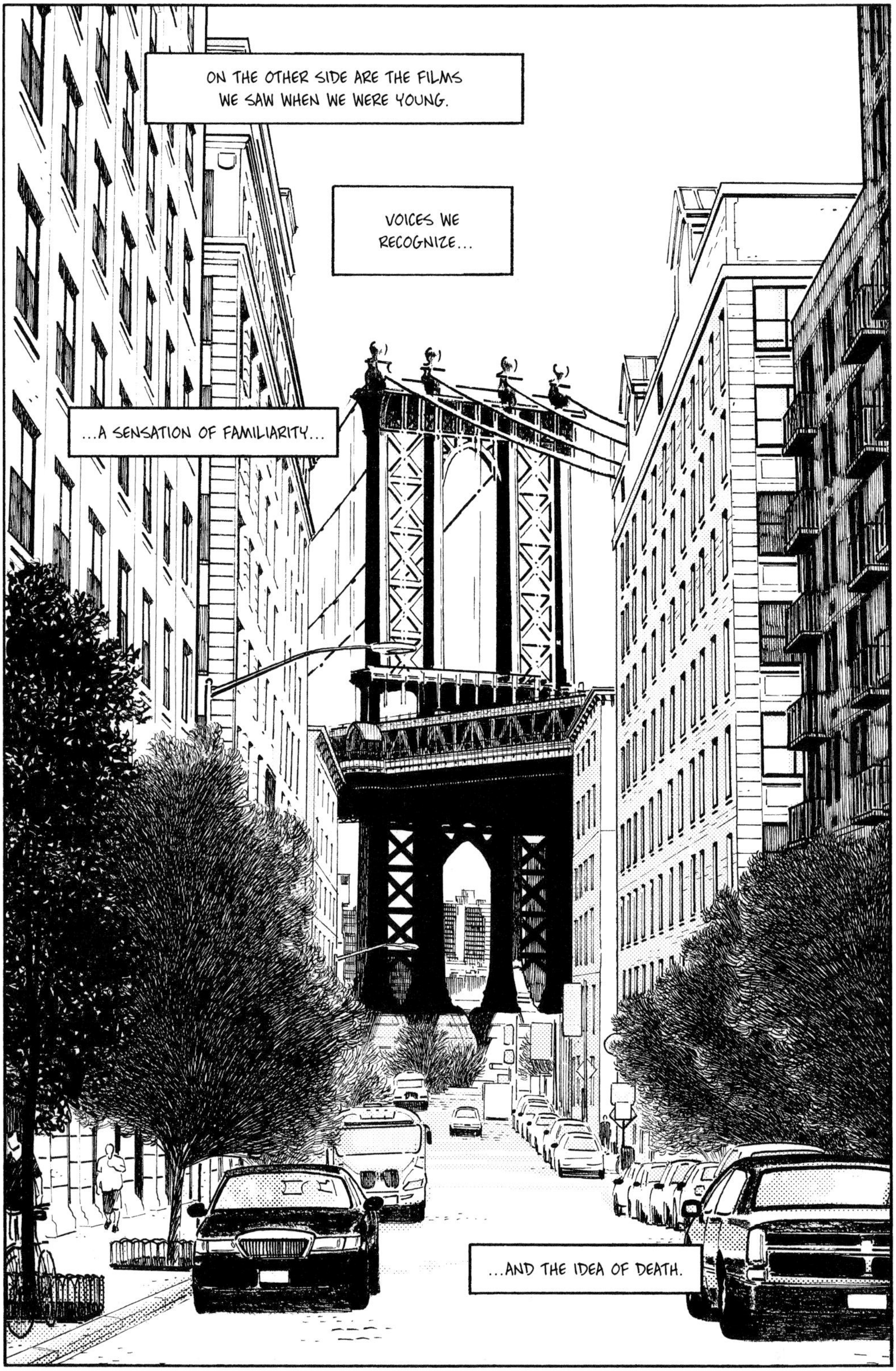
ON THE OTHER SIDE ARE THE FILMS WE SAW WHEN WE WERE YOUNG.
VOICES WE RECOGNIZE...
...A SENSATION OF FAMILIARITY...
...AND THE IDEA OF DEATH.

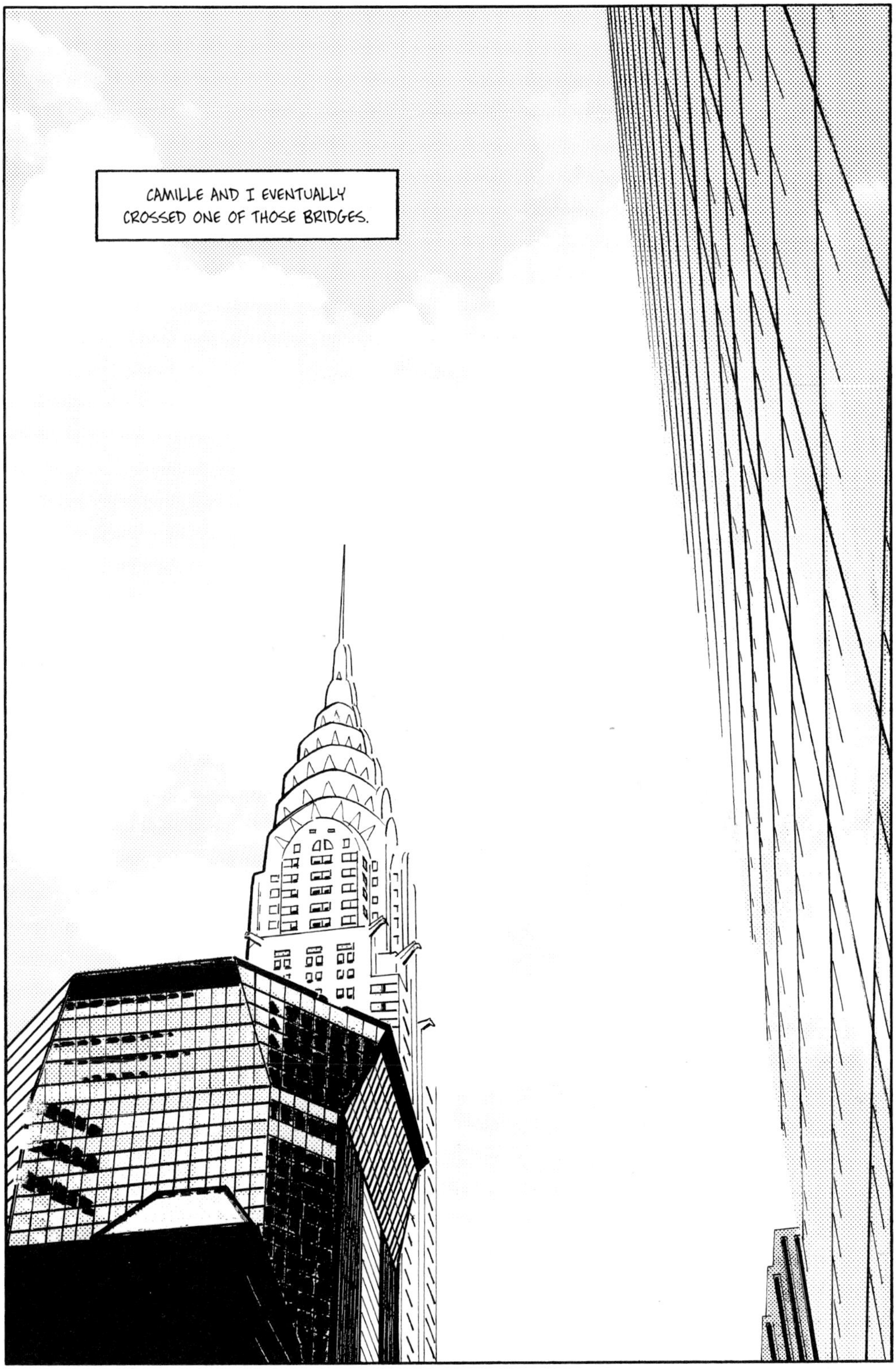
CAMILLE AND I EVENTUALLY CROSSED ONE OF THOSE BRIDGES.

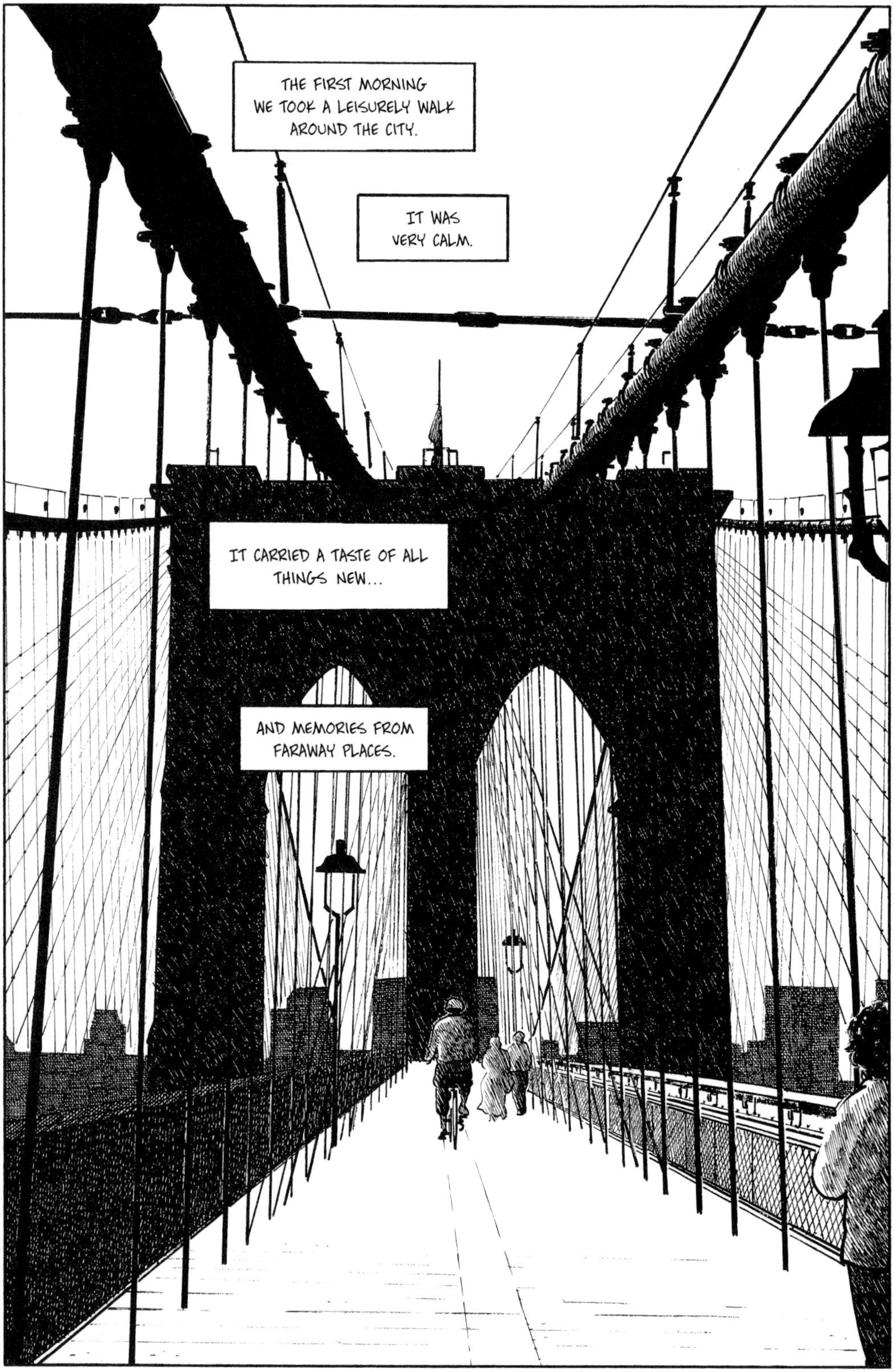
THE FIRST MORNING WE TOOK A LEISURELY WALK AROUND THE CITY.
IT WAS VERY CALM.
IT CARRIED A TASTE OF ALL THINGS NEW...
AND MEMORIES FROM FARAWAY PLACES.

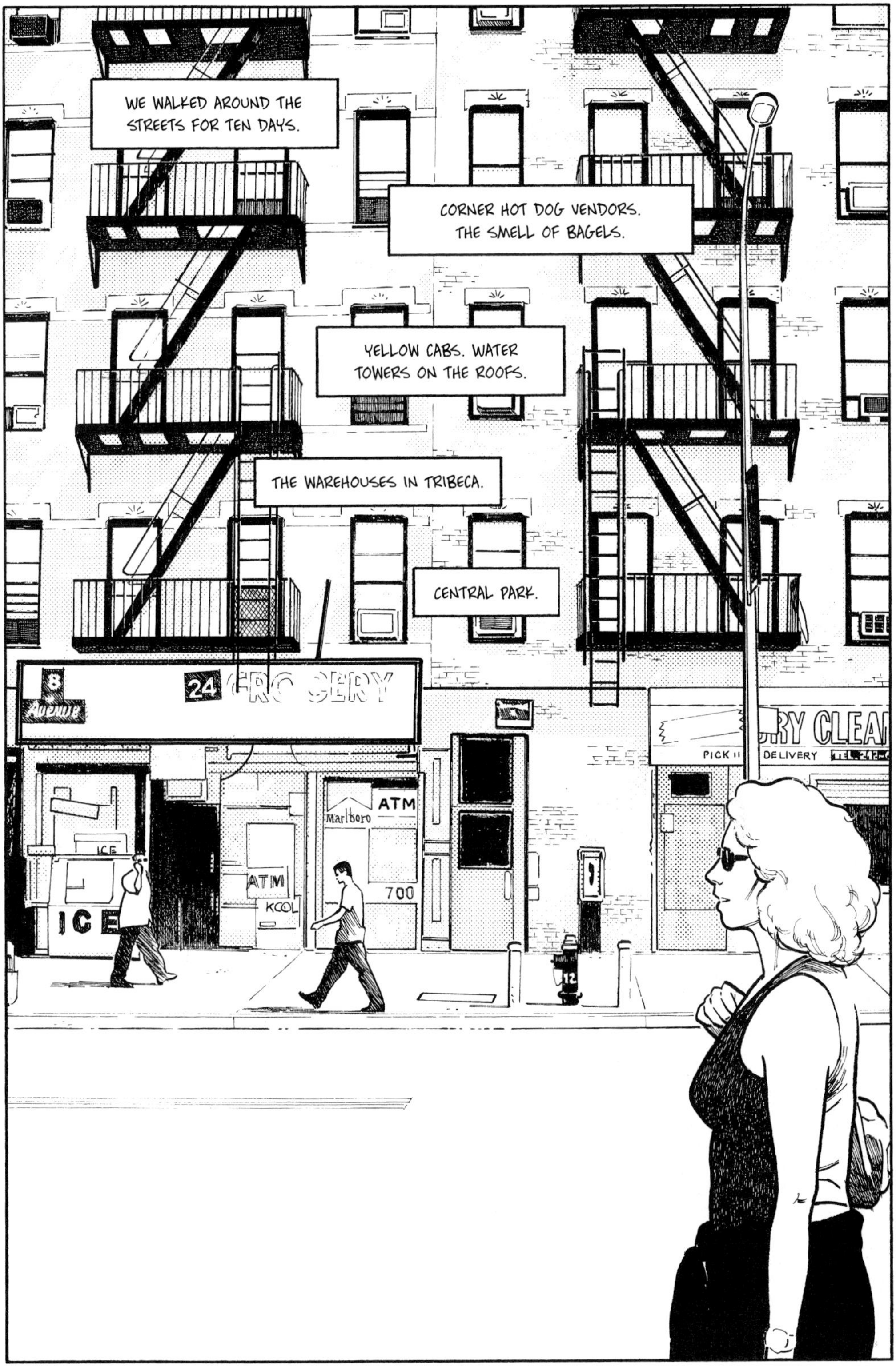
WE WALKED AROUND THE STREETS FOR TEN DAYS.
CORNER HOT DOG VENDORS. THE SMELL OF BAGELS.
YELLOW CABS. WATER TOWERS ON THE ROOFS.
THE WAREHOUSES IN TRIBECA.
CENTRAL PARK.
8 Avenue
24
ATM
Marlboro
ICE
ATM
KOOL
700
PICK
DELIVERY
12

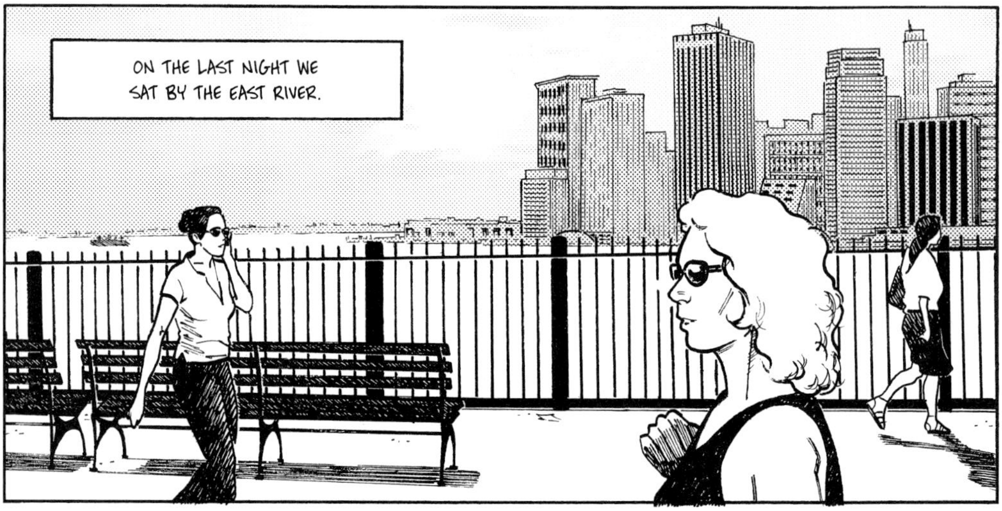
ON THE LAST NIGHT WE SAT BY THE EAST RIVER.

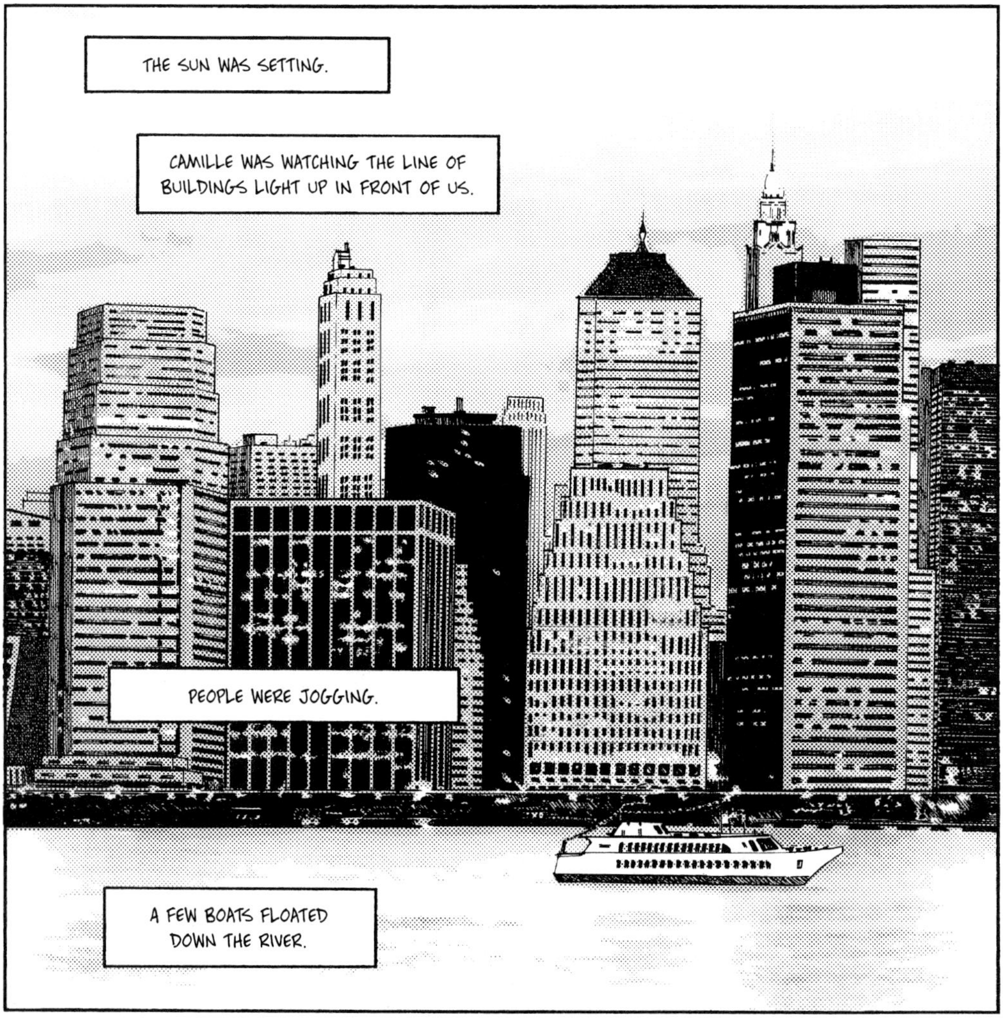
THE SUN WAS SETTING.
CAMILLE WAS WATCHING THE LINE OF BUILDINGS LIGHT UP IN FRONT OF US.
PEOPLE WERE JOGGING.
A FEW BOATS FLOATED DOWN THE RIVER.

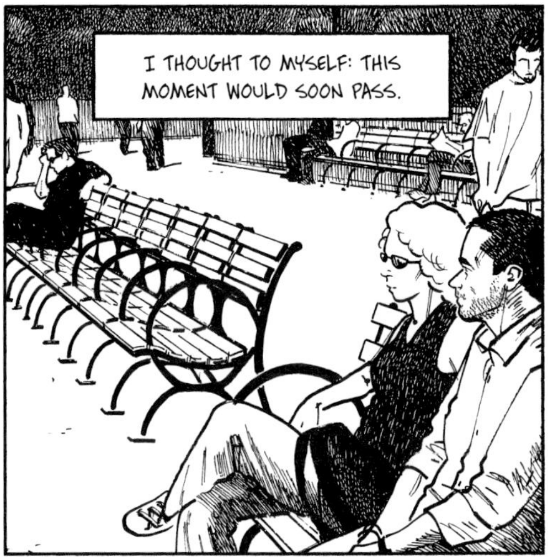
I THOUGHT TO MYSELF: THIS MOMENT WOULD SOON PASS.

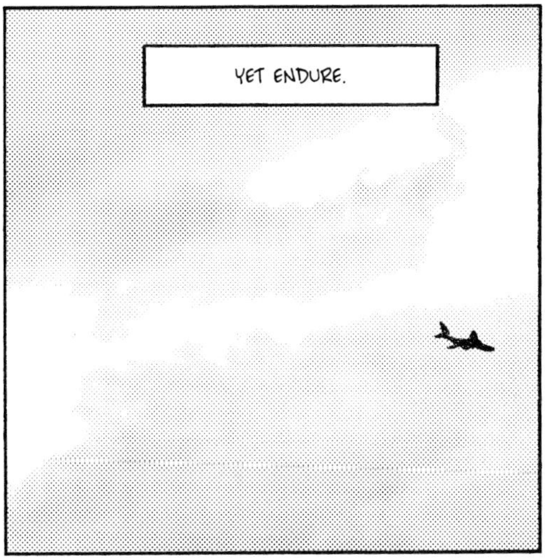
YET ENDURE.

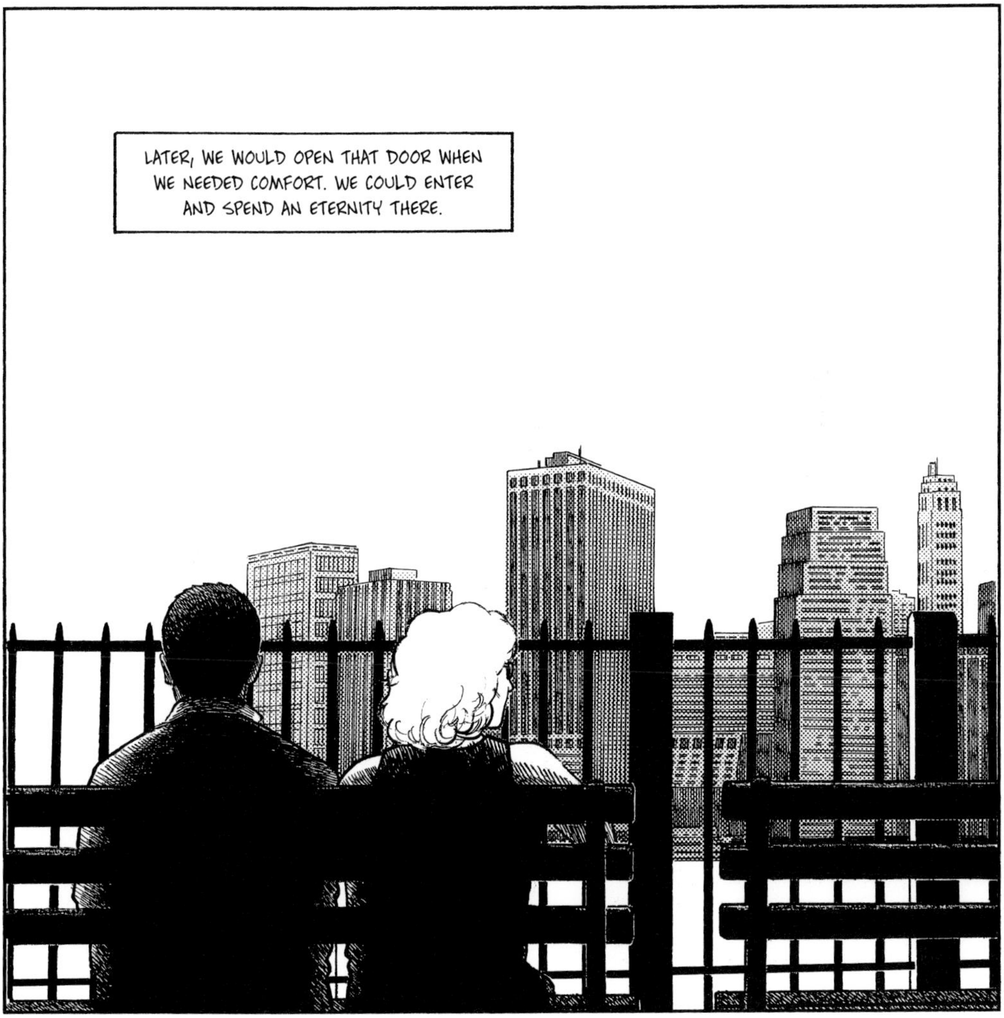
LATER, WE WOULD OPEN THAT DOOR WHEN WE NEEDED COMFORT. WE COULD ENTER AND SPEND AN ETERNITY THERE.

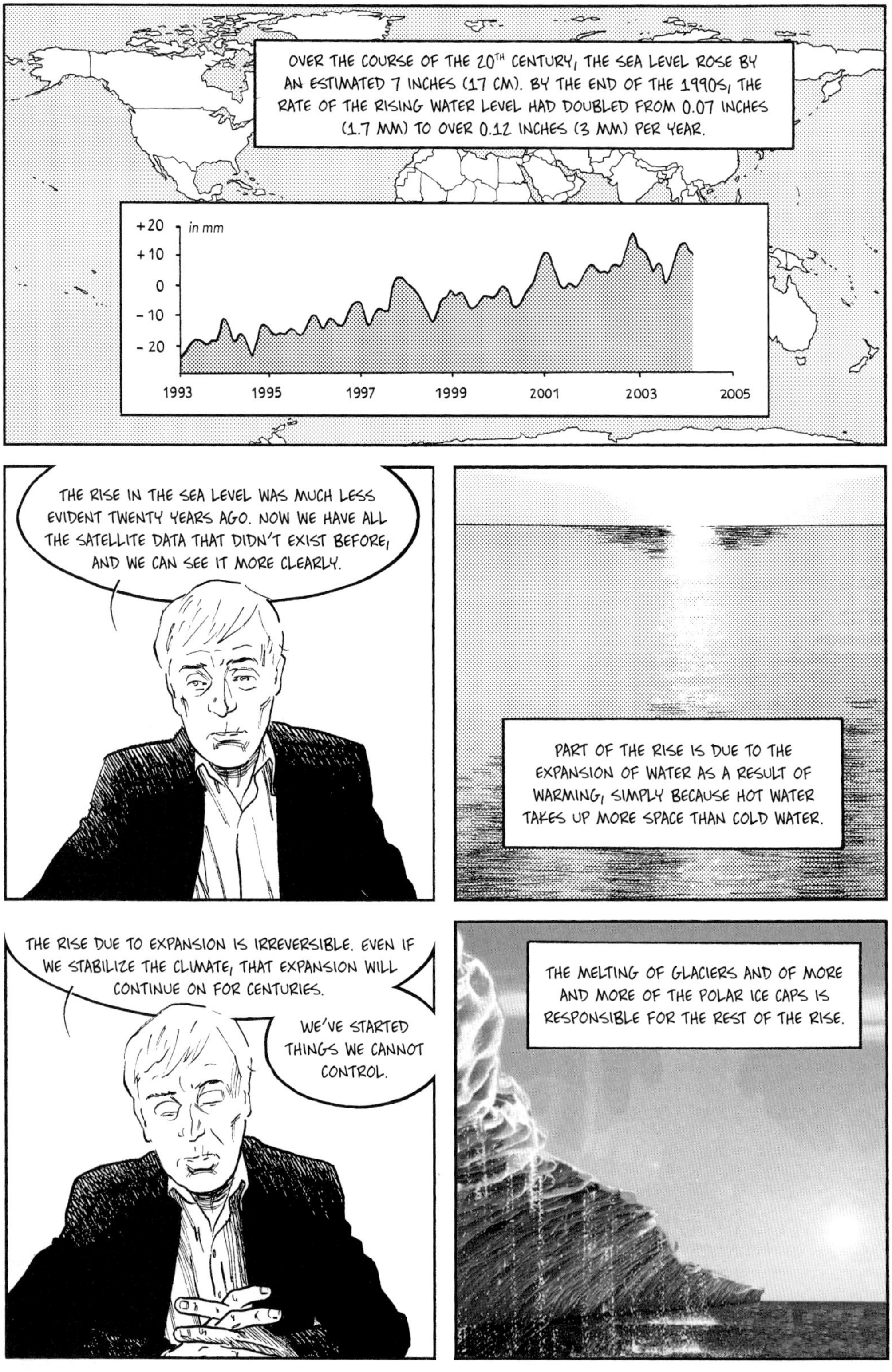
OVER THE COURSE OF THE 20TH CENTURY, THE SEA LEVEL ROSE BY AN ESTIMATED 7 INCHES (17 CM). BY THE END OF THE 1990S, THE RATE OF THE RISING WATER LEVEL HAD DOUBLED FROM 0.07 INCHES (1.7 MM) TO OVER 0.12 INCHES (3 MM) PER YEAR.
+20
+10
0
-10
-20
in mm
1993
1995
1997
1999
2001
2003
2005
THE RISE IN THE SEA LEVEL WAS MUCH LESS EVIDENT TWENTY YEARS AGO. NOW WE HAVE ALL THE SATELLITE DATA THAT DIDN'T EXIST BEFORE, AND WE CAN SEE IT MORE CLEARLY.
PART OF THE RISE IS DUE TO THE EXPANSION OF WATER AS A RESULT OF WARMING, SIMPLY BECAUSE HOT WATER TAKES UP MORE SPACE THAN COLD WATER.
THE RISE DUE TO EXPANSION IS IRREVERSIBLE. EVEN IF WE STABILIZE THE CLIMATE, THAT EXPANSION WILL CONTINUE ON FOR CENTURIES.
WE'VE STARTED THINGS WE CANNOT CONTROL.
THE MELTING OF GLACIERS AND OF MORE AND MORE OF THE POLAR ICE CAPS IS RESPONSIBLE FOR THE REST OF THE RISE.

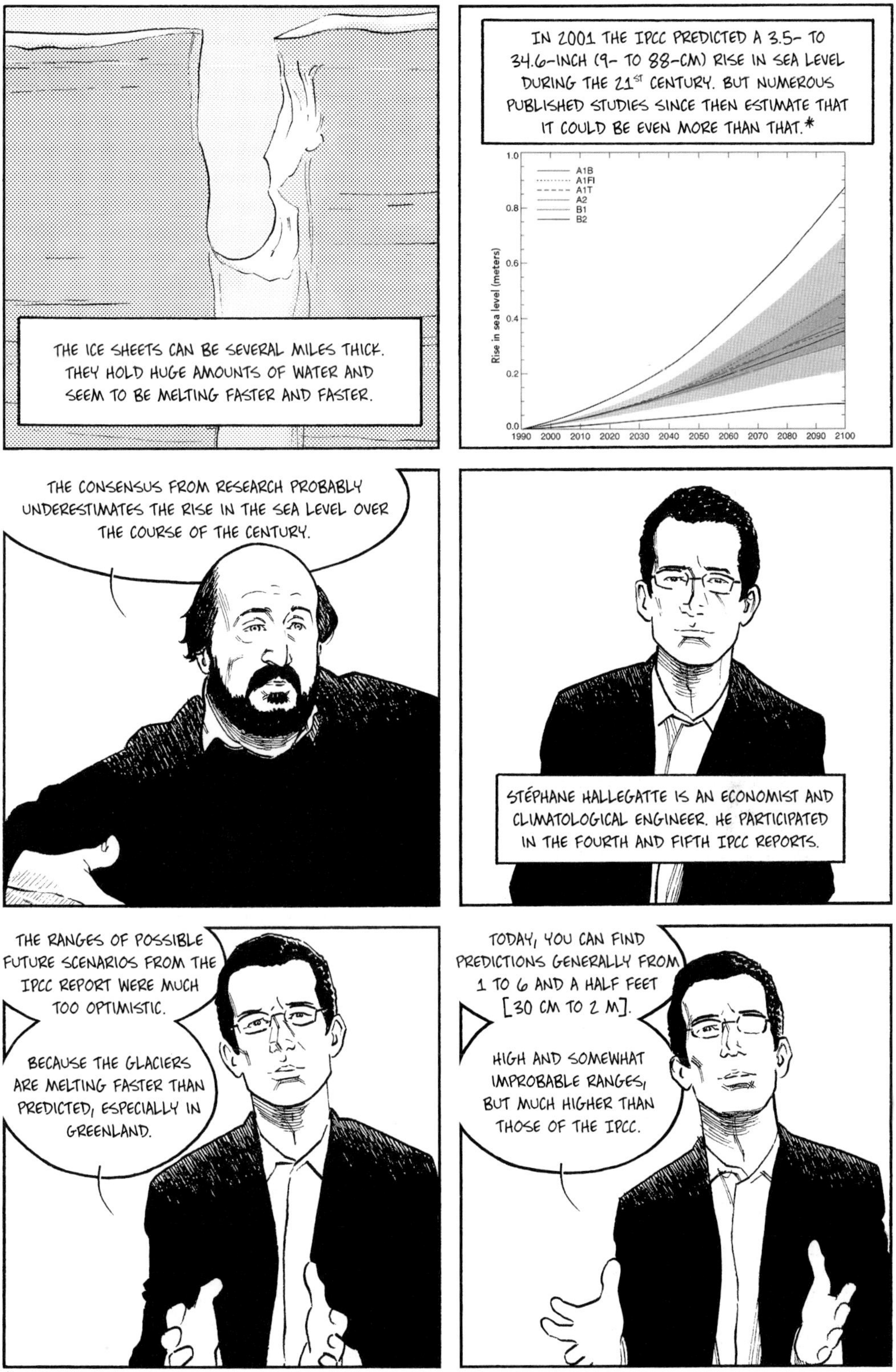

THE ICE SHEETS CAN BE SEVERAL MILES THICK. THEY HOLD HUGE AMOUNTS OF WATER AND SEEM TO BE MELTING FASTER AND FASTER.
IN 2001 THE IPCC PREDICTED A 3.5- TO 34.6-INCH (9- TO 88-CM) RISE IN SEA LEVEL DURING THE 21ST CENTURY. BUT NUMEROUS PUBLISHED STUDIES SINCE THEN ESTIMATE THAT IT COULD BE EVEN MORE THAN THAT.*
A1B
A1FI
A1T
A2
B1
B2
Rise in sea level (meters)
1.0
0.8
0.6
0.4
0.2
0.0
1990 2000 2010 2020 2030 2040 2050 2060 2070 2080 2090 2100
THE CONSENSUS FROM RESEARCH PROBABLY UNDERESTIMATES THE RISE IN THE SEA LEVEL OVER THE COURSE OF THE CENTURY.
STÉPHANE HALLEGATTE IS AN ECONOMIST AND CLIMATOLOGICAL ENGINEER. HE PARTICIPATED IN THE FOURTH AND FIFTH IPCC REPORTS.
THE RANGES OF POSSIBLE FUTURE SCENARIOS FROM THE IPCC REPORT WERE MUCH TOO OPTIMISTIC.
BECAUSE THE GLACIERS ARE MELTING FASTER THAN PREDICTED, ESPECIALLY IN GREENLAND.
TODAY, YOU CAN FIND PREDICTIONS GENERALLY FROM 1 TO 6 AND A HALF FEET [30 CM TO 2 M].
HIGH AND SOMEWHAT IMPROBABLE RANGES, BUT MUCH HIGHER THAN THOSE OF THE IPCC.

*IN 2013 THE IPCC REFINED THEIR PREDICTION TO A 10.2- TO 32.3-INCH (26- TO 82-CM) SEA-LEVEL RISE.

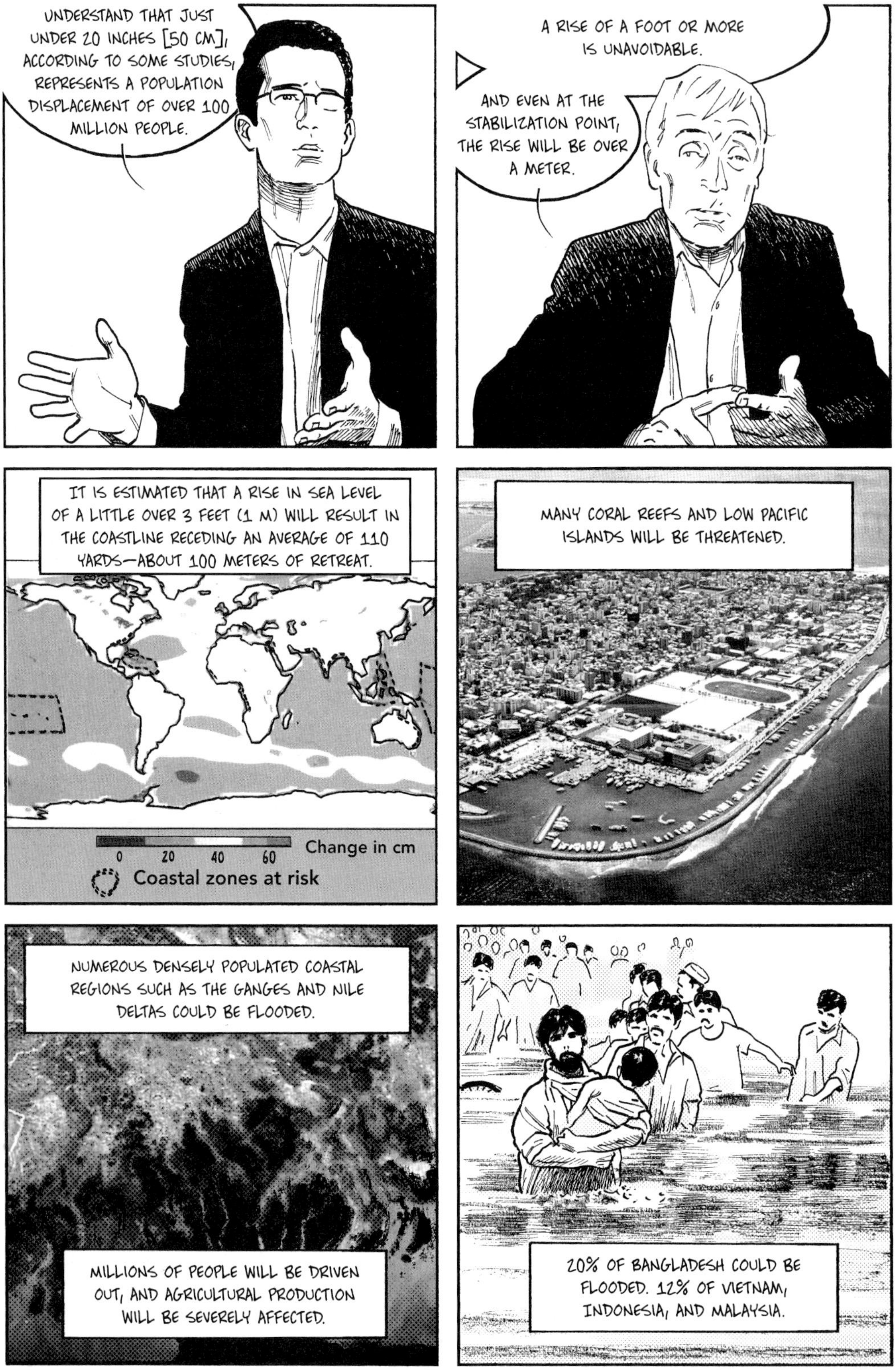

UNDERSTAND THAT JUST UNDER 20 INCHES [50 CM], ACCORDING TO SOME STUDIES, REPRESENTS A POPULATION DISPLACEMENT OF OVER 100 MILLION PEOPLE.
A RISE OF A FOOT OR MORE IS UNAVOIDABLE.
AND EVEN AT THE STABILIZATION POINT, THE RISE WILL BE OVER A METER.
IT IS ESTIMATED THAT A RISE IN SEA LEVEL OF A LITTLE OVER 3 FEET (1 M) WILL RESULT IN THE COASTLINE RECEDING AN AVERAGE OF 110 YARDS—ABOUT 100 METERS OF RETREAT.
0 20 40 60
Change in cm
Coastal zones at risk
MANY CORAL REEFS AND LOW PACIFIC ISLANDS WILL BE THREATENED.
NUMEROUS DENSELY POPULATED COASTAL REGIONS SUCH AS THE GANGES AND NILE DELTAS COULD BE FLOODED.
MILLIONS OF PEOPLE WILL BE DRIVEN OUT, AND AGRICULTURAL PRODUCTION WILL BE SEVERELY AFFECTED.
20% OF BANGLADESH COULD BE FLOODED. 12% OF VIETNAM, INDONESIA, AND MALAYSIA.

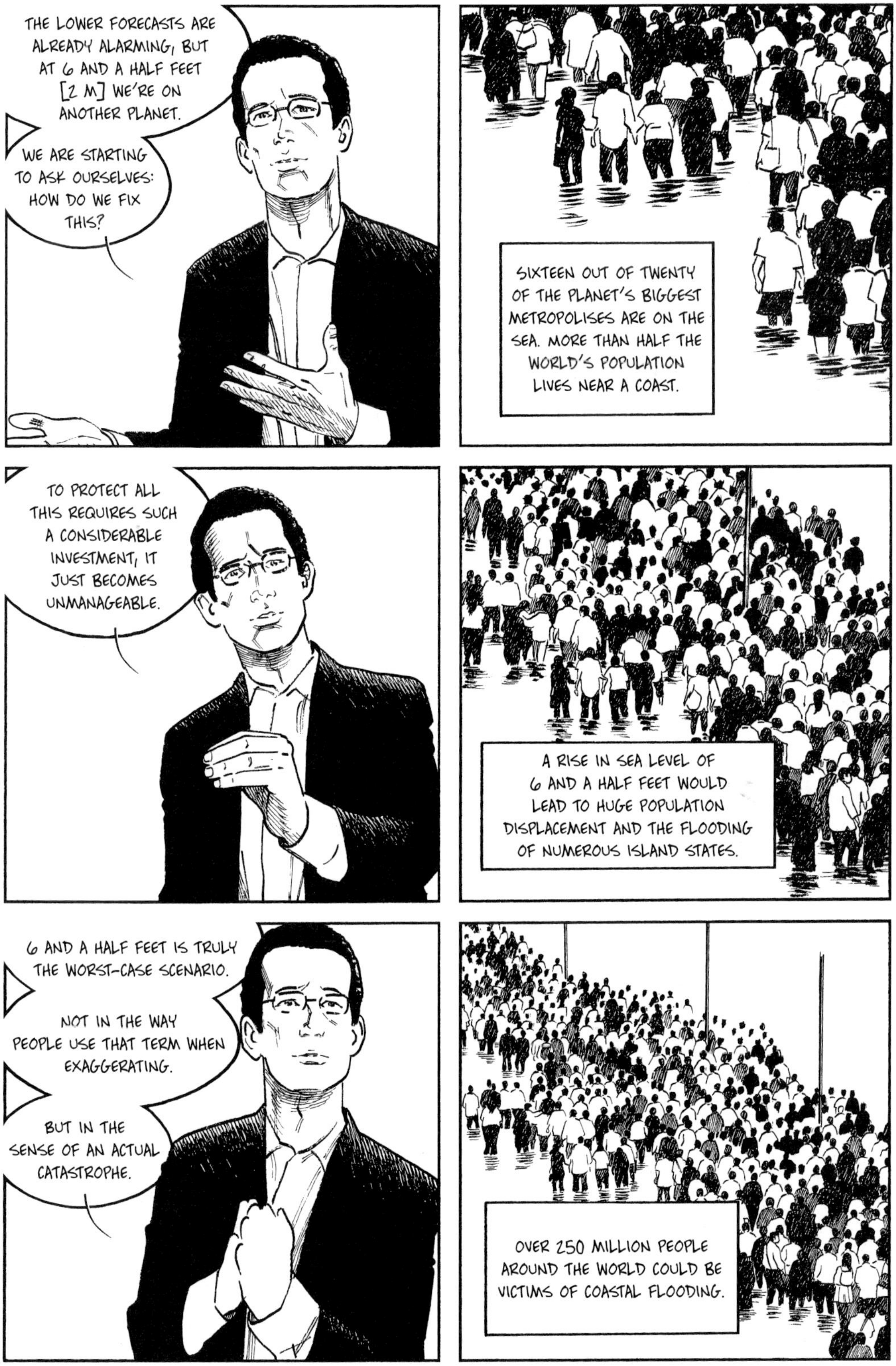
THE LOWER FORECASTS ARE ALREADY ALARMING, BUT AT 6 AND A HALF FEET [2 M] WE'RE ON ANOTHER PLANET.
WE ARE STARTING TO ASK OURSELVES: HOW DO WE FIX THIS?
SIXTEEN OUT OF TWENTY OF THE PLANET'S BIGGEST METROPOLISES ARE ON THE SEA. MORE THAN HALF THE WORLD'S POPULATION LIVES NEAR A COAST.
TO PROTECT ALL THIS REQUIRES SUCH A CONSIDERABLE INVESTMENT, IT JUST BECOMES UNMANAGEABLE.
A RISE IN SEA LEVEL OF 6 AND A HALF FEET WOULD LEAD TO HUGE POPULATION DISPLACEMENT AND THE FLOODING OF NUMEROUS ISLAND STATES.
6 AND A HALF FEET IS TRULY THE WORST-CASE SCENARIO.
NOT IN THE WAY PEOPLE USE THAT TERM WHEN EXAGGERATING.
BUT IN THE SENSE OF AN ACTUAL CATASTROPHE.
OVER 250 MILLION PEOPLE AROUND THE WORLD COULD BE VICTIMS OF COASTAL FLOODING.

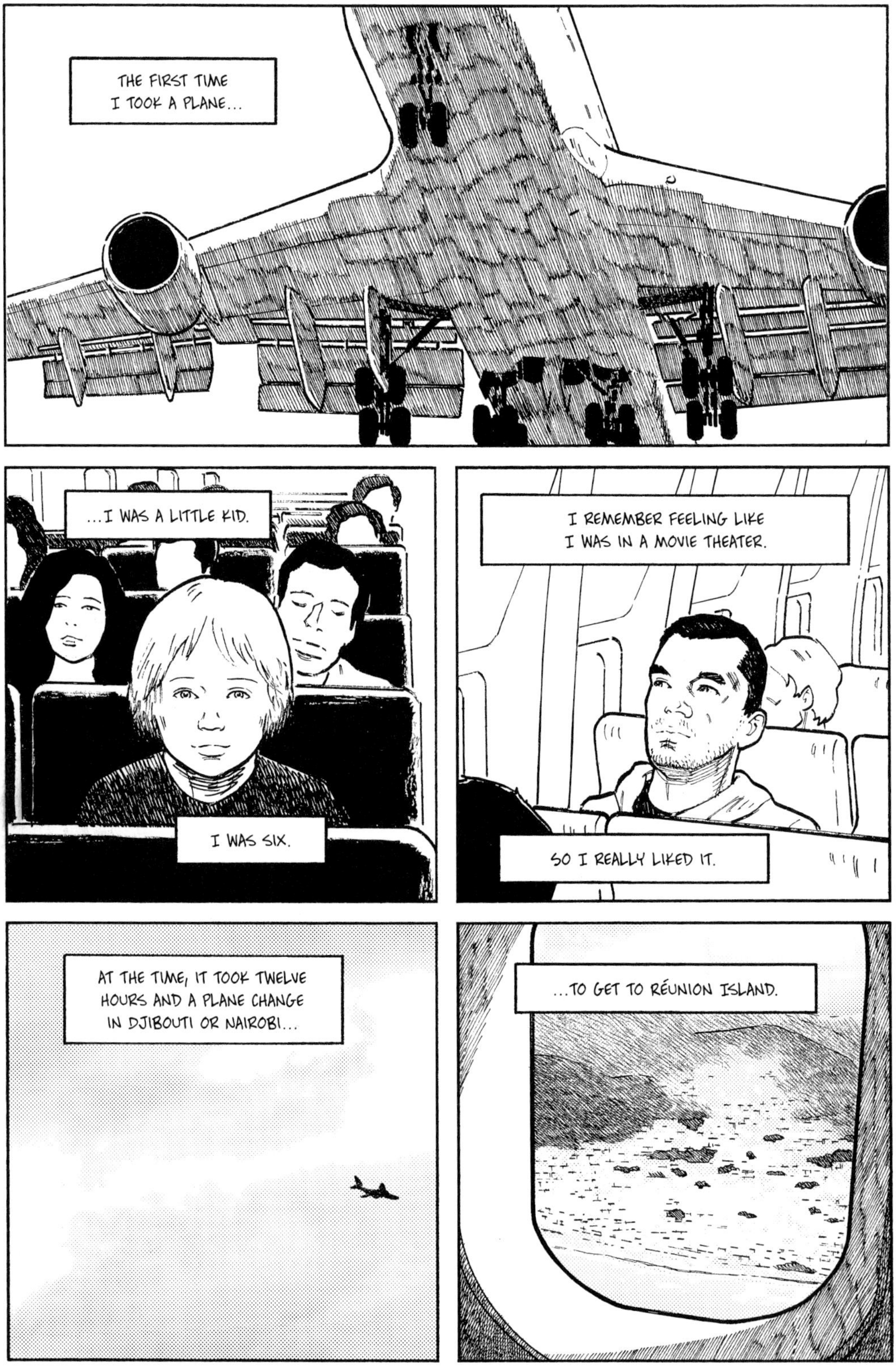
THE FIRST TIME I TOOK A PLANE...
...I WAS A LITTLE KID.
I WAS SIX.
I REMEMBER FEELING LIKE I WAS IN A MOVIE THEATER.
SO I REALLY LIKED IT.
AT THE TIME, IT TOOK TWELVE HOURS AND A PLANE CHANGE IN DJIBOUTI OR NAIROBI...
...TO GET TO RÉUNION ISLAND.

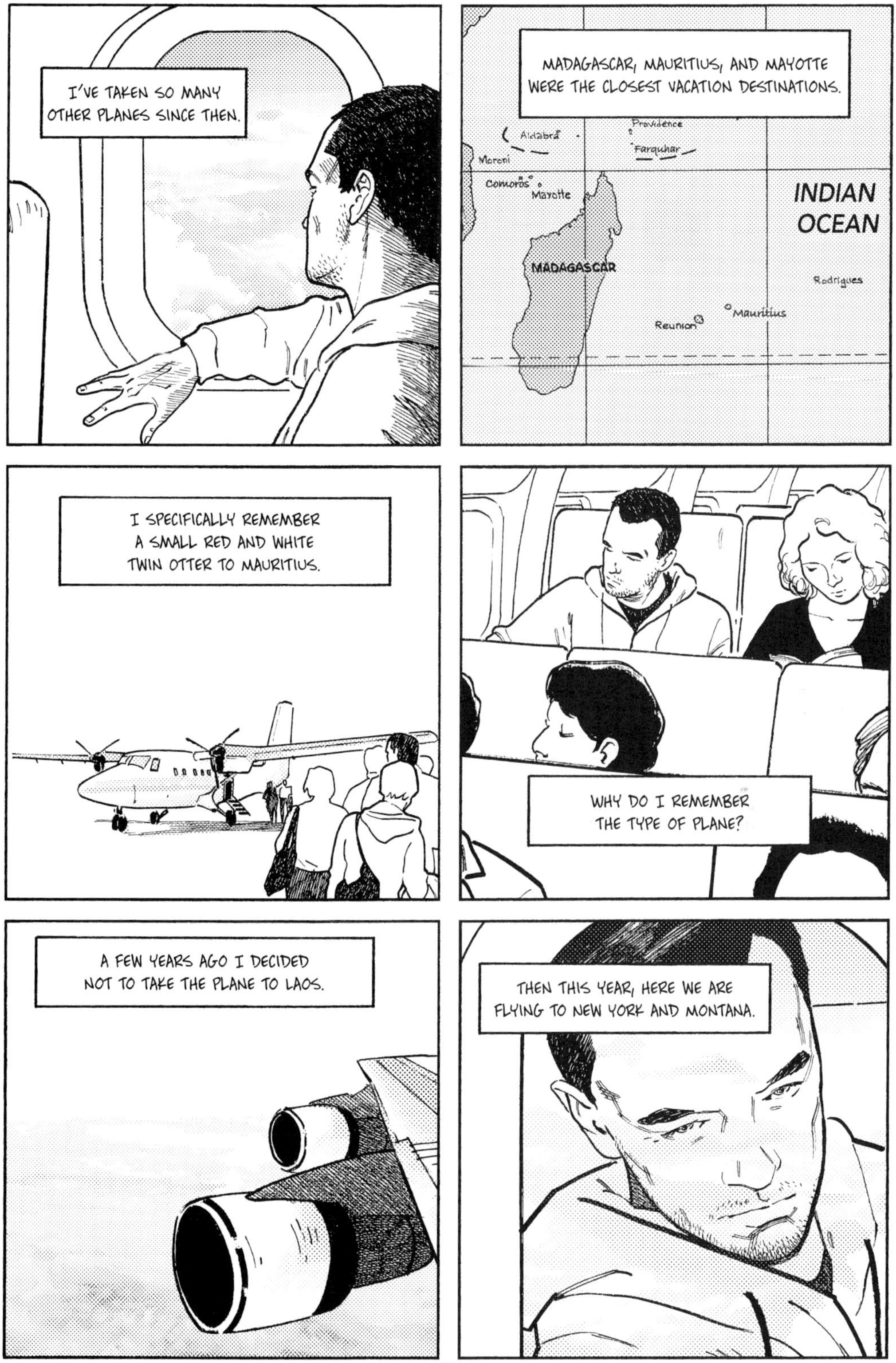
I'VE TAKEN SO MANY OTHER PLANES SINCE THEN.
MADAGASCAR, MAURITIUS, AND MAYOTTE WERE THE CLOSEST VACATION DESTINATIONS.
Providence
Aldabra
Farquhar
Moroni
Comoros
Mayotte
INDIAN OCEAN
MADAGASCAR
Rodrigues
Mauritius
Reunion
I SPECIFICALLY REMEMBER A SMALL RED AND WHITE TWIN OTTER TO MAURITIUS.
WHY DO I REMEMBER THE TYPE OF PLANE?
A FEW YEARS AGO I DECIDED NOT TO TAKE THE PLANE TO LAOS.
THEN THIS YEAR, HERE WE ARE FLYING TO NEW YORK AND MONTANA.

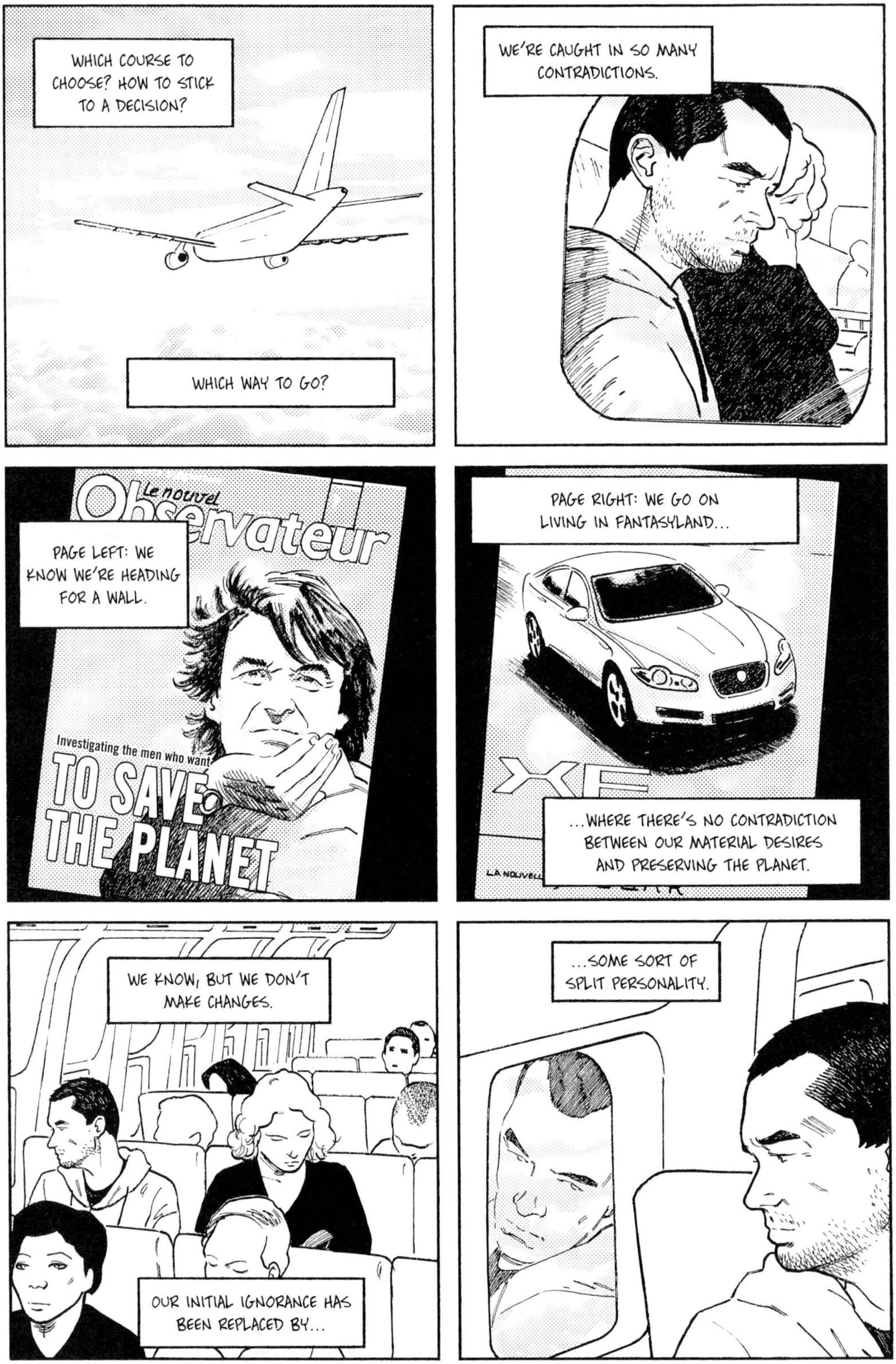
WHICH COURSE TO CHOOSE? HOW TO STICK TO A DECISION?
WHICH WAY TO GO?
WE'RE CAUGHT IN SO MANY CONTRADICTIONS.
Le nouvel Observateur
PAGE LEFT: WE KNOW WE'RE HEADING FOR A WALL.
Investigating the men who want
TO SAVE THE PLANET
PAGE RIGHT: WE GO ON LIVING IN FANTASYLAND...
XF
...WHERE THERE'S NO CONTRADICTION BETWEEN OUR MATERIAL DESIRES AND PRESERVING THE PLANET.
WE KNOW, BUT WE DON'T MAKE CHANGES.
OUR INITIAL IGNORANCE HAS BEEN REPLACED BY...
...SOME SORT OF SPLIT PERSONALITY.

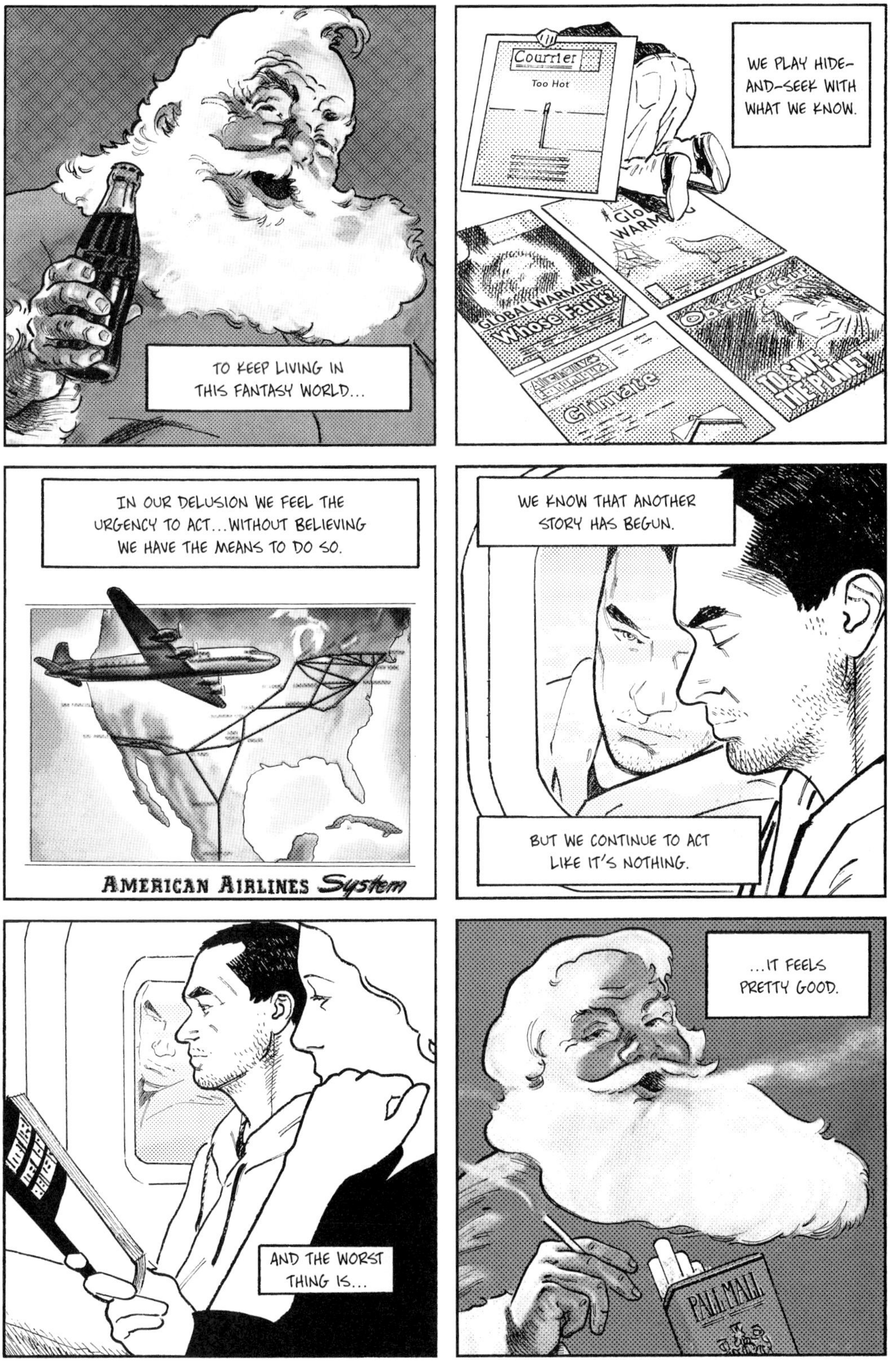
TO KEEP LIVING IN THIS FANTASY WORLD...
Courrier
Too Hot
WE PLAY HIDE-AND-SEEK WITH WHAT WE KNOW.
GLOBAL WARMING
Whose Fault?
Climate
TO SAVE THE PLANET
IN OUR DELUSION WE FEEL THE URGENCY TO ACT...WITHOUT BELIEVING WE HAVE THE MEANS TO DO SO.
AMERICAN AIRLINES System
WE KNOW THAT ANOTHER STORY HAS BEGUN.
BUT WE CONTINUE TO ACT LIKE IT'S NOTHING.
AND THE WORST THING IS...
...IT FEELS PRETTY GOOD.
PALL MALL

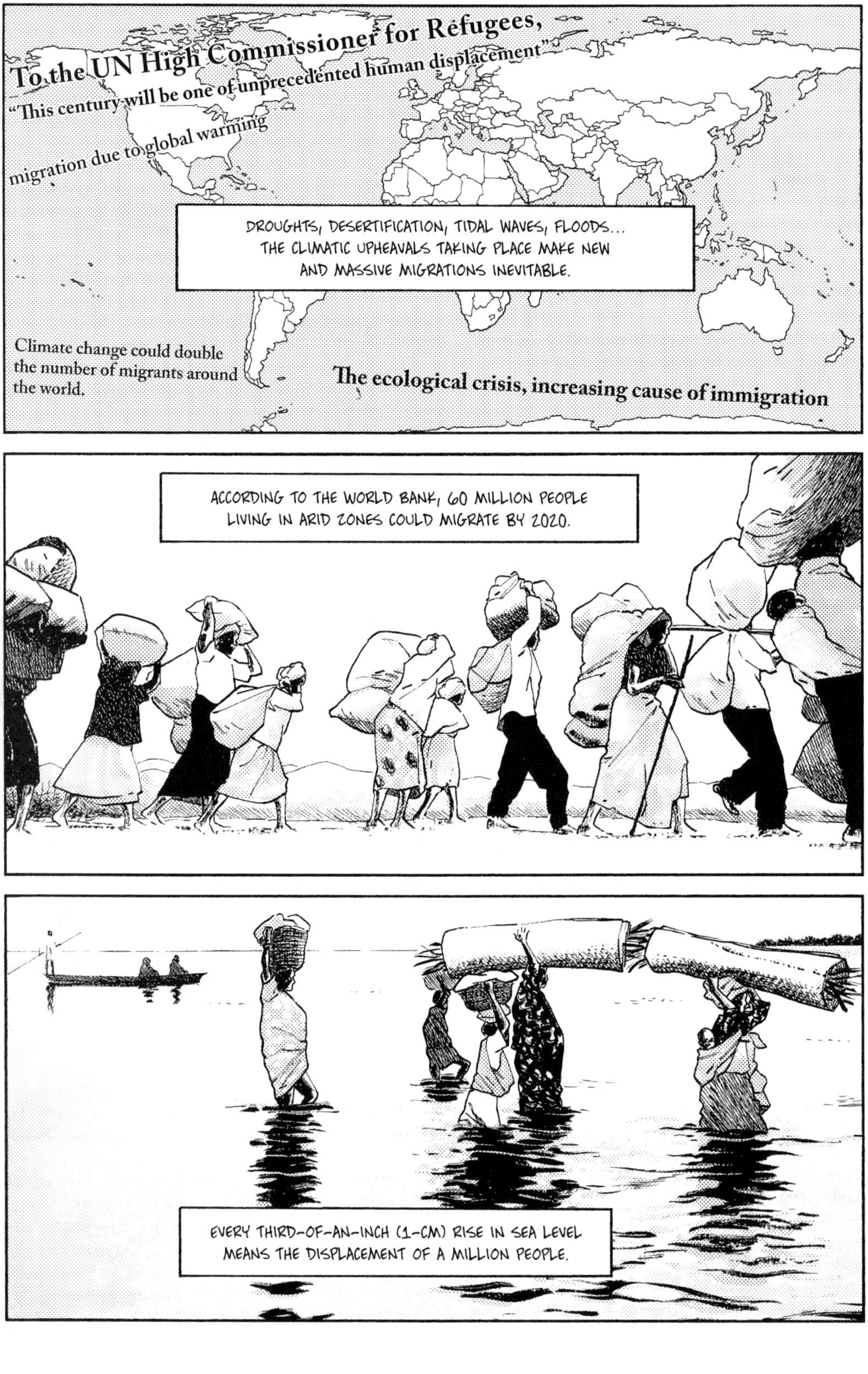
To the UN High Commissioner for Refugees,
"This century will be one of unprecedented human displacement"
migration due to global warming
DROUGHTS, DESERTIFICATION, TIDAL WAVES, FLOODS... THE CLIMATIC UPHEAVALS TAKING PLACE MAKE NEW AND MASSIVE MIGRATIONS INEVITABLE.
Climate change could double the number of migrants around the world.
The ecological crisis, increasing cause of immigration
ACCORDING TO THE WORLD BANK, 60 MILLION PEOPLE LIVING IN ARID ZONES COULD MIGRATE BY 2020.
EVERY THIRD-OF-AN-INCH (1-CM) RISE IN SEA LEVEL MEANS THE DISPLACEMENT OF A MILLION PEOPLE.

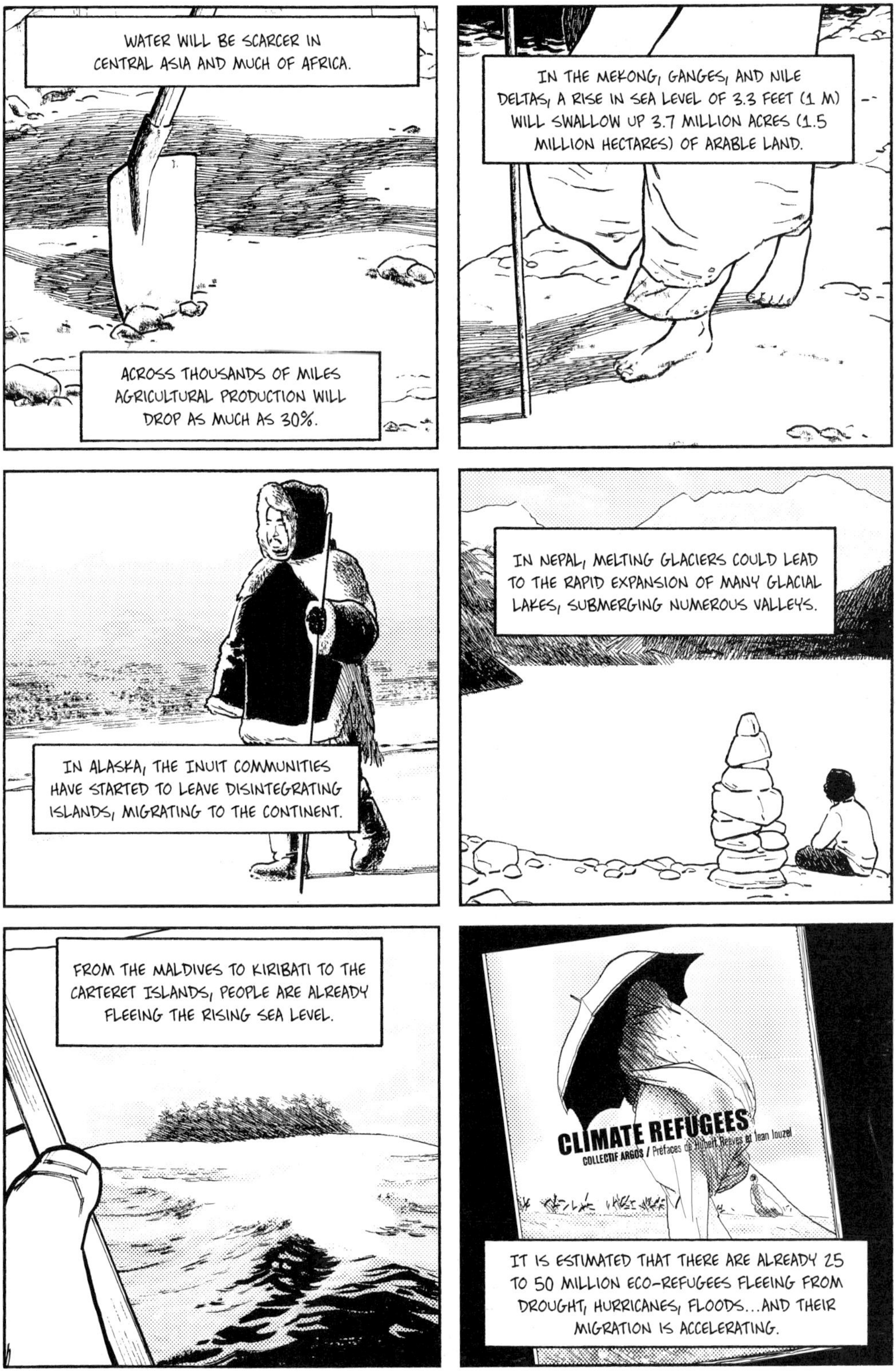
WATER WILL BE SCARCER IN CENTRAL ASIA AND MUCH OF AFRICA.
ACROSS THOUSANDS OF MILES AGRICULTURAL PRODUCTION WILL DROP AS MUCH AS 30%.
IN THE MEKONG, GANGES, AND NILE DELTAS, A RISE IN SEA LEVEL OF 3.3 FEET (1 M) WILL SWALLOW UP 3.7 MILLION ACRES (1.5 MILLION HECTARES) OF ARABLE LAND.
IN ALASKA, THE INUIT COMMUNITIES HAVE STARTED TO LEAVE DISINTEGRATING ISLANDS, MIGRATING TO THE CONTINENT.
IN NEPAL, MELTING GLACIERS COULD LEAD TO THE RAPID EXPANSION OF MANY GLACIAL LAKES, SUBMERGING NUMEROUS VALLEYS.
FROM THE MALDIVES TO KIRIBATI TO THE CARTERET ISLANDS, PEOPLE ARE ALREADY FLEEING THE RISING SEA LEVEL.
CLIMATE REFUGEES
COLLECTIF ARGOS / Préfaces de Hubert Reeves et Jean Jouzel
IT IS ESTIMATED THAT THERE ARE ALREADY 25 TO 50 MILLION ECO-REFUGEES FLEEING FROM DROUGHT, HURRICANES, FLOODS...AND THEIR MIGRATION IS ACCELERATING.

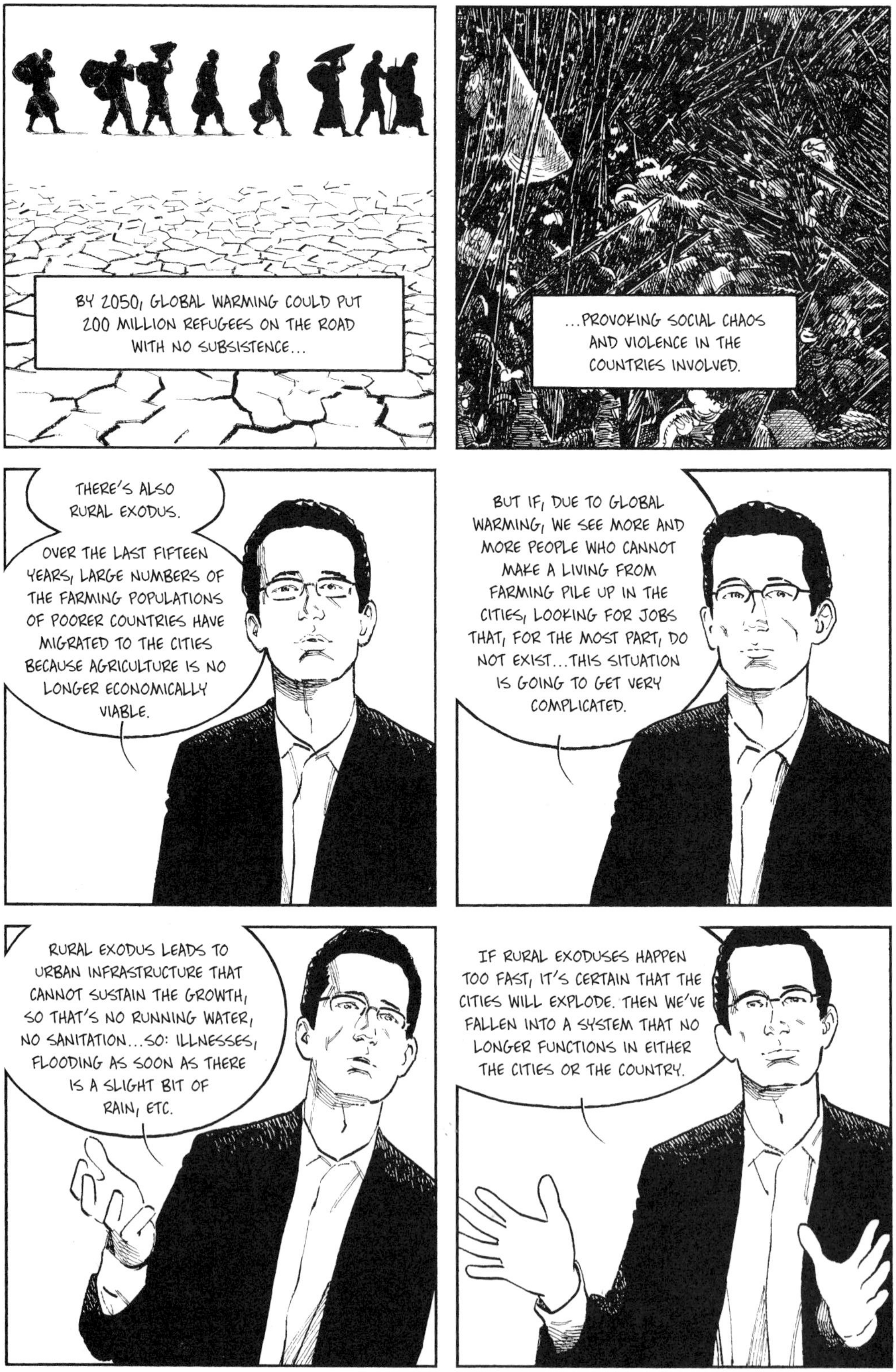
BY 2050, GLOBAL WARMING COULD PUT 200 MILLION REFUGEES ON THE ROAD WITH NO SUBSISTENCE...
...PROVOKING SOCIAL CHAOS AND VIOLENCE IN THE COUNTRIES INVOLVED.
THERE'S ALSO RURAL EXODUS.
OVER THE LAST FIFTEEN YEARS, LARGE NUMBERS OF THE FARMING POPULATIONS OF POORER COUNTRIES HAVE MIGRATED TO THE CITIES BECAUSE AGRICULTURE IS NO LONGER ECONOMICALLY VIABLE.
BUT IF, DUE TO GLOBAL WARMING, WE SEE MORE AND MORE PEOPLE WHO CANNOT MAKE A LIVING FROM FARMING PILE UP IN THE CITIES, LOOKING FOR JOBS THAT, FOR THE MOST PART, DO NOT EXIST...THIS SITUATION IS GOING TO GET VERY COMPLICATED.
RURAL EXODUS LEADS TO URBAN INFRASTRUCTURE THAT CANNOT SUSTAIN THE GROWTH, SO THAT'S NO RUNNING WATER, NO SANITATION...SO: ILLNESSES, FLOODING AS SOON AS THERE IS A SLIGHT BIT OF RAIN, ETC.
IF RURAL EXODUSES HAPPEN TOO FAST, IT'S CERTAIN THAT THE CITIES WILL EXPLODE. THEN WE'VE FALLEN INTO A SYSTEM THAT NO LONGER FUNCTIONS IN EITHER THE CITIES OR THE COUNTRY.

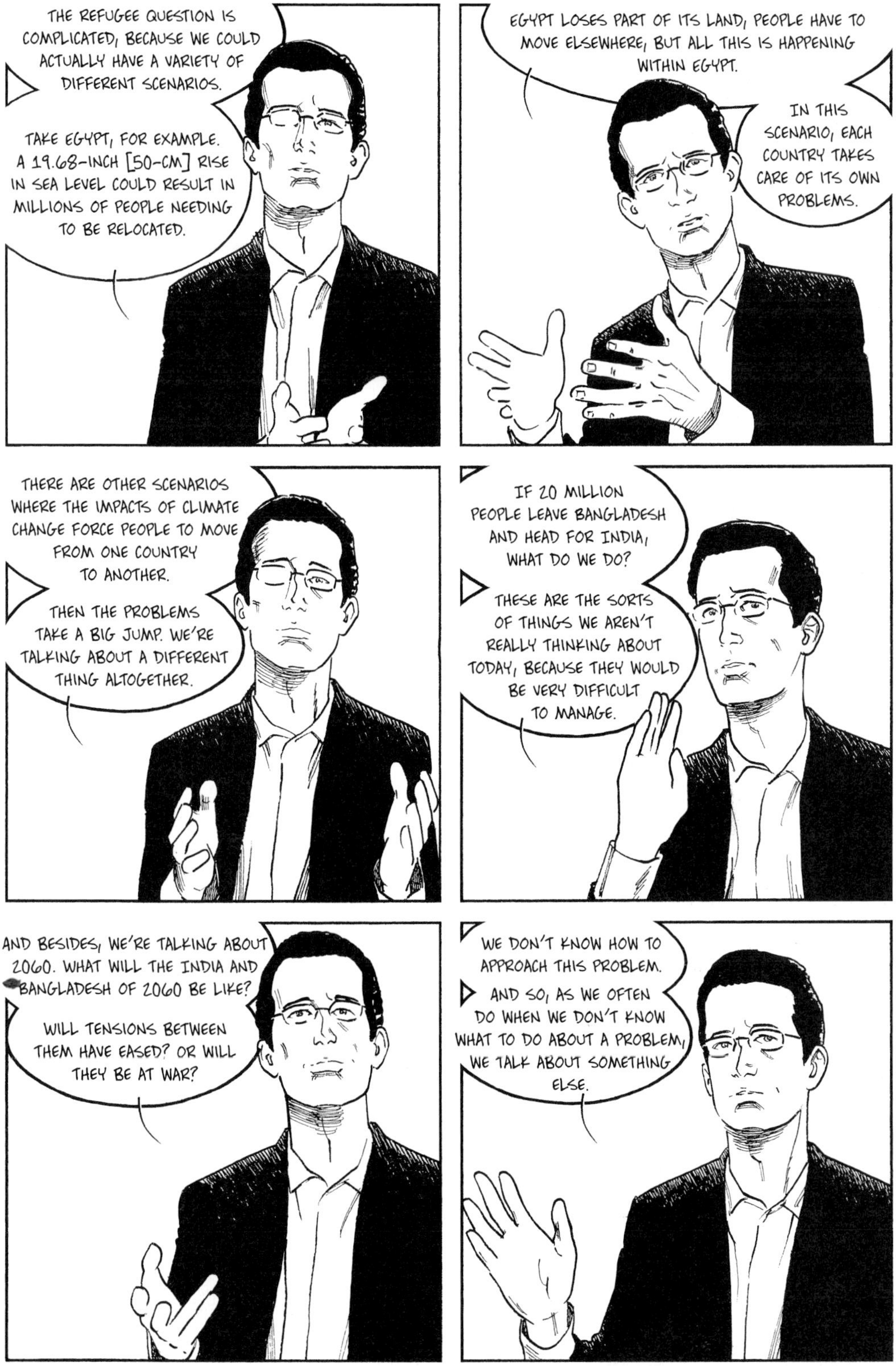
THE REFUGEE QUESTION IS COMPLICATED, BECAUSE WE COULD ACTUALLY HAVE A VARIETY OF DIFFERENT SCENARIOS.
TAKE EGYPT, FOR EXAMPLE. A 19.68-INCH [50-CM] RISE IN SEA LEVEL COULD RESULT IN MILLIONS OF PEOPLE NEEDING TO BE RELOCATED.
EGYPT LOSES PART OF ITS LAND, PEOPLE HAVE TO MOVE ELSEWHERE, BUT ALL THIS IS HAPPENING WITHIN EGYPT.
IN THIS SCENARIO, EACH COUNTRY TAKES CARE OF ITS OWN PROBLEMS.
THERE ARE OTHER SCENARIOS WHERE THE IMPACTS OF CLIMATE CHANGE FORCE PEOPLE TO MOVE FROM ONE COUNTRY TO ANOTHER.
THEN THE PROBLEMS TAKE A BIG JUMP. WE'RE TALKING ABOUT A DIFFERENT THING ALTOGETHER.
IF 20 MILLION PEOPLE LEAVE BANGLADESH AND HEAD FOR INDIA, WHAT DO WE DO?
THESE ARE THE SORTS OF THINGS WE AREN'T REALLY THINKING ABOUT TODAY, BECAUSE THEY WOULD BE VERY DIFFICULT TO MANAGE.
AND BESIDES, WE'RE TALKING ABOUT 2060. WHAT WILL THE INDIA AND BANGLADESH OF 2060 BE LIKE?
WILL TENSIONS BETWEEN THEM HAVE EASED? OR WILL THEY BE AT WAR?
WE DON'T KNOW HOW TO APPROACH THIS PROBLEM.
AND SO, AS WE OFTEN DO WHEN WE DON'T KNOW WHAT TO DO ABOUT A PROBLEM, WE TALK ABOUT SOMETHING ELSE.

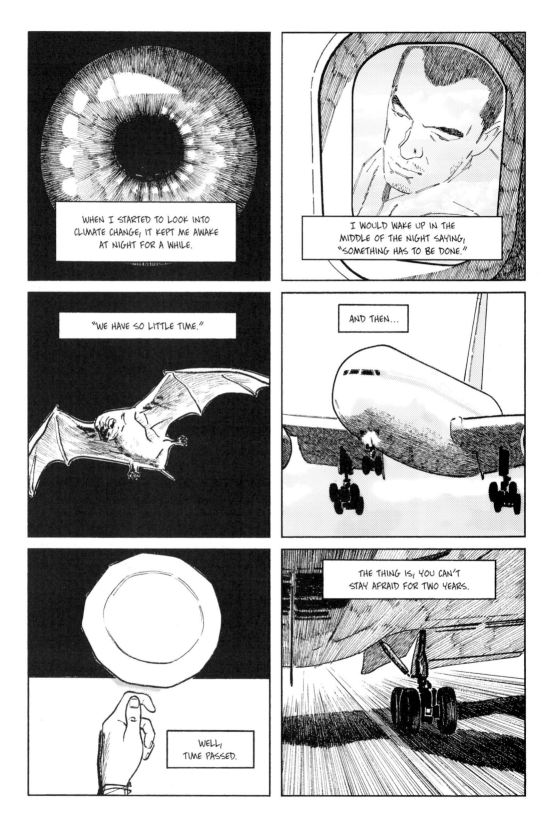

WHEN I STARTED TO LOOK INTO CLIMATE CHANGE, IT KEPT ME AWAKE AT NIGHT FOR A WHILE.

I WOULD WAKE UP IN THE MIDDLE OF THE NIGHT SAYING, "SOMETHING HAS TO BE DONE."

"WE HAVE SO LITTLE TIME."

AND THEN...

WELL, TIME PASSED.

THE THING IS, YOU CAN'T STAY AFRAID FOR TWO YEARS.

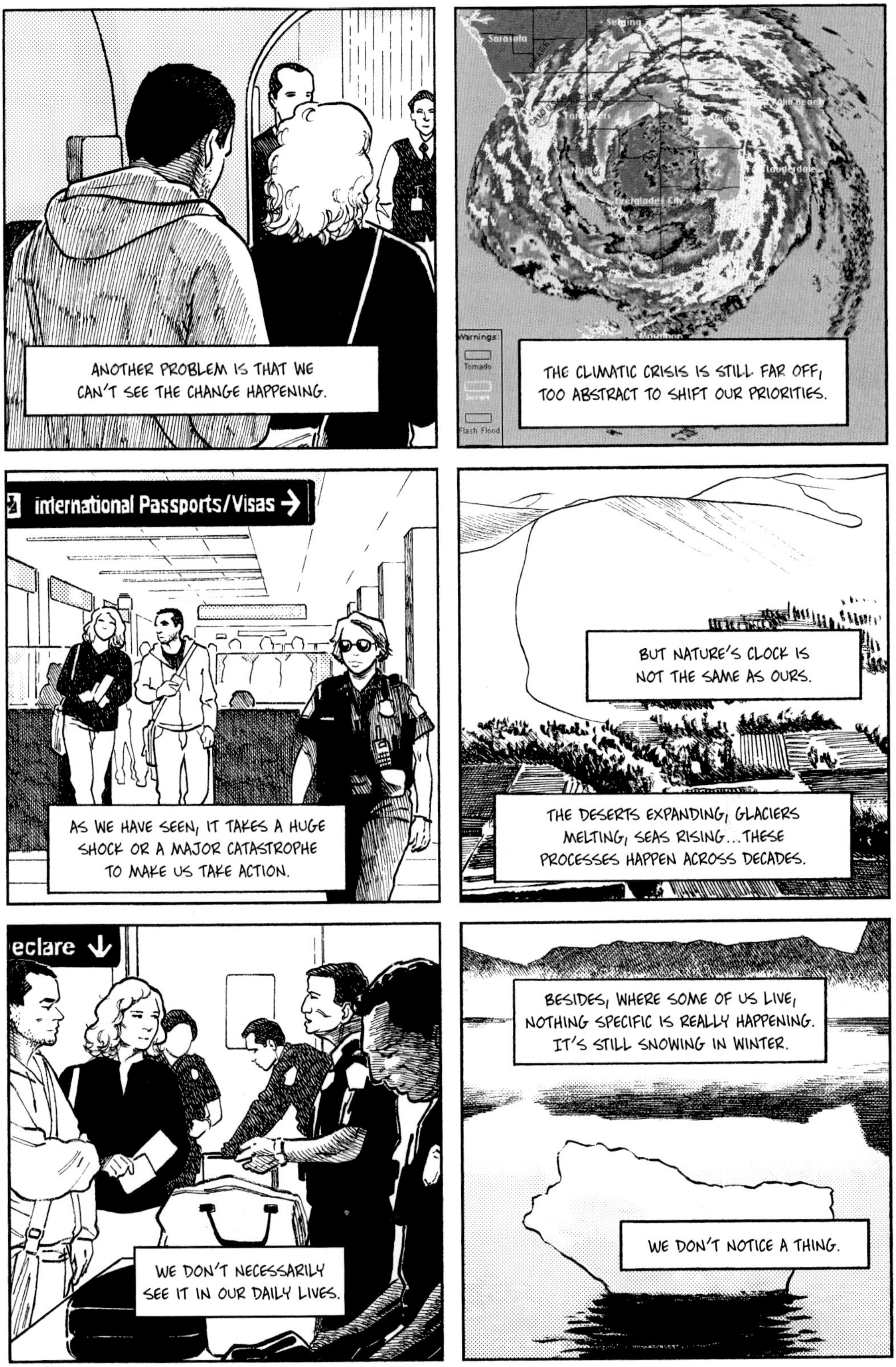
ANOTHER PROBLEM IS THAT WE CAN'T SEE THE CHANGE HAPPENING.
Sarasota
Warnings:
Tornado
Flash Flood
THE CLIMATIC CRISIS IS STILL FAR OFF, TOO ABSTRACT TO SHIFT OUR PRIORITIES.
international Passports/Visas →
AS WE HAVE SEEN, IT TAKES A HUGE SHOCK OR A MAJOR CATASTROPHE TO MAKE US TAKE ACTION.
BUT NATURE'S CLOCK IS NOT THE SAME AS OURS.
THE DESERTS EXPANDING, GLACIERS MELTING, SEAS RISING... THESE PROCESSES HAPPEN ACROSS DECADES.
eclare ↓
WE DON'T NECESSARILY SEE IT IN OUR DAILY LIVES.
BESIDES, WHERE SOME OF US LIVE, NOTHING SPECIFIC IS REALLY HAPPENING. IT'S STILL SNOWING IN WINTER.
WE DON'T NOTICE A THING.

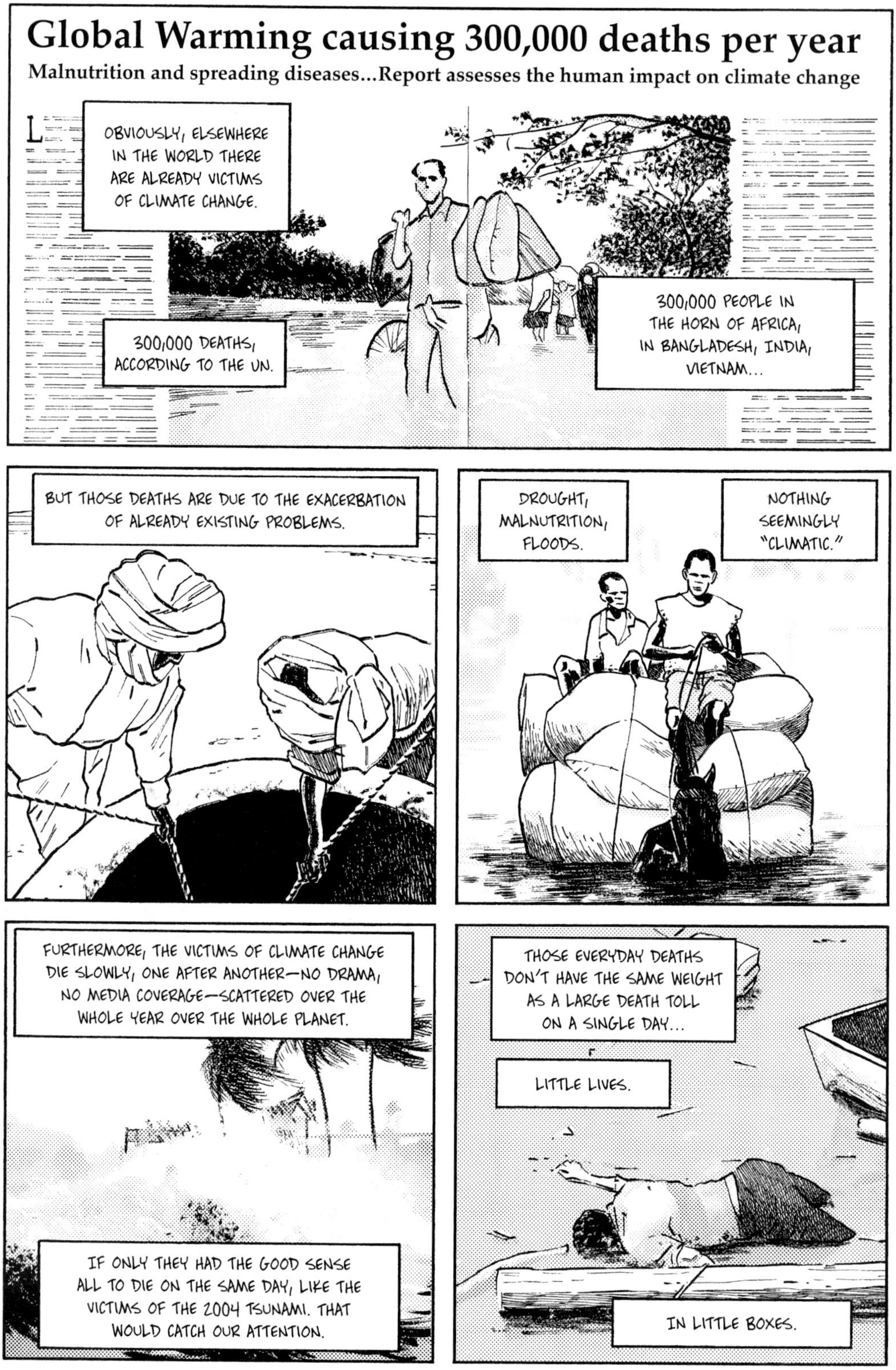

Global Warming causing 300,000 deaths per year
Malnutrition and spreading diseases...Report assesses the human impact on climate change
OBVIOUSLY, ELSEWHERE IN THE WORLD THERE ARE ALREADY VICTIMS OF CLIMATE CHANGE.
300,000 DEATHS, ACCORDING TO THE UN.
300,000 PEOPLE IN THE HORN OF AFRICA, IN BANGLADESH, INDIA, VIETNAM...
BUT THOSE DEATHS ARE DUE TO THE EXACERBATION OF ALREADY EXISTING PROBLEMS.
DROUGHT, MALNUTRITION, FLOODS.
NOTHING SEEMINGLY "CLIMATIC."
FURTHERMORE, THE VICTIMS OF CLIMATE CHANGE DIE SLOWLY, ONE AFTER ANOTHER—NO DRAMA, NO MEDIA COVERAGE—SCATTERED OVER THE WHOLE YEAR OVER THE WHOLE PLANET.
IF ONLY THEY HAD THE GOOD SENSE ALL TO DIE ON THE SAME DAY, LIKE THE VICTIMS OF THE 2004 TSUNAMI. THAT WOULD CATCH OUR ATTENTION.
THOSE EVERYDAY DEATHS DON'T HAVE THE SAME WEIGHT AS A LARGE DEATH TOLL ON A SINGLE DAY...
LITTLE LIVES.
IN LITTLE BOXES.

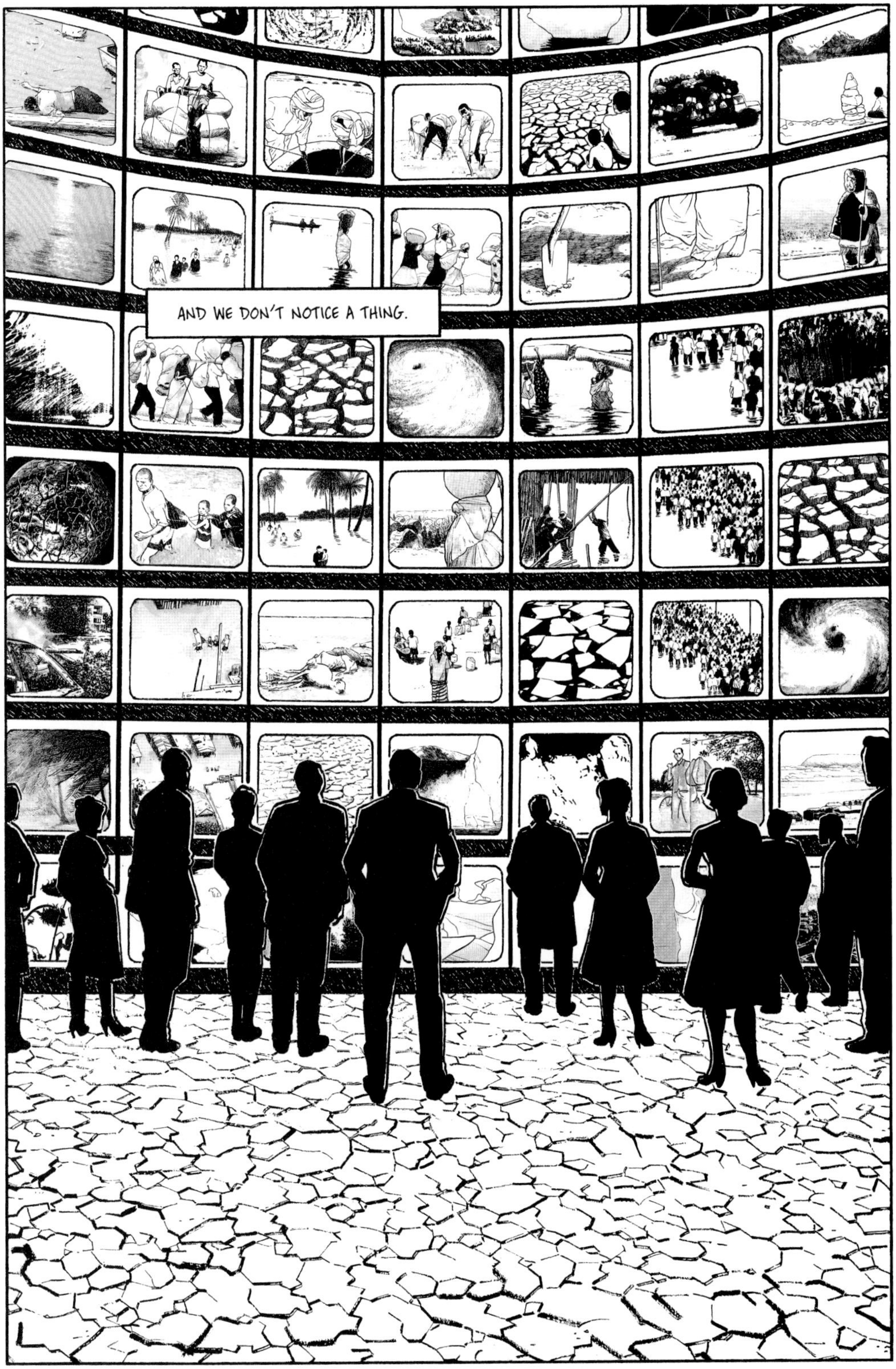
AND WE DON'T NOTICE A THING.

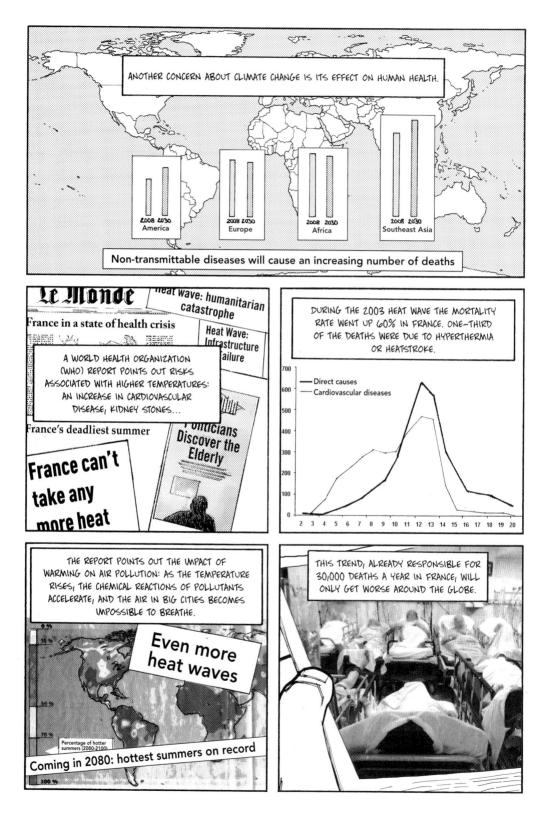

ANOTHER CONCERN ABOUT CLIMATE CHANGE IS ITS EFFECT ON HUMAN HEALTH.

2008 2030
America

2008 2030
Europe

2008 2030
Africa

2008 2030
Southeast Asia

Non-transmittable diseases will cause an increasing number of deaths

Le Monde

Heat wave: humanitarian catastrophe

France in a state of health crisis

Heat Wave: Infrastructure Failure

A WORLD HEALTH ORGANIZATION (WHO) REPORT POINTS OUT RISKS ASSOCIATED WITH HIGHER TEMPERATURES: AN INCREASE IN CARDIOVASCULAR DISEASE, KIDNEY STONES...

France's deadliest summer

Politicians Discover the Elderly

France can't take any more heat

DURING THE 2003 HEAT WAVE THE MORTALITY RATE WENT UP 60% IN FRANCE. ONE-THIRD OF THE DEATHS WERE DUE TO HYPERTHERMIA OR HEATSTROKE.

— Direct causes
— Cardiovascular diseases

700
600
500
400
300
200
100
0

2 3 4 5 6 7 8 9 10 11 12 13 14 15 16 17 18 19 20

THE REPORT POINTS OUT THE IMPACT OF WARMING ON AIR POLLUTION: AS THE TEMPERATURE RISES, THE CHEMICAL REACTIONS OF POLLUTANTS ACCELERATE, AND THE AIR IN BIG CITIES BECOMES IMPOSSIBLE TO BREATHE.

Even more heat waves

0 %
10 %
50 %
70 %

Percentage of hotter summers (2080-2100)

Coming in 2080: hottest summers on record

100 %

THIS TREND, ALREADY RESPONSIBLE FOR 30,000 DEATHS A YEAR IN FRANCE, WILL ONLY GET WORSE AROUND THE GLOBE.

252

HIGHER TEMPERATURES WILL ALSO EXPAND THE AREAS AFFECTED BY VECTOR-BORNE DISEASES—DISEASES TRANSMITTED BY ANIMALS.

CERTAIN DISEASE VECTORS—ANIMALS THAT THRIVE IN THE NEW TEMPERATURES—ARE MOVING INTO ZONES WHERE THE INHABITANTS ARE NOT IMMUNIZED.

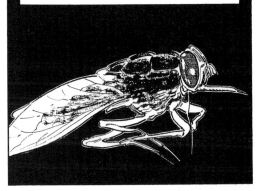

220 TO 400 MILLION MORE PEOPLE COULD BE EXPOSED TO MALARIA.

YELLOW FEVER, DENGUE FEVER, AND LYME DISEASE COULD BECOME ENDEMIC TO THE MIDDLE LATITUDES.

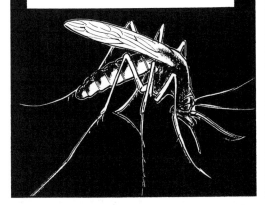

IN DEVELOPING COUNTRIES THE HEALTH REPERCUSSIONS OF WARMING WILL BE HARSHER ON POORER POPULATIONS.

ONCE AGAIN, CLIMATE CHANGE EXPOSES THE PROBLEM OF INCOME DISPARITY.

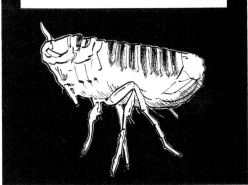

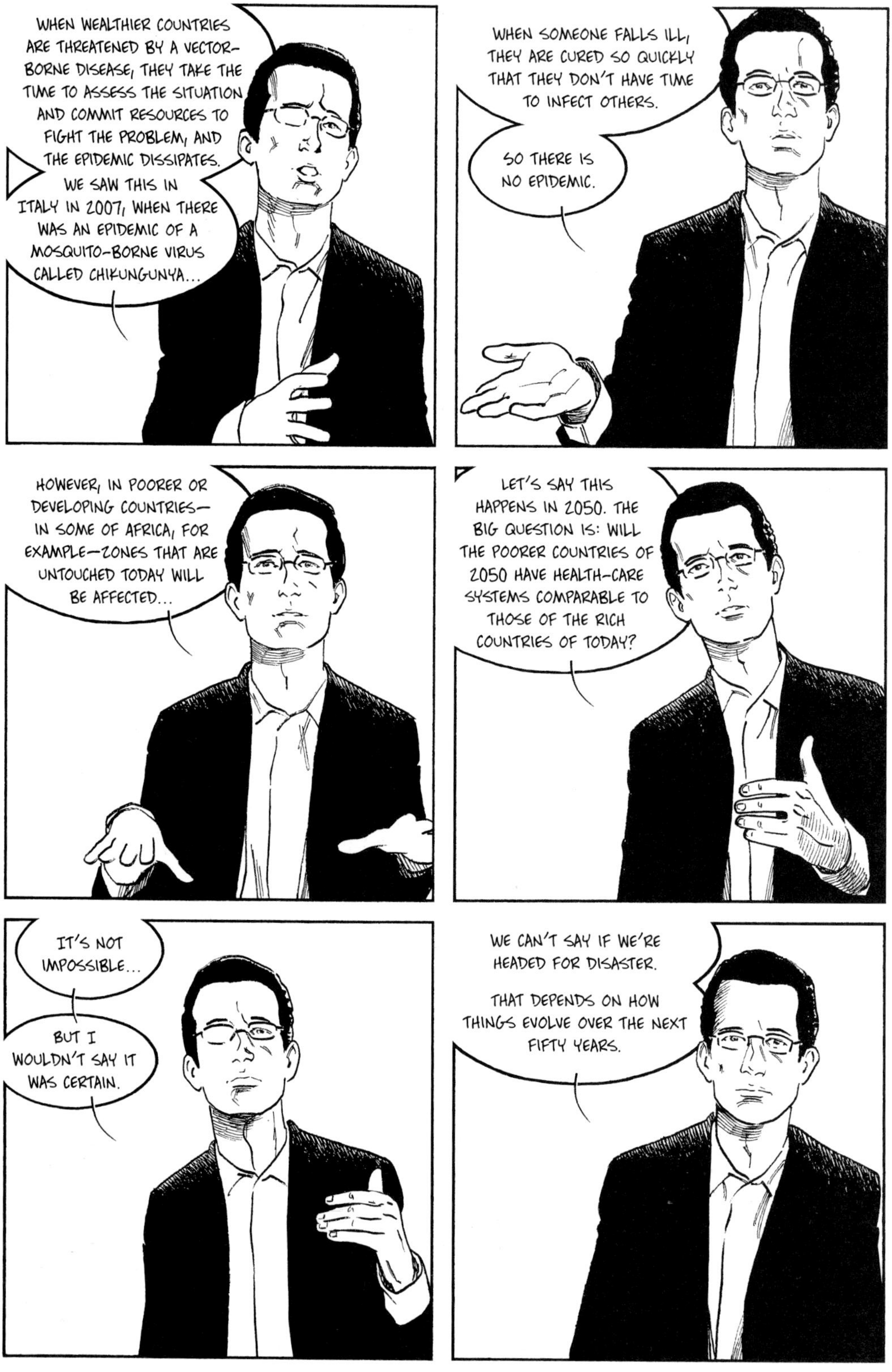
WHEN WEALTHIER COUNTRIES ARE THREATENED BY A VECTOR-BORNE DISEASE, THEY TAKE THE TIME TO ASSESS THE SITUATION AND COMMIT RESOURCES TO FIGHT THE PROBLEM, AND THE EPIDEMIC DISSIPATES.
WE SAW THIS IN ITALY IN 2007, WHEN THERE WAS AN EPIDEMIC OF A MOSQUITO-BORNE VIRUS CALLED CHIKUNGUNYA...
WHEN SOMEONE FALLS ILL, THEY ARE CURED SO QUICKLY THAT THEY DON'T HAVE TIME TO INFECT OTHERS.
SO THERE IS NO EPIDEMIC.
HOWEVER, IN POORER OR DEVELOPING COUNTRIES—IN SOME OF AFRICA, FOR EXAMPLE—ZONES THAT ARE UNTOUCHED TODAY WILL BE AFFECTED...
LET'S SAY THIS HAPPENS IN 2050. THE BIG QUESTION IS: WILL THE POORER COUNTRIES OF 2050 HAVE HEALTH-CARE SYSTEMS COMPARABLE TO THOSE OF THE RICH COUNTRIES OF TODAY?
IT'S NOT IMPOSSIBLE...
BUT I WOULDN'T SAY IT WAS CERTAIN.
WE CAN'T SAY IF WE'RE HEADED FOR DISASTER.
THAT DEPENDS ON HOW THINGS EVOLVE OVER THE NEXT FIFTY YEARS.

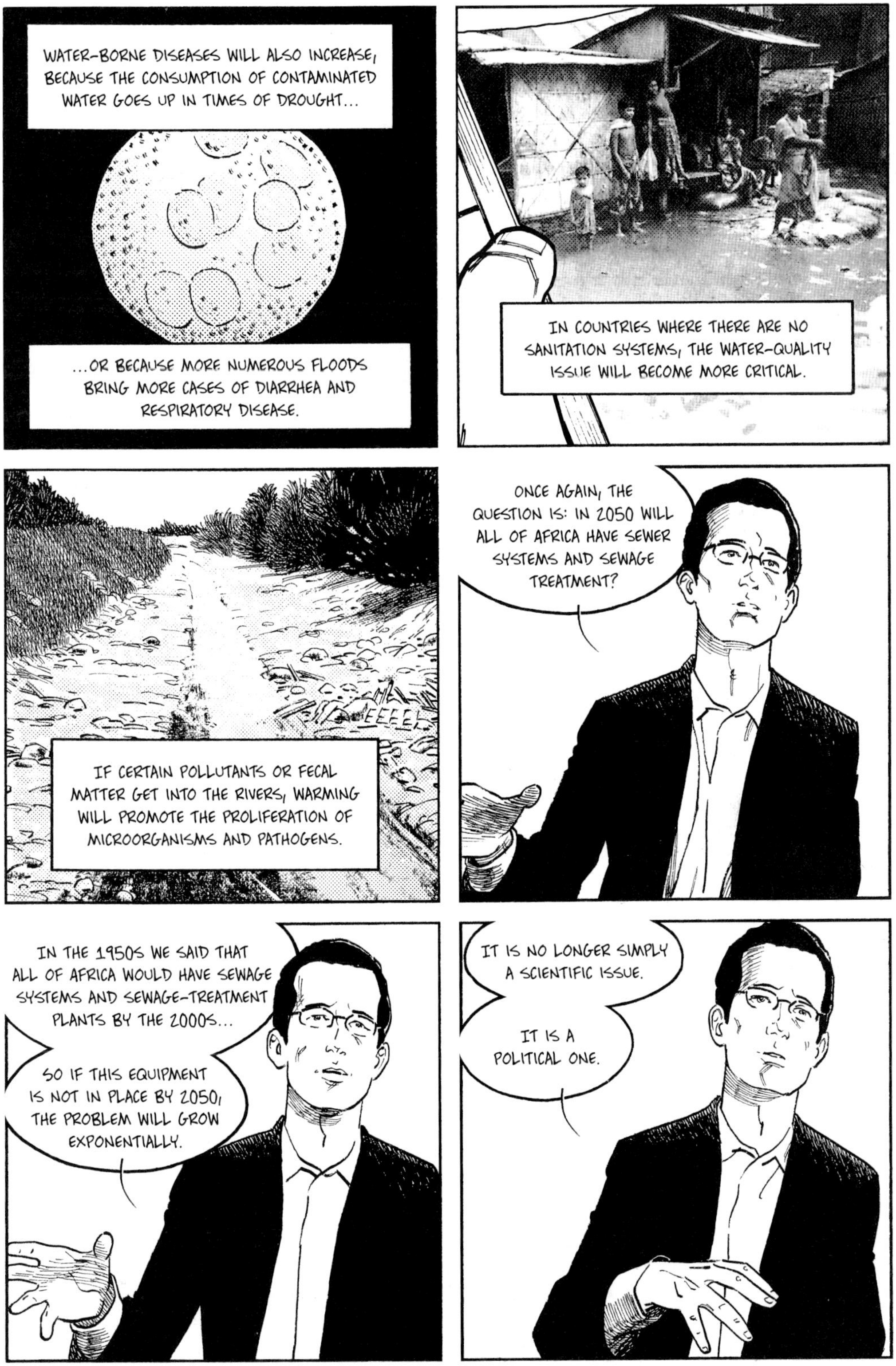
WATER-BORNE DISEASES WILL ALSO INCREASE, BECAUSE THE CONSUMPTION OF CONTAMINATED WATER GOES UP IN TIMES OF DROUGHT...
...OR BECAUSE MORE NUMEROUS FLOODS BRING MORE CASES OF DIARRHEA AND RESPIRATORY DISEASE.
IN COUNTRIES WHERE THERE ARE NO SANITATION SYSTEMS, THE WATER-QUALITY ISSUE WILL BECOME MORE CRITICAL.
IF CERTAIN POLLUTANTS OR FECAL MATTER GET INTO THE RIVERS, WARMING WILL PROMOTE THE PROLIFERATION OF MICROORGANISMS AND PATHOGENS.
ONCE AGAIN, THE QUESTION IS: IN 2050 WILL ALL OF AFRICA HAVE SEWER SYSTEMS AND SEWAGE TREATMENT?
IN THE 1950S WE SAID THAT ALL OF AFRICA WOULD HAVE SEWAGE SYSTEMS AND SEWAGE-TREATMENT PLANTS BY THE 2000S...
SO IF THIS EQUIPMENT IS NOT IN PLACE BY 2050, THE PROBLEM WILL GROW EXPONENTIALLY.
IT IS NO LONGER SIMPLY A SCIENTIFIC ISSUE.
IT IS A POLITICAL ONE.

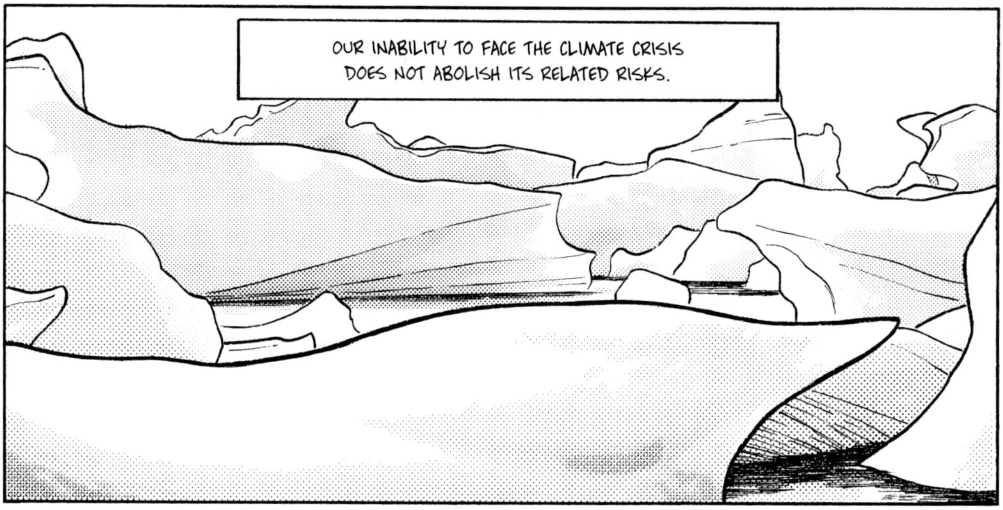
OUR INABILITY TO FACE THE CLIMATE CRISIS DOES NOT ABOLISH ITS RELATED RISKS.

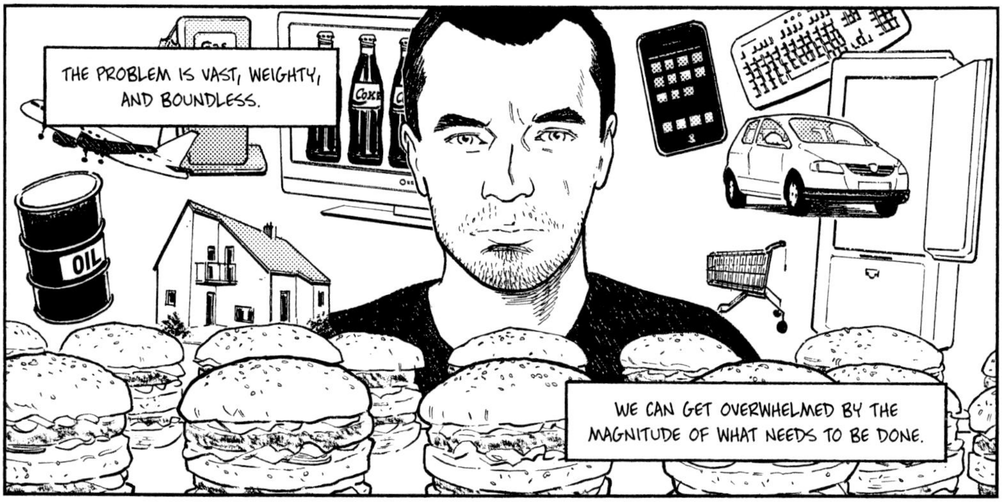
THE PROBLEM IS VAST, WEIGHTY, AND BOUNDLESS.
Coke
OIL
WE CAN GET OVERWHELMED BY THE MAGNITUDE OF WHAT NEEDS TO BE DONE.

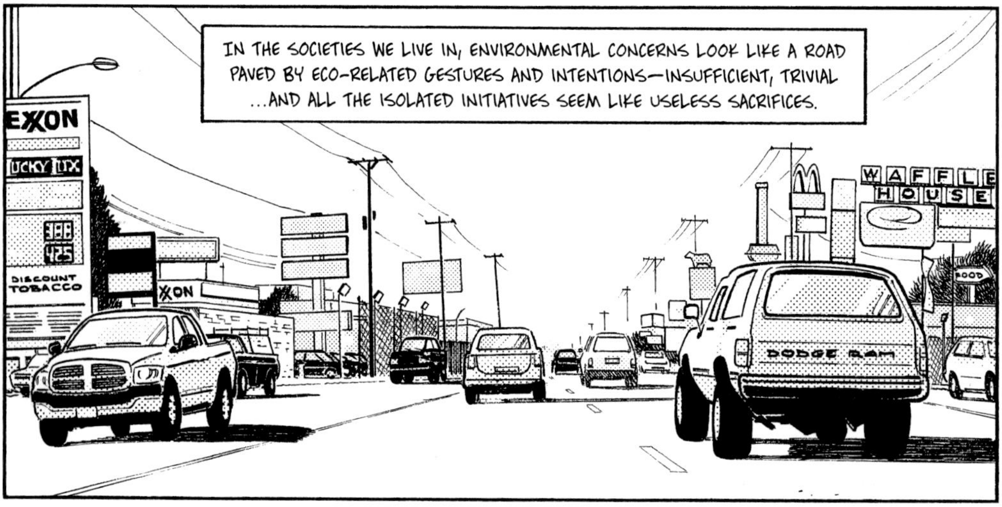
IN THE SOCIETIES WE LIVE IN, ENVIRONMENTAL CONCERNS LOOK LIKE A ROAD PAVED BY ECO-RELATED GESTURES AND INTENTIONS—INSUFFICIENT, TRIVIAL ...AND ALL THE ISOLATED INITIATIVES SEEM LIKE USELESS SACRIFICES.
EXXON
DISCOUNT TOBACCO
WAFFLE HOUSE
DODGE RAM

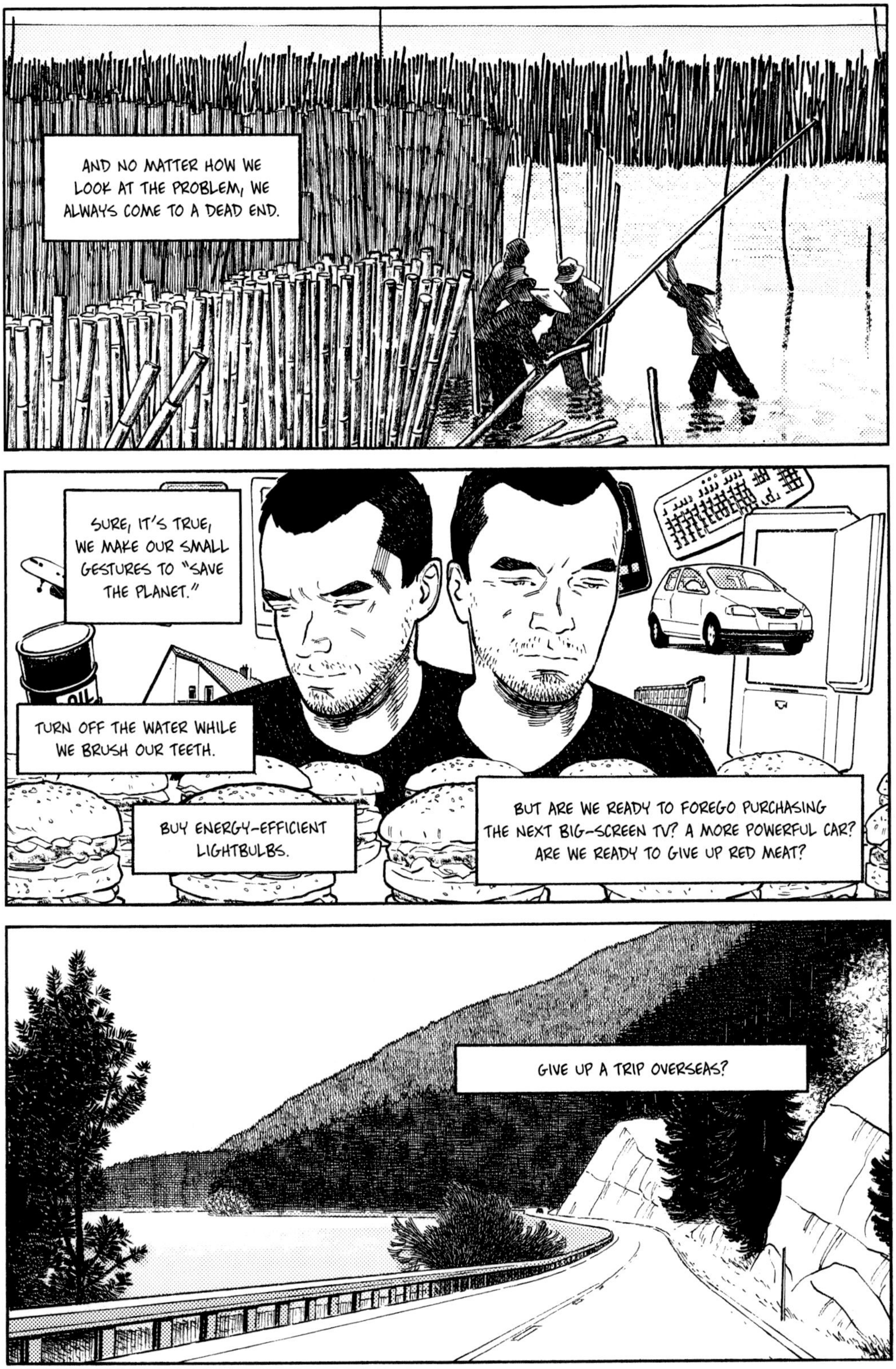
AND NO MATTER HOW WE LOOK AT THE PROBLEM, WE ALWAYS COME TO A DEAD END.
SURE, IT'S TRUE, WE MAKE OUR SMALL GESTURES TO "SAVE THE PLANET."
TURN OFF THE WATER WHILE WE BRUSH OUR TEETH.
BUY ENERGY-EFFICIENT LIGHTBULBS.
BUT ARE WE READY TO FOREGO PURCHASING THE NEXT BIG-SCREEN TV? A MORE POWERFUL CAR? ARE WE READY TO GIVE UP RED MEAT?
GIVE UP A TRIP OVERSEAS?

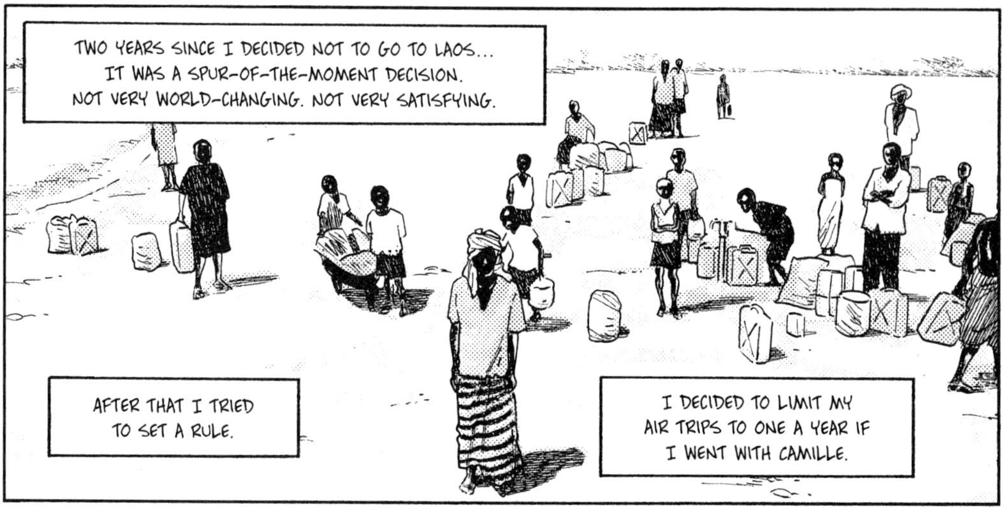
TWO YEARS SINCE I DECIDED NOT TO GO TO LAOS... IT WAS A SPUR-OF-THE-MOMENT DECISION. NOT VERY WORLD-CHANGING. NOT VERY SATISFYING.
AFTER THAT I TRIED TO SET A RULE.
I DECIDED TO LIMIT MY AIR TRIPS TO ONE A YEAR IF I WENT WITH CAMILLE.

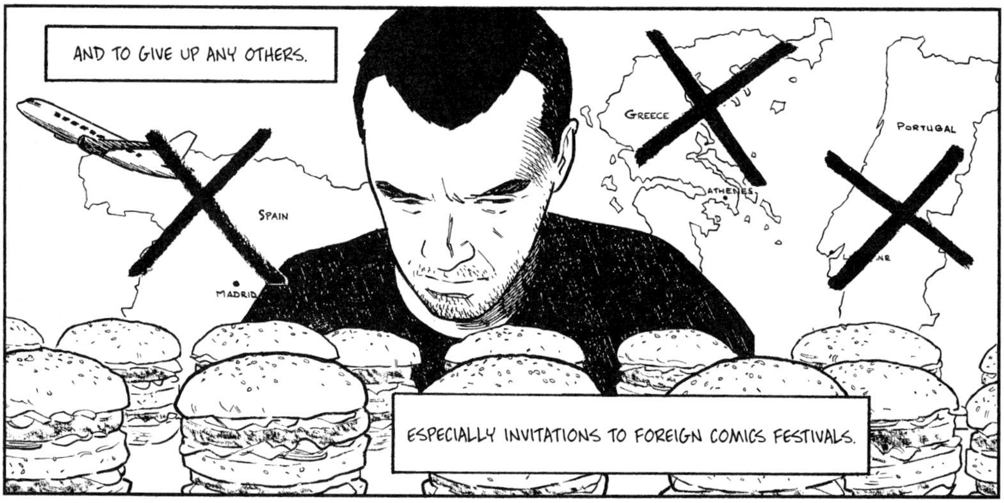
AND TO GIVE UP ANY OTHERS.
SPAIN
MADRID
GREECE
ATHENES
PORTUGAL
ESPECIALLY INVITATIONS TO FOREIGN COMICS FESTIVALS.

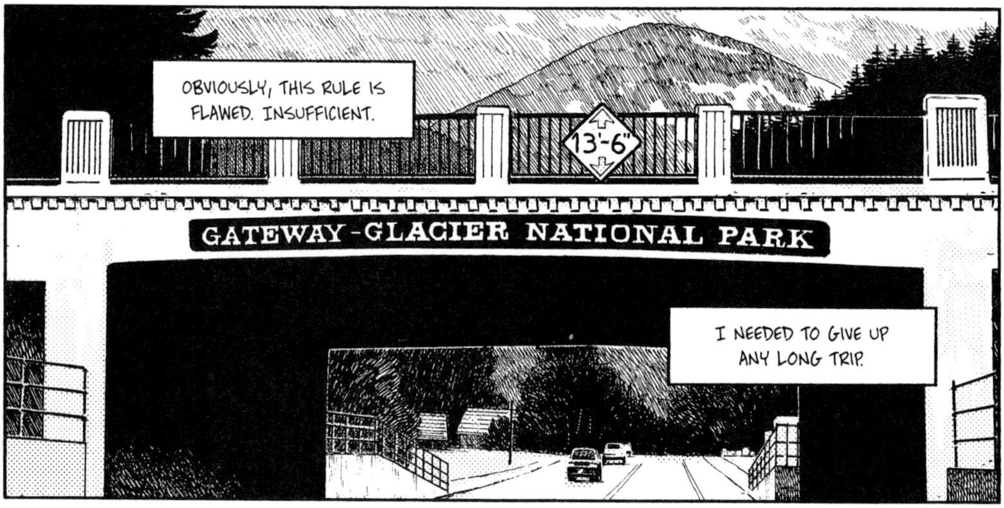
OBVIOUSLY, THIS RULE IS FLAWED. INSUFFICIENT.
13'-6"
GATEWAY-GLACIER NATIONAL PARK
I NEEDED TO GIVE UP ANY LONG TRIP.

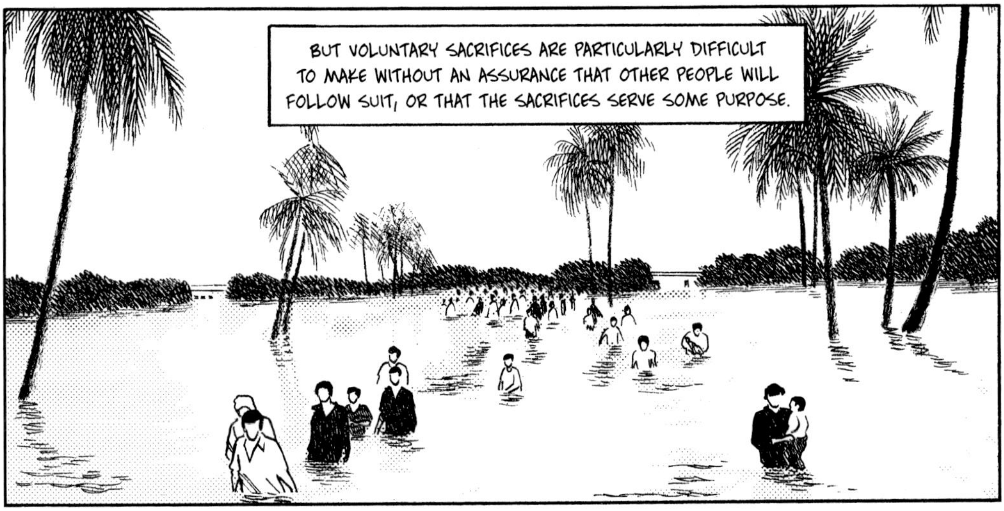
BUT VOLUNTARY SACRIFICES ARE PARTICULARLY DIFFICULT TO MAKE WITHOUT AN ASSURANCE THAT OTHER PEOPLE WILL FOLLOW SUIT, OR THAT THE SACRIFICES SERVE SOME PURPOSE.

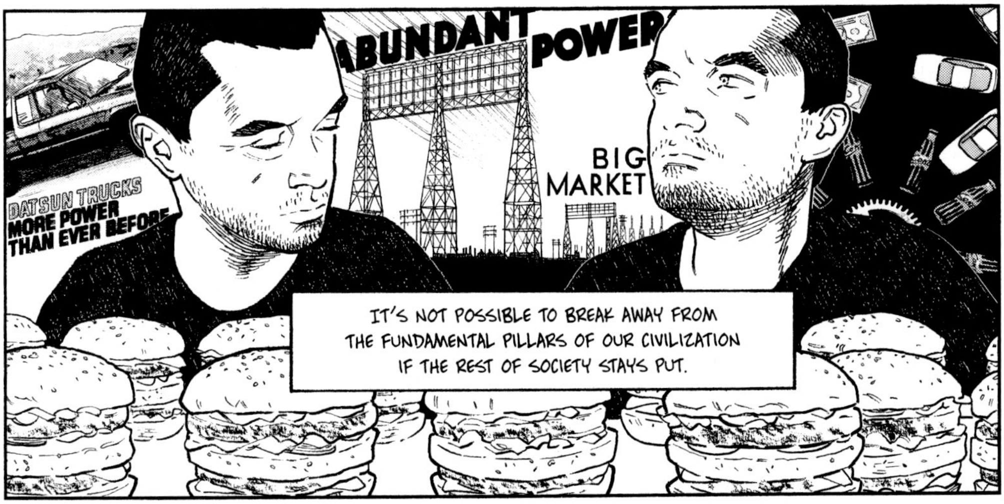
ABUNDANT POWER
BIG MARKET
DATSUN TRUCKS MORE POWER THAN EVER BEFORE
IT'S NOT POSSIBLE TO BREAK AWAY FROM THE FUNDAMENTAL PILLARS OF OUR CIVILIZATION IF THE REST OF SOCIETY STAYS PUT.

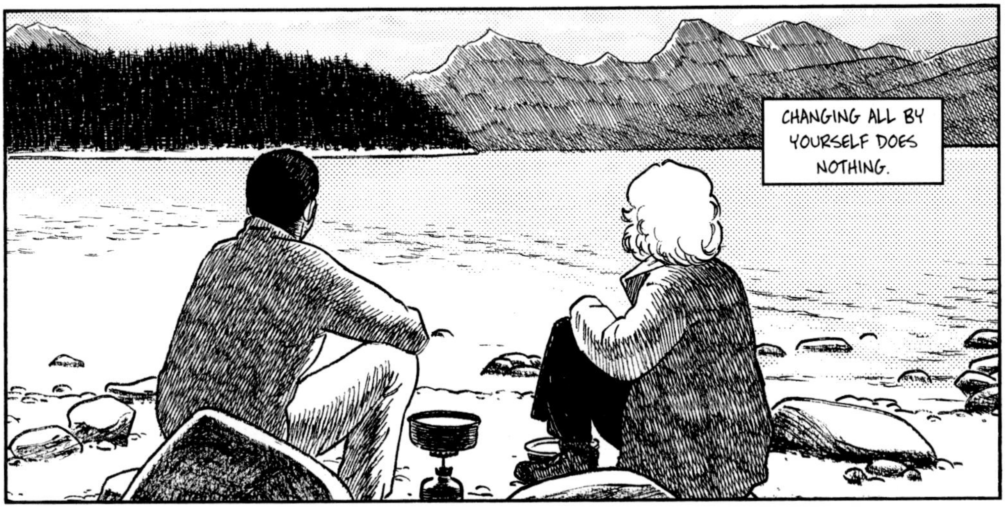
CHANGING ALL BY YOURSELF DOES NOTHING.

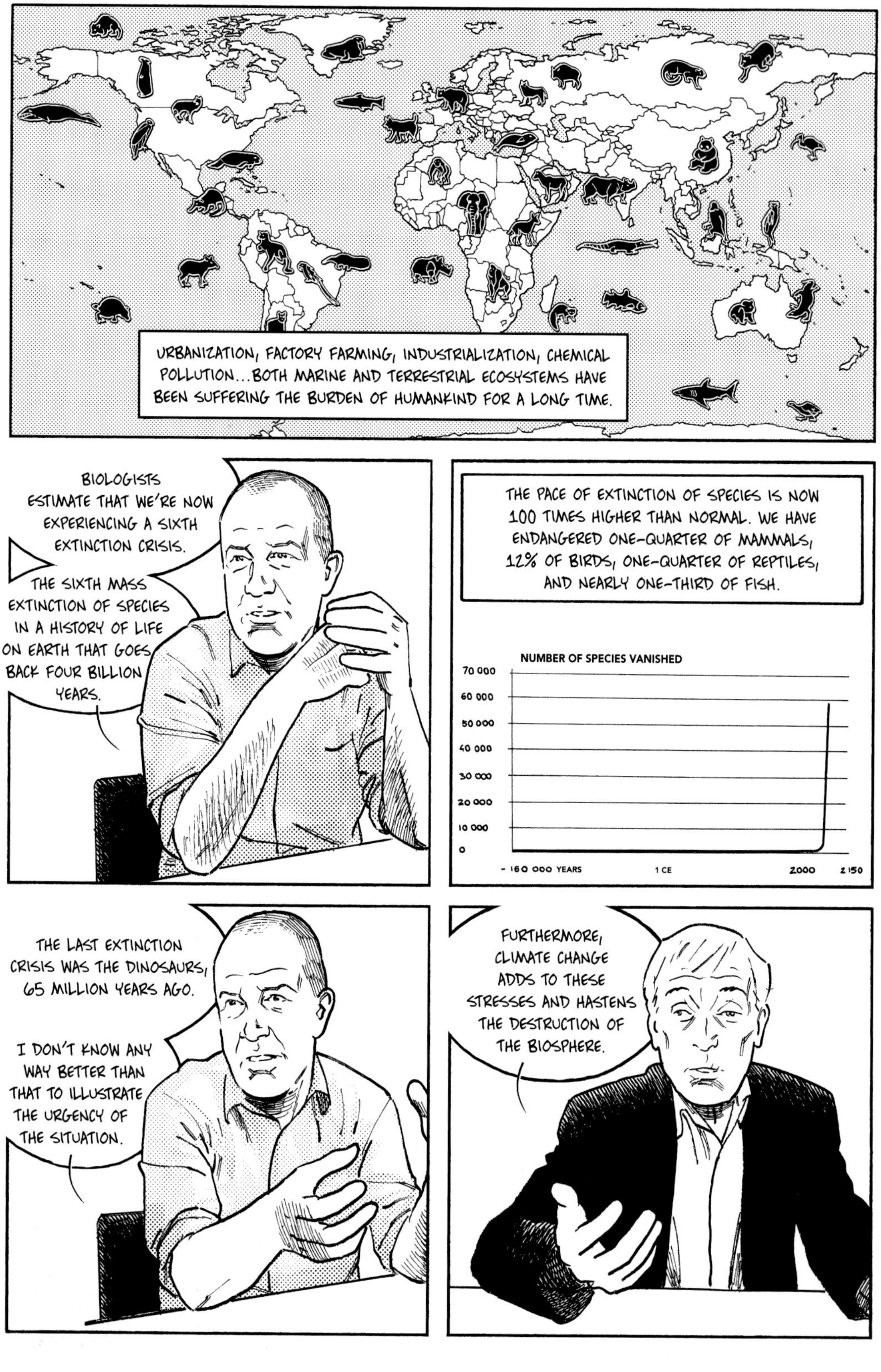
URBANIZATION, FACTORY FARMING, INDUSTRIALIZATION, CHEMICAL POLLUTION... BOTH MARINE AND TERRESTRIAL ECOSYSTEMS HAVE BEEN SUFFERING THE BURDEN OF HUMANKIND FOR A LONG TIME.
BIOLOGISTS ESTIMATE THAT WE'RE NOW EXPERIENCING A SIXTH EXTINCTION CRISIS.
THE SIXTH MASS EXTINCTION OF SPECIES IN A HISTORY OF LIFE ON EARTH THAT GOES BACK FOUR BILLION YEARS.
THE PACE OF EXTINCTION OF SPECIES IS NOW 100 TIMES HIGHER THAN NORMAL. WE HAVE ENDANGERED ONE-QUARTER OF MAMMALS, 12% OF BIRDS, ONE-QUARTER OF REPTILES, AND NEARLY ONE-THIRD OF FISH.
NUMBER OF SPECIES VANISHED
70 000
60 000
50 000
40 000
30 000
20 000
10 000
0
- 160 000 YEARS
1 CE
2000
2150
THE LAST EXTINCTION CRISIS WAS THE DINOSAURS, 65 MILLION YEARS AGO.
I DON'T KNOW ANY WAY BETTER THAN THAT TO ILLUSTRATE THE URGENCY OF THE SITUATION.
FURTHERMORE, CLIMATE CHANGE ADDS TO THESE STRESSES AND HASTENS THE DESTRUCTION OF THE BIOSPHERE.

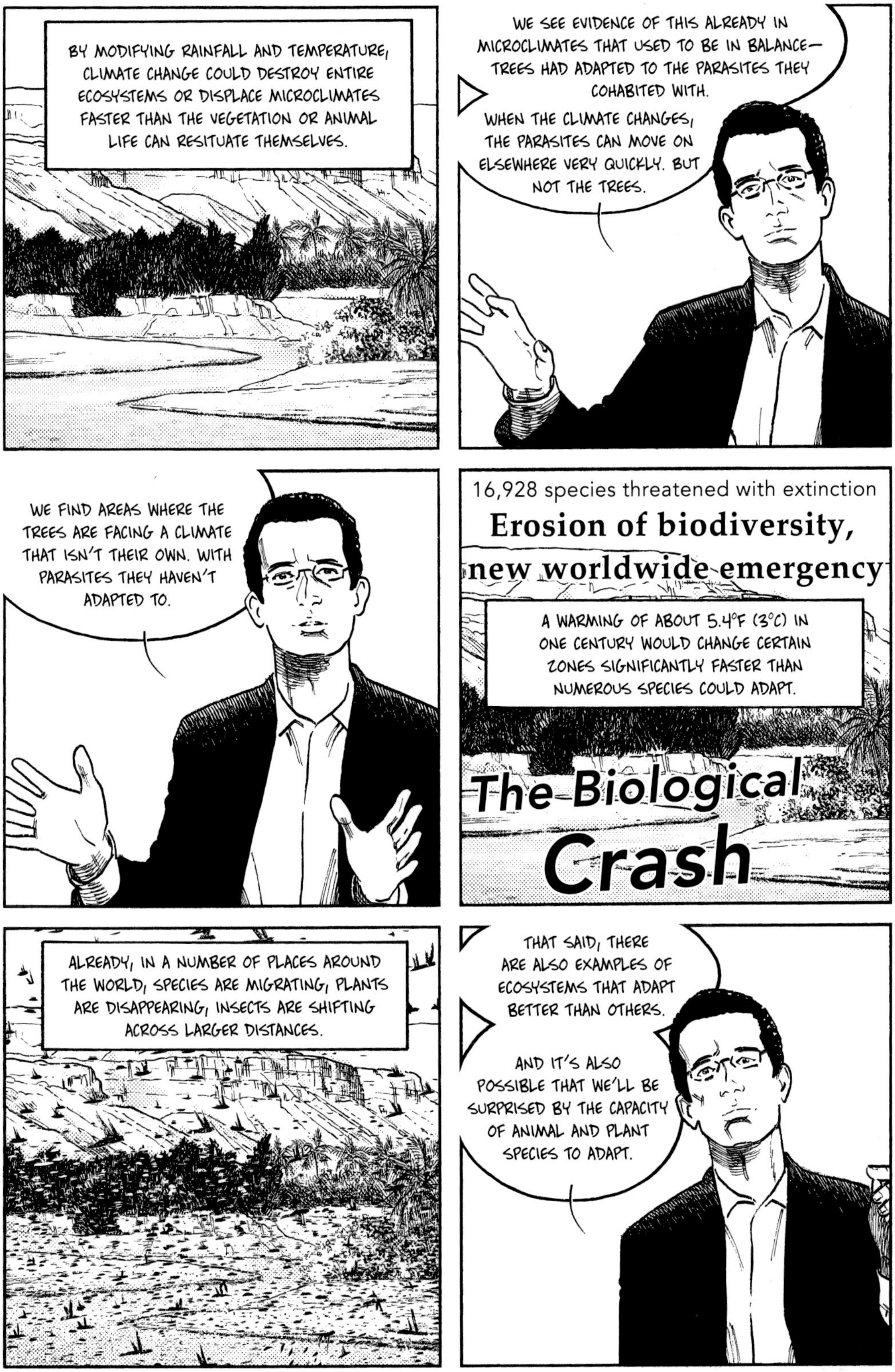
BY MODIFYING RAINFALL AND TEMPERATURE, CLIMATE CHANGE COULD DESTROY ENTIRE ECOSYSTEMS OR DISPLACE MICROCLIMATES FASTER THAN THE VEGETATION OR ANIMAL LIFE CAN RESITUATE THEMSELVES.
WE SEE EVIDENCE OF THIS ALREADY IN MICROCLIMATES THAT USED TO BE IN BALANCE—TREES HAD ADAPTED TO THE PARASITES THEY COHABITED WITH.
WHEN THE CLIMATE CHANGES, THE PARASITES CAN MOVE ON ELSEWHERE VERY QUICKLY. BUT NOT THE TREES.
WE FIND AREAS WHERE THE TREES ARE FACING A CLIMATE THAT ISN'T THEIR OWN. WITH PARASITES THEY HAVEN'T ADAPTED TO.
16,928 species threatened with extinction
Erosion of biodiversity, new worldwide emergency
A WARMING OF ABOUT 5.4°F (3°C) IN ONE CENTURY WOULD CHANGE CERTAIN ZONES SIGNIFICANTLY FASTER THAN NUMEROUS SPECIES COULD ADAPT.
The Biological Crash
ALREADY, IN A NUMBER OF PLACES AROUND THE WORLD, SPECIES ARE MIGRATING, PLANTS ARE DISAPPEARING, INSECTS ARE SHIFTING ACROSS LARGER DISTANCES.
THAT SAID, THERE ARE ALSO EXAMPLES OF ECOSYSTEMS THAT ADAPT BETTER THAN OTHERS.
AND IT'S ALSO POSSIBLE THAT WE'LL BE SURPRISED BY THE CAPACITY OF ANIMAL AND PLANT SPECIES TO ADAPT.

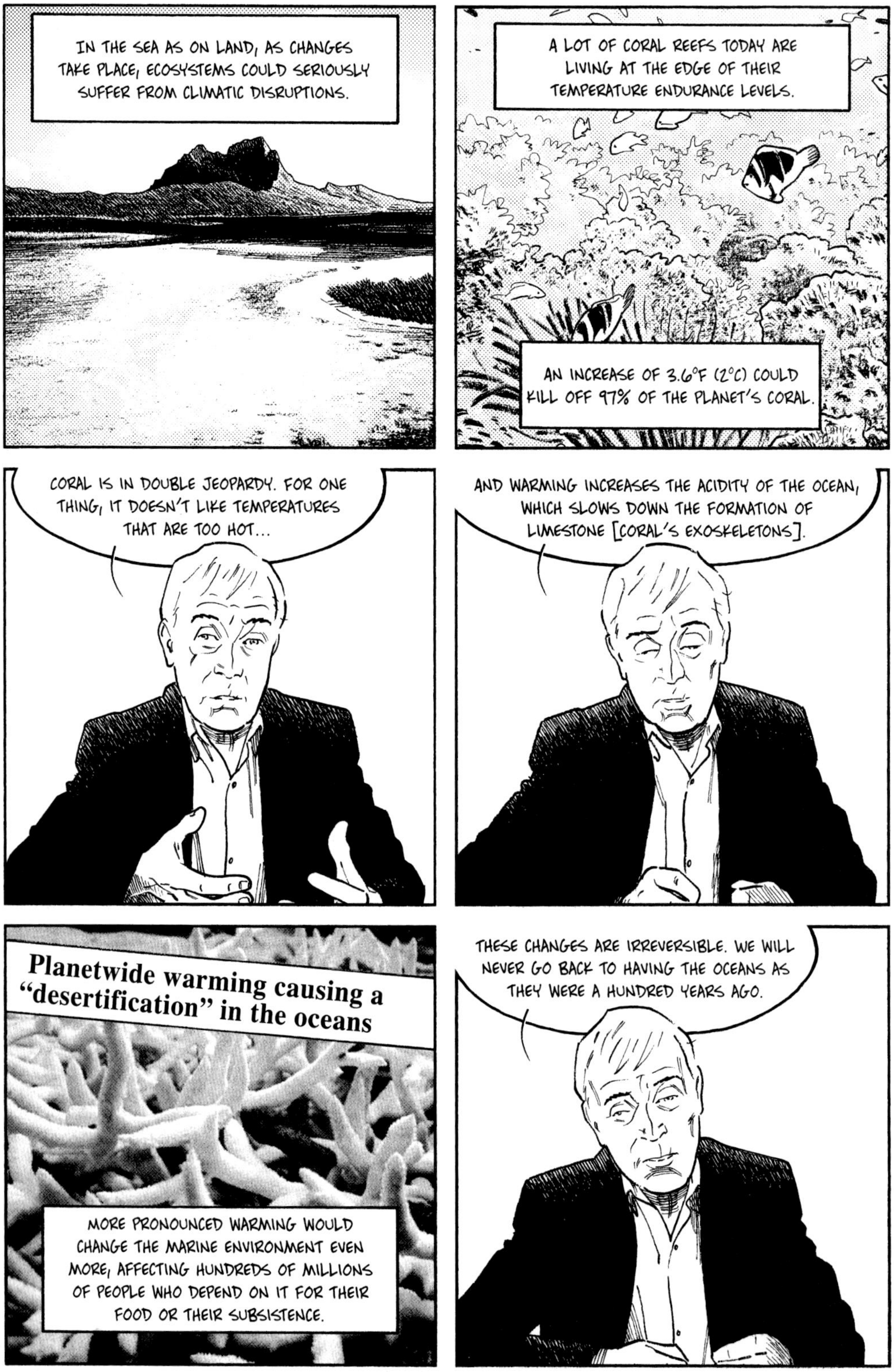
IN THE SEA AS ON LAND, AS CHANGES TAKE PLACE, ECOSYSTEMS COULD SERIOUSLY SUFFER FROM CLIMATIC DISRUPTIONS.
A LOT OF CORAL REEFS TODAY ARE LIVING AT THE EDGE OF THEIR TEMPERATURE ENDURANCE LEVELS.
AN INCREASE OF 3.6°F (2°C) COULD KILL OFF 97% OF THE PLANET'S CORAL.
CORAL IS IN DOUBLE JEOPARDY. FOR ONE THING, IT DOESN'T LIKE TEMPERATURES THAT ARE TOO HOT...
AND WARMING INCREASES THE ACIDITY OF THE OCEAN, WHICH SLOWS DOWN THE FORMATION OF LIMESTONE [CORAL'S EXOSKELETONS].
Planetwide warming causing a "desertification" in the oceans
MORE PRONOUNCED WARMING WOULD CHANGE THE MARINE ENVIRONMENT EVEN MORE, AFFECTING HUNDREDS OF MILLIONS OF PEOPLE WHO DEPEND ON IT FOR THEIR FOOD OR THEIR SUBSISTENCE.
THESE CHANGES ARE IRREVERSIBLE. WE WILL NEVER GO BACK TO HAVING THE OCEANS AS THEY WERE A HUNDRED YEARS AGO.

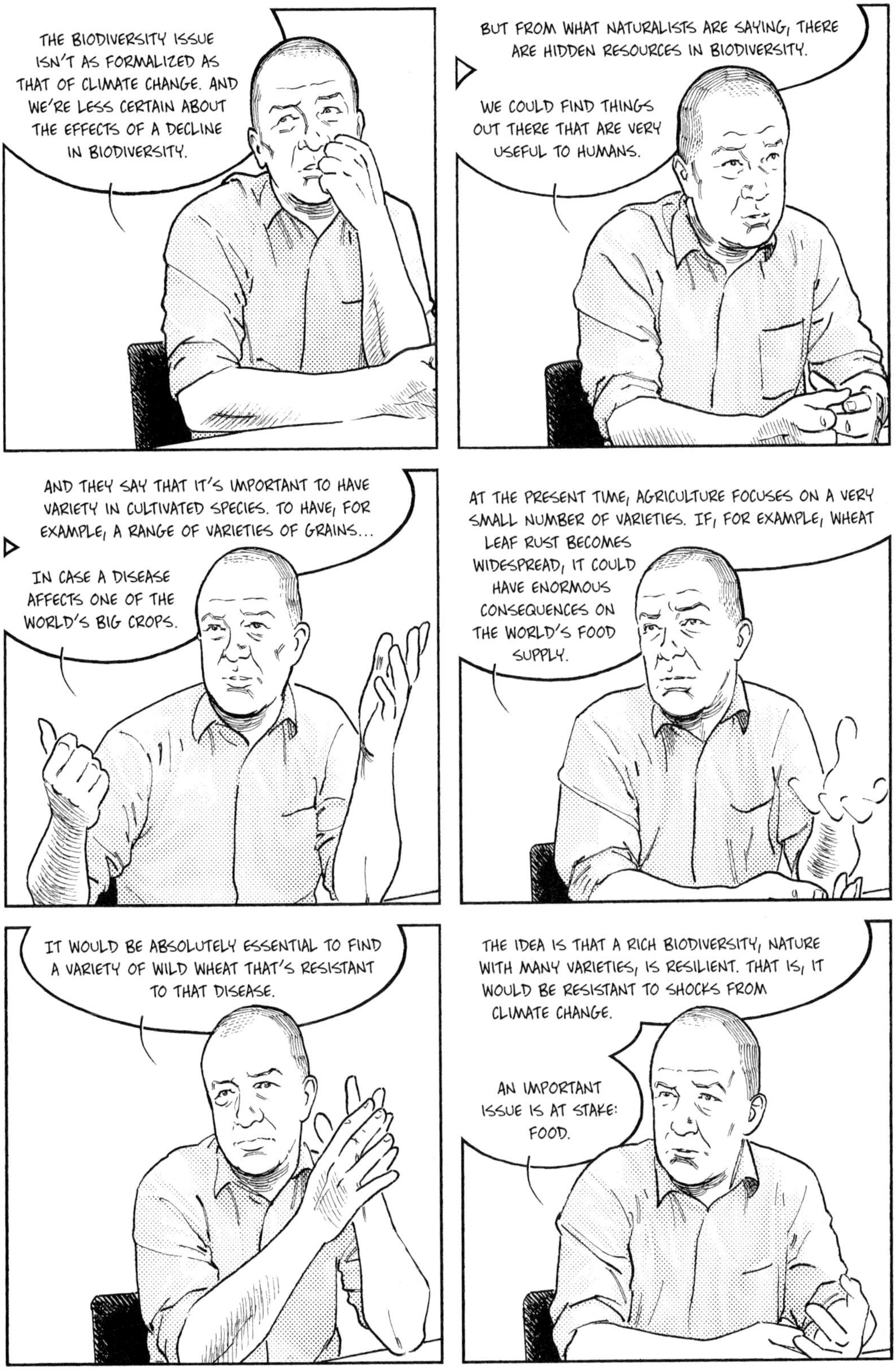
THE BIODIVERSITY ISSUE ISN'T AS FORMALIZED AS THAT OF CLIMATE CHANGE. AND WE'RE LESS CERTAIN ABOUT THE EFFECTS OF A DECLINE IN BIODIVERSITY.
BUT FROM WHAT NATURALISTS ARE SAYING, THERE ARE HIDDEN RESOURCES IN BIODIVERSITY.
WE COULD FIND THINGS OUT THERE THAT ARE VERY USEFUL TO HUMANS.
AND THEY SAY THAT IT'S IMPORTANT TO HAVE VARIETY IN CULTIVATED SPECIES. TO HAVE, FOR EXAMPLE, A RANGE OF VARIETIES OF GRAINS...
IN CASE A DISEASE AFFECTS ONE OF THE WORLD'S BIG CROPS.
AT THE PRESENT TIME, AGRICULTURE FOCUSES ON A VERY SMALL NUMBER OF VARIETIES. IF, FOR EXAMPLE, WHEAT LEAF RUST BECOMES WIDESPREAD, IT COULD HAVE ENORMOUS CONSEQUENCES ON THE WORLD'S FOOD SUPPLY.
IT WOULD BE ABSOLUTELY ESSENTIAL TO FIND A VARIETY OF WILD WHEAT THAT'S RESISTANT TO THAT DISEASE.
THE IDEA IS THAT A RICH BIODIVERSITY, NATURE WITH MANY VARIETIES, IS RESILIENT. THAT IS, IT WOULD BE RESISTANT TO SHOCKS FROM CLIMATE CHANGE.
AN IMPORTANT ISSUE IS AT STAKE: FOOD.

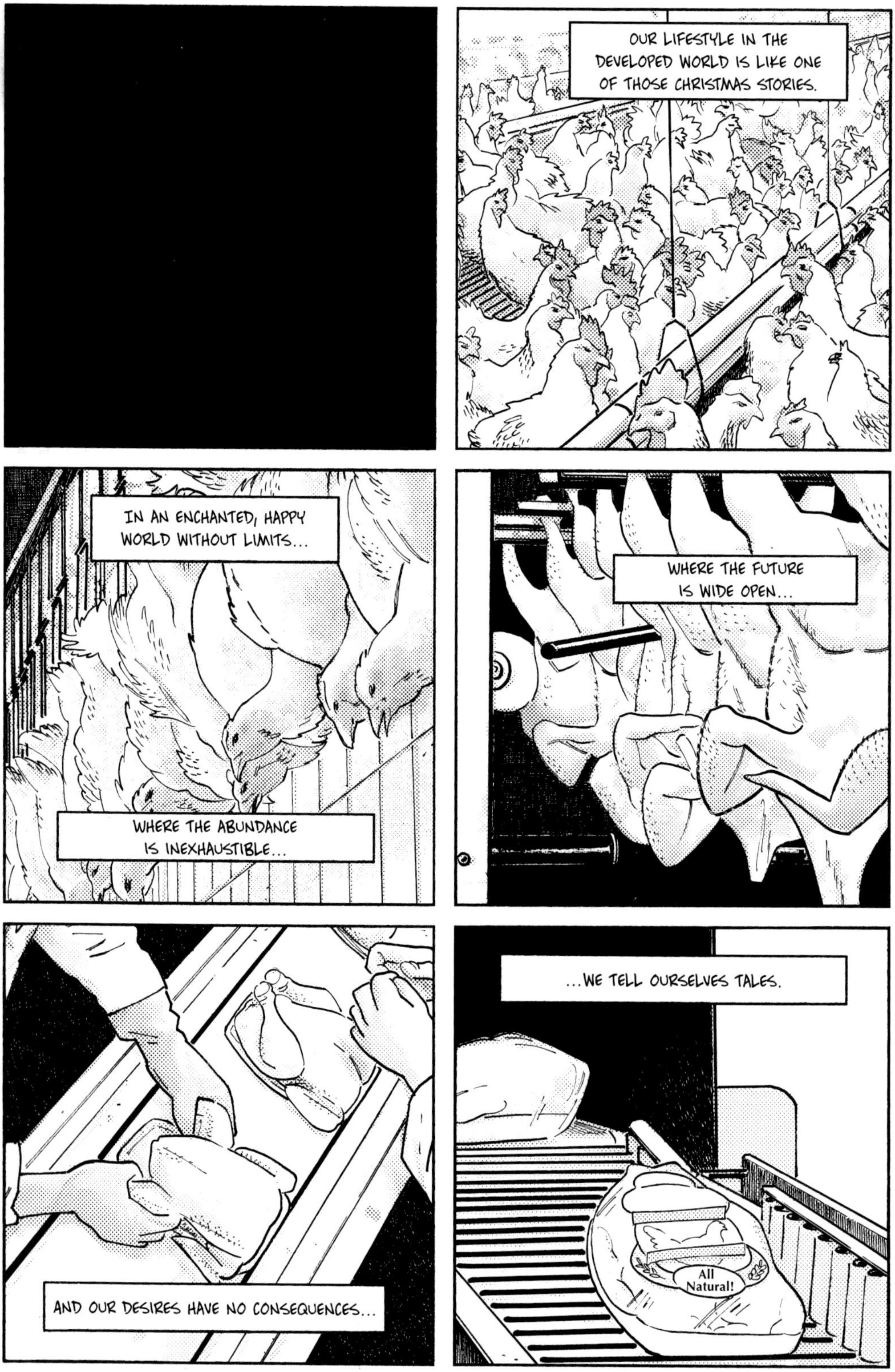
OUR LIFESTYLE IN THE DEVELOPED WORLD IS LIKE ONE OF THOSE CHRISTMAS STORIES.
IN AN ENCHANTED, HAPPY WORLD WITHOUT LIMITS...
WHERE THE ABUNDANCE IS INEXHAUSTIBLE...
WHERE THE FUTURE IS WIDE OPEN...
AND OUR DESIRES HAVE NO CONSEQUENCES...
...WE TELL OURSELVES TALES.
All Natural!

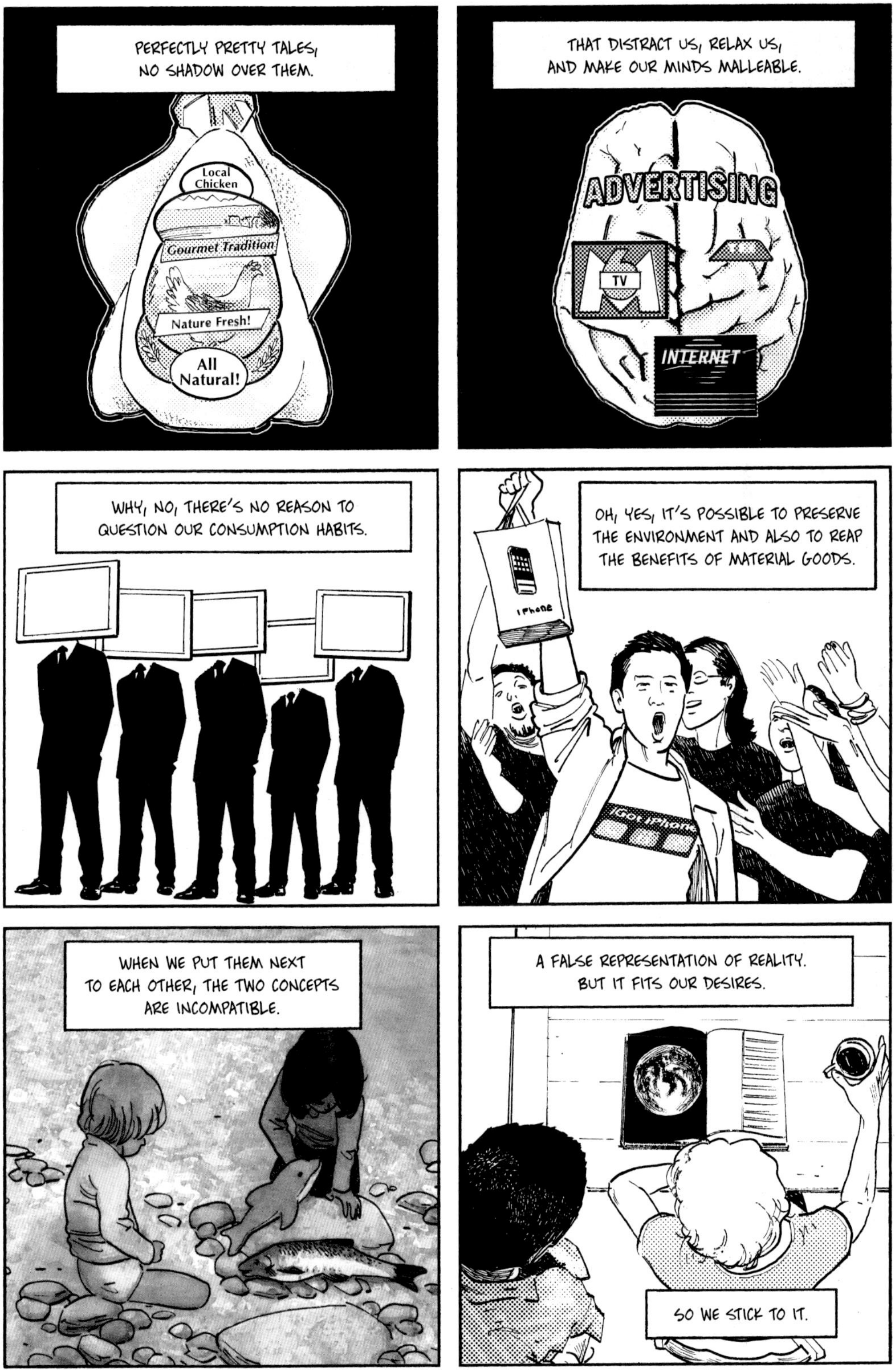
PERFECTLY PRETTY TALES, NO SHADOW OVER THEM.
Local Chicken
Gourmet Tradition
Nature Fresh!
All Natural!
THAT DISTRACT US, RELAX US, AND MAKE OUR MINDS MALLEABLE.
ADVERTISING
TV
INTERNET
WHY, NO, THERE'S NO REASON TO QUESTION OUR CONSUMPTION HABITS.
OH, YES, IT'S POSSIBLE TO PRESERVE THE ENVIRONMENT AND ALSO TO REAP THE BENEFITS OF MATERIAL GOODS.
iPhone
WHEN WE PUT THEM NEXT TO EACH OTHER, THE TWO CONCEPTS ARE INCOMPATIBLE.
A FALSE REPRESENTATION OF REALITY. BUT IT FITS OUR DESIRES.
SO WE STICK TO IT.

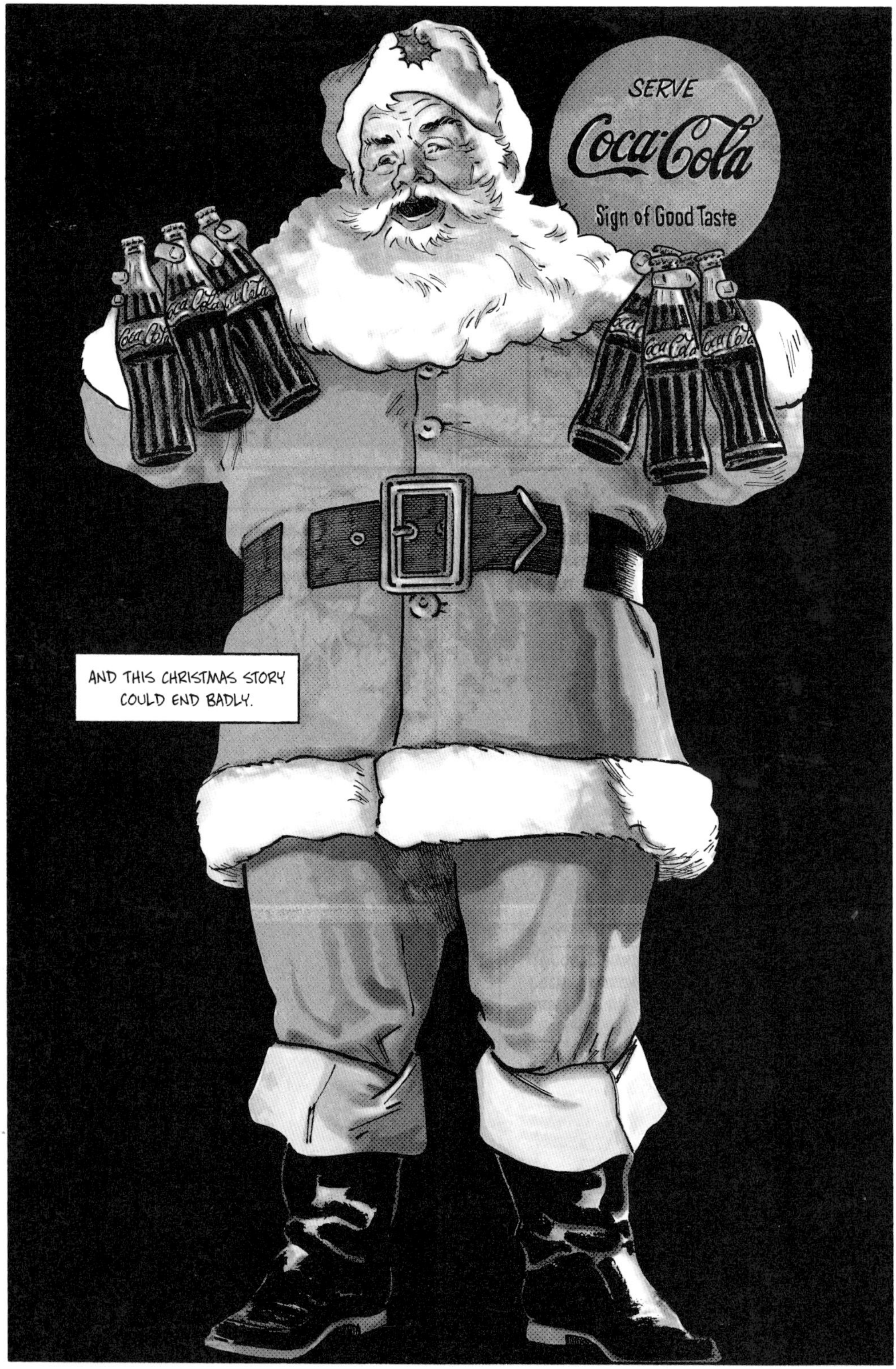
SERVE
Coca-Cola
Sign of Good Taste
AND THIS CHRISTMAS STORY COULD END BADLY.

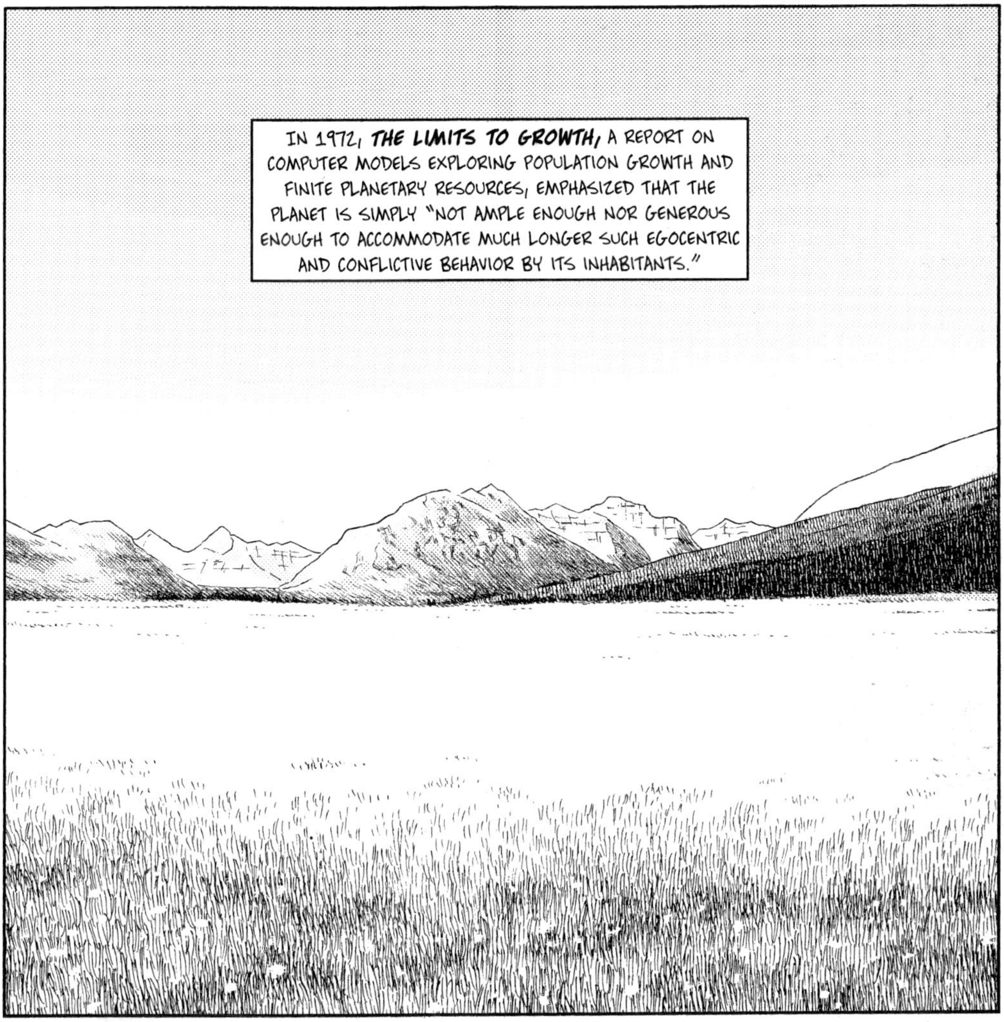
IN 1972, THE LIMITS TO GROWTH, A REPORT ON COMPUTER MODELS EXPLORING POPULATION GROWTH AND FINITE PLANETARY RESOURCES, EMPHASIZED THAT THE PLANET IS SIMPLY "NOT AMPLE ENOUGH NOR GENEROUS ENOUGH TO ACCOMMODATE MUCH LONGER SUCH EGOCENTRIC AND CONFLICTIVE BEHAVIOR BY ITS INHABITANTS."

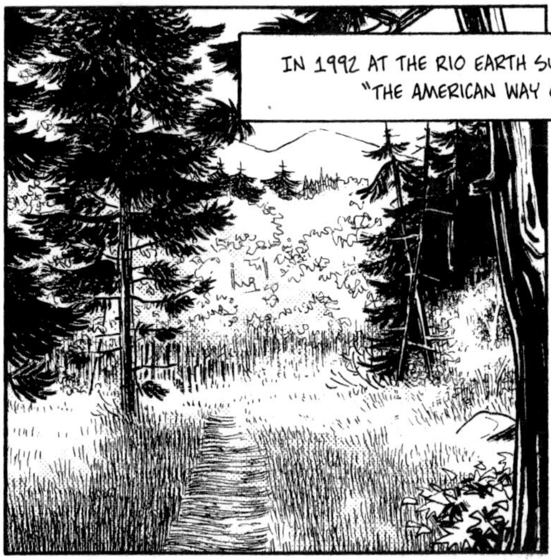
IN 1992 AT THE RIO EARTH SUMMIT, GEORGE H. W. BUSH DECLARED, "THE AMERICAN WAY OF LIFE IS NOT NEGOTIABLE."

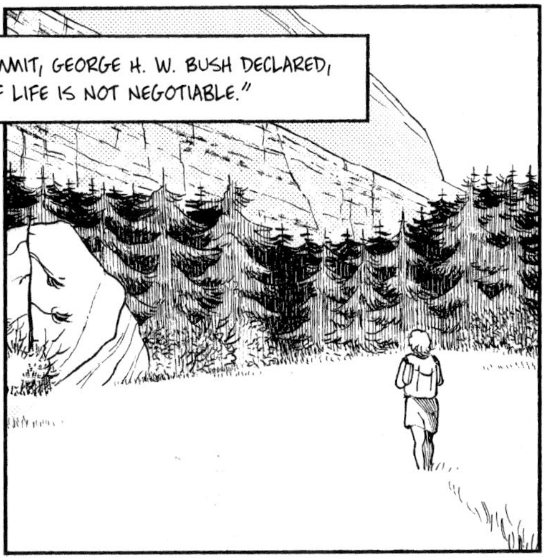

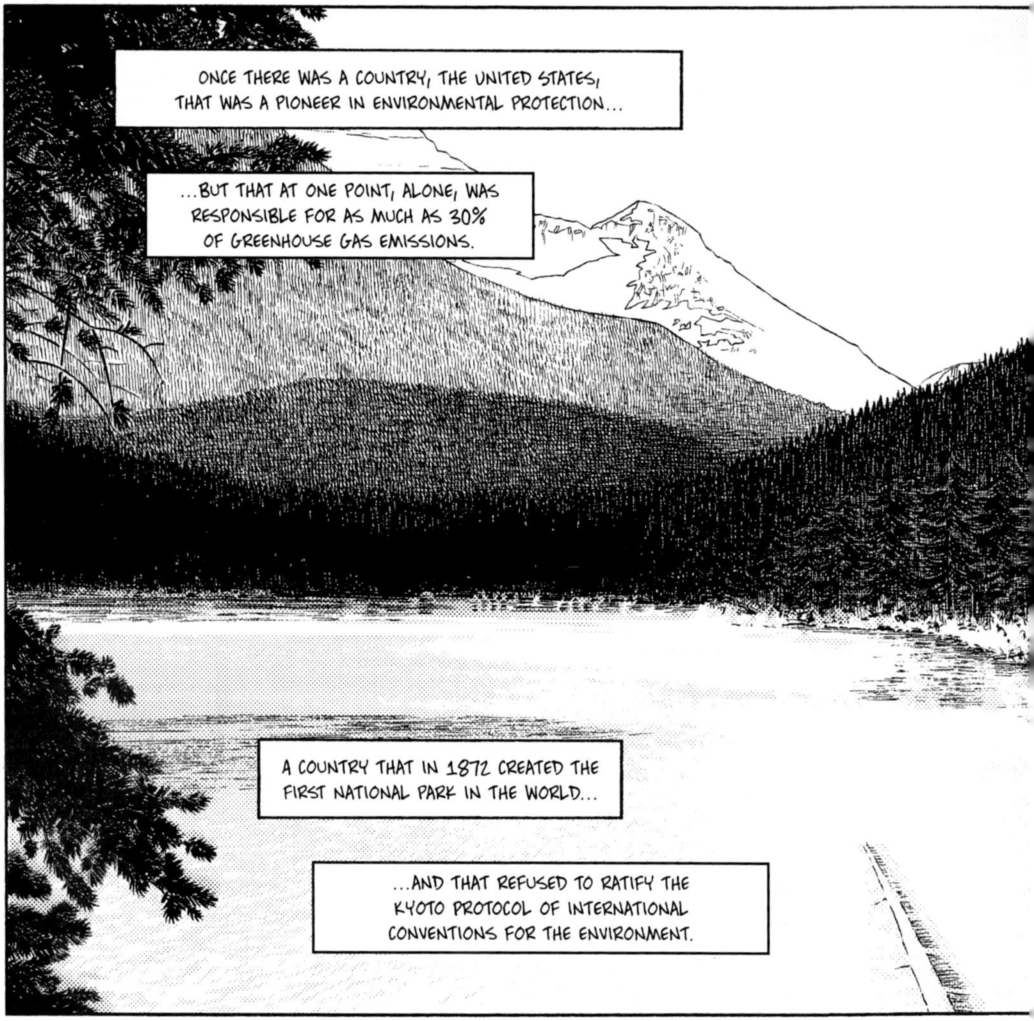
ONCE THERE WAS A COUNTRY, THE UNITED STATES, THAT WAS A PIONEER IN ENVIRONMENTAL PROTECTION...
...BUT THAT AT ONE POINT, ALONE, WAS RESPONSIBLE FOR AS MUCH AS 30% OF GREENHOUSE GAS EMISSIONS.
A COUNTRY THAT IN 1872 CREATED THE FIRST NATIONAL PARK IN THE WORLD...
...AND THAT REFUSED TO RATIFY THE KYOTO PROTOCOL OF INTERNATIONAL CONVENTIONS FOR THE ENVIRONMENT.

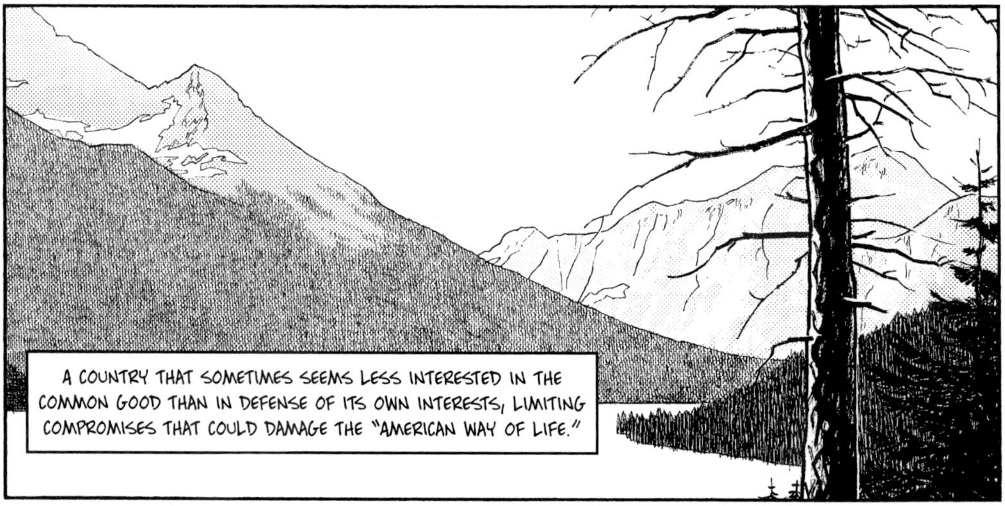
A COUNTRY THAT SOMETIMES SEEMS LESS INTERESTED IN THE COMMON GOOD THAN IN DEFENSE OF ITS OWN INTERESTS, LIMITING COMPROMISES THAT COULD DAMAGE THE "AMERICAN WAY OF LIFE."

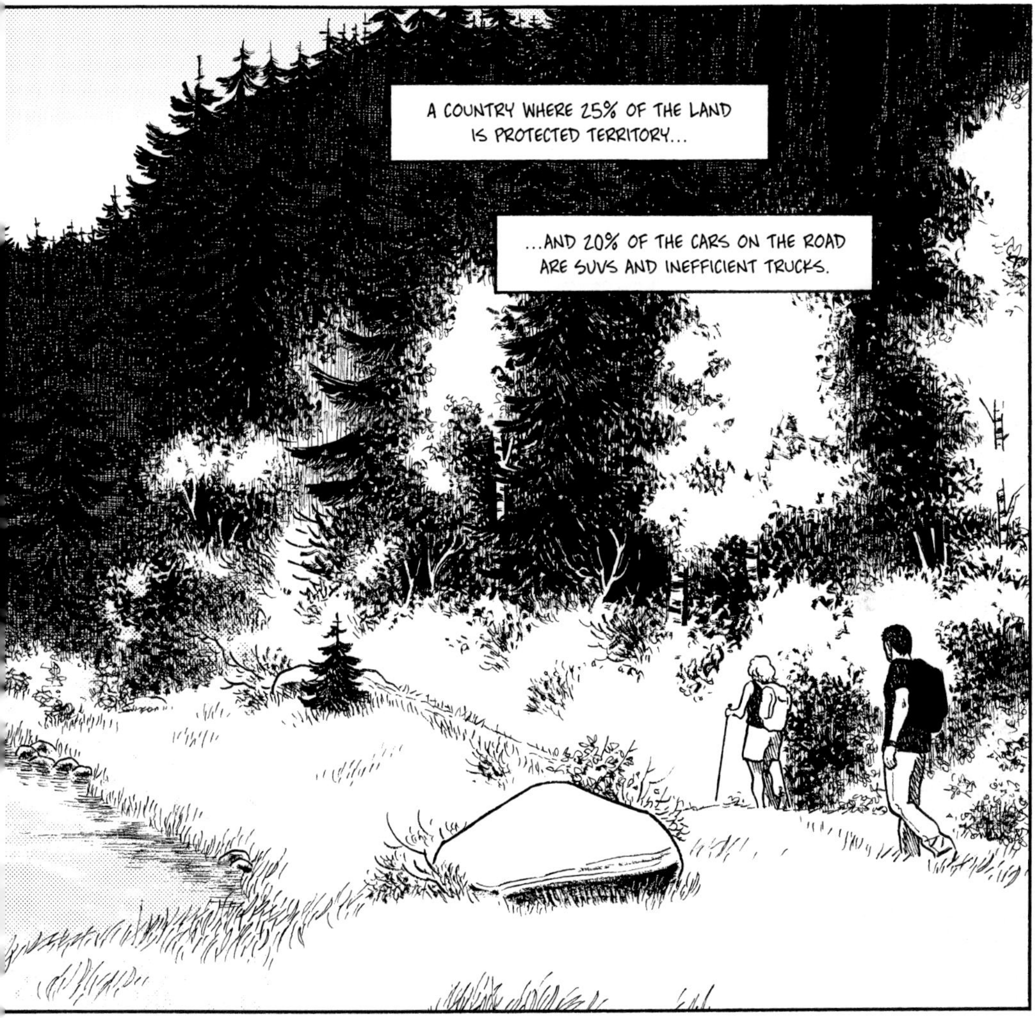
A COUNTRY WHERE 25% OF THE LAND IS PROTECTED TERRITORY...
...AND 20% OF THE CARS ON THE ROAD ARE SUVS AND INEFFICIENT TRUCKS.

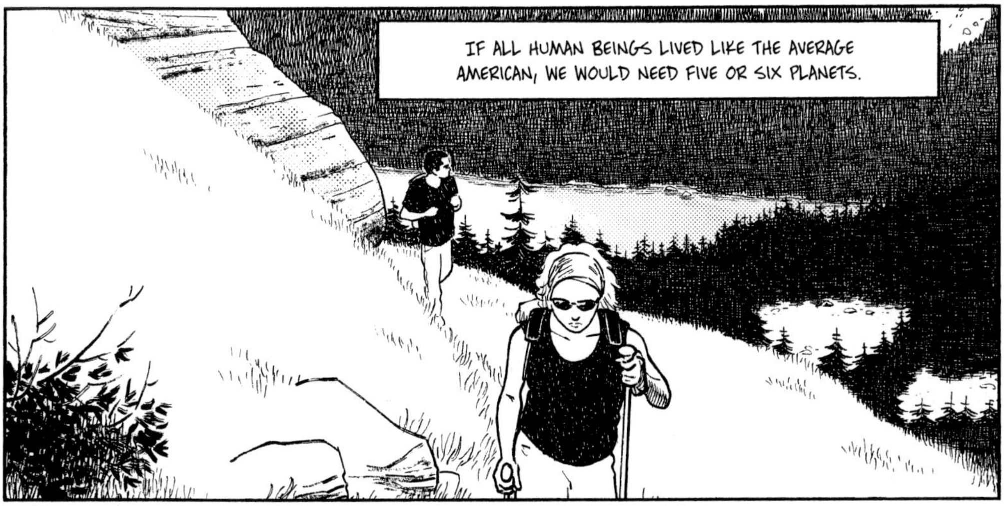
IF ALL HUMAN BEINGS LIVED LIKE THE AVERAGE AMERICAN, WE WOULD NEED FIVE OR SIX PLANETS.

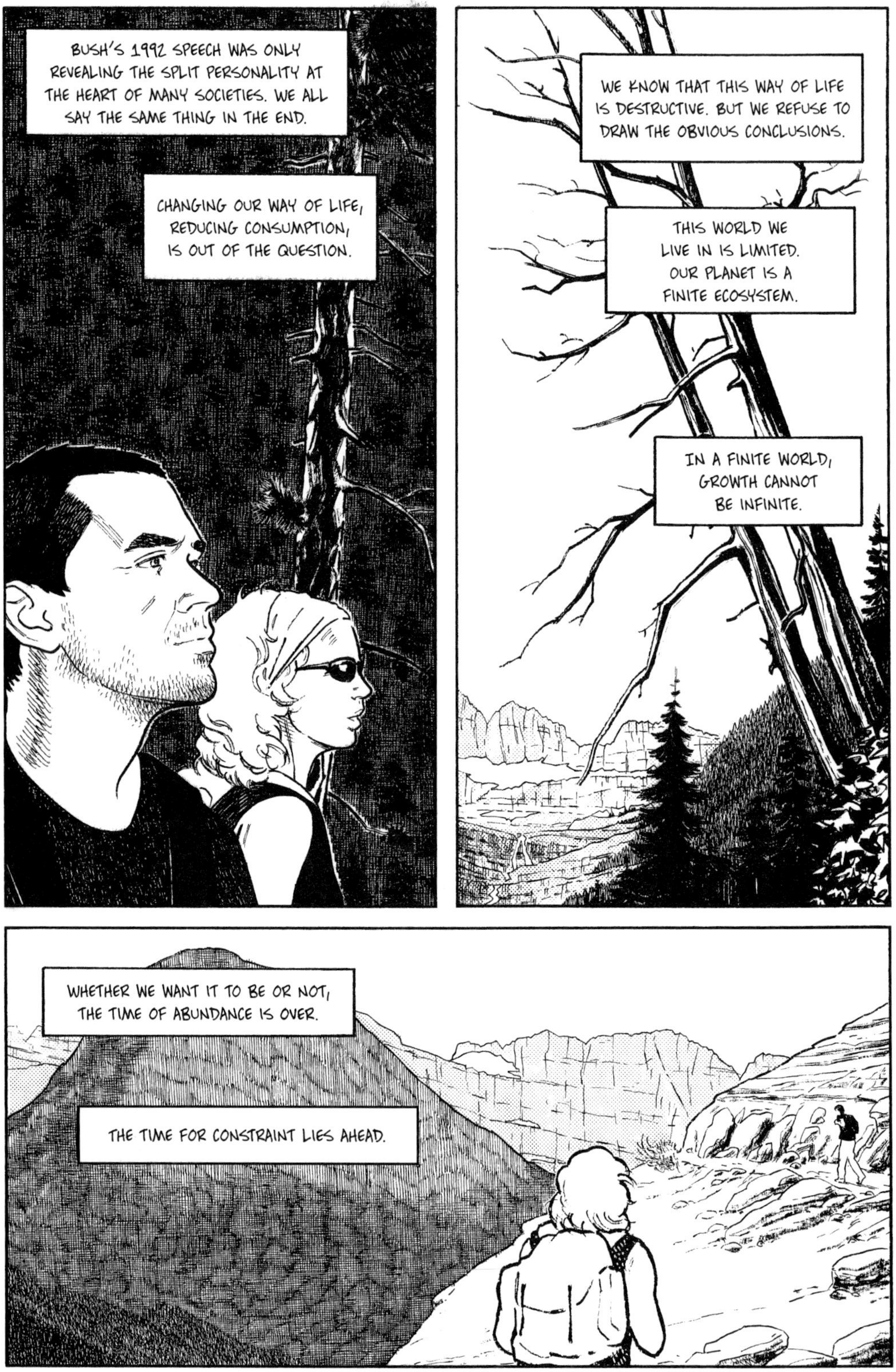
BUSH'S 1992 SPEECH WAS ONLY REVEALING THE SPLIT PERSONALITY AT THE HEART OF MANY SOCIETIES. WE ALL SAY THE SAME THING IN THE END.
CHANGING OUR WAY OF LIFE, REDUCING CONSUMPTION, IS OUT OF THE QUESTION.
WE KNOW THAT THIS WAY OF LIFE IS DESTRUCTIVE. BUT WE REFUSE TO DRAW THE OBVIOUS CONCLUSIONS.
THIS WORLD WE LIVE IN IS LIMITED. OUR PLANET IS A FINITE ECOSYSTEM.
IN A FINITE WORLD, GROWTH CANNOT BE INFINITE.
WHETHER WE WANT IT TO BE OR NOT, THE TIME OF ABUNDANCE IS OVER.
THE TIME FOR CONSTRAINT LIES AHEAD.

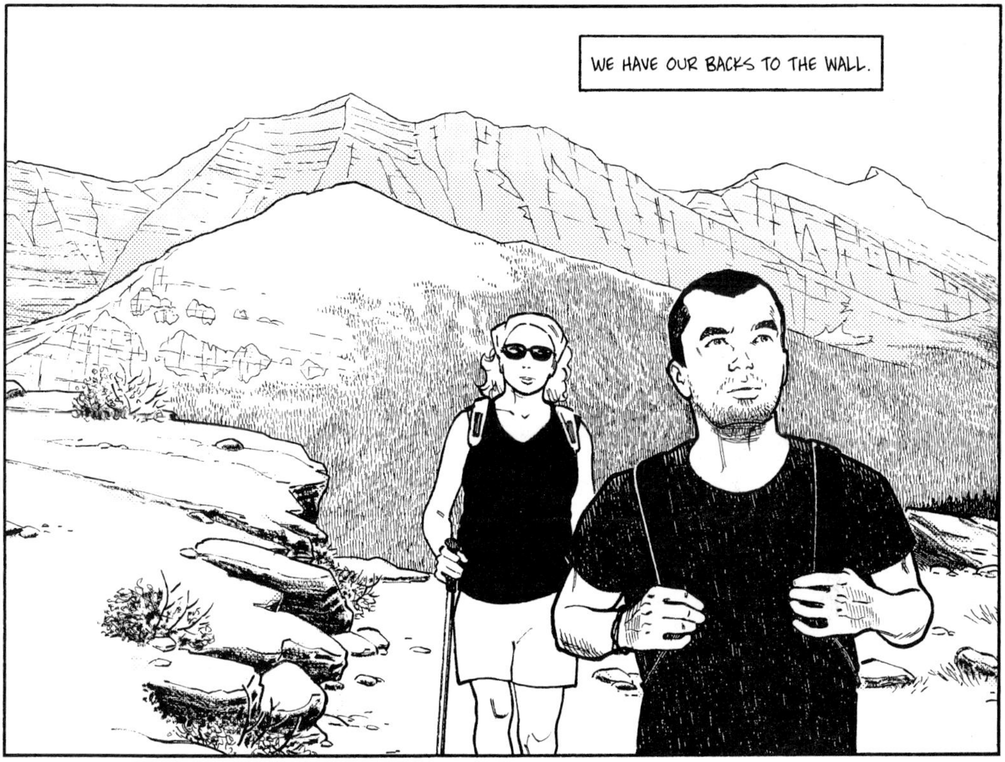
WE HAVE OUR BACKS TO THE WALL.

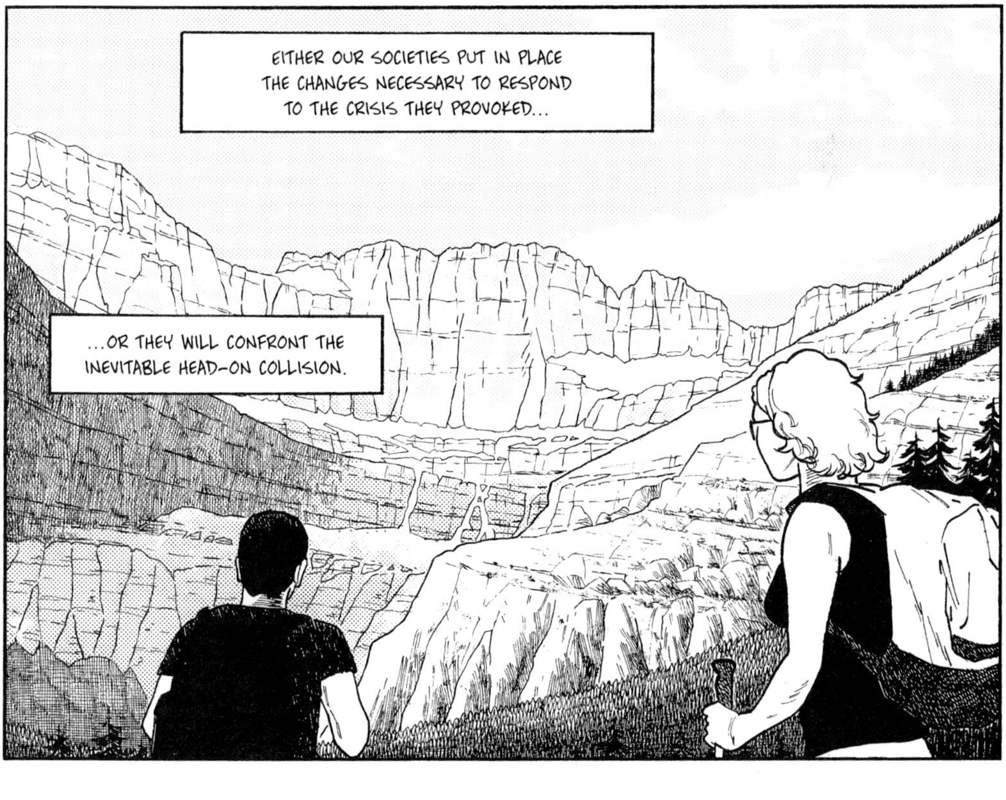
EITHER OUR SOCIETIES PUT IN PLACE THE CHANGES NECESSARY TO RESPOND TO THE CRISIS THEY PROVOKED...
...OR THEY WILL CONFRONT THE INEVITABLE HEAD-ON COLLISION.

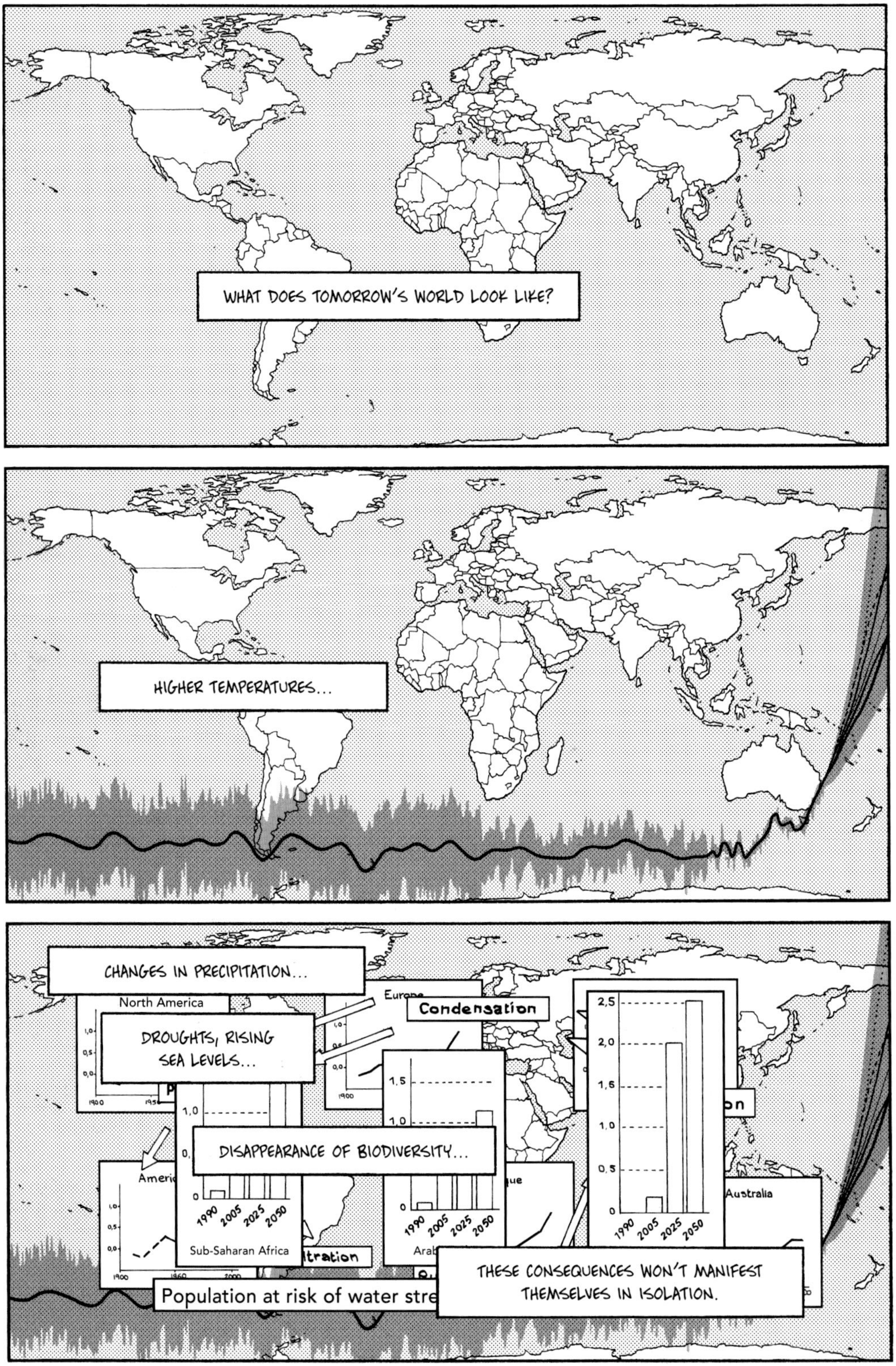
WHAT DOES TOMORROW'S WORLD LOOK LIKE?
HIGHER TEMPERATURES...
CHANGES IN PRECIPITATION...
North America
Europe
Condensation
DROUGHTS, RISING SEA LEVELS...
DISAPPEARANCE OF BIODIVERSITY...
Sub-Saharan Africa
Australia
Population at risk of water
THESE CONSEQUENCES WON'T MANIFEST THEMSELVES IN ISOLATION.

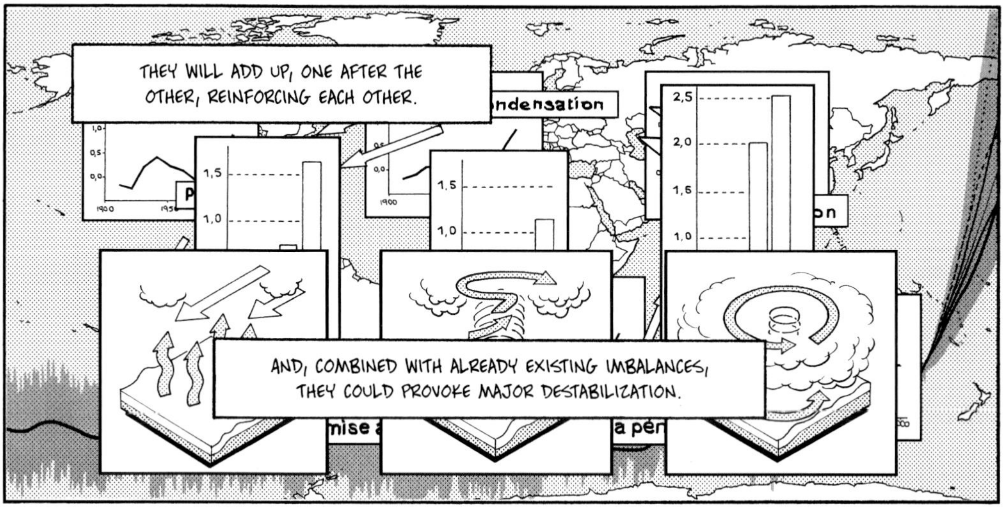
THEY WILL ADD UP, ONE AFTER THE OTHER, REINFORCING EACH OTHER.
AND, COMBINED WITH ALREADY EXISTING IMBALANCES, THEY COULD PROVOKE MAJOR DESTABILIZATION.

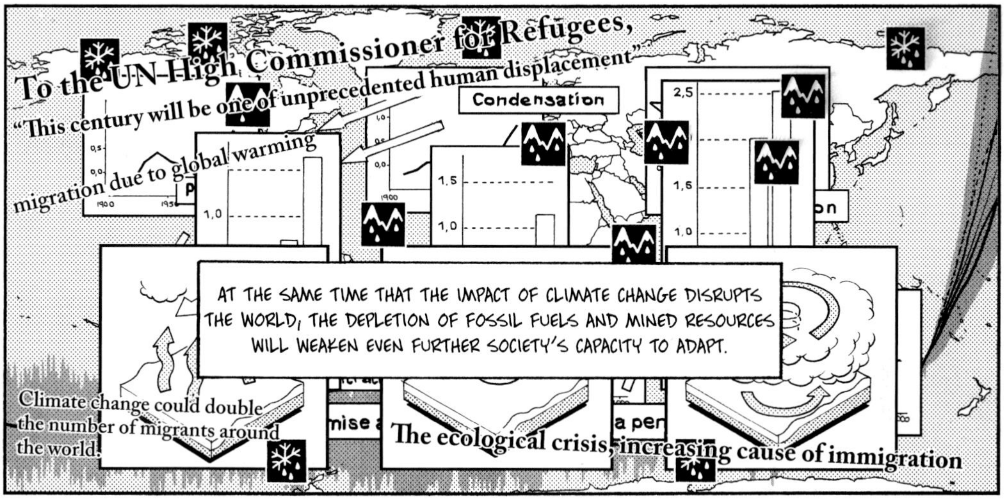
To the UN High Commissioner for Refugees,
"This century will be one of unprecedented human displacement"
migration due to global warming
Condensation
AT THE SAME TIME THAT THE IMPACT OF CLIMATE CHANGE DISRUPTS THE WORLD, THE DEPLETION OF FOSSIL FUELS AND MINED RESOURCES WILL WEAKEN EVEN FURTHER SOCIETY'S CAPACITY TO ADAPT.
Climate change could double the number of migrants around the world!
The ecological crisis, increasing cause of immigration

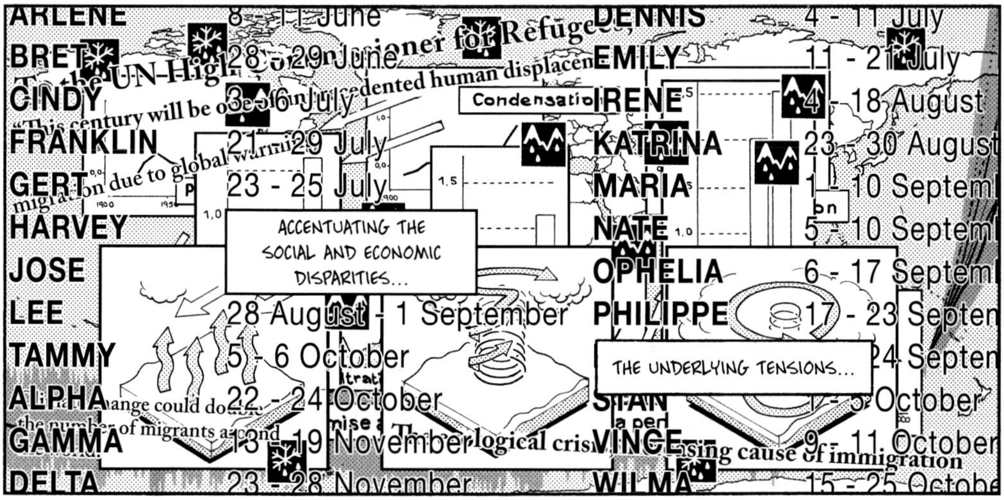
ARLENE 8 - 11 June
BRET 28 - 29 June
CINDY 3 - 6 July
FRANKLIN 21 - 29 July
GERT 23 - 25 July
HARVEY
JOSE
LEE 28 August - 1 September
TAMMY 5 - 6 October
ALPHA 22 - 24 October
GAMMA 13 - 19 November
DELTA 23 - 28 November
DENNIS 4 - 11 July
EMILY 11 - 21 July
IRENE 4 - 18 August
KATRINA 23 - 30 August
MARIA 1 - 10 Septem
NATE 5 - 10 Septem
OPHELIA 6 - 17 Septem
PHILIPPE 17 - 23 Septen
24 Septen
STAN October
VINCE 9 - 11 October
WILMA 15 - 25 Octobe
ACCENTUATING THE SOCIAL AND ECONOMIC DISPARITIES...
THE UNDERLYING TENSIONS...

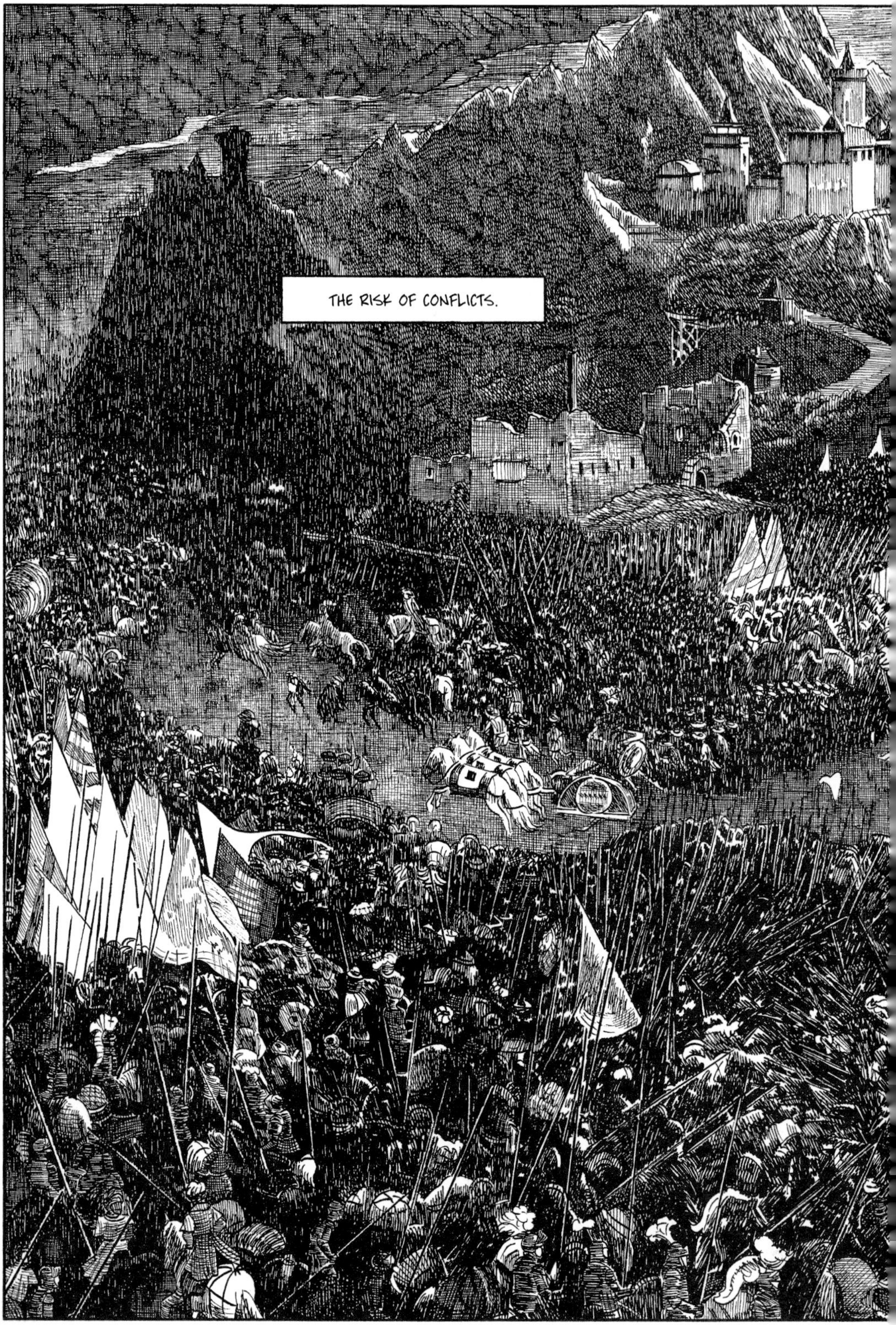
THE RISK OF CONFLICTS.

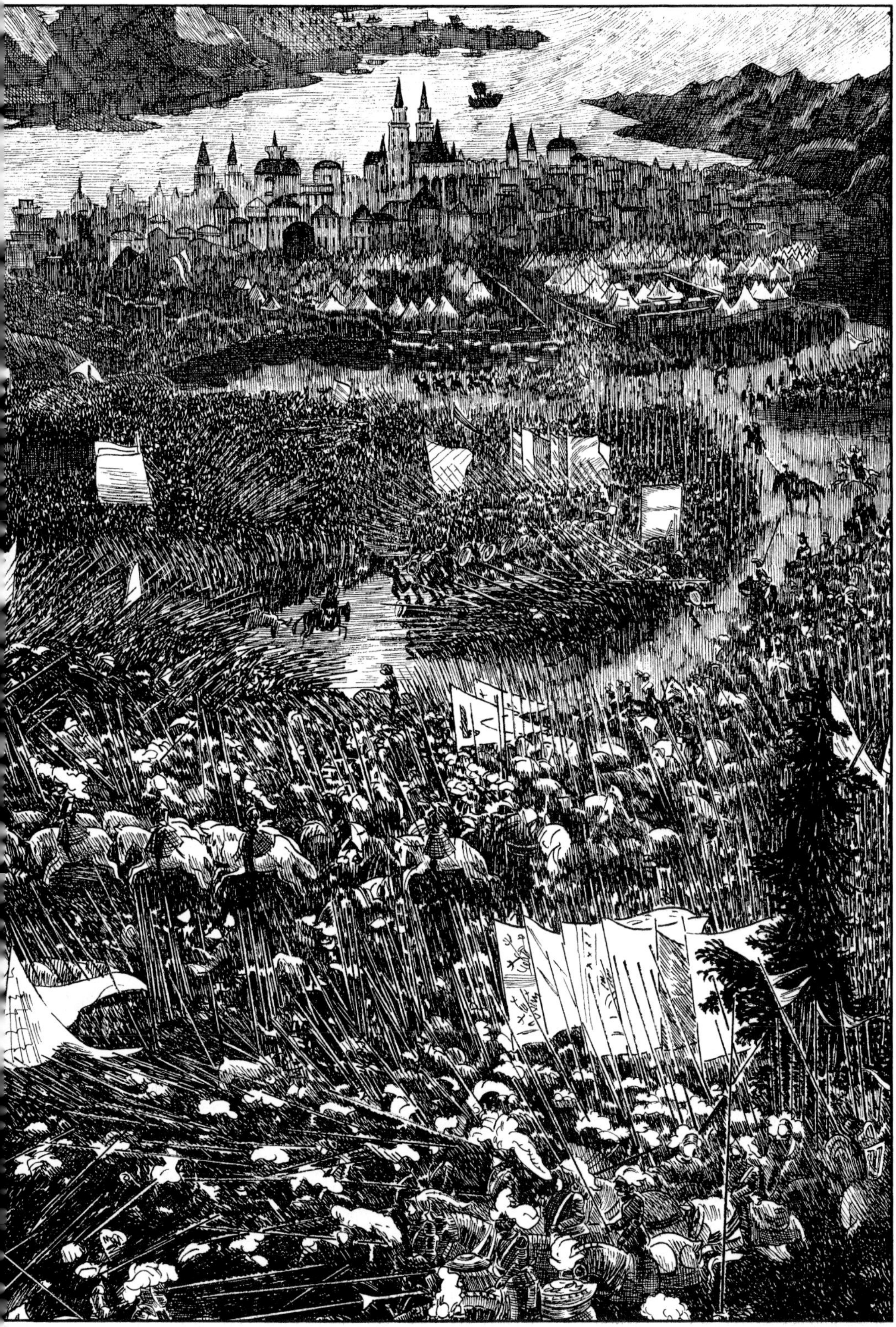

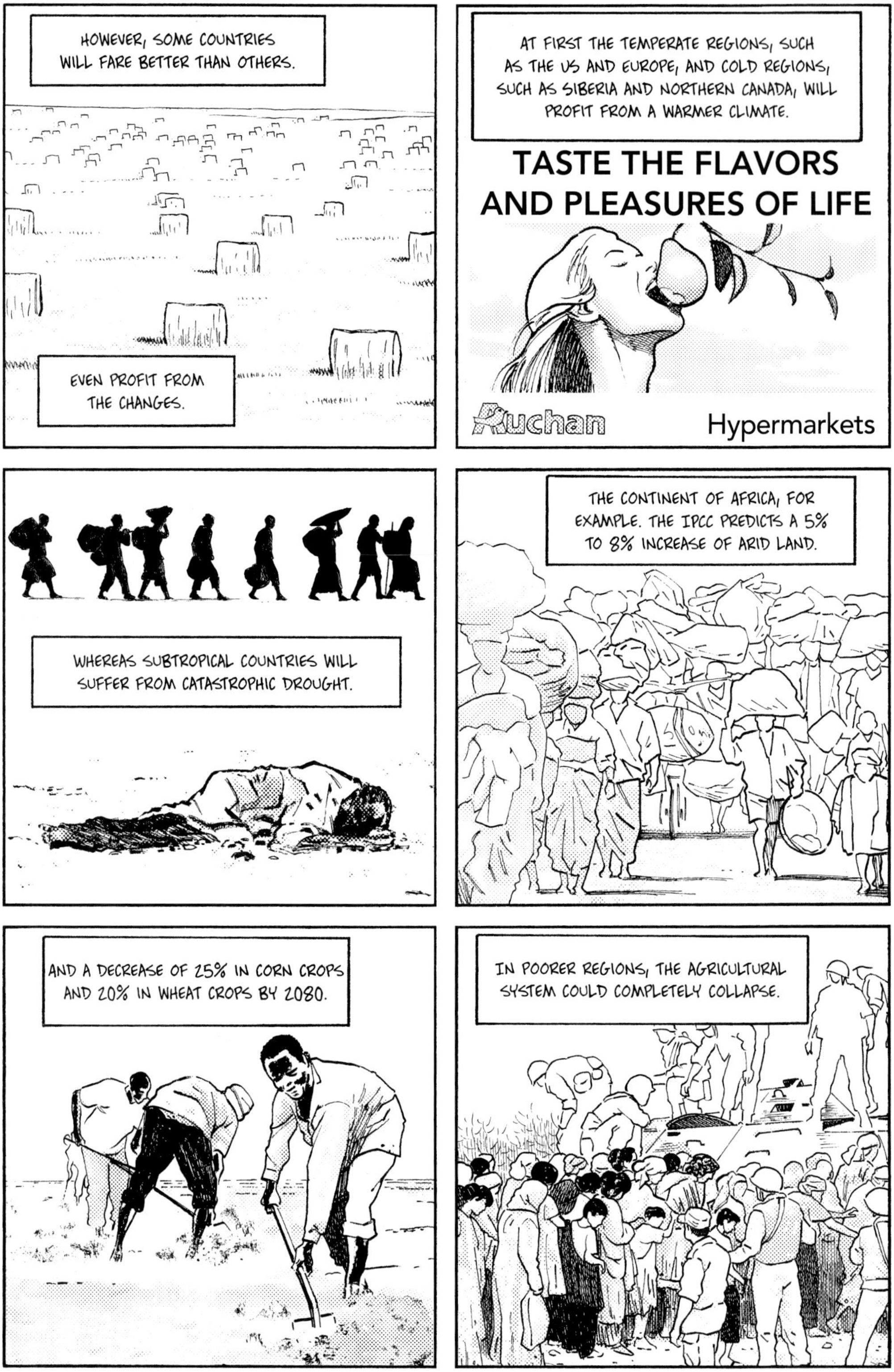
HOWEVER, SOME COUNTRIES WILL FARE BETTER THAN OTHERS.
EVEN PROFIT FROM THE CHANGES.
AT FIRST THE TEMPERATE REGIONS, SUCH AS THE US AND EUROPE, AND COLD REGIONS, SUCH AS SIBERIA AND NORTHERN CANADA, WILL PROFIT FROM A WARMER CLIMATE.
TASTE THE FLAVORS AND PLEASURES OF LIFE
Auchan
Hypermarkets
WHEREAS SUBTROPICAL COUNTRIES WILL SUFFER FROM CATASTROPHIC DROUGHT.
THE CONTINENT OF AFRICA, FOR EXAMPLE. THE IPCC PREDICTS A 5% TO 8% INCREASE OF ARID LAND.
AND A DECREASE OF 25% IN CORN CROPS AND 20% IN WHEAT CROPS BY 2080.
IN POORER REGIONS, THE AGRICULTURAL SYSTEM COULD COMPLETELY COLLAPSE.

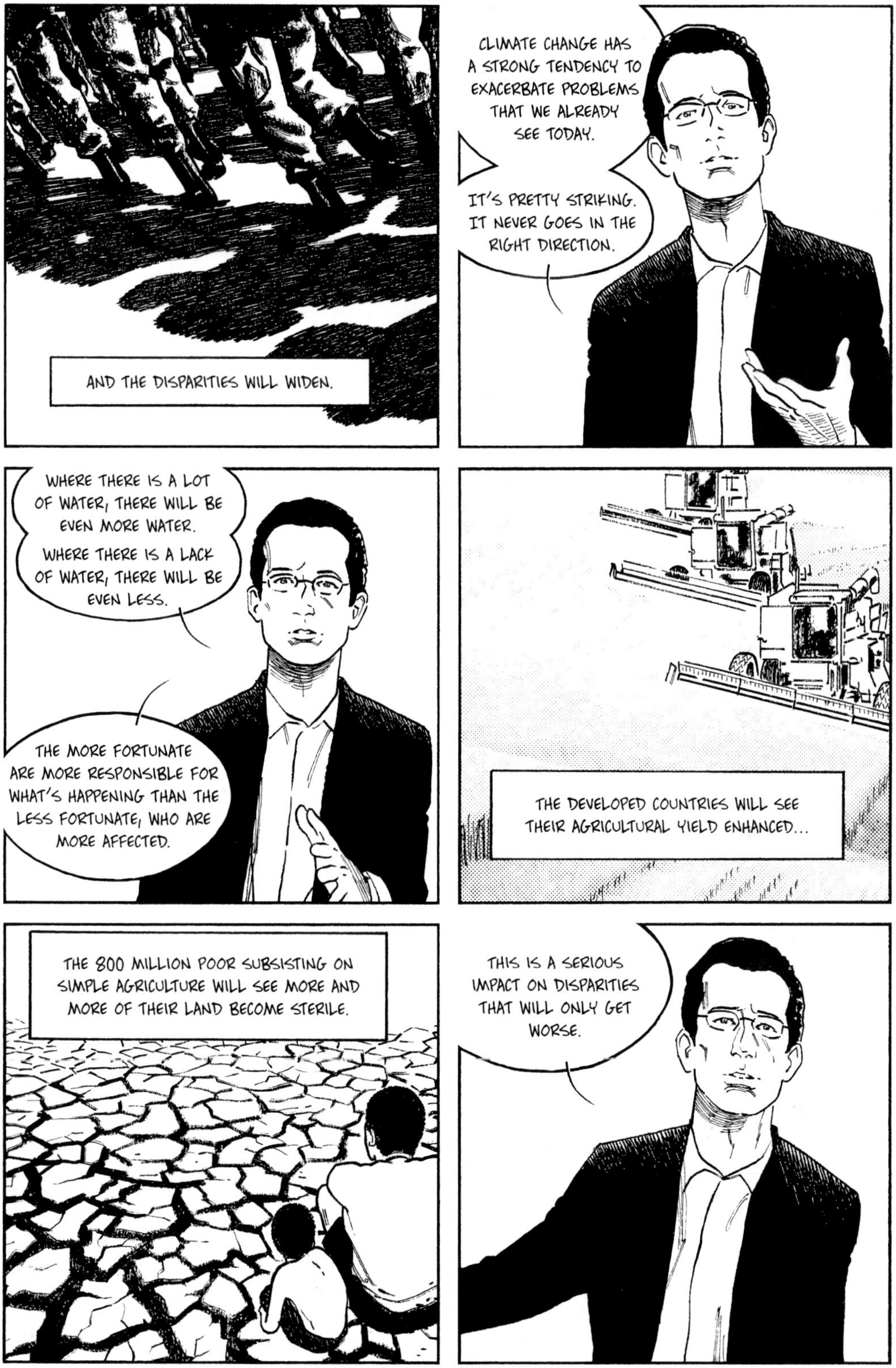

AND THE DISPARITIES WILL WIDEN.
CLIMATE CHANGE HAS A STRONG TENDENCY TO EXACERBATE PROBLEMS THAT WE ALREADY SEE TODAY.
IT'S PRETTY STRIKING. IT NEVER GOES IN THE RIGHT DIRECTION.
WHERE THERE IS A LOT OF WATER, THERE WILL BE EVEN MORE WATER.
WHERE THERE IS A LACK OF WATER, THERE WILL BE EVEN LESS.
THE MORE FORTUNATE ARE MORE RESPONSIBLE FOR WHAT'S HAPPENING THAN THE LESS FORTUNATE, WHO ARE MORE AFFECTED.
THE DEVELOPED COUNTRIES WILL SEE THEIR AGRICULTURAL YIELD ENHANCED...
THE 800 MILLION POOR SUBSISTING ON SIMPLE AGRICULTURE WILL SEE MORE AND MORE OF THEIR LAND BECOME STERILE.
THIS IS A SERIOUS IMPACT ON DISPARITIES THAT WILL ONLY GET WORSE.

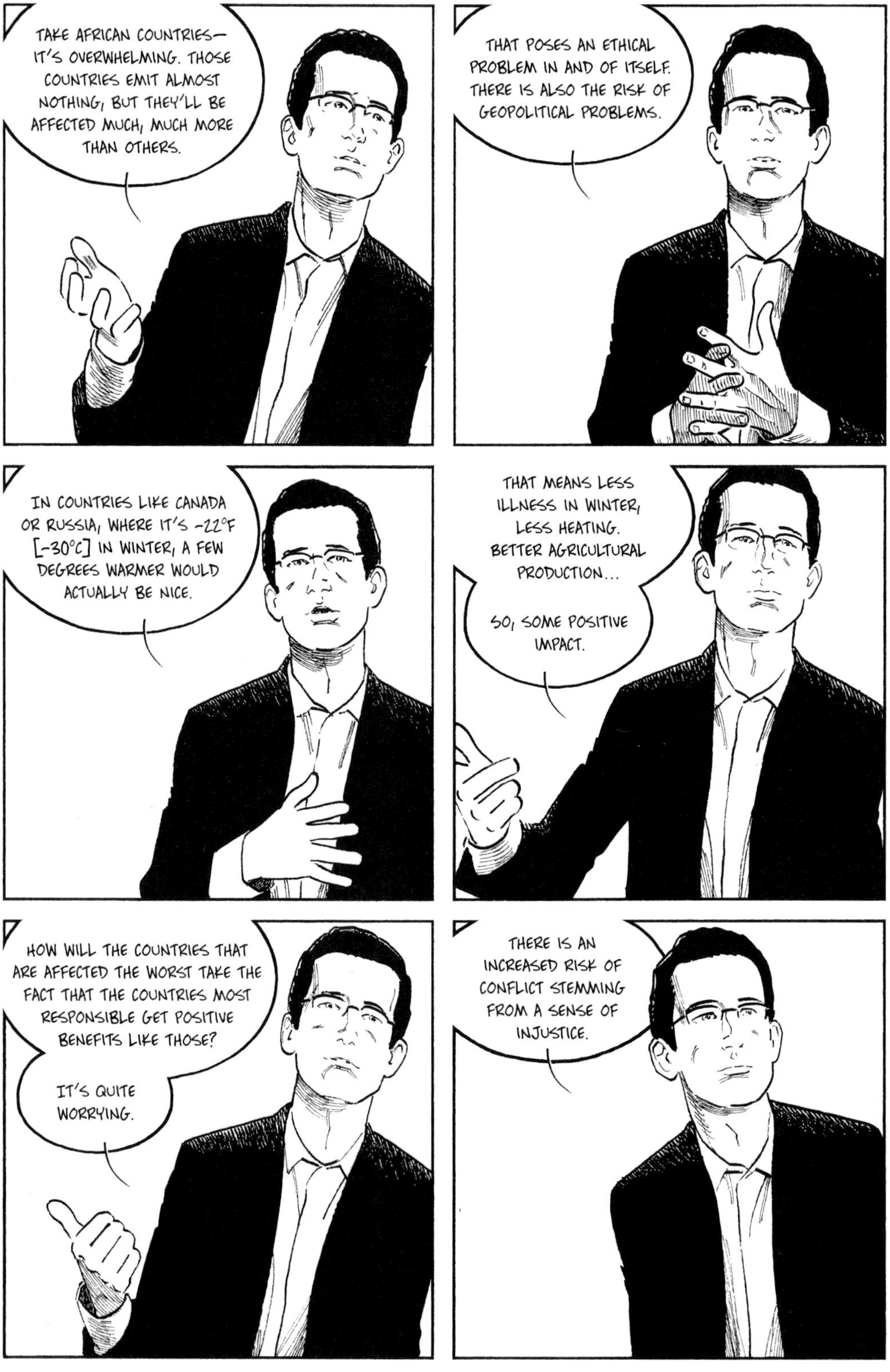
TAKE AFRICAN COUNTRIES— IT'S OVERWHELMING. THOSE COUNTRIES EMIT ALMOST NOTHING, BUT THEY'LL BE AFFECTED MUCH, MUCH MORE THAN OTHERS.
THAT POSES AN ETHICAL PROBLEM IN AND OF ITSELF. THERE IS ALSO THE RISK OF GEOPOLITICAL PROBLEMS.
IN COUNTRIES LIKE CANADA OR RUSSIA, WHERE IT'S -22°F [-30°C] IN WINTER, A FEW DEGREES WARMER WOULD ACTUALLY BE NICE.
THAT MEANS LESS ILLNESS IN WINTER, LESS HEATING. BETTER AGRICULTURAL PRODUCTION...
SO, SOME POSITIVE IMPACT.
HOW WILL THE COUNTRIES THAT ARE AFFECTED THE WORST TAKE THE FACT THAT THE COUNTRIES MOST RESPONSIBLE GET POSITIVE BENEFITS LIKE THOSE?
IT'S QUITE WORRYING.
THERE IS AN INCREASED RISK OF CONFLICT STEMMING FROM A SENSE OF INJUSTICE.

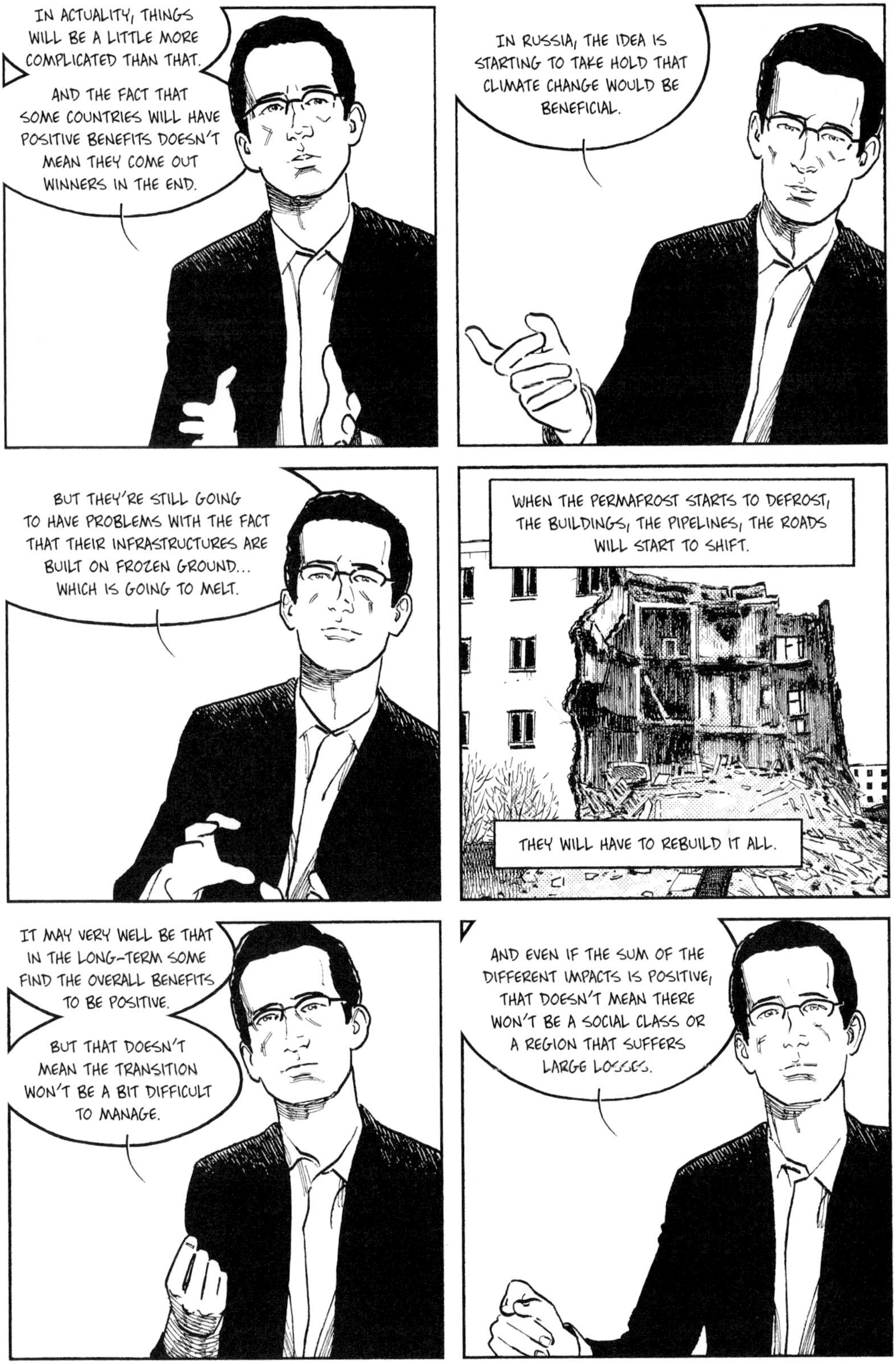
IN ACTUALITY, THINGS WILL BE A LITTLE MORE COMPLICATED THAN THAT.
AND THE FACT THAT SOME COUNTRIES WILL HAVE POSITIVE BENEFITS DOESN'T MEAN THEY COME OUT WINNERS IN THE END.
IN RUSSIA, THE IDEA IS STARTING TO TAKE HOLD THAT CLIMATE CHANGE WOULD BE BENEFICIAL.
BUT THEY'RE STILL GOING TO HAVE PROBLEMS WITH THE FACT THAT THEIR INFRASTRUCTURES ARE BUILT ON FROZEN GROUND... WHICH IS GOING TO MELT.
WHEN THE PERMAFROST STARTS TO DEFROST, THE BUILDINGS, THE PIPELINES, THE ROADS WILL START TO SHIFT.
THEY WILL HAVE TO REBUILD IT ALL.
IT MAY VERY WELL BE THAT IN THE LONG-TERM SOME FIND THE OVERALL BENEFITS TO BE POSITIVE.
BUT THAT DOESN'T MEAN THE TRANSITION WON'T BE A BIT DIFFICULT TO MANAGE.
AND EVEN IF THE SUM OF THE DIFFERENT IMPACTS IS POSITIVE, THAT DOESN'T MEAN THERE WON'T BE A SOCIAL CLASS OR A REGION THAT SUFFERS LARGE LOSSES.

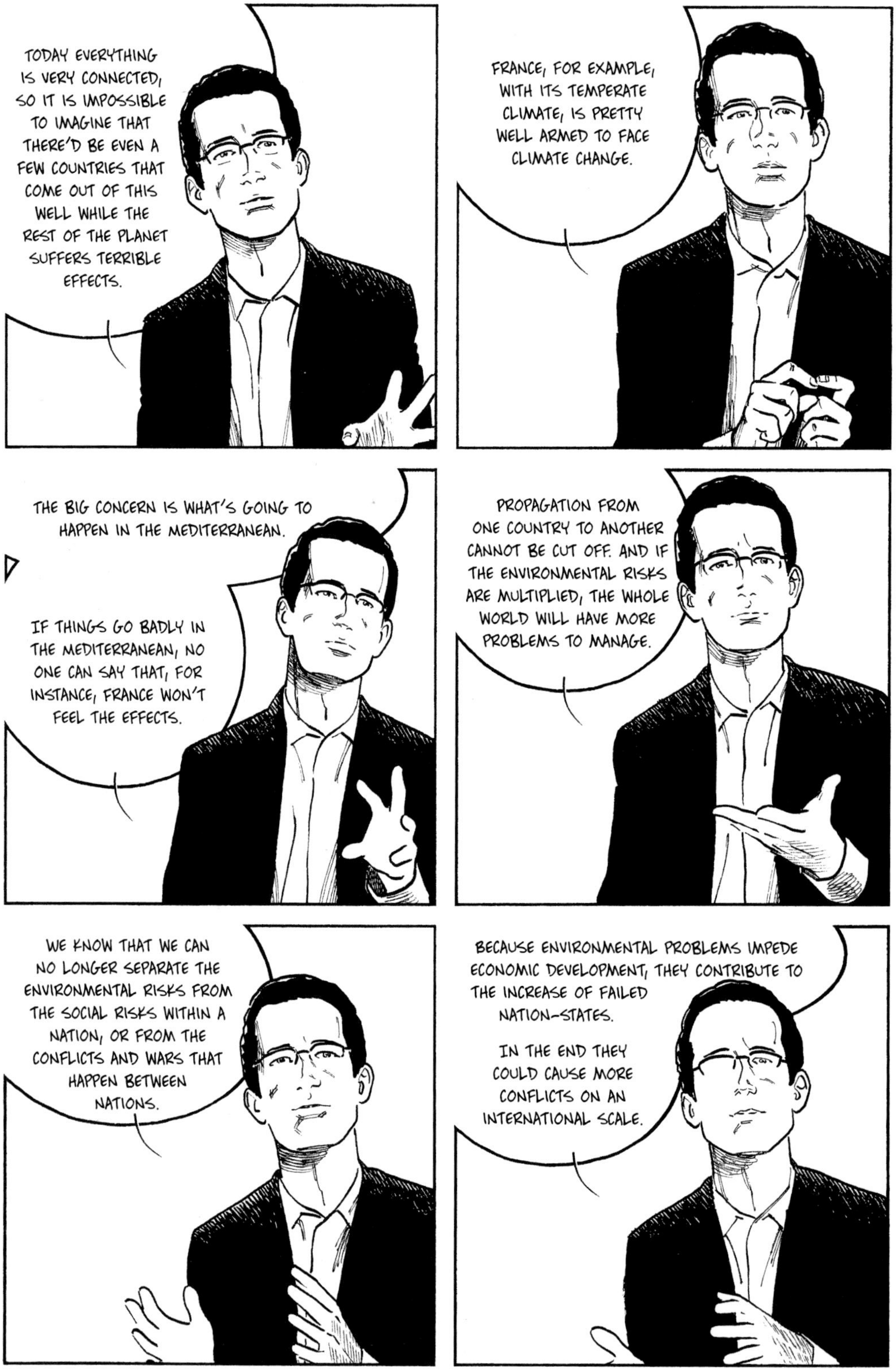
TODAY EVERYTHING IS VERY CONNECTED, SO IT IS IMPOSSIBLE TO IMAGINE THAT THERE'D BE EVEN A FEW COUNTRIES THAT COME OUT OF THIS WELL WHILE THE REST OF THE PLANET SUFFERS TERRIBLE EFFECTS.
FRANCE, FOR EXAMPLE, WITH ITS TEMPERATE CLIMATE, IS PRETTY WELL ARMED TO FACE CLIMATE CHANGE.
THE BIG CONCERN IS WHAT'S GOING TO HAPPEN IN THE MEDITERRANEAN.
IF THINGS GO BADLY IN THE MEDITERRANEAN, NO ONE CAN SAY THAT, FOR INSTANCE, FRANCE WON'T FEEL THE EFFECTS.
PROPAGATION FROM ONE COUNTRY TO ANOTHER CANNOT BE CUT OFF. AND IF THE ENVIRONMENTAL RISKS ARE MULTIPLIED, THE WHOLE WORLD WILL HAVE MORE PROBLEMS TO MANAGE.
WE KNOW THAT WE CAN NO LONGER SEPARATE THE ENVIRONMENTAL RISKS FROM THE SOCIAL RISKS WITHIN A NATION, OR FROM THE CONFLICTS AND WARS THAT HAPPEN BETWEEN NATIONS.
BECAUSE ENVIRONMENTAL PROBLEMS IMPEDE ECONOMIC DEVELOPMENT, THEY CONTRIBUTE TO THE INCREASE OF FAILED NATION-STATES.
IN THE END THEY COULD CAUSE MORE CONFLICTS ON AN INTERNATIONAL SCALE.

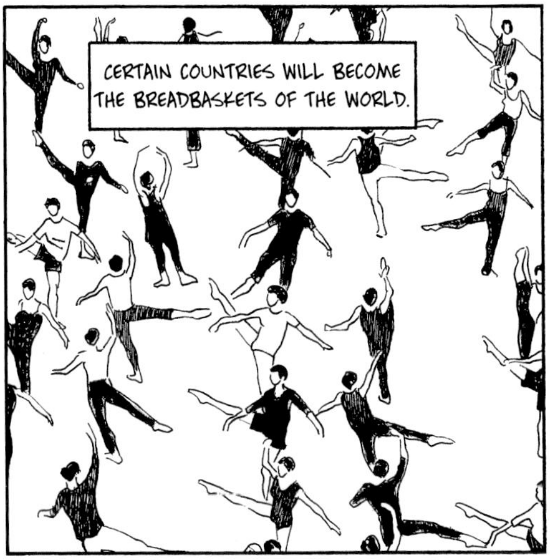
CERTAIN COUNTRIES WILL BECOME THE BREADBASKETS OF THE WORLD.

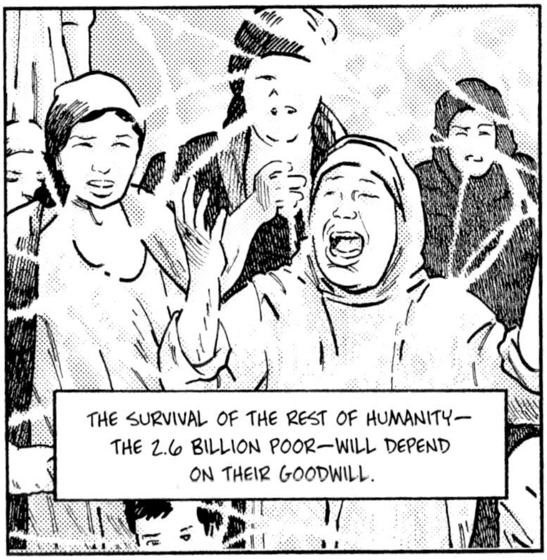
THE SURVIVAL OF THE REST OF HUMANITY—THE 2.6 BILLION POOR—WILL DEPEND ON THEIR GOODWILL.

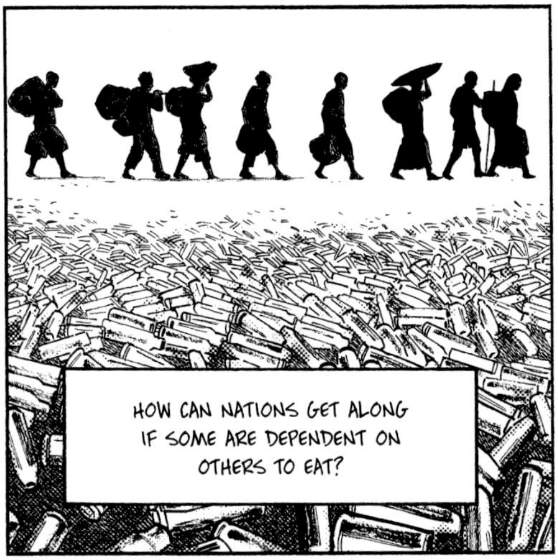
HOW CAN NATIONS GET ALONG IF SOME ARE DEPENDENT ON OTHERS TO EAT?

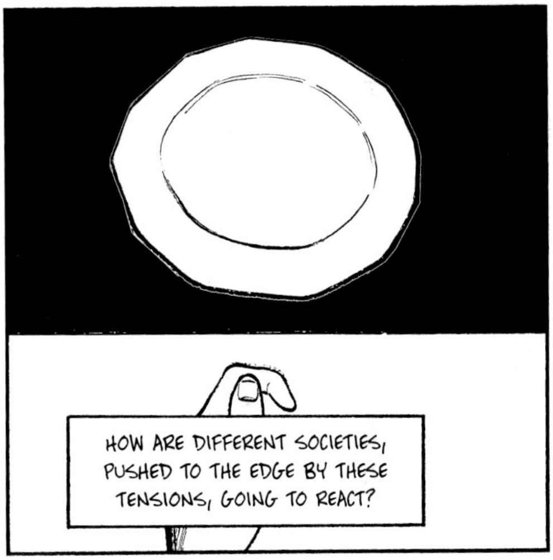
HOW ARE DIFFERENT SOCIETIES, PUSHED TO THE EDGE BY THESE TENSIONS, GOING TO REACT?

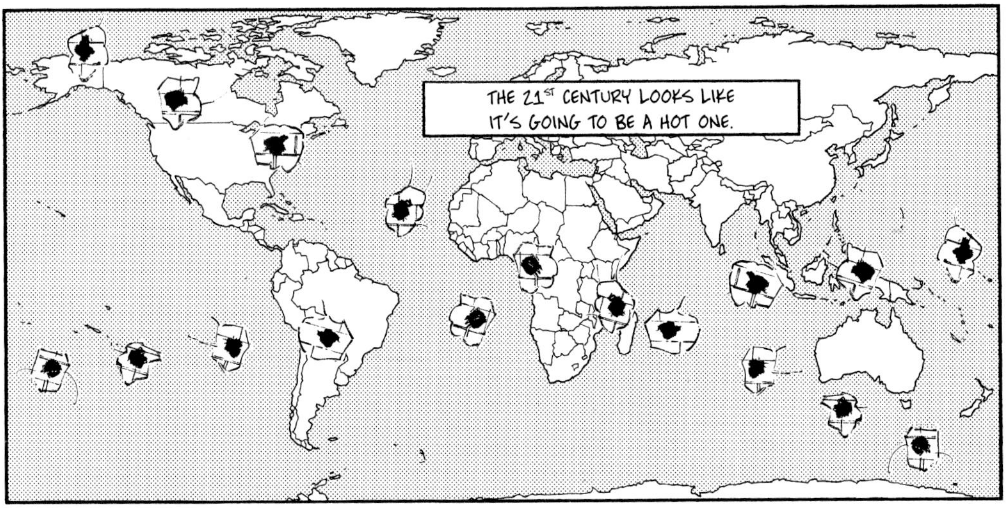
THE 21ST CENTURY LOOKS LIKE IT'S GOING TO BE A HOT ONE.

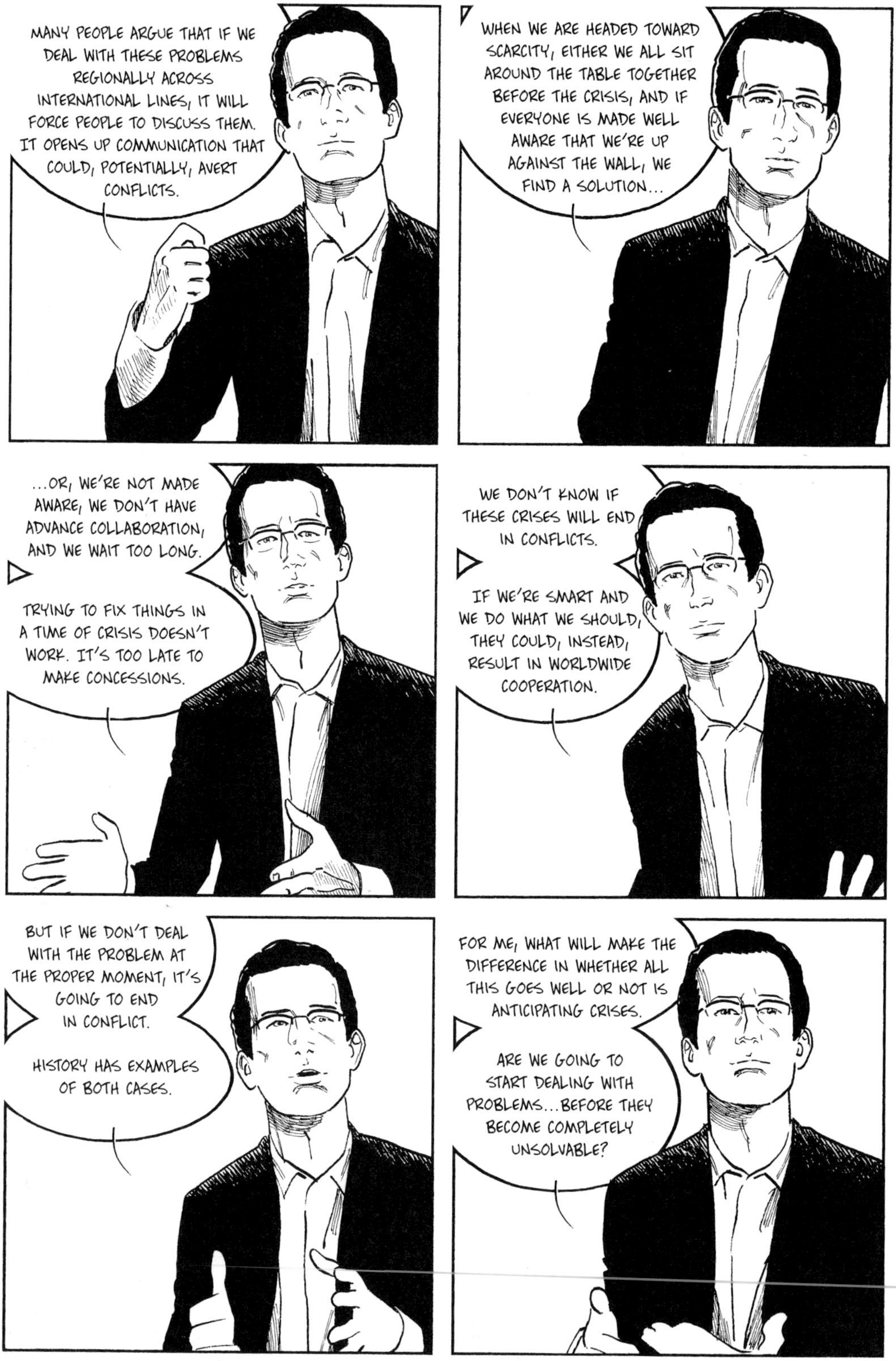
MANY PEOPLE ARGUE THAT IF WE DEAL WITH THESE PROBLEMS REGIONALLY ACROSS INTERNATIONAL LINES, IT WILL FORCE PEOPLE TO DISCUSS THEM. IT OPENS UP COMMUNICATION THAT COULD, POTENTIALLY, AVERT CONFLICTS.
WHEN WE ARE HEADED TOWARD SCARCITY, EITHER WE ALL SIT AROUND THE TABLE TOGETHER BEFORE THE CRISIS, AND IF EVERYONE IS MADE WELL AWARE THAT WE'RE UP AGAINST THE WALL, WE FIND A SOLUTION...
...OR, WE'RE NOT MADE AWARE, WE DON'T HAVE ADVANCE COLLABORATION, AND WE WAIT TOO LONG.
TRYING TO FIX THINGS IN A TIME OF CRISIS DOESN'T WORK. IT'S TOO LATE TO MAKE CONCESSIONS.
WE DON'T KNOW IF THESE CRISES WILL END IN CONFLICTS.
IF WE'RE SMART AND WE DO WHAT WE SHOULD, THEY COULD, INSTEAD, RESULT IN WORLDWIDE COOPERATION.
BUT IF WE DON'T DEAL WITH THE PROBLEM AT THE PROPER MOMENT, IT'S GOING TO END IN CONFLICT.
HISTORY HAS EXAMPLES OF BOTH CASES.
FOR ME, WHAT WILL MAKE THE DIFFERENCE IN WHETHER ALL THIS GOES WELL OR NOT IS ANTICIPATING CRISES.
ARE WE GOING TO START DEALING WITH PROBLEMS...BEFORE THEY BECOME COMPLETELY UNSOLVABLE?

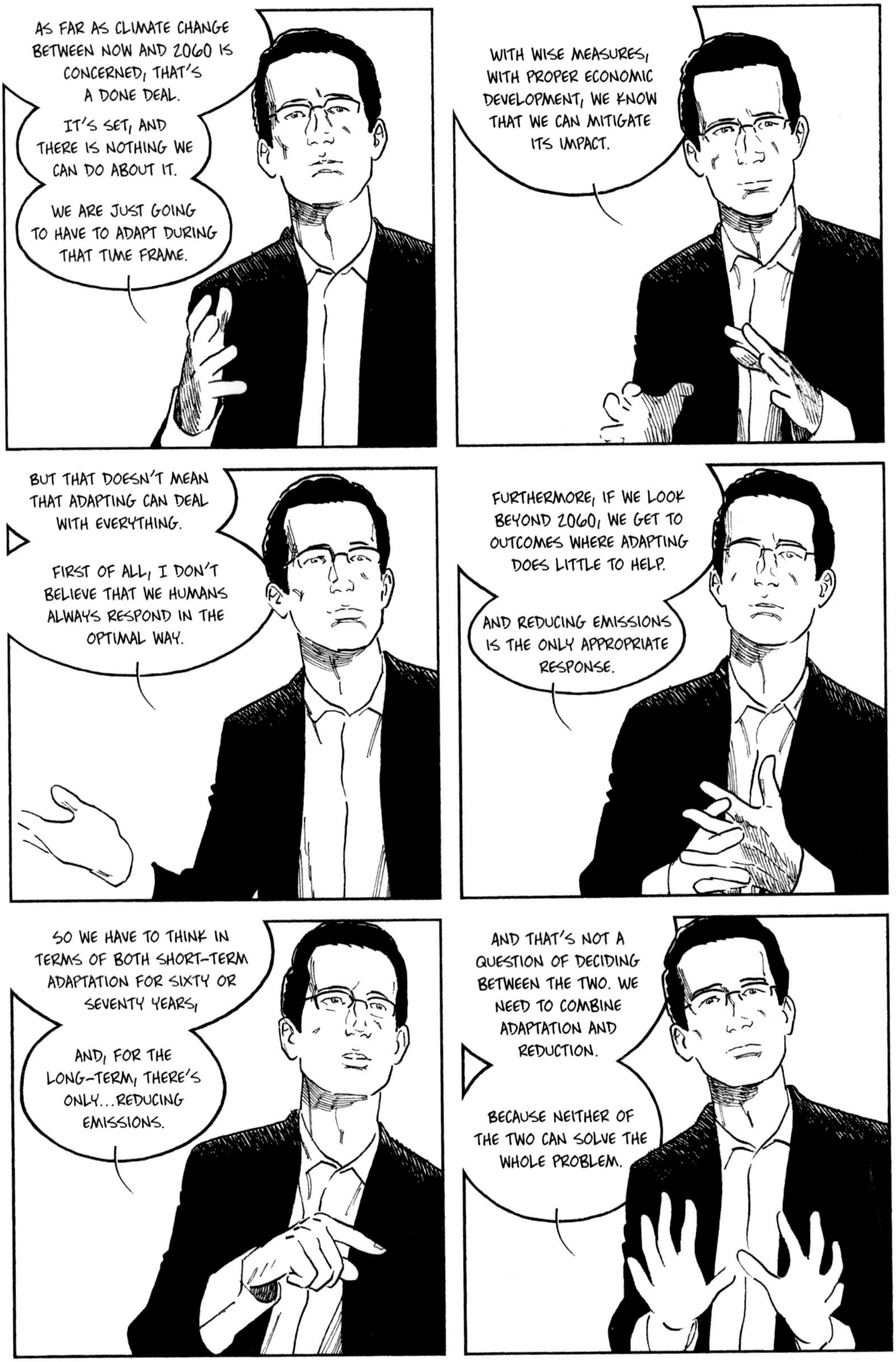
AS FAR AS CLIMATE CHANGE BETWEEN NOW AND 2060 IS CONCERNED, THAT'S A DONE DEAL.
IT'S SET, AND THERE IS NOTHING WE CAN DO ABOUT IT.
WE ARE JUST GOING TO HAVE TO ADAPT DURING THAT TIME FRAME.
WITH WISE MEASURES, WITH PROPER ECONOMIC DEVELOPMENT, WE KNOW THAT WE CAN MITIGATE ITS IMPACT.
BUT THAT DOESN'T MEAN THAT ADAPTING CAN DEAL WITH EVERYTHING.
FIRST OF ALL, I DON'T BELIEVE THAT WE HUMANS ALWAYS RESPOND IN THE OPTIMAL WAY.
FURTHERMORE, IF WE LOOK BEYOND 2060, WE GET TO OUTCOMES WHERE ADAPTING DOES LITTLE TO HELP.
AND REDUCING EMISSIONS IS THE ONLY APPROPRIATE RESPONSE.
SO WE HAVE TO THINK IN TERMS OF BOTH SHORT-TERM ADAPTATION FOR SIXTY OR SEVENTY YEARS,
AND, FOR THE LONG-TERM, THERE'S ONLY...REDUCING EMISSIONS.
AND THAT'S NOT A QUESTION OF DECIDING BETWEEN THE TWO. WE NEED TO COMBINE ADAPTATION AND REDUCTION.
BECAUSE NEITHER OF THE TWO CAN SOLVE THE WHOLE PROBLEM.

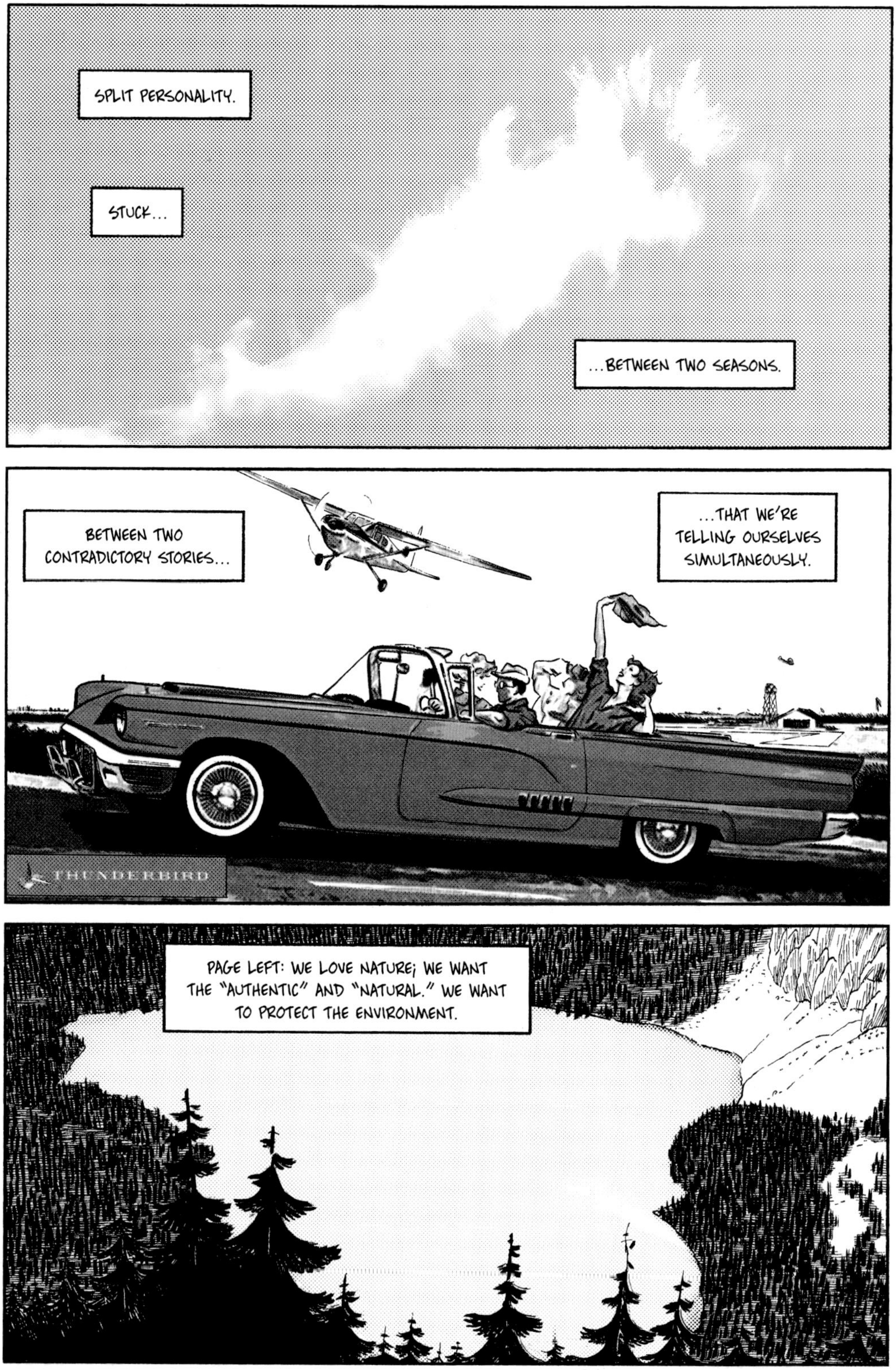
SPLIT PERSONALITY.
STUCK...
...BETWEEN TWO SEASONS.
BETWEEN TWO CONTRADICTORY STORIES...
...THAT WE'RE TELLING OURSELVES SIMULTANEOUSLY.
THUNDERBIRD
PAGE LEFT: WE LOVE NATURE; WE WANT THE "AUTHENTIC" AND "NATURAL." WE WANT TO PROTECT THE ENVIRONMENT.

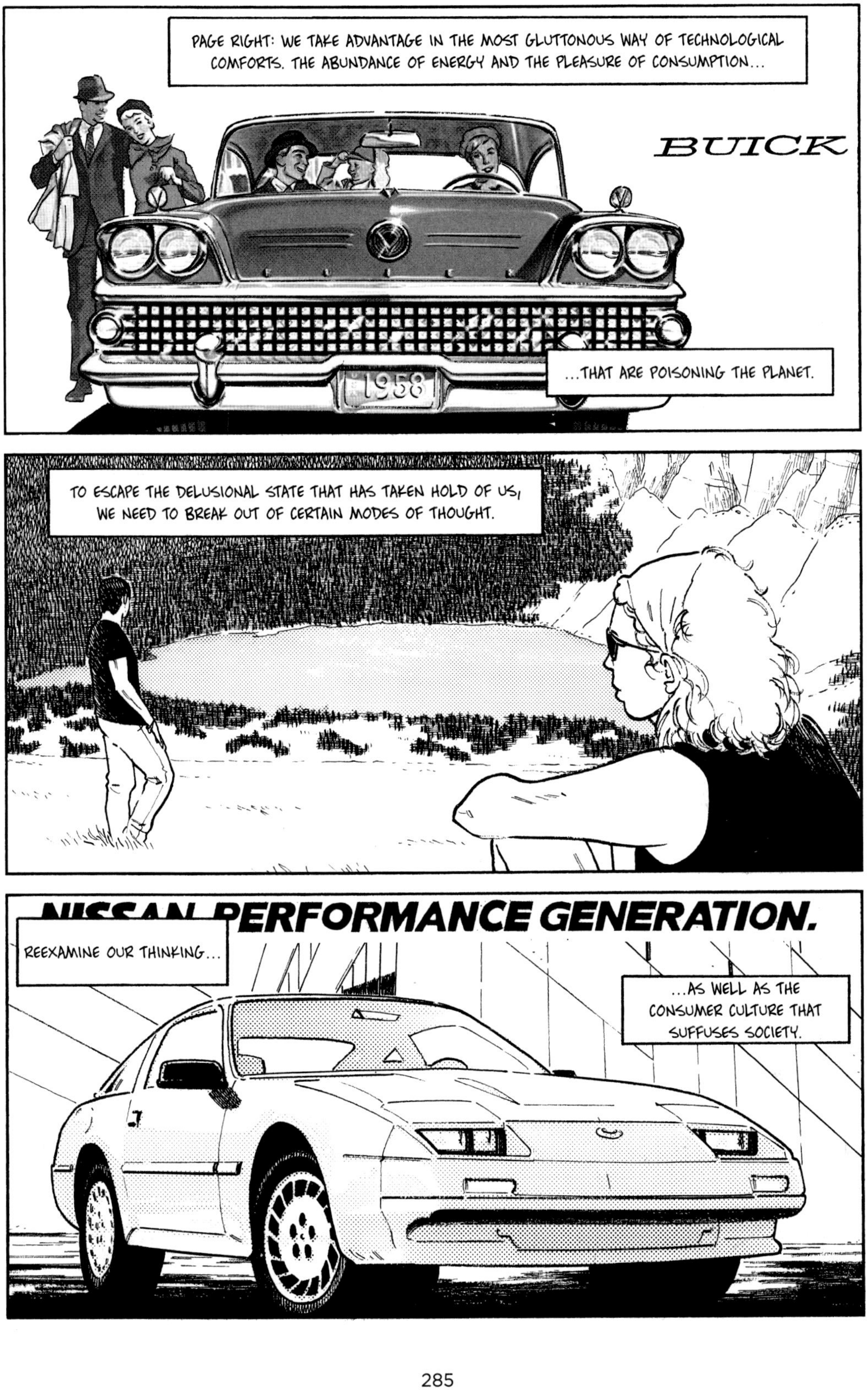
PAGE RIGHT: WE TAKE ADVANTAGE IN THE MOST GLUTTONOUS WAY OF TECHNOLOGICAL COMFORTS. THE ABUNDANCE OF ENERGY AND THE PLEASURE OF CONSUMPTION...
BUICK
1958
...THAT ARE POISONING THE PLANET.
TO ESCAPE THE DELUSIONAL STATE THAT HAS TAKEN HOLD OF US, WE NEED TO BREAK OUT OF CERTAIN MODES OF THOUGHT.
PERFORMANCE GENERATION.
REEXAMINE OUR THINKING...
...AS WELL AS THE CONSUMER CULTURE THAT SUFFUSES SOCIETY.

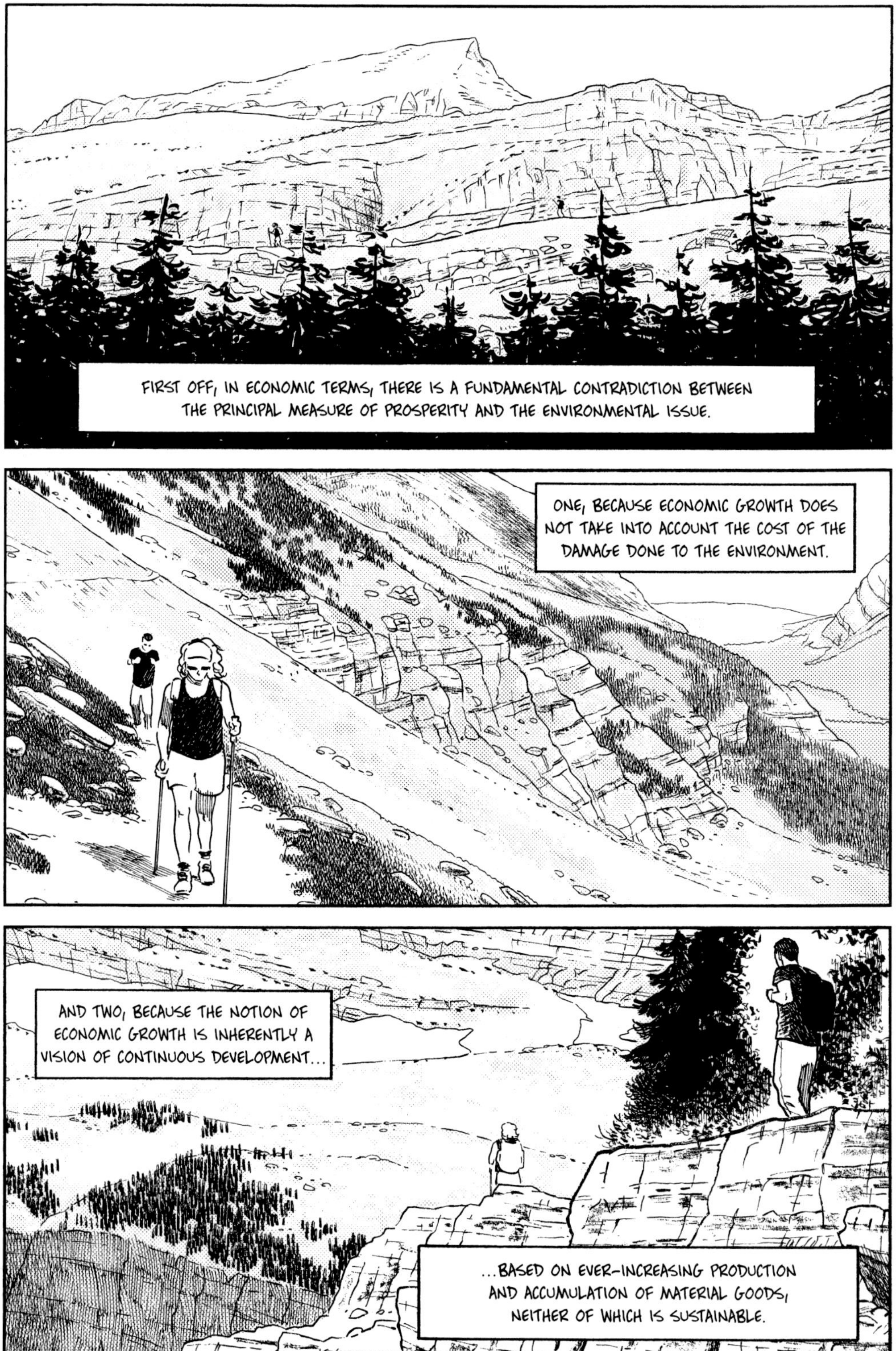
FIRST OFF, IN ECONOMIC TERMS, THERE IS A FUNDAMENTAL CONTRADICTION BETWEEN THE PRINCIPAL MEASURE OF PROSPERITY AND THE ENVIRONMENTAL ISSUE.
ONE, BECAUSE ECONOMIC GROWTH DOES NOT TAKE INTO ACCOUNT THE COST OF THE DAMAGE DONE TO THE ENVIRONMENT.
AND TWO, BECAUSE THE NOTION OF ECONOMIC GROWTH IS INHERENTLY A VISION OF CONTINUOUS DEVELOPMENT...
...BASED ON EVER-INCREASING PRODUCTION AND ACCUMULATION OF MATERIAL GOODS, NEITHER OF WHICH IS SUSTAINABLE.

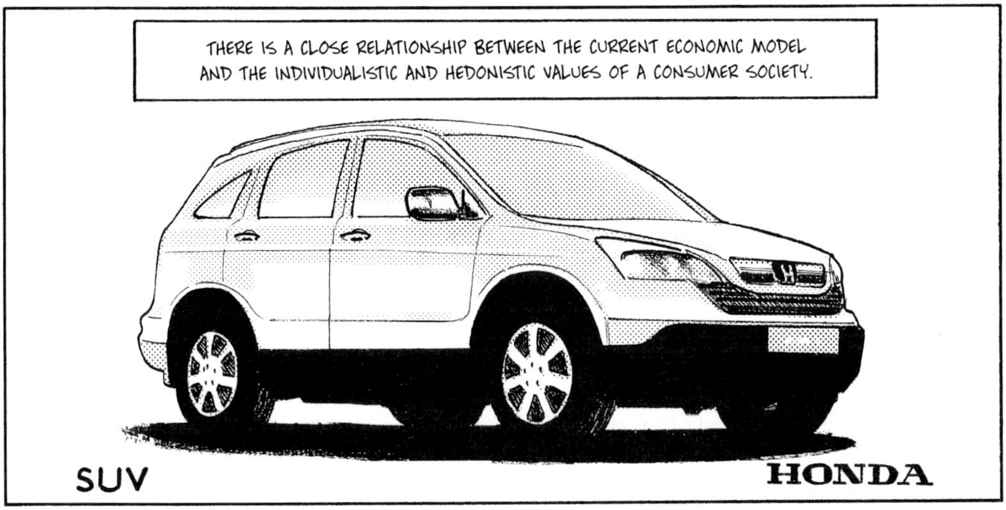
THERE IS A CLOSE RELATIONSHIP BETWEEN THE CURRENT ECONOMIC MODEL AND THE INDIVIDUALISTIC AND HEDONISTIC VALUES OF A CONSUMER SOCIETY.
SUV
HONDA

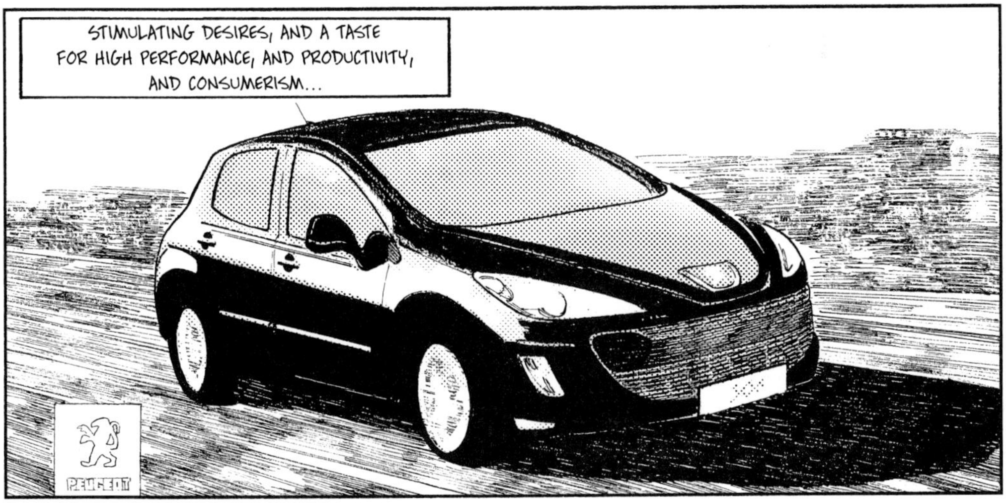
STIMULATING DESIRES, AND A TASTE FOR HIGH PERFORMANCE, AND PRODUCTIVITY, AND CONSUMERISM...
PEUGEOT

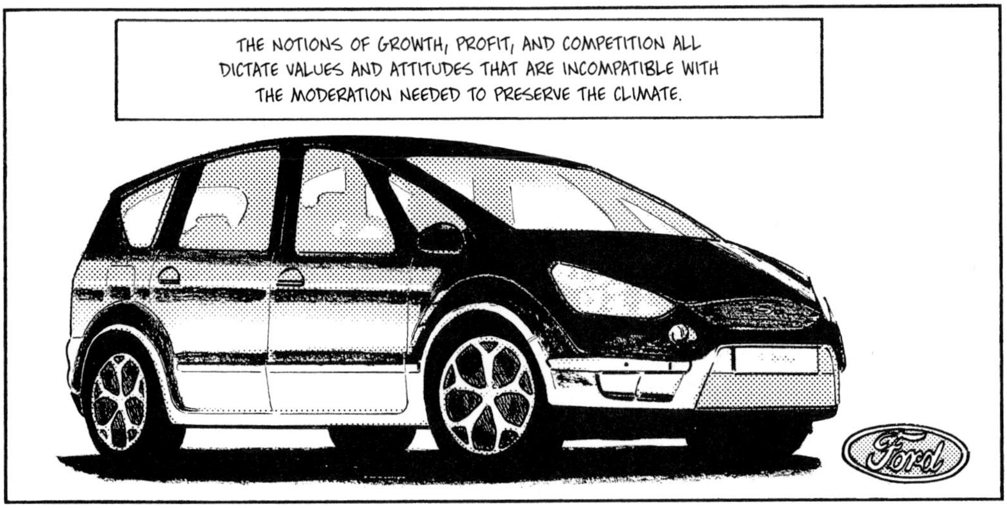
THE NOTIONS OF GROWTH, PROFIT, AND COMPETITION ALL DICTATE VALUES AND ATTITUDES THAT ARE INCOMPATIBLE WITH THE MODERATION NEEDED TO PRESERVE THE CLIMATE.
Ford

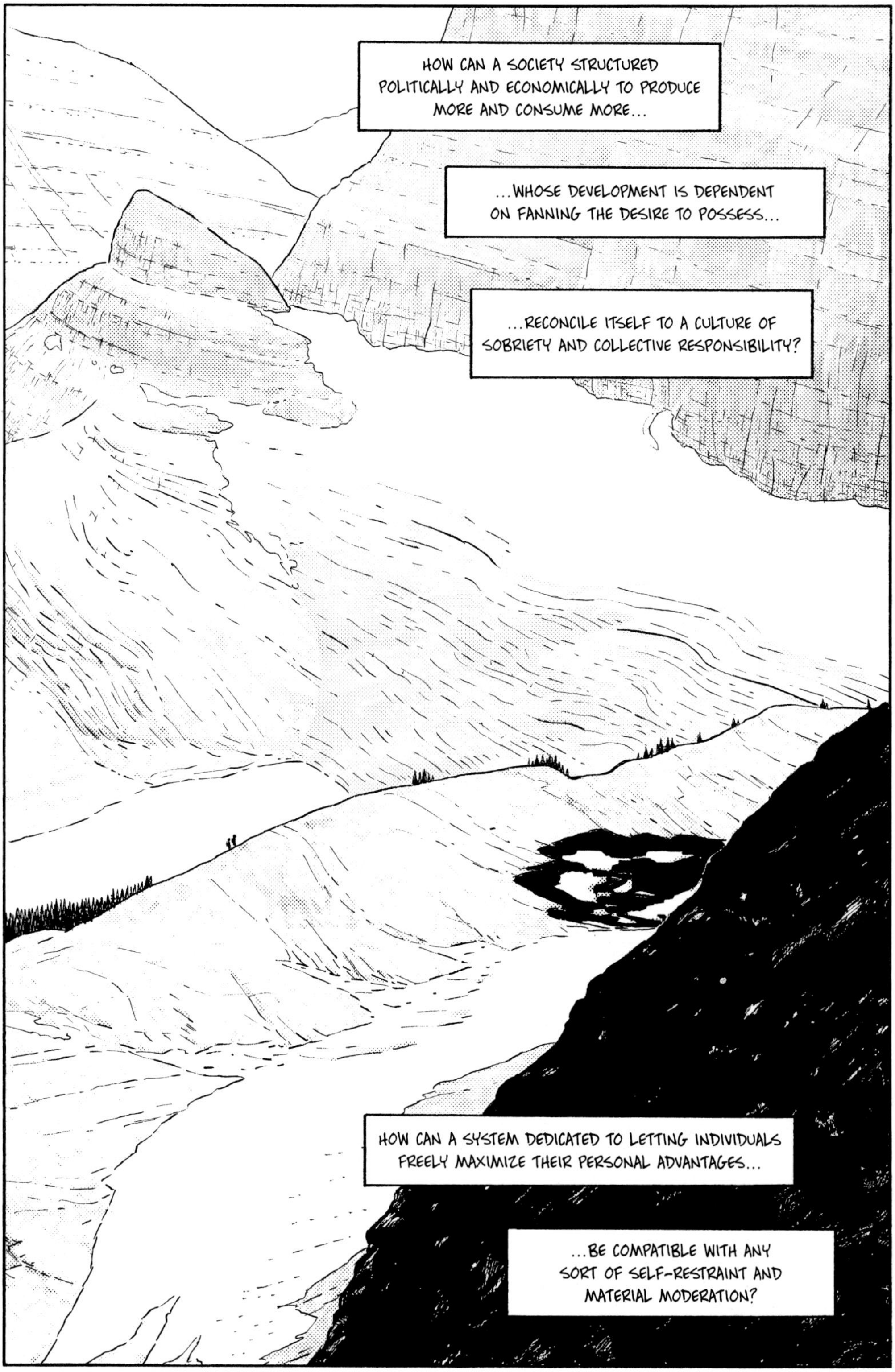
HOW CAN A SOCIETY STRUCTURED POLITICALLY AND ECONOMICALLY TO PRODUCE MORE AND CONSUME MORE...
...WHOSE DEVELOPMENT IS DEPENDENT ON FANNING THE DESIRE TO POSSESS...
...RECONCILE ITSELF TO A CULTURE OF SOBRIETY AND COLLECTIVE RESPONSIBILITY?
HOW CAN A SYSTEM DEDICATED TO LETTING INDIVIDUALS FREELY MAXIMIZE THEIR PERSONAL ADVANTAGES...
...BE COMPATIBLE WITH ANY SORT OF SELF-RESTRAINT AND MATERIAL MODERATION?

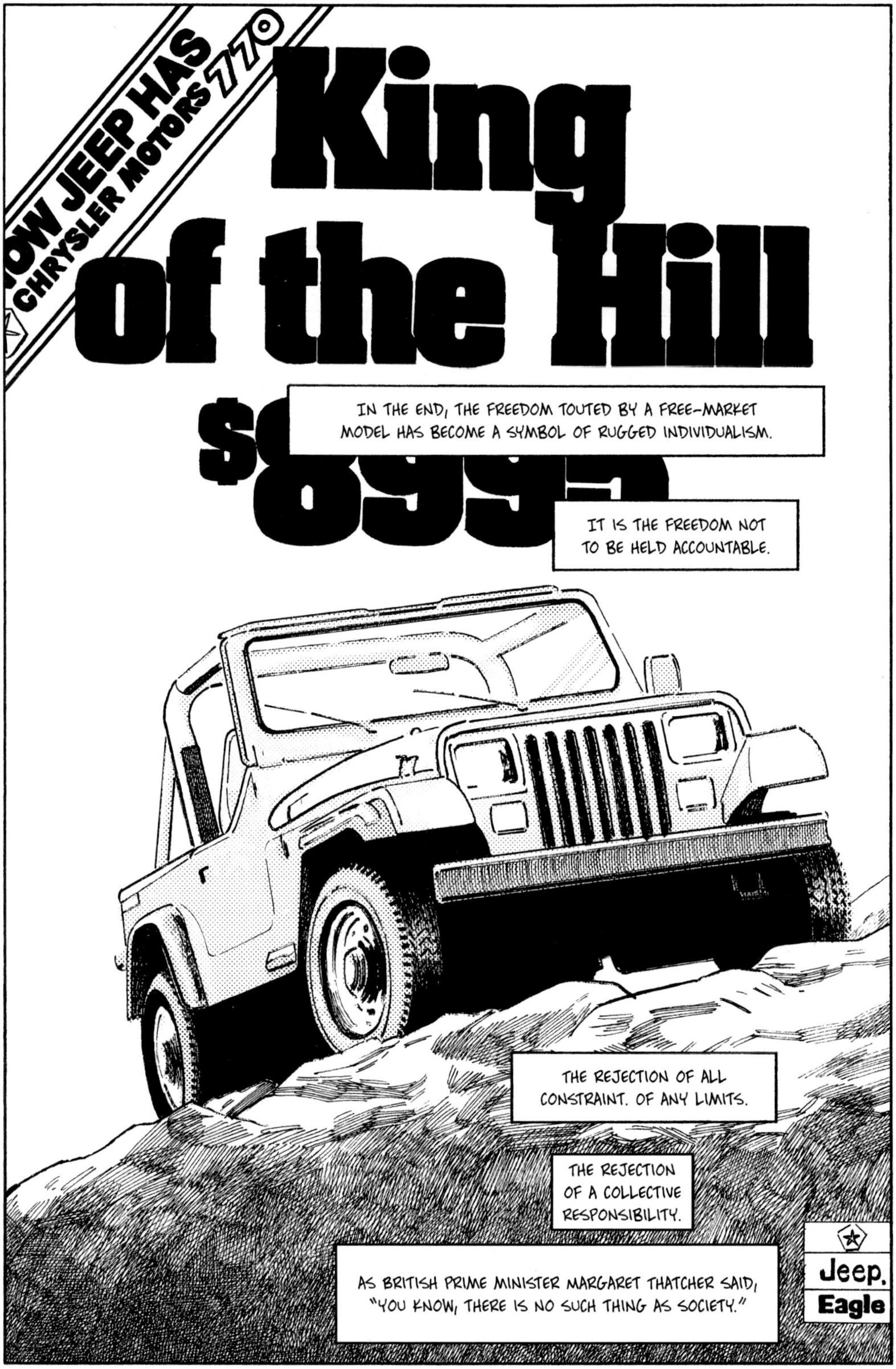
NOW JEEP HAS
CHRYSLER MOTORS 7/70
King of the Hill
IN THE END, THE FREEDOM TOUTED BY A FREE-MARKET MODEL HAS BECOME A SYMBOL OF RUGGED INDIVIDUALISM.
IT IS THE FREEDOM NOT TO BE HELD ACCOUNTABLE.
THE REJECTION OF ALL CONSTRAINT. OF ANY LIMITS.
THE REJECTION OF A COLLECTIVE RESPONSIBILITY.
AS BRITISH PRIME MINISTER MARGARET THATCHER SAID, "YOU KNOW, THERE IS NO SUCH THING AS SOCIETY."
Jeep.
Eagle

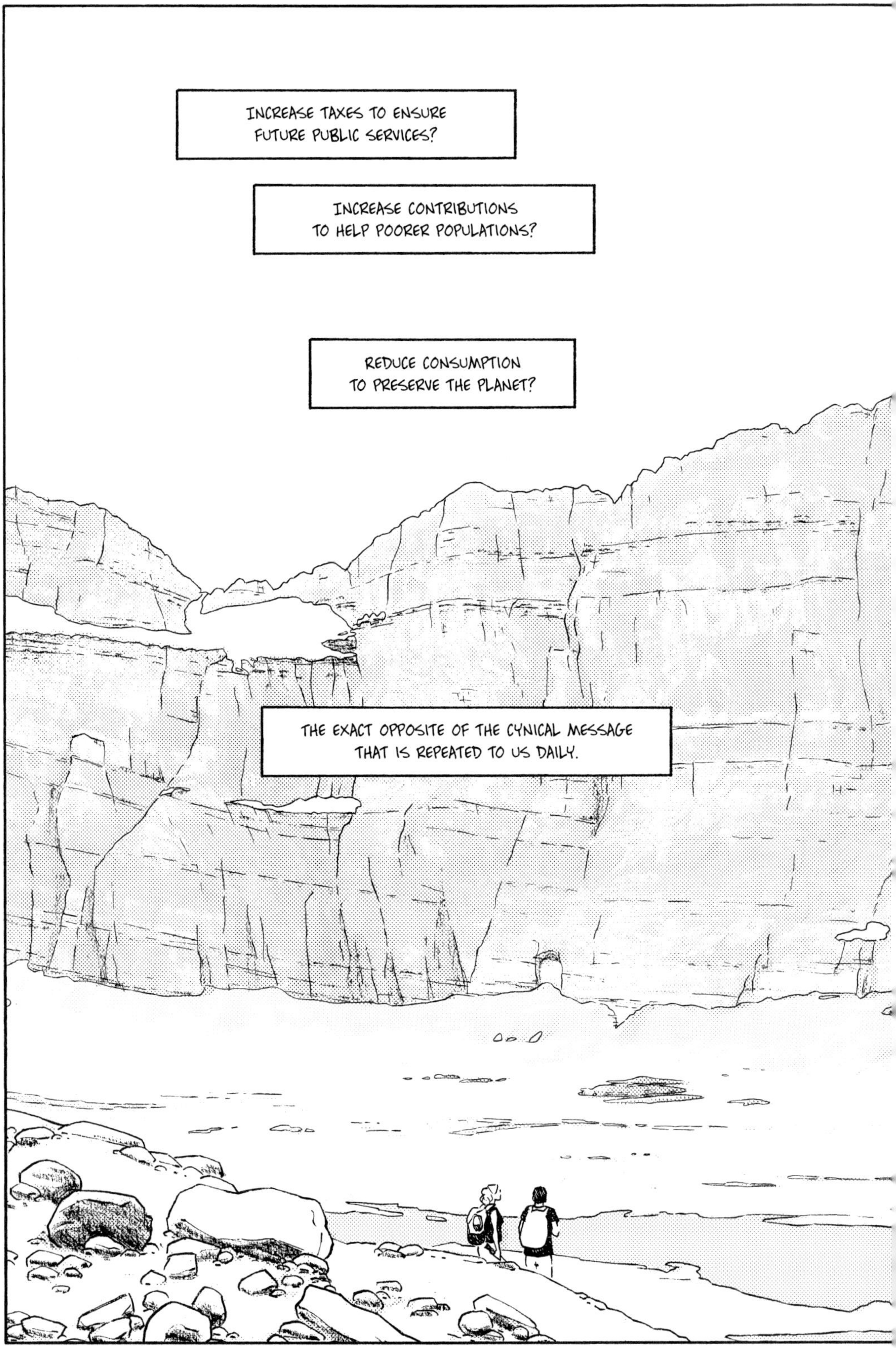
INCREASE TAXES TO ENSURE FUTURE PUBLIC SERVICES?
INCREASE CONTRIBUTIONS TO HELP POORER POPULATIONS?
REDUCE CONSUMPTION TO PRESERVE THE PLANET?
THE EXACT OPPOSITE OF THE CYNICAL MESSAGE THAT IS REPEATED TO US DAILY.

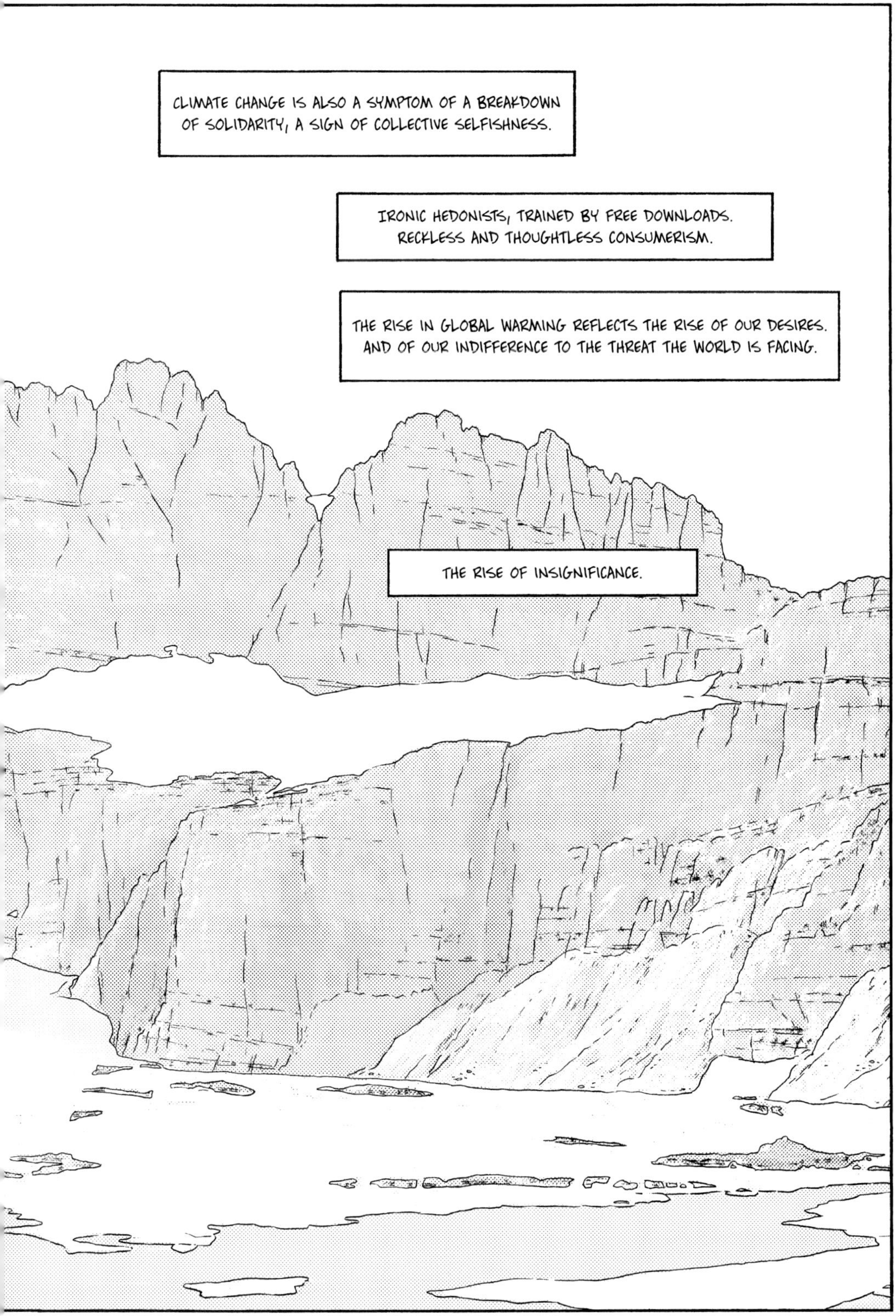
CLIMATE CHANGE IS ALSO A SYMPTOM OF A BREAKDOWN OF SOLIDARITY, A SIGN OF COLLECTIVE SELFISHNESS.
IRONIC HEDONISTS, TRAINED BY FREE DOWNLOADS. RECKLESS AND THOUGHTLESS CONSUMERISM.
THE RISE IN GLOBAL WARMING REFLECTS THE RISE OF OUR DESIRES. AND OF OUR INDIFFERENCE TO THE THREAT THE WORLD IS FACING.
THE RISE OF INSIGNIFICANCE.

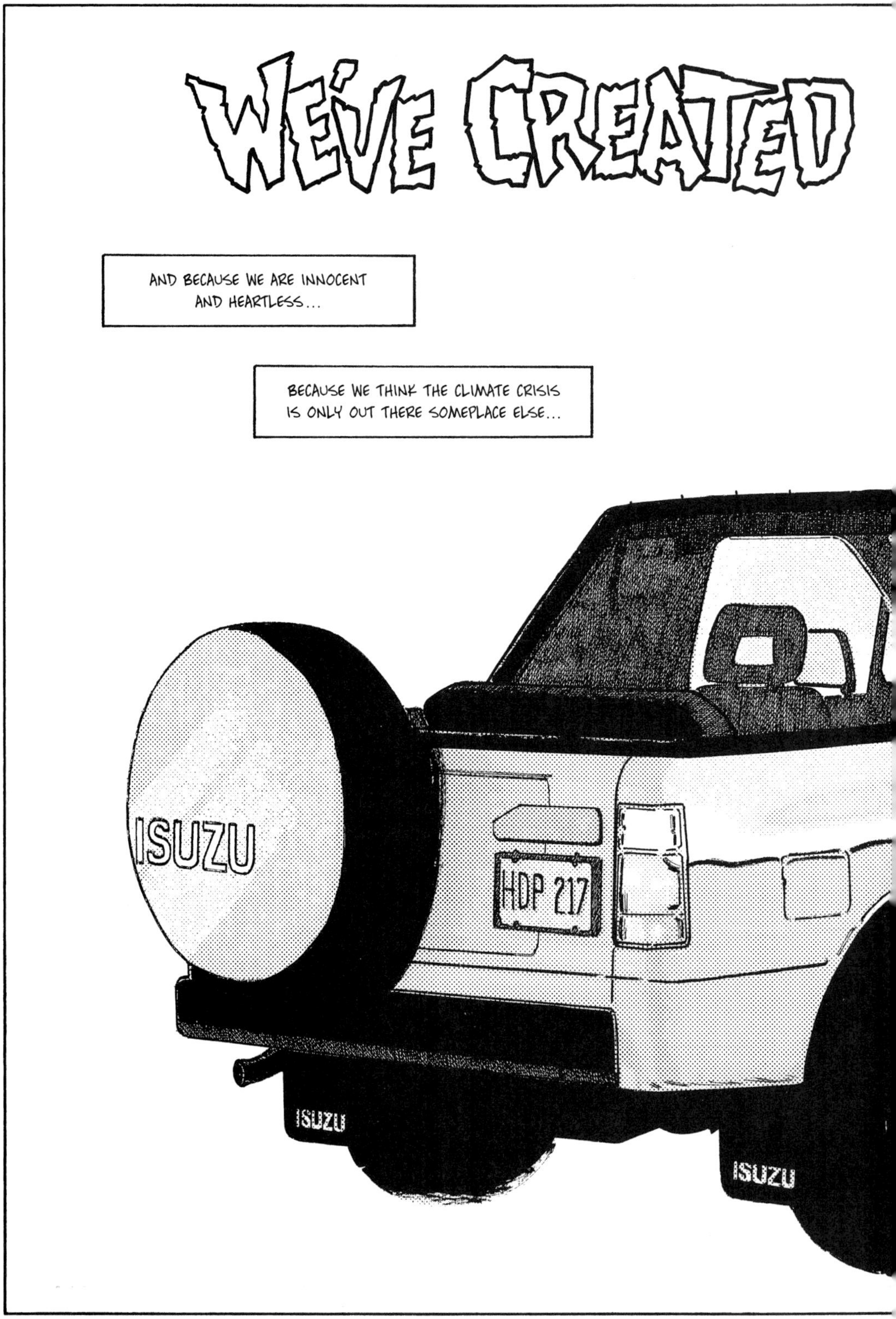
WE'VE CREATED
AND BECAUSE WE ARE INNOCENT AND HEARTLESS...
BECAUSE WE THINK THE CLIMATE CRISIS IS ONLY OUT THERE SOMEPLACE ELSE...
ISUZU
HDP 217
ISUZU
ISUZU

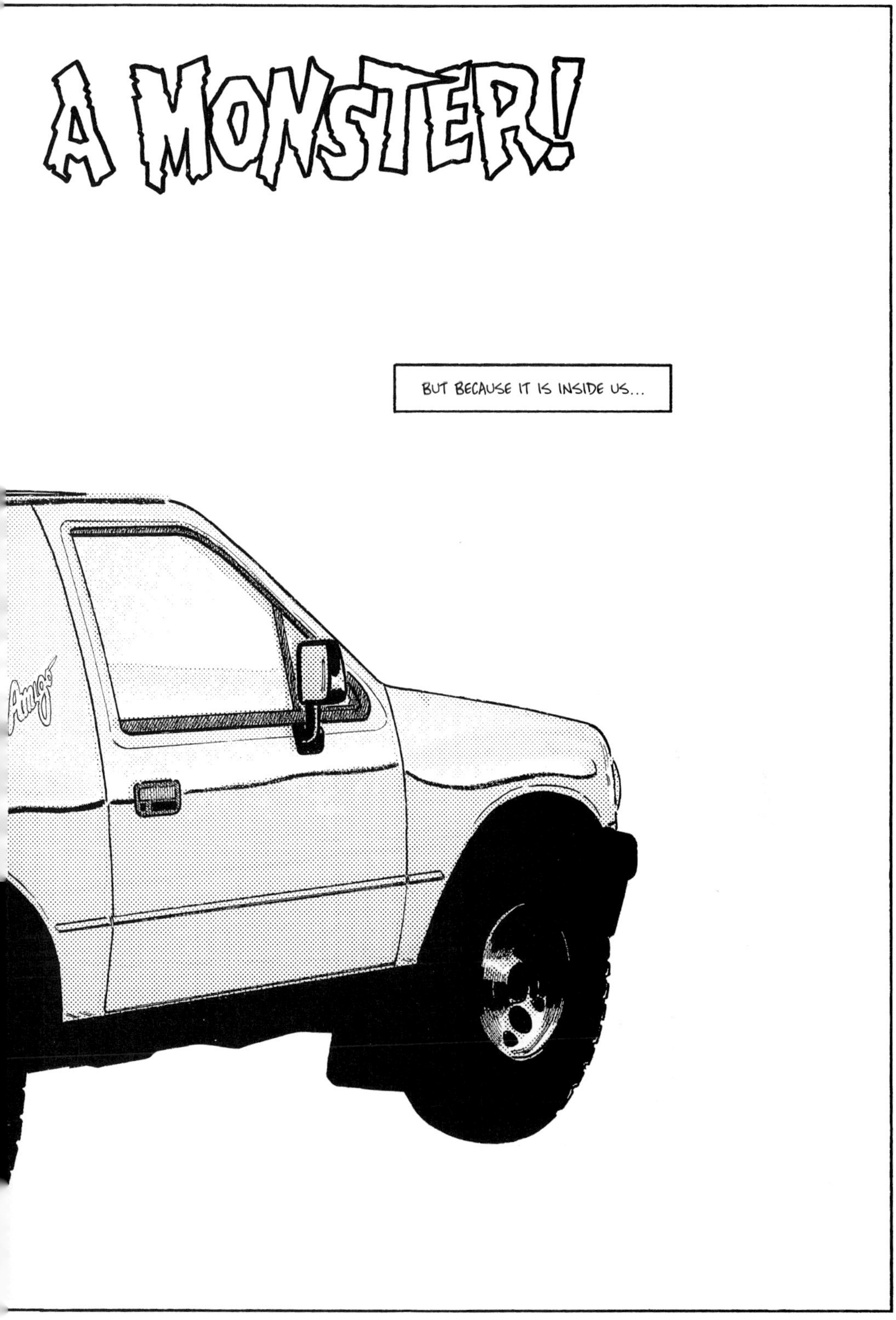
A MONSTER!
BUT BECAUSE IT IS INSIDE US...
Amigo

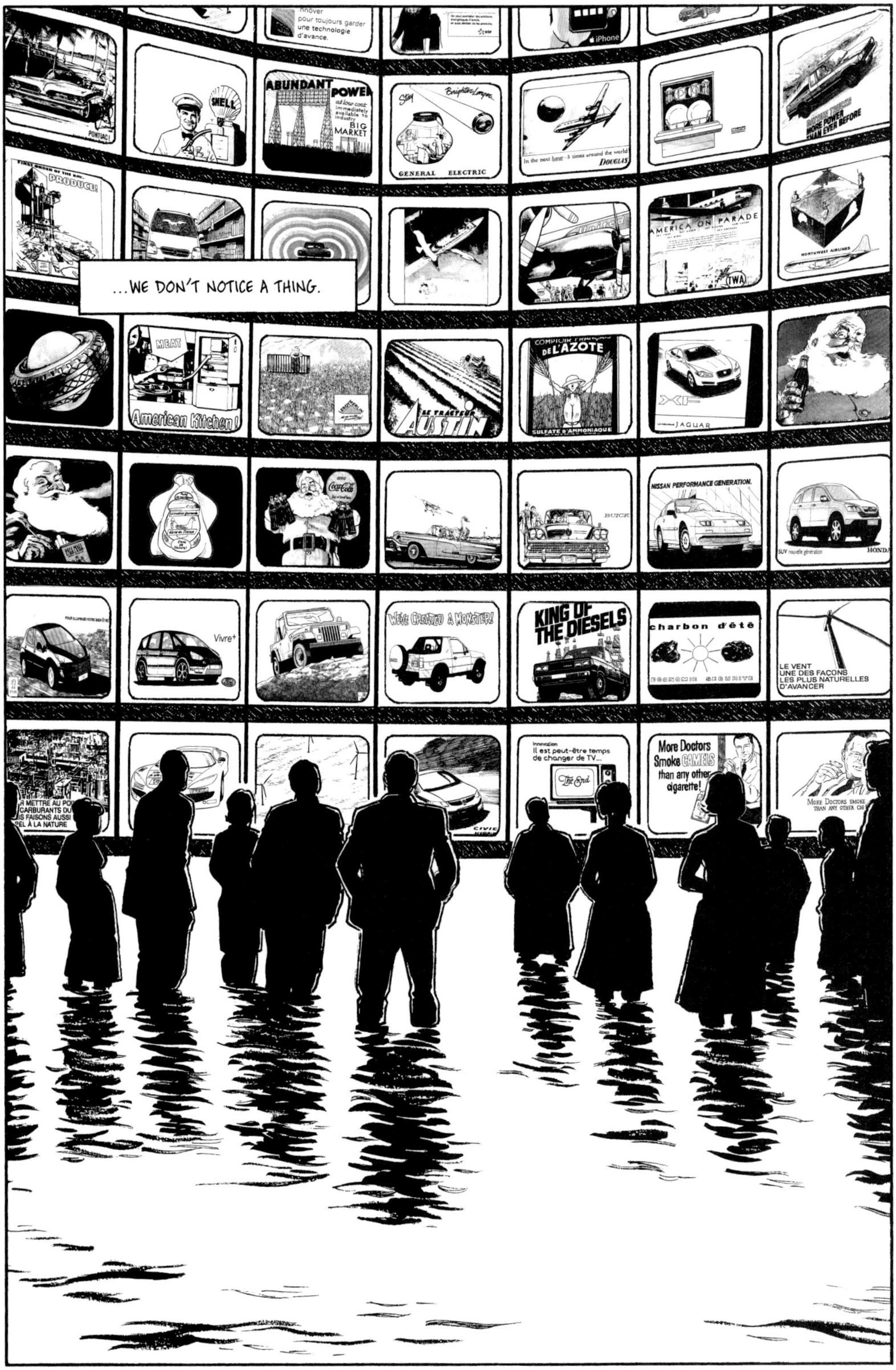
...WE DON'T NOTICE A THING.
ABUNDANT POWER
at low cost immediately available to industry
BIG MARKET
GENERAL ELECTRIC
DOUGLAS
PRODUCE!
AMERICA ON PARADE
NORTHWEST AIRLINES
TWA
MEAT
American Kitchen!
LE TRACTEUR AUSTIN
COMPTOIR FRANÇAIS DE L'AZOTE
SULFATE D'AMMONIAQUE
XF
JAGUAR
Coca-Cola
BUICK
NISSAN PERFORMANCE GENERATION.
SUV nouvelle génération
HONDA
Vivre+
WE'VE CREATED A MONSTER!
KING OF THE DIESELS
charbon d'été
LE VENT
UNE DES FACONS
LES PLUS NATURELLES
D'AVANCER
Innovation
Il est peut-être temps
de changer de TV...
The End
More Doctors
Smoke CAMELS
than any other
cigarette!
MORE DOCTORS SMOKE

HOW DO YOU END A GRAPHIC NOVEL?

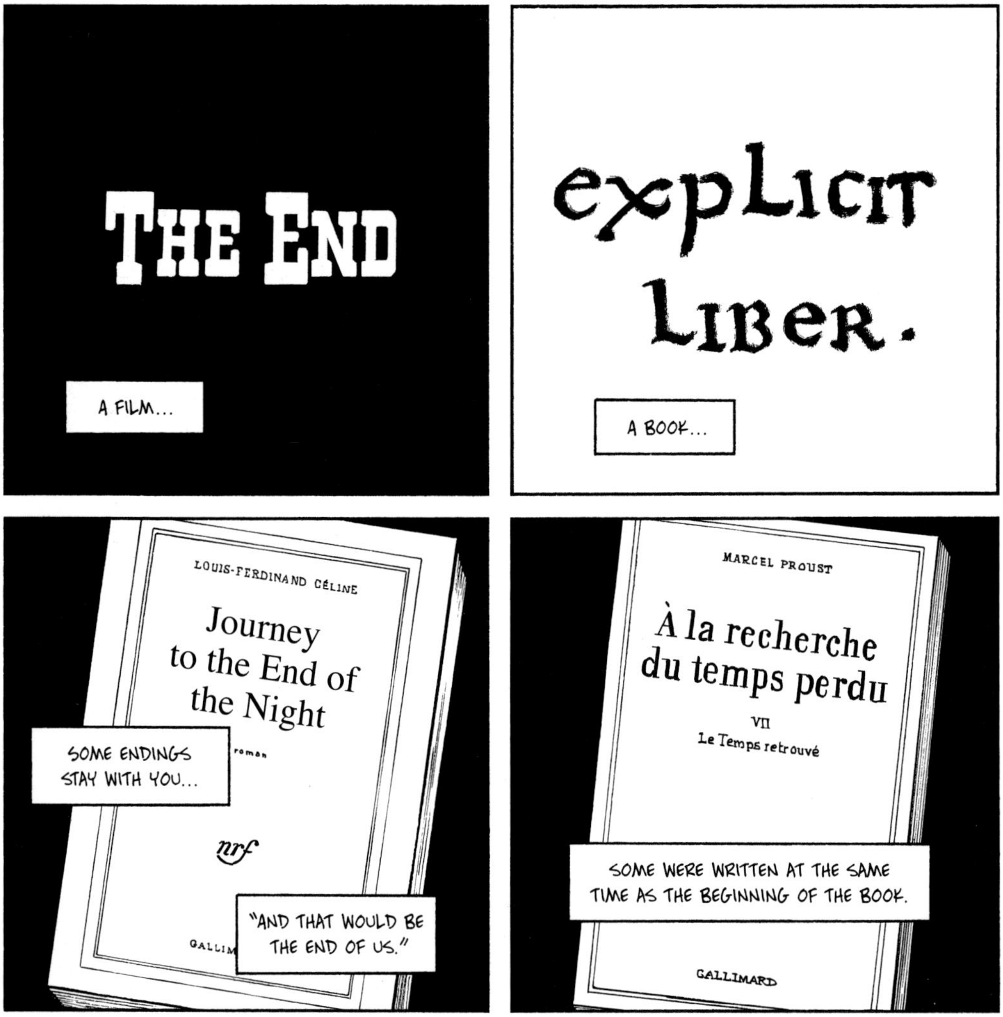
THE END
A FILM...
explicit
liber.
A BOOK...
LOUIS-FERDINAND CÉLINE
Journey to the End of the Night
roman
SOME ENDINGS STAY WITH YOU...
nrf
"AND THAT WOULD BE THE END OF US."
MARCEL PROUST
À la recherche du temps perdu
VII
Le Temps retrouvé
SOME WERE WRITTEN AT THE SAME TIME AS THE BEGINNING OF THE BOOK.
GALLIMARD

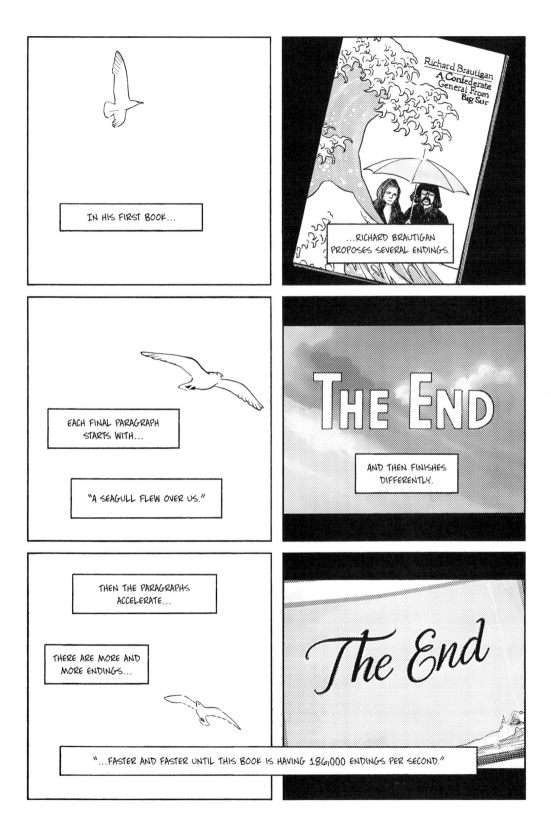

IN HIS FIRST BOOK...

Richard Brautigan
A Confederate General From Big Sur

...RICHARD BRAUTIGAN PROPOSES SEVERAL ENDINGS.

EACH FINAL PARAGRAPH STARTS WITH...

"A SEAGULL FLEW OVER US."

THE END

AND THEN FINISHES DIFFERENTLY.

THEN THE PARAGRAPHS ACCELERATE...

THERE ARE MORE AND MORE ENDINGS...

The End

"...FASTER AND FASTER UNTIL THIS BOOK IS HAVING 186,000 ENDINGS PER SECOND."

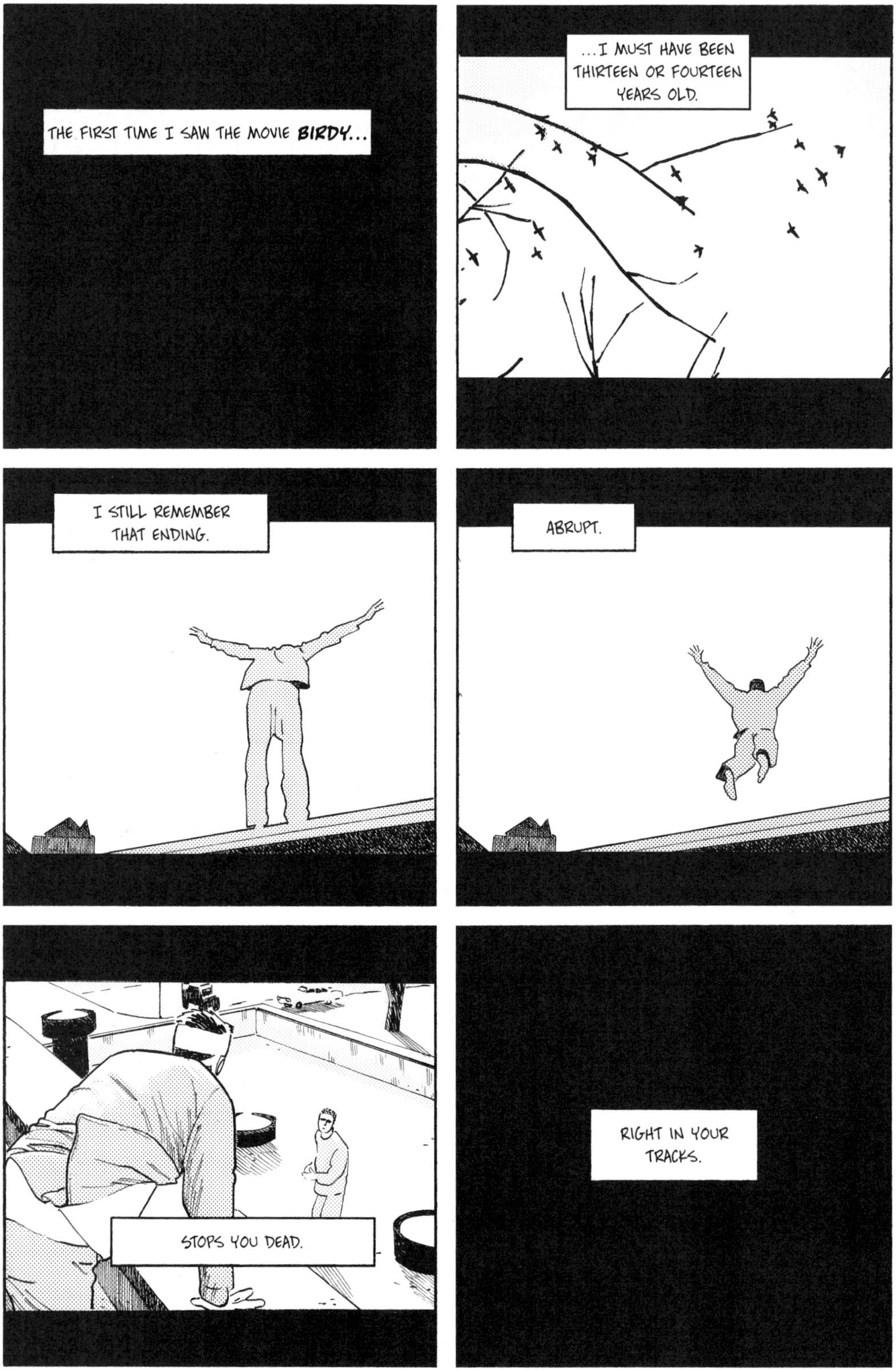
THE FIRST TIME I SAW THE MOVIE **BIRDY**...
...I MUST HAVE BEEN THIRTEEN OR FOURTEEN YEARS OLD.
I STILL REMEMBER THAT ENDING.
ABRUPT.
STOPS YOU DEAD.
RIGHT IN YOUR TRACKS.

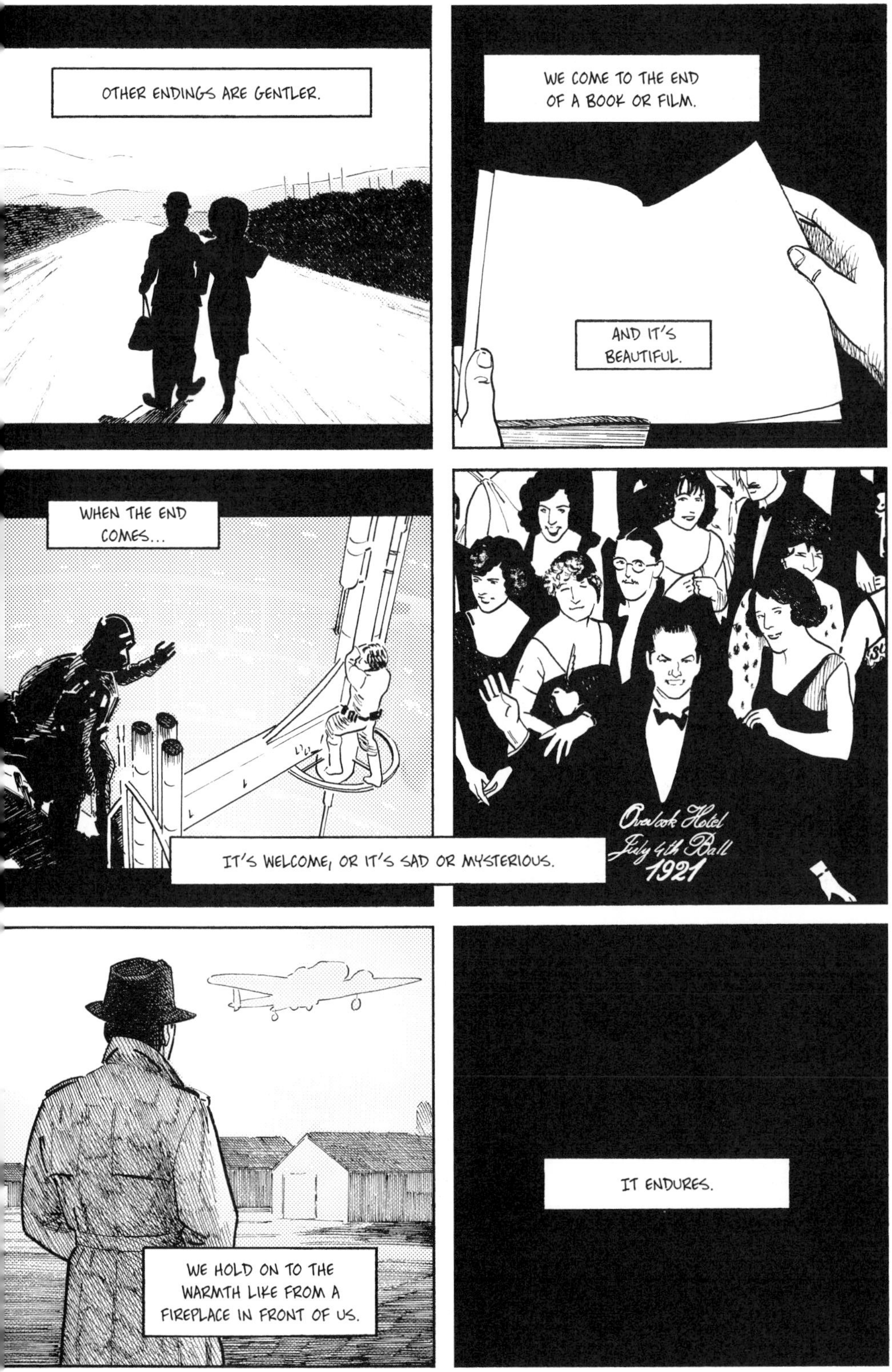
OTHER ENDINGS ARE GENTLER.
WE COME TO THE END OF A BOOK OR FILM.
AND IT'S BEAUTIFUL.
WHEN THE END COMES...
Overlook Hotel July 4th Ball 1921
IT'S WELCOME, OR IT'S SAD OR MYSTERIOUS.
WE HOLD ON TO THE WARMTH LIKE FROM A FIREPLACE IN FRONT OF US.
IT ENDURES.

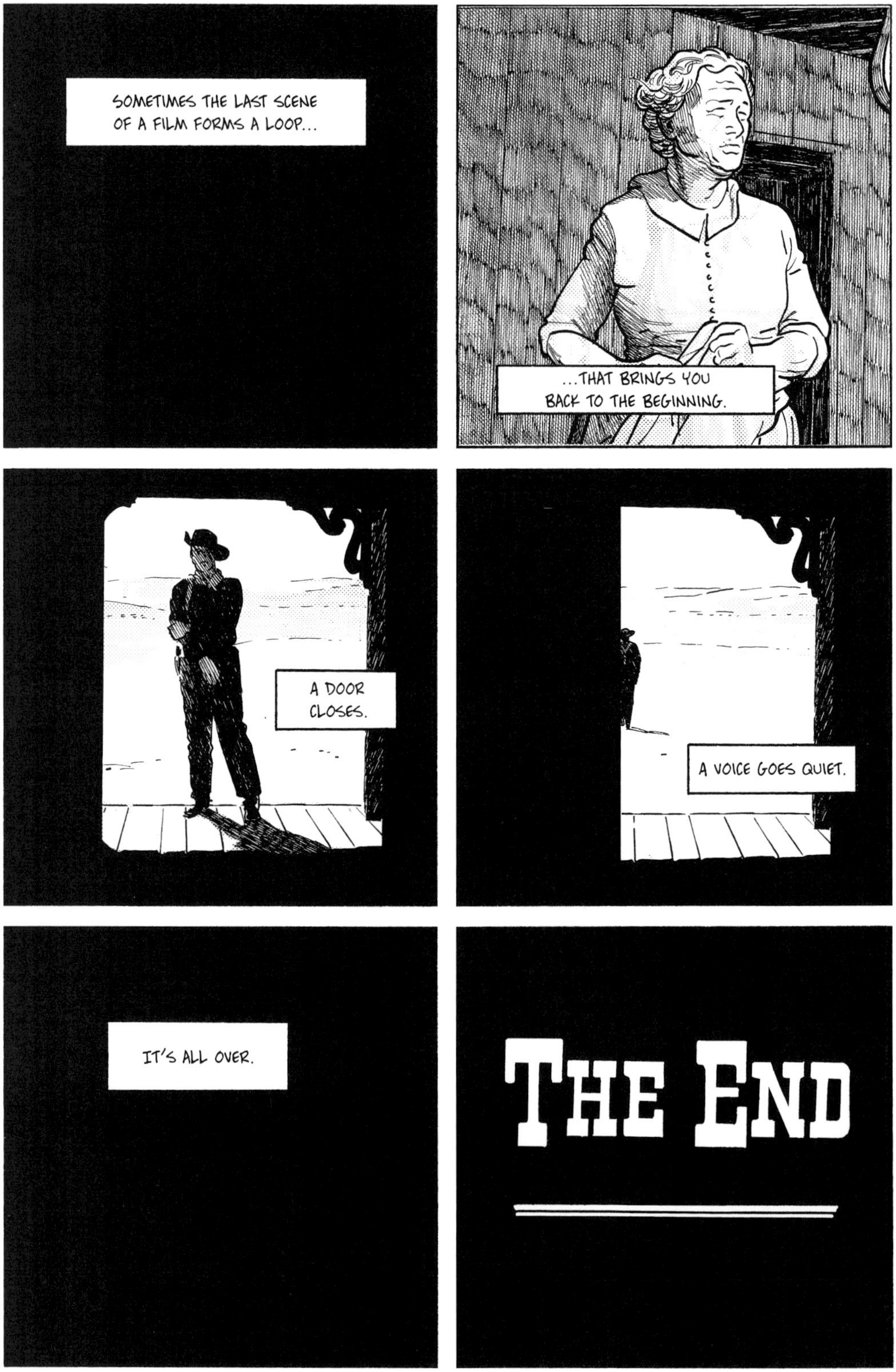
SOMETIMES THE LAST SCENE OF A FILM FORMS A LOOP...
...THAT BRINGS YOU BACK TO THE BEGINNING.
A DOOR CLOSES.
A VOICE GOES QUIET.
IT'S ALL OVER.
THE END

THE

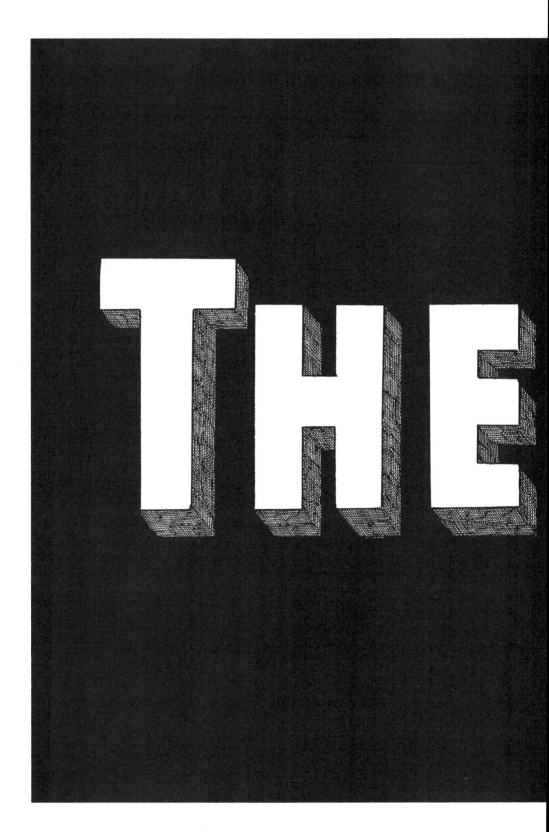

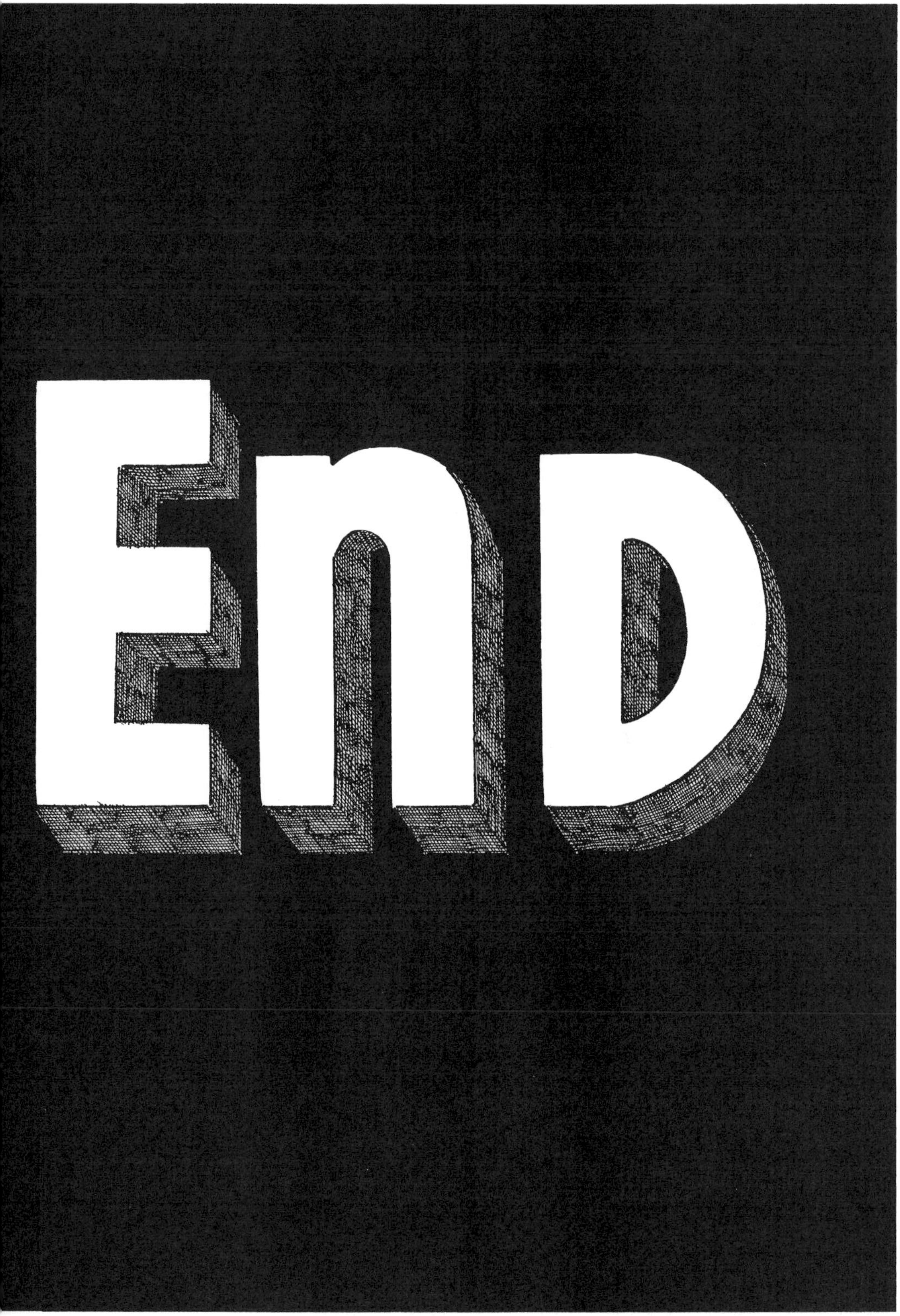
End

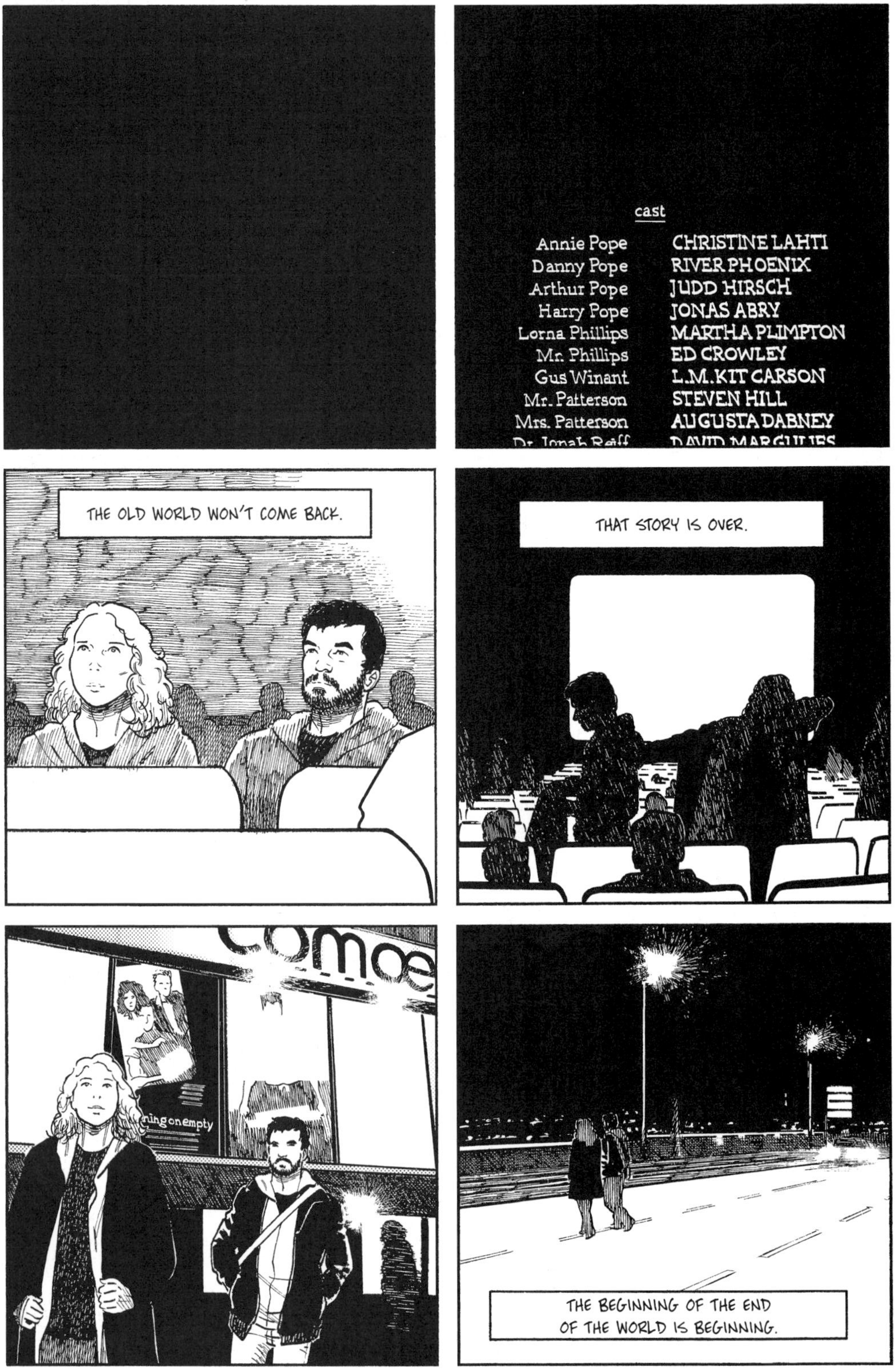
cast
Annie Pope CHRISTINE LAHTI
Danny Pope RIVER PHOENIX
Arthur Pope JUDD HIRSCH
Harry Pope JONAS ABRY
Lorna Phillips MARTHA PLIMPTON
Mr. Phillips ED CROWLEY
Gus Winant L.M. KIT CARSON
Mr. Patterson STEVEN HILL
Mrs. Patterson AUGUSTA DABNEY
THE OLD WORLD WON'T COME BACK.
THAT STORY IS OVER.
THE BEGINNING OF THE END
OF THE WORLD IS BEGINNING.

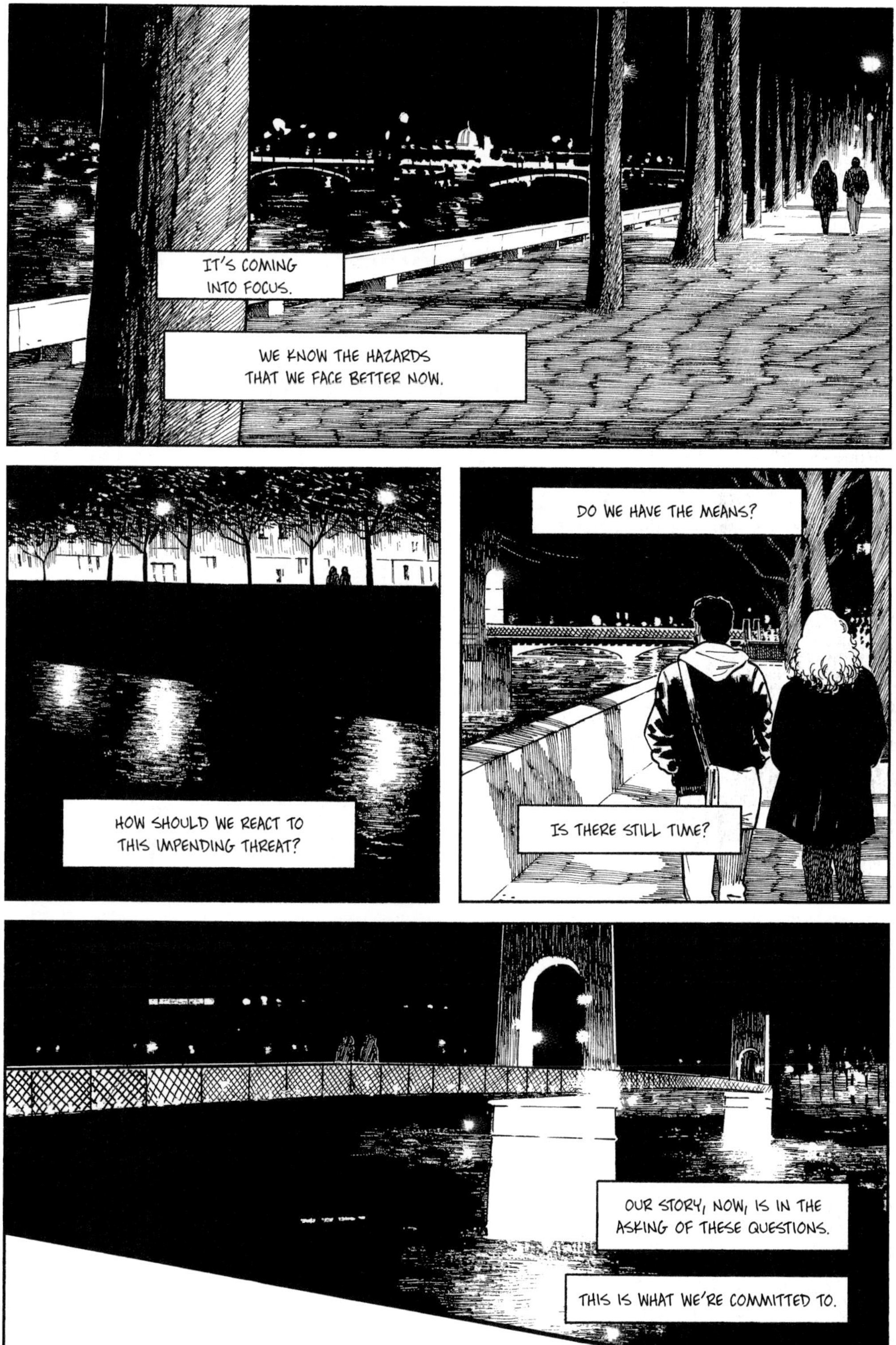
IT'S COMING INTO FOCUS.
WE KNOW THE HAZARDS THAT WE FACE BETTER NOW.
HOW SHOULD WE REACT TO THIS IMPENDING THREAT?
DO WE HAVE THE MEANS?
IS THERE STILL TIME?
OUR STORY, NOW, IS IN THE ASKING OF THESE QUESTIONS.
THIS IS WHAT WE'RE COMMITTED TO.

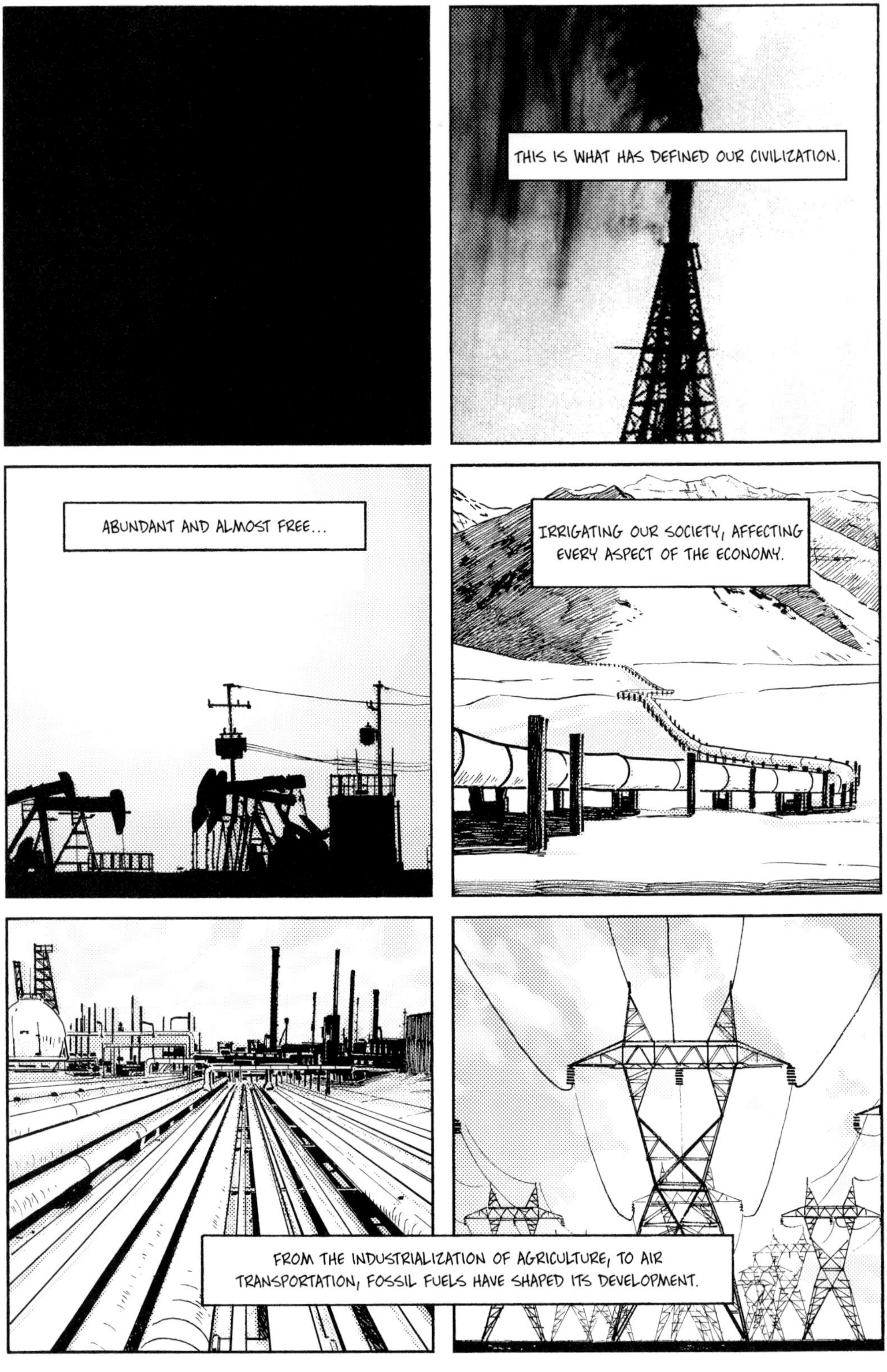
THIS IS WHAT HAS DEFINED OUR CIVILIZATION.
ABUNDANT AND ALMOST FREE...
IRRIGATING OUR SOCIETY, AFFECTING EVERY ASPECT OF THE ECONOMY.
FROM THE INDUSTRIALIZATION OF AGRICULTURE, TO AIR TRANSPORTATION, FOSSIL FUELS HAVE SHAPED ITS DEVELOPMENT.

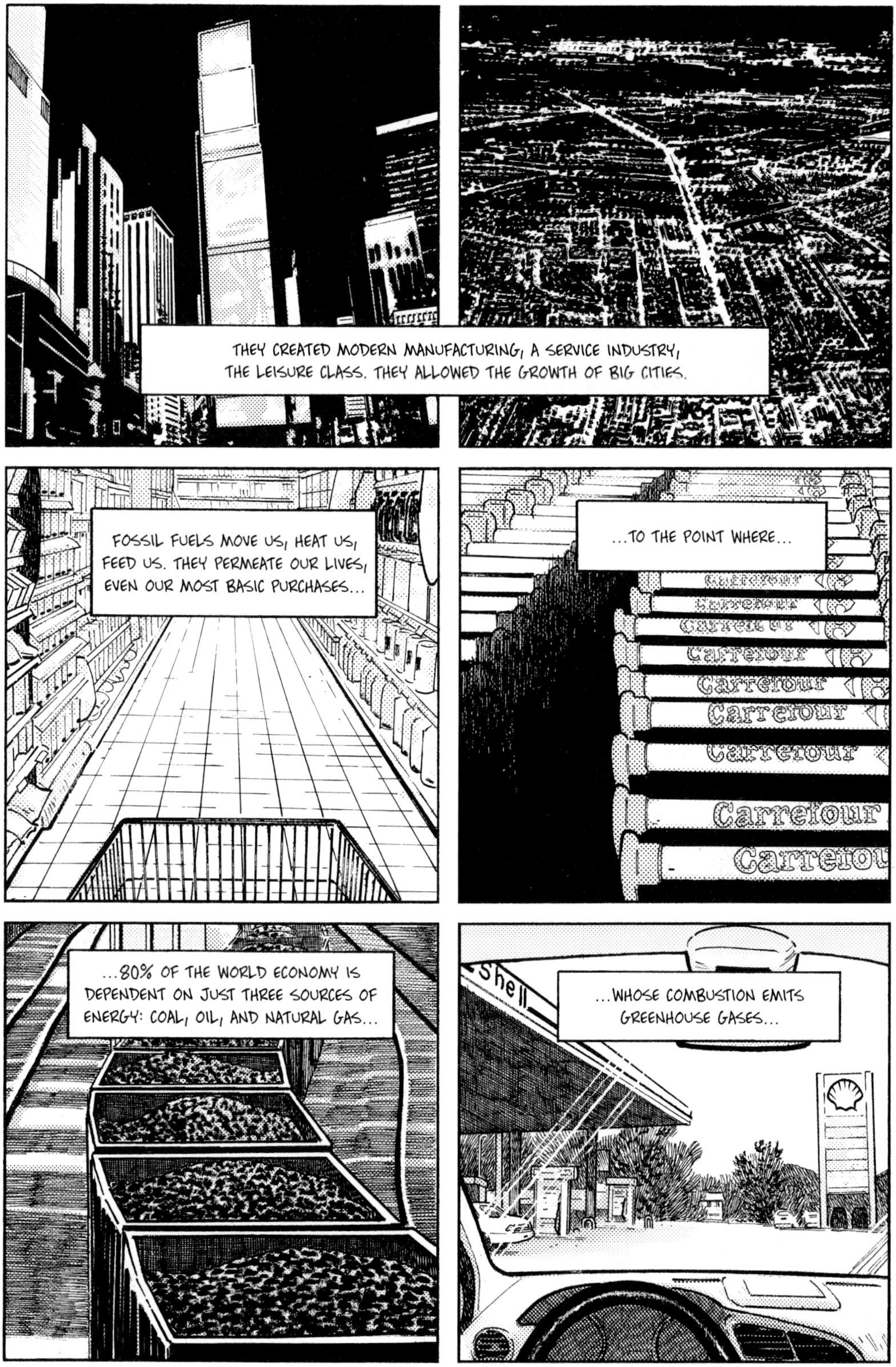
THEY CREATED MODERN MANUFACTURING, A SERVICE INDUSTRY, THE LEISURE CLASS. THEY ALLOWED THE GROWTH OF BIG CITIES.
FOSSIL FUELS MOVE US, HEAT US, FEED US. THEY PERMEATE OUR LIVES, EVEN OUR MOST BASIC PURCHASES...
...TO THE POINT WHERE...
Carrefour
...80% OF THE WORLD ECONOMY IS DEPENDENT ON JUST THREE SOURCES OF ENERGY: COAL, OIL, AND NATURAL GAS...
Shell
...WHOSE COMBUSTION EMITS GREENHOUSE GASES...

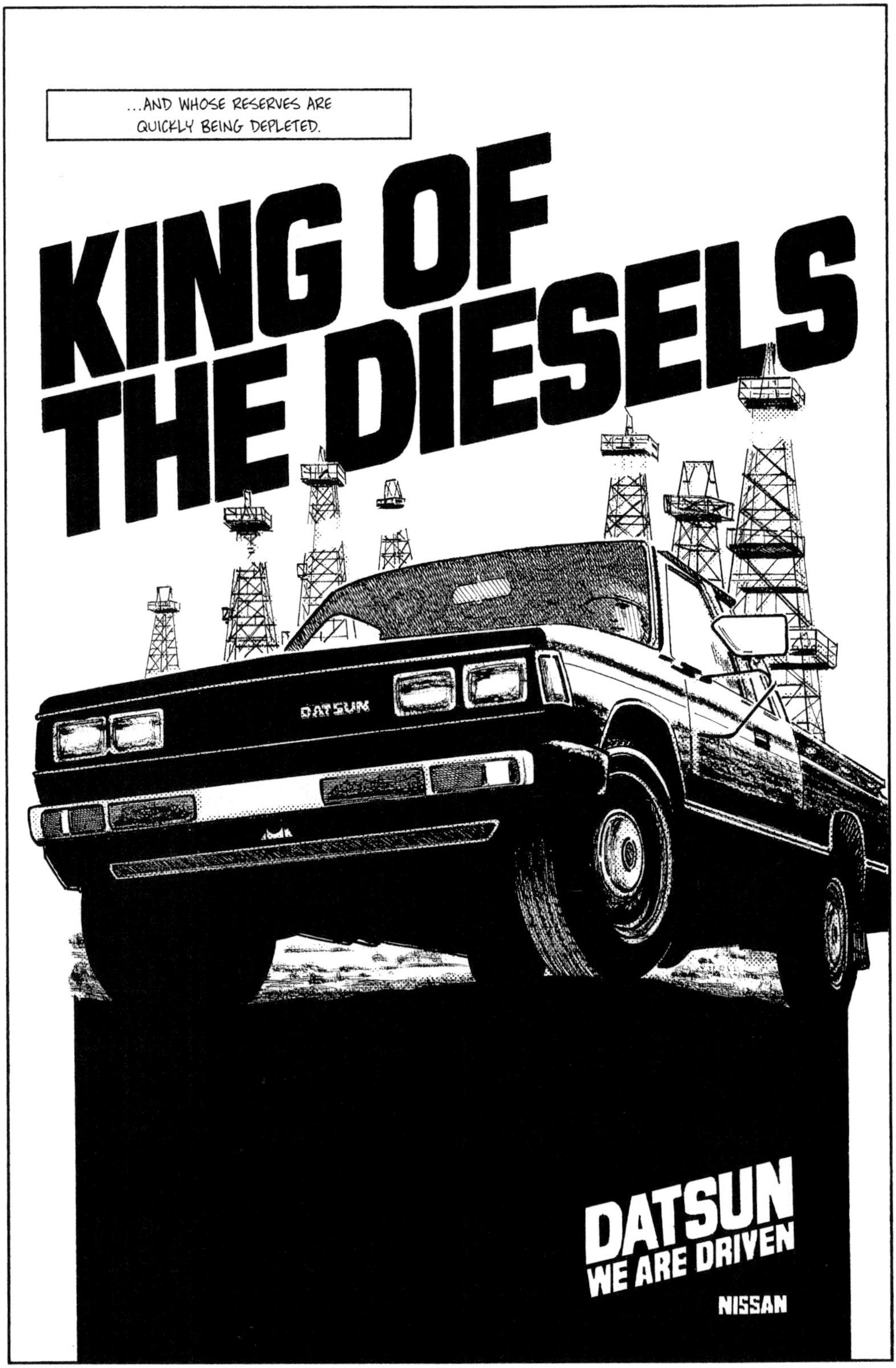
...AND WHOSE RESERVES ARE QUICKLY BEING DEPLETED.
KING OF THE DIESELS
DATSUN
DATSUN
WE ARE DRIVEN
NISSAN

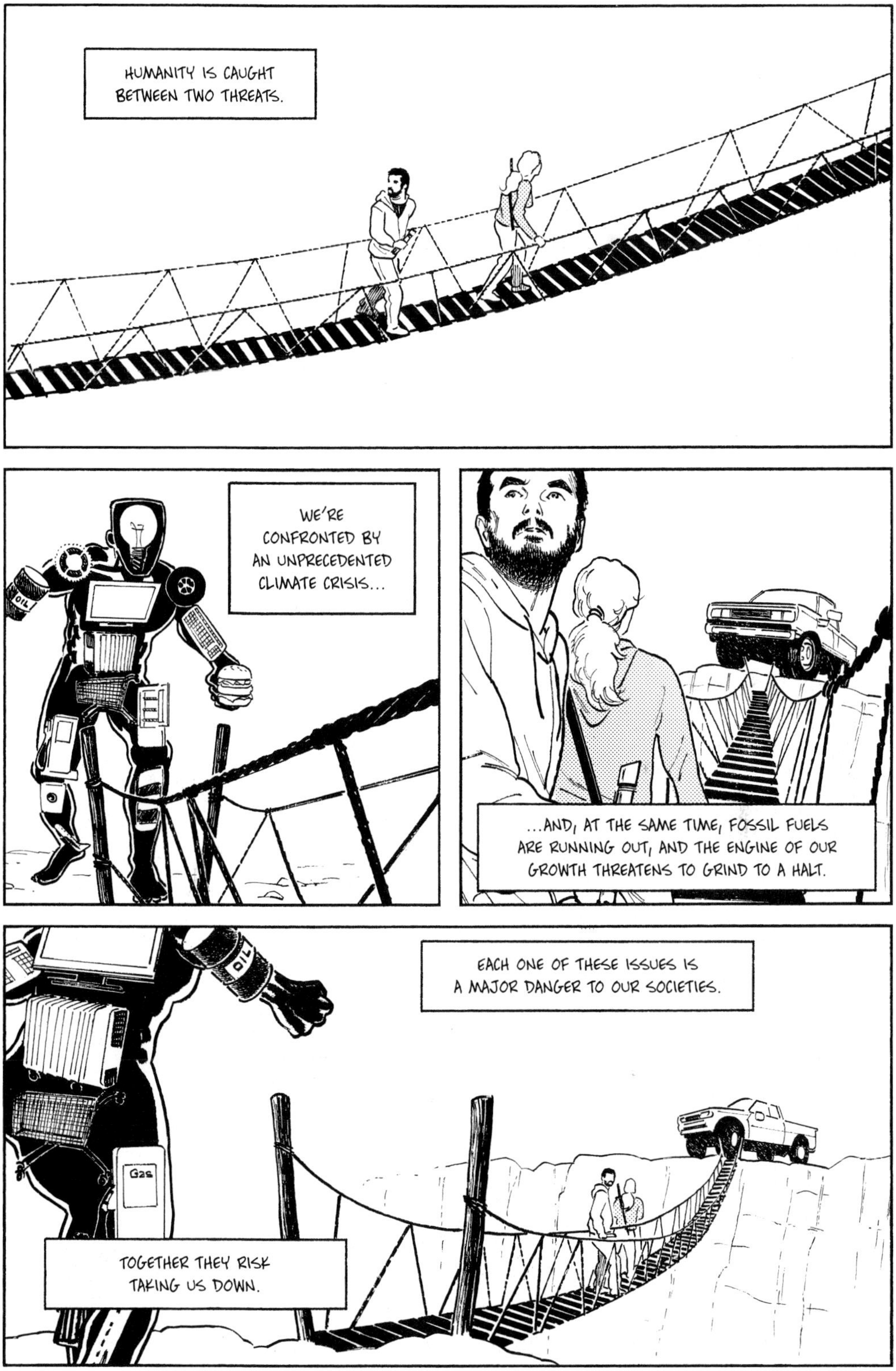
HUMANITY IS CAUGHT BETWEEN TWO THREATS.
WE'RE CONFRONTED BY AN UNPRECEDENTED CLIMATE CRISIS...
OIL
...AND, AT THE SAME TIME, FOSSIL FUELS ARE RUNNING OUT, AND THE ENGINE OF OUR GROWTH THREATENS TO GRIND TO A HALT.
EACH ONE OF THESE ISSUES IS A MAJOR DANGER TO OUR SOCIETIES.
OIL
Gas
TOGETHER THEY RISK TAKING US DOWN.

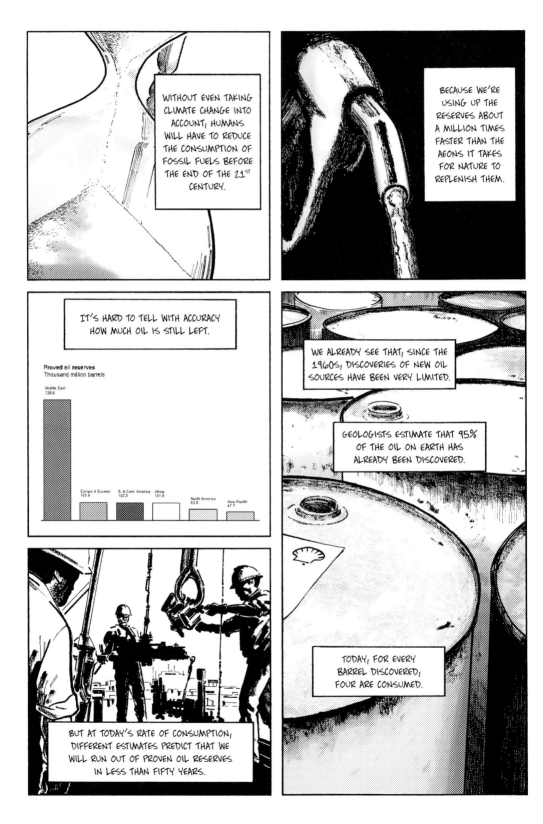

WITHOUT EVEN TAKING CLIMATE CHANGE INTO ACCOUNT, HUMANS WILL HAVE TO REDUCE THE CONSUMPTION OF FOSSIL FUELS BEFORE THE END OF THE 21ST CENTURY.

BECAUSE WE'RE USING UP THE RESERVES ABOUT A MILLION TIMES FASTER THAN THE AEONS IT TAKES FOR NATURE TO REPLENISH THEM.

IT'S HARD TO TELL WITH ACCURACY HOW MUCH OIL IS STILL LEFT.

Proved oil reserves
Thousand million barrels

Middle East
726.6

Europe & Eurasia
105.9

S. & Cent. America
102.2

Africa
101.8

North America
63.8

Asia Pacific
47.7

WE ALREADY SEE THAT, SINCE THE 1960s, DISCOVERIES OF NEW OIL SOURCES HAVE BEEN VERY LIMITED.

GEOLOGISTS ESTIMATE THAT 95% OF THE OIL ON EARTH HAS ALREADY BEEN DISCOVERED.

TODAY, FOR EVERY BARREL DISCOVERED, FOUR ARE CONSUMED.

BUT AT TODAY'S RATE OF CONSUMPTION, DIFFERENT ESTIMATES PREDICT THAT WE WILL RUN OUT OF PROVEN OIL RESERVES IN LESS THAN FIFTY YEARS.

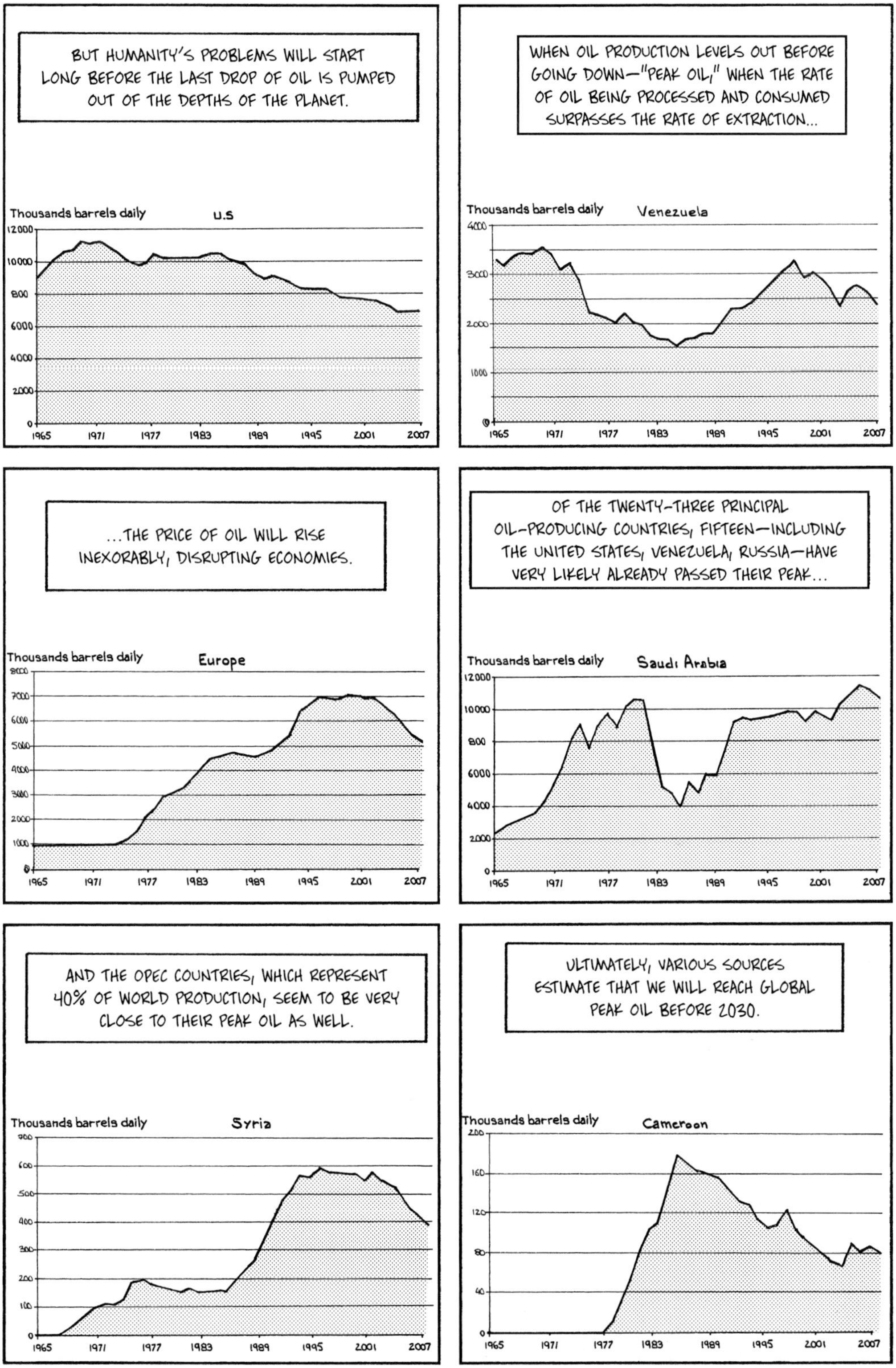
BUT HUMANITY'S PROBLEMS WILL START LONG BEFORE THE LAST DROP OF OIL IS PUMPED OUT OF THE DEPTHS OF THE PLANET.
Thousands barrels daily
U.S
12000
10000
800
6000
4000
2000
0
1965
1971
1977
1983
1989
1995
2001
2007
WHEN OIL PRODUCTION LEVELS OUT BEFORE GOING DOWN—"PEAK OIL," WHEN THE RATE OF OIL BEING PROCESSED AND CONSUMED SURPASSES THE RATE OF EXTRACTION...
Thousands barrels daily
Venezuela
4000
3000
2000
1000
0
1965
1971
1977
1983
1989
1995
2001
2007
...THE PRICE OF OIL WILL RISE INEXORABLY, DISRUPTING ECONOMIES.
Thousands barrels daily
Europe
8000
7000
6000
5000
4000
3000
2000
1000
0
1965
1971
1977
1983
1989
1995
2001
2007
OF THE TWENTY-THREE PRINCIPAL OIL-PRODUCING COUNTRIES, FIFTEEN—INCLUDING THE UNITED STATES, VENEZUELA, RUSSIA—HAVE VERY LIKELY ALREADY PASSED THEIR PEAK...
Thousands barrels daily
Saudi Arabia
12000
10000
800
6000
4000
2000
0
1965
1971
1977
1983
1989
1995
2001
2007
AND THE OPEC COUNTRIES, WHICH REPRESENT 40% OF WORLD PRODUCTION, SEEM TO BE VERY CLOSE TO THEIR PEAK OIL AS WELL.
Thousands barrels daily
Syria
700
600
500
400
300
200
100
0
1965
1971
1977
1983
1989
1995
2001
2007
ULTIMATELY, VARIOUS SOURCES ESTIMATE THAT WE WILL REACH GLOBAL PEAK OIL BEFORE 2030.
Thousands barrels daily
Cameroon
200
160
120
80
40
0
1965
1971
1977
1983
1989
1995
2001
2007

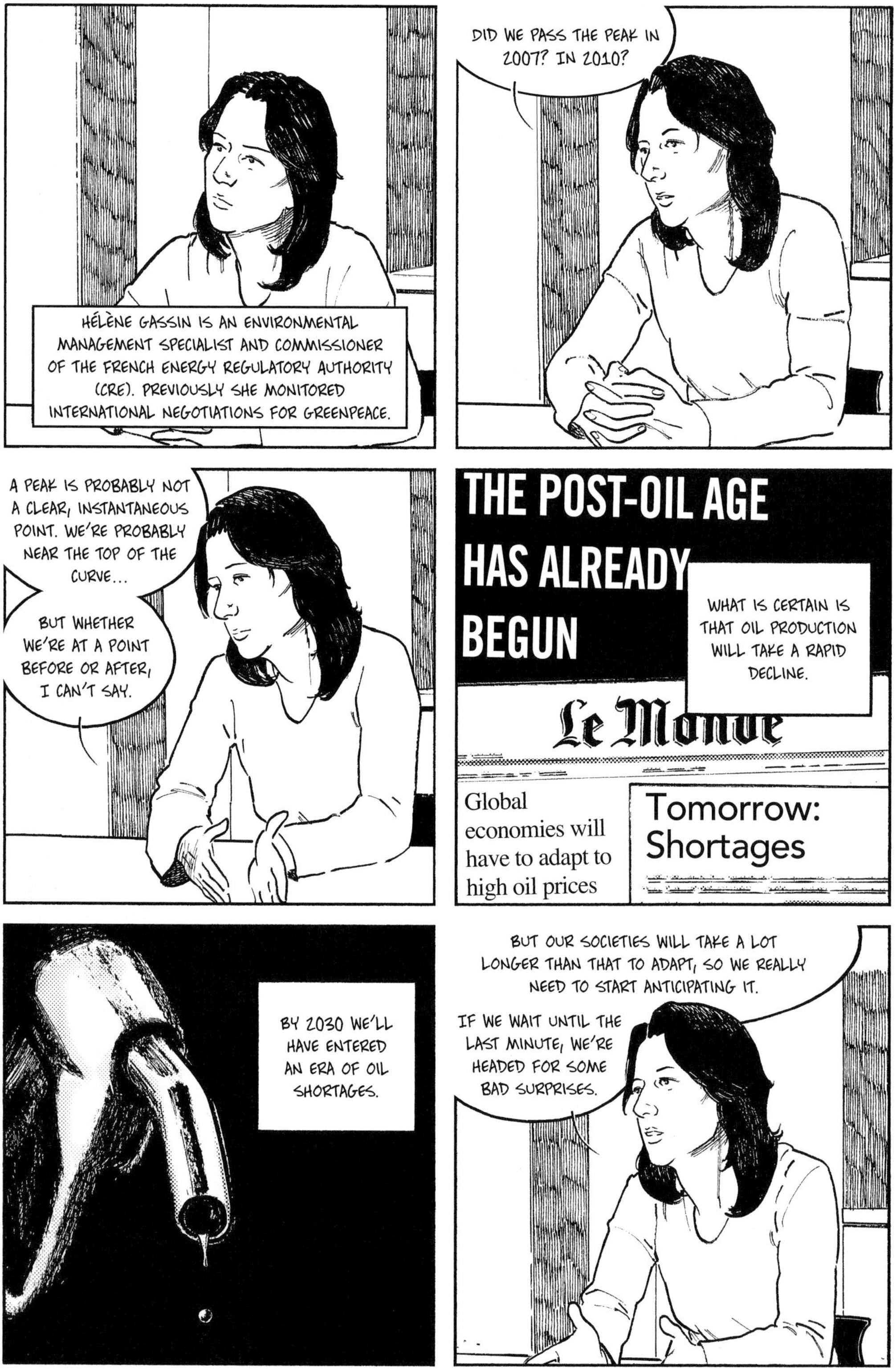
HÉLÈNE GASSIN IS AN ENVIRONMENTAL MANAGEMENT SPECIALIST AND COMMISSIONER OF THE FRENCH ENERGY REGULATORY AUTHORITY (CRE). PREVIOUSLY SHE MONITORED INTERNATIONAL NEGOTIATIONS FOR GREENPEACE.
DID WE PASS THE PEAK IN 2007? IN 2010?
A PEAK IS PROBABLY NOT A CLEAR, INSTANTANEOUS POINT. WE'RE PROBABLY NEAR THE TOP OF THE CURVE...
BUT WHETHER WE'RE AT A POINT BEFORE OR AFTER, I CAN'T SAY.
THE POST-OIL AGE HAS ALREADY BEGUN
WHAT IS CERTAIN IS THAT OIL PRODUCTION WILL TAKE A RAPID DECLINE.
Le Monde
Global economies will have to adapt to high oil prices
Tomorrow: Shortages
BY 2030 WE'LL HAVE ENTERED AN ERA OF OIL SHORTAGES.
BUT OUR SOCIETIES WILL TAKE A LOT LONGER THAN THAT TO ADAPT, SO WE REALLY NEED TO START ANTICIPATING IT.
IF WE WAIT UNTIL THE LAST MINUTE, WE'RE HEADED FOR SOME BAD SURPRISES.

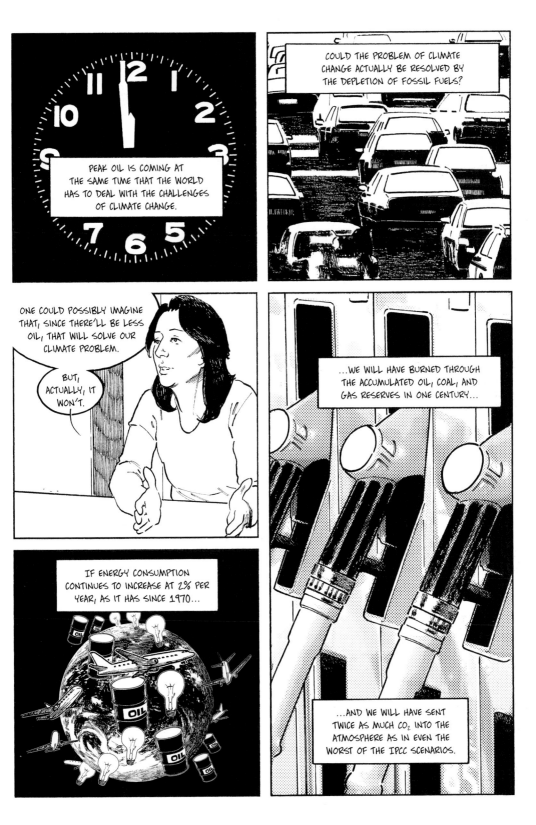

PEAK OIL IS COMING AT THE SAME TIME THAT THE WORLD HAS TO DEAL WITH THE CHALLENGES OF CLIMATE CHANGE.

COULD THE PROBLEM OF CLIMATE CHANGE ACTUALLY BE RESOLVED BY THE DEPLETION OF FOSSIL FUELS?

ONE COULD POSSIBLY IMAGINE THAT, SINCE THERE'LL BE LESS OIL, THAT WILL SOLVE OUR CLIMATE PROBLEM.

BUT, ACTUALLY, IT WON'T.

...WE WILL HAVE BURNED THROUGH THE ACCUMULATED OIL, COAL, AND GAS RESERVES IN ONE CENTURY...

IF ENERGY CONSUMPTION CONTINUES TO INCREASE AT 2% PER YEAR, AS IT HAS SINCE 1970...

...AND WE WILL HAVE SENT TWICE AS MUCH CO_2 INTO THE ATMOSPHERE AS IN EVEN THE WORST OF THE IPCC SCENARIOS.

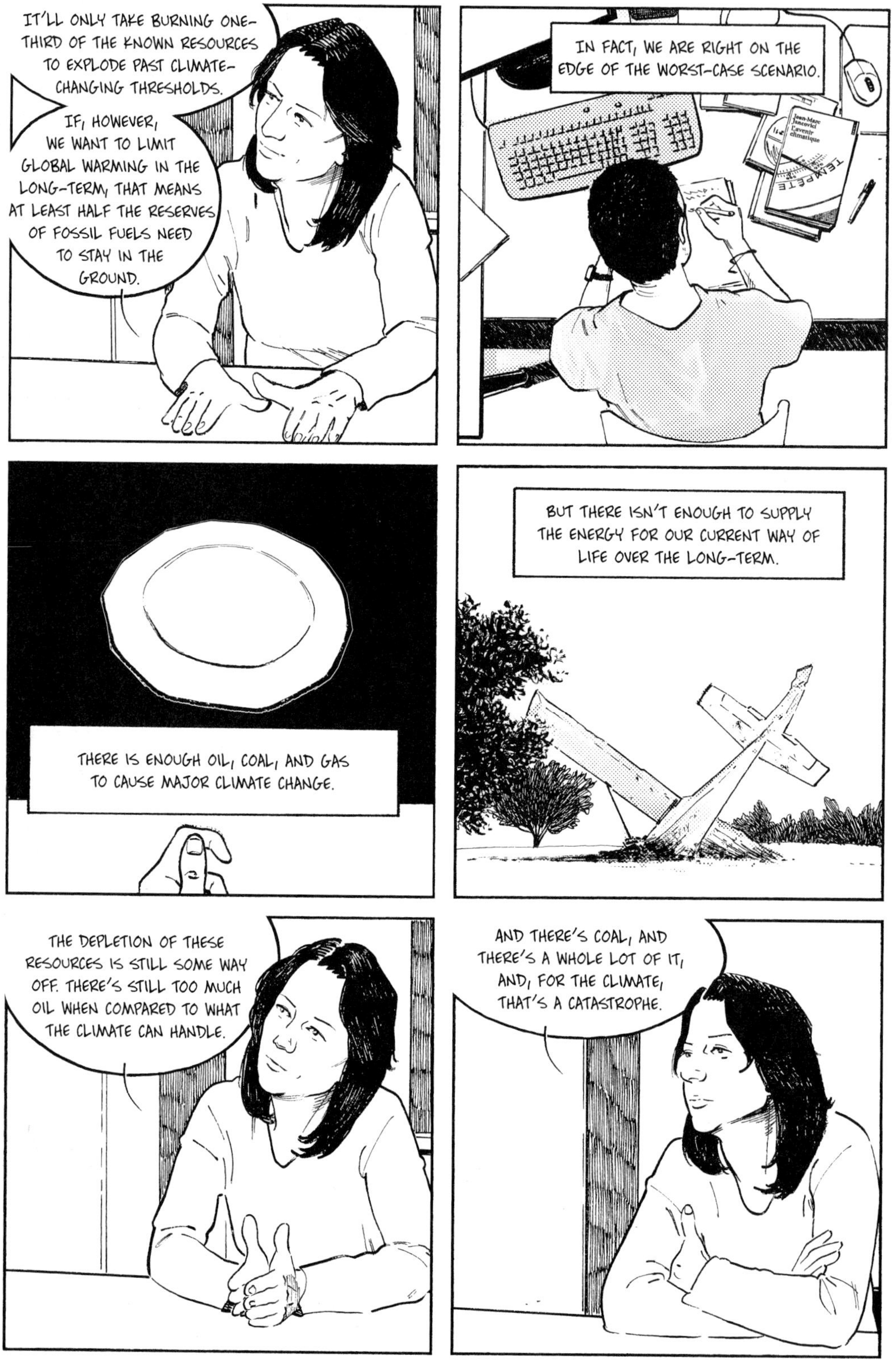
IT'LL ONLY TAKE BURNING ONE-THIRD OF THE KNOWN RESOURCES TO EXPLODE PAST CLIMATE-CHANGING THRESHOLDS.
IF, HOWEVER, WE WANT TO LIMIT GLOBAL WARMING IN THE LONG-TERM, THAT MEANS AT LEAST HALF THE RESERVES OF FOSSIL FUELS NEED TO STAY IN THE GROUND.
IN FACT, WE ARE RIGHT ON THE EDGE OF THE WORST-CASE SCENARIO.
THERE IS ENOUGH OIL, COAL, AND GAS TO CAUSE MAJOR CLIMATE CHANGE.
BUT THERE ISN'T ENOUGH TO SUPPLY THE ENERGY FOR OUR CURRENT WAY OF LIFE OVER THE LONG-TERM.
THE DEPLETION OF THESE RESOURCES IS STILL SOME WAY OFF. THERE'S STILL TOO MUCH OIL WHEN COMPARED TO WHAT THE CLIMATE CAN HANDLE.
AND THERE'S COAL, AND THERE'S A WHOLE LOT OF IT, AND, FOR THE CLIMATE, THAT'S A CATASTROPHE.

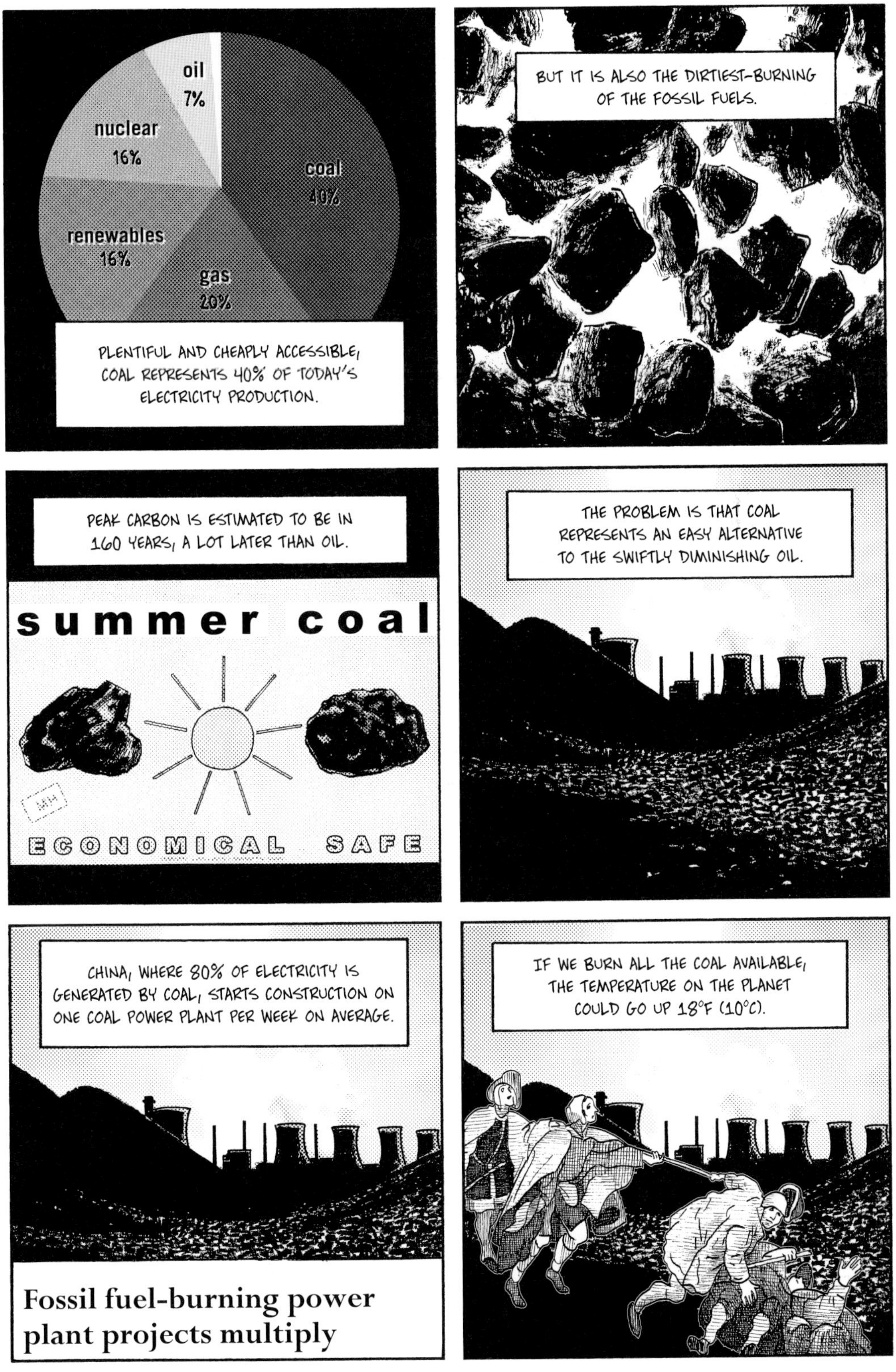
oil
7%
nuclear
16%
coal
40%
renewables
16%
gas
20%
PLENTIFUL AND CHEAPLY ACCESSIBLE, COAL REPRESENTS 40% OF TODAY'S ELECTRICITY PRODUCTION.
BUT IT IS ALSO THE DIRTIEST-BURNING OF THE FOSSIL FUELS.
PEAK CARBON IS ESTIMATED TO BE IN 160 YEARS, A LOT LATER THAN OIL.
summer coal
ECONOMICAL SAFE
THE PROBLEM IS THAT COAL REPRESENTS AN EASY ALTERNATIVE TO THE SWIFTLY DIMINISHING OIL.
CHINA, WHERE 80% OF ELECTRICITY IS GENERATED BY COAL, STARTS CONSTRUCTION ON ONE COAL POWER PLANT PER WEEK ON AVERAGE.
Fossil fuel-burning power plant projects multiply
IF WE BURN ALL THE COAL AVAILABLE, THE TEMPERATURE ON THE PLANET COULD GO UP 18°F (10°C).

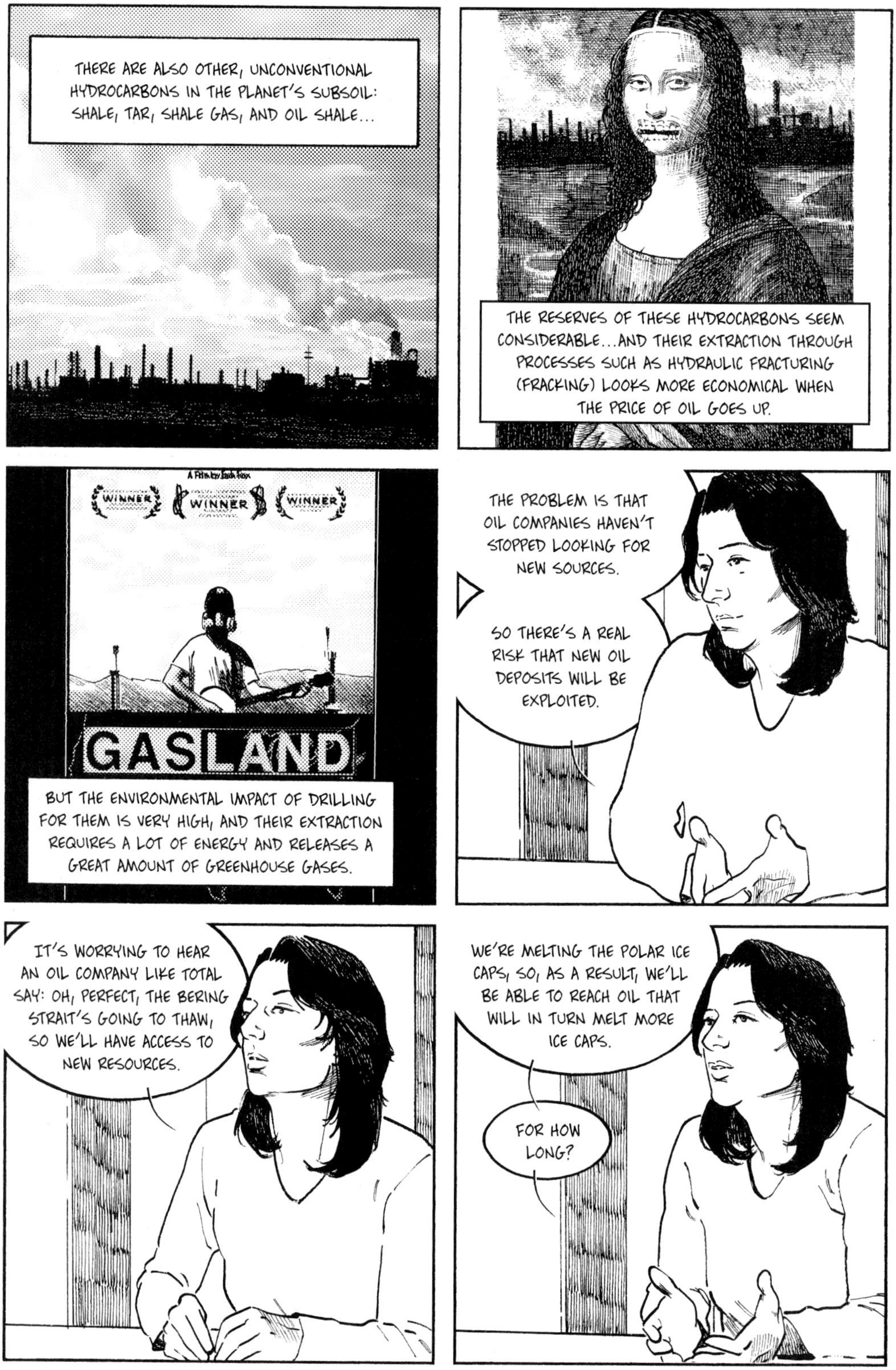

THERE ARE ALSO OTHER, UNCONVENTIONAL HYDROCARBONS IN THE PLANET'S SUBSOIL: SHALE, TAR, SHALE GAS, AND OIL SHALE...
THE RESERVES OF THESE HYDROCARBONS SEEM CONSIDERABLE...AND THEIR EXTRACTION THROUGH PROCESSES SUCH AS HYDRAULIC FRACTURING (FRACKING) LOOKS MORE ECONOMICAL WHEN THE PRICE OF OIL GOES UP.
WINNER
WINNER
WINNER
GASLAND
BUT THE ENVIRONMENTAL IMPACT OF DRILLING FOR THEM IS VERY HIGH, AND THEIR EXTRACTION REQUIRES A LOT OF ENERGY AND RELEASES A GREAT AMOUNT OF GREENHOUSE GASES.
THE PROBLEM IS THAT OIL COMPANIES HAVEN'T STOPPED LOOKING FOR NEW SOURCES.
SO THERE'S A REAL RISK THAT NEW OIL DEPOSITS WILL BE EXPLOITED.
IT'S WORRYING TO HEAR AN OIL COMPANY LIKE TOTAL SAY: OH, PERFECT, THE BERING STRAIT'S GOING TO THAW, SO WE'LL HAVE ACCESS TO NEW RESOURCES.
WE'RE MELTING THE POLAR ICE CAPS, SO, AS A RESULT, WE'LL BE ABLE TO REACH OIL THAT WILL IN TURN MELT MORE ICE CAPS.
FOR HOW LONG?

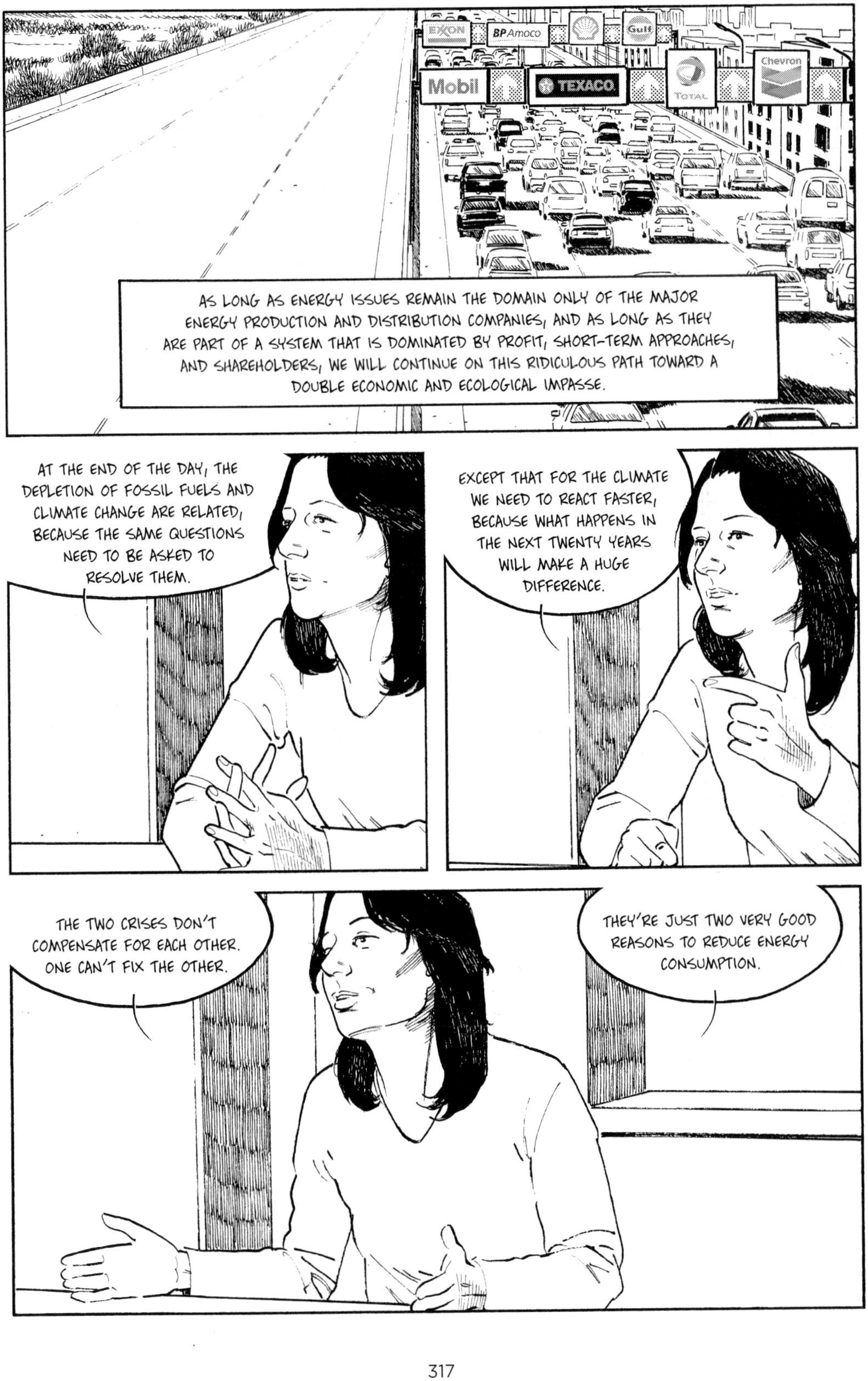
EXXON
BP Amoco
Gulf
Chevron
Mobil
TEXACO
TOTAL
AS LONG AS ENERGY ISSUES REMAIN THE DOMAIN ONLY OF THE MAJOR ENERGY PRODUCTION AND DISTRIBUTION COMPANIES, AND AS LONG AS THEY ARE PART OF A SYSTEM THAT IS DOMINATED BY PROFIT, SHORT-TERM APPROACHES, AND SHAREHOLDERS, WE WILL CONTINUE ON THIS RIDICULOUS PATH TOWARD A DOUBLE ECONOMIC AND ECOLOGICAL IMPASSE.
AT THE END OF THE DAY, THE DEPLETION OF FOSSIL FUELS AND CLIMATE CHANGE ARE RELATED, BECAUSE THE SAME QUESTIONS NEED TO BE ASKED TO RESOLVE THEM.
EXCEPT THAT FOR THE CLIMATE WE NEED TO REACT FASTER, BECAUSE WHAT HAPPENS IN THE NEXT TWENTY YEARS WILL MAKE A HUGE DIFFERENCE.
THE TWO CRISES DON'T COMPENSATE FOR EACH OTHER. ONE CAN'T FIX THE OTHER.
THEY'RE JUST TWO VERY GOOD REASONS TO REDUCE ENERGY CONSUMPTION.

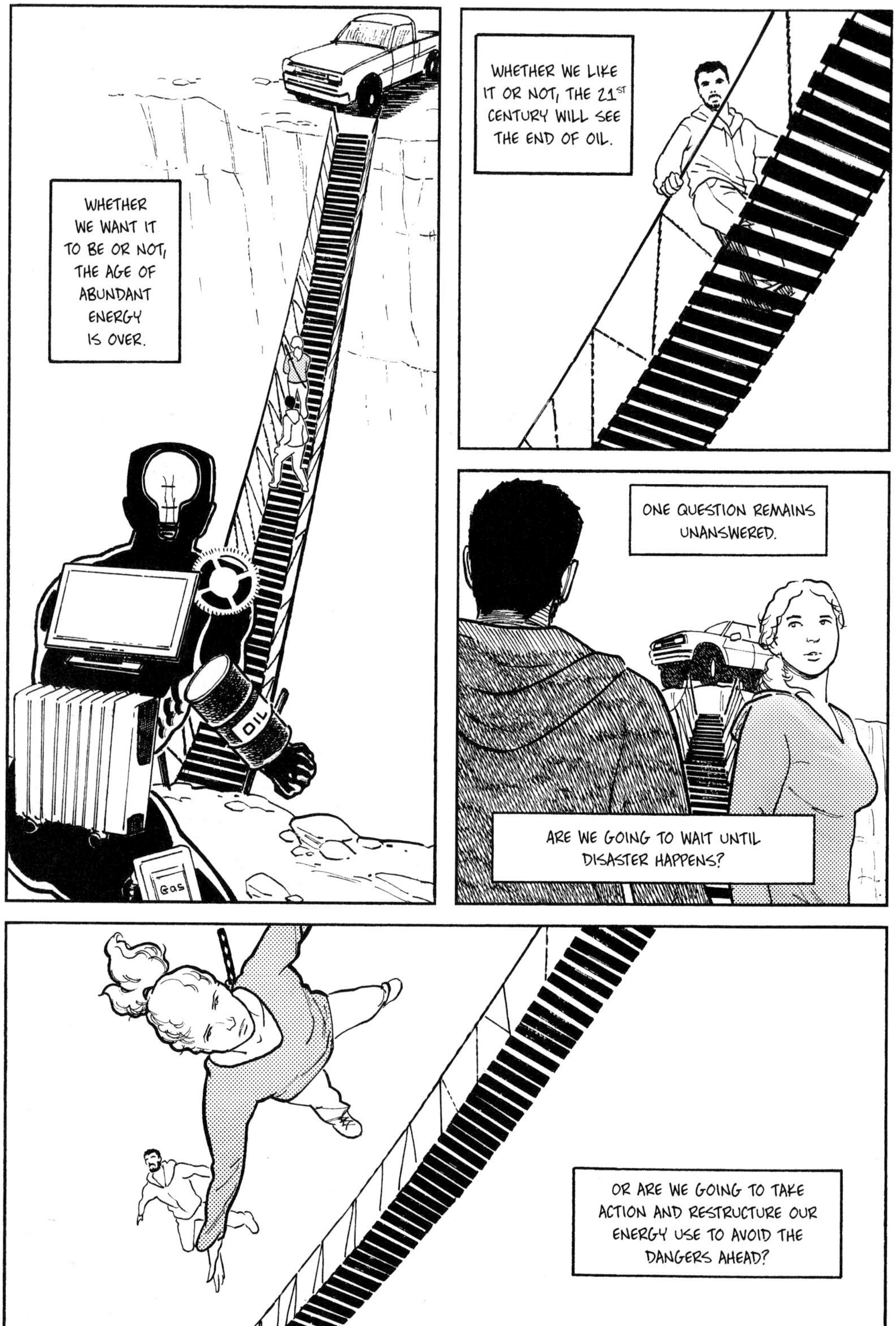
WHETHER WE WANT IT TO BE OR NOT, THE AGE OF ABUNDANT ENERGY IS OVER.
OIL
Gas
WHETHER WE LIKE IT OR NOT, THE 21ST CENTURY WILL SEE THE END OF OIL.
ONE QUESTION REMAINS UNANSWERED.
ARE WE GOING TO WAIT UNTIL DISASTER HAPPENS?
OR ARE WE GOING TO TAKE ACTION AND RESTRUCTURE OUR ENERGY USE TO AVOID THE DANGERS AHEAD?

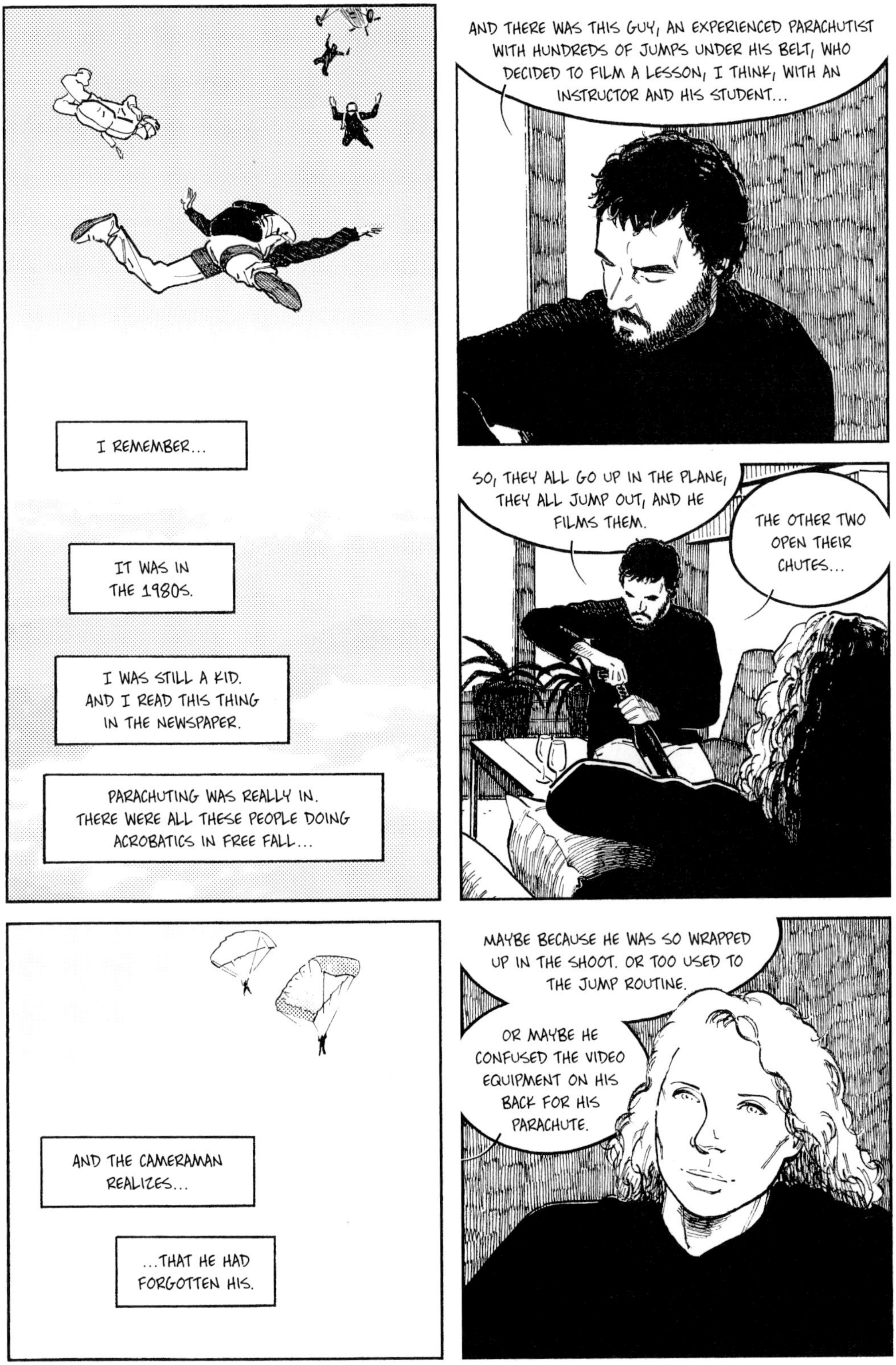
I REMEMBER...
IT WAS IN THE 1980S.
I WAS STILL A KID. AND I READ THIS THING IN THE NEWSPAPER.
PARACHUTING WAS REALLY IN. THERE WERE ALL THESE PEOPLE DOING ACROBATICS IN FREE FALL...
AND THERE WAS THIS GUY, AN EXPERIENCED PARACHUTIST WITH HUNDREDS OF JUMPS UNDER HIS BELT, WHO DECIDED TO FILM A LESSON, I THINK, WITH AN INSTRUCTOR AND HIS STUDENT...
SO, THEY ALL GO UP IN THE PLANE, THEY ALL JUMP OUT, AND HE FILMS THEM.
THE OTHER TWO OPEN THEIR CHUTES...
AND THE CAMERAMAN REALIZES...
...THAT HE HAD FORGOTTEN HIS.
MAYBE BECAUSE HE WAS SO WRAPPED UP IN THE SHOOT. OR TOO USED TO THE JUMP ROUTINE.
OR MAYBE HE CONFUSED THE VIDEO EQUIPMENT ON HIS BACK FOR HIS PARACHUTE.

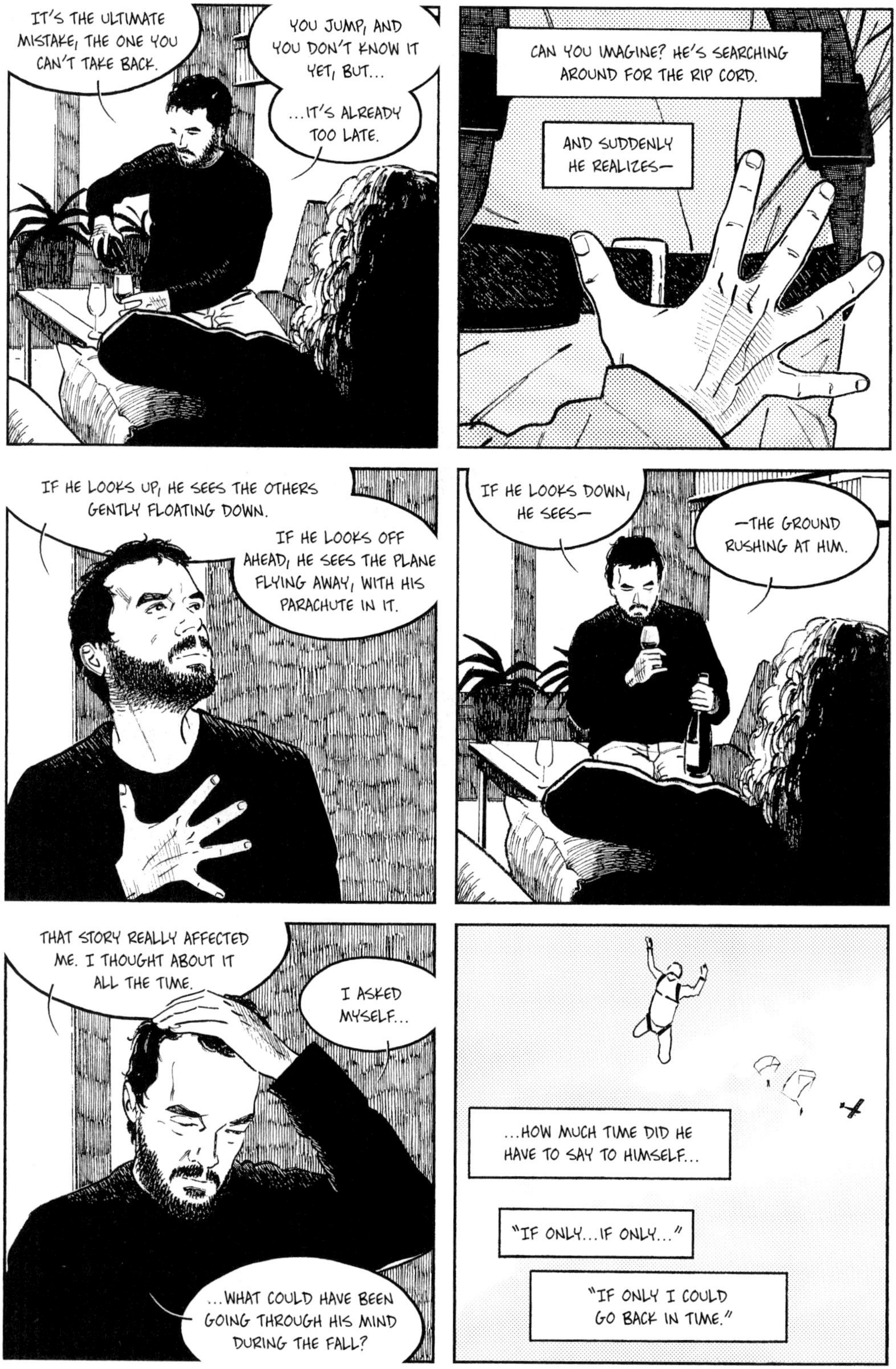
IT'S THE ULTIMATE MISTAKE, THE ONE YOU CAN'T TAKE BACK.
YOU JUMP, AND YOU DON'T KNOW IT YET, BUT...
...IT'S ALREADY TOO LATE.
CAN YOU IMAGINE? HE'S SEARCHING AROUND FOR THE RIP CORD.
AND SUDDENLY HE REALIZES—
IF HE LOOKS UP, HE SEES THE OTHERS GENTLY FLOATING DOWN.
IF HE LOOKS OFF AHEAD, HE SEES THE PLANE FLYING AWAY, WITH HIS PARACHUTE IN IT.
IF HE LOOKS DOWN, HE SEES—
—THE GROUND RUSHING AT HIM.
THAT STORY REALLY AFFECTED ME. I THOUGHT ABOUT IT ALL THE TIME.
I ASKED MYSELF...
...WHAT COULD HAVE BEEN GOING THROUGH HIS MIND DURING THE FALL?
...HOW MUCH TIME DID HE HAVE TO SAY TO HIMSELF...
"IF ONLY...IF ONLY..."
"IF ONLY I COULD GO BACK IN TIME."

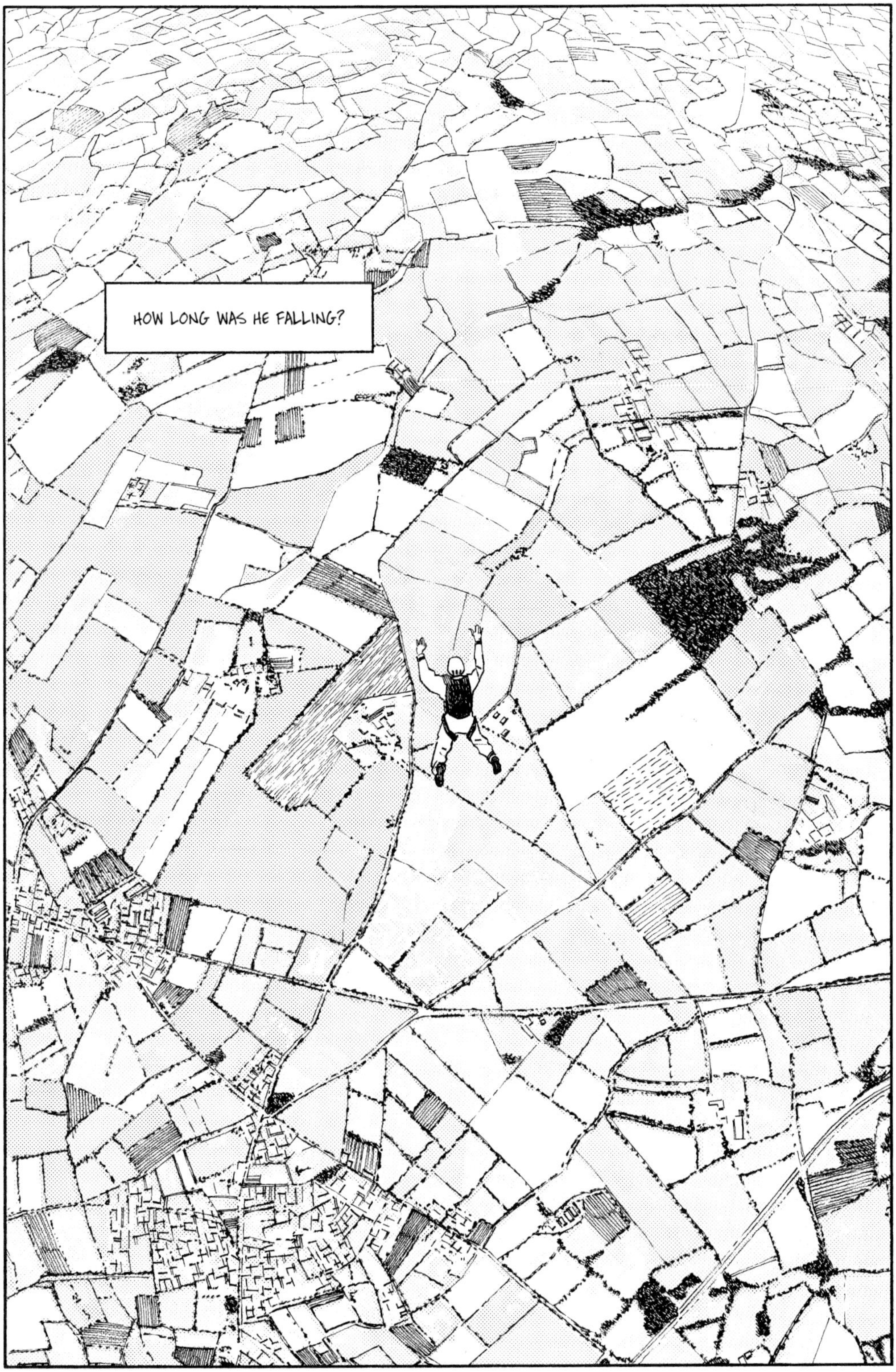
HOW LONG WAS HE FALLING?

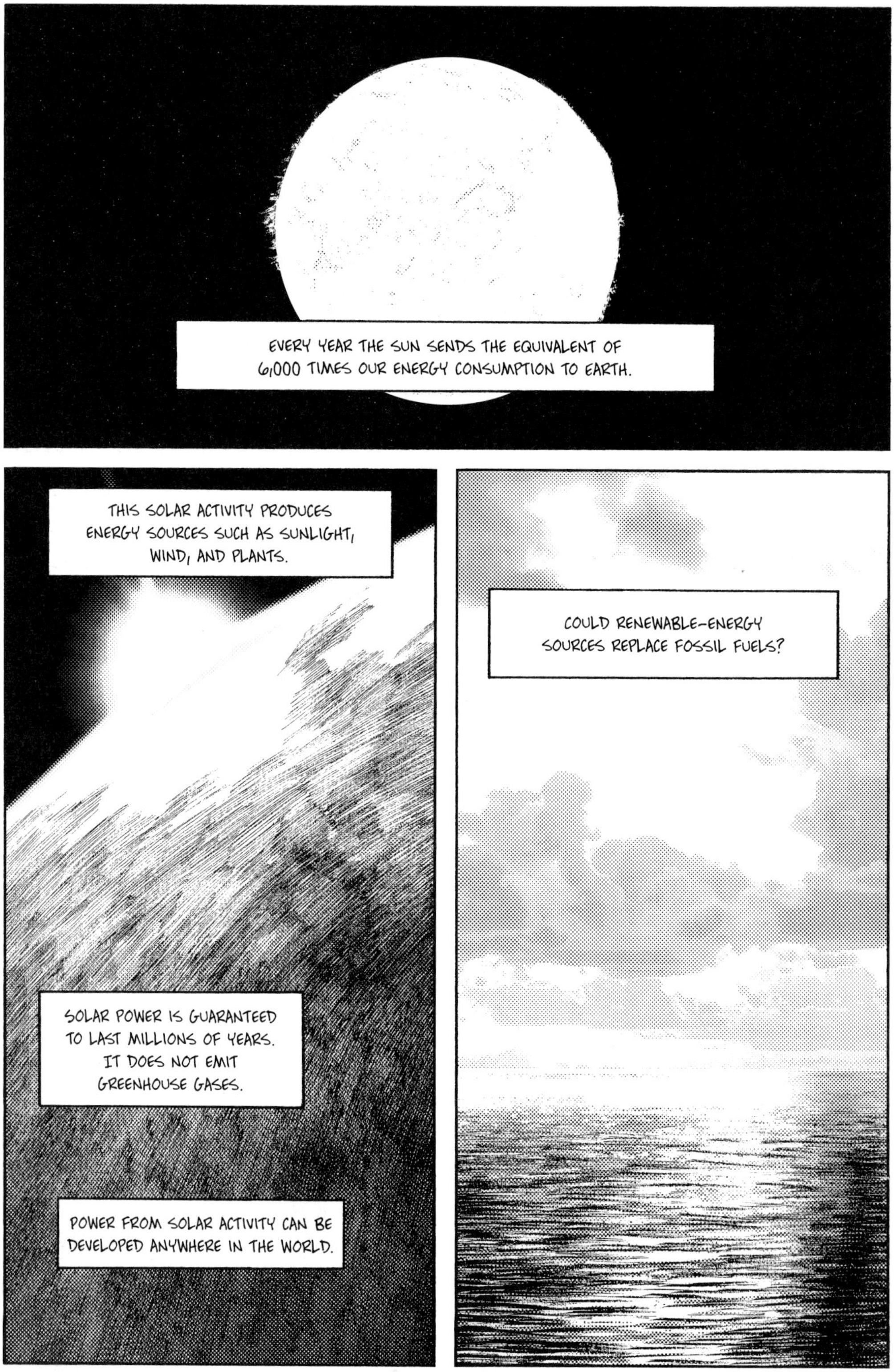
EVERY YEAR THE SUN SENDS THE EQUIVALENT OF 6,000 TIMES OUR ENERGY CONSUMPTION TO EARTH.
THIS SOLAR ACTIVITY PRODUCES ENERGY SOURCES SUCH AS SUNLIGHT, WIND, AND PLANTS.
SOLAR POWER IS GUARANTEED TO LAST MILLIONS OF YEARS. IT DOES NOT EMIT GREENHOUSE GASES.
POWER FROM SOLAR ACTIVITY CAN BE DEVELOPED ANYWHERE IN THE WORLD.
COULD RENEWABLE-ENERGY SOURCES REPLACE FOSSIL FUELS?

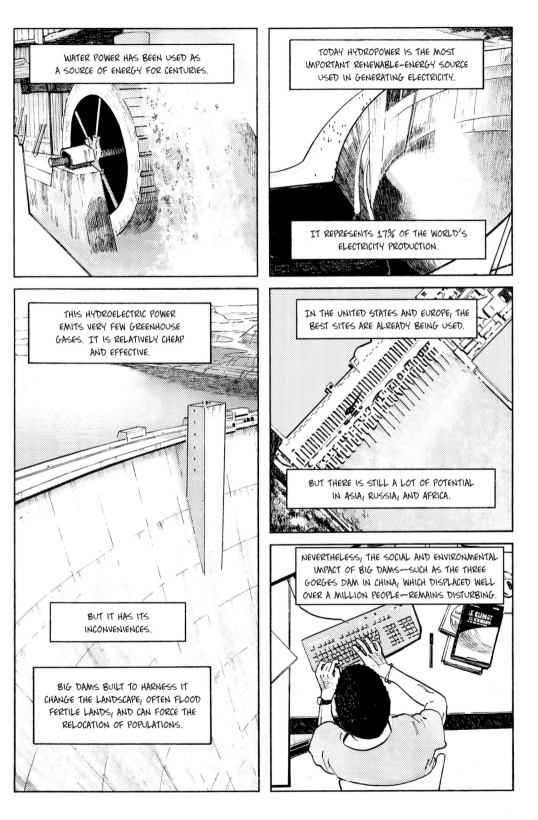

WATER POWER HAS BEEN USED AS A SOURCE OF ENERGY FOR CENTURIES.

TODAY HYDROPOWER IS THE MOST IMPORTANT RENEWABLE-ENERGY SOURCE USED IN GENERATING ELECTRICITY.

IT REPRESENTS 17% OF THE WORLD'S ELECTRICITY PRODUCTION.

THIS HYDROELECTRIC POWER EMITS VERY FEW GREENHOUSE GASES. IT IS RELATIVELY CHEAP AND EFFECTIVE.

IN THE UNITED STATES AND EUROPE, THE BEST SITES ARE ALREADY BEING USED.

BUT THERE IS STILL A LOT OF POTENTIAL IN ASIA, RUSSIA, AND AFRICA.

NEVERTHELESS, THE SOCIAL AND ENVIRONMENTAL IMPACT OF BIG DAMS—SUCH AS THE THREE GORGES DAM IN CHINA, WHICH DISPLACED WELL OVER A MILLION PEOPLE—REMAINS DISTURBING.

BUT IT HAS ITS INCONVENIENCES.

BIG DAMS BUILT TO HARNESS IT CHANGE THE LANDSCAPE, OFTEN FLOOD FERTILE LANDS, AND CAN FORCE THE RELOCATION OF POPULATIONS.

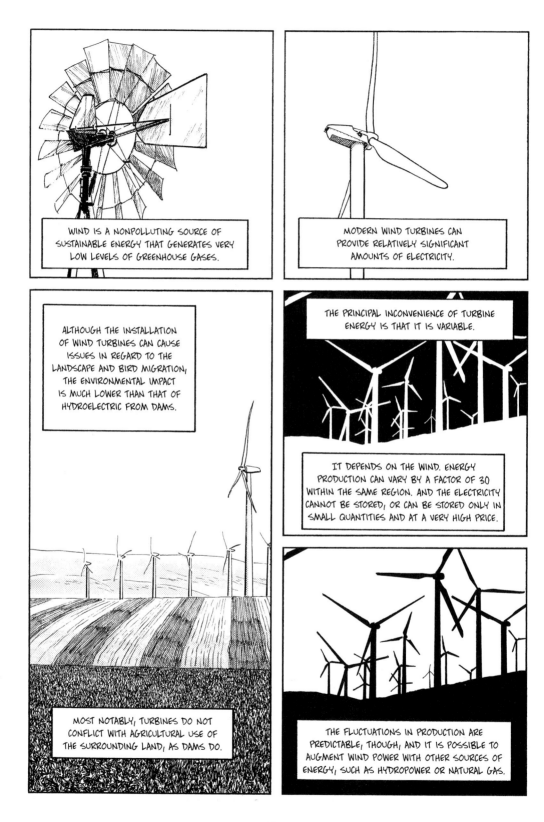

WIND IS A NONPOLLUTING SOURCE OF SUSTAINABLE ENERGY THAT GENERATES VERY LOW LEVELS OF GREENHOUSE GASES.

MODERN WIND TURBINES CAN PROVIDE RELATIVELY SIGNIFICANT AMOUNTS OF ELECTRICITY.

ALTHOUGH THE INSTALLATION OF WIND TURBINES CAN CAUSE ISSUES IN REGARD TO THE LANDSCAPE AND BIRD MIGRATION, THE ENVIRONMENTAL IMPACT IS MUCH LOWER THAN THAT OF HYDROELECTRIC FROM DAMS.

THE PRINCIPAL INCONVENIENCE OF TURBINE ENERGY IS THAT IT IS VARIABLE.

IT DEPENDS ON THE WIND. ENERGY PRODUCTION CAN VARY BY A FACTOR OF 30 WITHIN THE SAME REGION. AND THE ELECTRICITY CANNOT BE STORED, OR CAN BE STORED ONLY IN SMALL QUANTITIES AND AT A VERY HIGH PRICE.

MOST NOTABLY, TURBINES DO NOT CONFLICT WITH AGRICULTURAL USE OF THE SURROUNDING LAND, AS DAMS DO.

THE FLUCTUATIONS IN PRODUCTION ARE PREDICTABLE, THOUGH, AND IT IS POSSIBLE TO AUGMENT WIND POWER WITH OTHER SOURCES OF ENERGY, SUCH AS HYDROPOWER OR NATURAL GAS.

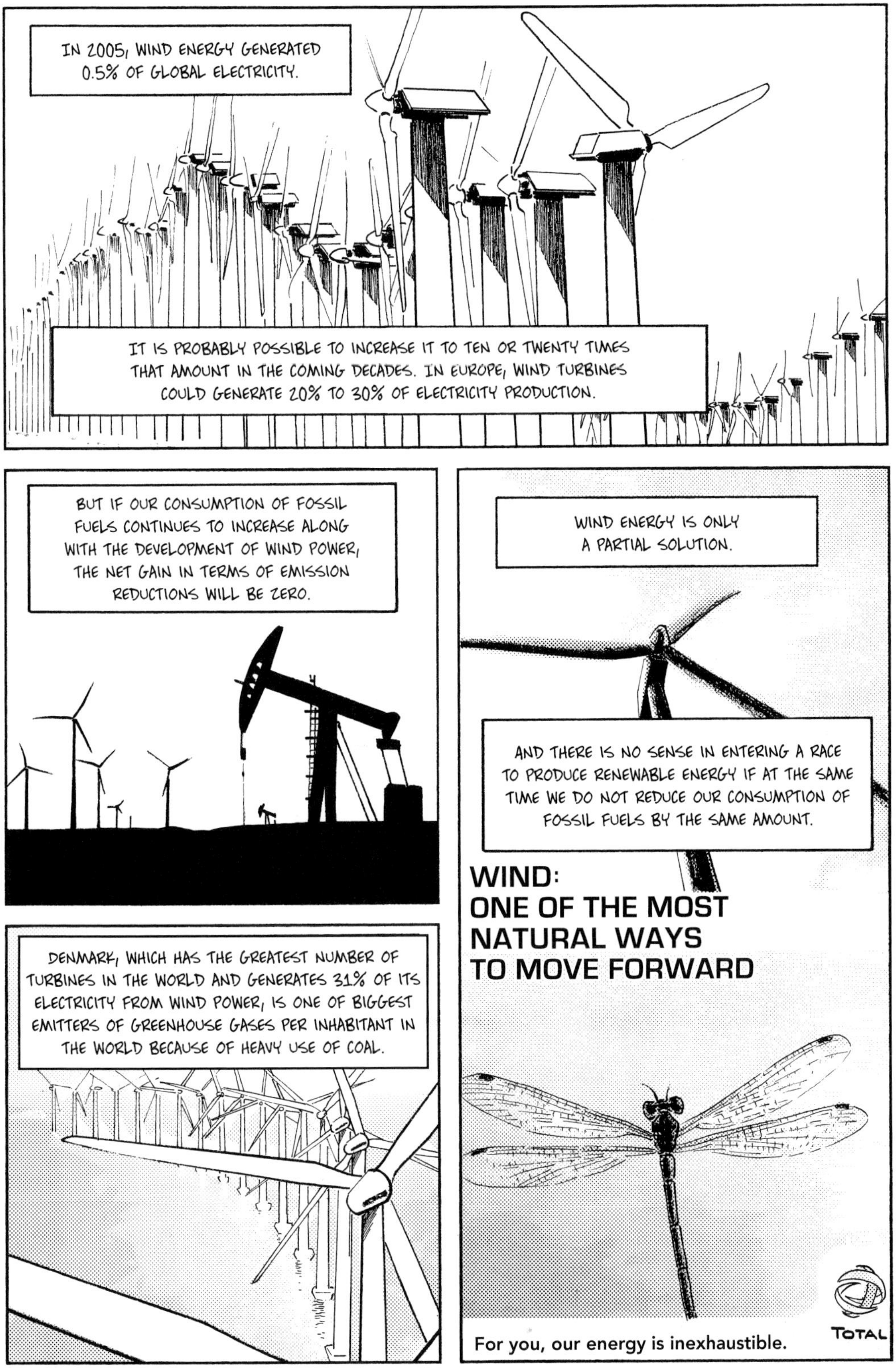
IN 2005, WIND ENERGY GENERATED 0.5% OF GLOBAL ELECTRICITY.
IT IS PROBABLY POSSIBLE TO INCREASE IT TO TEN OR TWENTY TIMES THAT AMOUNT IN THE COMING DECADES. IN EUROPE, WIND TURBINES COULD GENERATE 20% TO 30% OF ELECTRICITY PRODUCTION.
BUT IF OUR CONSUMPTION OF FOSSIL FUELS CONTINUES TO INCREASE ALONG WITH THE DEVELOPMENT OF WIND POWER, THE NET GAIN IN TERMS OF EMISSION REDUCTIONS WILL BE ZERO.
DENMARK, WHICH HAS THE GREATEST NUMBER OF TURBINES IN THE WORLD AND GENERATES 31% OF ITS ELECTRICITY FROM WIND POWER, IS ONE OF BIGGEST EMITTERS OF GREENHOUSE GASES PER INHABITANT IN THE WORLD BECAUSE OF HEAVY USE OF COAL.
WIND ENERGY IS ONLY A PARTIAL SOLUTION.
AND THERE IS NO SENSE IN ENTERING A RACE TO PRODUCE RENEWABLE ENERGY IF AT THE SAME TIME WE DO NOT REDUCE OUR CONSUMPTION OF FOSSIL FUELS BY THE SAME AMOUNT.
WIND:
ONE OF THE MOST NATURAL WAYS TO MOVE FORWARD
TOTAL
For you, our energy is inexhaustible.

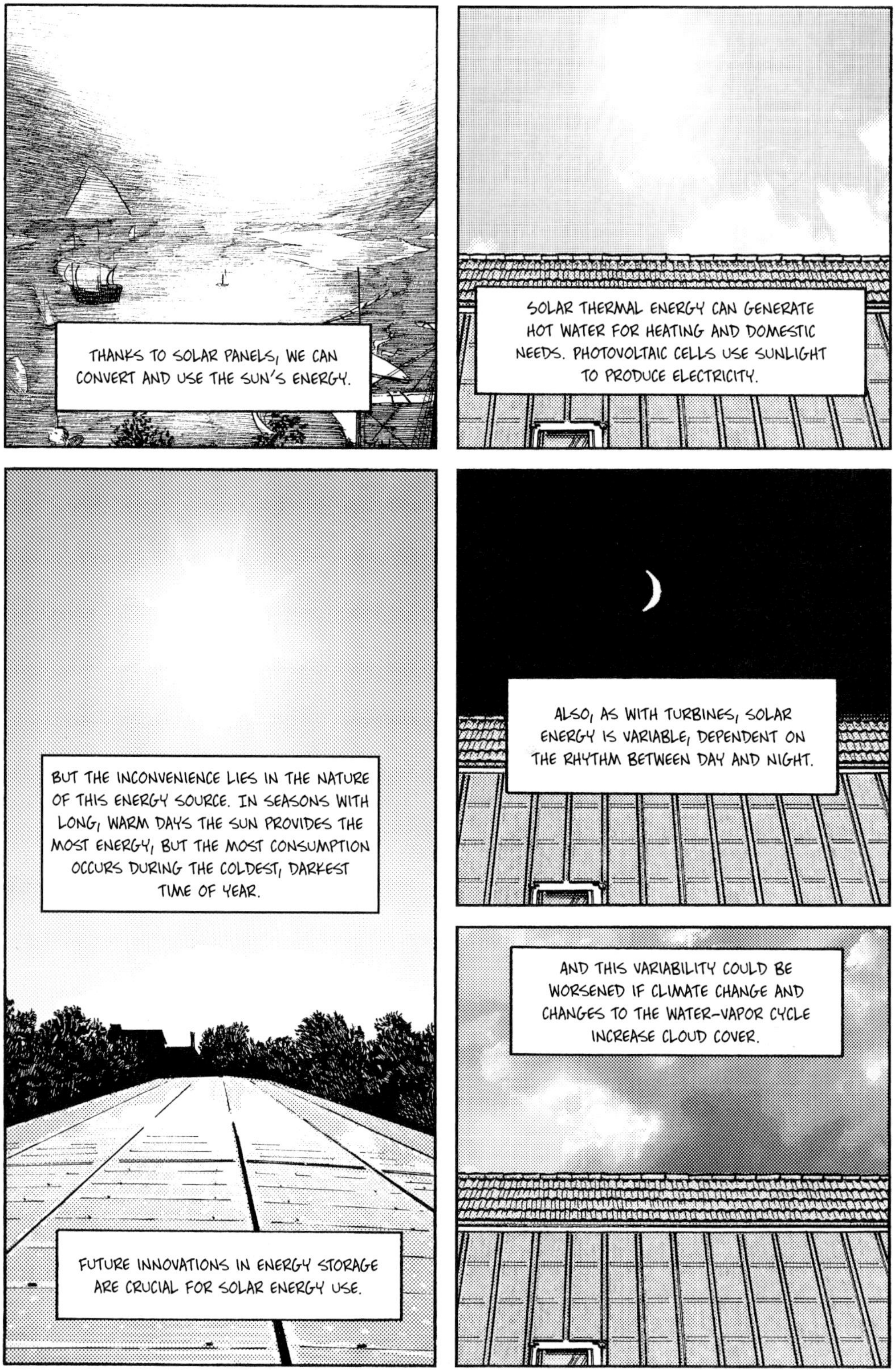
THANKS TO SOLAR PANELS, WE CAN CONVERT AND USE THE SUN'S ENERGY.
SOLAR THERMAL ENERGY CAN GENERATE HOT WATER FOR HEATING AND DOMESTIC NEEDS. PHOTOVOLTAIC CELLS USE SUNLIGHT TO PRODUCE ELECTRICITY.
BUT THE INCONVENIENCE LIES IN THE NATURE OF THIS ENERGY SOURCE. IN SEASONS WITH LONG, WARM DAYS THE SUN PROVIDES THE MOST ENERGY, BUT THE MOST CONSUMPTION OCCURS DURING THE COLDEST, DARKEST TIME OF YEAR.
FUTURE INNOVATIONS IN ENERGY STORAGE ARE CRUCIAL FOR SOLAR ENERGY USE.
ALSO, AS WITH TURBINES, SOLAR ENERGY IS VARIABLE, DEPENDENT ON THE RHYTHM BETWEEN DAY AND NIGHT.
AND THIS VARIABILITY COULD BE WORSENED IF CLIMATE CHANGE AND CHANGES TO THE WATER-VAPOR CYCLE INCREASE CLOUD COVER.

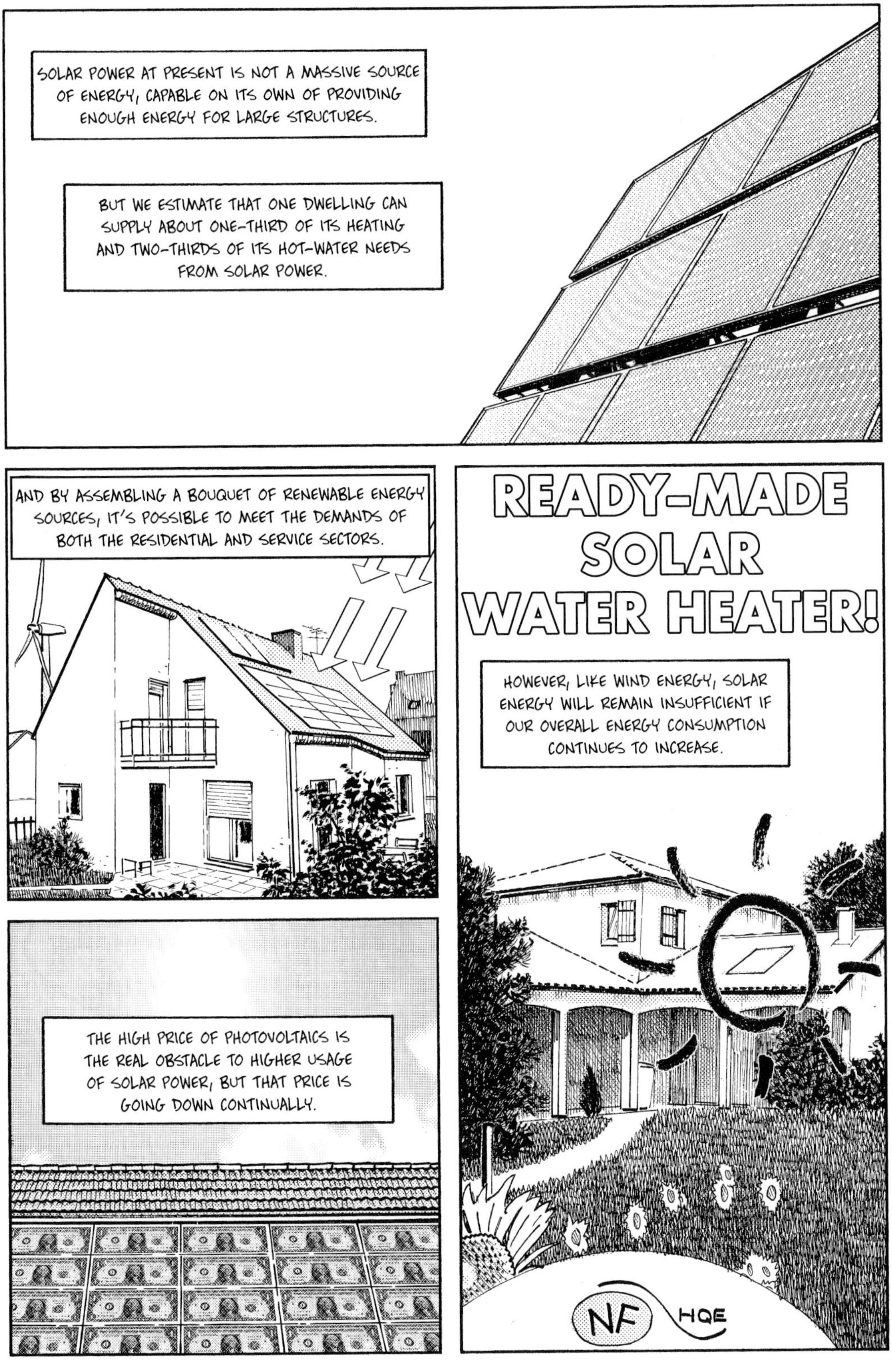
SOLAR POWER AT PRESENT IS NOT A MASSIVE SOURCE OF ENERGY, CAPABLE ON ITS OWN OF PROVIDING ENOUGH ENERGY FOR LARGE STRUCTURES.
BUT WE ESTIMATE THAT ONE DWELLING CAN SUPPLY ABOUT ONE-THIRD OF ITS HEATING AND TWO-THIRDS OF ITS HOT-WATER NEEDS FROM SOLAR POWER.
AND BY ASSEMBLING A BOUQUET OF RENEWABLE ENERGY SOURCES, IT'S POSSIBLE TO MEET THE DEMANDS OF BOTH THE RESIDENTIAL AND SERVICE SECTORS.
THE HIGH PRICE OF PHOTOVOLTAICS IS THE REAL OBSTACLE TO HIGHER USAGE OF SOLAR POWER, BUT THAT PRICE IS GOING DOWN CONTINUALLY.
READY-MADE SOLAR WATER HEATER!
HOWEVER, LIKE WIND ENERGY, SOLAR ENERGY WILL REMAIN INSUFFICIENT IF OUR OVERALL ENERGY CONSUMPTION CONTINUES TO INCREASE.
NF
HQE

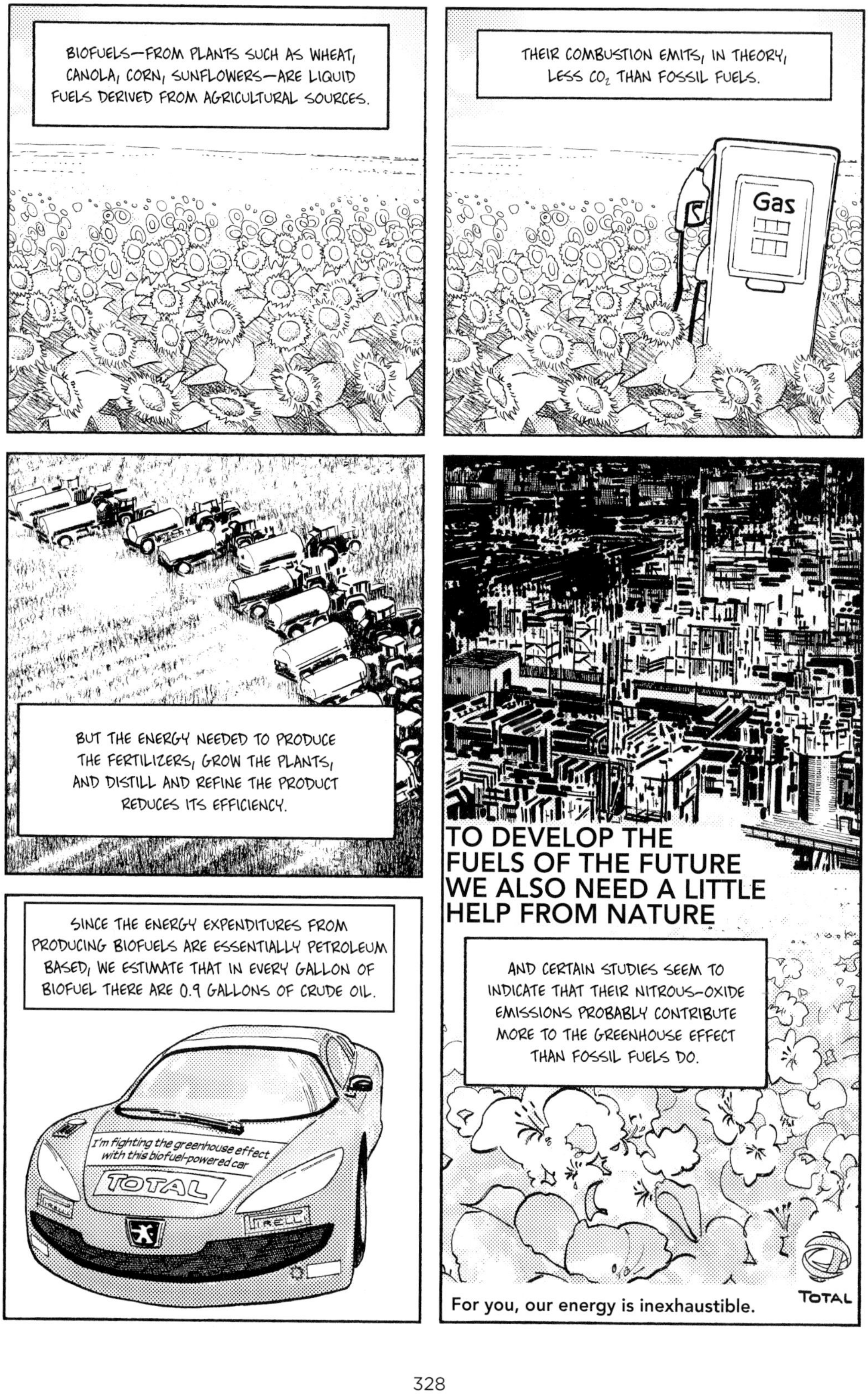
BIOFUELS—FROM PLANTS SUCH AS WHEAT, CANOLA, CORN, SUNFLOWERS—ARE LIQUID FUELS DERIVED FROM AGRICULTURAL SOURCES.
THEIR COMBUSTION EMITS, IN THEORY, LESS CO_2 THAN FOSSIL FUELS.
Gas
BUT THE ENERGY NEEDED TO PRODUCE THE FERTILIZERS, GROW THE PLANTS, AND DISTILL AND REFINE THE PRODUCT REDUCES ITS EFFICIENCY.
TO DEVELOP THE FUELS OF THE FUTURE WE ALSO NEED A LITTLE HELP FROM NATURE
SINCE THE ENERGY EXPENDITURES FROM PRODUCING BIOFUELS ARE ESSENTIALLY PETROLEUM BASED, WE ESTIMATE THAT IN EVERY GALLON OF BIOFUEL THERE ARE 0.9 GALLONS OF CRUDE OIL.
I'm fighting the greenhouse effect with this biofuel-powered car
TOTAL
AND CERTAIN STUDIES SEEM TO INDICATE THAT THEIR NITROUS-OXIDE EMISSIONS PROBABLY CONTRIBUTE MORE TO THE GREENHOUSE EFFECT THAN FOSSIL FUELS DO.
TOTAL
For you, our energy is inexhaustible.

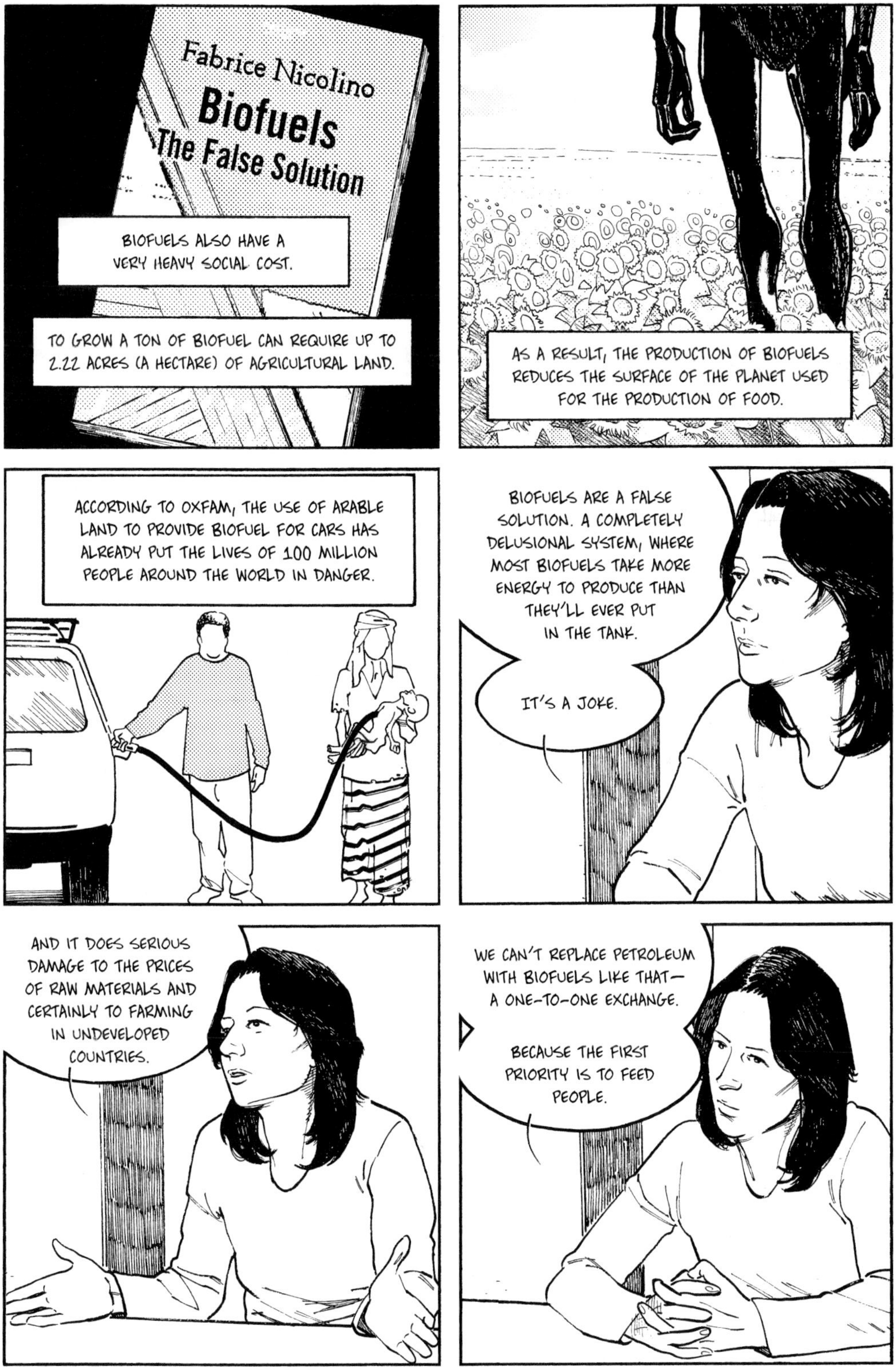
Fabrice Nicolino
Biofuels
The False Solution
BIOFUELS ALSO HAVE A VERY HEAVY SOCIAL COST.
TO GROW A TON OF BIOFUEL CAN REQUIRE UP TO 2.22 ACRES (A HECTARE) OF AGRICULTURAL LAND.
AS A RESULT, THE PRODUCTION OF BIOFUELS REDUCES THE SURFACE OF THE PLANET USED FOR THE PRODUCTION OF FOOD.
ACCORDING TO OXFAM, THE USE OF ARABLE LAND TO PROVIDE BIOFUEL FOR CARS HAS ALREADY PUT THE LIVES OF 100 MILLION PEOPLE AROUND THE WORLD IN DANGER.
BIOFUELS ARE A FALSE SOLUTION. A COMPLETELY DELUSIONAL SYSTEM, WHERE MOST BIOFUELS TAKE MORE ENERGY TO PRODUCE THAN THEY'LL EVER PUT IN THE TANK.
IT'S A JOKE.
AND IT DOES SERIOUS DAMAGE TO THE PRICES OF RAW MATERIALS AND CERTAINLY TO FARMING IN UNDEVELOPED COUNTRIES.
WE CAN'T REPLACE PETROLEUM WITH BIOFUELS LIKE THAT—A ONE-TO-ONE EXCHANGE.
BECAUSE THE FIRST PRIORITY IS TO FEED PEOPLE.

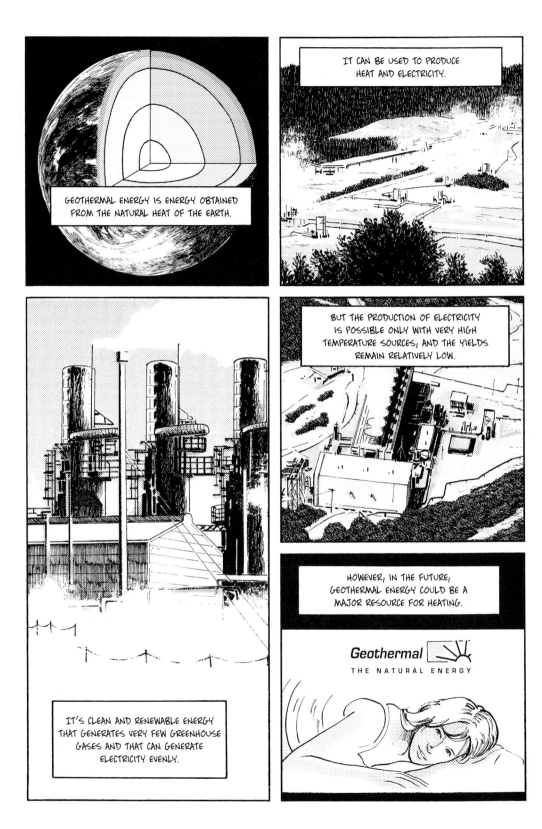

GEOTHERMAL ENERGY IS ENERGY OBTAINED FROM THE NATURAL HEAT OF THE EARTH.

IT CAN BE USED TO PRODUCE HEAT AND ELECTRICITY.

BUT THE PRODUCTION OF ELECTRICITY IS POSSIBLE ONLY WITH VERY HIGH TEMPERATURE SOURCES, AND THE YIELDS REMAIN RELATIVELY LOW.

HOWEVER, IN THE FUTURE, GEOTHERMAL ENERGY COULD BE A MAJOR RESOURCE FOR HEATING.

Geothermal
THE NATURAL ENERGY

IT'S CLEAN AND RENEWABLE ENERGY THAT GENERATES VERY FEW GREENHOUSE GASES AND THAT CAN GENERATE ELECTRICITY EVENLY.

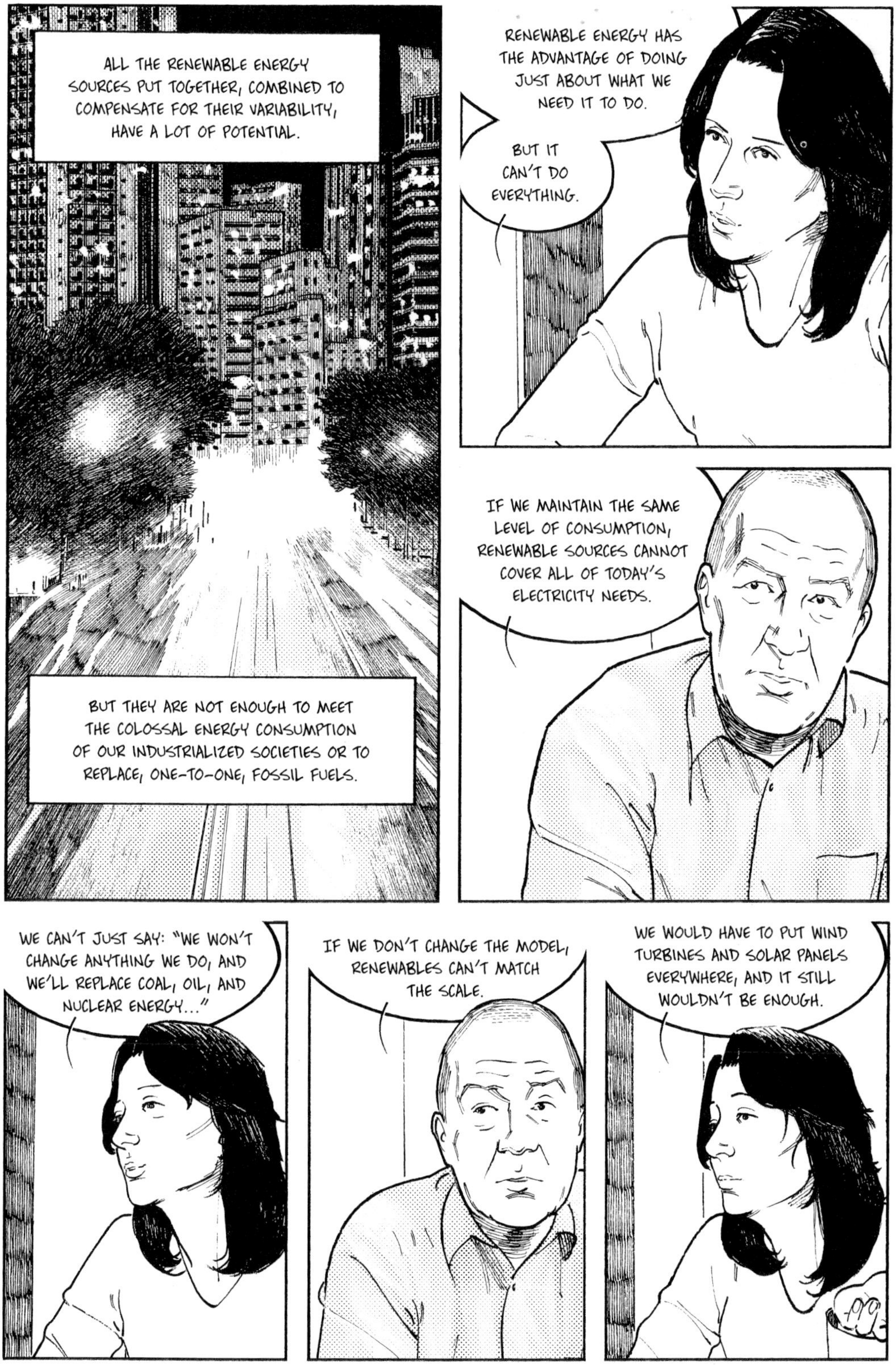
ALL THE RENEWABLE ENERGY SOURCES PUT TOGETHER, COMBINED TO COMPENSATE FOR THEIR VARIABILITY, HAVE A LOT OF POTENTIAL.
BUT THEY ARE NOT ENOUGH TO MEET THE COLOSSAL ENERGY CONSUMPTION OF OUR INDUSTRIALIZED SOCIETIES OR TO REPLACE, ONE-TO-ONE, FOSSIL FUELS.
RENEWABLE ENERGY HAS THE ADVANTAGE OF DOING JUST ABOUT WHAT WE NEED IT TO DO.
BUT IT CAN'T DO EVERYTHING.
IF WE MAINTAIN THE SAME LEVEL OF CONSUMPTION, RENEWABLE SOURCES CANNOT COVER ALL OF TODAY'S ELECTRICITY NEEDS.
WE CAN'T JUST SAY: "WE WON'T CHANGE ANYTHING WE DO, AND WE'LL REPLACE COAL, OIL, AND NUCLEAR ENERGY..."
IF WE DON'T CHANGE THE MODEL, RENEWABLES CAN'T MATCH THE SCALE.
WE WOULD HAVE TO PUT WIND TURBINES AND SOLAR PANELS EVERYWHERE, AND IT STILL WOULDN'T BE ENOUGH.

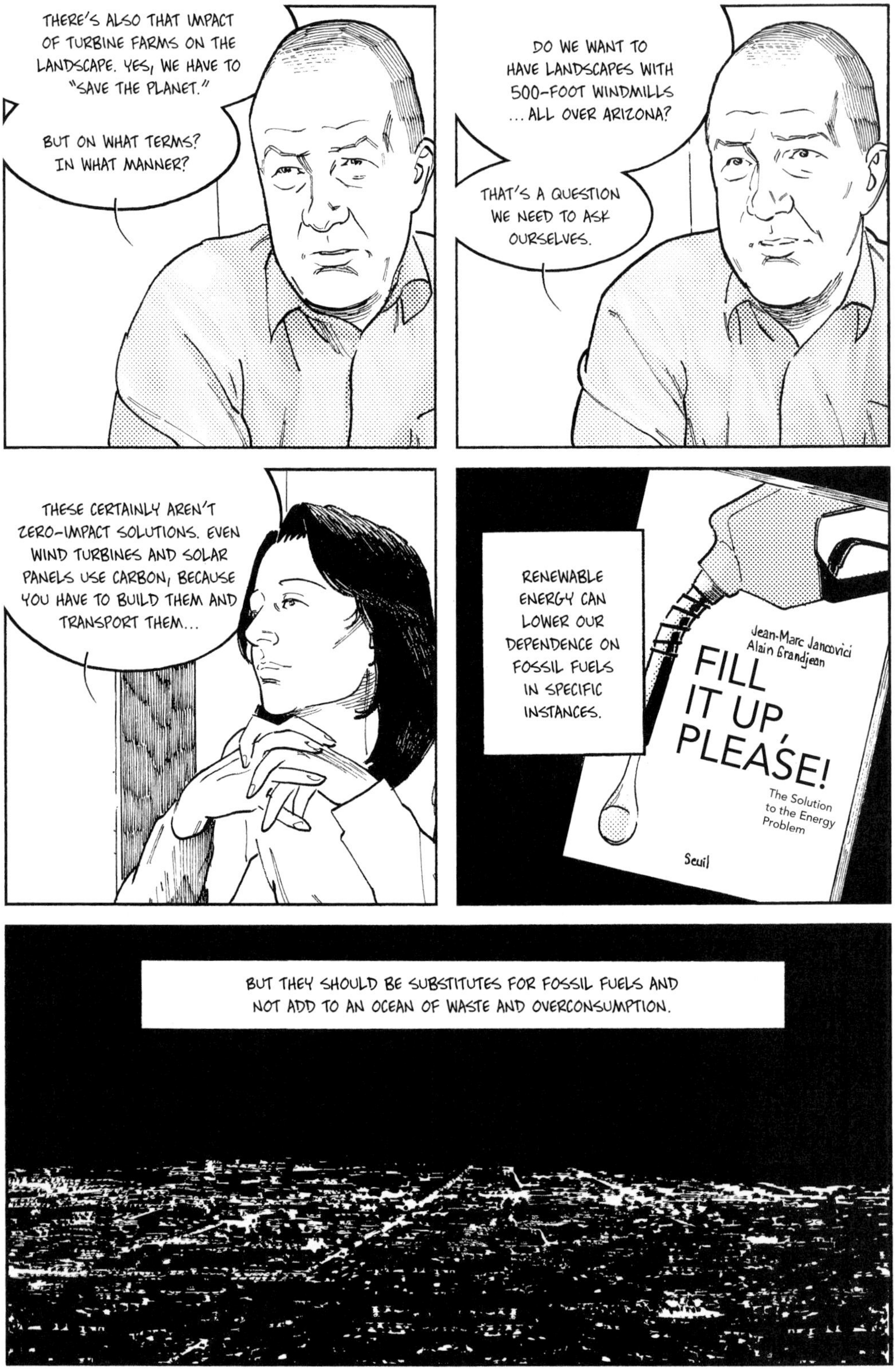
THERE'S ALSO THAT IMPACT OF TURBINE FARMS ON THE LANDSCAPE. YES, WE HAVE TO "SAVE THE PLANET."
BUT ON WHAT TERMS? IN WHAT MANNER?
DO WE WANT TO HAVE LANDSCAPES WITH 500-FOOT WINDMILLS ... ALL OVER ARIZONA?
THAT'S A QUESTION WE NEED TO ASK OURSELVES.
THESE CERTAINLY AREN'T ZERO-IMPACT SOLUTIONS. EVEN WIND TURBINES AND SOLAR PANELS USE CARBON, BECAUSE YOU HAVE TO BUILD THEM AND TRANSPORT THEM...
RENEWABLE ENERGY CAN LOWER OUR DEPENDENCE ON FOSSIL FUELS IN SPECIFIC INSTANCES.
Jean-Marc Jancovici
Alain Grandjean
FILL IT UP, PLEASE!
The Solution to the Energy Problem
Seuil
BUT THEY SHOULD BE SUBSTITUTES FOR FOSSIL FUELS AND NOT ADD TO AN OCEAN OF WASTE AND OVERCONSUMPTION.

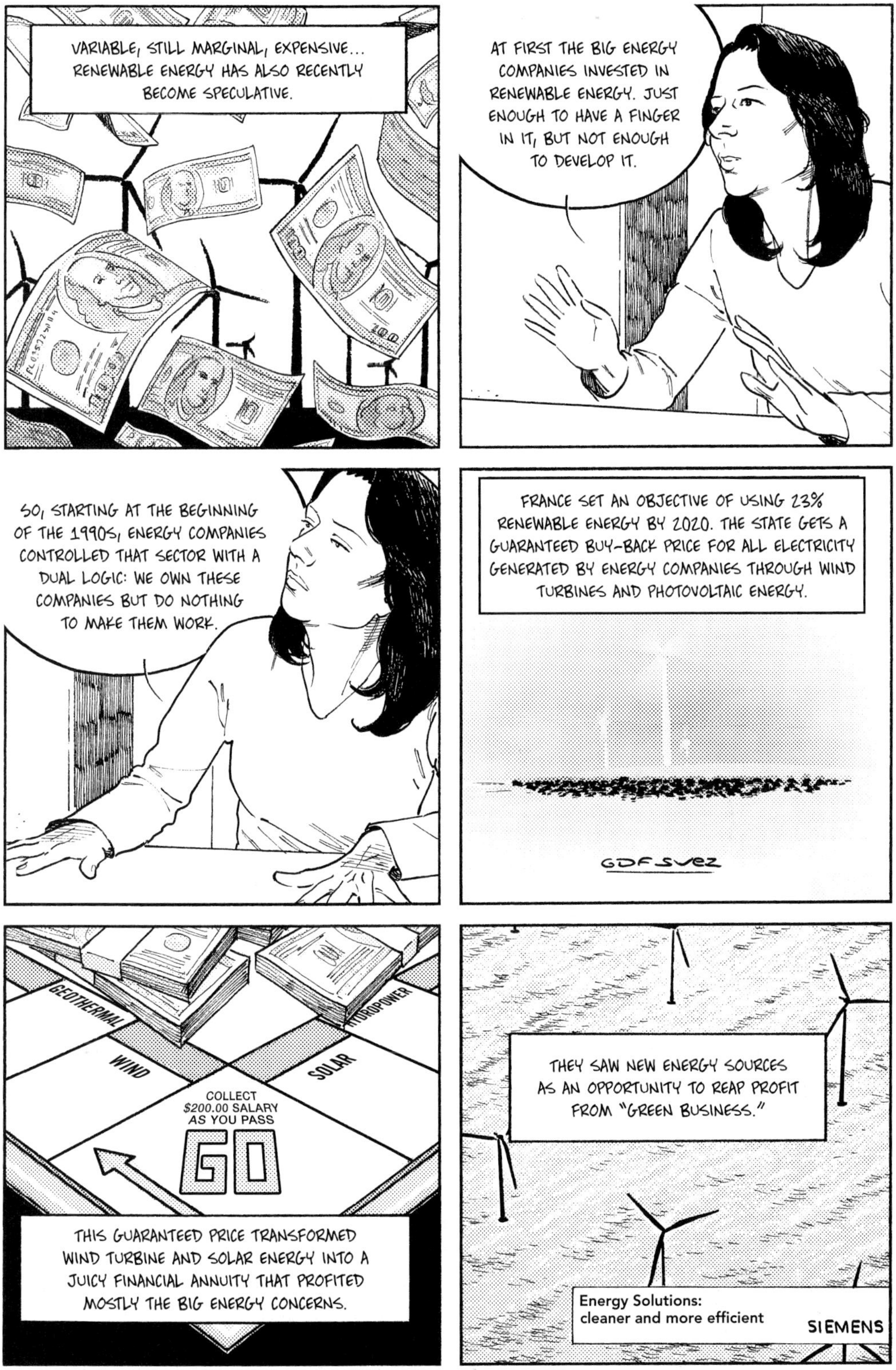
VARIABLE, STILL MARGINAL, EXPENSIVE... RENEWABLE ENERGY HAS ALSO RECENTLY BECOME SPECULATIVE.
AT FIRST THE BIG ENERGY COMPANIES INVESTED IN RENEWABLE ENERGY. JUST ENOUGH TO HAVE A FINGER IN IT, BUT NOT ENOUGH TO DEVELOP IT.
SO, STARTING AT THE BEGINNING OF THE 1990S, ENERGY COMPANIES CONTROLLED THAT SECTOR WITH A DUAL LOGIC: WE OWN THESE COMPANIES BUT DO NOTHING TO MAKE THEM WORK.
FRANCE SET AN OBJECTIVE OF USING 23% RENEWABLE ENERGY BY 2020. THE STATE GETS A GUARANTEED BUY-BACK PRICE FOR ALL ELECTRICITY GENERATED BY ENERGY COMPANIES THROUGH WIND TURBINES AND PHOTOVOLTAIC ENERGY.
GDF SUEZ
GEOTHERMAL
HYDROPOWER
WIND
SOLAR
COLLECT $200.00 SALARY AS YOU PASS
GO
THIS GUARANTEED PRICE TRANSFORMED WIND TURBINE AND SOLAR ENERGY INTO A JUICY FINANCIAL ANNUITY THAT PROFITED MOSTLY THE BIG ENERGY CONCERNS.
THEY SAW NEW ENERGY SOURCES AS AN OPPORTUNITY TO REAP PROFIT FROM "GREEN BUSINESS."
Energy Solutions:
cleaner and more efficient
SIEMENS

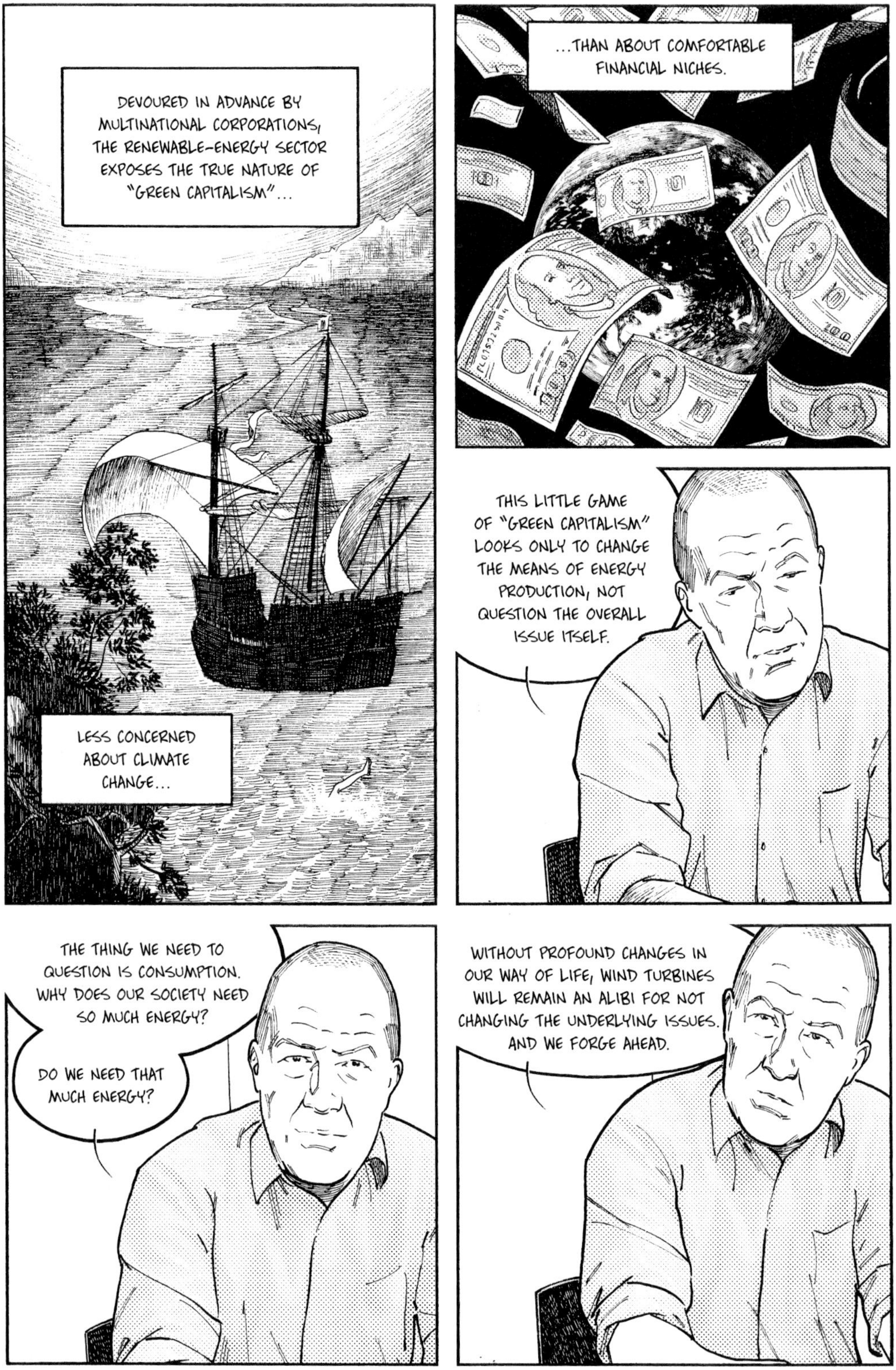
DEVOURED IN ADVANCE BY MULTINATIONAL CORPORATIONS, THE RENEWABLE-ENERGY SECTOR EXPOSES THE TRUE NATURE OF "GREEN CAPITALISM"...
LESS CONCERNED ABOUT CLIMATE CHANGE...
...THAN ABOUT COMFORTABLE FINANCIAL NICHES.
THIS LITTLE GAME OF "GREEN CAPITALISM" LOOKS ONLY TO CHANGE THE MEANS OF ENERGY PRODUCTION, NOT QUESTION THE OVERALL ISSUE ITSELF.
THE THING WE NEED TO QUESTION IS CONSUMPTION. WHY DOES OUR SOCIETY NEED SO MUCH ENERGY?
DO WE NEED THAT MUCH ENERGY?
WITHOUT PROFOUND CHANGES IN OUR WAY OF LIFE, WIND TURBINES WILL REMAIN AN ALIBI FOR NOT CHANGING THE UNDERLYING ISSUES. AND WE FORGE AHEAD.

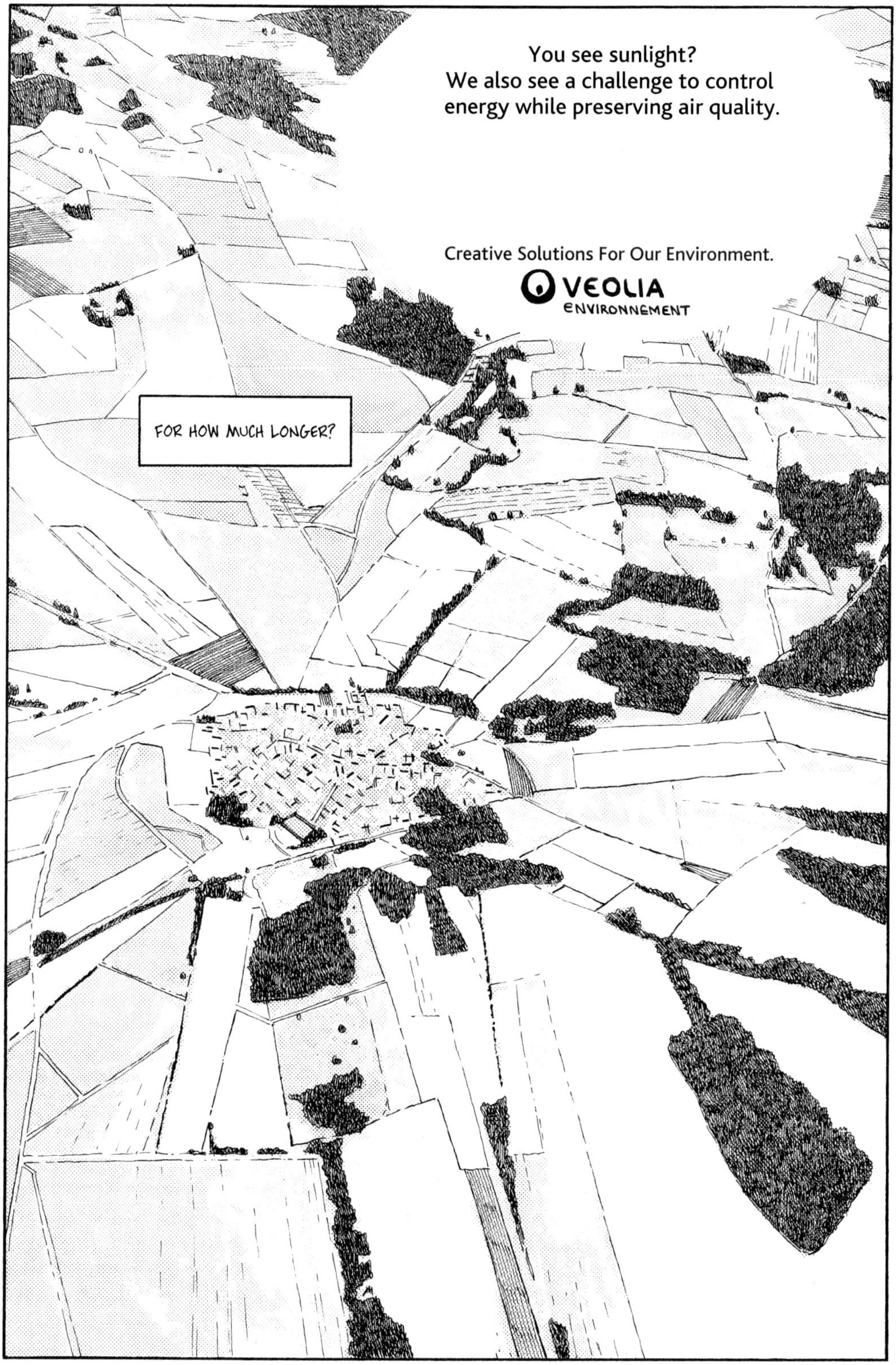
You see sunlight?
We also see a challenge to control energy while preserving air quality.
Creative Solutions For Our Environment.
VEOLIA ENVIRONNEMENT
FOR HOW MUCH LONGER?

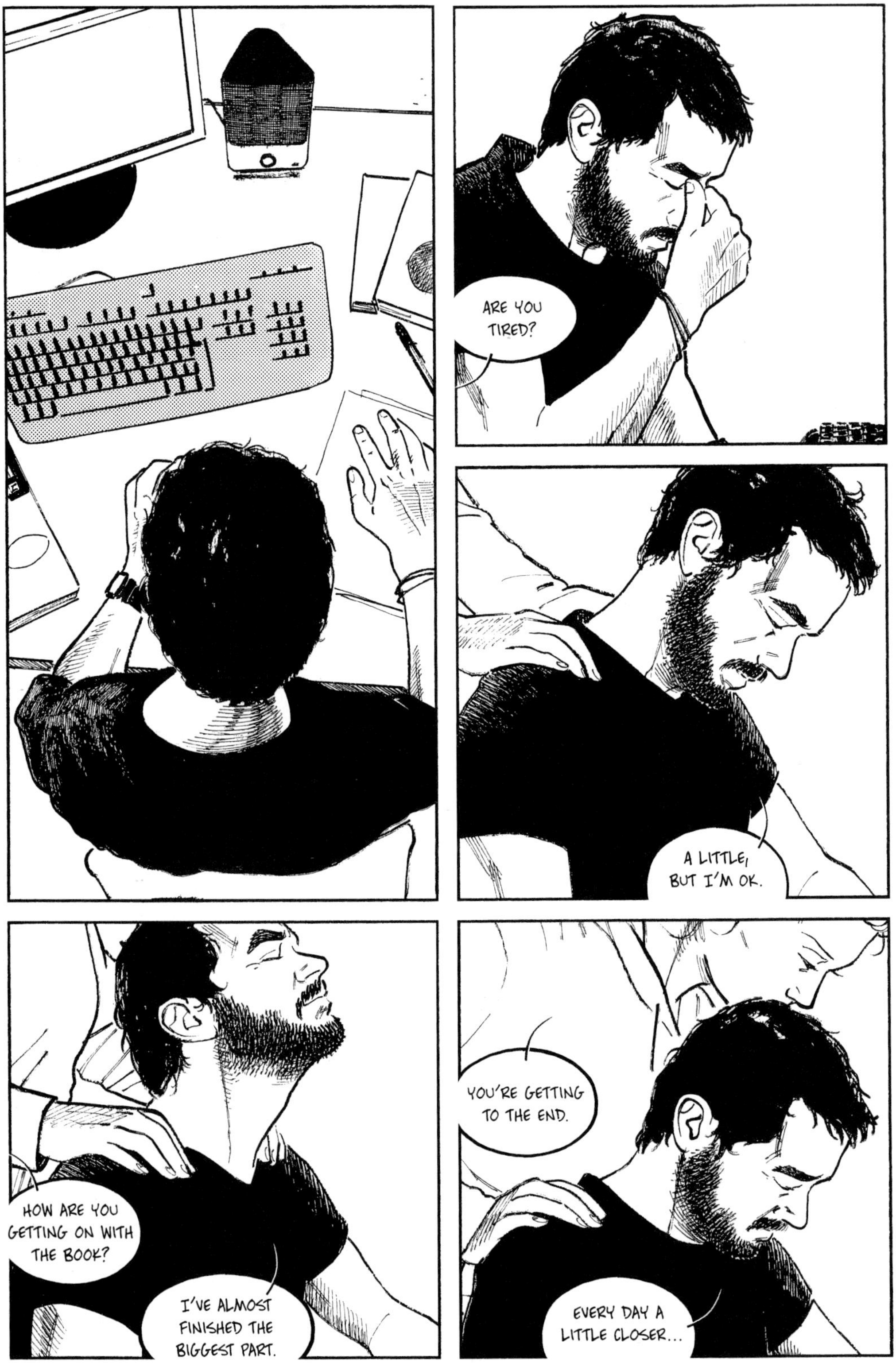
ARE YOU TIRED?
A LITTLE, BUT I'M OK.
HOW ARE YOU GETTING ON WITH THE BOOK?
I'VE ALMOST FINISHED THE BIGGEST PART.
YOU'RE GETTING TO THE END.
EVERY DAY A LITTLE CLOSER...

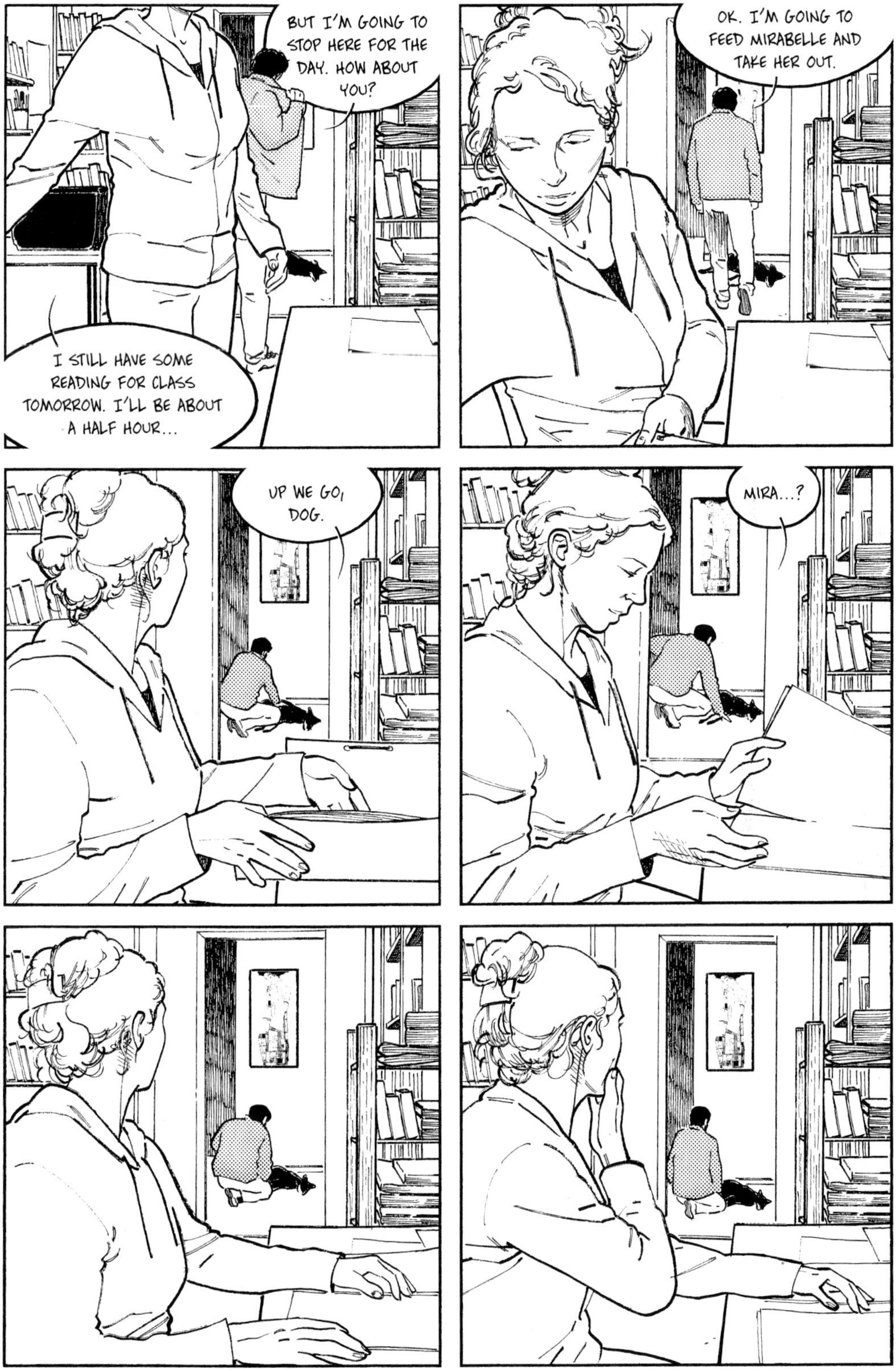
BUT I'M GOING TO STOP HERE FOR THE DAY. HOW ABOUT YOU?
I STILL HAVE SOME READING FOR CLASS TOMORROW. I'LL BE ABOUT A HALF HOUR...
OK. I'M GOING TO FEED MIRABELLE AND TAKE HER OUT.
UP WE GO, DOG.
MIRA...?

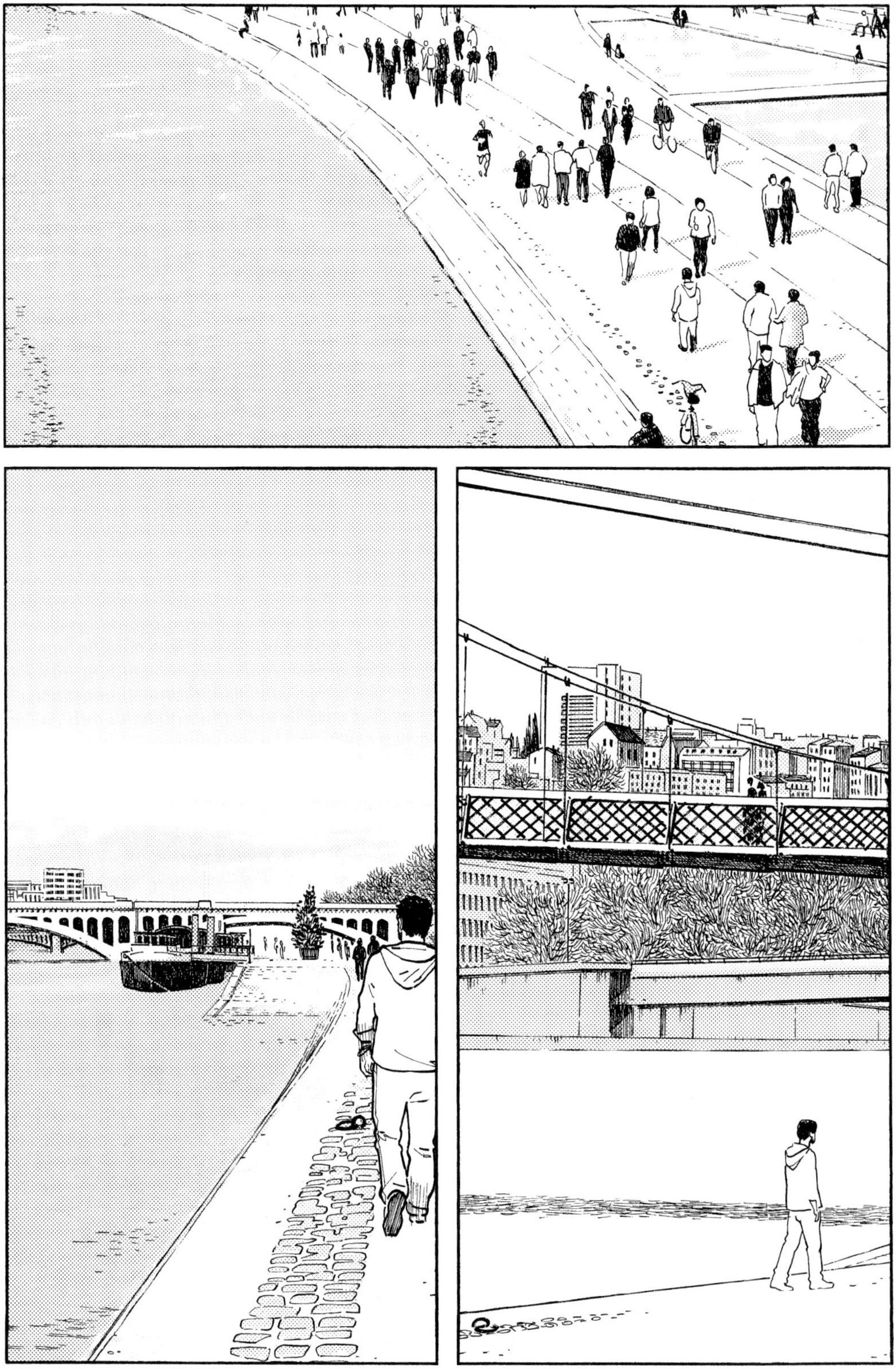

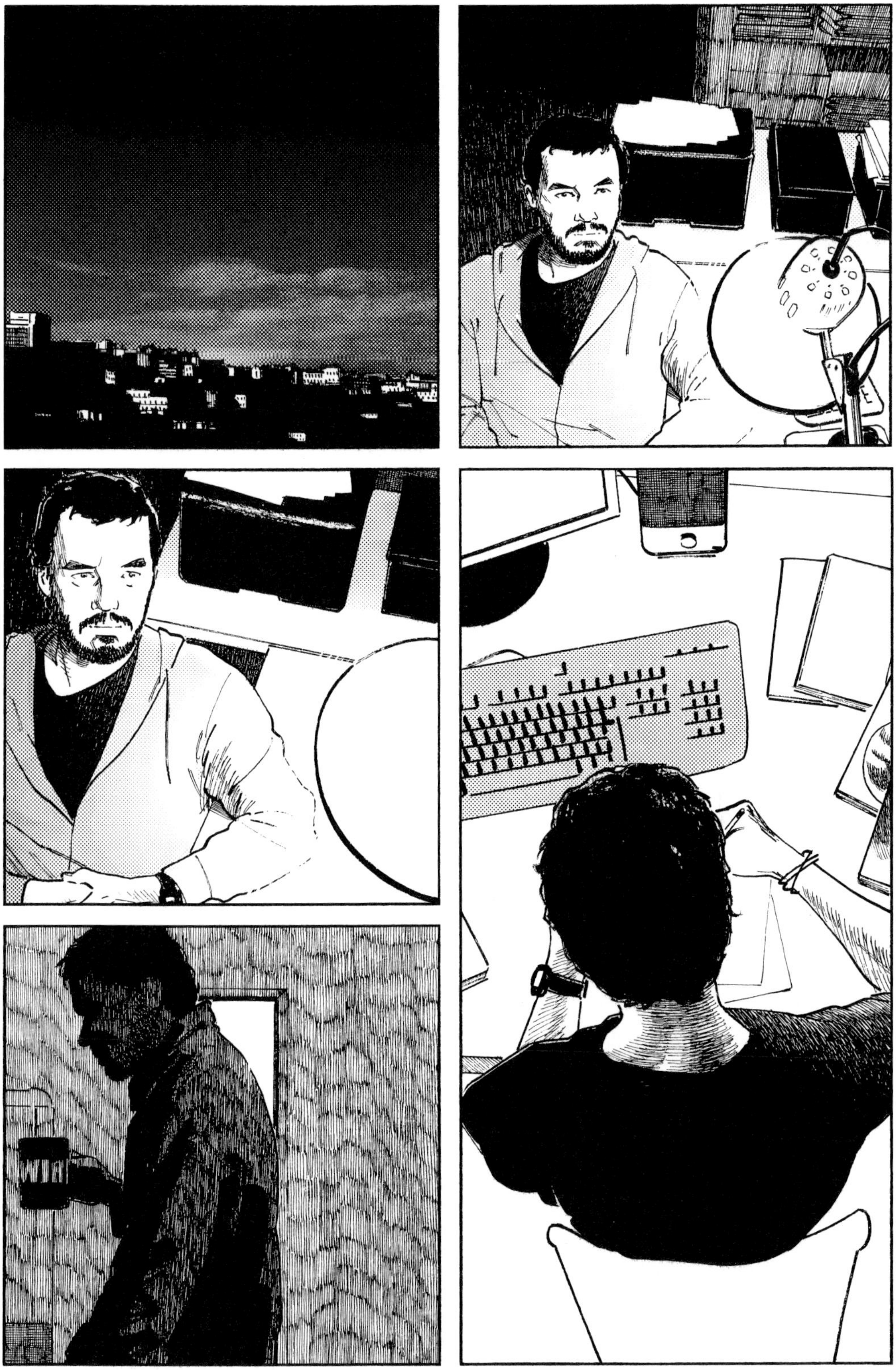

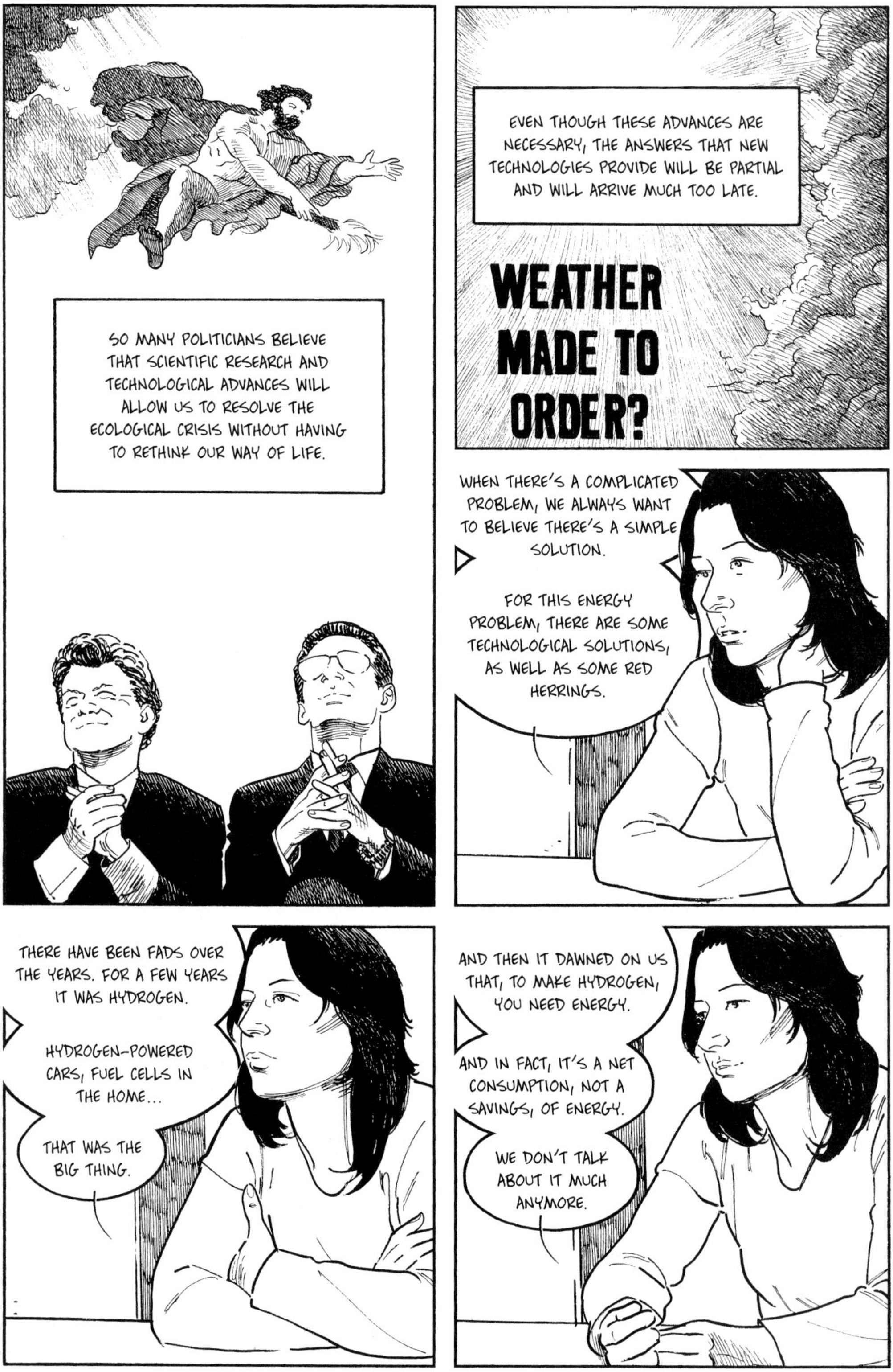
SO MANY POLITICIANS BELIEVE THAT SCIENTIFIC RESEARCH AND TECHNOLOGICAL ADVANCES WILL ALLOW US TO RESOLVE THE ECOLOGICAL CRISIS WITHOUT HAVING TO RETHINK OUR WAY OF LIFE.
EVEN THOUGH THESE ADVANCES ARE NECESSARY, THE ANSWERS THAT NEW TECHNOLOGIES PROVIDE WILL BE PARTIAL AND WILL ARRIVE MUCH TOO LATE.
WEATHER MADE TO ORDER?
WHEN THERE'S A COMPLICATED PROBLEM, WE ALWAYS WANT TO BELIEVE THERE'S A SIMPLE SOLUTION.
FOR THIS ENERGY PROBLEM, THERE ARE SOME TECHNOLOGICAL SOLUTIONS, AS WELL AS SOME RED HERRINGS.
THERE HAVE BEEN FADS OVER THE YEARS. FOR A FEW YEARS IT WAS HYDROGEN.
HYDROGEN-POWERED CARS, FUEL CELLS IN THE HOME...
THAT WAS THE BIG THING.
AND THEN IT DAWNED ON US THAT, TO MAKE HYDROGEN, YOU NEED ENERGY.
AND IN FACT, IT'S A NET CONSUMPTION, NOT A SAVINGS, OF ENERGY.
WE DON'T TALK ABOUT IT MUCH ANYMORE.

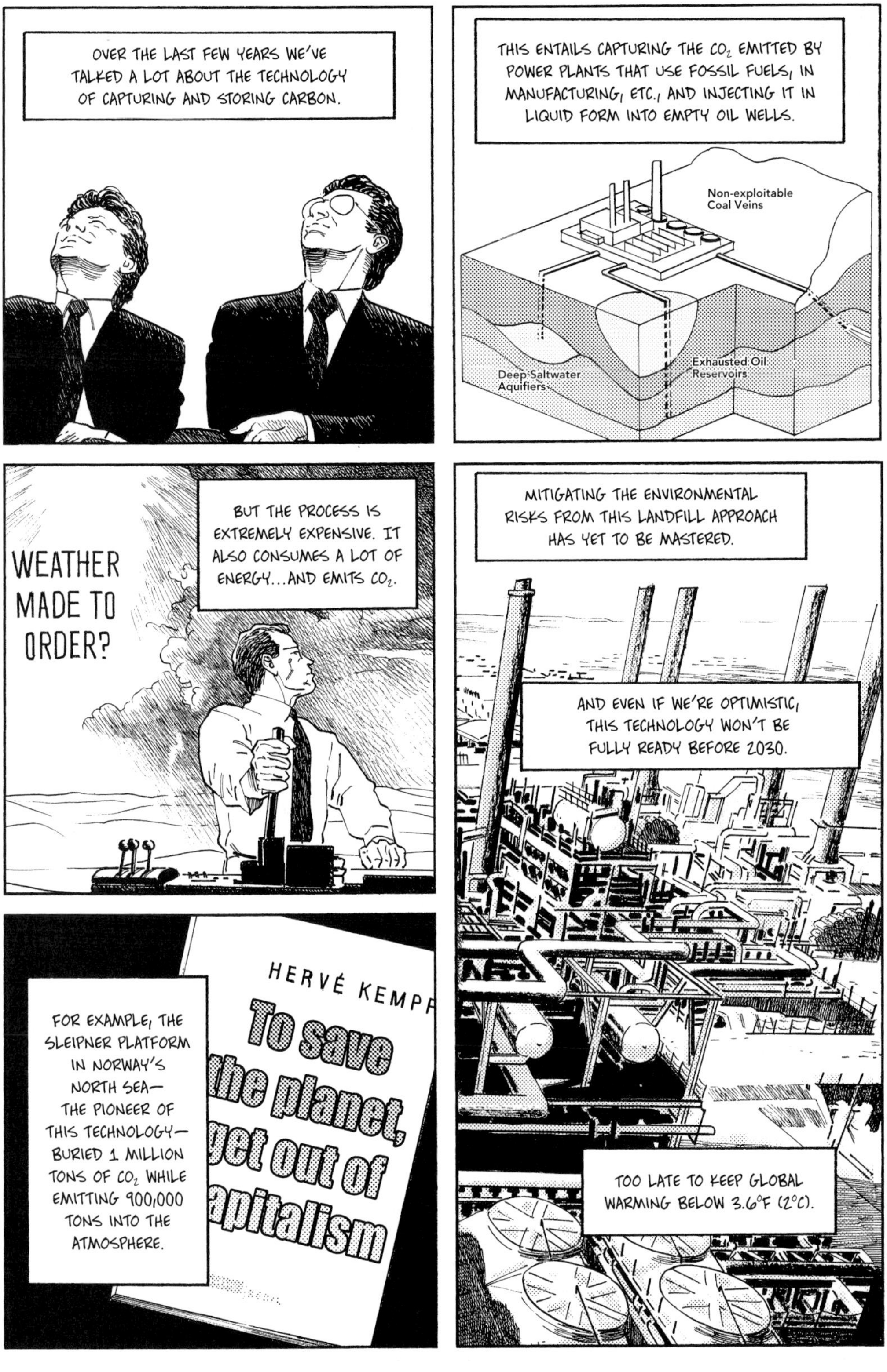
OVER THE LAST FEW YEARS WE'VE TALKED A LOT ABOUT THE TECHNOLOGY OF CAPTURING AND STORING CARBON.
THIS ENTAILS CAPTURING THE CO_2 EMITTED BY POWER PLANTS THAT USE FOSSIL FUELS, IN MANUFACTURING, ETC., AND INJECTING IT IN LIQUID FORM INTO EMPTY OIL WELLS.
Non-exploitable Coal Veins
Deep Saltwater Aquifiers
Exhausted Oil Reservoirs
WEATHER MADE TO ORDER?
BUT THE PROCESS IS EXTREMELY EXPENSIVE. IT ALSO CONSUMES A LOT OF ENERGY...AND EMITS CO_2.
MITIGATING THE ENVIRONMENTAL RISKS FROM THIS LANDFILL APPROACH HAS YET TO BE MASTERED.
AND EVEN IF WE'RE OPTIMISTIC, THIS TECHNOLOGY WON'T BE FULLY READY BEFORE 2030.
FOR EXAMPLE, THE SLEIPNER PLATFORM IN NORWAY'S NORTH SEA—THE PIONEER OF THIS TECHNOLOGY—BURIED 1 MILLION TONS OF CO_2 WHILE EMITTING 900,000 TONS INTO THE ATMOSPHERE.
HERVÉ KEMPF
To save the planet, get out of apitalism
TOO LATE TO KEEP GLOBAL WARMING BELOW 3.6°F (2°C).

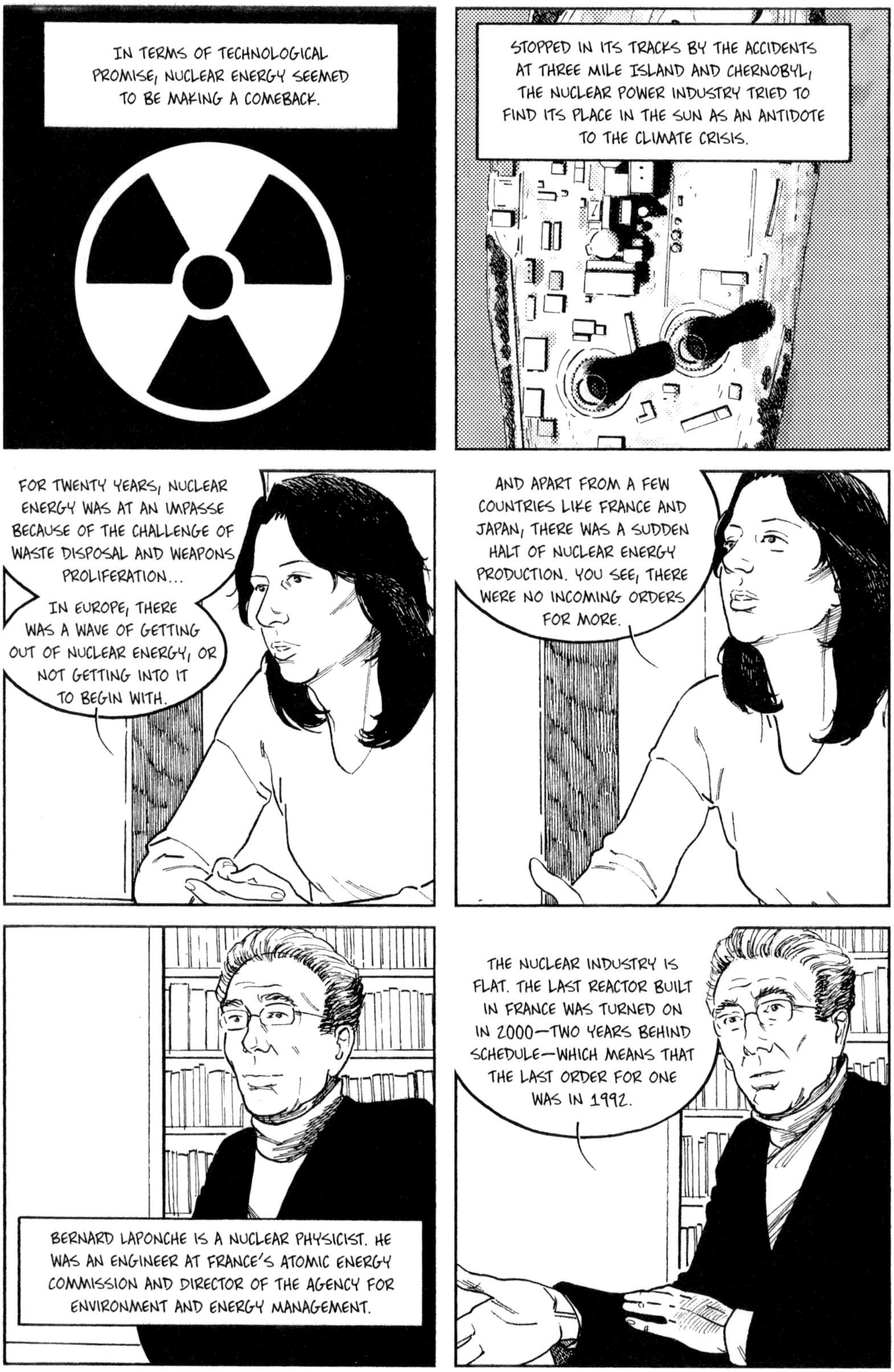
IN TERMS OF TECHNOLOGICAL PROMISE, NUCLEAR ENERGY SEEMED TO BE MAKING A COMEBACK.
STOPPED IN ITS TRACKS BY THE ACCIDENTS AT THREE MILE ISLAND AND CHERNOBYL, THE NUCLEAR POWER INDUSTRY TRIED TO FIND ITS PLACE IN THE SUN AS AN ANTIDOTE TO THE CLIMATE CRISIS.
FOR TWENTY YEARS, NUCLEAR ENERGY WAS AT AN IMPASSE BECAUSE OF THE CHALLENGE OF WASTE DISPOSAL AND WEAPONS PROLIFERATION...
IN EUROPE, THERE WAS A WAVE OF GETTING OUT OF NUCLEAR ENERGY, OR NOT GETTING INTO IT TO BEGIN WITH.
AND APART FROM A FEW COUNTRIES LIKE FRANCE AND JAPAN, THERE WAS A SUDDEN HALT OF NUCLEAR ENERGY PRODUCTION. YOU SEE, THERE WERE NO INCOMING ORDERS FOR MORE.
BERNARD LAPONCHE IS A NUCLEAR PHYSICIST. HE WAS AN ENGINEER AT FRANCE'S ATOMIC ENERGY COMMISSION AND DIRECTOR OF THE AGENCY FOR ENVIRONMENT AND ENERGY MANAGEMENT.
THE NUCLEAR INDUSTRY IS FLAT. THE LAST REACTOR BUILT IN FRANCE WAS TURNED ON IN 2000—TWO YEARS BEHIND SCHEDULE—WHICH MEANS THAT THE LAST ORDER FOR ONE WAS IN 1992.

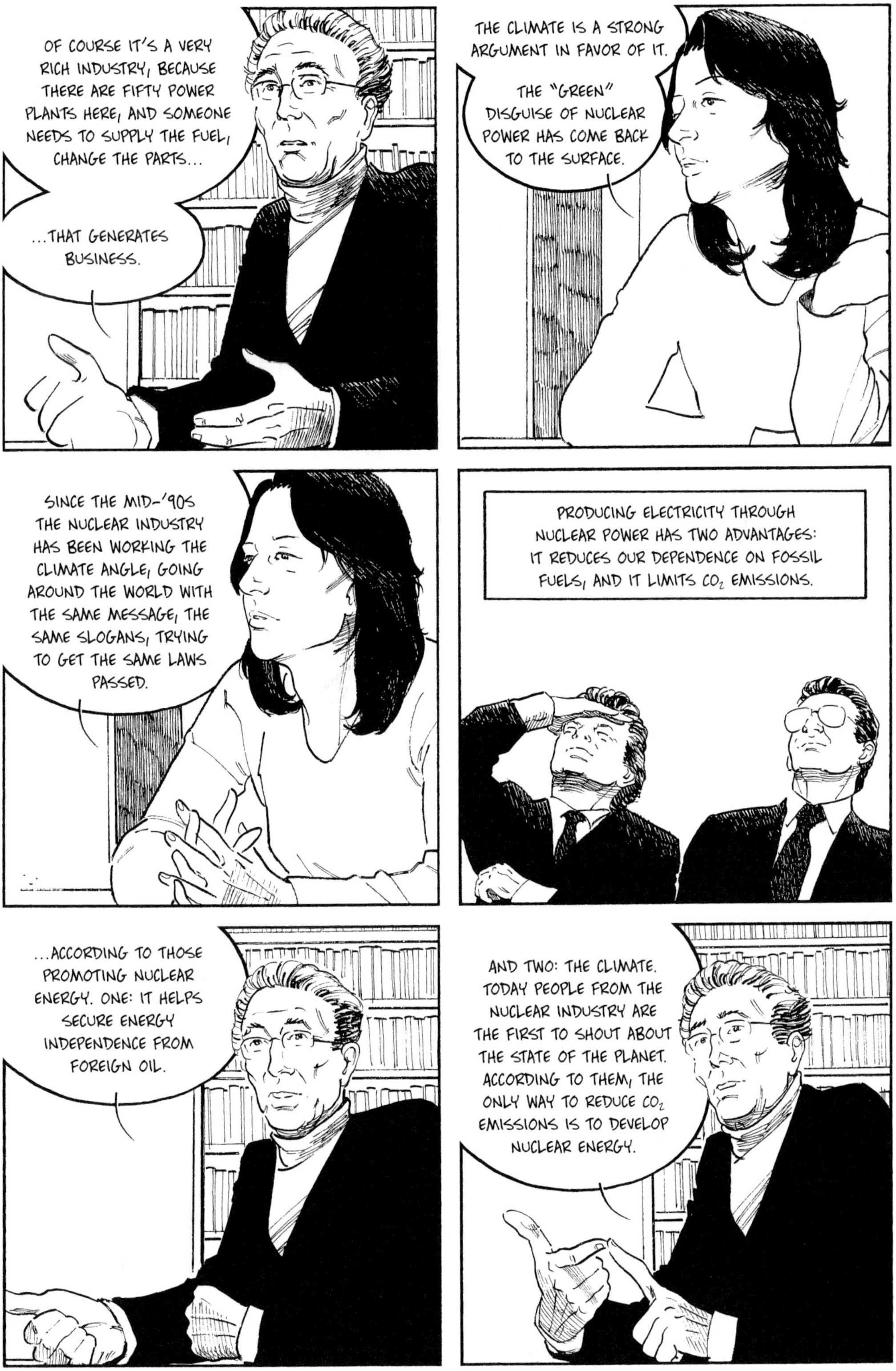
OF COURSE IT'S A VERY RICH INDUSTRY, BECAUSE THERE ARE FIFTY POWER PLANTS HERE, AND SOMEONE NEEDS TO SUPPLY THE FUEL, CHANGE THE PARTS...
...THAT GENERATES BUSINESS.
THE CLIMATE IS A STRONG ARGUMENT IN FAVOR OF IT.
THE "GREEN" DISGUISE OF NUCLEAR POWER HAS COME BACK TO THE SURFACE.
SINCE THE MID-'90S THE NUCLEAR INDUSTRY HAS BEEN WORKING THE CLIMATE ANGLE, GOING AROUND THE WORLD WITH THE SAME MESSAGE, THE SAME SLOGANS, TRYING TO GET THE SAME LAWS PASSED.
PRODUCING ELECTRICITY THROUGH NUCLEAR POWER HAS TWO ADVANTAGES: IT REDUCES OUR DEPENDENCE ON FOSSIL FUELS, AND IT LIMITS CO_2 EMISSIONS.
...ACCORDING TO THOSE PROMOTING NUCLEAR ENERGY. ONE: IT HELPS SECURE ENERGY INDEPENDENCE FROM FOREIGN OIL.
AND TWO: THE CLIMATE. TODAY PEOPLE FROM THE NUCLEAR INDUSTRY ARE THE FIRST TO SHOUT ABOUT THE STATE OF THE PLANET. ACCORDING TO THEM, THE ONLY WAY TO REDUCE CO_2 EMISSIONS IS TO DEVELOP NUCLEAR ENERGY.

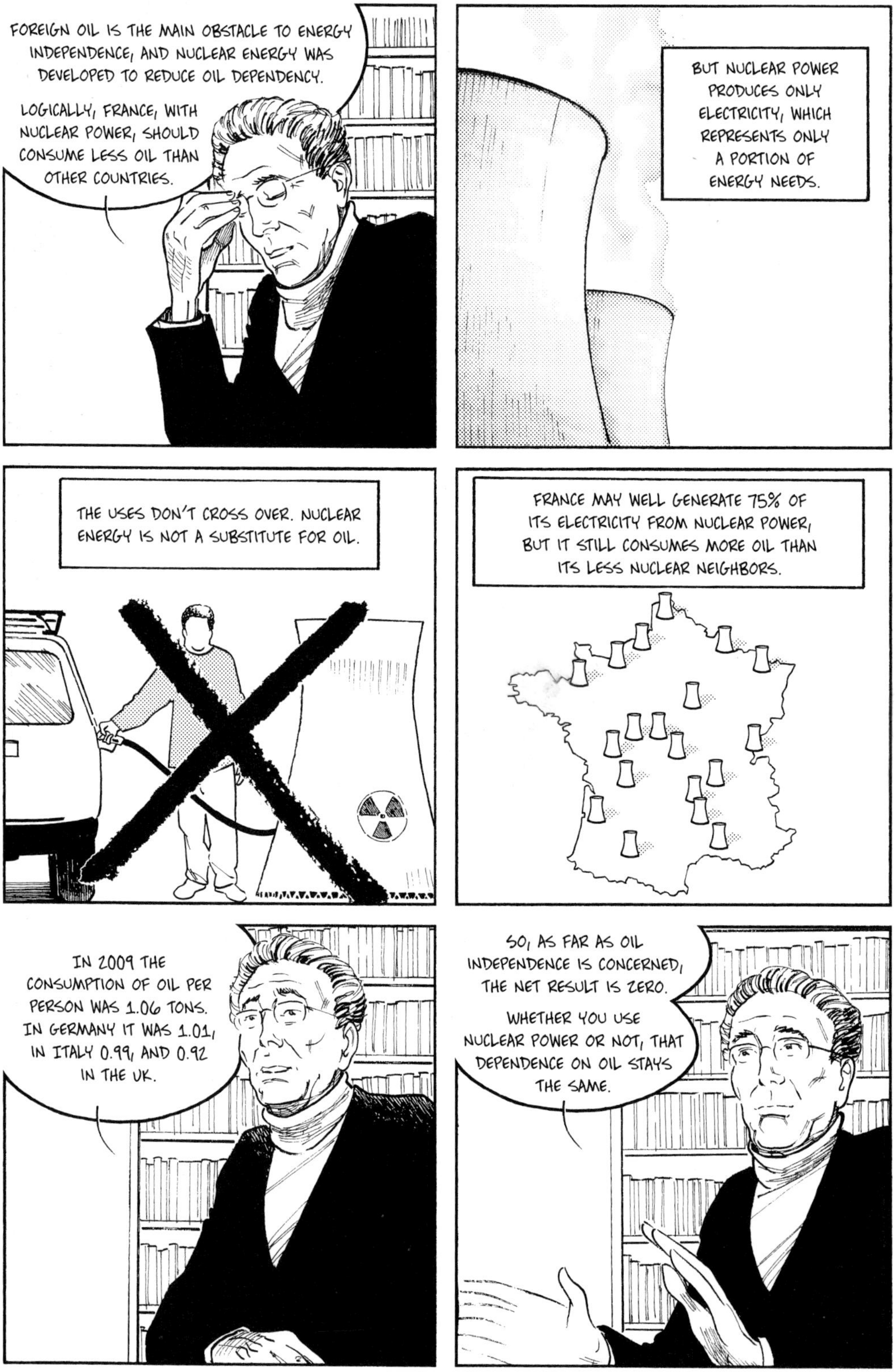
FOREIGN OIL IS THE MAIN OBSTACLE TO ENERGY INDEPENDENCE, AND NUCLEAR ENERGY WAS DEVELOPED TO REDUCE OIL DEPENDENCY.
LOGICALLY, FRANCE, WITH NUCLEAR POWER, SHOULD CONSUME LESS OIL THAN OTHER COUNTRIES.
BUT NUCLEAR POWER PRODUCES ONLY ELECTRICITY, WHICH REPRESENTS ONLY A PORTION OF ENERGY NEEDS.
THE USES DON'T CROSS OVER. NUCLEAR ENERGY IS NOT A SUBSTITUTE FOR OIL.
FRANCE MAY WELL GENERATE 75% OF ITS ELECTRICITY FROM NUCLEAR POWER, BUT IT STILL CONSUMES MORE OIL THAN ITS LESS NUCLEAR NEIGHBORS.
IN 2009 THE CONSUMPTION OF OIL PER PERSON WAS 1.06 TONS. IN GERMANY IT WAS 1.01, IN ITALY 0.99, AND 0.92 IN THE UK.
SO, AS FAR AS OIL INDEPENDENCE IS CONCERNED, THE NET RESULT IS ZERO.
WHETHER YOU USE NUCLEAR POWER OR NOT, THAT DEPENDENCE ON OIL STAYS THE SAME.

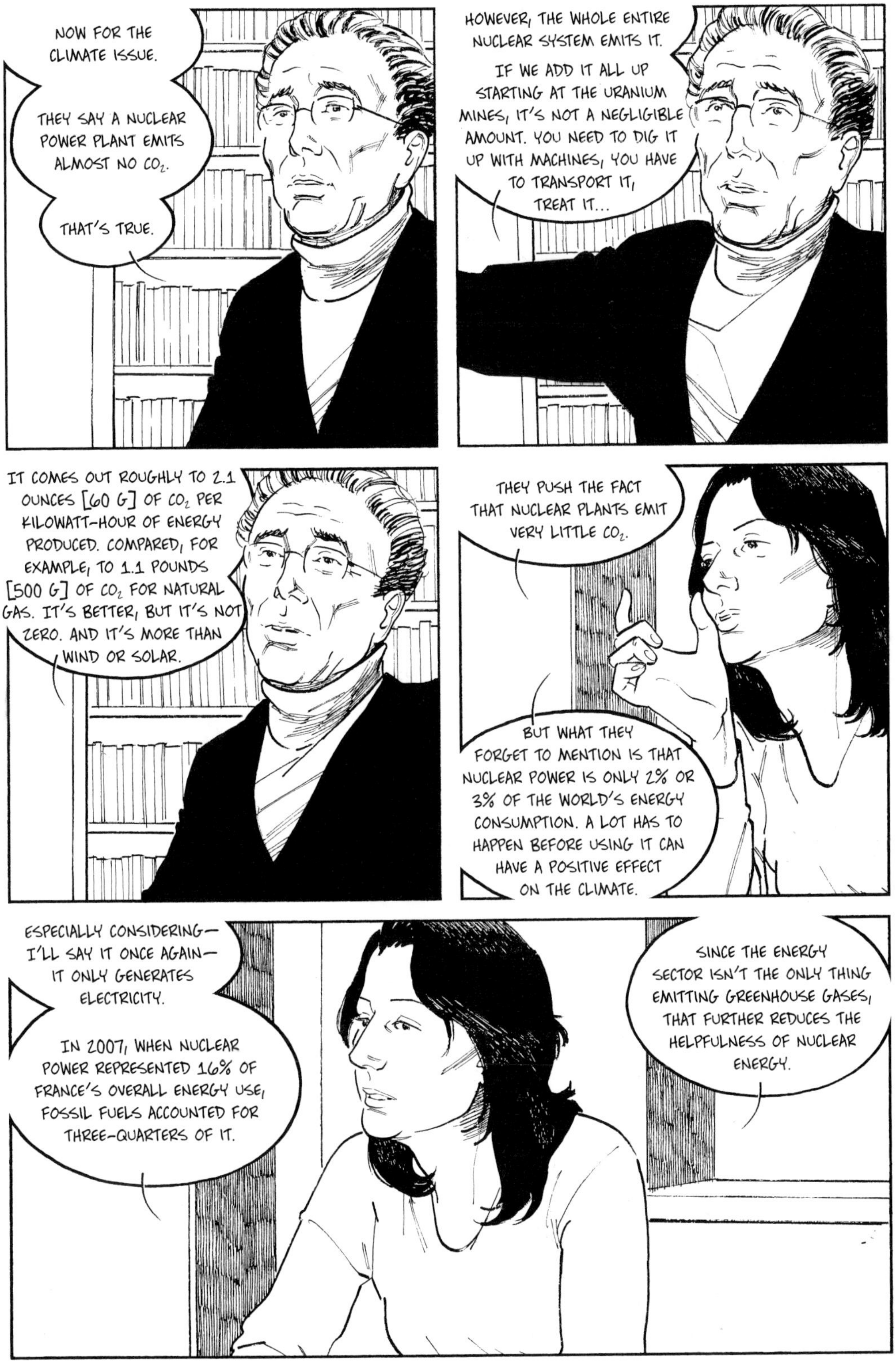
NOW FOR THE CLIMATE ISSUE.
THEY SAY A NUCLEAR POWER PLANT EMITS ALMOST NO CO_2.
THAT'S TRUE.
HOWEVER, THE WHOLE ENTIRE NUCLEAR SYSTEM EMITS IT.
IF WE ADD IT ALL UP STARTING AT THE URANIUM MINES, IT'S NOT A NEGLIGIBLE AMOUNT. YOU NEED TO DIG IT UP WITH MACHINES, YOU HAVE TO TRANSPORT IT, TREAT IT...
IT COMES OUT ROUGHLY TO 2.1 OUNCES [60 G] OF CO_2 PER KILOWATT-HOUR OF ENERGY PRODUCED. COMPARED, FOR EXAMPLE, TO 1.1 POUNDS [500 G] OF CO_2 FOR NATURAL GAS. IT'S BETTER, BUT IT'S NOT ZERO. AND IT'S MORE THAN WIND OR SOLAR.
THEY PUSH THE FACT THAT NUCLEAR PLANTS EMIT VERY LITTLE CO_2.
BUT WHAT THEY FORGET TO MENTION IS THAT NUCLEAR POWER IS ONLY 2% OR 3% OF THE WORLD'S ENERGY CONSUMPTION. A LOT HAS TO HAPPEN BEFORE USING IT CAN HAVE A POSITIVE EFFECT ON THE CLIMATE.
ESPECIALLY CONSIDERING— I'LL SAY IT ONCE AGAIN— IT ONLY GENERATES ELECTRICITY.
IN 2007, WHEN NUCLEAR POWER REPRESENTED 16% OF FRANCE'S OVERALL ENERGY USE, FOSSIL FUELS ACCOUNTED FOR THREE-QUARTERS OF IT.
SINCE THE ENERGY SECTOR ISN'T THE ONLY THING EMITTING GREENHOUSE GASES, THAT FURTHER REDUCES THE HELPFULNESS OF NUCLEAR ENERGY.

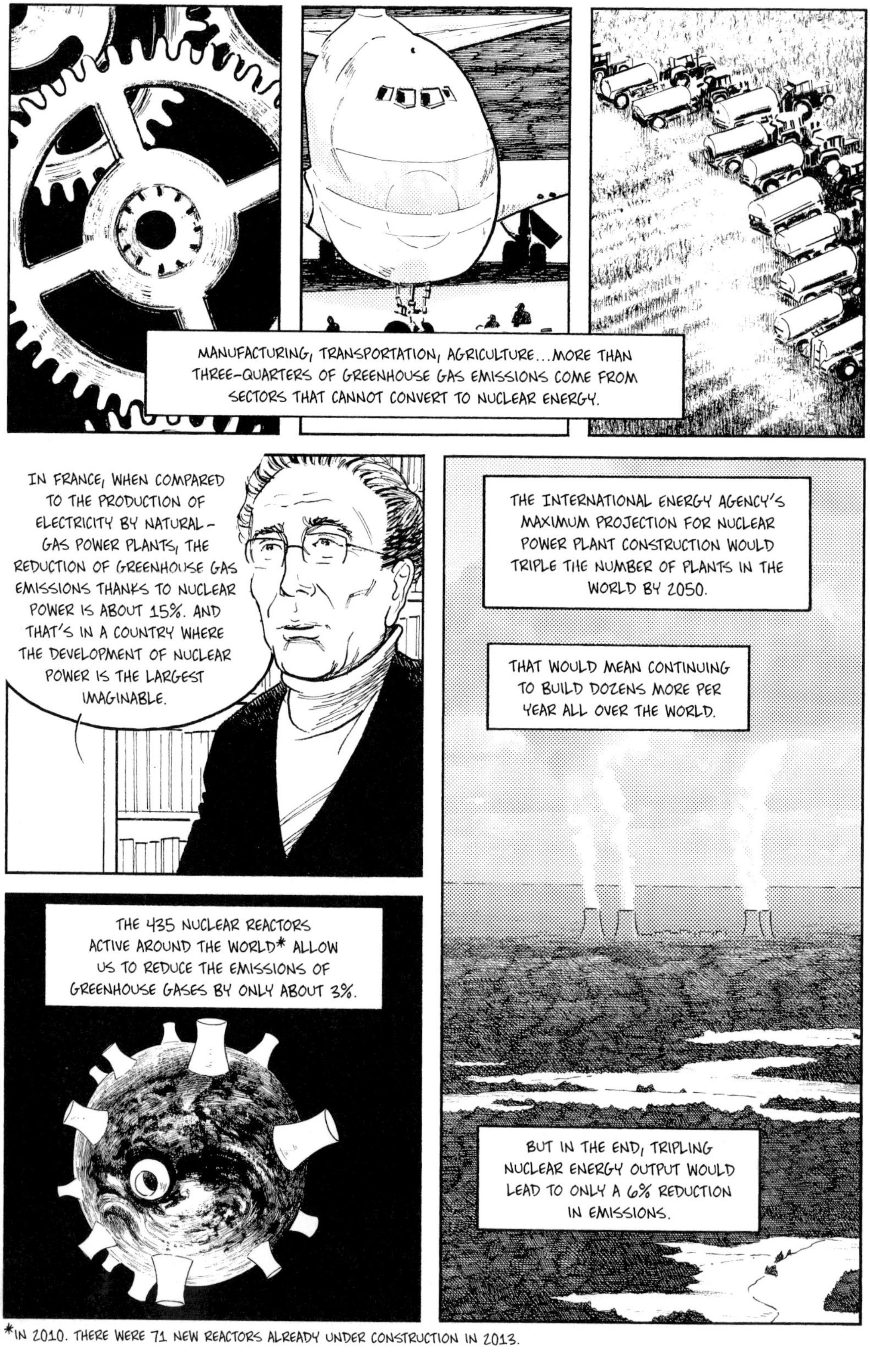
MANUFACTURING, TRANSPORTATION, AGRICULTURE... MORE THAN THREE-QUARTERS OF GREENHOUSE GAS EMISSIONS COME FROM SECTORS THAT CANNOT CONVERT TO NUCLEAR ENERGY.
IN FRANCE, WHEN COMPARED TO THE PRODUCTION OF ELECTRICITY BY NATURAL-GAS POWER PLANTS, THE REDUCTION OF GREENHOUSE GAS EMISSIONS THANKS TO NUCLEAR POWER IS ABOUT 15%. AND THAT'S IN A COUNTRY WHERE THE DEVELOPMENT OF NUCLEAR POWER IS THE LARGEST IMAGINABLE.
THE INTERNATIONAL ENERGY AGENCY'S MAXIMUM PROJECTION FOR NUCLEAR POWER PLANT CONSTRUCTION WOULD TRIPLE THE NUMBER OF PLANTS IN THE WORLD BY 2050.
THAT WOULD MEAN CONTINUING TO BUILD DOZENS MORE PER YEAR ALL OVER THE WORLD.
THE 435 NUCLEAR REACTORS ACTIVE AROUND THE WORLD* ALLOW US TO REDUCE THE EMISSIONS OF GREENHOUSE GASES BY ONLY ABOUT 3%.
BUT IN THE END, TRIPLING NUCLEAR ENERGY OUTPUT WOULD LEAD TO ONLY A 6% REDUCTION IN EMISSIONS.

*IN 2010. THERE WERE 71 NEW REACTORS ALREADY UNDER CONSTRUCTION IN 2013.

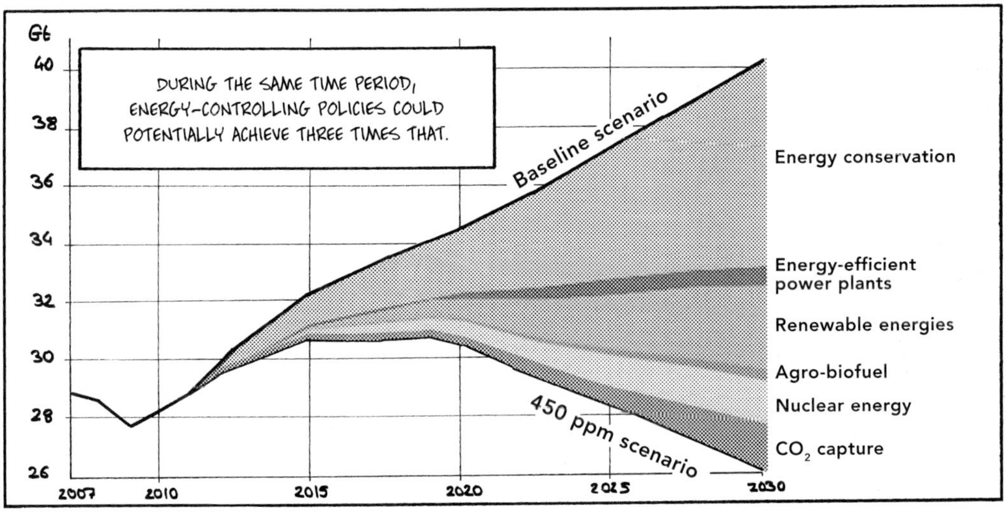
Gt
40
38
36
34
32
30
28
26
2007
2010
2015
2020
2025
2030
DURING THE SAME TIME PERIOD, ENERGY-CONTROLLING POLICIES COULD POTENTIALLY ACHIEVE THREE TIMES THAT.
Baseline scenario
450 ppm scenario
Energy conservation
Energy-efficient power plants
Renewable energies
Agro-biofuel
Nuclear energy
CO_2 capture

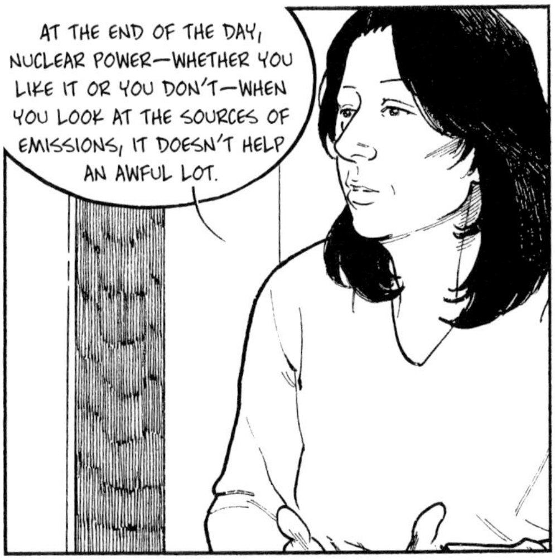
AT THE END OF THE DAY, NUCLEAR POWER—WHETHER YOU LIKE IT OR YOU DON'T—WHEN YOU LOOK AT THE SOURCES OF EMISSIONS, IT DOESN'T HELP AN AWFUL LOT.

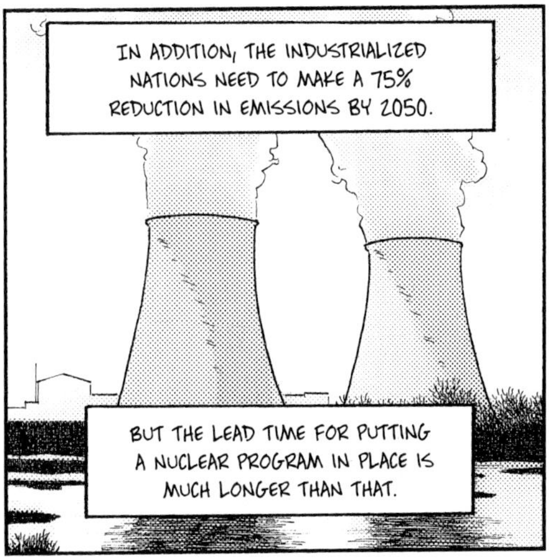
IN ADDITION, THE INDUSTRIALIZED NATIONS NEED TO MAKE A 75% REDUCTION IN EMISSIONS BY 2050.
BUT THE LEAD TIME FOR PUTTING A NUCLEAR PROGRAM IN PLACE IS MUCH LONGER THAN THAT.

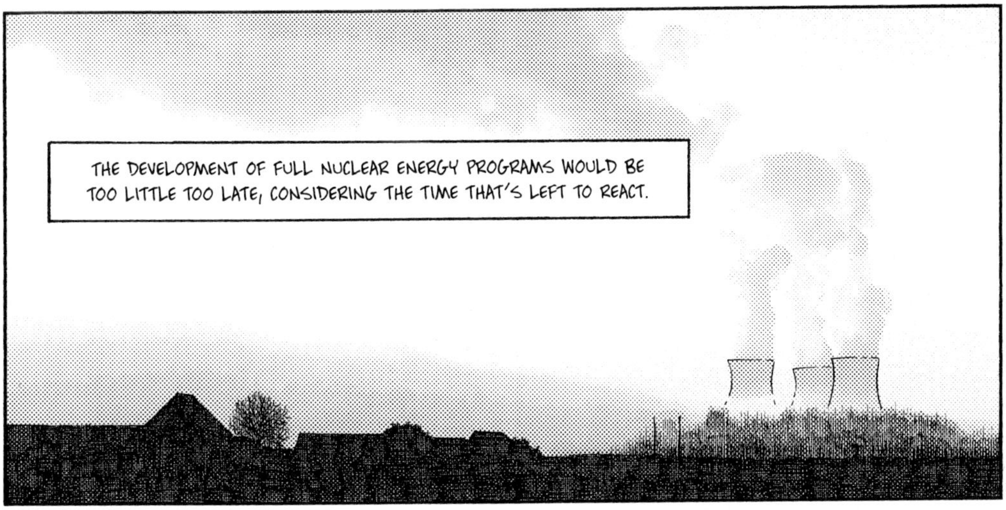
THE DEVELOPMENT OF FULL NUCLEAR ENERGY PROGRAMS WOULD BE TOO LITTLE TOO LATE, CONSIDERING THE TIME THAT'S LEFT TO REACT.

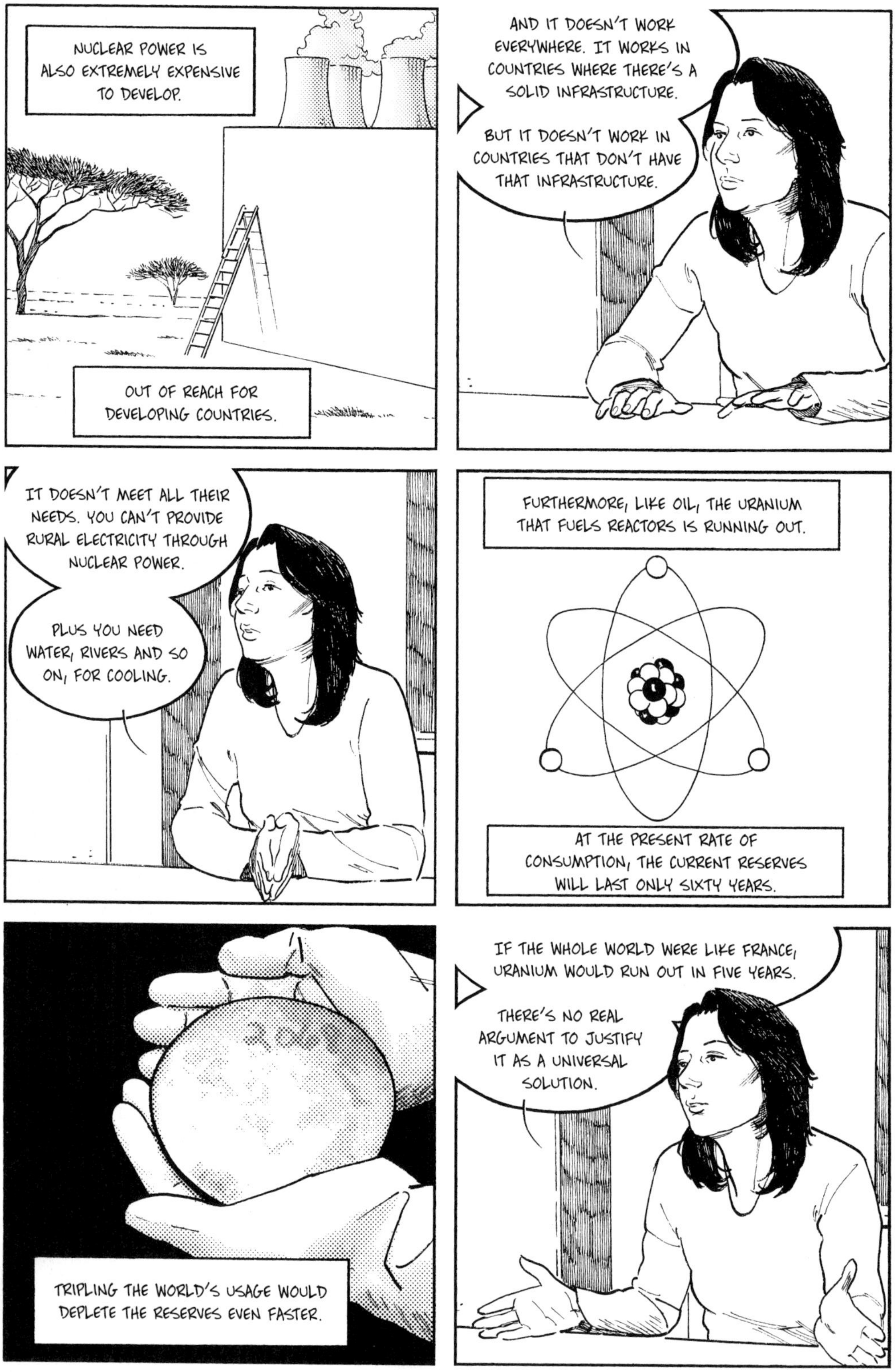
NUCLEAR POWER IS ALSO EXTREMELY EXPENSIVE TO DEVELOP.
OUT OF REACH FOR DEVELOPING COUNTRIES.
AND IT DOESN'T WORK EVERYWHERE. IT WORKS IN COUNTRIES WHERE THERE'S A SOLID INFRASTRUCTURE.
BUT IT DOESN'T WORK IN COUNTRIES THAT DON'T HAVE THAT INFRASTRUCTURE.
IT DOESN'T MEET ALL THEIR NEEDS. YOU CAN'T PROVIDE RURAL ELECTRICITY THROUGH NUCLEAR POWER.
PLUS YOU NEED WATER, RIVERS AND SO ON, FOR COOLING.
FURTHERMORE, LIKE OIL, THE URANIUM THAT FUELS REACTORS IS RUNNING OUT.
AT THE PRESENT RATE OF CONSUMPTION, THE CURRENT RESERVES WILL LAST ONLY SIXTY YEARS.
TRIPLING THE WORLD'S USAGE WOULD DEPLETE THE RESERVES EVEN FASTER.
IF THE WHOLE WORLD WERE LIKE FRANCE, URANIUM WOULD RUN OUT IN FIVE YEARS.
THERE'S NO REAL ARGUMENT TO JUSTIFY IT AS A UNIVERSAL SOLUTION.

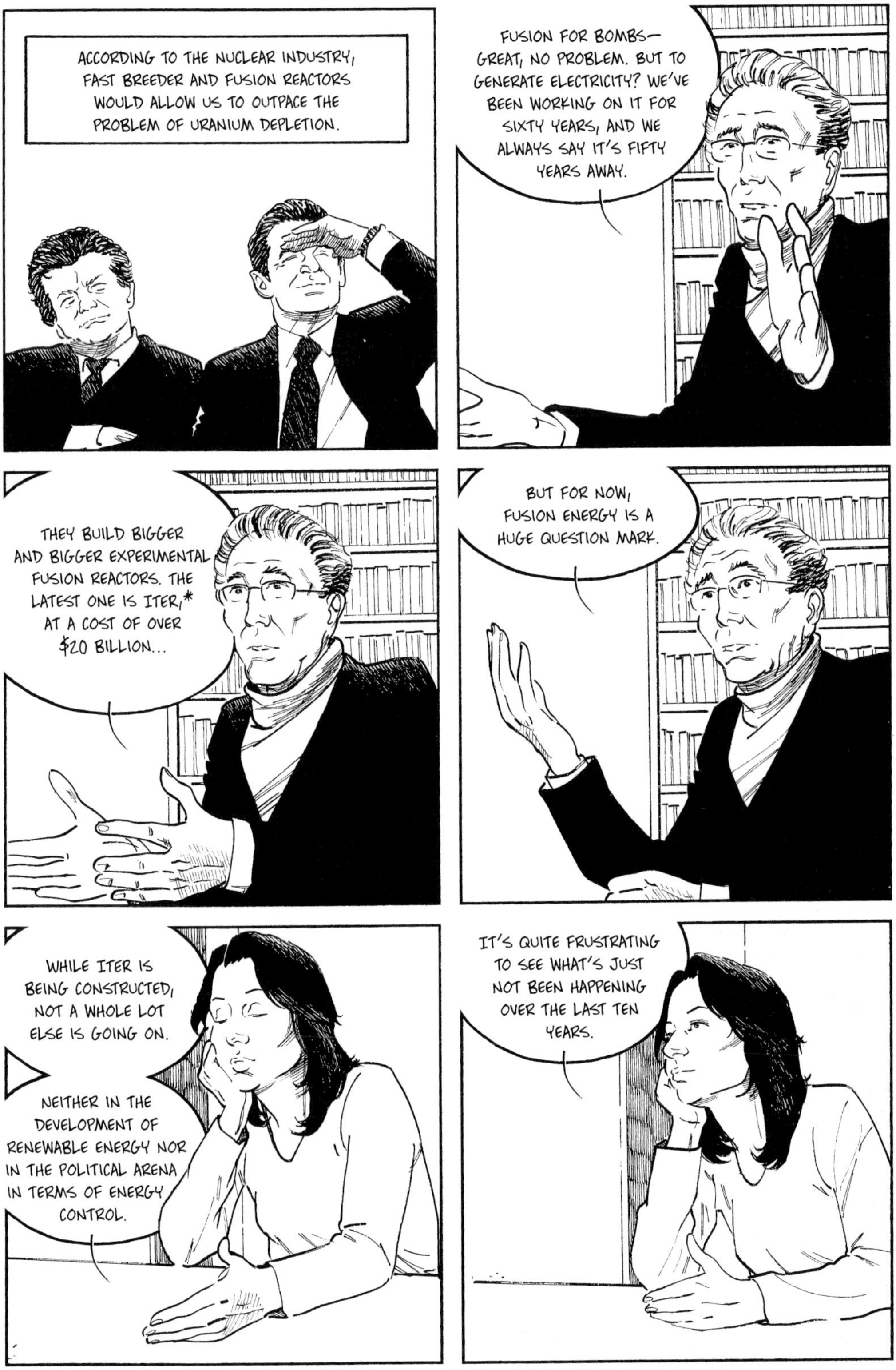

ACCORDING TO THE NUCLEAR INDUSTRY, FAST BREEDER AND FUSION REACTORS WOULD ALLOW US TO OUTPACE THE PROBLEM OF URANIUM DEPLETION.
FUSION FOR BOMBS—GREAT, NO PROBLEM. BUT TO GENERATE ELECTRICITY? WE'VE BEEN WORKING ON IT FOR SIXTY YEARS, AND WE ALWAYS SAY IT'S FIFTY YEARS AWAY.
THEY BUILD BIGGER AND BIGGER EXPERIMENTAL FUSION REACTORS. THE LATEST ONE IS ITER,* AT A COST OF OVER $20 BILLION...
BUT FOR NOW, FUSION ENERGY IS A HUGE QUESTION MARK.
WHILE ITER IS BEING CONSTRUCTED, NOT A WHOLE LOT ELSE IS GOING ON.
NEITHER IN THE DEVELOPMENT OF RENEWABLE ENERGY NOR IN THE POLITICAL ARENA IN TERMS OF ENERGY CONTROL.
IT'S QUITE FRUSTRATING TO SEE WHAT'S JUST NOT BEEN HAPPENING OVER THE LAST TEN YEARS.

*THE INTERNATIONAL THERMONUCLEAR EXPERIMENTAL REACTOR. THE GOAL IS FOR ITER TO GO ONLINE IN 2020.

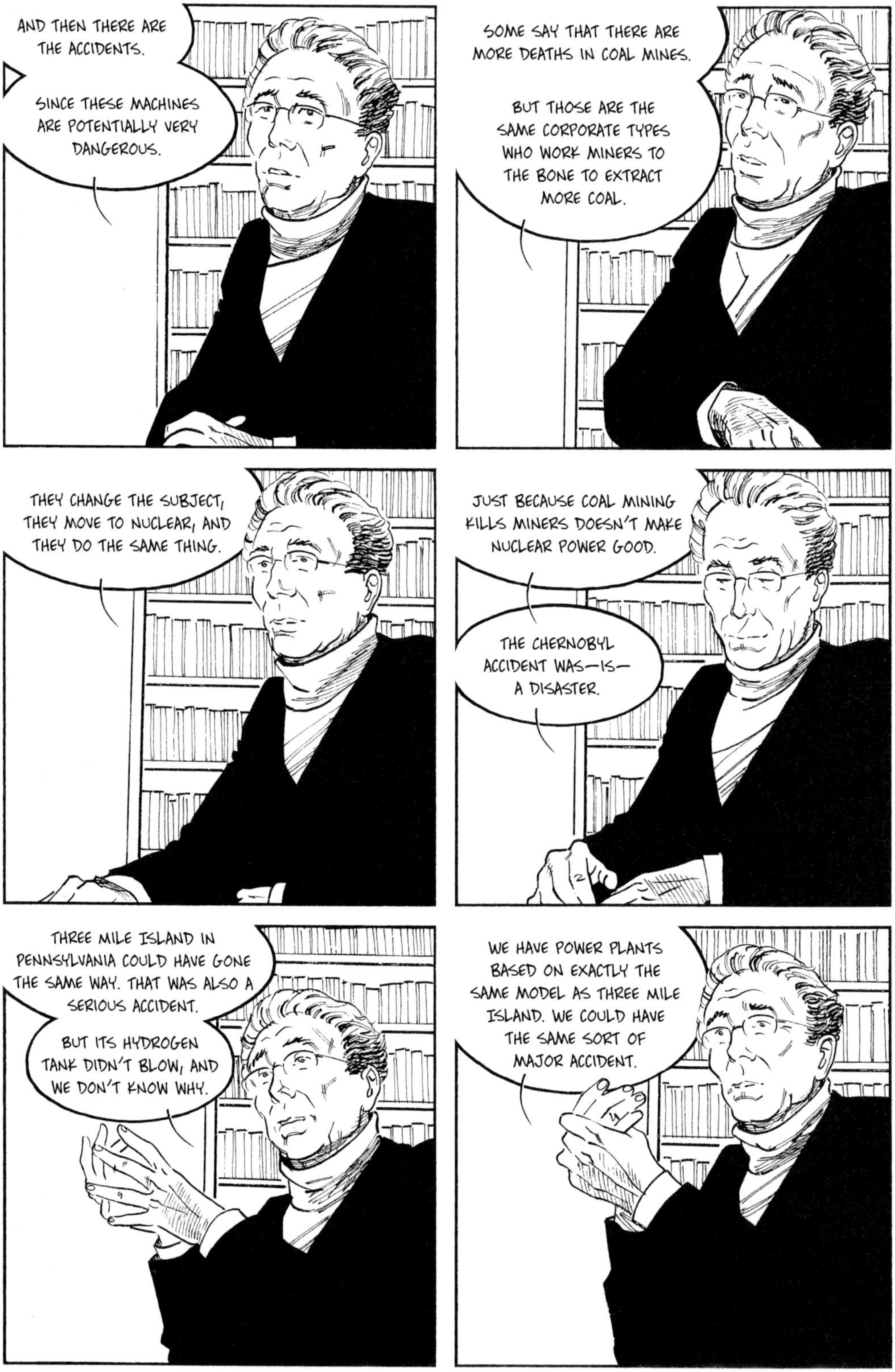
AND THEN THERE ARE THE ACCIDENTS.
SINCE THESE MACHINES ARE POTENTIALLY VERY DANGEROUS.
SOME SAY THAT THERE ARE MORE DEATHS IN COAL MINES.
BUT THOSE ARE THE SAME CORPORATE TYPES WHO WORK MINERS TO THE BONE TO EXTRACT MORE COAL.
THEY CHANGE THE SUBJECT, THEY MOVE TO NUCLEAR, AND THEY DO THE SAME THING.
JUST BECAUSE COAL MINING KILLS MINERS DOESN'T MAKE NUCLEAR POWER GOOD.
THE CHERNOBYL ACCIDENT WAS—IS—A DISASTER.
THREE MILE ISLAND IN PENNSYLVANIA COULD HAVE GONE THE SAME WAY. THAT WAS ALSO A SERIOUS ACCIDENT.
BUT ITS HYDROGEN TANK DIDN'T BLOW, AND WE DON'T KNOW WHY.
WE HAVE POWER PLANTS BASED ON EXACTLY THE SAME MODEL AS THREE MILE ISLAND. WE COULD HAVE THE SAME SORT OF MAJOR ACCIDENT.

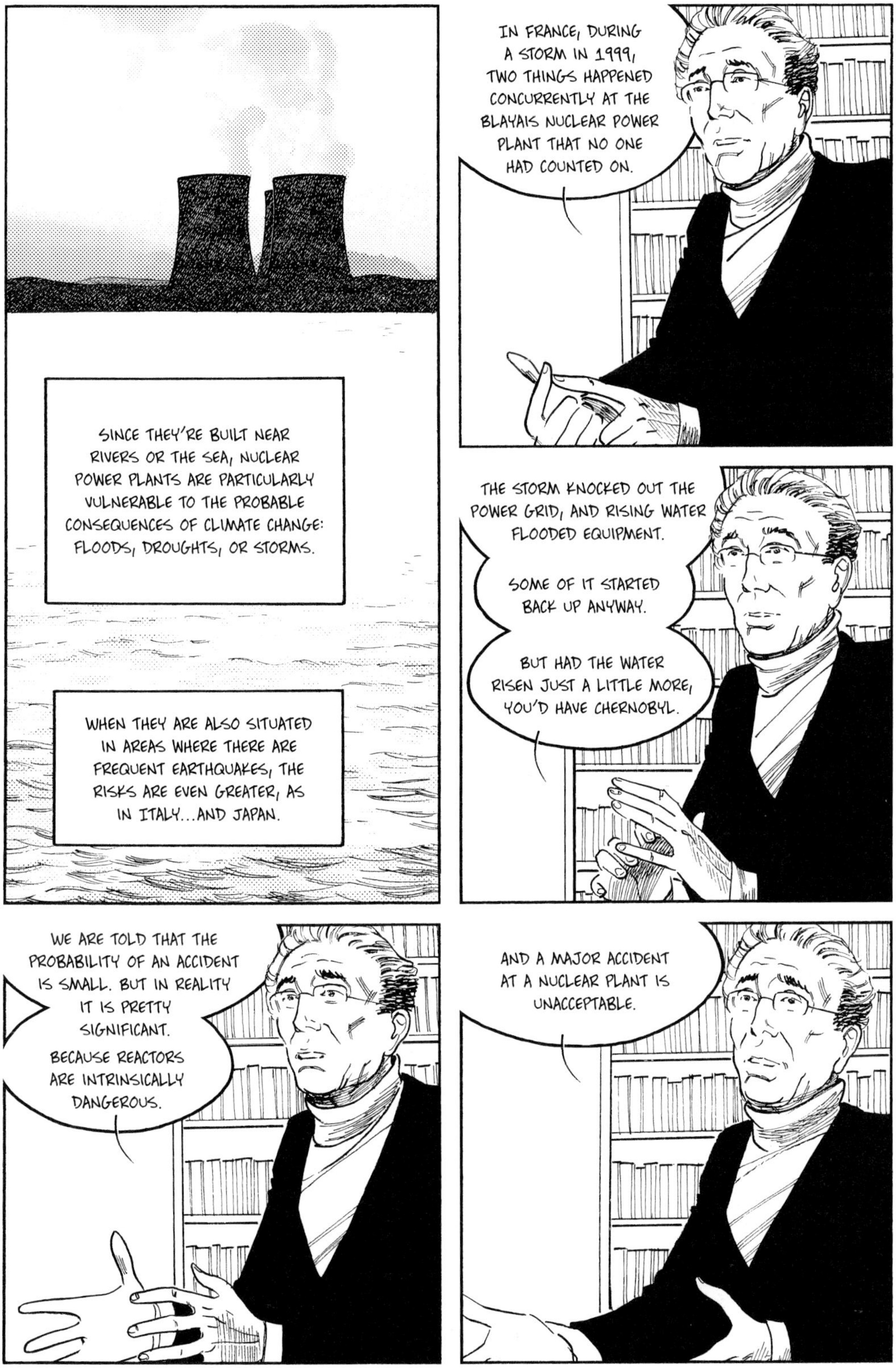
SINCE THEY'RE BUILT NEAR RIVERS OR THE SEA, NUCLEAR POWER PLANTS ARE PARTICULARLY VULNERABLE TO THE PROBABLE CONSEQUENCES OF CLIMATE CHANGE: FLOODS, DROUGHTS, OR STORMS.
WHEN THEY ARE ALSO SITUATED IN AREAS WHERE THERE ARE FREQUENT EARTHQUAKES, THE RISKS ARE EVEN GREATER, AS IN ITALY...AND JAPAN.
IN FRANCE, DURING A STORM IN 1999, TWO THINGS HAPPENED CONCURRENTLY AT THE BLAYAIS NUCLEAR POWER PLANT THAT NO ONE HAD COUNTED ON.
THE STORM KNOCKED OUT THE POWER GRID, AND RISING WATER FLOODED EQUIPMENT.
SOME OF IT STARTED BACK UP ANYWAY.
BUT HAD THE WATER RISEN JUST A LITTLE MORE, YOU'D HAVE CHERNOBYL.
WE ARE TOLD THAT THE PROBABILITY OF AN ACCIDENT IS SMALL. BUT IN REALITY IT IS PRETTY SIGNIFICANT.
BECAUSE REACTORS ARE INTRINSICALLY DANGEROUS.
AND A MAJOR ACCIDENT AT A NUCLEAR PLANT IS UNACCEPTABLE.

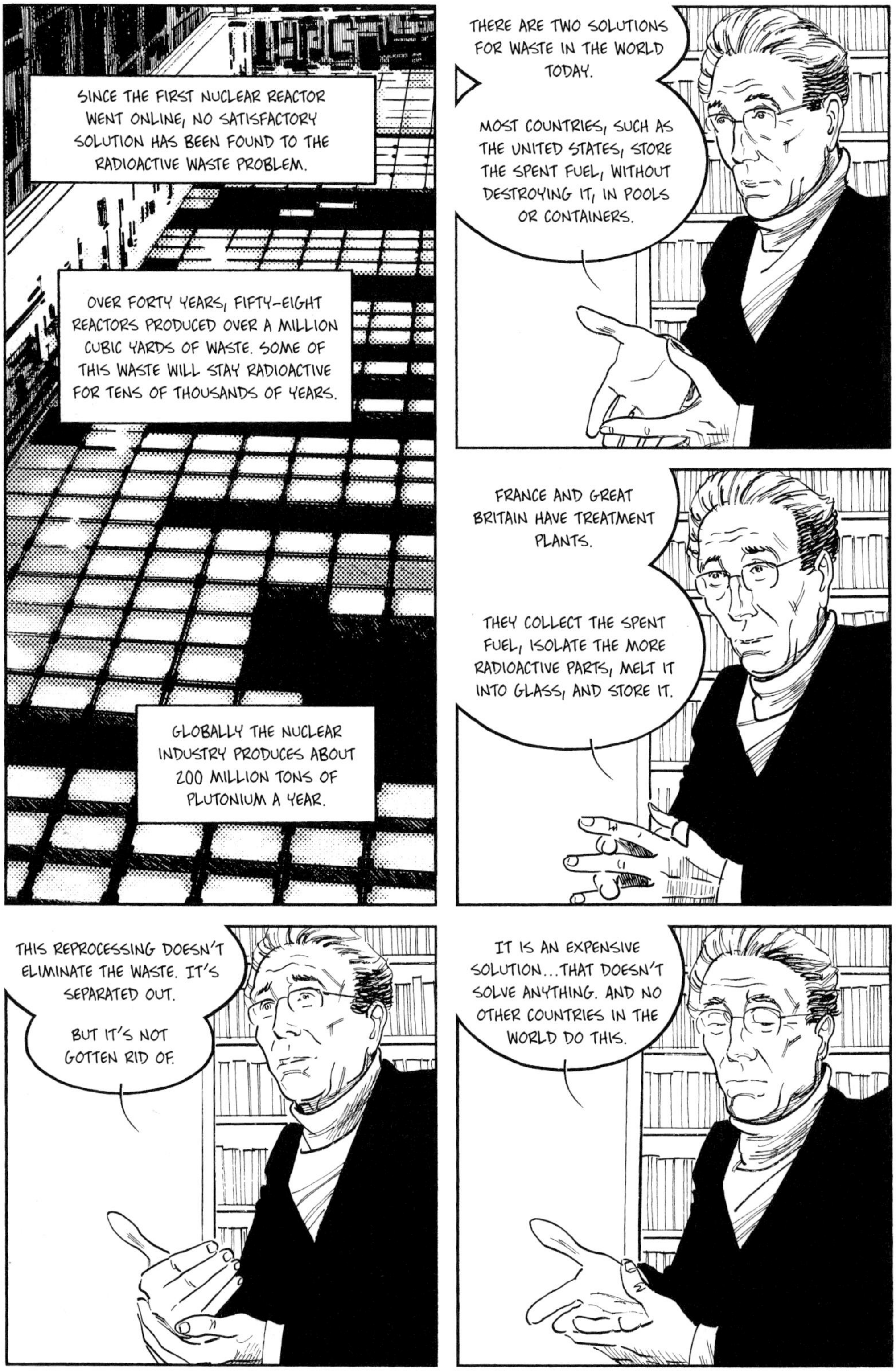
SINCE THE FIRST NUCLEAR REACTOR WENT ONLINE, NO SATISFACTORY SOLUTION HAS BEEN FOUND TO THE RADIOACTIVE WASTE PROBLEM.
OVER FORTY YEARS, FIFTY-EIGHT REACTORS PRODUCED OVER A MILLION CUBIC YARDS OF WASTE. SOME OF THIS WASTE WILL STAY RADIOACTIVE FOR TENS OF THOUSANDS OF YEARS.
GLOBALLY THE NUCLEAR INDUSTRY PRODUCES ABOUT 200 MILLION TONS OF PLUTONIUM A YEAR.
THERE ARE TWO SOLUTIONS FOR WASTE IN THE WORLD TODAY.
MOST COUNTRIES, SUCH AS THE UNITED STATES, STORE THE SPENT FUEL, WITHOUT DESTROYING IT, IN POOLS OR CONTAINERS.
FRANCE AND GREAT BRITAIN HAVE TREATMENT PLANTS.
THEY COLLECT THE SPENT FUEL, ISOLATE THE MORE RADIOACTIVE PARTS, MELT IT INTO GLASS, AND STORE IT.
THIS REPROCESSING DOESN'T ELIMINATE THE WASTE. IT'S SEPARATED OUT.
BUT IT'S NOT GOTTEN RID OF.
IT IS AN EXPENSIVE SOLUTION...THAT DOESN'T SOLVE ANYTHING. AND NO OTHER COUNTRIES IN THE WORLD DO THIS.

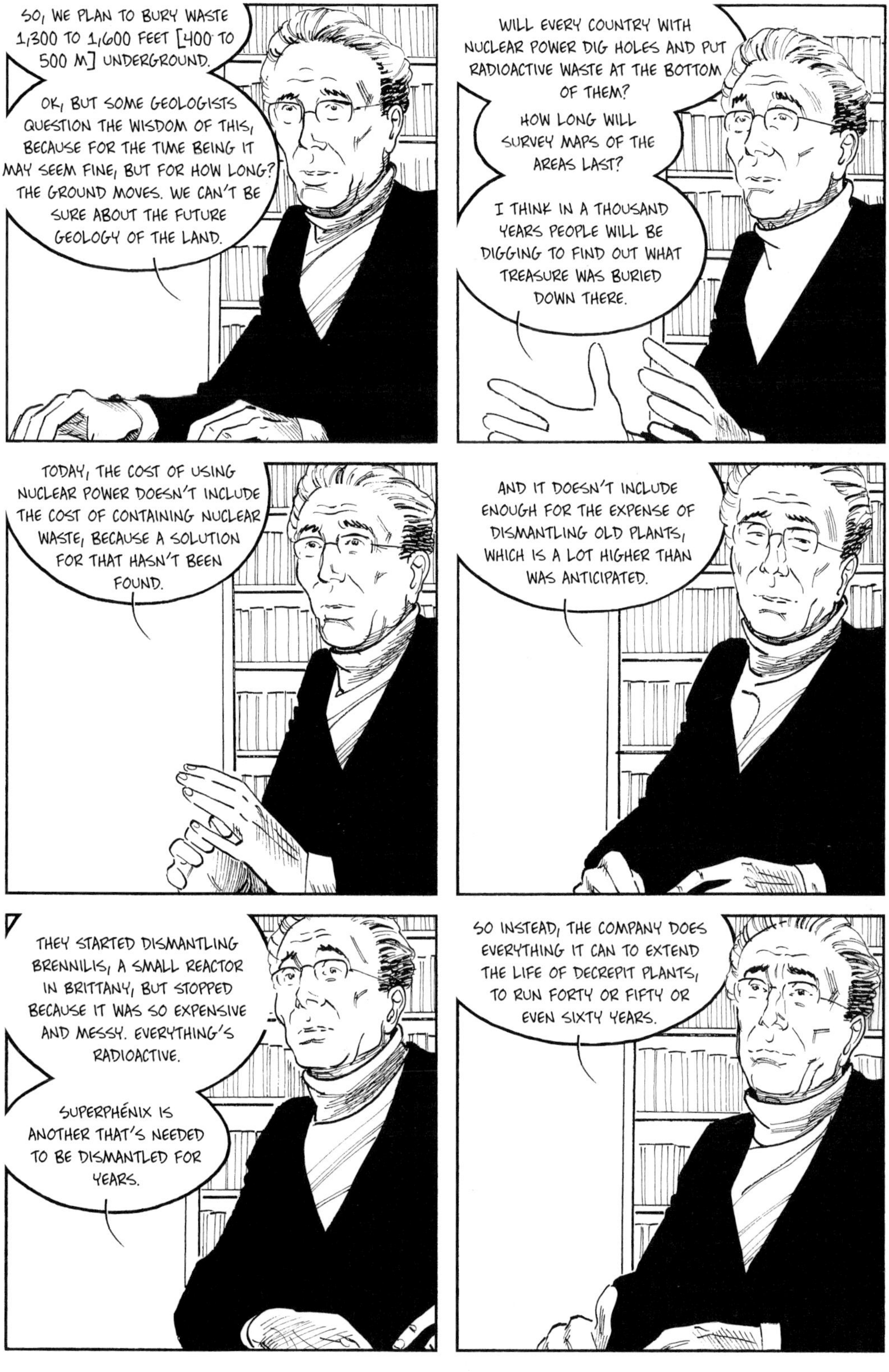
SO, WE PLAN TO BURY WASTE 1,300 TO 1,600 FEET [400 TO 500 M] UNDERGROUND.
OK, BUT SOME GEOLOGISTS QUESTION THE WISDOM OF THIS, BECAUSE FOR THE TIME BEING IT MAY SEEM FINE, BUT FOR HOW LONG? THE GROUND MOVES. WE CAN'T BE SURE ABOUT THE FUTURE GEOLOGY OF THE LAND.
WILL EVERY COUNTRY WITH NUCLEAR POWER DIG HOLES AND PUT RADIOACTIVE WASTE AT THE BOTTOM OF THEM?
HOW LONG WILL SURVEY MAPS OF THE AREAS LAST?
I THINK IN A THOUSAND YEARS PEOPLE WILL BE DIGGING TO FIND OUT WHAT TREASURE WAS BURIED DOWN THERE.
TODAY, THE COST OF USING NUCLEAR POWER DOESN'T INCLUDE THE COST OF CONTAINING NUCLEAR WASTE, BECAUSE A SOLUTION FOR THAT HASN'T BEEN FOUND.
AND IT DOESN'T INCLUDE ENOUGH FOR THE EXPENSE OF DISMANTLING OLD PLANTS, WHICH IS A LOT HIGHER THAN WAS ANTICIPATED.
THEY STARTED DISMANTLING BRENNILIS, A SMALL REACTOR IN BRITTANY, BUT STOPPED BECAUSE IT WAS SO EXPENSIVE AND MESSY. EVERYTHING'S RADIOACTIVE.
SUPERPHÉNIX IS ANOTHER THAT'S NEEDED TO BE DISMANTLED FOR YEARS.
SO INSTEAD, THE COMPANY DOES EVERYTHING IT CAN TO EXTEND THE LIFE OF DECREPIT PLANTS, TO RUN FORTY OR FIFTY OR EVEN SIXTY YEARS.

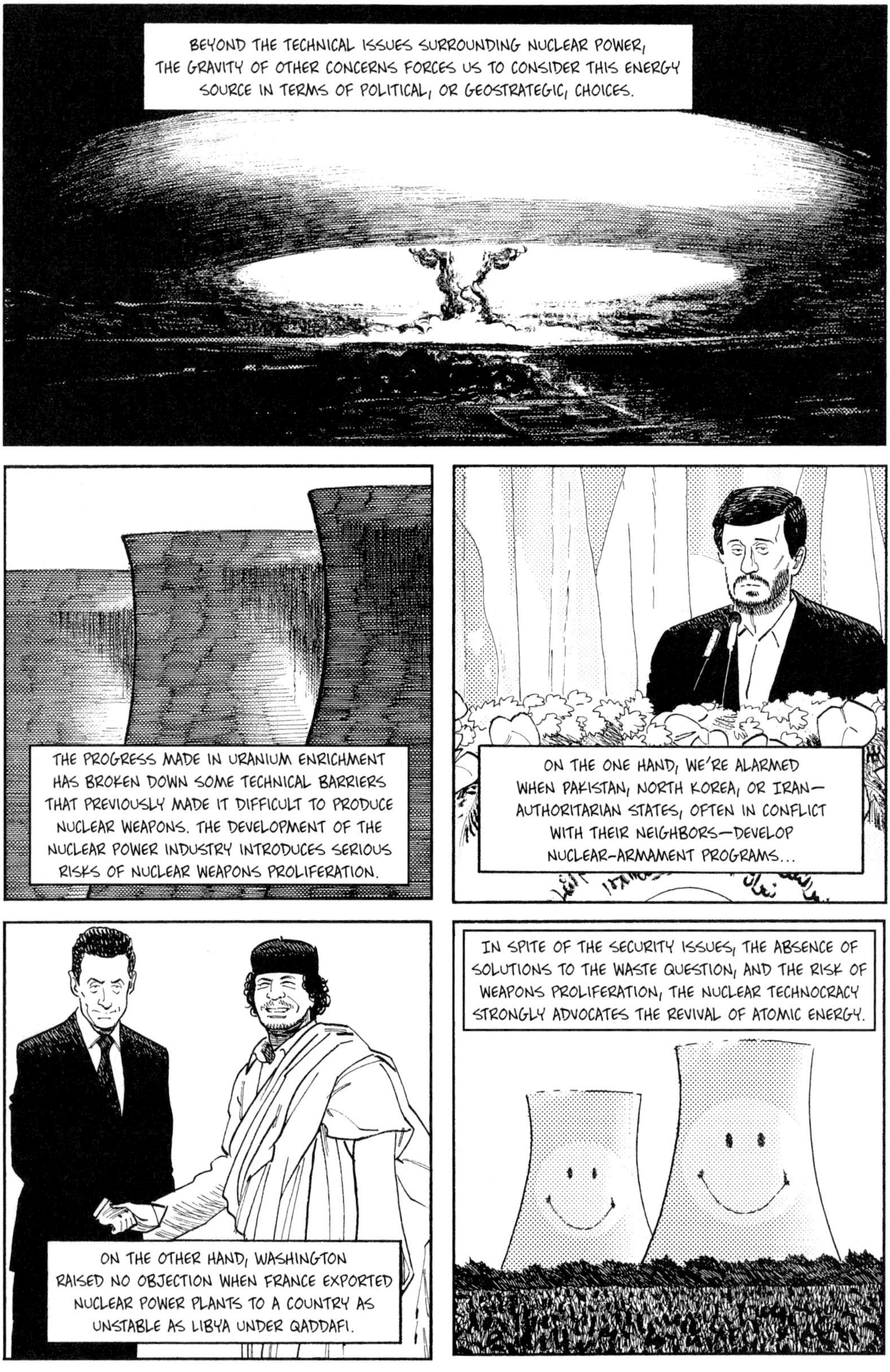
BEYOND THE TECHNICAL ISSUES SURROUNDING NUCLEAR POWER, THE GRAVITY OF OTHER CONCERNS FORCES US TO CONSIDER THIS ENERGY SOURCE IN TERMS OF POLITICAL, OR GEOSTRATEGIC, CHOICES.
THE PROGRESS MADE IN URANIUM ENRICHMENT HAS BROKEN DOWN SOME TECHNICAL BARRIERS THAT PREVIOUSLY MADE IT DIFFICULT TO PRODUCE NUCLEAR WEAPONS. THE DEVELOPMENT OF THE NUCLEAR POWER INDUSTRY INTRODUCES SERIOUS RISKS OF NUCLEAR WEAPONS PROLIFERATION.
ON THE ONE HAND, WE'RE ALARMED WHEN PAKISTAN, NORTH KOREA, OR IRAN—AUTHORITARIAN STATES, OFTEN IN CONFLICT WITH THEIR NEIGHBORS—DEVELOP NUCLEAR-ARMAMENT PROGRAMS...
ON THE OTHER HAND, WASHINGTON RAISED NO OBJECTION WHEN FRANCE EXPORTED NUCLEAR POWER PLANTS TO A COUNTRY AS UNSTABLE AS LIBYA UNDER QADDAFI.
IN SPITE OF THE SECURITY ISSUES, THE ABSENCE OF SOLUTIONS TO THE WASTE QUESTION, AND THE RISK OF WEAPONS PROLIFERATION, THE NUCLEAR TECHNOCRACY STRONGLY ADVOCATES THE REVIVAL OF ATOMIC ENERGY.

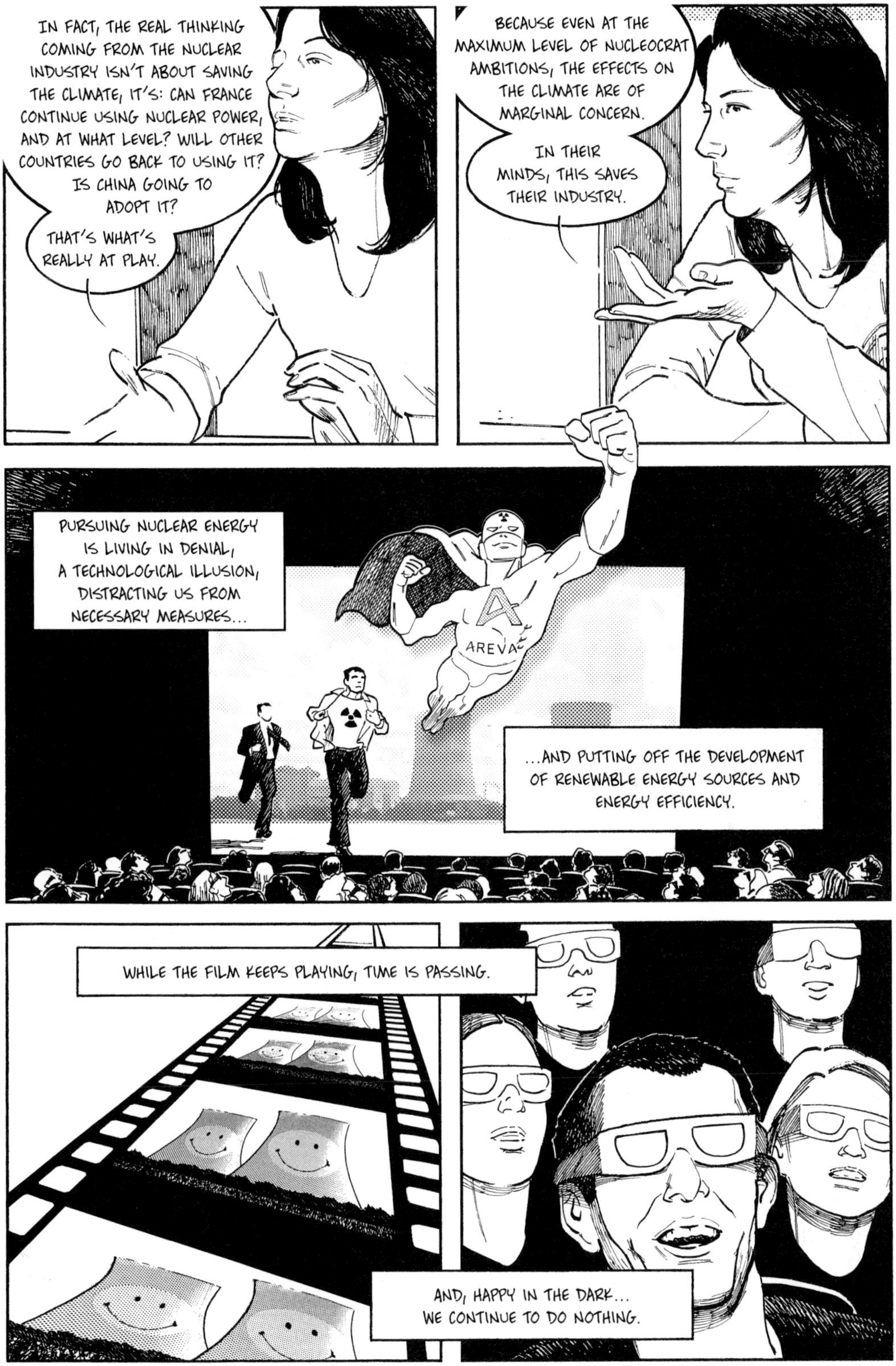
IN FACT, THE REAL THINKING COMING FROM THE NUCLEAR INDUSTRY ISN'T ABOUT SAVING THE CLIMATE, IT'S: CAN FRANCE CONTINUE USING NUCLEAR POWER, AND AT WHAT LEVEL? WILL OTHER COUNTRIES GO BACK TO USING IT? IS CHINA GOING TO ADOPT IT?
THAT'S WHAT'S REALLY AT PLAY.
BECAUSE EVEN AT THE MAXIMUM LEVEL OF NUCLEOCRAT AMBITIONS, THE EFFECTS ON THE CLIMATE ARE OF MARGINAL CONCERN.
IN THEIR MINDS, THIS SAVES THEIR INDUSTRY.
PURSUING NUCLEAR ENERGY IS LIVING IN DENIAL, A TECHNOLOGICAL ILLUSION, DISTRACTING US FROM NECESSARY MEASURES...
AREVA
...AND PUTTING OFF THE DEVELOPMENT OF RENEWABLE ENERGY SOURCES AND ENERGY EFFICIENCY.
WHILE THE FILM KEEPS PLAYING, TIME IS PASSING.
AND, HAPPY IN THE DARK... WE CONTINUE TO DO NOTHING.

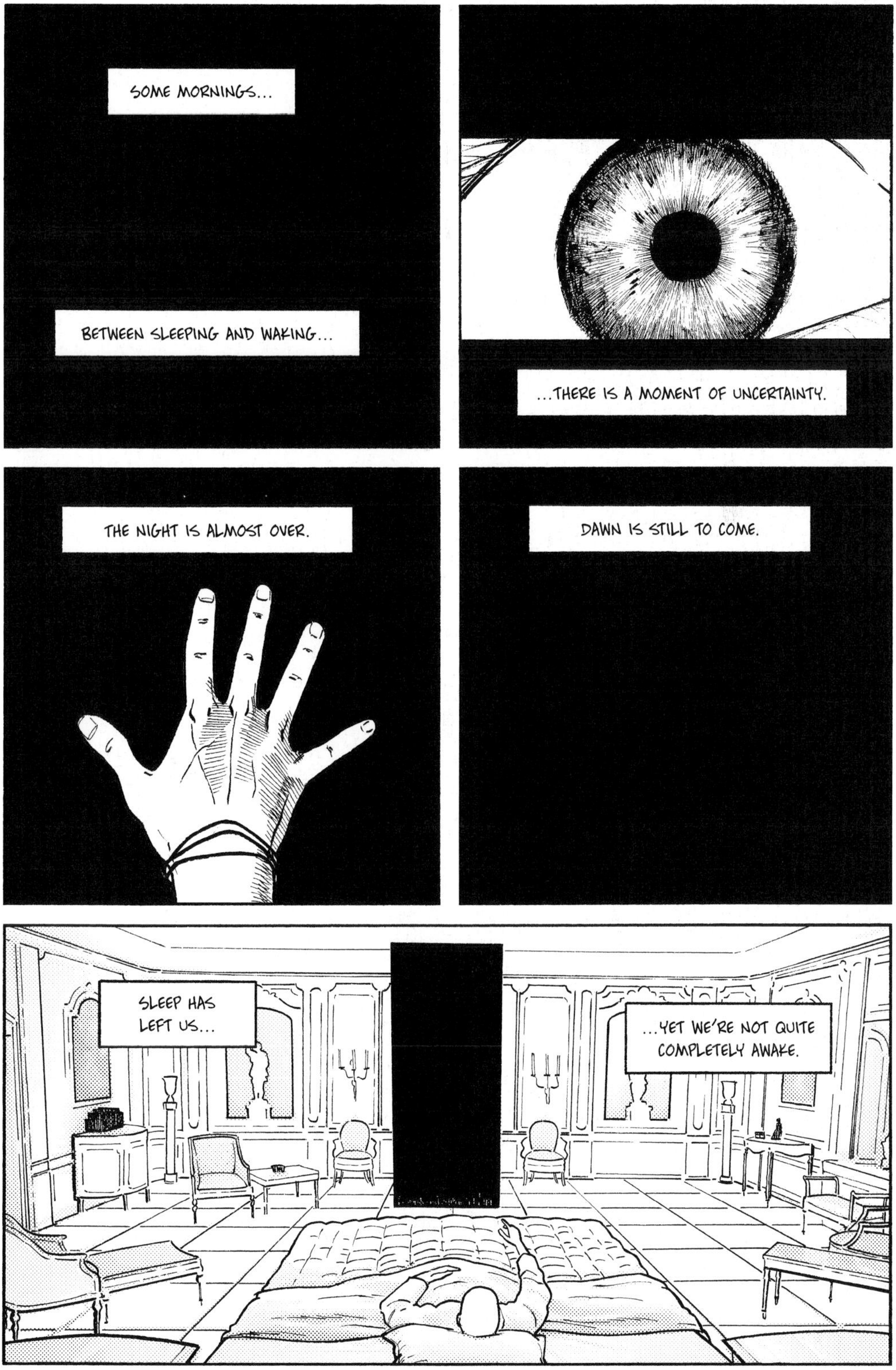
SOME MORNINGS...
BETWEEN SLEEPING AND WAKING...
...THERE IS A MOMENT OF UNCERTAINTY.
THE NIGHT IS ALMOST OVER.
DAWN IS STILL TO COME.
SLEEP HAS LEFT US...
...YET WE'RE NOT QUITE COMPLETELY AWAKE.

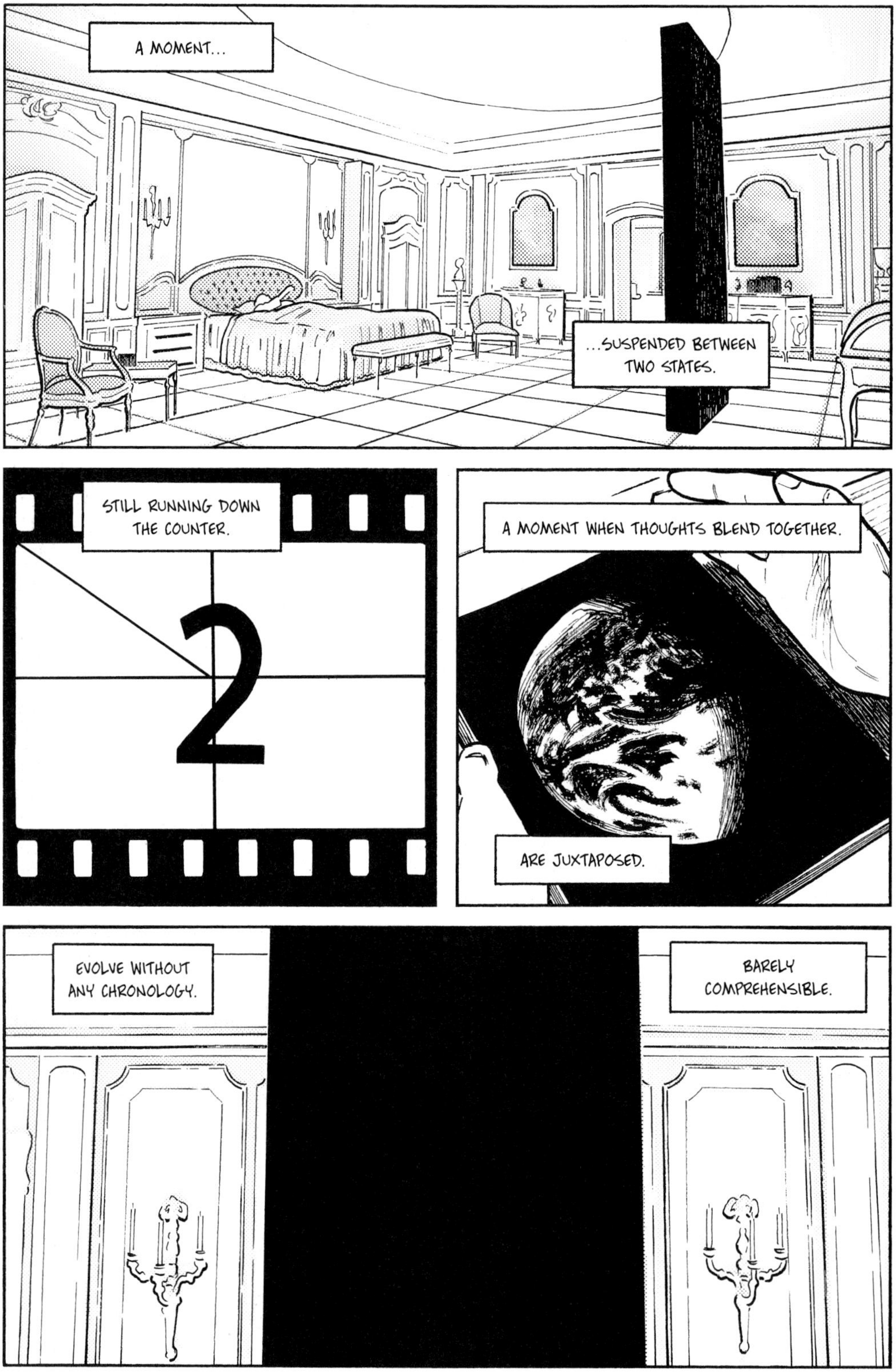
A MOMENT...
...SUSPENDED BETWEEN TWO STATES.
STILL RUNNING DOWN THE COUNTER.
2
A MOMENT WHEN THOUGHTS BLEND TOGETHER.
ARE JUXTAPOSED.
EVOLVE WITHOUT ANY CHRONOLOGY.
BARELY COMPREHENSIBLE.

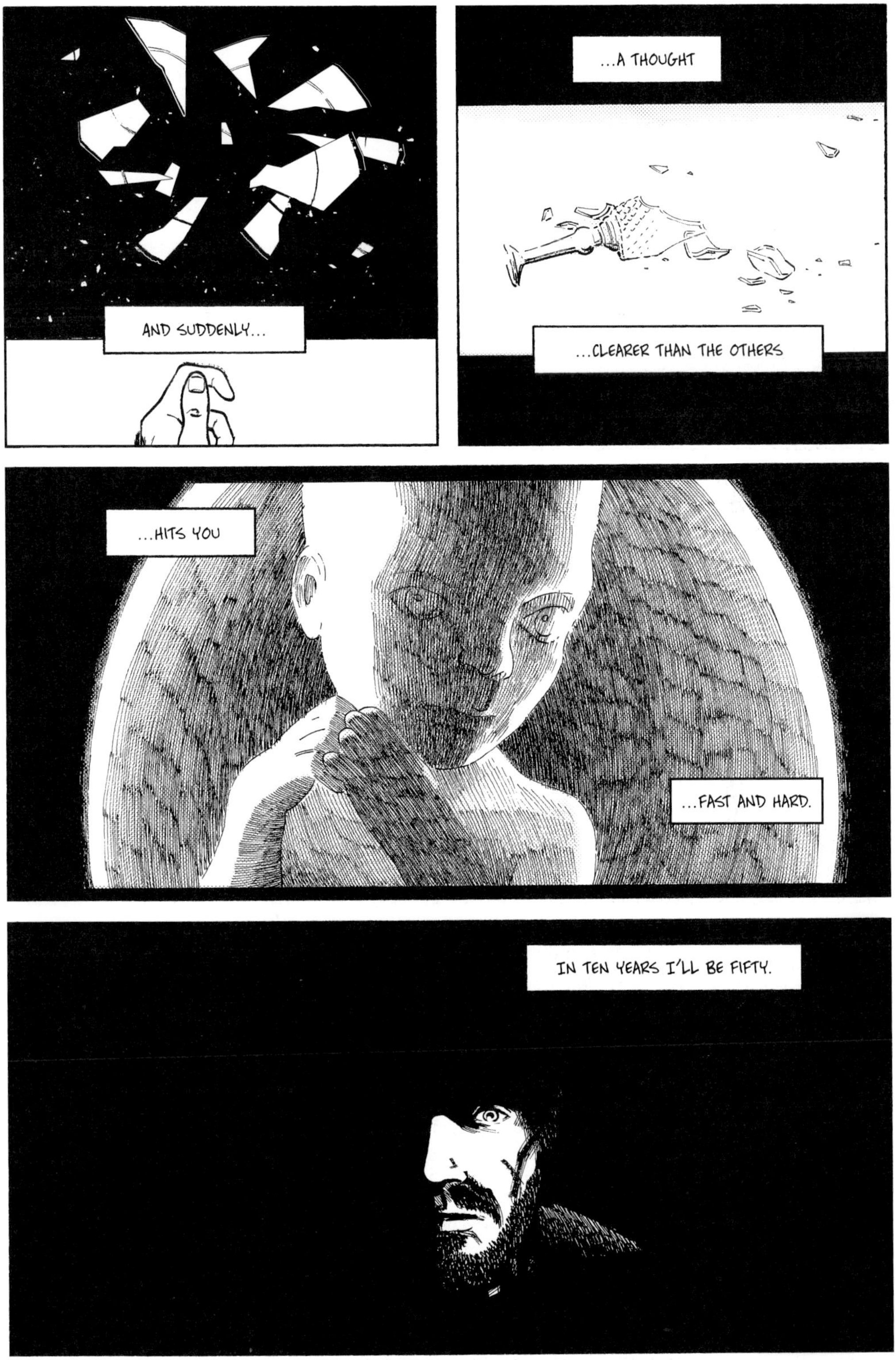
AND SUDDENLY...
...A THOUGHT
...CLEARER THAN THE OTHERS
...HITS YOU
...FAST AND HARD.
IN TEN YEARS I'LL BE FIFTY.

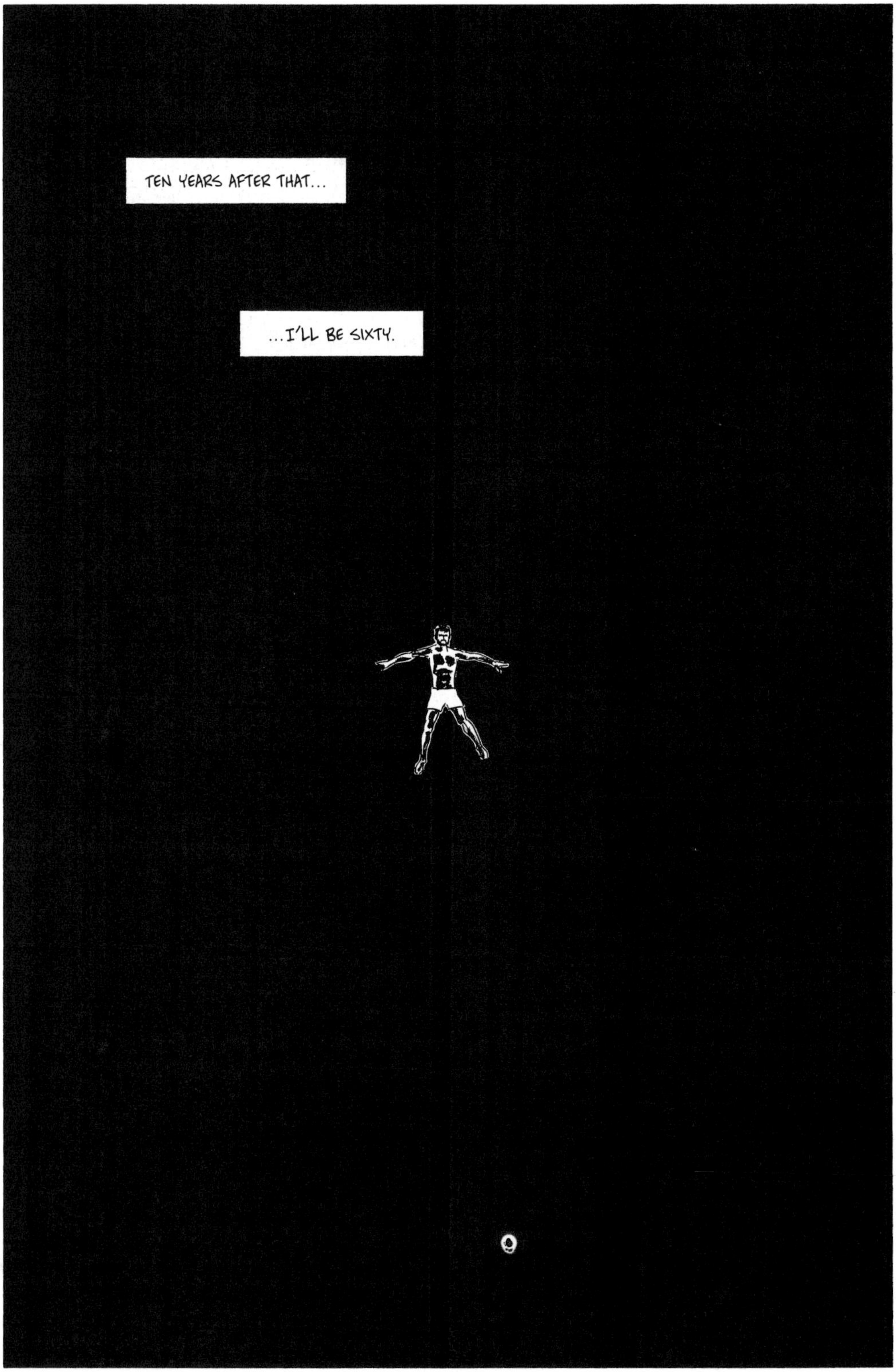
TEN YEARS AFTER THAT...
...I'LL BE SIXTY.

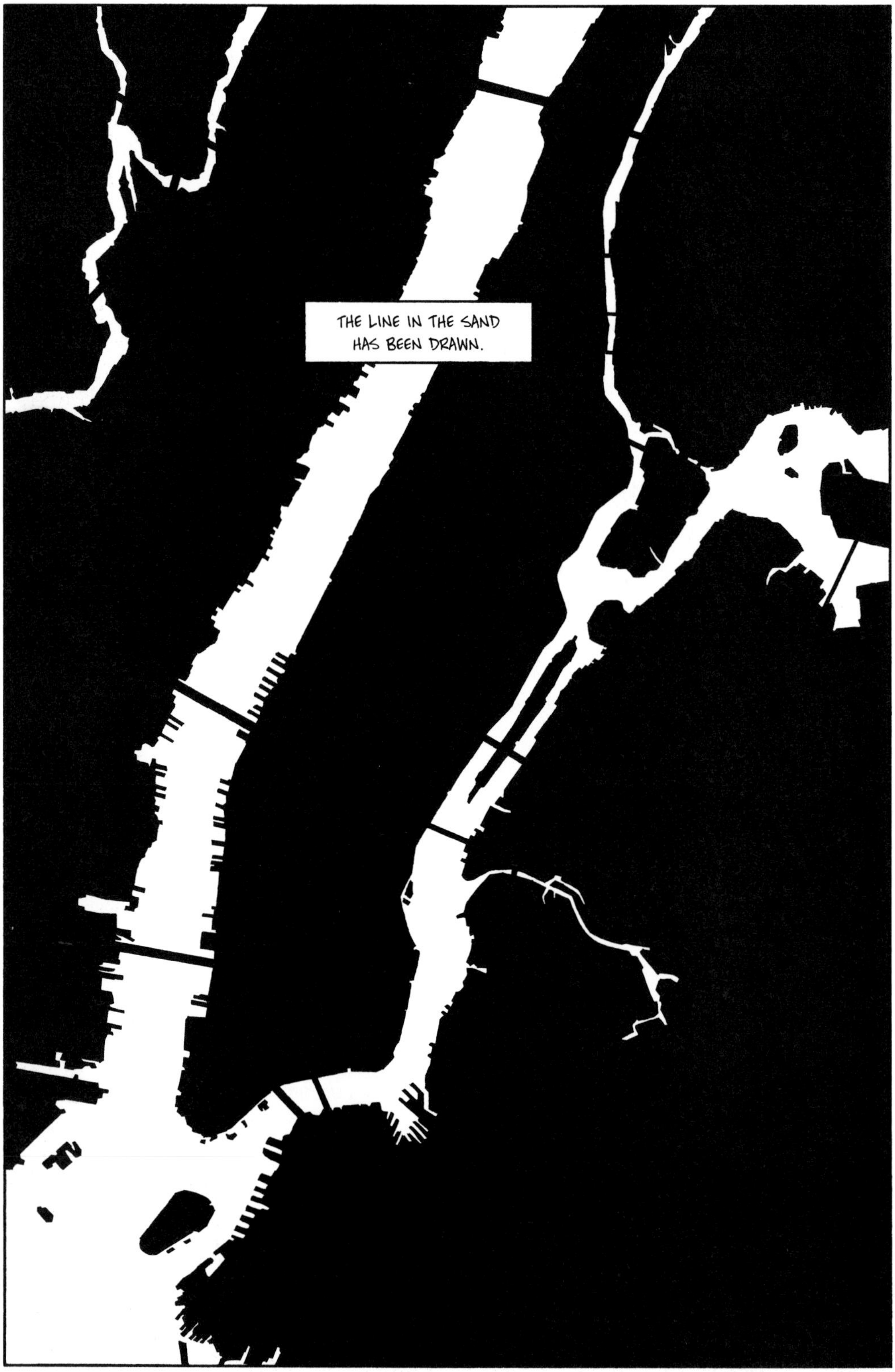
THE LINE IN THE SAND HAS BEEN DRAWN.

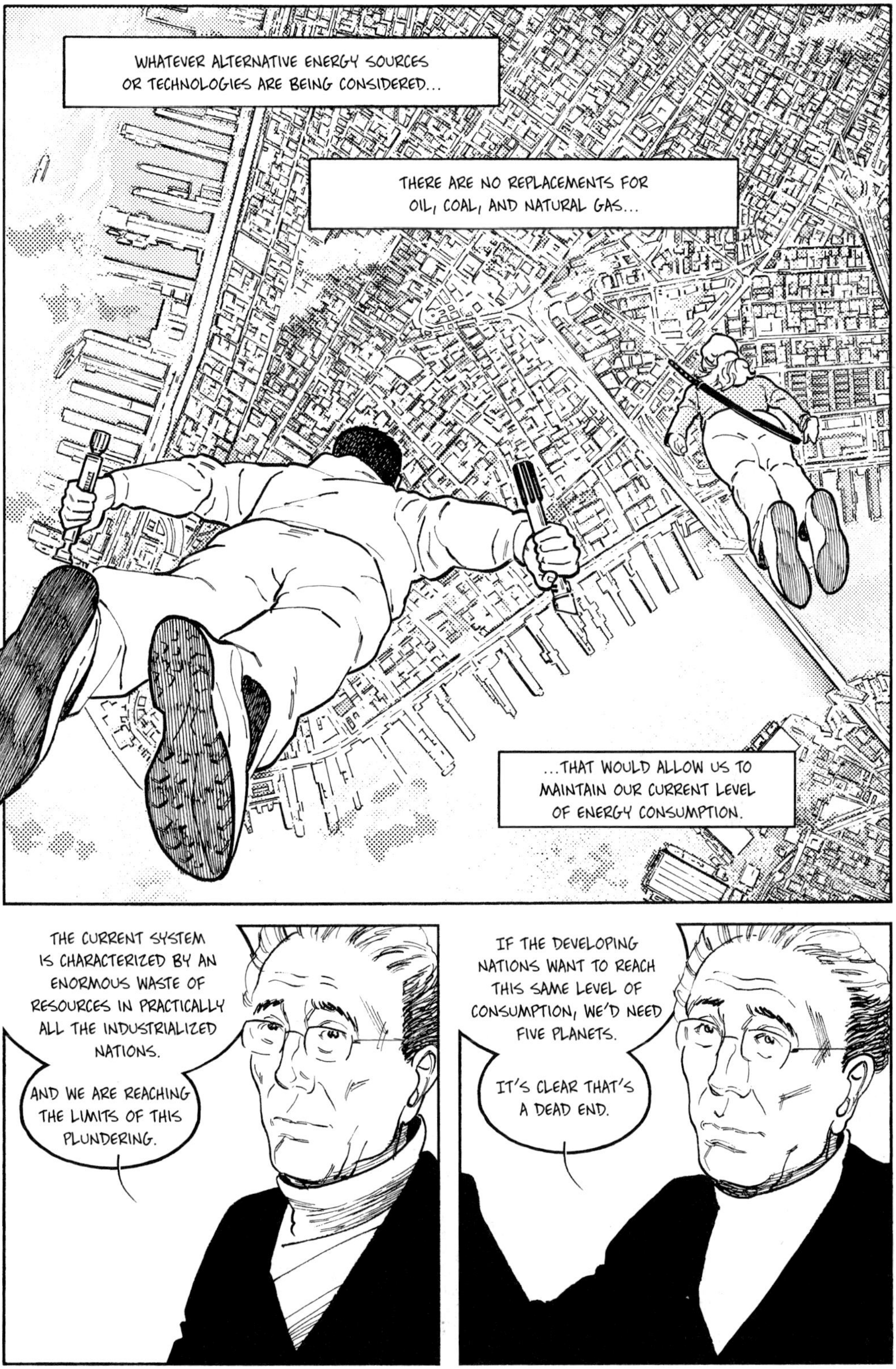
WHATEVER ALTERNATIVE ENERGY SOURCES OR TECHNOLOGIES ARE BEING CONSIDERED...
THERE ARE NO REPLACEMENTS FOR OIL, COAL, AND NATURAL GAS...
...THAT WOULD ALLOW US TO MAINTAIN OUR CURRENT LEVEL OF ENERGY CONSUMPTION.
THE CURRENT SYSTEM IS CHARACTERIZED BY AN ENORMOUS WASTE OF RESOURCES IN PRACTICALLY ALL THE INDUSTRIALIZED NATIONS.
AND WE ARE REACHING THE LIMITS OF THIS PLUNDERING.
IF THE DEVELOPING NATIONS WANT TO REACH THIS SAME LEVEL OF CONSUMPTION, WE'D NEED FIVE PLANETS.
IT'S CLEAR THAT'S A DEAD END.

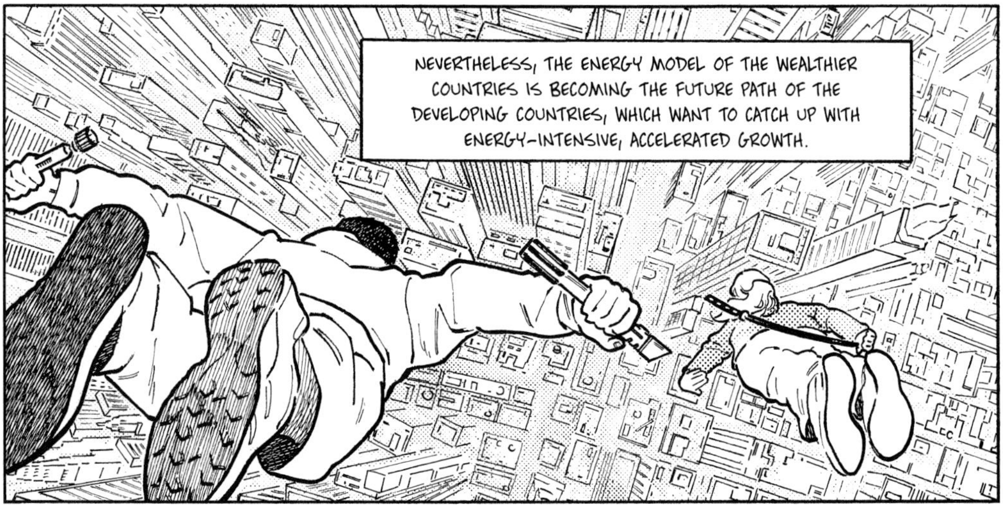
NEVERTHELESS, THE ENERGY MODEL OF THE WEALTHIER COUNTRIES IS BECOMING THE FUTURE PATH OF THE DEVELOPING COUNTRIES, WHICH WANT TO CATCH UP WITH ENERGY-INTENSIVE, ACCELERATED GROWTH.

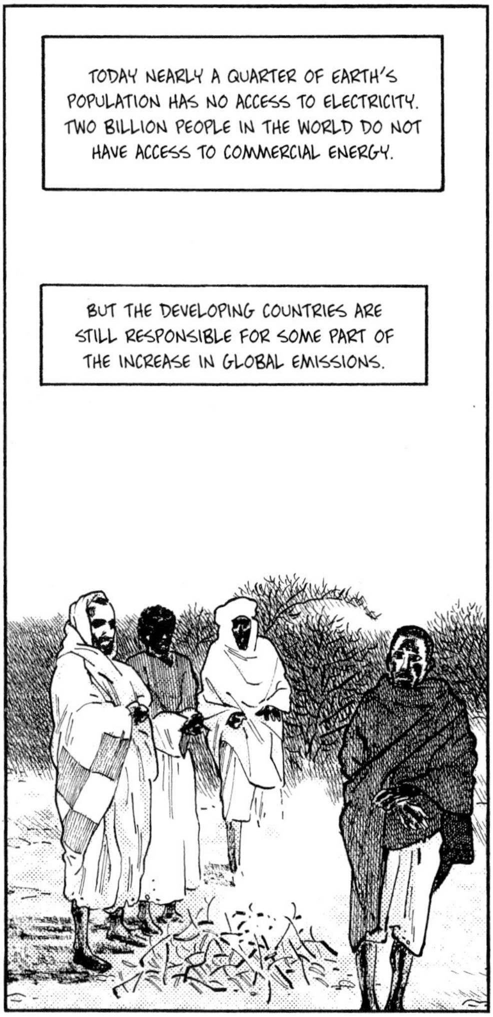
TODAY NEARLY A QUARTER OF EARTH'S POPULATION HAS NO ACCESS TO ELECTRICITY. TWO BILLION PEOPLE IN THE WORLD DO NOT HAVE ACCESS TO COMMERCIAL ENERGY.
BUT THE DEVELOPING COUNTRIES ARE STILL RESPONSIBLE FOR SOME PART OF THE INCREASE IN GLOBAL EMISSIONS.

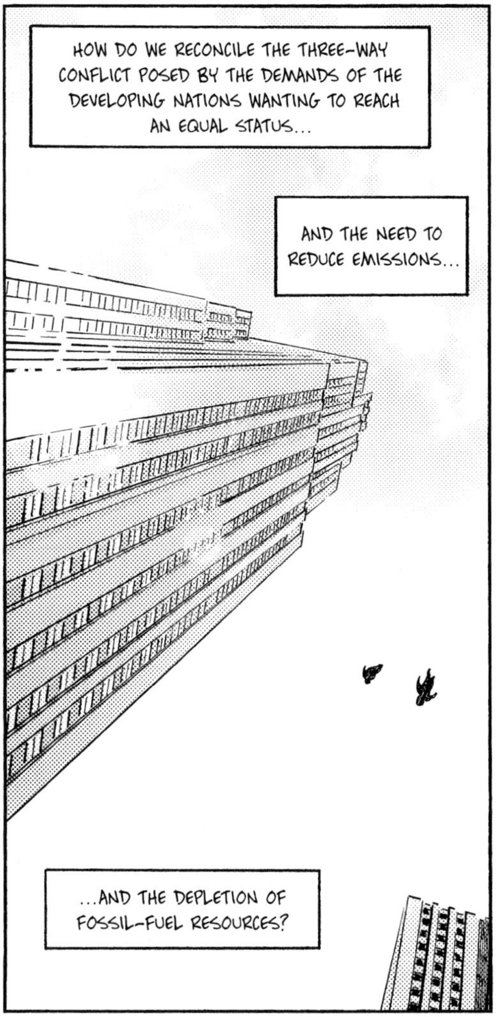
HOW DO WE RECONCILE THE THREE-WAY CONFLICT POSED BY THE DEMANDS OF THE DEVELOPING NATIONS WANTING TO REACH AN EQUAL STATUS...
AND THE NEED TO REDUCE EMISSIONS...
...AND THE DEPLETION OF FOSSIL-FUEL RESOURCES?

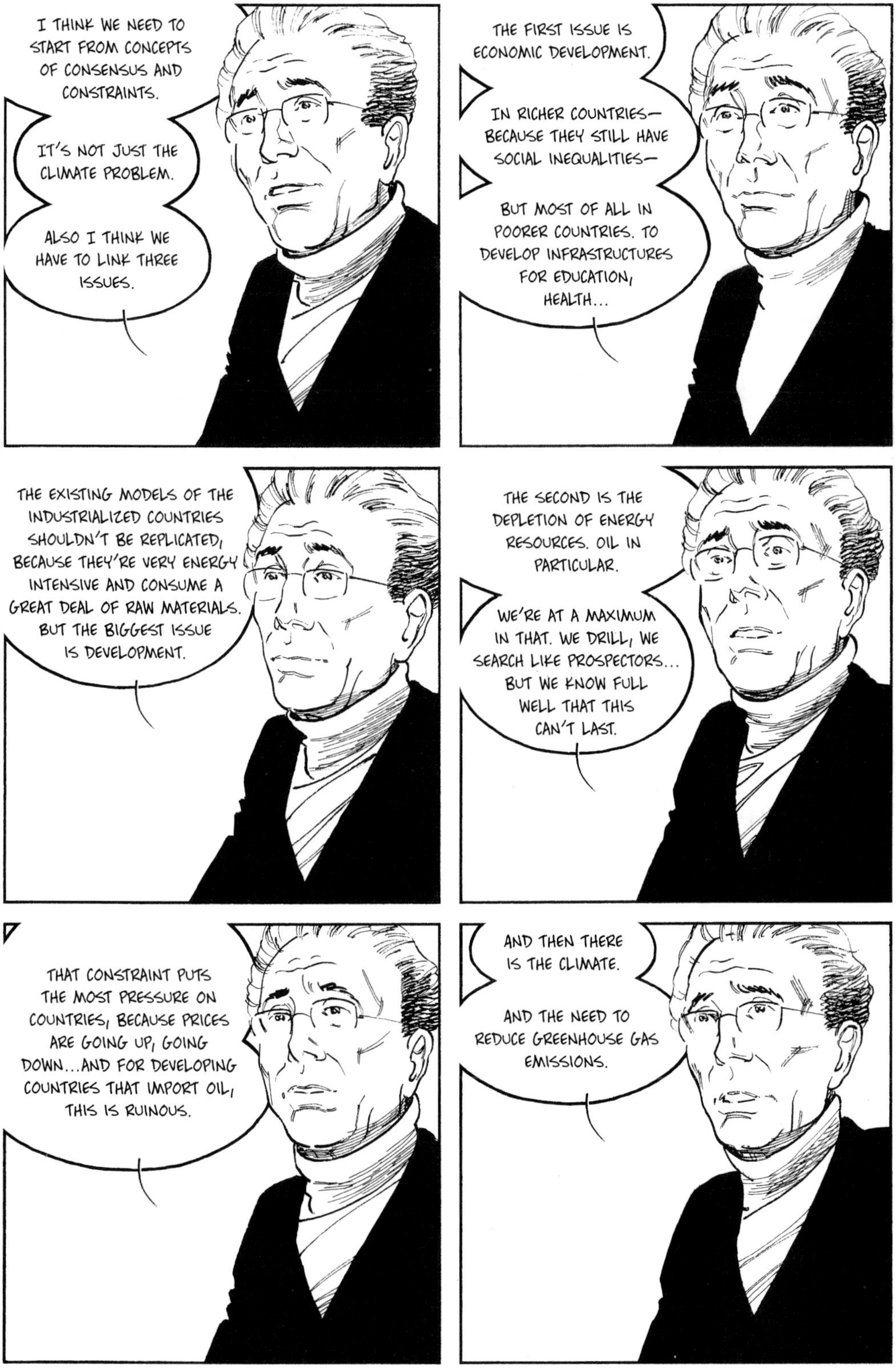
I THINK WE NEED TO START FROM CONCEPTS OF CONSENSUS AND CONSTRAINTS.
IT'S NOT JUST THE CLIMATE PROBLEM.
ALSO I THINK WE HAVE TO LINK THREE ISSUES.
THE FIRST ISSUE IS ECONOMIC DEVELOPMENT.
IN RICHER COUNTRIES—BECAUSE THEY STILL HAVE SOCIAL INEQUALITIES—
BUT MOST OF ALL IN POORER COUNTRIES. TO DEVELOP INFRASTRUCTURES FOR EDUCATION, HEALTH...
THE EXISTING MODELS OF THE INDUSTRIALIZED COUNTRIES SHOULDN'T BE REPLICATED, BECAUSE THEY'RE VERY ENERGY INTENSIVE AND CONSUME A GREAT DEAL OF RAW MATERIALS. BUT THE BIGGEST ISSUE IS DEVELOPMENT.
THE SECOND IS THE DEPLETION OF ENERGY RESOURCES. OIL IN PARTICULAR.
WE'RE AT A MAXIMUM IN THAT. WE DRILL, WE SEARCH LIKE PROSPECTORS... BUT WE KNOW FULL WELL THAT THIS CAN'T LAST.
THAT CONSTRAINT PUTS THE MOST PRESSURE ON COUNTRIES, BECAUSE PRICES ARE GOING UP, GOING DOWN...AND FOR DEVELOPING COUNTRIES THAT IMPORT OIL, THIS IS RUINOUS.
AND THEN THERE IS THE CLIMATE.
AND THE NEED TO REDUCE GREENHOUSE GAS EMISSIONS.

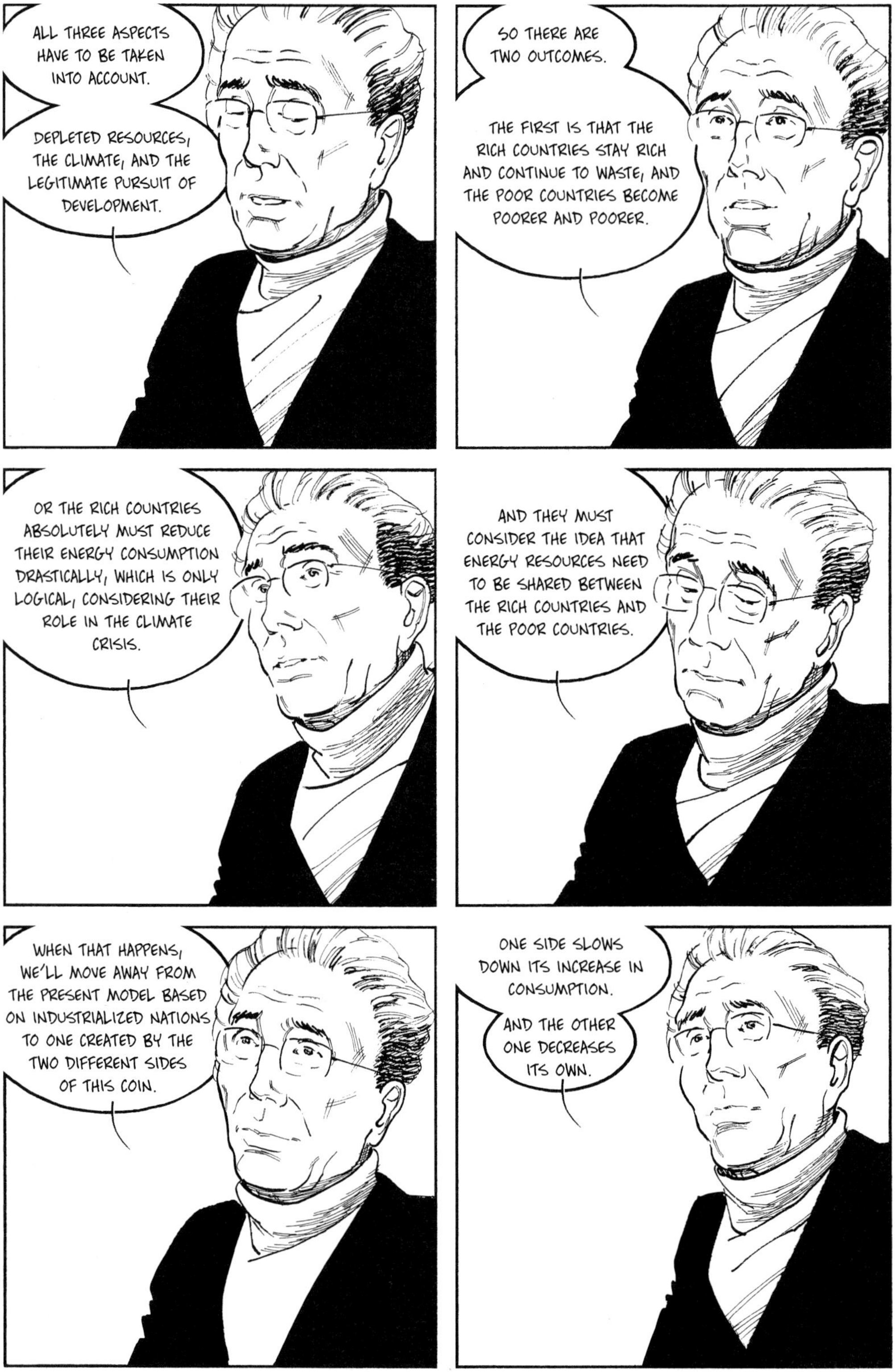
ALL THREE ASPECTS HAVE TO BE TAKEN INTO ACCOUNT.
DEPLETED RESOURCES, THE CLIMATE, AND THE LEGITIMATE PURSUIT OF DEVELOPMENT.
SO THERE ARE TWO OUTCOMES.
THE FIRST IS THAT THE RICH COUNTRIES STAY RICH AND CONTINUE TO WASTE, AND THE POOR COUNTRIES BECOME POORER AND POORER.
OR THE RICH COUNTRIES ABSOLUTELY MUST REDUCE THEIR ENERGY CONSUMPTION DRASTICALLY, WHICH IS ONLY LOGICAL, CONSIDERING THEIR ROLE IN THE CLIMATE CRISIS.
AND THEY MUST CONSIDER THE IDEA THAT ENERGY RESOURCES NEED TO BE SHARED BETWEEN THE RICH COUNTRIES AND THE POOR COUNTRIES.
WHEN THAT HAPPENS, WE'LL MOVE AWAY FROM THE PRESENT MODEL BASED ON INDUSTRIALIZED NATIONS TO ONE CREATED BY THE TWO DIFFERENT SIDES OF THIS COIN.
ONE SIDE SLOWS DOWN ITS INCREASE IN CONSUMPTION.
AND THE OTHER ONE DECREASES ITS OWN.

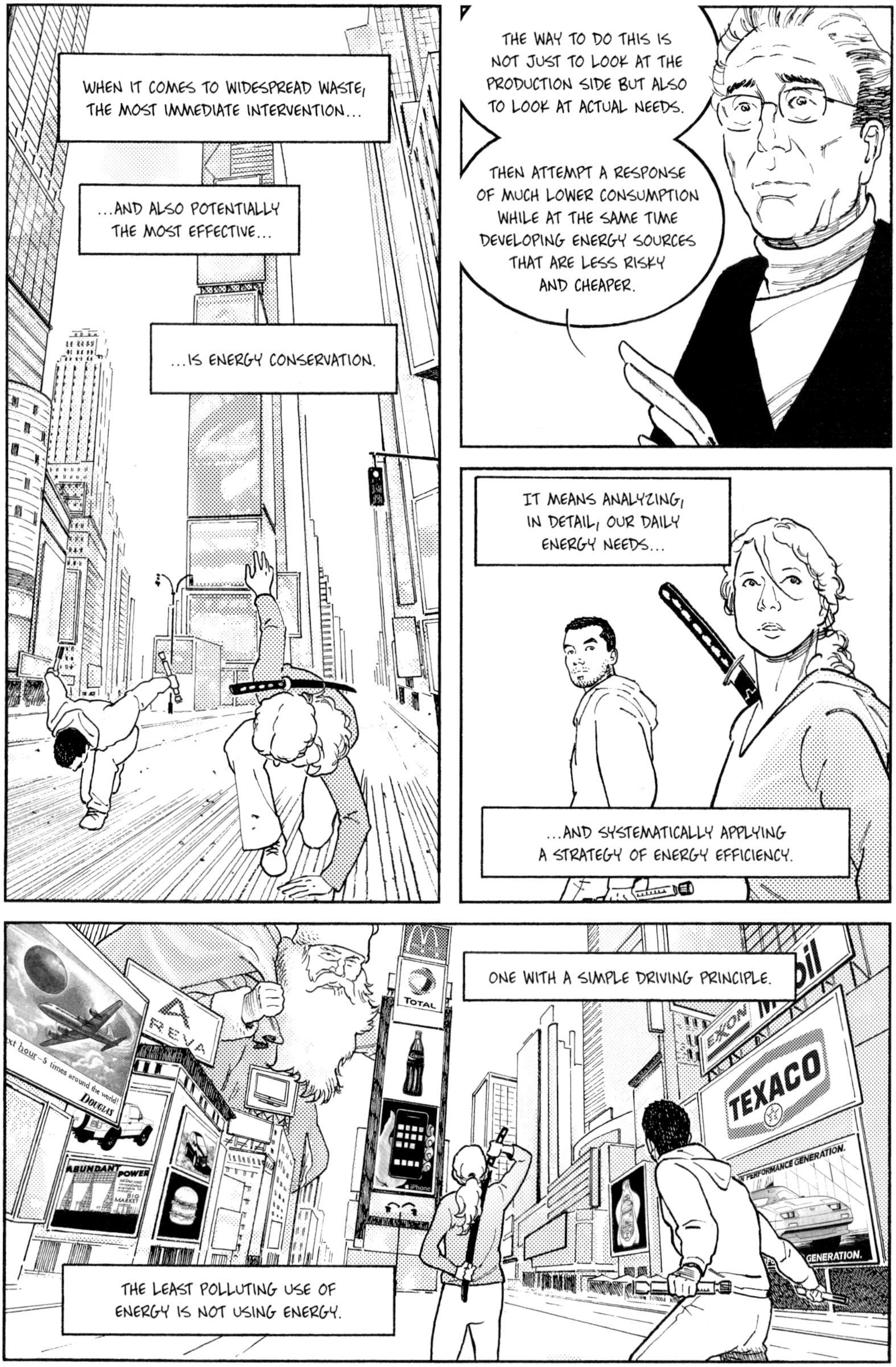

WHEN IT COMES TO WIDESPREAD WASTE, THE MOST IMMEDIATE INTERVENTION...
...AND ALSO POTENTIALLY THE MOST EFFECTIVE...
...IS ENERGY CONSERVATION.
THE WAY TO DO THIS IS NOT JUST TO LOOK AT THE PRODUCTION SIDE BUT ALSO TO LOOK AT ACTUAL NEEDS.
THEN ATTEMPT A RESPONSE OF MUCH LOWER CONSUMPTION WHILE AT THE SAME TIME DEVELOPING ENERGY SOURCES THAT ARE LESS RISKY AND CHEAPER.
IT MEANS ANALYZING, IN DETAIL, OUR DAILY ENERGY NEEDS...
...AND SYSTEMATICALLY APPLYING A STRATEGY OF ENERGY EFFICIENCY.
ONE WITH A SIMPLE DRIVING PRINCIPLE.
DOUGLAS
ABUNDANT POWER
TOTAL
iPhone
TEXACO
PERFORMANCE GENERATION.
GENERATION.
THE LEAST POLLUTING USE OF ENERGY IS NOT USING ENERGY.

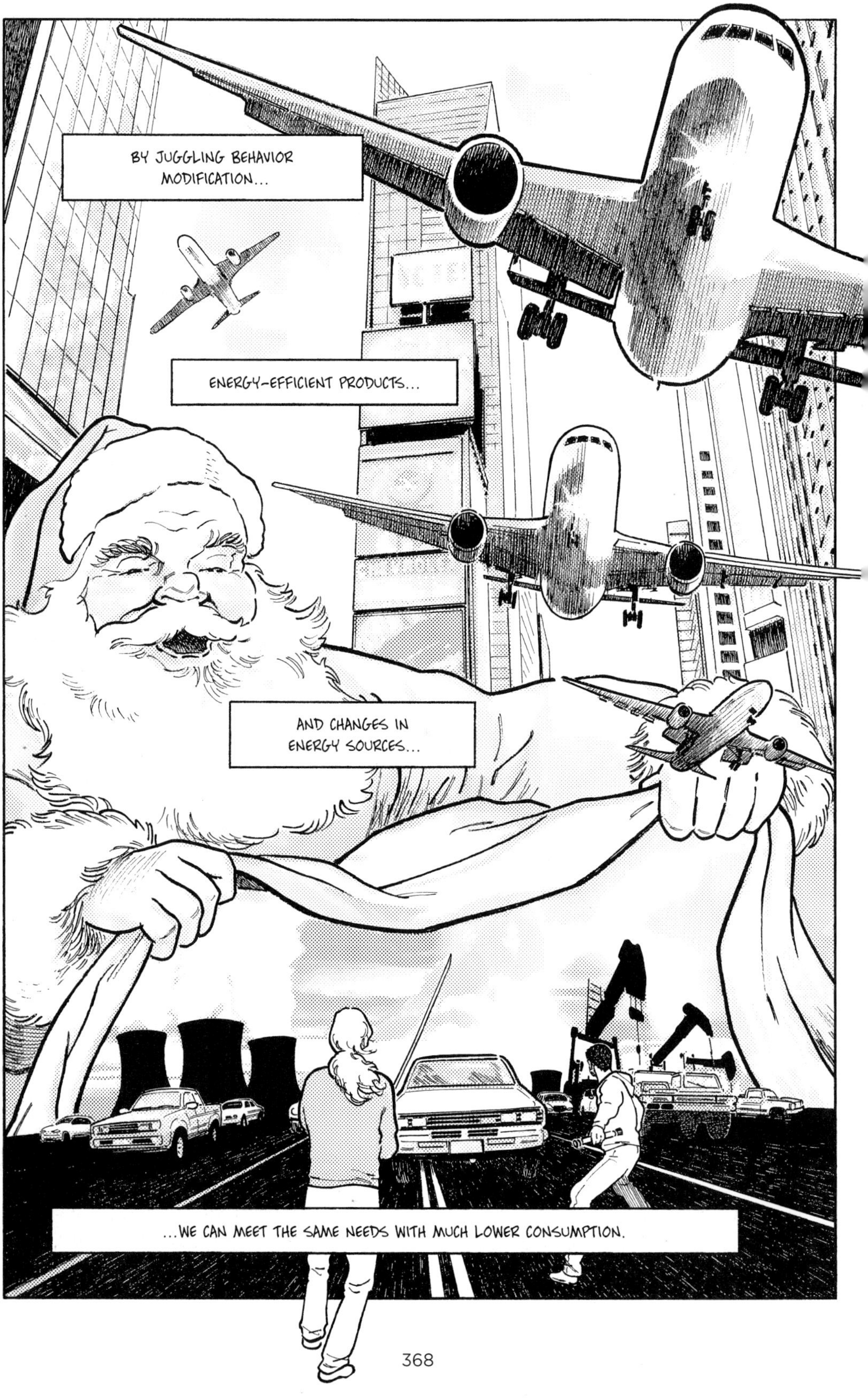
BY JUGGLING BEHAVIOR MODIFICATION...
ENERGY-EFFICIENT PRODUCTS...
AND CHANGES IN ENERGY SOURCES...
...WE CAN MEET THE SAME NEEDS WITH MUCH LOWER CONSUMPTION.

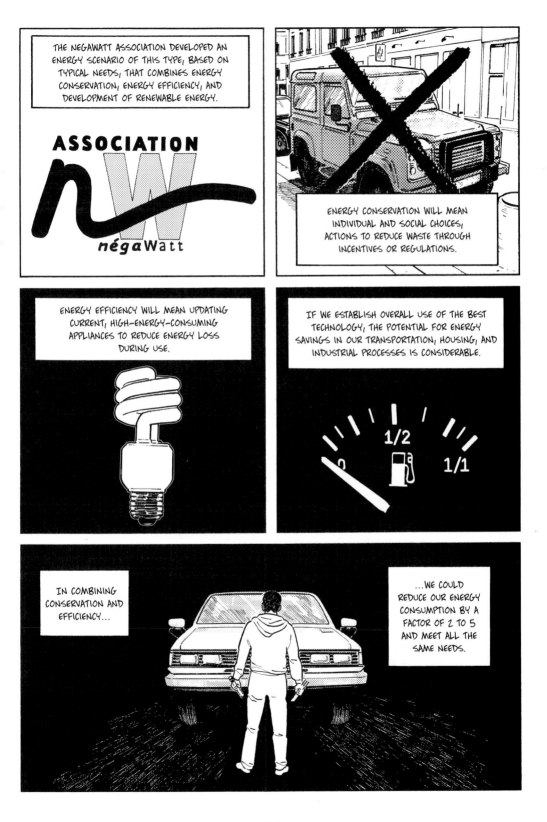

THE NEGAWATT ASSOCIATION DEVELOPED AN ENERGY SCENARIO OF THIS TYPE, BASED ON TYPICAL NEEDS, THAT COMBINES ENERGY CONSERVATION, ENERGY EFFICIENCY, AND DEVELOPMENT OF RENEWABLE ENERGY.

ASSOCIATION
négaWatt

ENERGY CONSERVATION WILL MEAN INDIVIDUAL AND SOCIAL CHOICES, ACTIONS TO REDUCE WASTE THROUGH INCENTIVES OR REGULATIONS.

ENERGY EFFICIENCY WILL MEAN UPDATING CURRENT, HIGH-ENERGY-CONSUMING APPLIANCES TO REDUCE ENERGY LOSS DURING USE.

IF WE ESTABLISH OVERALL USE OF THE BEST TECHNOLOGY, THE POTENTIAL FOR ENERGY SAVINGS IN OUR TRANSPORTATION, HOUSING, AND INDUSTRIAL PROCESSES IS CONSIDERABLE.

1/2
1/1

IN COMBINING CONSERVATION AND EFFICIENCY...

...WE COULD REDUCE OUR ENERGY CONSUMPTION BY A FACTOR OF 2 TO 5 AND MEET ALL THE SAME NEEDS.

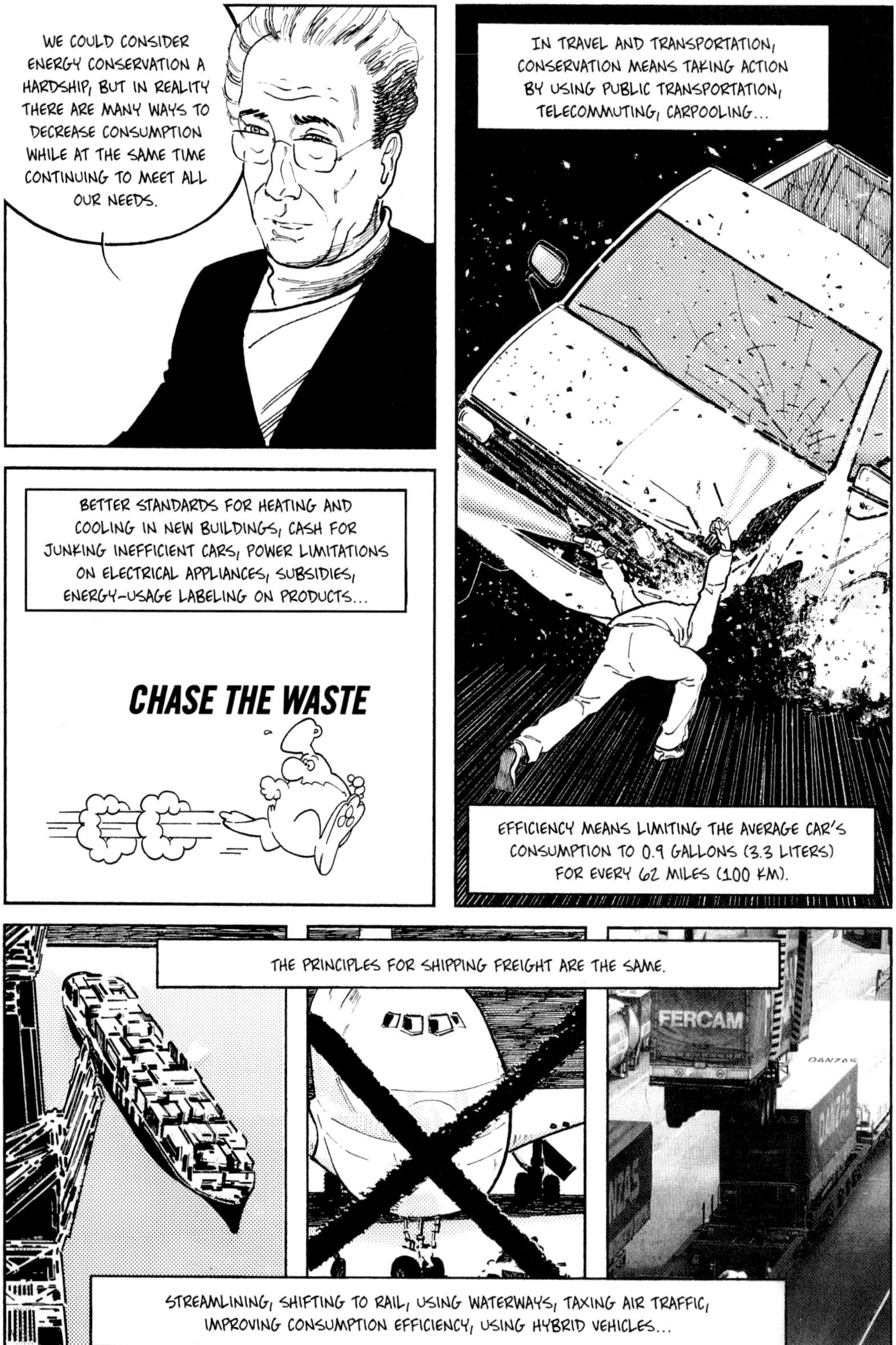
WE COULD CONSIDER ENERGY CONSERVATION A HARDSHIP, BUT IN REALITY THERE ARE MANY WAYS TO DECREASE CONSUMPTION WHILE AT THE SAME TIME CONTINUING TO MEET ALL OUR NEEDS.
IN TRAVEL AND TRANSPORTATION, CONSERVATION MEANS TAKING ACTION BY USING PUBLIC TRANSPORTATION, TELECOMMUTING, CARPOOLING...
BETTER STANDARDS FOR HEATING AND COOLING IN NEW BUILDINGS, CASH FOR JUNKING INEFFICIENT CARS, POWER LIMITATIONS ON ELECTRICAL APPLIANCES, SUBSIDIES, ENERGY-USAGE LABELING ON PRODUCTS...
CHASE THE WASTE
EFFICIENCY MEANS LIMITING THE AVERAGE CAR'S CONSUMPTION TO 0.9 GALLONS (3.3 LITERS) FOR EVERY 62 MILES (100 KM).
THE PRINCIPLES FOR SHIPPING FREIGHT ARE THE SAME.
FERCAM
DANZAS
STREAMLINING, SHIFTING TO RAIL, USING WATERWAYS, TAXING AIR TRAFFIC, IMPROVING CONSUMPTION EFFICIENCY, USING HYBRID VEHICLES...

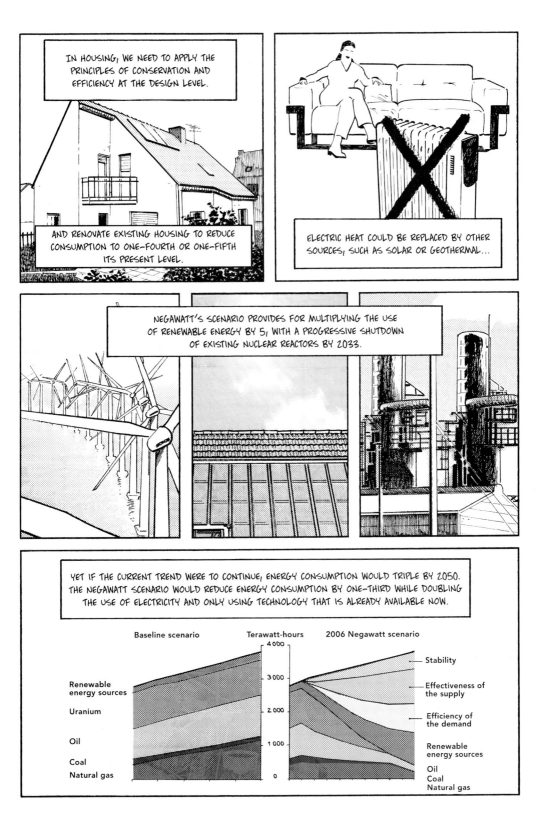

IN HOUSING, WE NEED TO APPLY THE PRINCIPLES OF CONSERVATION AND EFFICIENCY AT THE DESIGN LEVEL.

AND RENOVATE EXISTING HOUSING TO REDUCE CONSUMPTION TO ONE-FOURTH OR ONE-FIFTH ITS PRESENT LEVEL.

ELECTRIC HEAT COULD BE REPLACED BY OTHER SOURCES, SUCH AS SOLAR OR GEOTHERMAL...

NEGAWATT'S SCENARIO PROVIDES FOR MULTIPLYING THE USE OF RENEWABLE ENERGY BY 5, WITH A PROGRESSIVE SHUTDOWN OF EXISTING NUCLEAR REACTORS BY 2033.

YET IF THE CURRENT TREND WERE TO CONTINUE, ENERGY CONSUMPTION WOULD TRIPLE BY 2050. THE NEGAWATT SCENARIO WOULD REDUCE ENERGY CONSUMPTION BY ONE-THIRD WHILE DOUBLING THE USE OF ELECTRICITY AND ONLY USING TECHNOLOGY THAT IS ALREADY AVAILABLE NOW.

Baseline scenario   Terawatt-hours   2006 Negawatt scenario
4 000

Renewable energy sources   3 000   Stability

Uranium   Effectiveness of the supply

2 000   Efficiency of the demand

Oil   1 000

Coal   Renewable energy sources

Natural gas   0   Oil
                                                         Coal
                                                         Natural gas

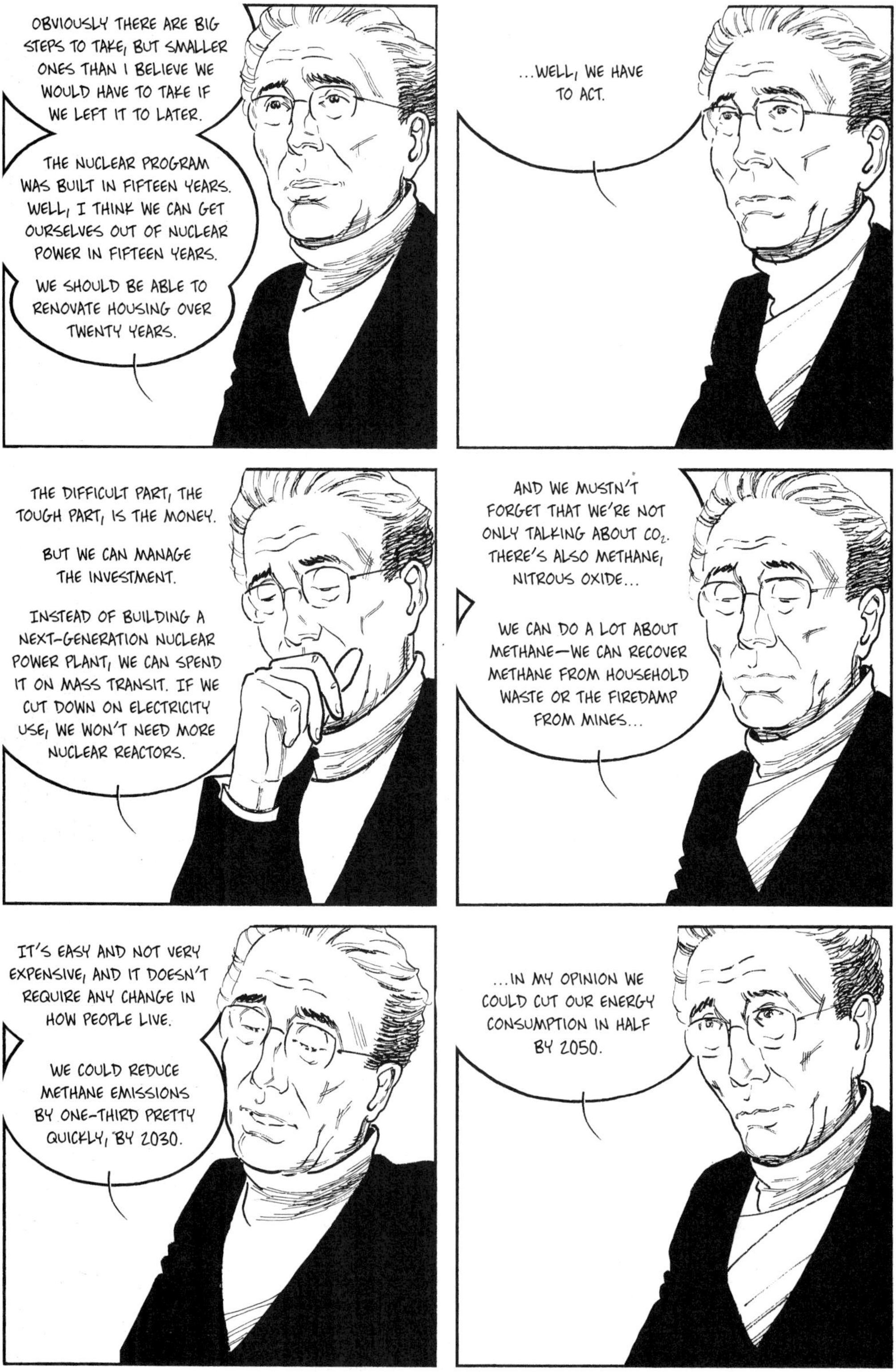
OBVIOUSLY THERE ARE BIG STEPS TO TAKE, BUT SMALLER ONES THAN I BELIEVE WE WOULD HAVE TO TAKE IF WE LEFT IT TO LATER.
THE NUCLEAR PROGRAM WAS BUILT IN FIFTEEN YEARS. WELL, I THINK WE CAN GET OURSELVES OUT OF NUCLEAR POWER IN FIFTEEN YEARS.
WE SHOULD BE ABLE TO RENOVATE HOUSING OVER TWENTY YEARS.
...WELL, WE HAVE TO ACT.
THE DIFFICULT PART, THE TOUGH PART, IS THE MONEY.
BUT WE CAN MANAGE THE INVESTMENT.
INSTEAD OF BUILDING A NEXT-GENERATION NUCLEAR POWER PLANT, WE CAN SPEND IT ON MASS TRANSIT. IF WE CUT DOWN ON ELECTRICITY USE, WE WON'T NEED MORE NUCLEAR REACTORS.
AND WE MUSTN'T FORGET THAT WE'RE NOT ONLY TALKING ABOUT CO_2. THERE'S ALSO METHANE, NITROUS OXIDE...
WE CAN DO A LOT ABOUT METHANE—WE CAN RECOVER METHANE FROM HOUSEHOLD WASTE OR THE FIREDAMP FROM MINES...
IT'S EASY AND NOT VERY EXPENSIVE, AND IT DOESN'T REQUIRE ANY CHANGE IN HOW PEOPLE LIVE.
WE COULD REDUCE METHANE EMISSIONS BY ONE-THIRD PRETTY QUICKLY, BY 2030.
...IN MY OPINION WE COULD CUT OUR ENERGY CONSUMPTION IN HALF BY 2050.

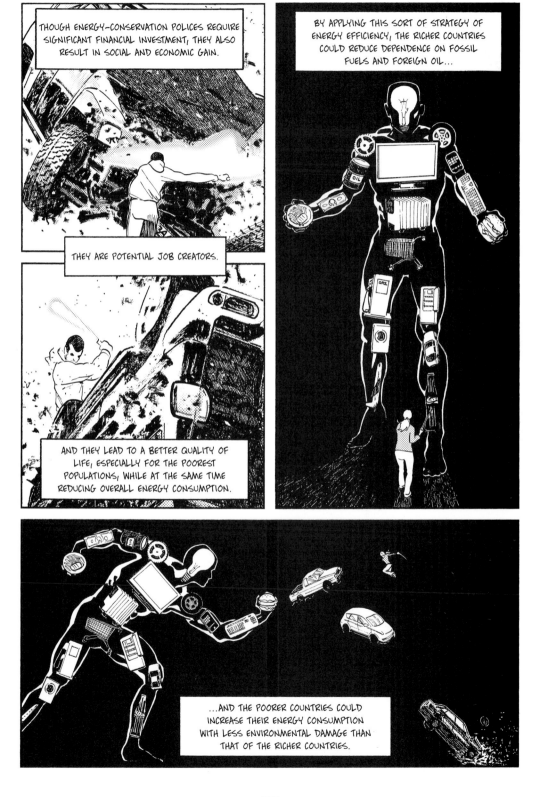

THOUGH ENERGY-CONSERVATION POLICES REQUIRE SIGNIFICANT FINANCIAL INVESTMENT, THEY ALSO RESULT IN SOCIAL AND ECONOMIC GAIN.

THEY ARE POTENTIAL JOB CREATORS.

AND THEY LEAD TO A BETTER QUALITY OF LIFE, ESPECIALLY FOR THE POOREST POPULATIONS, WHILE AT THE SAME TIME REDUCING OVERALL ENERGY CONSUMPTION.

BY APPLYING THIS SORT OF STRATEGY OF ENERGY EFFICIENCY, THE RICHER COUNTRIES COULD REDUCE DEPENDENCE ON FOSSIL FUELS AND FOREIGN OIL...

...AND THE POORER COUNTRIES COULD INCREASE THEIR ENERGY CONSUMPTION WITH LESS ENVIRONMENTAL DAMAGE THAN THAT OF THE RICHER COUNTRIES.

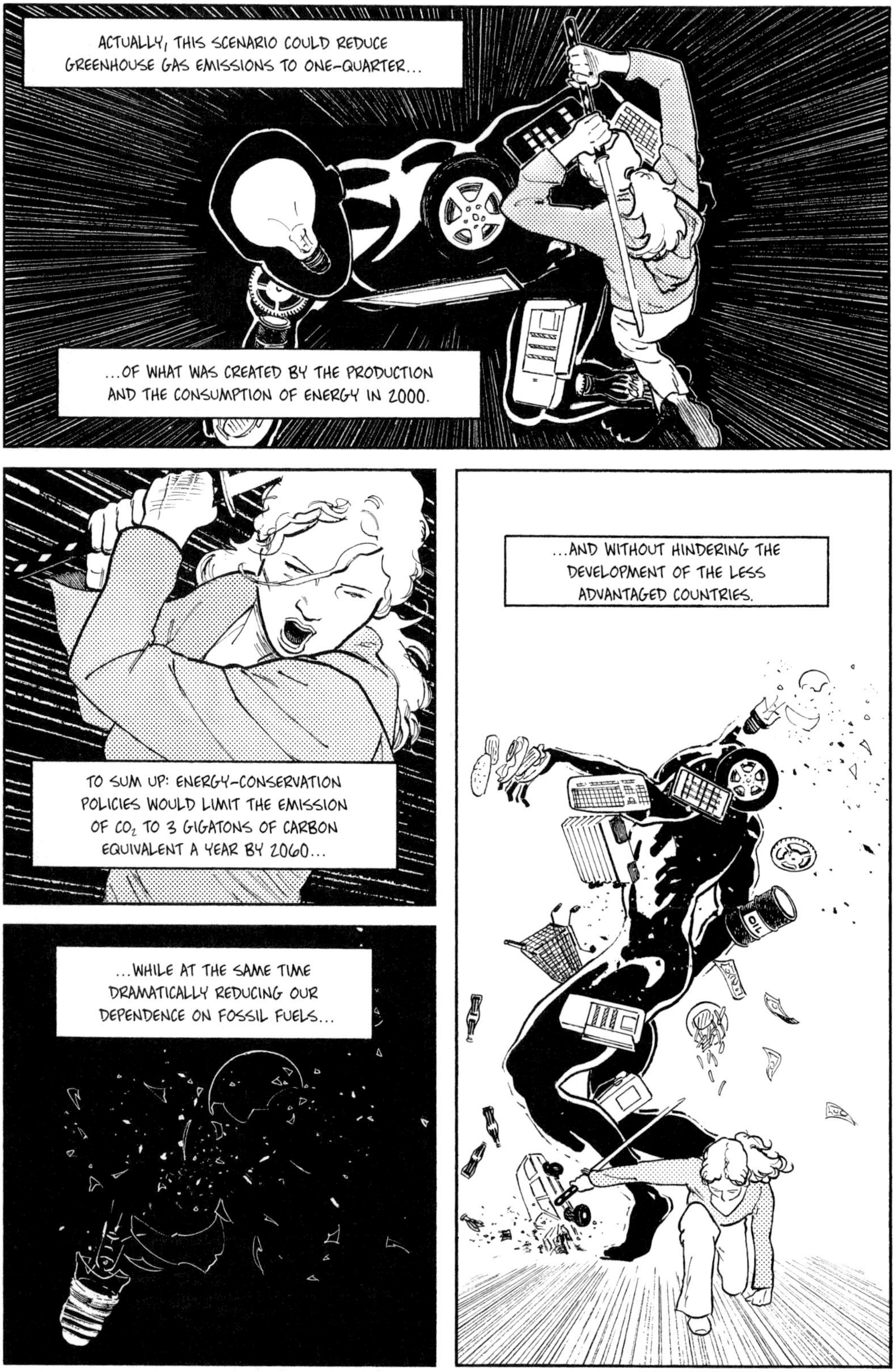
ACTUALLY, THIS SCENARIO COULD REDUCE GREENHOUSE GAS EMISSIONS TO ONE-QUARTER...
...OF WHAT WAS CREATED BY THE PRODUCTION AND THE CONSUMPTION OF ENERGY IN 2000.
TO SUM UP: ENERGY-CONSERVATION POLICIES WOULD LIMIT THE EMISSION OF CO_2 TO 3 GIGATONS OF CARBON EQUIVALENT A YEAR BY 2060...
...WHILE AT THE SAME TIME DRAMATICALLY REDUCING OUR DEPENDENCE ON FOSSIL FUELS...
...AND WITHOUT HINDERING THE DEVELOPMENT OF THE LESS ADVANTAGED COUNTRIES.
OIL

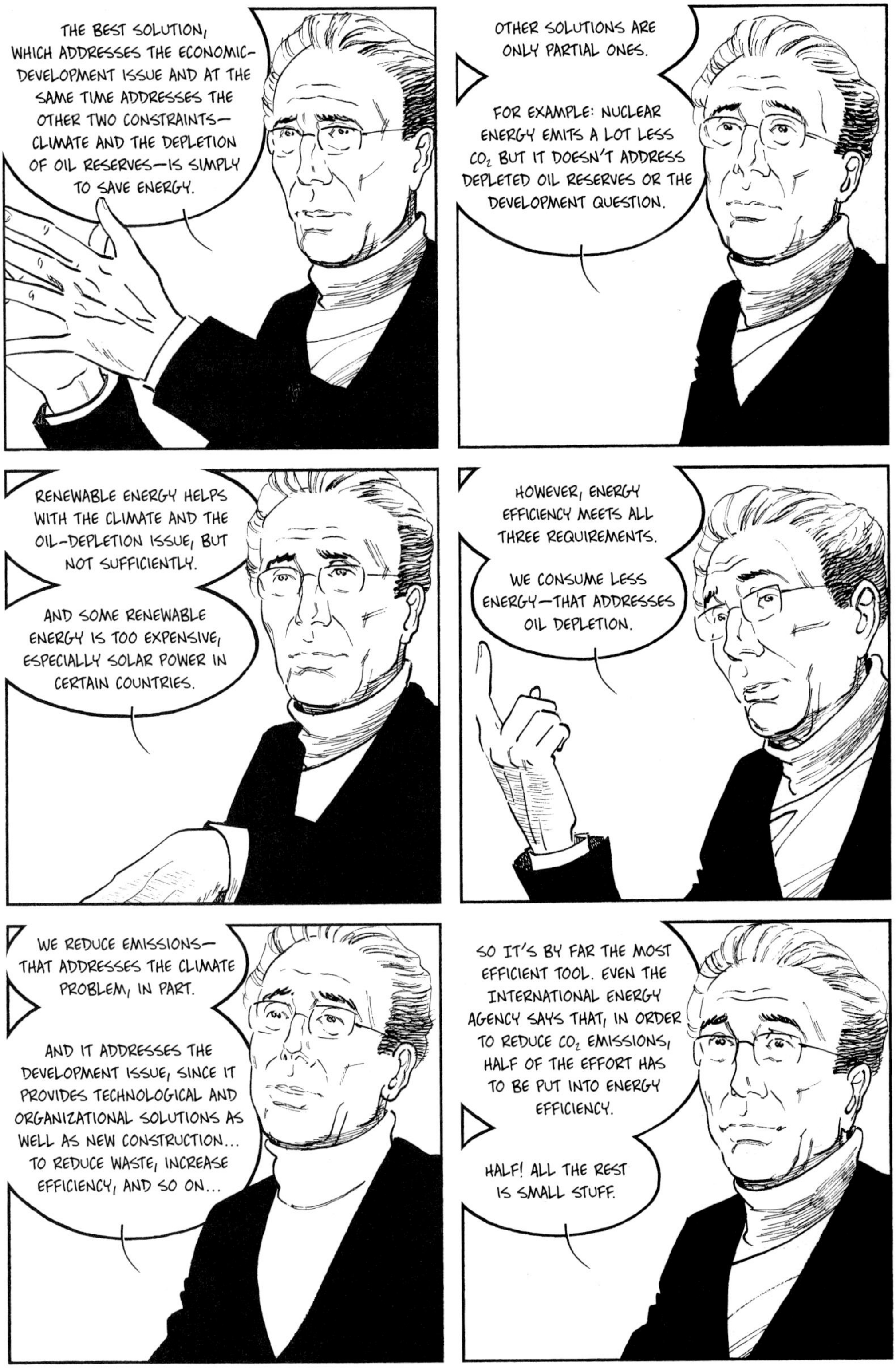
THE BEST SOLUTION, WHICH ADDRESSES THE ECONOMIC-DEVELOPMENT ISSUE AND AT THE SAME TIME ADDRESSES THE OTHER TWO CONSTRAINTS—CLIMATE AND THE DEPLETION OF OIL RESERVES—IS SIMPLY TO SAVE ENERGY.
OTHER SOLUTIONS ARE ONLY PARTIAL ONES.
FOR EXAMPLE: NUCLEAR ENERGY EMITS A LOT LESS CO_2 BUT IT DOESN'T ADDRESS DEPLETED OIL RESERVES OR THE DEVELOPMENT QUESTION.
RENEWABLE ENERGY HELPS WITH THE CLIMATE AND THE OIL-DEPLETION ISSUE, BUT NOT SUFFICIENTLY.
AND SOME RENEWABLE ENERGY IS TOO EXPENSIVE, ESPECIALLY SOLAR POWER IN CERTAIN COUNTRIES.
HOWEVER, ENERGY EFFICIENCY MEETS ALL THREE REQUIREMENTS.
WE CONSUME LESS ENERGY—THAT ADDRESSES OIL DEPLETION.
WE REDUCE EMISSIONS—THAT ADDRESSES THE CLIMATE PROBLEM, IN PART.
AND IT ADDRESSES THE DEVELOPMENT ISSUE, SINCE IT PROVIDES TECHNOLOGICAL AND ORGANIZATIONAL SOLUTIONS AS WELL AS NEW CONSTRUCTION... TO REDUCE WASTE, INCREASE EFFICIENCY, AND SO ON...
SO IT'S BY FAR THE MOST EFFICIENT TOOL. EVEN THE INTERNATIONAL ENERGY AGENCY SAYS THAT, IN ORDER TO REDUCE CO_2 EMISSIONS, HALF OF THE EFFORT HAS TO BE PUT INTO ENERGY EFFICIENCY.
HALF! ALL THE REST IS SMALL STUFF.

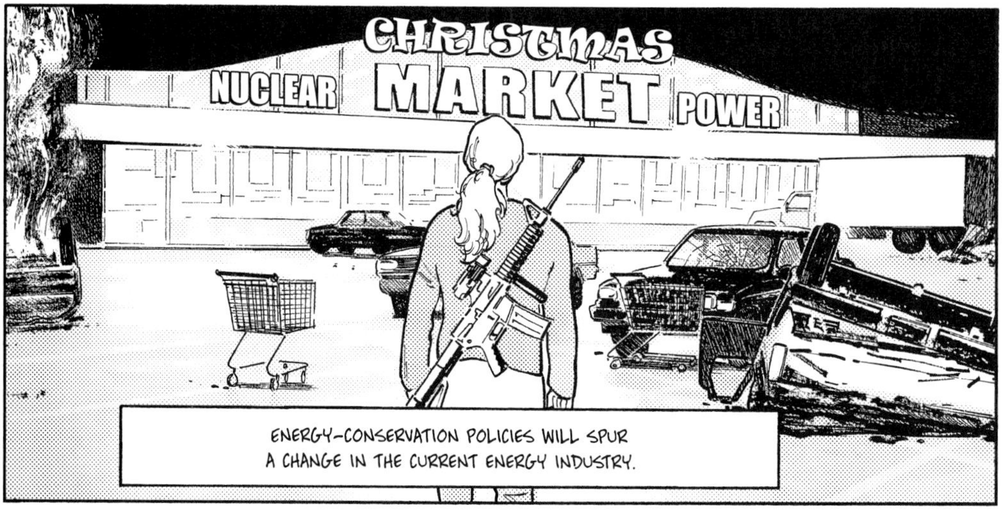
CHRISTMAS
NUCLEAR MARKET POWER
ENERGY-CONSERVATION POLICIES WILL SPUR A CHANGE IN THE CURRENT ENERGY INDUSTRY.

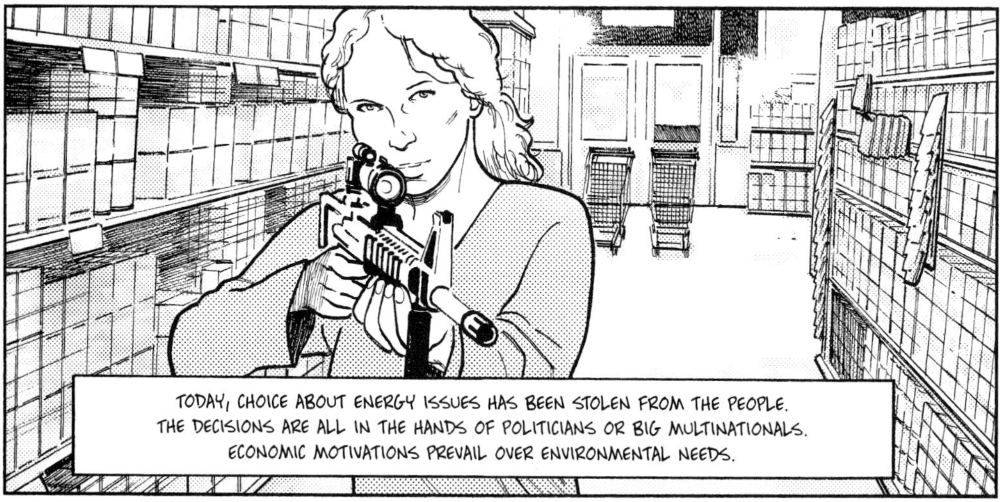
TODAY, CHOICE ABOUT ENERGY ISSUES HAS BEEN STOLEN FROM THE PEOPLE. THE DECISIONS ARE ALL IN THE HANDS OF POLITICIANS OR BIG MULTINATIONALS. ECONOMIC MOTIVATIONS PREVAIL OVER ENVIRONMENTAL NEEDS.

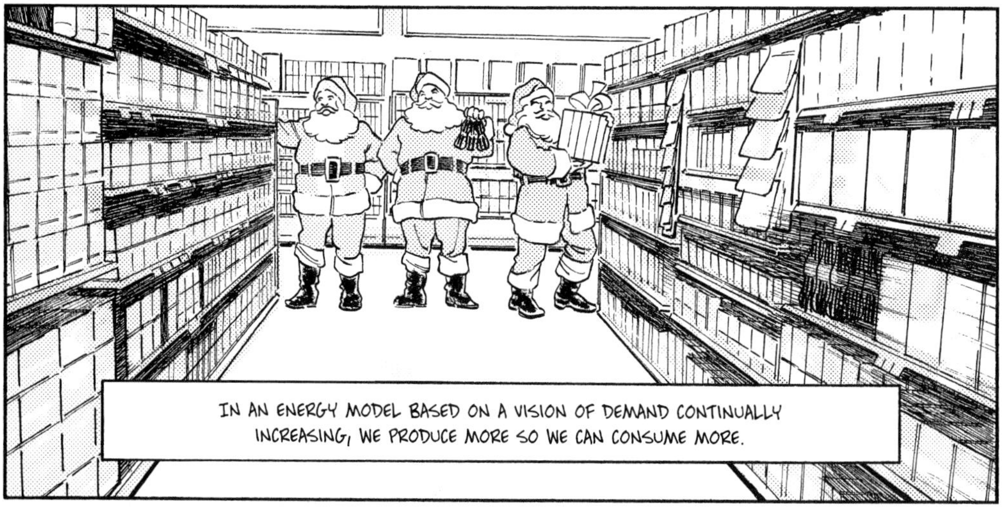
IN AN ENERGY MODEL BASED ON A VISION OF DEMAND CONTINUALLY INCREASING, WE PRODUCE MORE SO WE CAN CONSUME MORE.

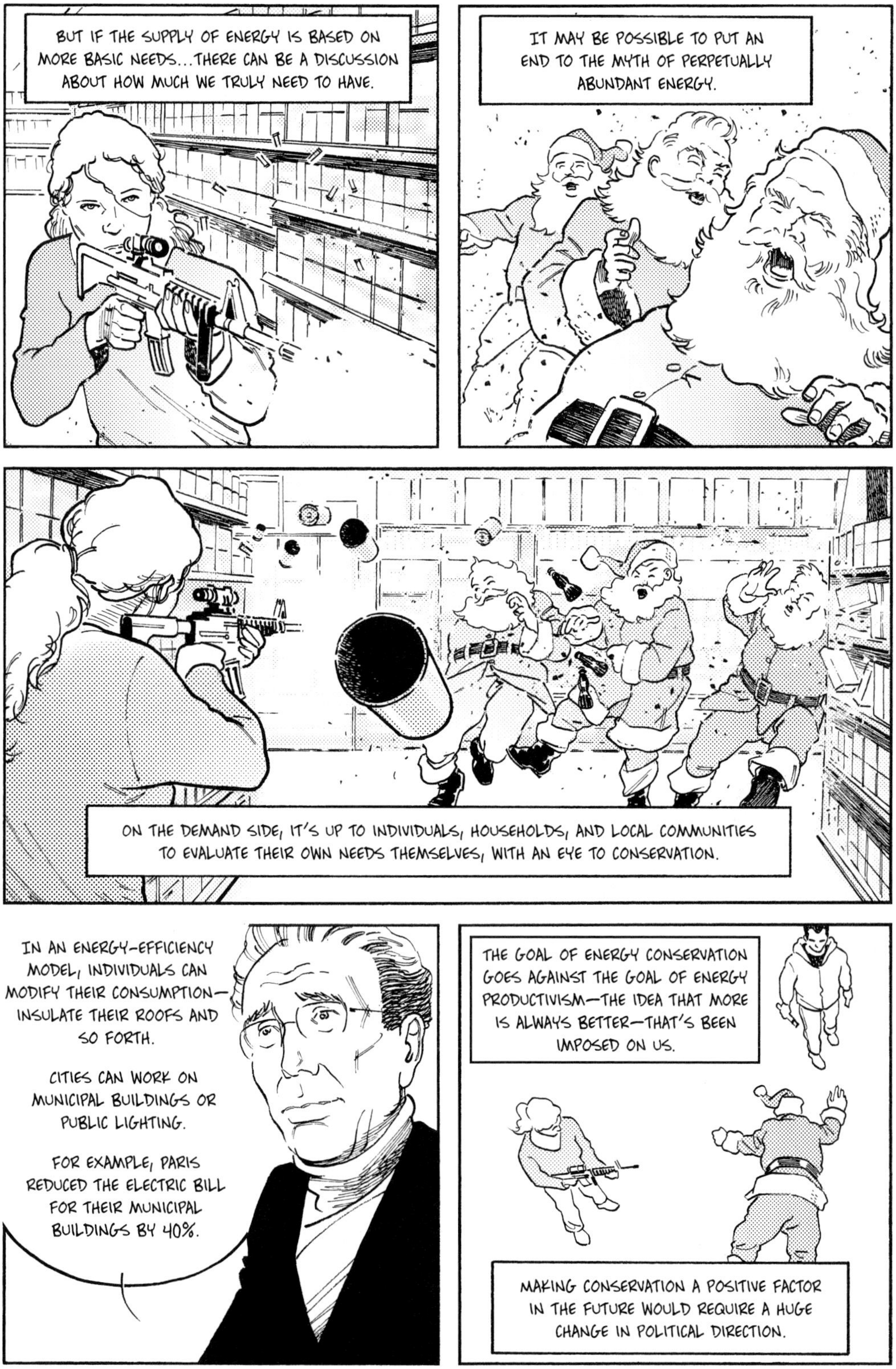
BUT IF THE SUPPLY OF ENERGY IS BASED ON MORE BASIC NEEDS...THERE CAN BE A DISCUSSION ABOUT HOW MUCH WE TRULY NEED TO HAVE.
IT MAY BE POSSIBLE TO PUT AN END TO THE MYTH OF PERPETUALLY ABUNDANT ENERGY.
ON THE DEMAND SIDE, IT'S UP TO INDIVIDUALS, HOUSEHOLDS, AND LOCAL COMMUNITIES TO EVALUATE THEIR OWN NEEDS THEMSELVES, WITH AN EYE TO CONSERVATION.
IN AN ENERGY-EFFICIENCY MODEL, INDIVIDUALS CAN MODIFY THEIR CONSUMPTION—INSULATE THEIR ROOFS AND SO FORTH.
CITIES CAN WORK ON MUNICIPAL BUILDINGS OR PUBLIC LIGHTING.
FOR EXAMPLE, PARIS REDUCED THE ELECTRIC BILL FOR THEIR MUNICIPAL BUILDINGS BY 40%.
THE GOAL OF ENERGY CONSERVATION GOES AGAINST THE GOAL OF ENERGY PRODUCTIVISM—THE IDEA THAT MORE IS ALWAYS BETTER—THAT'S BEEN IMPOSED ON US.
MAKING CONSERVATION A POSITIVE FACTOR IN THE FUTURE WOULD REQUIRE A HUGE CHANGE IN POLITICAL DIRECTION.

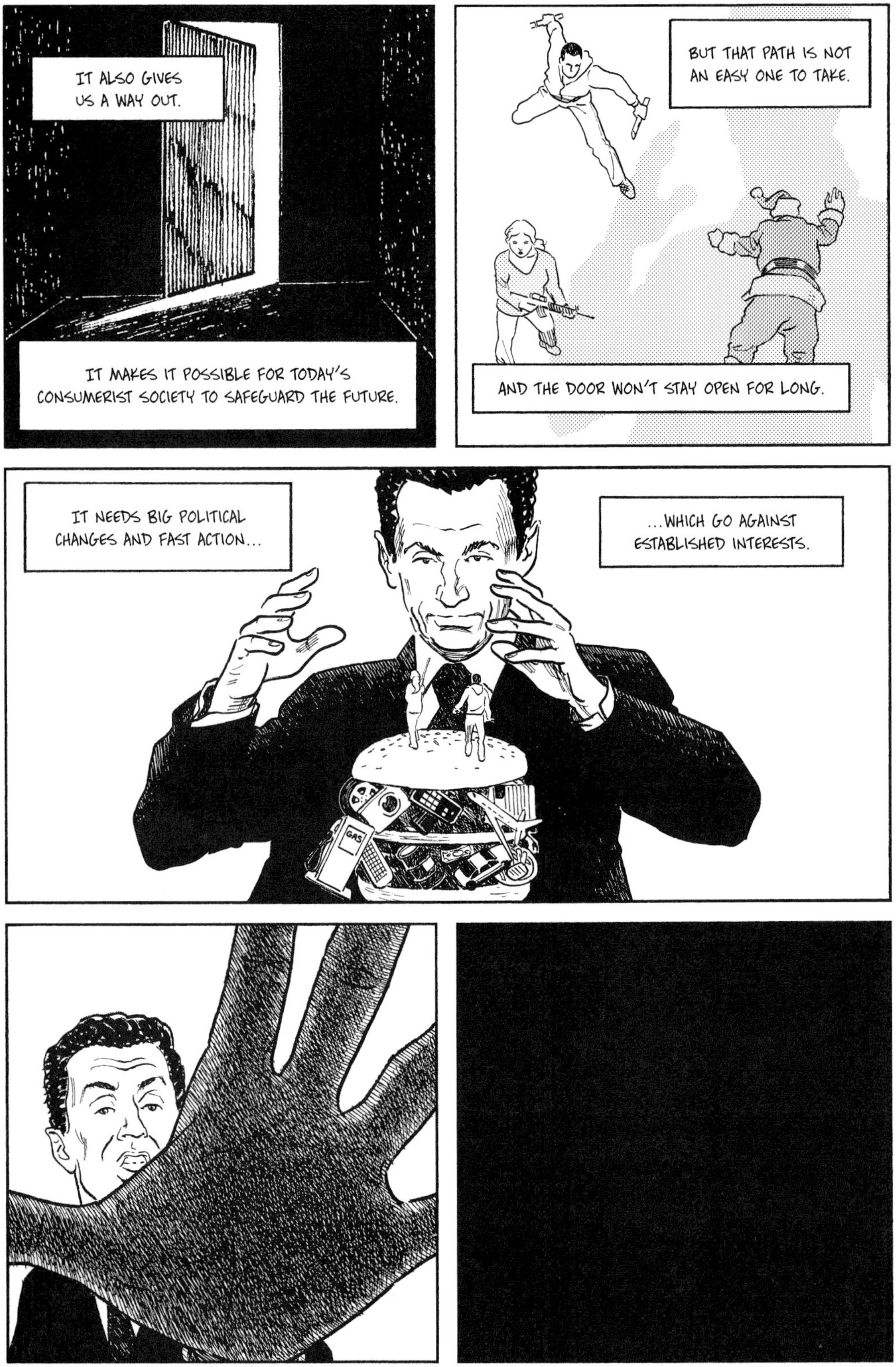
IT ALSO GIVES US A WAY OUT.
IT MAKES IT POSSIBLE FOR TODAY'S CONSUMERIST SOCIETY TO SAFEGUARD THE FUTURE.
BUT THAT PATH IS NOT AN EASY ONE TO TAKE.
AND THE DOOR WON'T STAY OPEN FOR LONG.
IT NEEDS BIG POLITICAL CHANGES AND FAST ACTION...
...WHICH GO AGAINST ESTABLISHED INTERESTS.
GAS

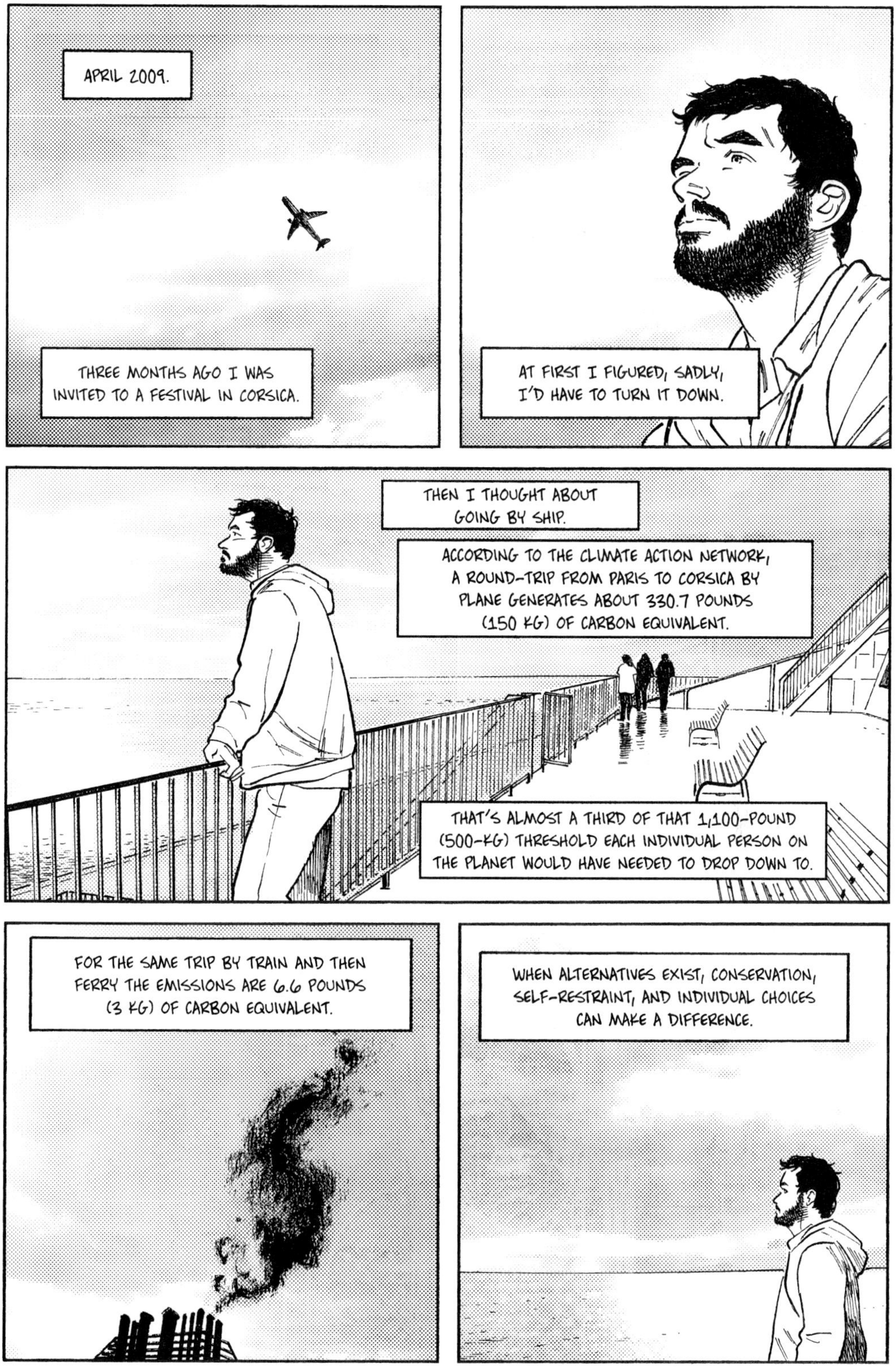
APRIL 2009.
THREE MONTHS AGO I WAS INVITED TO A FESTIVAL IN CORSICA.
AT FIRST I FIGURED, SADLY, I'D HAVE TO TURN IT DOWN.
THEN I THOUGHT ABOUT GOING BY SHIP.
ACCORDING TO THE CLIMATE ACTION NETWORK, A ROUND-TRIP FROM PARIS TO CORSICA BY PLANE GENERATES ABOUT 330.7 POUNDS (150 KG) OF CARBON EQUIVALENT.
THAT'S ALMOST A THIRD OF THAT 1,100-POUND (500-KG) THRESHOLD EACH INDIVIDUAL PERSON ON THE PLANET WOULD HAVE NEEDED TO DROP DOWN TO.
FOR THE SAME TRIP BY TRAIN AND THEN FERRY THE EMISSIONS ARE 6.6 POUNDS (3 KG) OF CARBON EQUIVALENT.
WHEN ALTERNATIVES EXIST, CONSERVATION, SELF-RESTRAINT, AND INDIVIDUAL CHOICES CAN MAKE A DIFFERENCE.

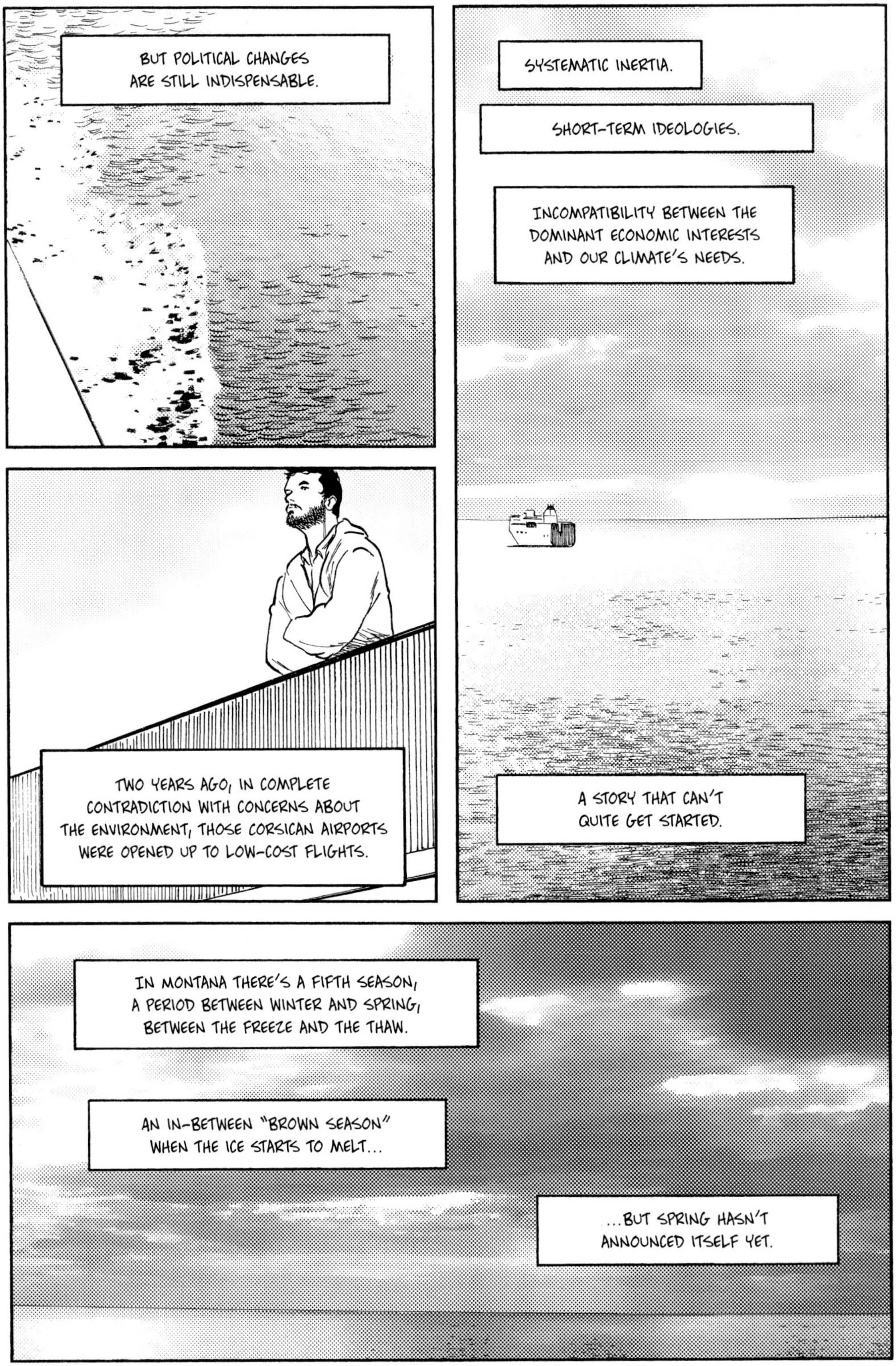
BUT POLITICAL CHANGES ARE STILL INDISPENSABLE.
TWO YEARS AGO, IN COMPLETE CONTRADICTION WITH CONCERNS ABOUT THE ENVIRONMENT, THOSE CORSICAN AIRPORTS WERE OPENED UP TO LOW-COST FLIGHTS.
SYSTEMATIC INERTIA.
SHORT-TERM IDEOLOGIES.
INCOMPATIBILITY BETWEEN THE DOMINANT ECONOMIC INTERESTS AND OUR CLIMATE'S NEEDS.
A STORY THAT CAN'T QUITE GET STARTED.
IN MONTANA THERE'S A FIFTH SEASON, A PERIOD BETWEEN WINTER AND SPRING, BETWEEN THE FREEZE AND THE THAW.
AN IN-BETWEEN "BROWN SEASON" WHEN THE ICE STARTS TO MELT...
...BUT SPRING HASN'T ANNOUNCED ITSELF YET.

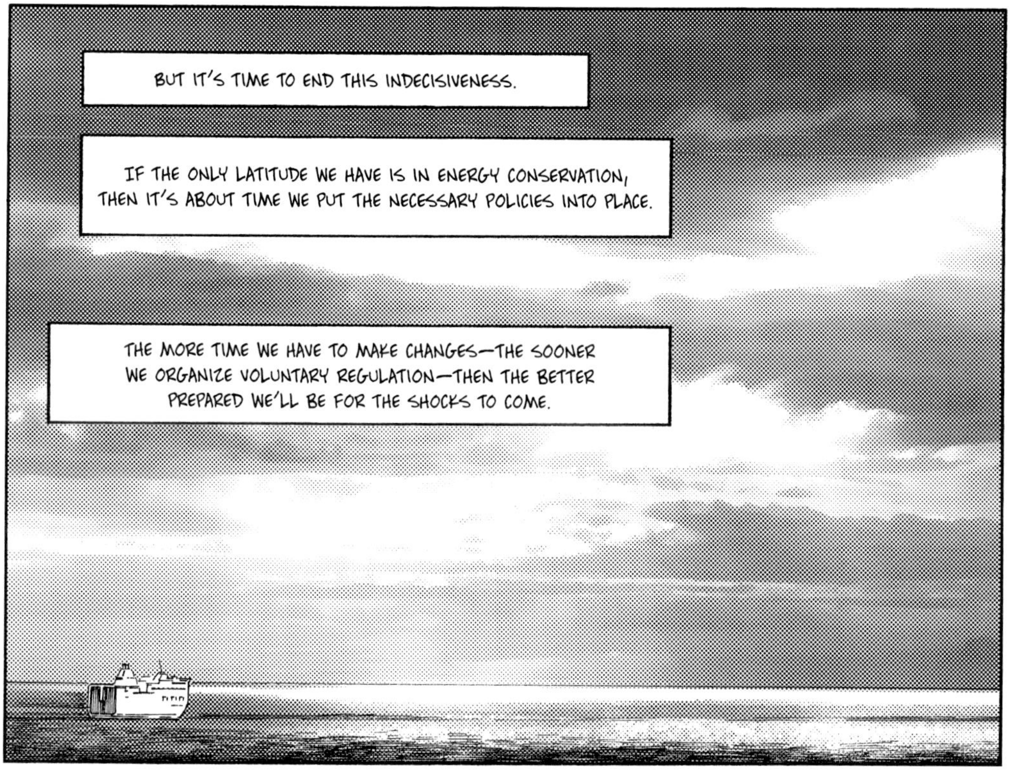
BUT IT'S TIME TO END THIS INDECISIVENESS.
IF THE ONLY LATITUDE WE HAVE IS IN ENERGY CONSERVATION, THEN IT'S ABOUT TIME WE PUT THE NECESSARY POLICIES INTO PLACE.
THE MORE TIME WE HAVE TO MAKE CHANGES—THE SOONER WE ORGANIZE VOLUNTARY REGULATION—THEN THE BETTER PREPARED WE'LL BE FOR THE SHOCKS TO COME.

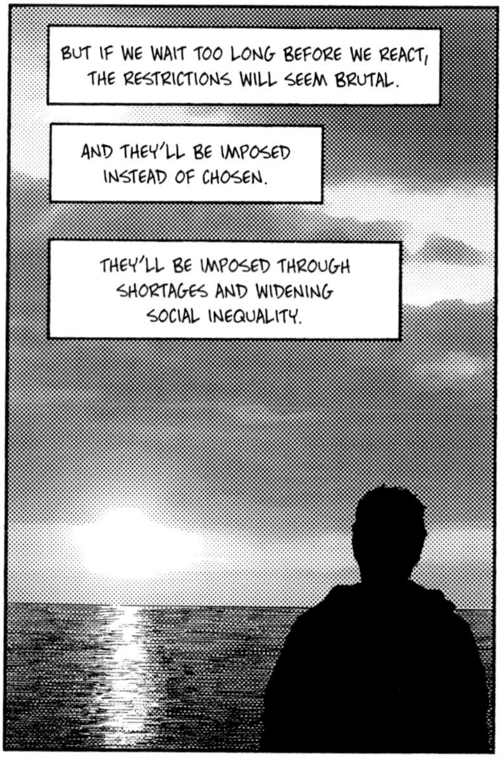
BUT IF WE WAIT TOO LONG BEFORE WE REACT, THE RESTRICTIONS WILL SEEM BRUTAL.
AND THEY'LL BE IMPOSED INSTEAD OF CHOSEN.
THEY'LL BE IMPOSED THROUGH SHORTAGES AND WIDENING SOCIAL INEQUALITY.

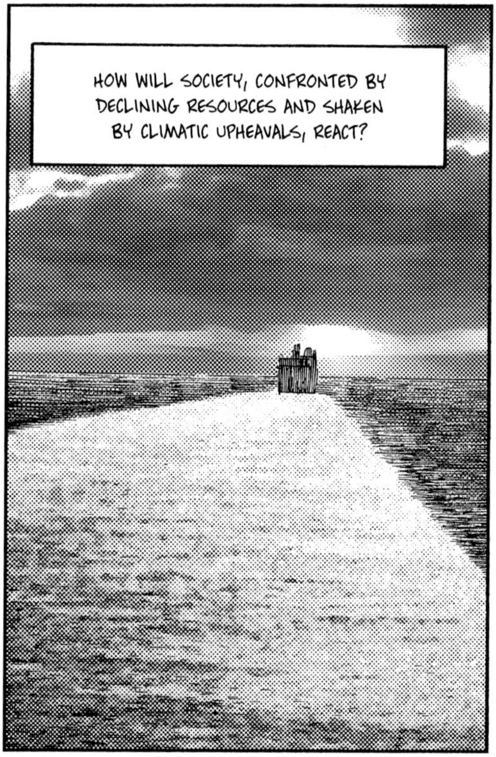
HOW WILL SOCIETY, CONFRONTED BY DECLINING RESOURCES AND SHAKEN BY CLIMATIC UPHEAVALS, REACT?

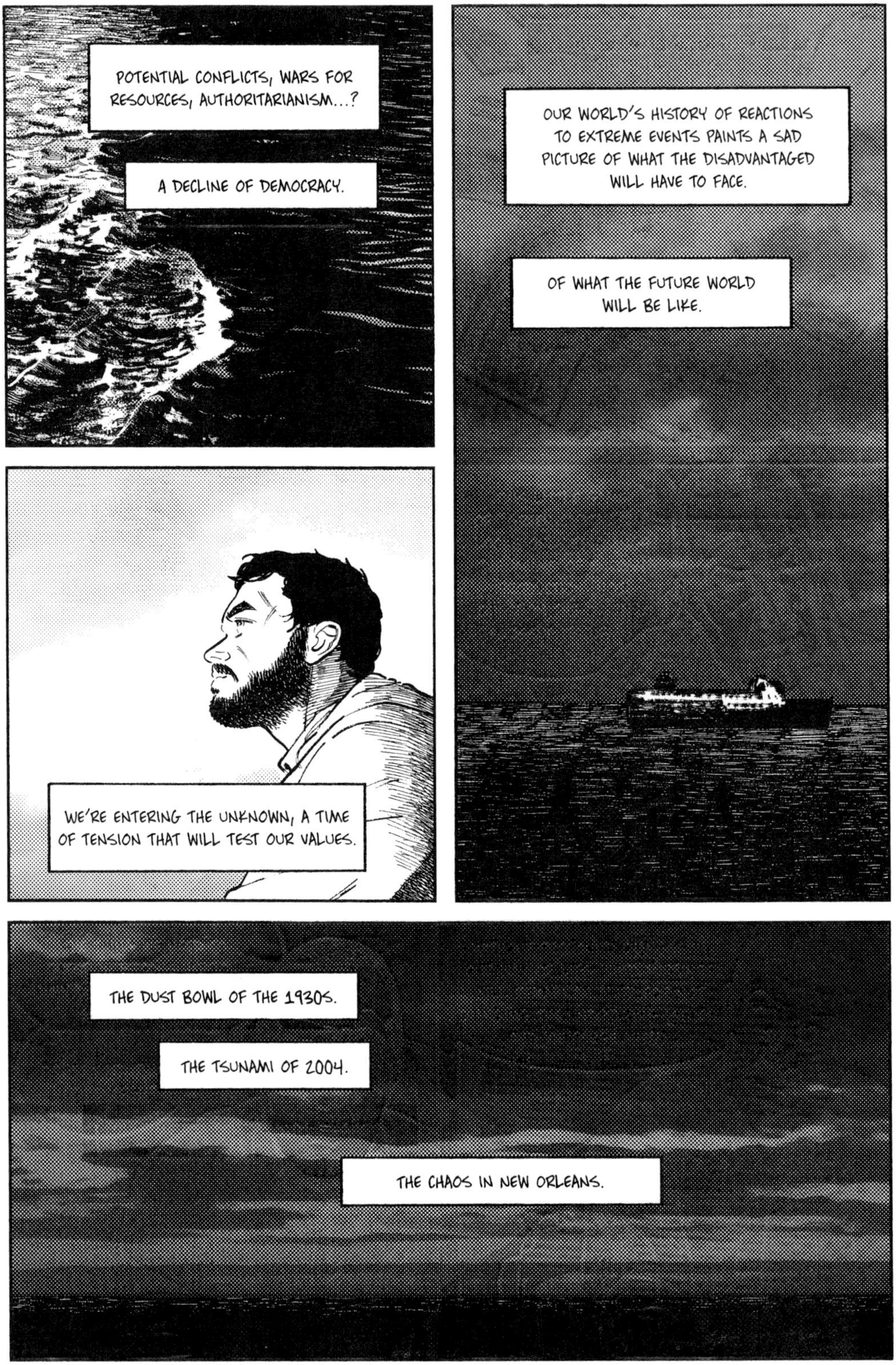
POTENTIAL CONFLICTS, WARS FOR RESOURCES, AUTHORITARIANISM...?
A DECLINE OF DEMOCRACY.
WE'RE ENTERING THE UNKNOWN, A TIME OF TENSION THAT WILL TEST OUR VALUES.
OUR WORLD'S HISTORY OF REACTIONS TO EXTREME EVENTS PAINTS A SAD PICTURE OF WHAT THE DISADVANTAGED WILL HAVE TO FACE.
OF WHAT THE FUTURE WORLD WILL BE LIKE.
THE DUST BOWL OF THE 1930S.
THE TSUNAMI OF 2004.
THE CHAOS IN NEW ORLEANS.

SOMETIMES TOWARD THE END...

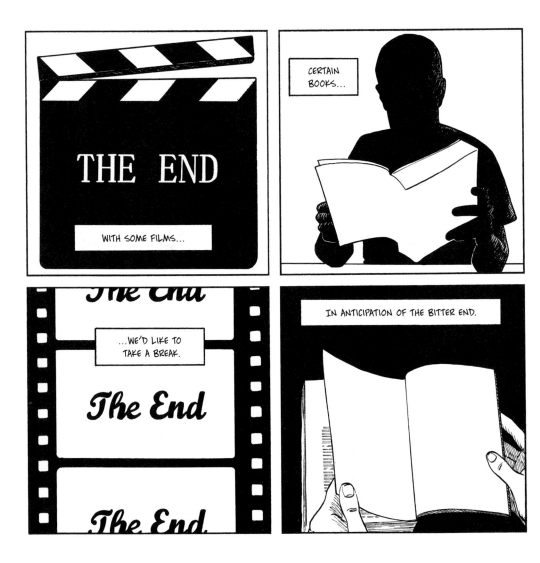

THE END

WITH SOME FILMS...

CERTAIN BOOKS...

...WE'D LIKE TO TAKE A BREAK.

The End

The End

The End.

IN ANTICIPATION OF THE BITTER END.

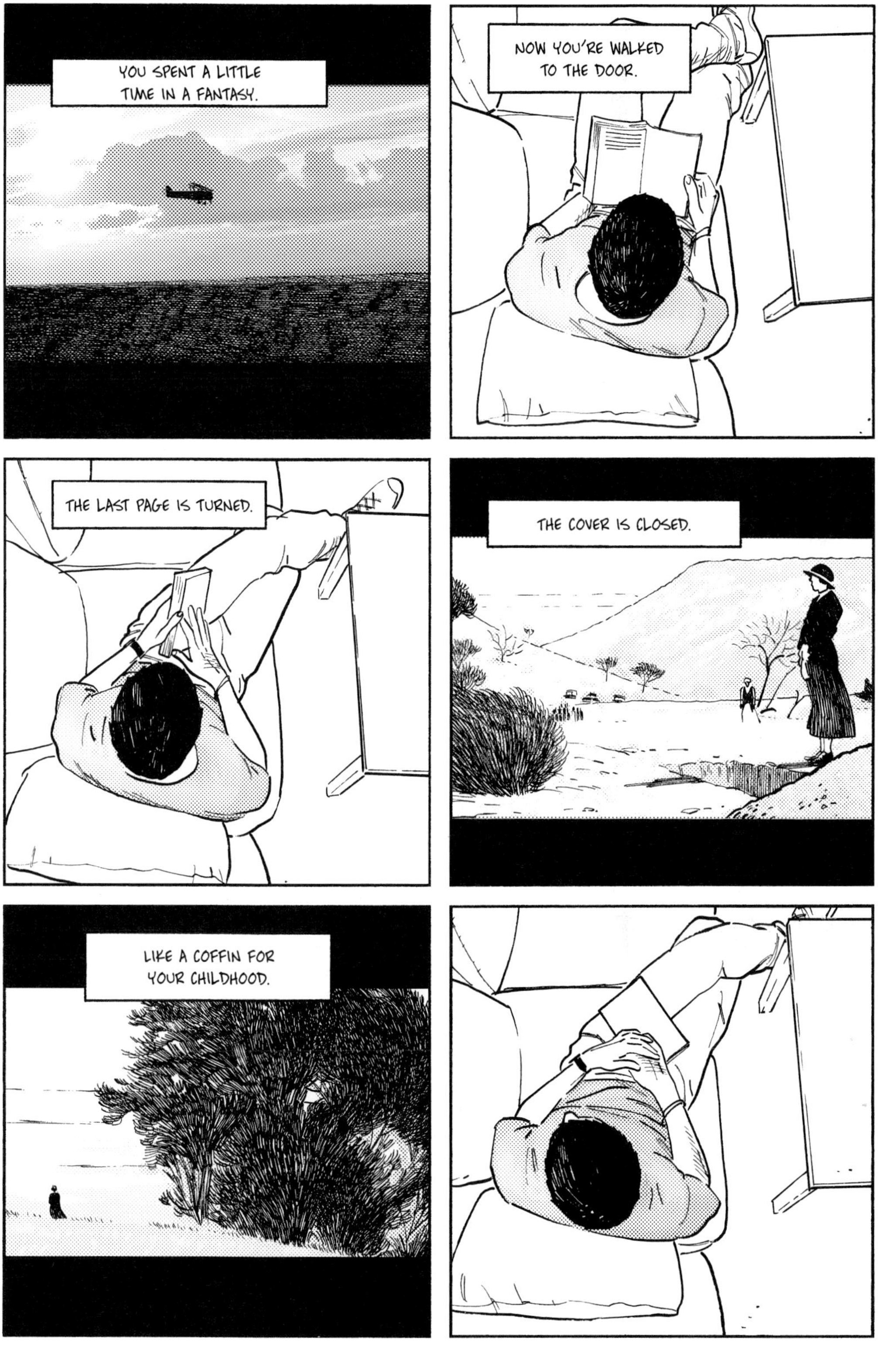
YOU SPENT A LITTLE TIME IN A FANTASY.
NOW YOU'RE WALKED TO THE DOOR.
THE LAST PAGE IS TURNED.
THE COVER IS CLOSED.
LIKE A COFFIN FOR YOUR CHILDHOOD.

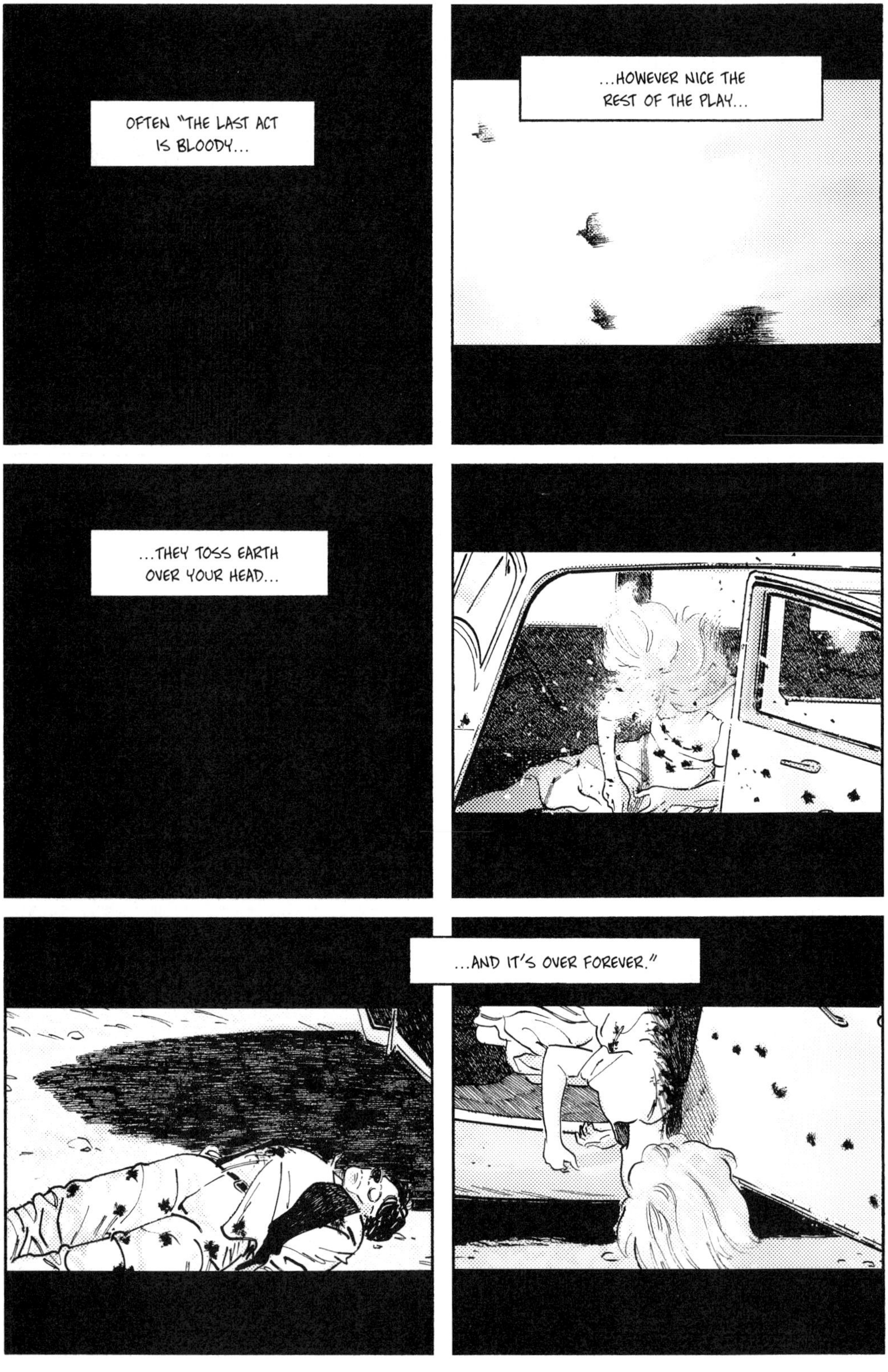
OFTEN "THE LAST ACT IS BLOODY...
...HOWEVER NICE THE REST OF THE PLAY...
...THEY TOSS EARTH OVER YOUR HEAD...
...AND IT'S OVER FOREVER."

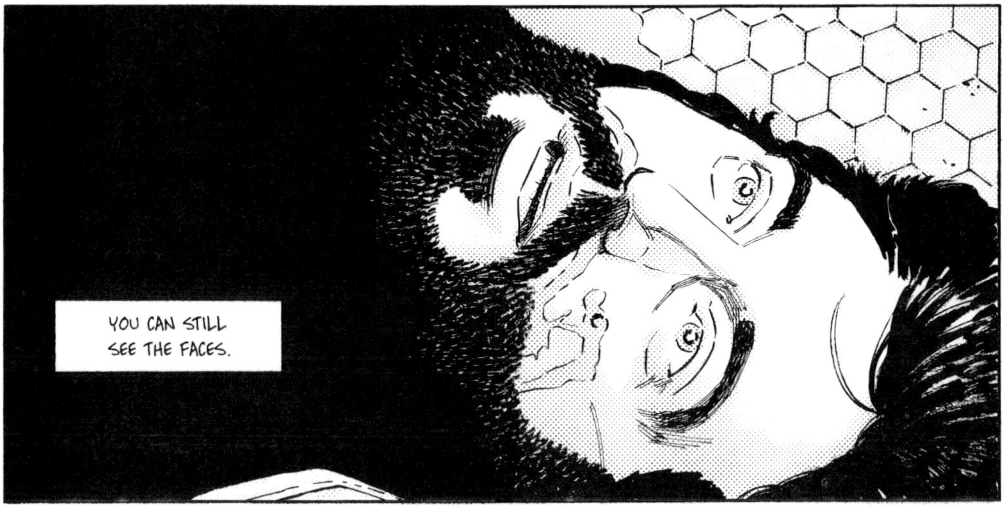
YOU CAN STILL SEE THE FACES.

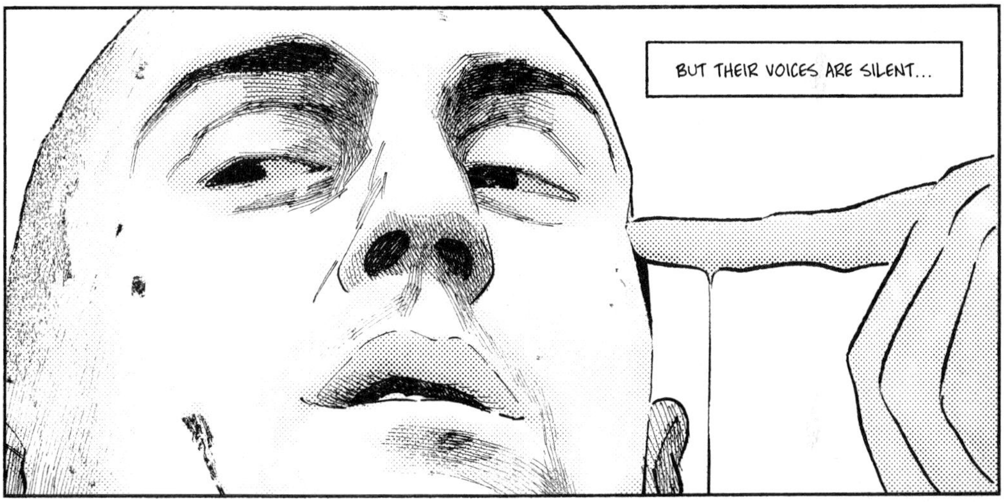
BUT THEIR VOICES ARE SILENT...

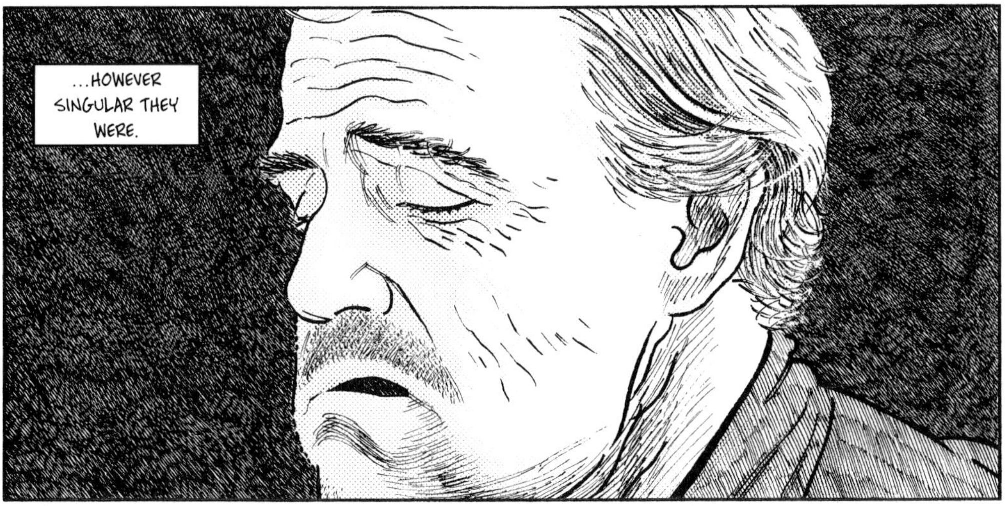
...HOWEVER SINGULAR THEY WERE.

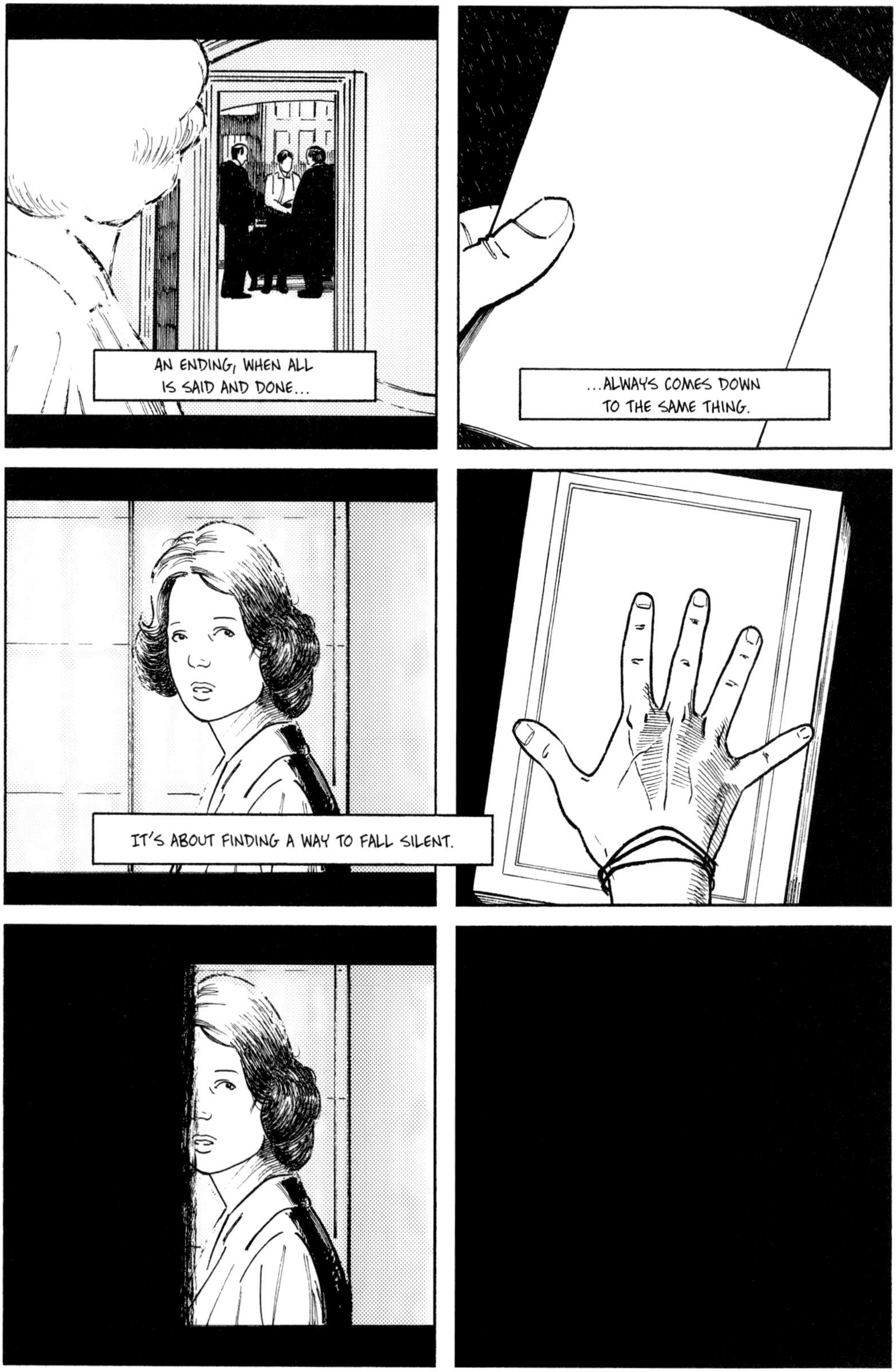
AN ENDING, WHEN ALL IS SAID AND DONE...
...ALWAYS COMES DOWN TO THE SAME THING.
IT'S ABOUT FINDING A WAY TO FALL SILENT.

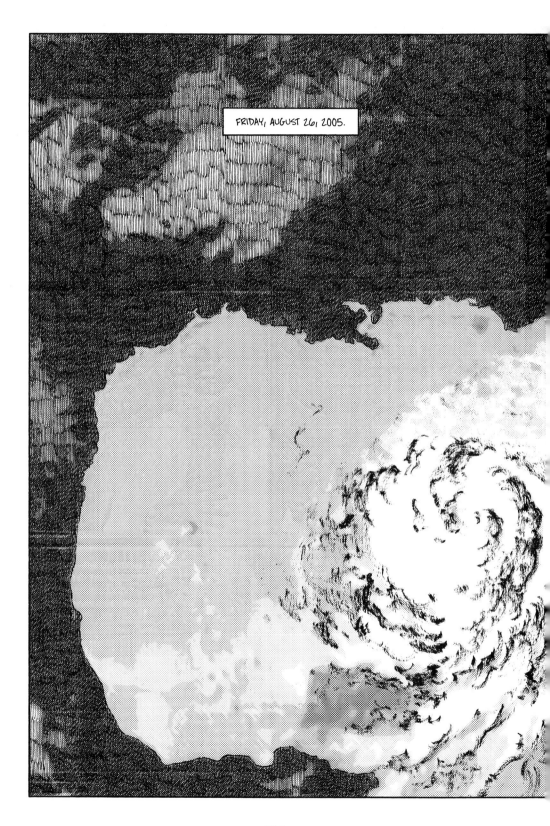

FRIDAY, AUGUST 26, 2005.

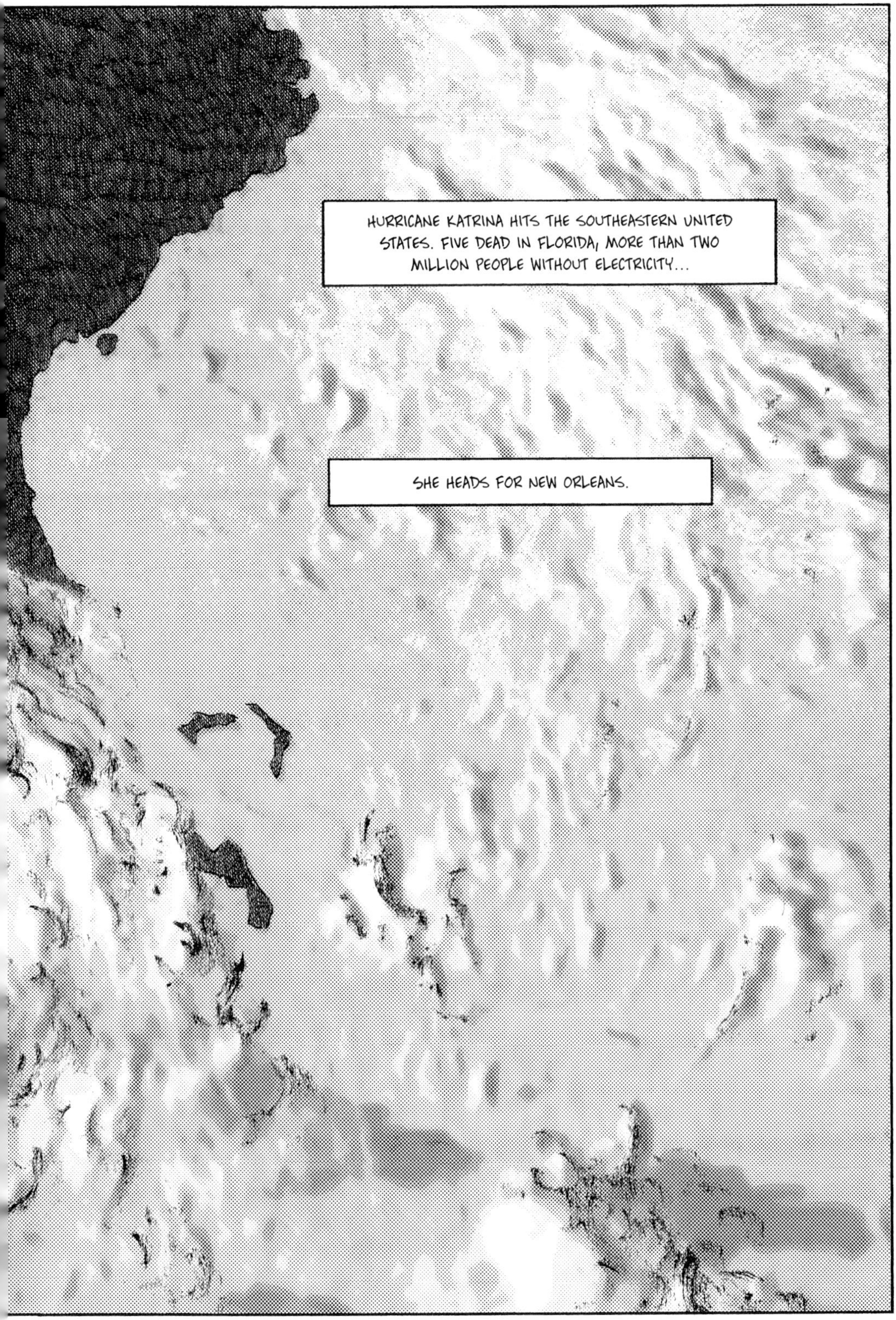
HURRICANE KATRINA HITS THE SOUTHEASTERN UNITED STATES. FIVE DEAD IN FLORIDA, MORE THAN TWO MILLION PEOPLE WITHOUT ELECTRICITY...
SHE HEADS FOR NEW ORLEANS.

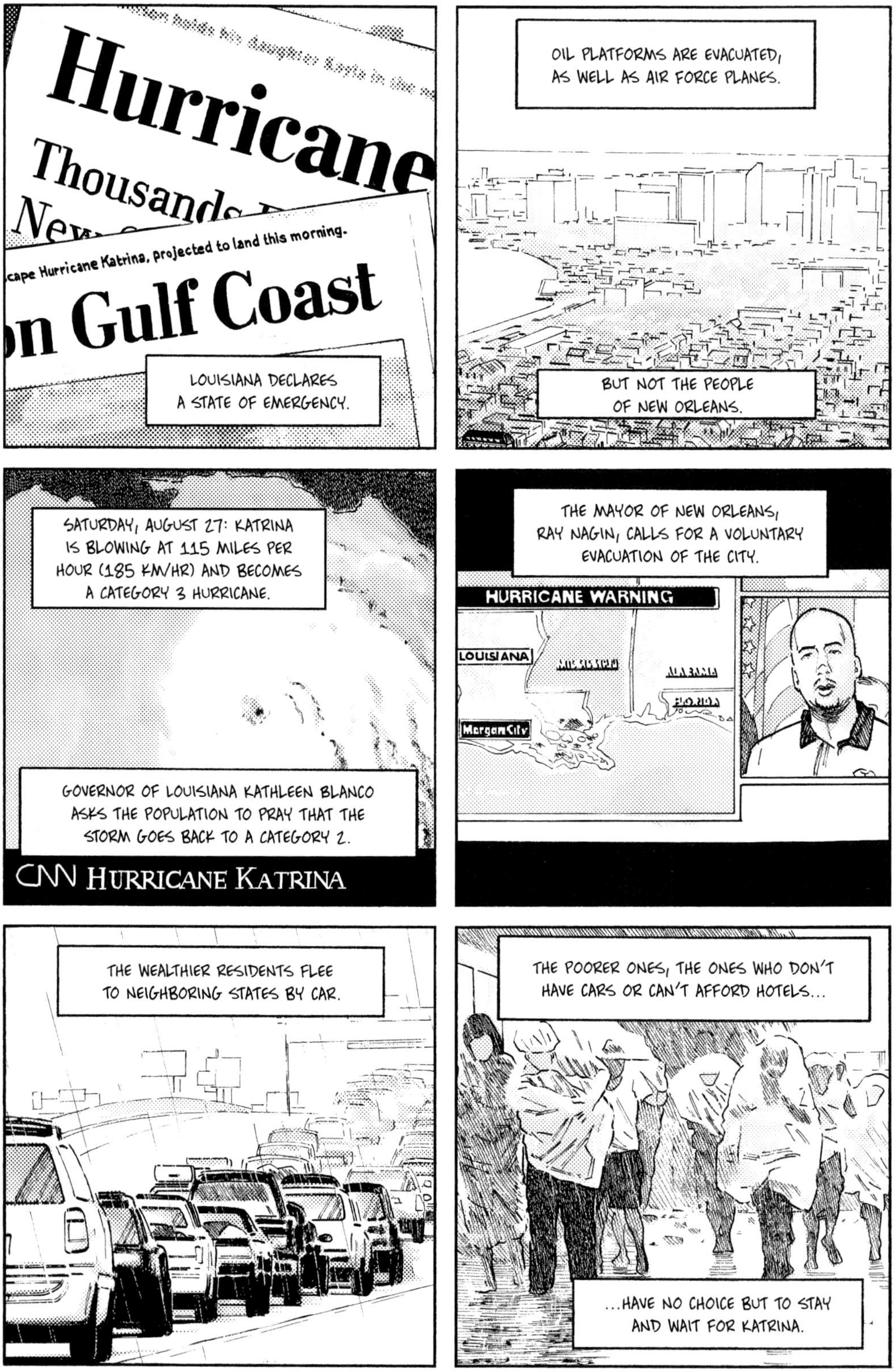
Hurricane
Thousands
cape Hurricane Katrina, projected to land this morning.
on Gulf Coast
LOUISIANA DECLARES A STATE OF EMERGENCY.
OIL PLATFORMS ARE EVACUATED, AS WELL AS AIR FORCE PLANES.
BUT NOT THE PEOPLE OF NEW ORLEANS.
SATURDAY, AUGUST 27: KATRINA IS BLOWING AT 115 MILES PER HOUR (185 KM/HR) AND BECOMES A CATEGORY 3 HURRICANE.
GOVERNOR OF LOUISIANA KATHLEEN BLANCO ASKS THE POPULATION TO PRAY THAT THE STORM GOES BACK TO A CATEGORY 2.
CNN HURRICANE KATRINA
THE MAYOR OF NEW ORLEANS, RAY NAGIN, CALLS FOR A VOLUNTARY EVACUATION OF THE CITY.
HURRICANE WARNING
LOUISIANA
Morgan City
THE WEALTHIER RESIDENTS FLEE TO NEIGHBORING STATES BY CAR.
THE POORER ONES, THE ONES WHO DON'T HAVE CARS OR CAN'T AFFORD HOTELS...
...HAVE NO CHOICE BUT TO STAY AND WAIT FOR KATRINA.

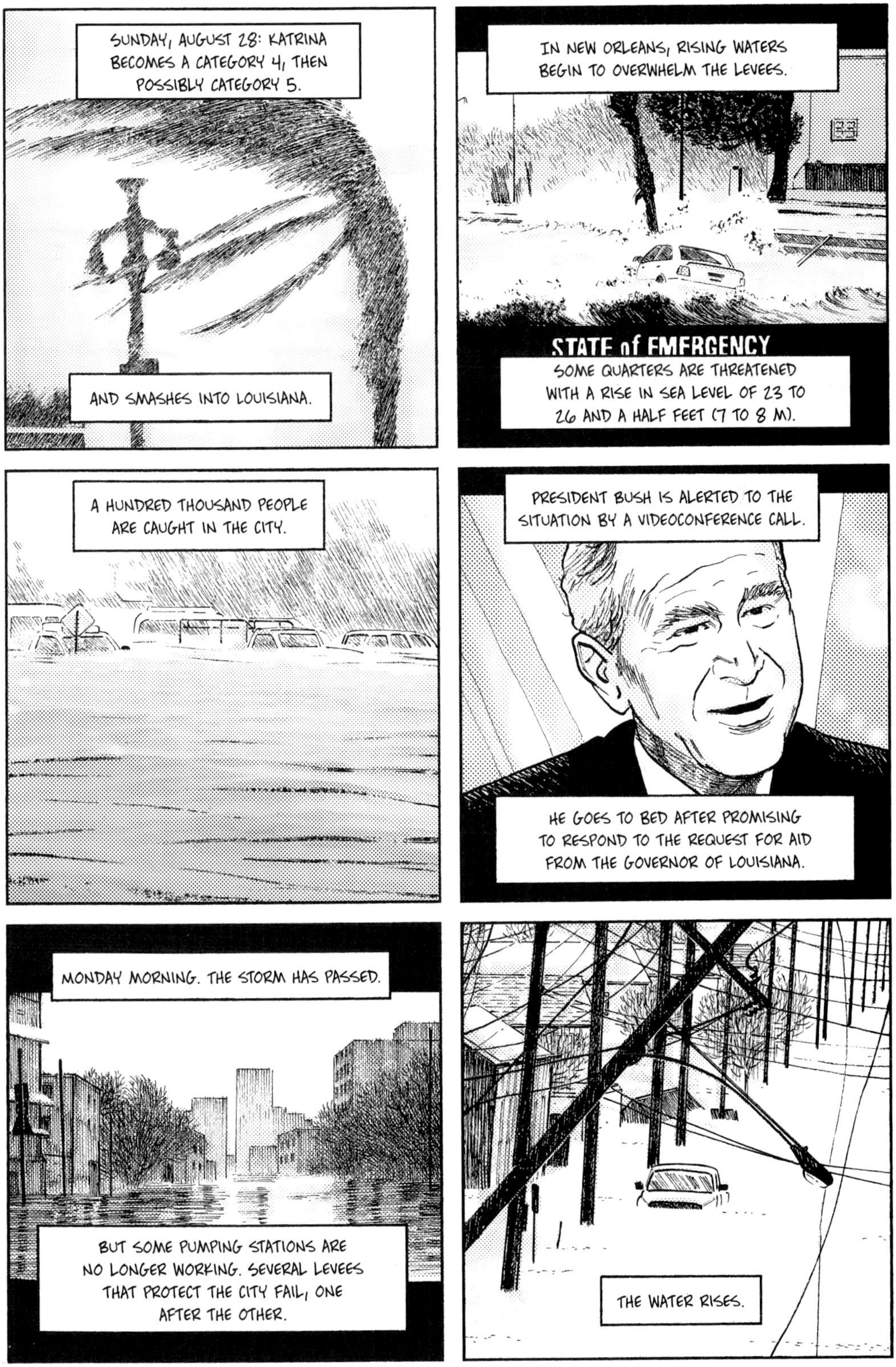
SUNDAY, AUGUST 28: KATRINA BECOMES A CATEGORY 4, THEN POSSIBLY CATEGORY 5.
AND SMASHES INTO LOUISIANA.
IN NEW ORLEANS, RISING WATERS BEGIN TO OVERWHELM THE LEVEES.
STATE of EMERGENCY
SOME QUARTERS ARE THREATENED WITH A RISE IN SEA LEVEL OF 23 TO 26 AND A HALF FEET (7 TO 8 M).
A HUNDRED THOUSAND PEOPLE ARE CAUGHT IN THE CITY.
PRESIDENT BUSH IS ALERTED TO THE SITUATION BY A VIDEOCONFERENCE CALL.
HE GOES TO BED AFTER PROMISING TO RESPOND TO THE REQUEST FOR AID FROM THE GOVERNOR OF LOUISIANA.
MONDAY MORNING. THE STORM HAS PASSED.
BUT SOME PUMPING STATIONS ARE NO LONGER WORKING. SEVERAL LEVEES THAT PROTECT THE CITY FAIL, ONE AFTER THE OTHER.
THE WATER RISES.

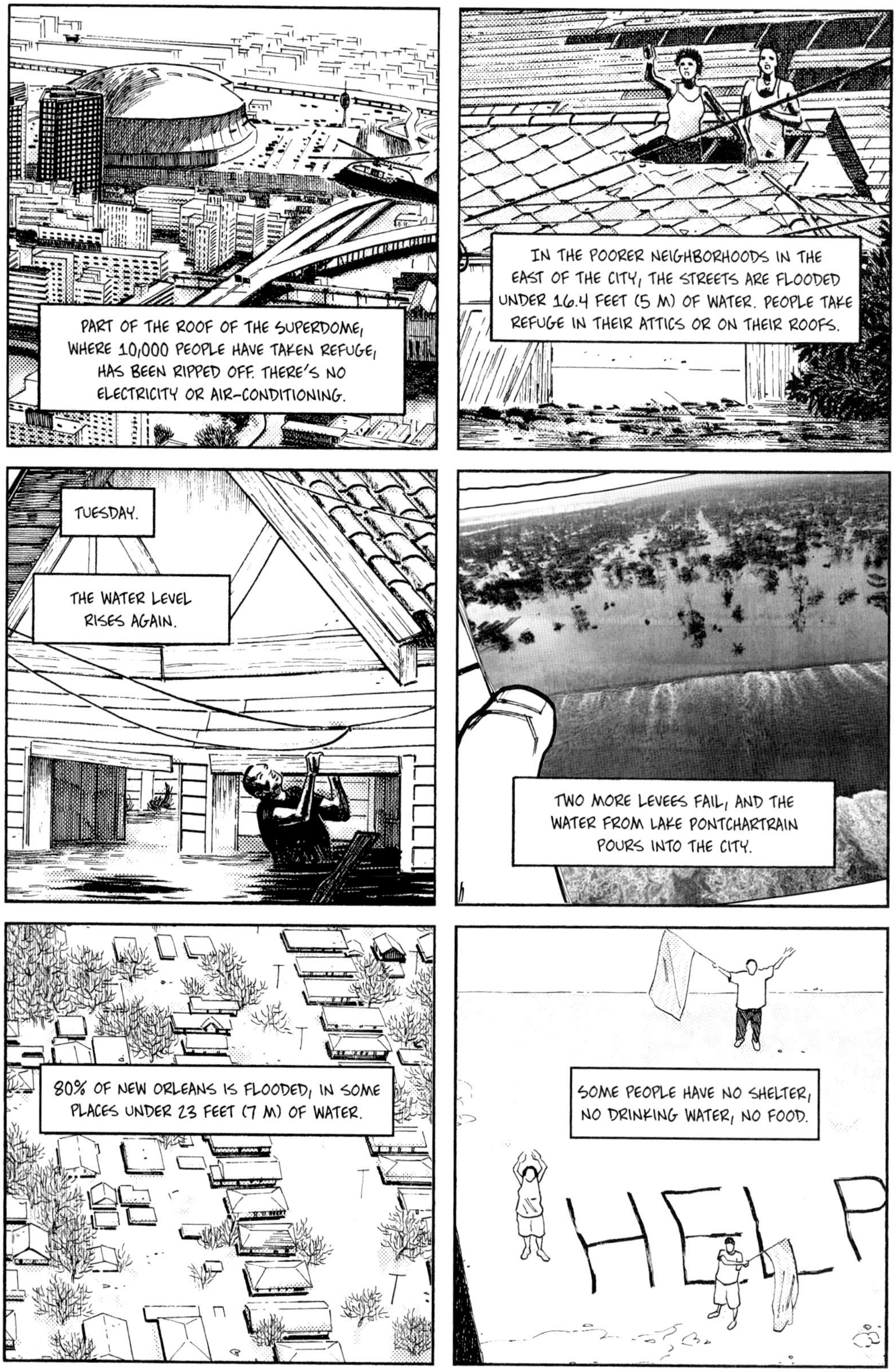
PART OF THE ROOF OF THE SUPERDOME, WHERE 10,000 PEOPLE HAVE TAKEN REFUGE, HAS BEEN RIPPED OFF. THERE'S NO ELECTRICITY OR AIR-CONDITIONING.
IN THE POORER NEIGHBORHOODS IN THE EAST OF THE CITY, THE STREETS ARE FLOODED UNDER 16.4 FEET (5 M) OF WATER. PEOPLE TAKE REFUGE IN THEIR ATTICS OR ON THEIR ROOFS.
TUESDAY.
THE WATER LEVEL RISES AGAIN.
TWO MORE LEVEES FAIL, AND THE WATER FROM LAKE PONTCHARTRAIN POURS INTO THE CITY.
80% OF NEW ORLEANS IS FLOODED, IN SOME PLACES UNDER 23 FEET (7 M) OF WATER.
SOME PEOPLE HAVE NO SHELTER, NO DRINKING WATER, NO FOOD.
HELP

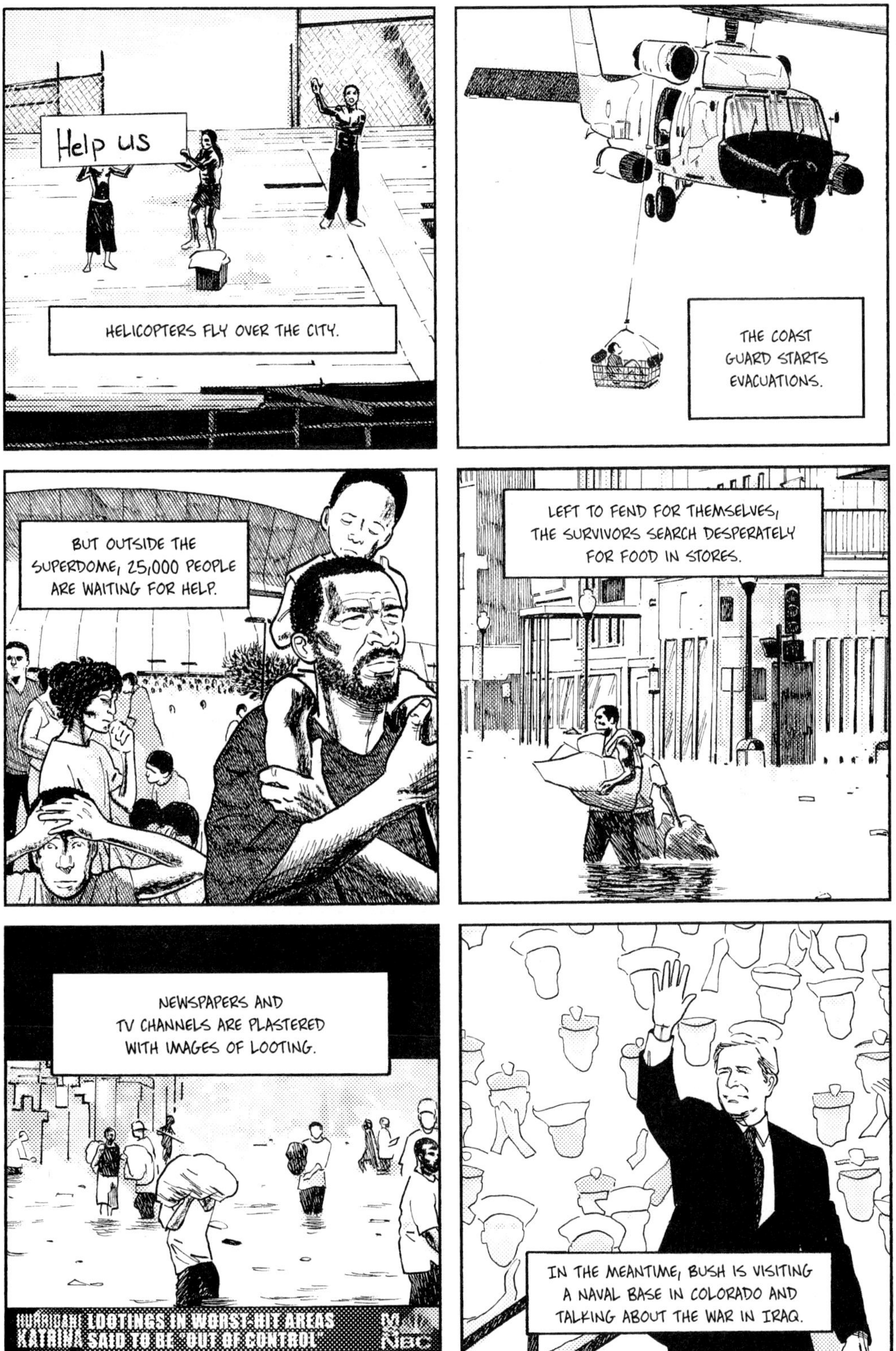
Help us
HELICOPTERS FLY OVER THE CITY.
THE COAST GUARD STARTS EVACUATIONS.
BUT OUTSIDE THE SUPERDOME, 25,000 PEOPLE ARE WAITING FOR HELP.
LEFT TO FEND FOR THEMSELVES, THE SURVIVORS SEARCH DESPERATELY FOR FOOD IN STORES.
NEWSPAPERS AND TV CHANNELS ARE PLASTERED WITH IMAGES OF LOOTING.
HURRICANE KATRINA
LOOTINGS IN WORST-HIT AREAS SAID TO BE "OUT OF CONTROL"
IN THE MEANTIME, BUSH IS VISITING A NAVAL BASE IN COLORADO AND TALKING ABOUT THE WAR IN IRAQ.

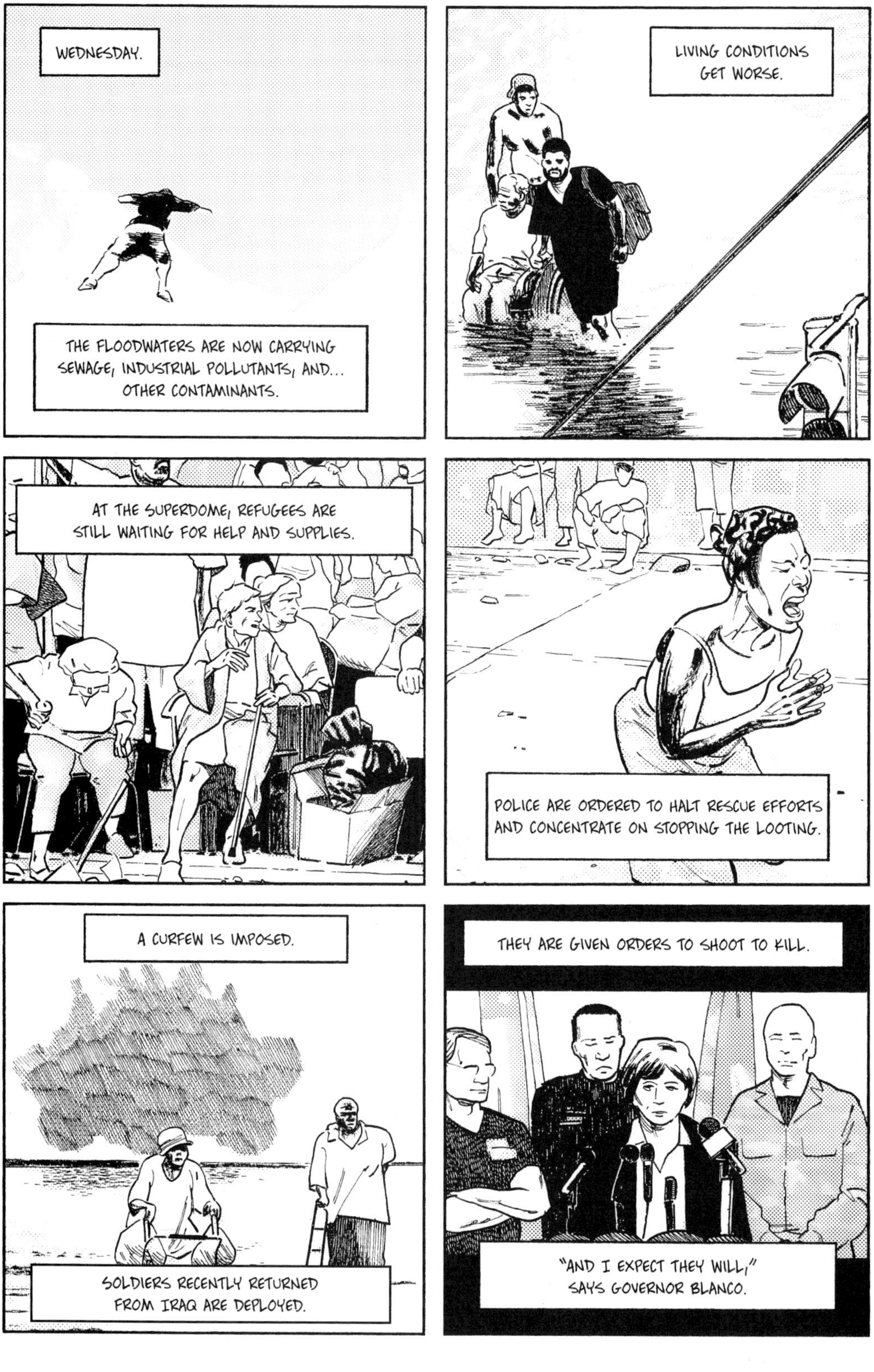
WEDNESDAY.
THE FLOODWATERS ARE NOW CARRYING SEWAGE, INDUSTRIAL POLLUTANTS, AND... OTHER CONTAMINANTS.
LIVING CONDITIONS GET WORSE.
AT THE SUPERDOME, REFUGEES ARE STILL WAITING FOR HELP AND SUPPLIES.
POLICE ARE ORDERED TO HALT RESCUE EFFORTS AND CONCENTRATE ON STOPPING THE LOOTING.
A CURFEW IS IMPOSED.
SOLDIERS RECENTLY RETURNED FROM IRAQ ARE DEPLOYED.
THEY ARE GIVEN ORDERS TO SHOOT TO KILL.
"AND I EXPECT THEY WILL," SAYS GOVERNOR BLANCO.

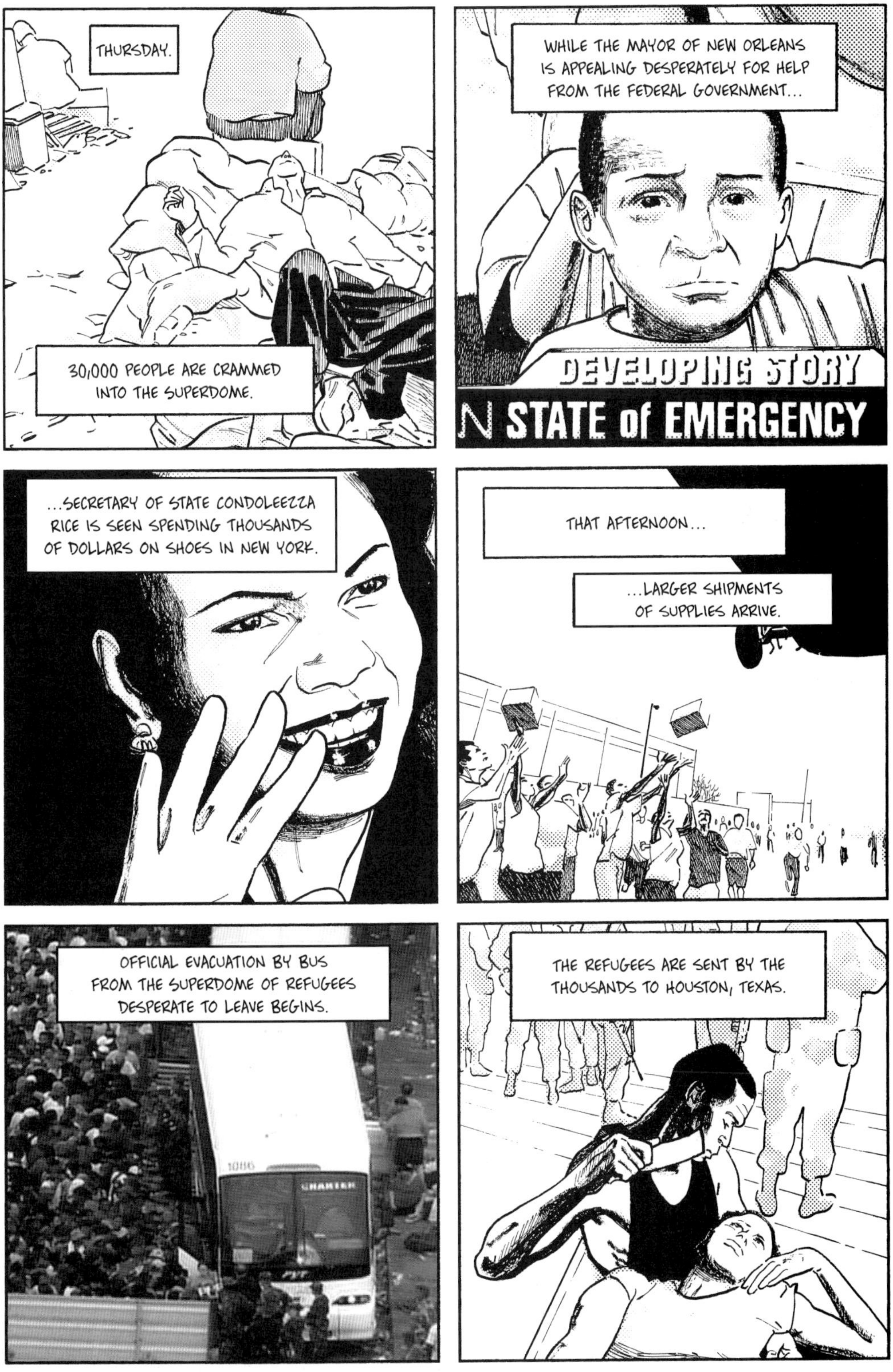
THURSDAY.
30,000 PEOPLE ARE CRAMMED INTO THE SUPERDOME.
WHILE THE MAYOR OF NEW ORLEANS IS APPEALING DESPERATELY FOR HELP FROM THE FEDERAL GOVERNMENT...
DEVELOPING STORY
N STATE of EMERGENCY
...SECRETARY OF STATE CONDOLEEZZA RICE IS SEEN SPENDING THOUSANDS OF DOLLARS ON SHOES IN NEW YORK.
THAT AFTERNOON...
...LARGER SHIPMENTS OF SUPPLIES ARRIVE.
OFFICIAL EVACUATION BY BUS FROM THE SUPERDOME OF REFUGEES DESPERATE TO LEAVE BEGINS.
CHARTER
THE REFUGEES ARE SENT BY THE THOUSANDS TO HOUSTON, TEXAS.

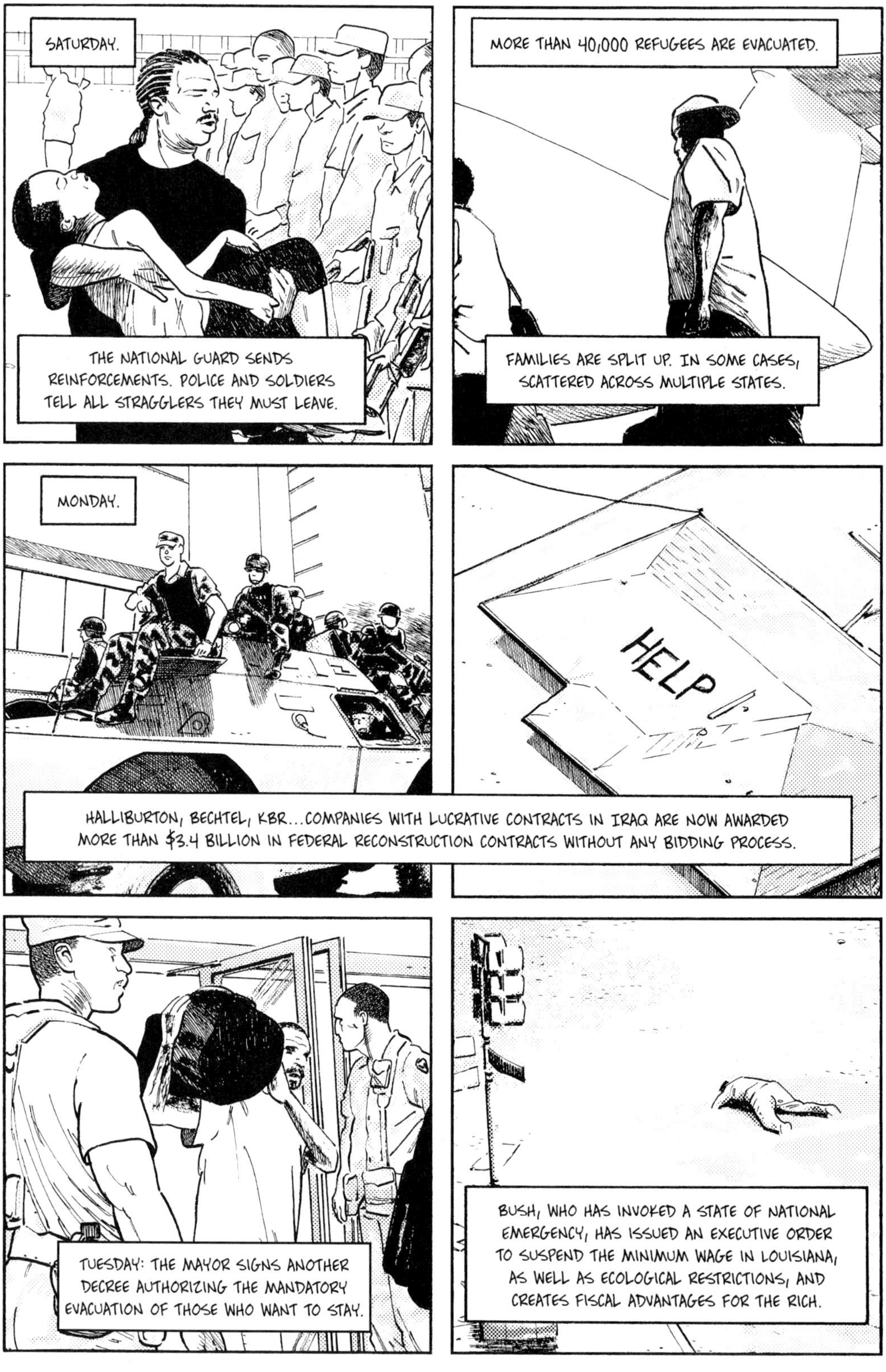
SATURDAY.
THE NATIONAL GUARD SENDS REINFORCEMENTS. POLICE AND SOLDIERS TELL ALL STRAGGLERS THEY MUST LEAVE.
MORE THAN 40,000 REFUGEES ARE EVACUATED.
FAMILIES ARE SPLIT UP. IN SOME CASES, SCATTERED ACROSS MULTIPLE STATES.
MONDAY.
HELP!
HALLIBURTON, BECHTEL, KBR...COMPANIES WITH LUCRATIVE CONTRACTS IN IRAQ ARE NOW AWARDED MORE THAN $3.4 BILLION IN FEDERAL RECONSTRUCTION CONTRACTS WITHOUT ANY BIDDING PROCESS.
TUESDAY: THE MAYOR SIGNS ANOTHER DECREE AUTHORIZING THE MANDATORY EVACUATION OF THOSE WHO WANT TO STAY.
BUSH, WHO HAS INVOKED A STATE OF NATIONAL EMERGENCY, HAS ISSUED AN EXECUTIVE ORDER TO SUSPEND THE MINIMUM WAGE IN LOUISIANA, AS WELL AS ECOLOGICAL RESTRICTIONS, AND CREATES FISCAL ADVANTAGES FOR THE RICH.

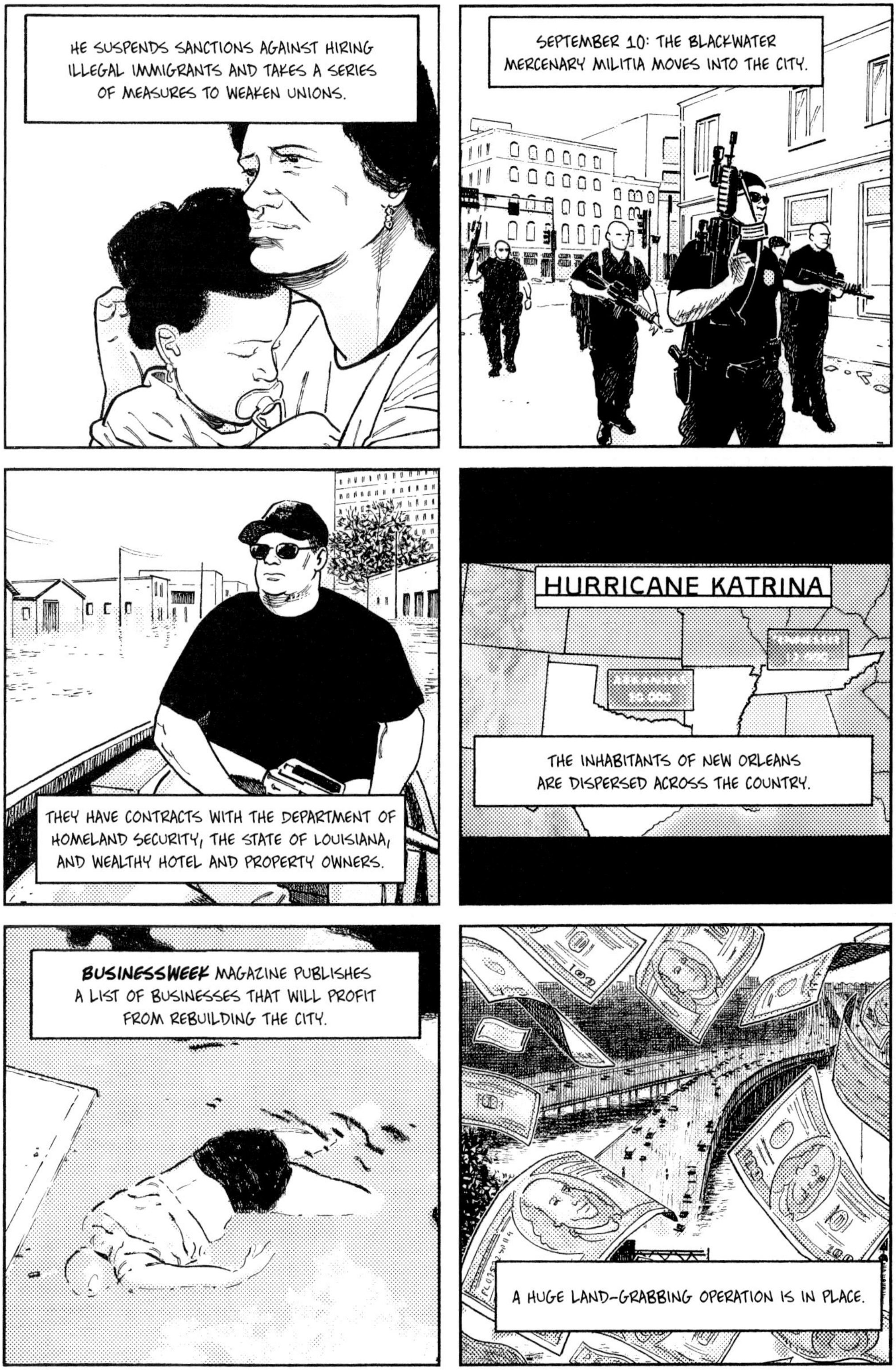
HE SUSPENDS SANCTIONS AGAINST HIRING ILLEGAL IMMIGRANTS AND TAKES A SERIES OF MEASURES TO WEAKEN UNIONS.
SEPTEMBER 10: THE BLACKWATER MERCENARY MILITIA MOVES INTO THE CITY.
THEY HAVE CONTRACTS WITH THE DEPARTMENT OF HOMELAND SECURITY, THE STATE OF LOUISIANA, AND WEALTHY HOTEL AND PROPERTY OWNERS.
HURRICANE KATRINA
THE INHABITANTS OF NEW ORLEANS ARE DISPERSED ACROSS THE COUNTRY.
BUSINESSWEEK MAGAZINE PUBLISHES A LIST OF BUSINESSES THAT WILL PROFIT FROM REBUILDING THE CITY.
A HUGE LAND-GRABBING OPERATION IS IN PLACE.

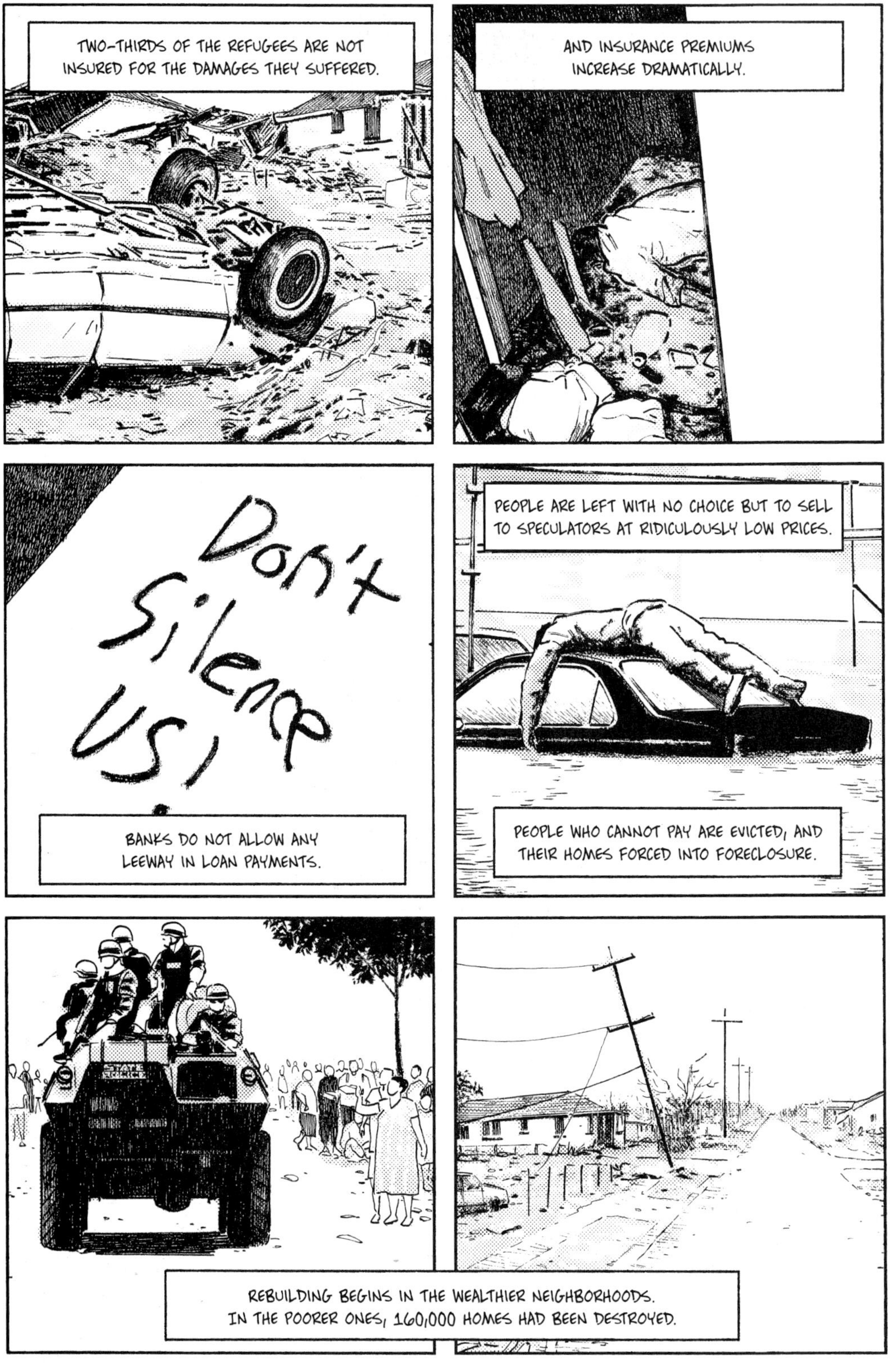
TWO-THIRDS OF THE REFUGEES ARE NOT INSURED FOR THE DAMAGES THEY SUFFERED.
AND INSURANCE PREMIUMS INCREASE DRAMATICALLY.
Don't Silence Us!
BANKS DO NOT ALLOW ANY LEEWAY IN LOAN PAYMENTS.
PEOPLE ARE LEFT WITH NO CHOICE BUT TO SELL TO SPECULATORS AT RIDICULOUSLY LOW PRICES.
PEOPLE WHO CANNOT PAY ARE EVICTED, AND THEIR HOMES FORCED INTO FORECLOSURE.
STATE
REBUILDING BEGINS IN THE WEALTHIER NEIGHBORHOODS. IN THE POORER ONES, 160,000 HOMES HAD BEEN DESTROYED.

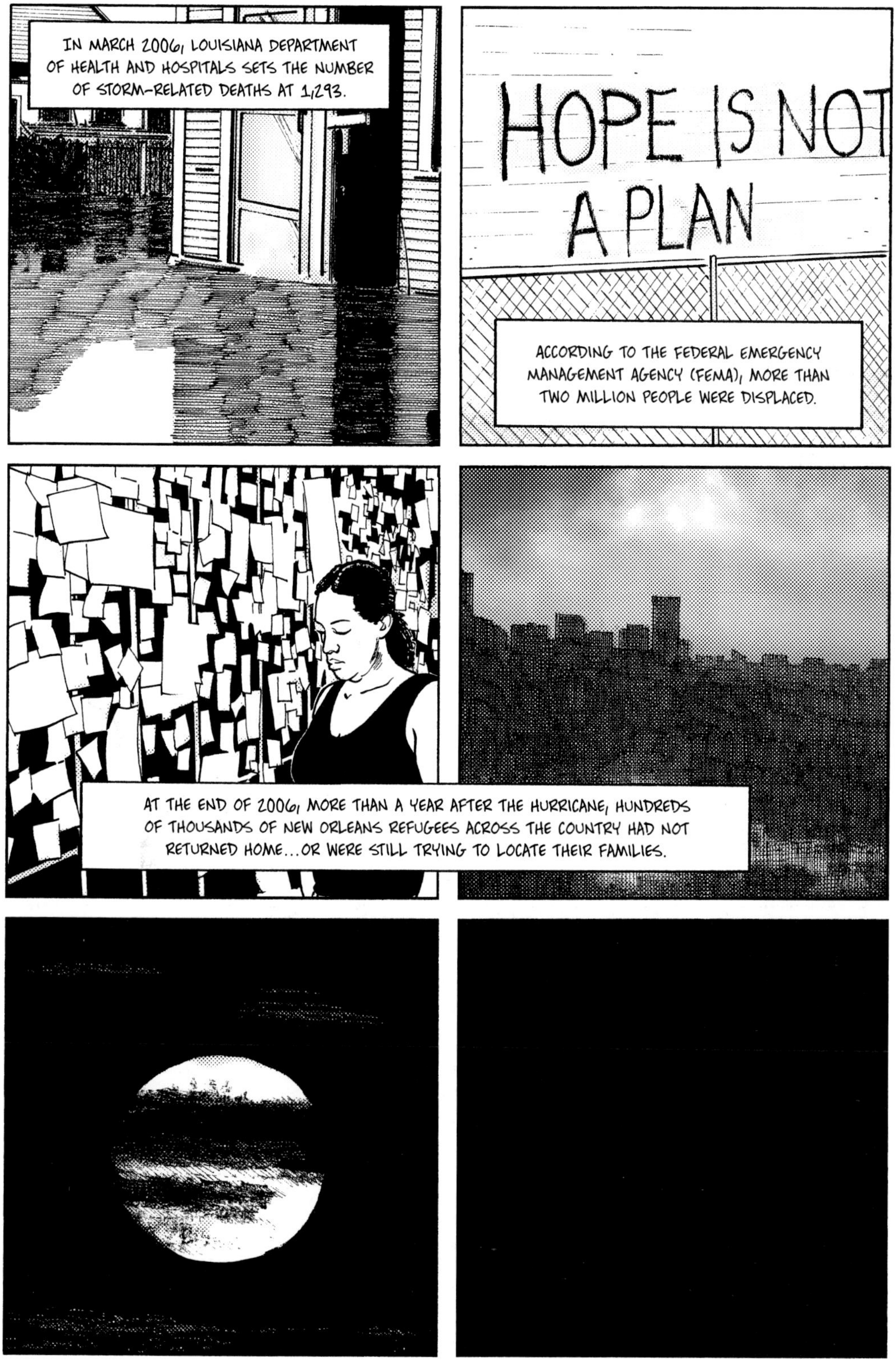
IN MARCH 2006, LOUISIANA DEPARTMENT OF HEALTH AND HOSPITALS SETS THE NUMBER OF STORM-RELATED DEATHS AT 1,293.
HOPE IS NOT A PLAN
ACCORDING TO THE FEDERAL EMERGENCY MANAGEMENT AGENCY (FEMA), MORE THAN TWO MILLION PEOPLE WERE DISPLACED.
AT THE END OF 2006, MORE THAN A YEAR AFTER THE HURRICANE, HUNDREDS OF THOUSANDS OF NEW ORLEANS REFUGEES ACROSS THE COUNTRY HAD NOT RETURNED HOME...OR WERE STILL TRYING TO LOCATE THEIR FAMILIES.

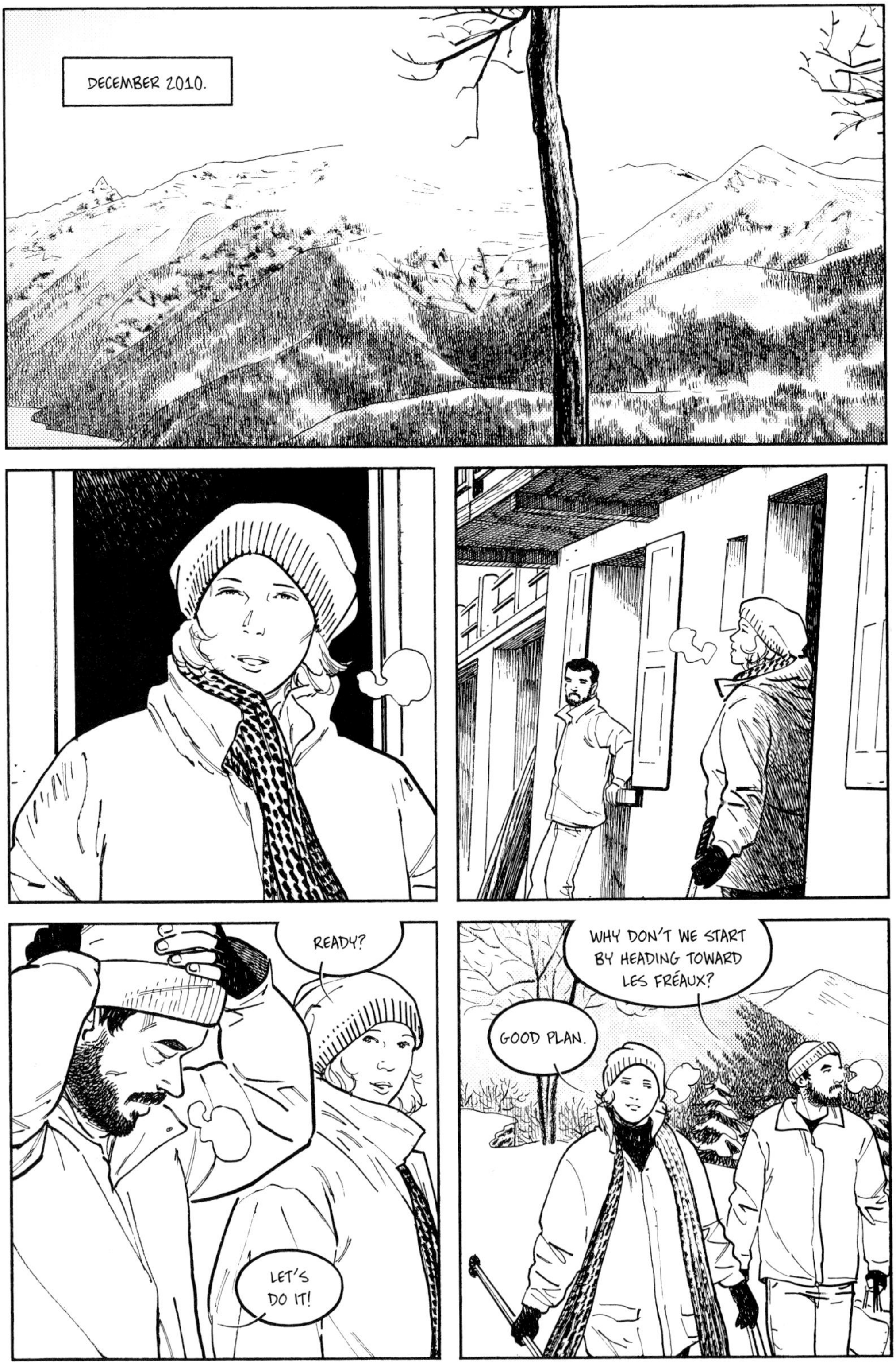
DECEMBER 2010.
READY?
LET'S DO IT!
WHY DON'T WE START BY HEADING TOWARD LES FRÉAUX?
GOOD PLAN.

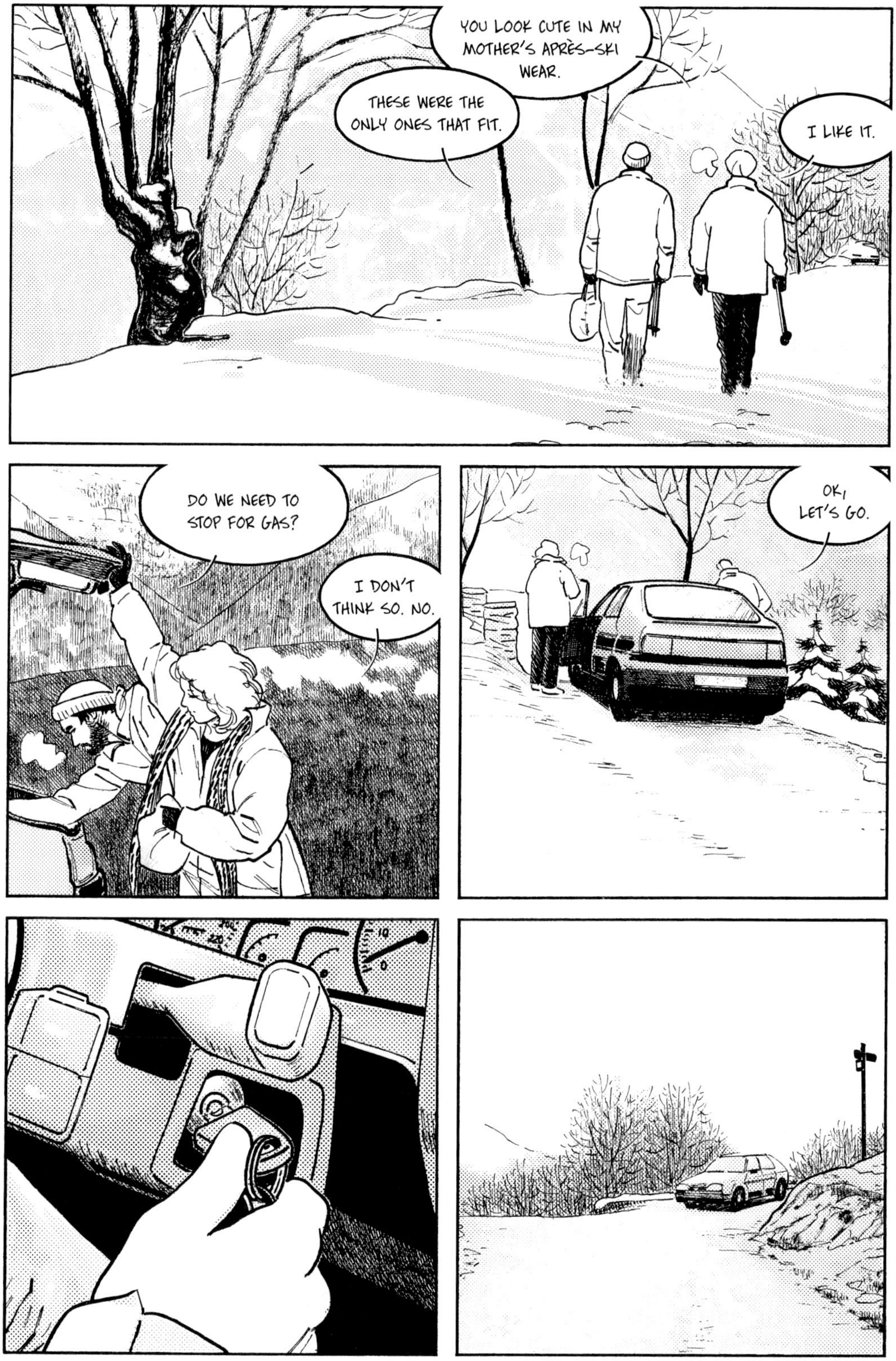
YOU LOOK CUTE IN MY MOTHER'S APRÈS-SKI WEAR.
THESE WERE THE ONLY ONES THAT FIT.
I LIKE IT.
DO WE NEED TO STOP FOR GAS?
I DON'T THINK SO. NO.
OK, LET'S GO.

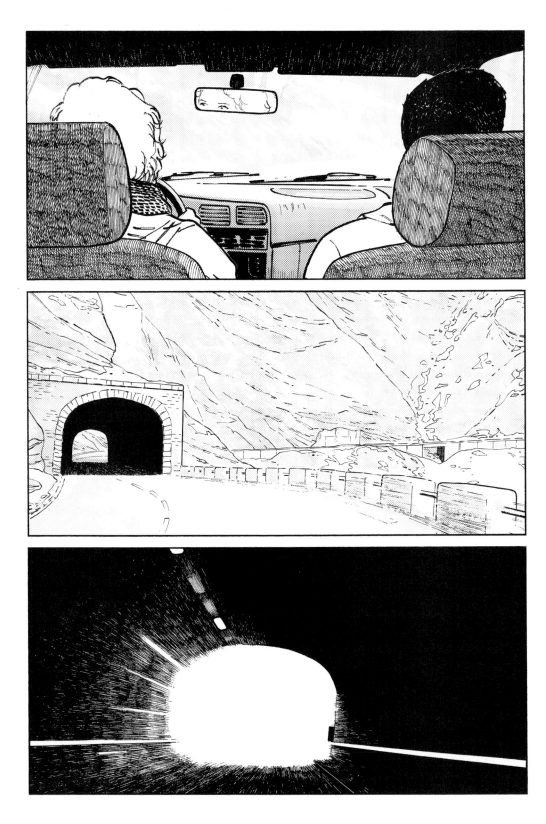

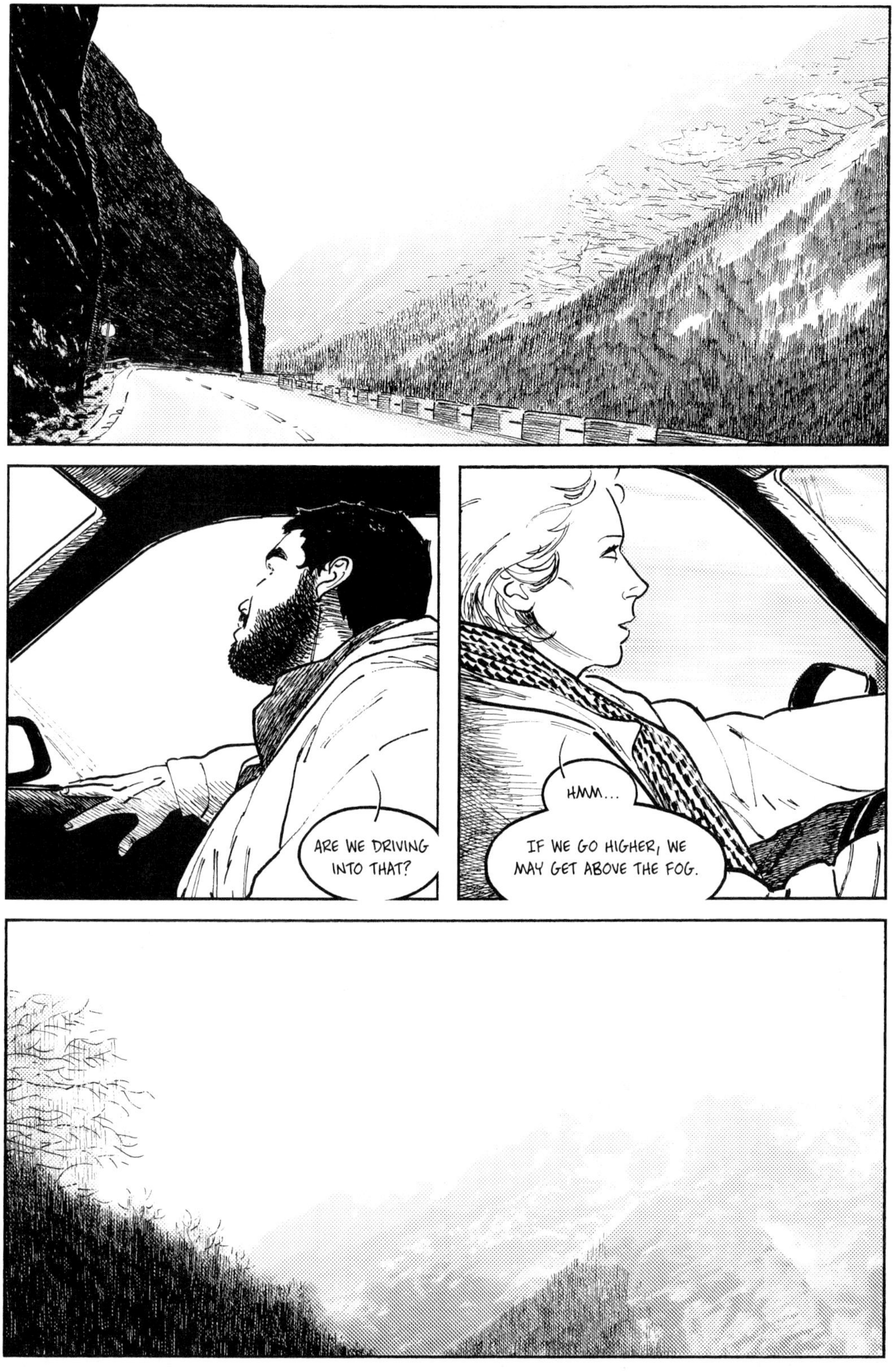
ARE WE DRIVING INTO THAT?
HMM...
IF WE GO HIGHER, WE MAY GET ABOVE THE FOG.

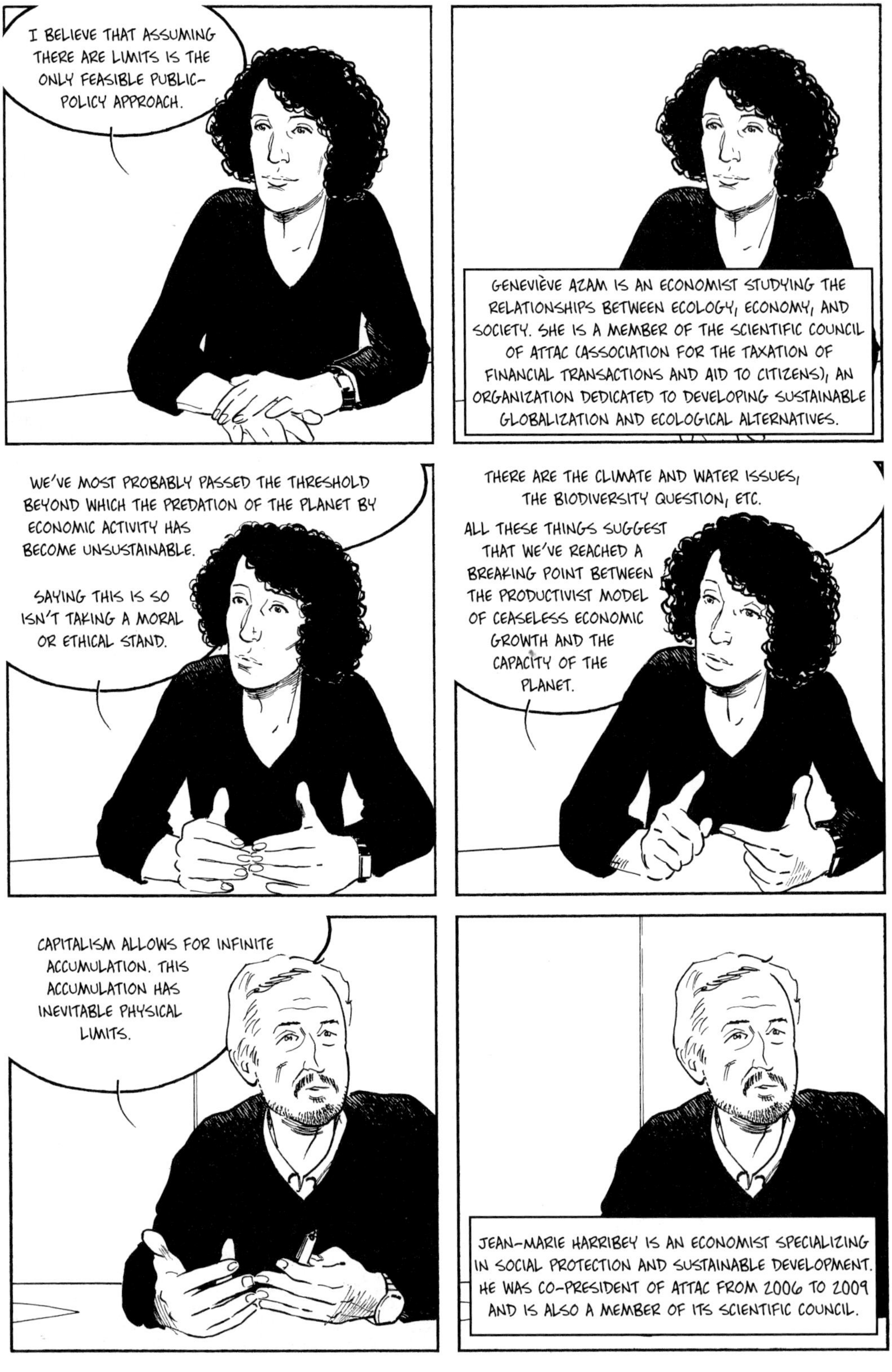
I BELIEVE THAT ASSUMING THERE ARE LIMITS IS THE ONLY FEASIBLE PUBLIC-POLICY APPROACH.
GENEVIÈVE AZAM IS AN ECONOMIST STUDYING THE RELATIONSHIPS BETWEEN ECOLOGY, ECONOMY, AND SOCIETY. SHE IS A MEMBER OF THE SCIENTIFIC COUNCIL OF ATTAC (ASSOCIATION FOR THE TAXATION OF FINANCIAL TRANSACTIONS AND AID TO CITIZENS), AN ORGANIZATION DEDICATED TO DEVELOPING SUSTAINABLE GLOBALIZATION AND ECOLOGICAL ALTERNATIVES.
WE'VE MOST PROBABLY PASSED THE THRESHOLD BEYOND WHICH THE PREDATION OF THE PLANET BY ECONOMIC ACTIVITY HAS BECOME UNSUSTAINABLE.
SAYING THIS IS SO ISN'T TAKING A MORAL OR ETHICAL STAND.
THERE ARE THE CLIMATE AND WATER ISSUES, THE BIODIVERSITY QUESTION, ETC.
ALL THESE THINGS SUGGEST THAT WE'VE REACHED A BREAKING POINT BETWEEN THE PRODUCTIVIST MODEL OF CEASELESS ECONOMIC GROWTH AND THE CAPACITY OF THE PLANET.
CAPITALISM ALLOWS FOR INFINITE ACCUMULATION. THIS ACCUMULATION HAS INEVITABLE PHYSICAL LIMITS.
JEAN-MARIE HARRIBEY IS AN ECONOMIST SPECIALIZING IN SOCIAL PROTECTION AND SUSTAINABLE DEVELOPMENT. HE WAS CO-PRESIDENT OF ATTAC FROM 2006 TO 2009 AND IS ALSO A MEMBER OF ITS SCIENTIFIC COUNCIL.

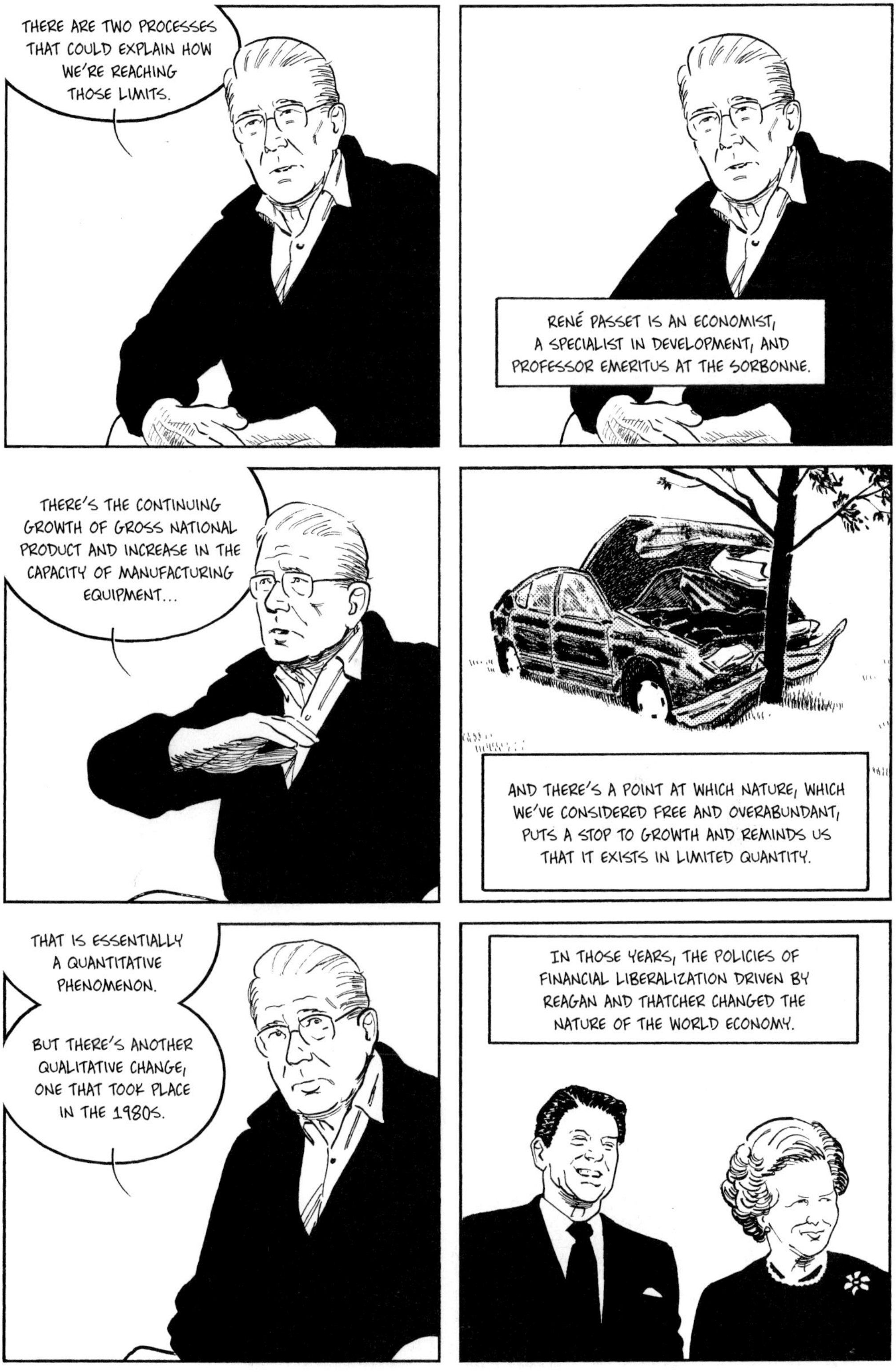
THERE ARE TWO PROCESSES THAT COULD EXPLAIN HOW WE'RE REACHING THOSE LIMITS.
RENÉ PASSET IS AN ECONOMIST, A SPECIALIST IN DEVELOPMENT, AND PROFESSOR EMERITUS AT THE SORBONNE.
THERE'S THE CONTINUING GROWTH OF GROSS NATIONAL PRODUCT AND INCREASE IN THE CAPACITY OF MANUFACTURING EQUIPMENT...
AND THERE'S A POINT AT WHICH NATURE, WHICH WE'VE CONSIDERED FREE AND OVERABUNDANT, PUTS A STOP TO GROWTH AND REMINDS US THAT IT EXISTS IN LIMITED QUANTITY.
THAT IS ESSENTIALLY A QUANTITATIVE PHENOMENON.
BUT THERE'S ANOTHER QUALITATIVE CHANGE, ONE THAT TOOK PLACE IN THE 1980S.
IN THOSE YEARS, THE POLICIES OF FINANCIAL LIBERALIZATION DRIVEN BY REAGAN AND THATCHER CHANGED THE NATURE OF THE WORLD ECONOMY.

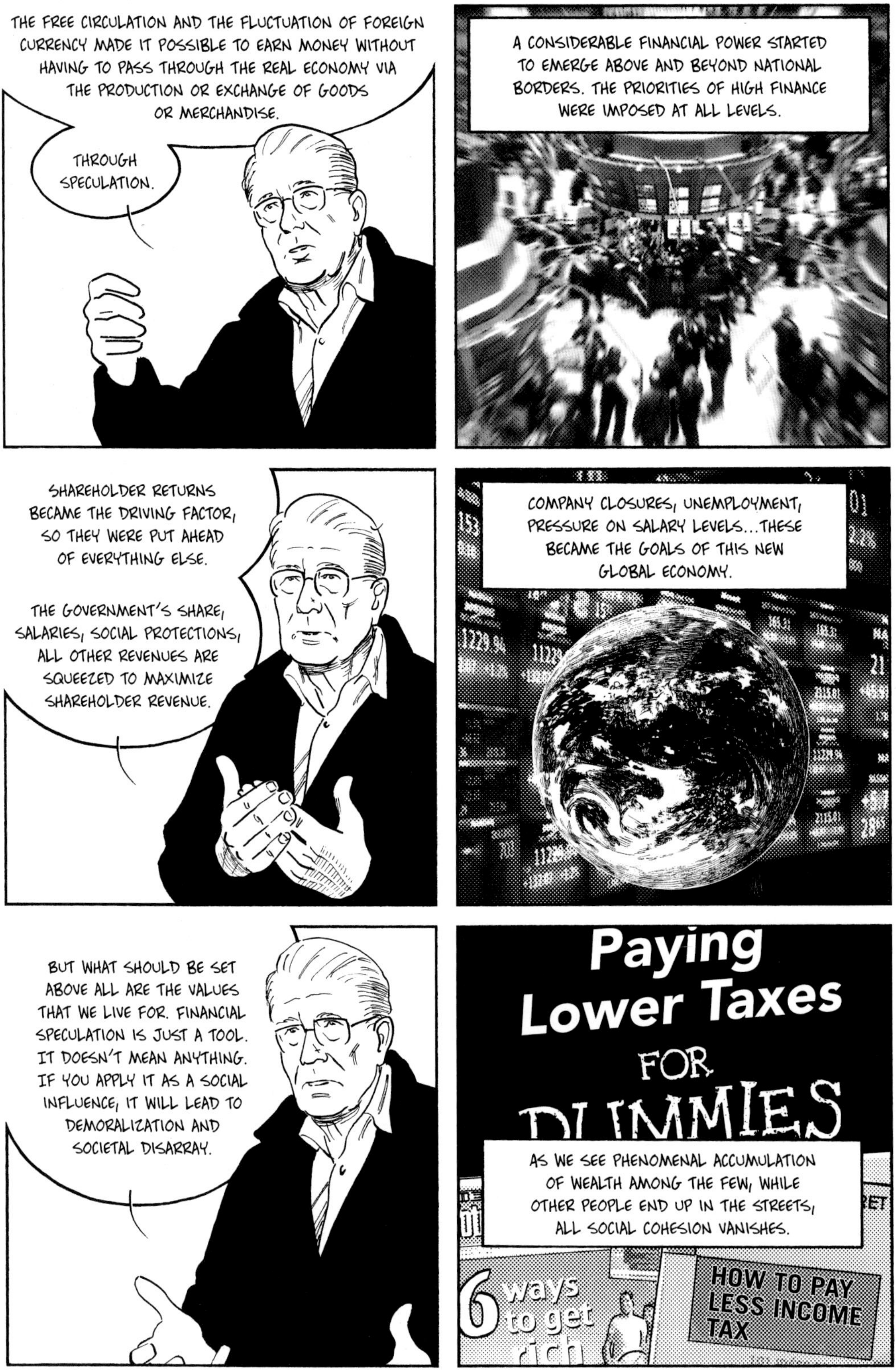
THE FREE CIRCULATION AND THE FLUCTUATION OF FOREIGN CURRENCY MADE IT POSSIBLE TO EARN MONEY WITHOUT HAVING TO PASS THROUGH THE REAL ECONOMY VIA THE PRODUCTION OR EXCHANGE OF GOODS OR MERCHANDISE.
THROUGH SPECULATION.
A CONSIDERABLE FINANCIAL POWER STARTED TO EMERGE ABOVE AND BEYOND NATIONAL BORDERS. THE PRIORITIES OF HIGH FINANCE WERE IMPOSED AT ALL LEVELS.
SHAREHOLDER RETURNS BECAME THE DRIVING FACTOR, SO THEY WERE PUT AHEAD OF EVERYTHING ELSE.
THE GOVERNMENT'S SHARE, SALARIES, SOCIAL PROTECTIONS, ALL OTHER REVENUES ARE SQUEEZED TO MAXIMIZE SHAREHOLDER REVENUE.
COMPANY CLOSURES, UNEMPLOYMENT, PRESSURE ON SALARY LEVELS...THESE BECAME THE GOALS OF THIS NEW GLOBAL ECONOMY.
BUT WHAT SHOULD BE SET ABOVE ALL ARE THE VALUES THAT WE LIVE FOR. FINANCIAL SPECULATION IS JUST A TOOL. IT DOESN'T MEAN ANYTHING. IF YOU APPLY IT AS A SOCIAL INFLUENCE, IT WILL LEAD TO DEMORALIZATION AND SOCIETAL DISARRAY.
Paying Lower Taxes
FOR
AS WE SEE PHENOMENAL ACCUMULATION OF WEALTH AMONG THE FEW, WHILE OTHER PEOPLE END UP IN THE STREETS, ALL SOCIAL COHESION VANISHES.
6 ways to get
HOW TO PAY LESS INCOME TAX

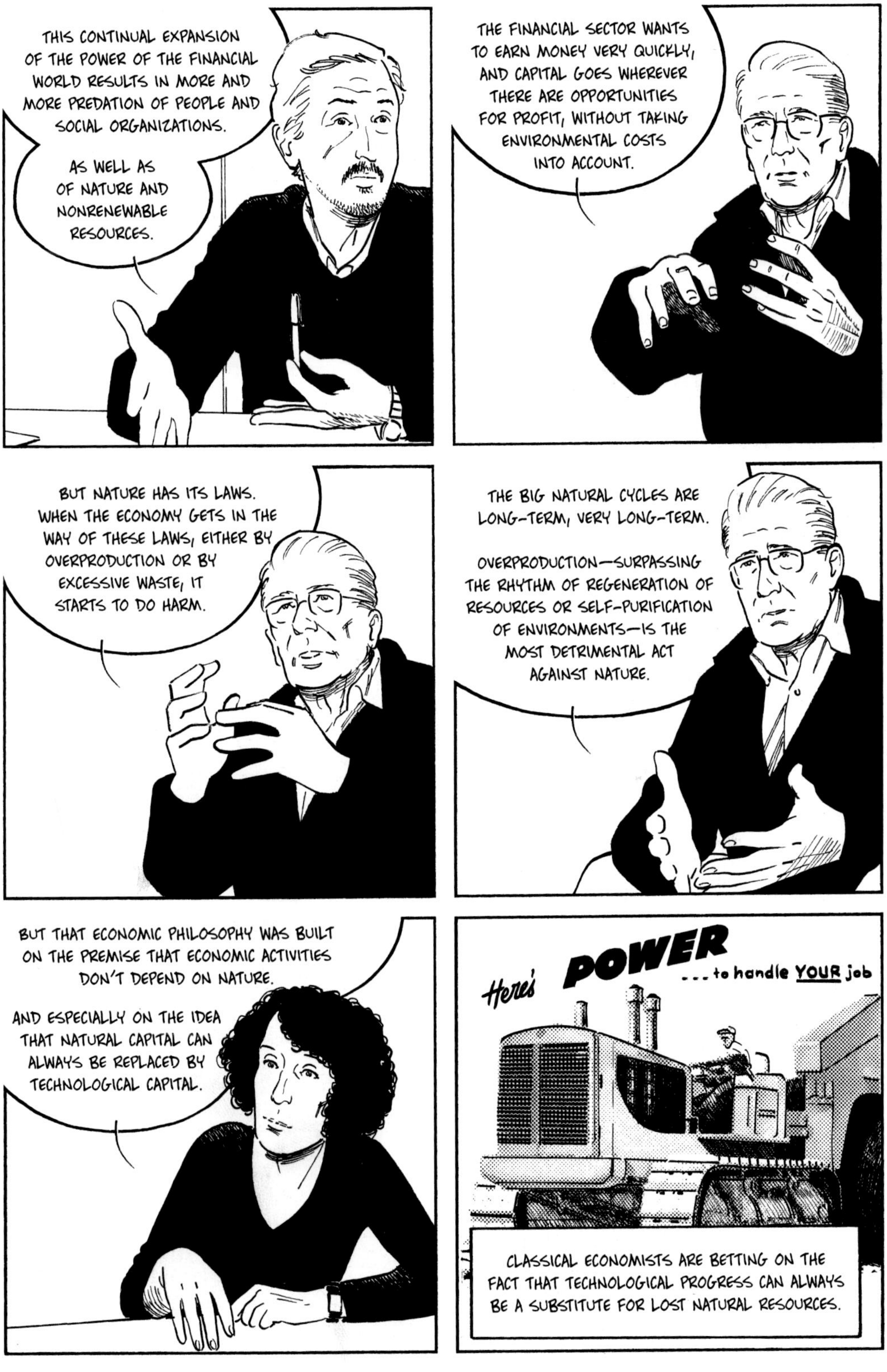
THIS CONTINUAL EXPANSION OF THE POWER OF THE FINANCIAL WORLD RESULTS IN MORE AND MORE PREDATION OF PEOPLE AND SOCIAL ORGANIZATIONS.
AS WELL AS OF NATURE AND NONRENEWABLE RESOURCES.
THE FINANCIAL SECTOR WANTS TO EARN MONEY VERY QUICKLY, AND CAPITAL GOES WHEREVER THERE ARE OPPORTUNITIES FOR PROFIT, WITHOUT TAKING ENVIRONMENTAL COSTS INTO ACCOUNT.
BUT NATURE HAS ITS LAWS. WHEN THE ECONOMY GETS IN THE WAY OF THESE LAWS, EITHER BY OVERPRODUCTION OR BY EXCESSIVE WASTE, IT STARTS TO DO HARM.
THE BIG NATURAL CYCLES ARE LONG-TERM, VERY LONG-TERM.
OVERPRODUCTION—SURPASSING THE RHYTHM OF REGENERATION OF RESOURCES OR SELF-PURIFICATION OF ENVIRONMENTS—IS THE MOST DETRIMENTAL ACT AGAINST NATURE.
BUT THAT ECONOMIC PHILOSOPHY WAS BUILT ON THE PREMISE THAT ECONOMIC ACTIVITIES DON'T DEPEND ON NATURE.
AND ESPECIALLY ON THE IDEA THAT NATURAL CAPITAL CAN ALWAYS BE REPLACED BY TECHNOLOGICAL CAPITAL.
Here's POWER ... to handle YOUR job
CLASSICAL ECONOMISTS ARE BETTING ON THE FACT THAT TECHNOLOGICAL PROGRESS CAN ALWAYS BE A SUBSTITUTE FOR LOST NATURAL RESOURCES.

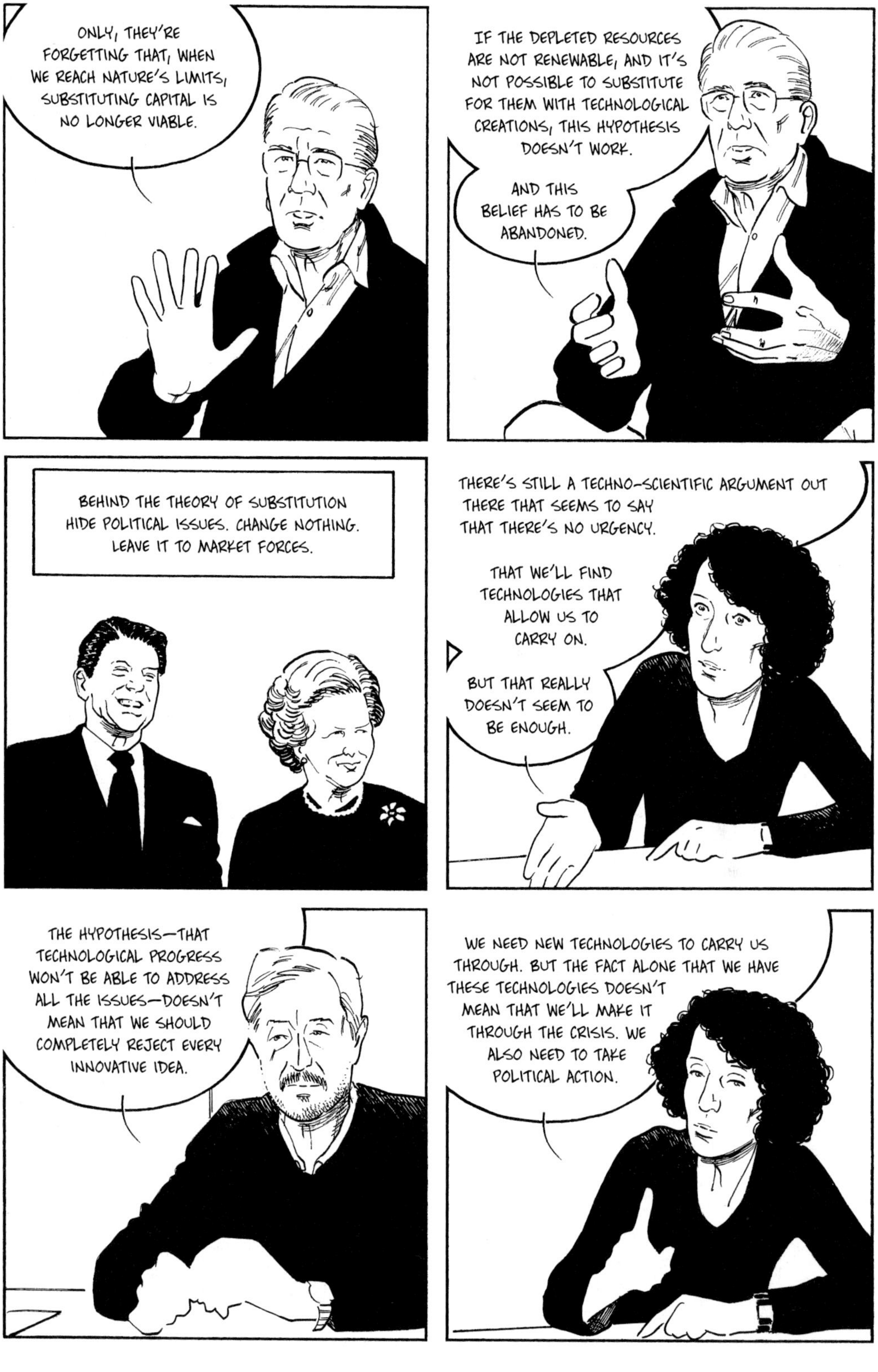
ONLY, THEY'RE FORGETTING THAT, WHEN WE REACH NATURE'S LIMITS, SUBSTITUTING CAPITAL IS NO LONGER VIABLE.
IF THE DEPLETED RESOURCES ARE NOT RENEWABLE, AND IT'S NOT POSSIBLE TO SUBSTITUTE FOR THEM WITH TECHNOLOGICAL CREATIONS, THIS HYPOTHESIS DOESN'T WORK.
AND THIS BELIEF HAS TO BE ABANDONED.
BEHIND THE THEORY OF SUBSTITUTION HIDE POLITICAL ISSUES. CHANGE NOTHING. LEAVE IT TO MARKET FORCES.
THERE'S STILL A TECHNO-SCIENTIFIC ARGUMENT OUT THERE THAT SEEMS TO SAY THAT THERE'S NO URGENCY.
THAT WE'LL FIND TECHNOLOGIES THAT ALLOW US TO CARRY ON.
BUT THAT REALLY DOESN'T SEEM TO BE ENOUGH.
THE HYPOTHESIS—THAT TECHNOLOGICAL PROGRESS WON'T BE ABLE TO ADDRESS ALL THE ISSUES—DOESN'T MEAN THAT WE SHOULD COMPLETELY REJECT EVERY INNOVATIVE IDEA.
WE NEED NEW TECHNOLOGIES TO CARRY US THROUGH. BUT THE FACT ALONE THAT WE HAVE THESE TECHNOLOGIES DOESN'T MEAN THAT WE'LL MAKE IT THROUGH THE CRISIS. WE ALSO NEED TO TAKE POLITICAL ACTION.

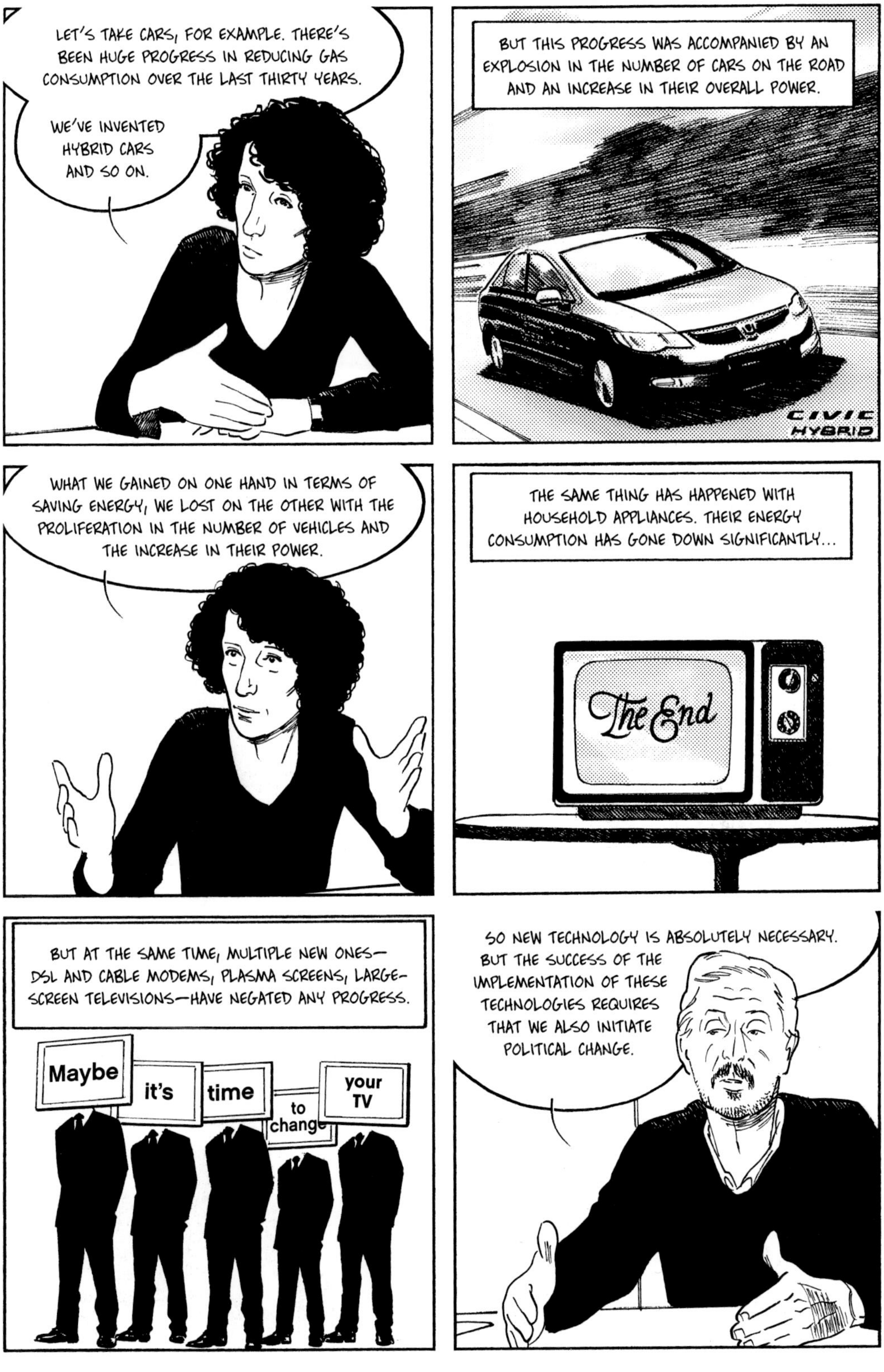
LET'S TAKE CARS, FOR EXAMPLE. THERE'S BEEN HUGE PROGRESS IN REDUCING GAS CONSUMPTION OVER THE LAST THIRTY YEARS.
WE'VE INVENTED HYBRID CARS AND SO ON.
BUT THIS PROGRESS WAS ACCOMPANIED BY AN EXPLOSION IN THE NUMBER OF CARS ON THE ROAD AND AN INCREASE IN THEIR OVERALL POWER.
CIVIC HYBRID
WHAT WE GAINED ON ONE HAND IN TERMS OF SAVING ENERGY, WE LOST ON THE OTHER WITH THE PROLIFERATION IN THE NUMBER OF VEHICLES AND THE INCREASE IN THEIR POWER.
THE SAME THING HAS HAPPENED WITH HOUSEHOLD APPLIANCES. THEIR ENERGY CONSUMPTION HAS GONE DOWN SIGNIFICANTLY...
The End
BUT AT THE SAME TIME, MULTIPLE NEW ONES—DSL AND CABLE MODEMS, PLASMA SCREENS, LARGE-SCREEN TELEVISIONS—HAVE NEGATED ANY PROGRESS.
Maybe it's time to change your TV
SO NEW TECHNOLOGY IS ABSOLUTELY NECESSARY.
BUT THE SUCCESS OF THE IMPLEMENTATION OF THESE TECHNOLOGIES REQUIRES THAT WE ALSO INITIATE POLITICAL CHANGE.

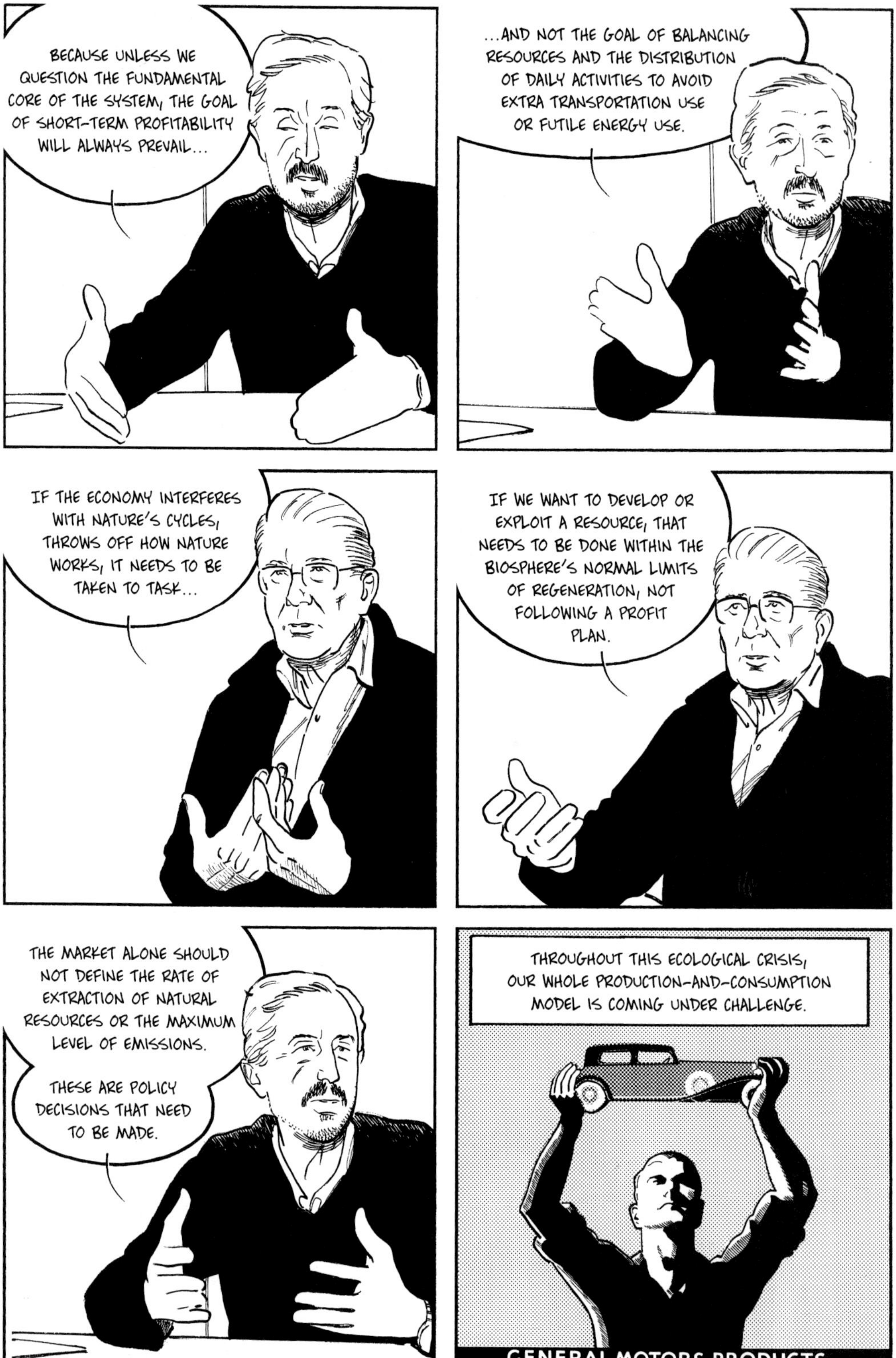
BECAUSE UNLESS WE QUESTION THE FUNDAMENTAL CORE OF THE SYSTEM, THE GOAL OF SHORT-TERM PROFITABILITY WILL ALWAYS PREVAIL...
...AND NOT THE GOAL OF BALANCING RESOURCES AND THE DISTRIBUTION OF DAILY ACTIVITIES TO AVOID EXTRA TRANSPORTATION USE OR FUTILE ENERGY USE.
IF THE ECONOMY INTERFERES WITH NATURE'S CYCLES, THROWS OFF HOW NATURE WORKS, IT NEEDS TO BE TAKEN TO TASK...
IF WE WANT TO DEVELOP OR EXPLOIT A RESOURCE, THAT NEEDS TO BE DONE WITHIN THE BIOSPHERE'S NORMAL LIMITS OF REGENERATION, NOT FOLLOWING A PROFIT PLAN.
THE MARKET ALONE SHOULD NOT DEFINE THE RATE OF EXTRACTION OF NATURAL RESOURCES OR THE MAXIMUM LEVEL OF EMISSIONS.
THESE ARE POLICY DECISIONS THAT NEED TO BE MADE.
THROUGHOUT THIS ECOLOGICAL CRISIS, OUR WHOLE PRODUCTION-AND-CONSUMPTION MODEL IS COMING UNDER CHALLENGE.
GENERAL MOTORS PRODUCTS

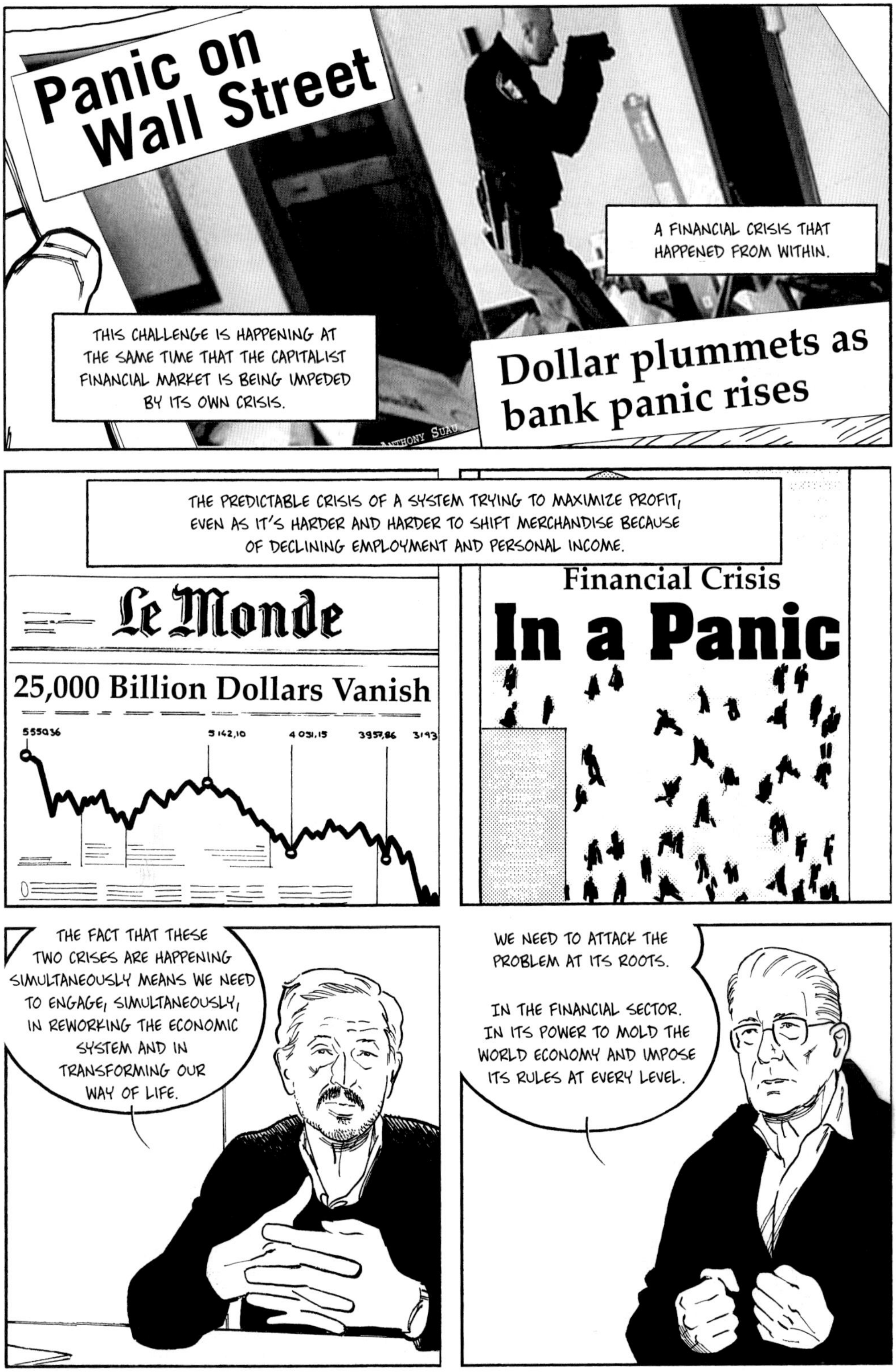
Panic on Wall Street
A FINANCIAL CRISIS THAT HAPPENED FROM WITHIN.
THIS CHALLENGE IS HAPPENING AT THE SAME TIME THAT THE CAPITALIST FINANCIAL MARKET IS BEING IMPEDED BY ITS OWN CRISIS.
ANTHONY SUAU
Dollar plummets as bank panic rises
THE PREDICTABLE CRISIS OF A SYSTEM TRYING TO MAXIMIZE PROFIT, EVEN AS IT'S HARDER AND HARDER TO SHIFT MERCHANDISE BECAUSE OF DECLINING EMPLOYMENT AND PERSONAL INCOME.
Le Monde
25,000 Billion Dollars Vanish
5142,10
4 051,15
3957,86
Financial Crisis
In a Panic
THE FACT THAT THESE TWO CRISES ARE HAPPENING SIMULTANEOUSLY MEANS WE NEED TO ENGAGE, SIMULTANEOUSLY, IN REWORKING THE ECONOMIC SYSTEM AND IN TRANSFORMING OUR WAY OF LIFE.
WE NEED TO ATTACK THE PROBLEM AT ITS ROOTS.
IN THE FINANCIAL SECTOR. IN ITS POWER TO MOLD THE WORLD ECONOMY AND IMPOSE ITS RULES AT EVERY LEVEL.

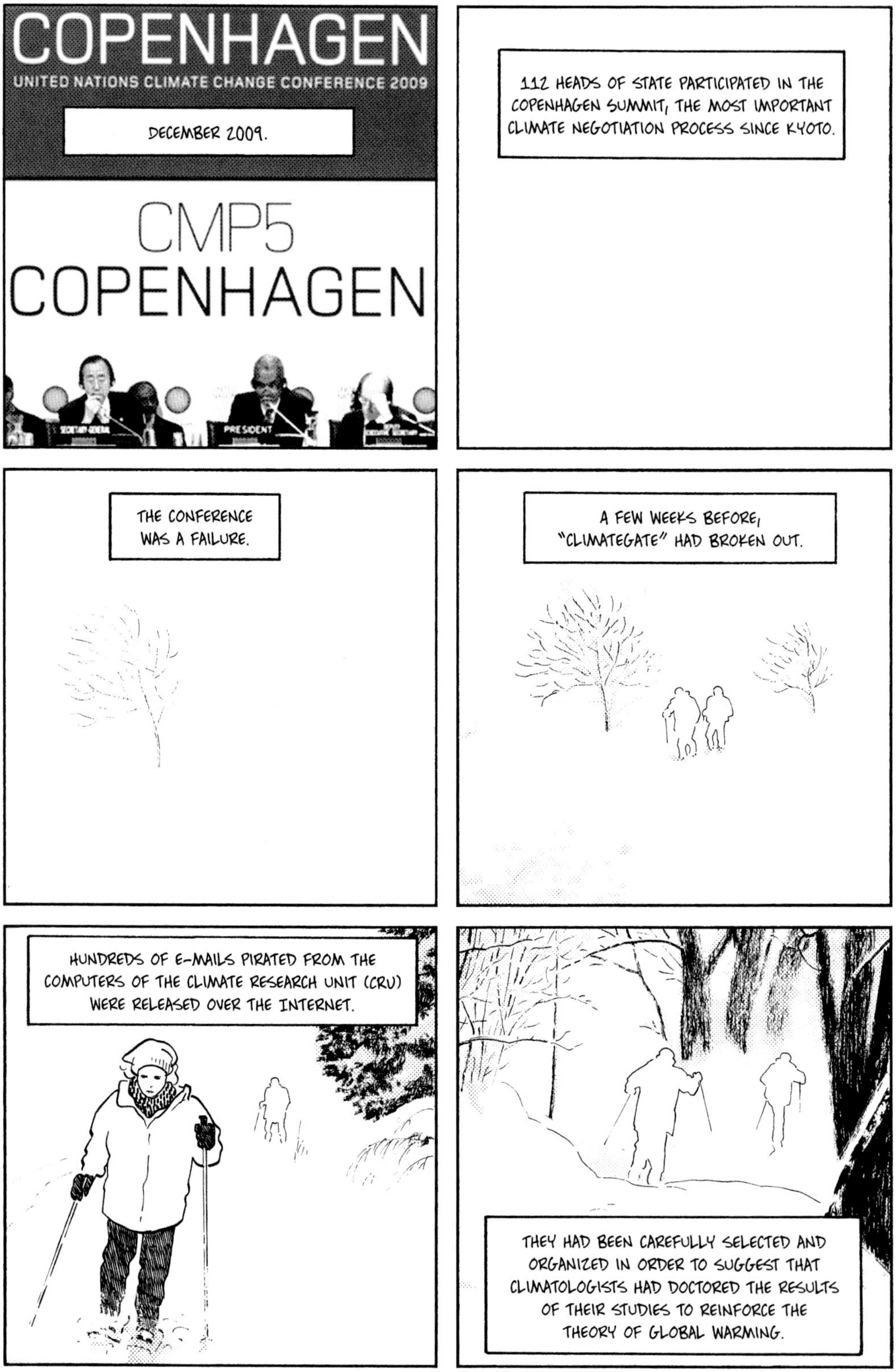
COPENHAGEN
UNITED NATIONS CLIMATE CHANGE CONFERENCE 2009
DECEMBER 2009.
CMP5
COPENHAGEN
SECRETARY-GENERAL
PRESIDENT
112 HEADS OF STATE PARTICIPATED IN THE COPENHAGEN SUMMIT, THE MOST IMPORTANT CLIMATE NEGOTIATION PROCESS SINCE KYOTO.
THE CONFERENCE WAS A FAILURE.
A FEW WEEKS BEFORE, "CLIMATEGATE" HAD BROKEN OUT.
HUNDREDS OF E-MAILS PIRATED FROM THE COMPUTERS OF THE CLIMATE RESEARCH UNIT (CRU) WERE RELEASED OVER THE INTERNET.
THEY HAD BEEN CAREFULLY SELECTED AND ORGANIZED IN ORDER TO SUGGEST THAT CLIMATOLOGISTS HAD DOCTORED THE RESULTS OF THEIR STUDIES TO REINFORCE THE THEORY OF GLOBAL WARMING.

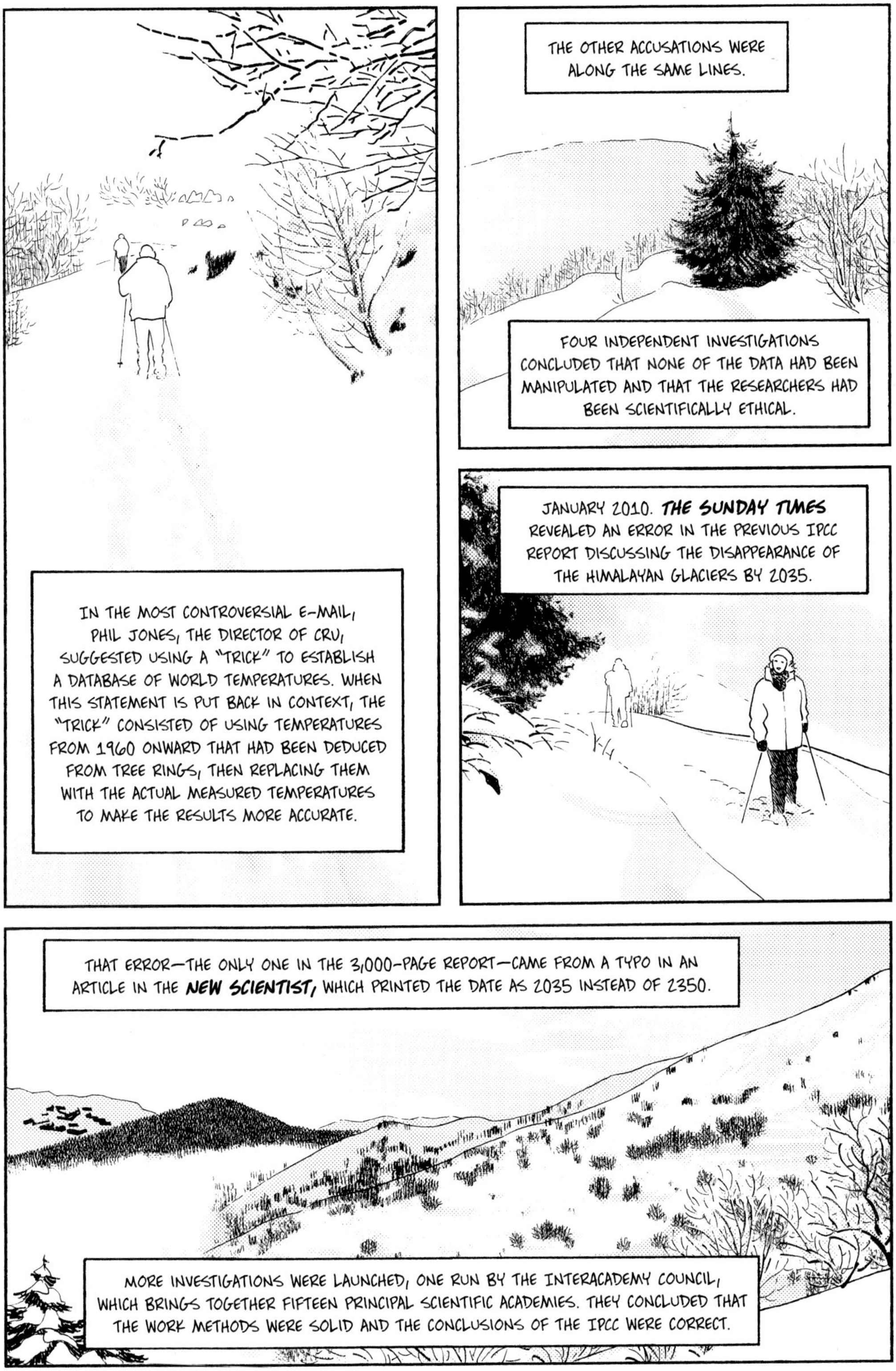
IN THE MOST CONTROVERSIAL E-MAIL, PHIL JONES, THE DIRECTOR OF CRU, SUGGESTED USING A "TRICK" TO ESTABLISH A DATABASE OF WORLD TEMPERATURES. WHEN THIS STATEMENT IS PUT BACK IN CONTEXT, THE "TRICK" CONSISTED OF USING TEMPERATURES FROM 1960 ONWARD THAT HAD BEEN DEDUCED FROM TREE RINGS, THEN REPLACING THEM WITH THE ACTUAL MEASURED TEMPERATURES TO MAKE THE RESULTS MORE ACCURATE.
THE OTHER ACCUSATIONS WERE ALONG THE SAME LINES.
FOUR INDEPENDENT INVESTIGATIONS CONCLUDED THAT NONE OF THE DATA HAD BEEN MANIPULATED AND THAT THE RESEARCHERS HAD BEEN SCIENTIFICALLY ETHICAL.
JANUARY 2010. **THE SUNDAY TIMES** REVEALED AN ERROR IN THE PREVIOUS IPCC REPORT DISCUSSING THE DISAPPEARANCE OF THE HIMALAYAN GLACIERS BY 2035.
THAT ERROR—THE ONLY ONE IN THE 3,000-PAGE REPORT—CAME FROM A TYPO IN AN ARTICLE IN THE **NEW SCIENTIST,** WHICH PRINTED THE DATE AS 2035 INSTEAD OF 2350.
MORE INVESTIGATIONS WERE LAUNCHED, ONE RUN BY THE INTERACADEMY COUNCIL, WHICH BRINGS TOGETHER FIFTEEN PRINCIPAL SCIENTIFIC ACADEMIES. THEY CONCLUDED THAT THE WORK METHODS WERE SOLID AND THE CONCLUSIONS OF THE IPCC WERE CORRECT.

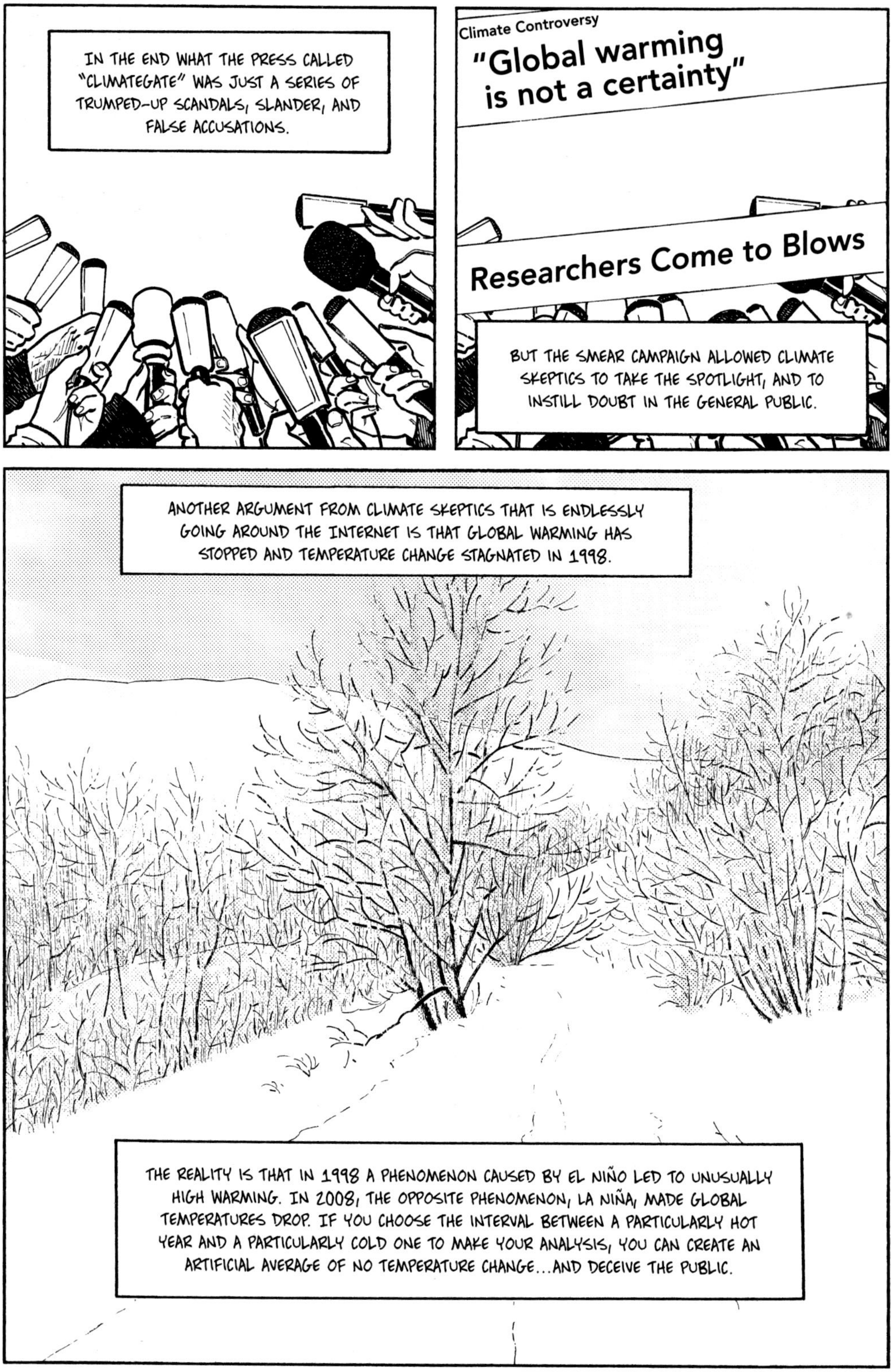
IN THE END WHAT THE PRESS CALLED "CLIMATEGATE" WAS JUST A SERIES OF TRUMPED-UP SCANDALS, SLANDER, AND FALSE ACCUSATIONS.
Climate Controversy
"Global warming is not a certainty"
Researchers Come to Blows
BUT THE SMEAR CAMPAIGN ALLOWED CLIMATE SKEPTICS TO TAKE THE SPOTLIGHT, AND TO INSTILL DOUBT IN THE GENERAL PUBLIC.
ANOTHER ARGUMENT FROM CLIMATE SKEPTICS THAT IS ENDLESSLY GOING AROUND THE INTERNET IS THAT GLOBAL WARMING HAS STOPPED AND TEMPERATURE CHANGE STAGNATED IN 1998.
THE REALITY IS THAT IN 1998 A PHENOMENON CAUSED BY EL NIÑO LED TO UNUSUALLY HIGH WARMING. IN 2008, THE OPPOSITE PHENOMENON, LA NIÑA, MADE GLOBAL TEMPERATURES DROP. IF YOU CHOOSE THE INTERVAL BETWEEN A PARTICULARLY HOT YEAR AND A PARTICULARLY COLD ONE TO MAKE YOUR ANALYSIS, YOU CAN CREATE AN ARTIFICIAL AVERAGE OF NO TEMPERATURE CHANGE...AND DECEIVE THE PUBLIC.

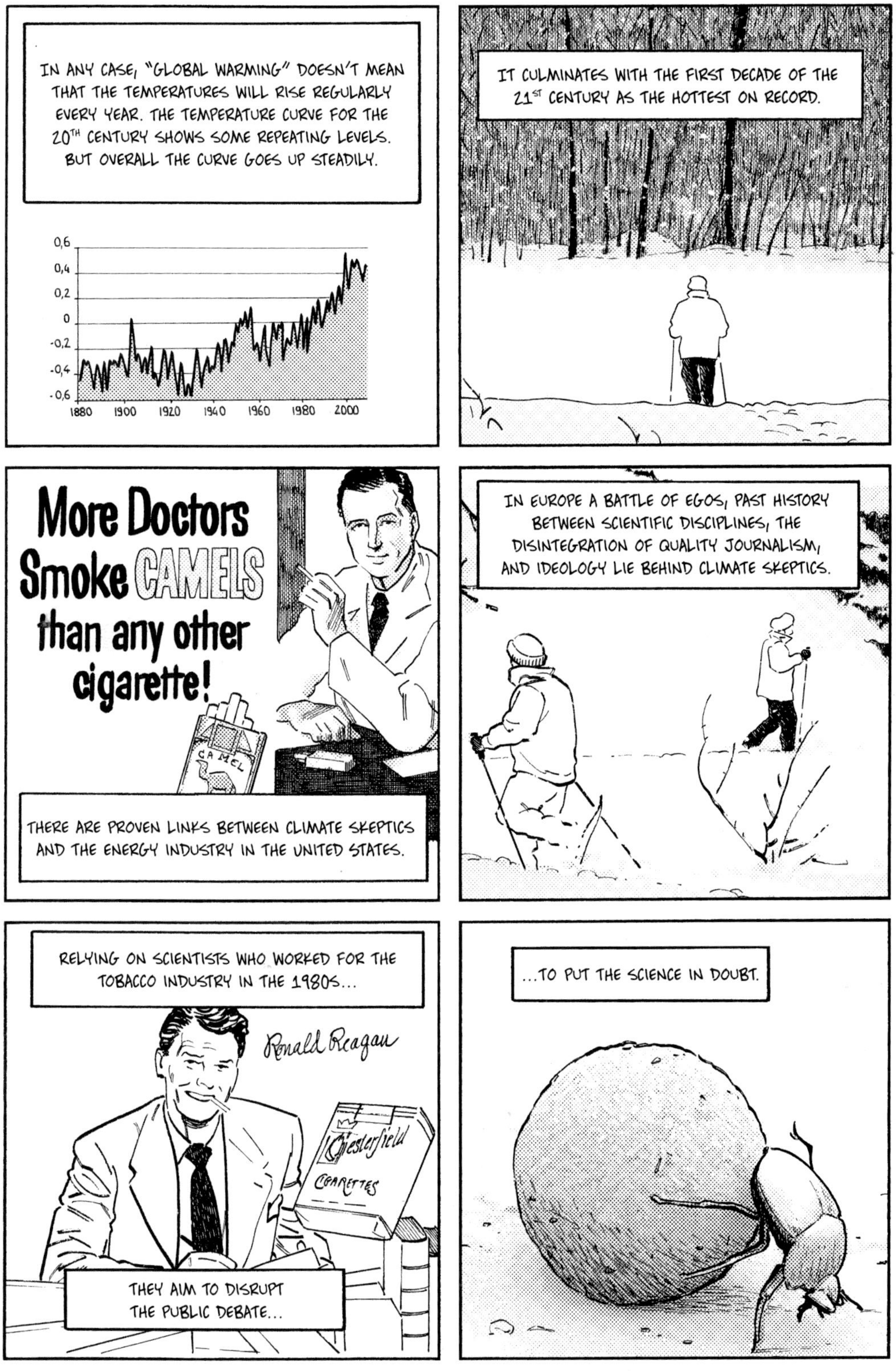
IN ANY CASE, "GLOBAL WARMING" DOESN'T MEAN THAT THE TEMPERATURES WILL RISE REGULARLY EVERY YEAR. THE TEMPERATURE CURVE FOR THE 20TH CENTURY SHOWS SOME REPEATING LEVELS. BUT OVERALL THE CURVE GOES UP STEADILY.
0,6
0,4
0,2
0
-0,2
-0,4
-0,6
1880
1900
1920
1940
1960
1980
2000
IT CULMINATES WITH THE FIRST DECADE OF THE 21ST CENTURY AS THE HOTTEST ON RECORD.
More Doctors Smoke CAMELS than any other cigarette!
CAMEL
THERE ARE PROVEN LINKS BETWEEN CLIMATE SKEPTICS AND THE ENERGY INDUSTRY IN THE UNITED STATES.
IN EUROPE A BATTLE OF EGOS, PAST HISTORY BETWEEN SCIENTIFIC DISCIPLINES, THE DISINTEGRATION OF QUALITY JOURNALISM, AND IDEOLOGY LIE BEHIND CLIMATE SKEPTICS.
RELYING ON SCIENTISTS WHO WORKED FOR THE TOBACCO INDUSTRY IN THE 1980S...
Ronald Reagan
Chesterfield
CIGARETTES
THEY AIM TO DISRUPT THE PUBLIC DEBATE...
...TO PUT THE SCIENCE IN DOUBT.

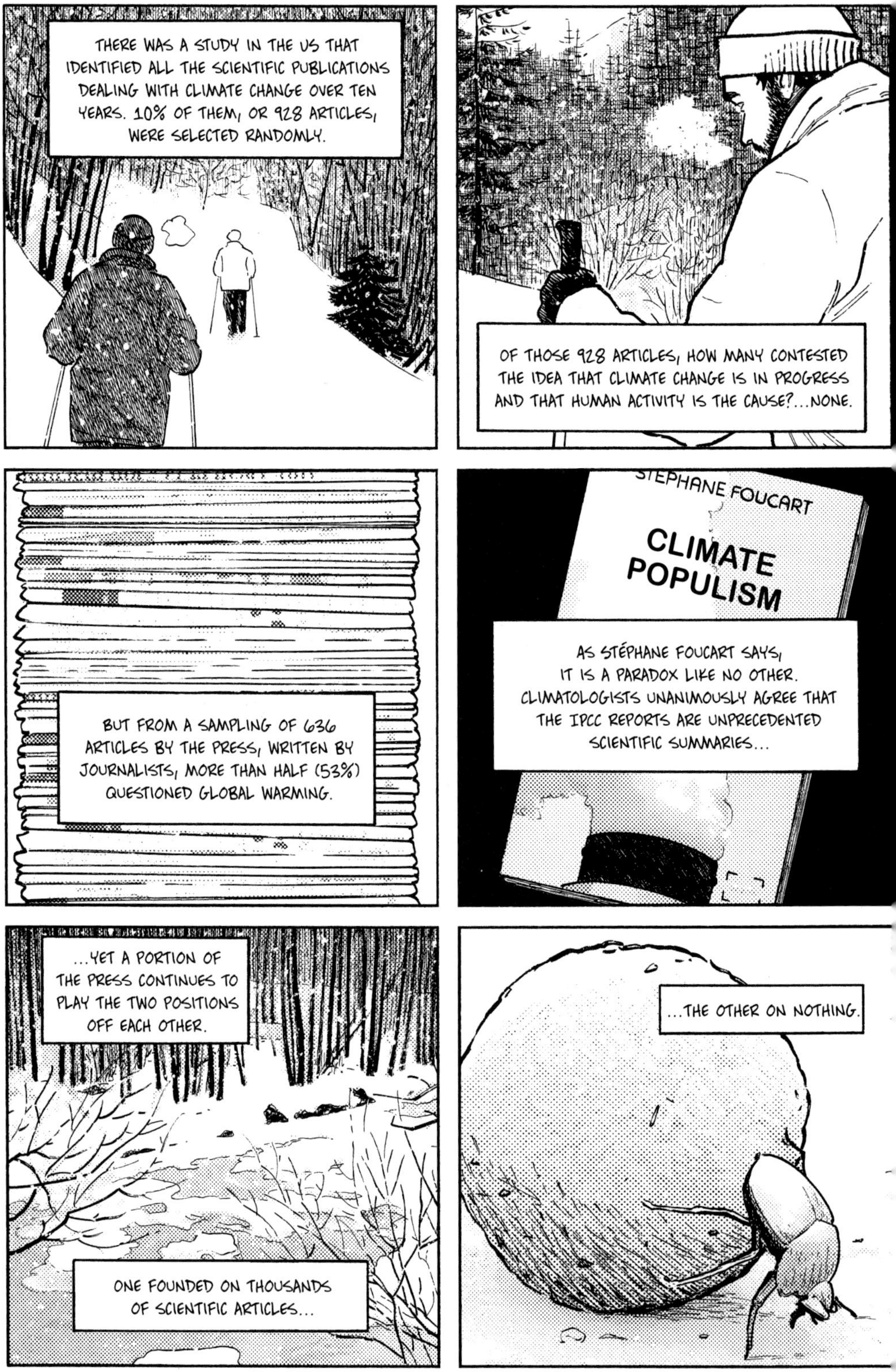
THERE WAS A STUDY IN THE US THAT IDENTIFIED ALL THE SCIENTIFIC PUBLICATIONS DEALING WITH CLIMATE CHANGE OVER TEN YEARS. 10% OF THEM, OR 928 ARTICLES, WERE SELECTED RANDOMLY.
OF THOSE 928 ARTICLES, HOW MANY CONTESTED THE IDEA THAT CLIMATE CHANGE IS IN PROGRESS AND THAT HUMAN ACTIVITY IS THE CAUSE?...NONE.
BUT FROM A SAMPLING OF 636 ARTICLES BY THE PRESS, WRITTEN BY JOURNALISTS, MORE THAN HALF (53%) QUESTIONED GLOBAL WARMING.
STÉPHANE FOUCART
CLIMATE POPULISM
AS STÉPHANE FOUCART SAYS, IT IS A PARADOX LIKE NO OTHER. CLIMATOLOGISTS UNANIMOUSLY AGREE THAT THE IPCC REPORTS ARE UNPRECEDENTED SCIENTIFIC SUMMARIES...
...YET A PORTION OF THE PRESS CONTINUES TO PLAY THE TWO POSITIONS OFF EACH OTHER.
ONE FOUNDED ON THOUSANDS OF SCIENTIFIC ARTICLES...
...THE OTHER ON NOTHING.

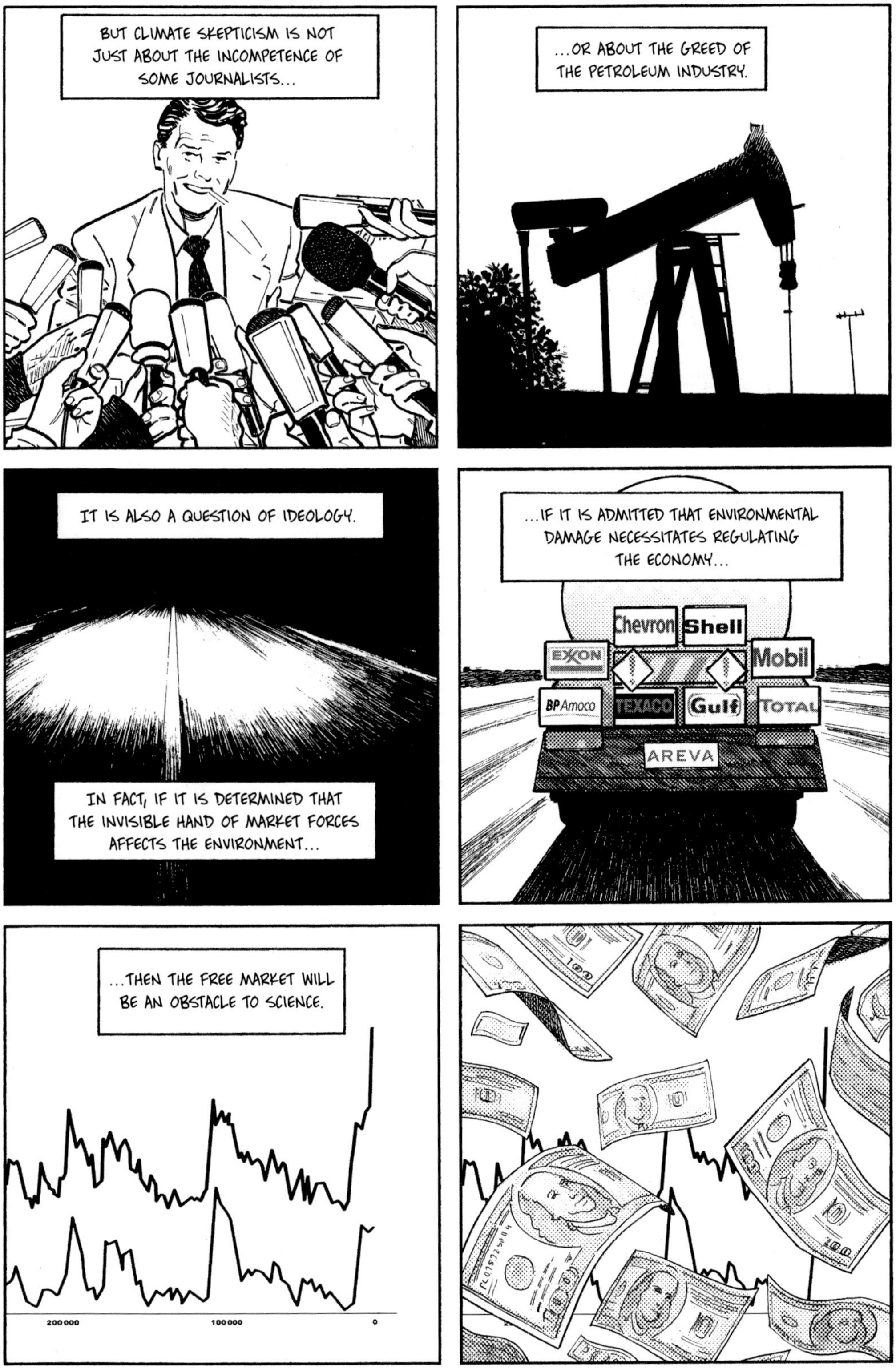
BUT CLIMATE SKEPTICISM IS NOT JUST ABOUT THE INCOMPETENCE OF SOME JOURNALISTS...
...OR ABOUT THE GREED OF THE PETROLEUM INDUSTRY.
IT IS ALSO A QUESTION OF IDEOLOGY.
IN FACT, IF IT IS DETERMINED THAT THE INVISIBLE HAND OF MARKET FORCES AFFECTS THE ENVIRONMENT...
...IF IT IS ADMITTED THAT ENVIRONMENTAL DAMAGE NECESSITATES REGULATING THE ECONOMY...
Chevron
Shell
EXXON
Mobil
BP Amoco
TEXACO
Gulf
TOTAL
AREVA
...THEN THE FREE MARKET WILL BE AN OBSTACLE TO SCIENCE.
200000
100000
0

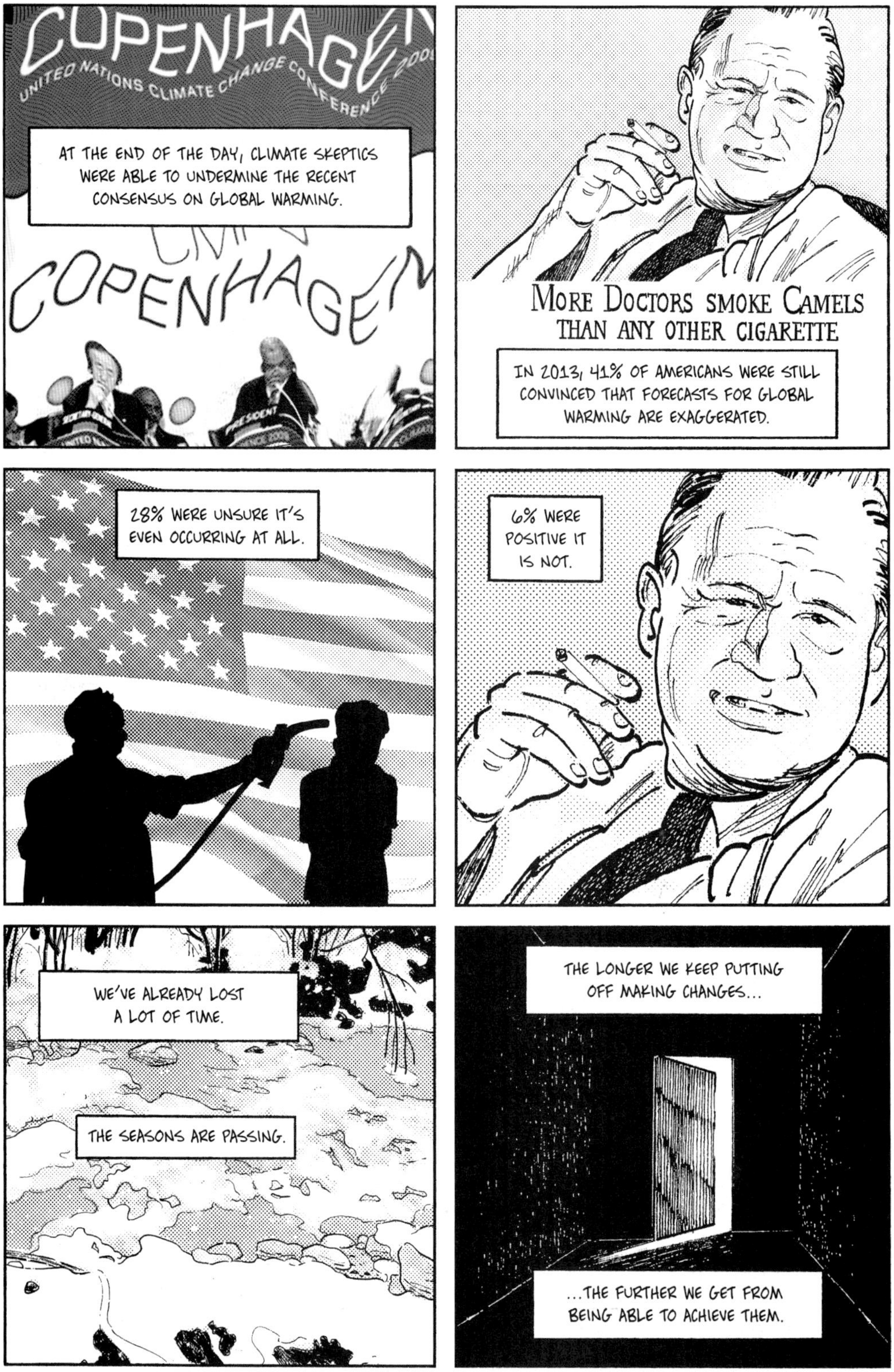
COPENHAGEN
UNITED NATIONS CLIMATE CHANGE CONFERENCE 2009
AT THE END OF THE DAY, CLIMATE SKEPTICS WERE ABLE TO UNDERMINE THE RECENT CONSENSUS ON GLOBAL WARMING.
COPENHAGEN
PRESIDENT
MORE DOCTORS SMOKE CAMELS THAN ANY OTHER CIGARETTE
IN 2013, 41% OF AMERICANS WERE STILL CONVINCED THAT FORECASTS FOR GLOBAL WARMING ARE EXAGGERATED.
28% WERE UNSURE IT'S EVEN OCCURRING AT ALL.
6% WERE POSITIVE IT IS NOT.
WE'VE ALREADY LOST A LOT OF TIME.
THE SEASONS ARE PASSING.
THE LONGER WE KEEP PUTTING OFF MAKING CHANGES...
...THE FURTHER WE GET FROM BEING ABLE TO ACHIEVE THEM.

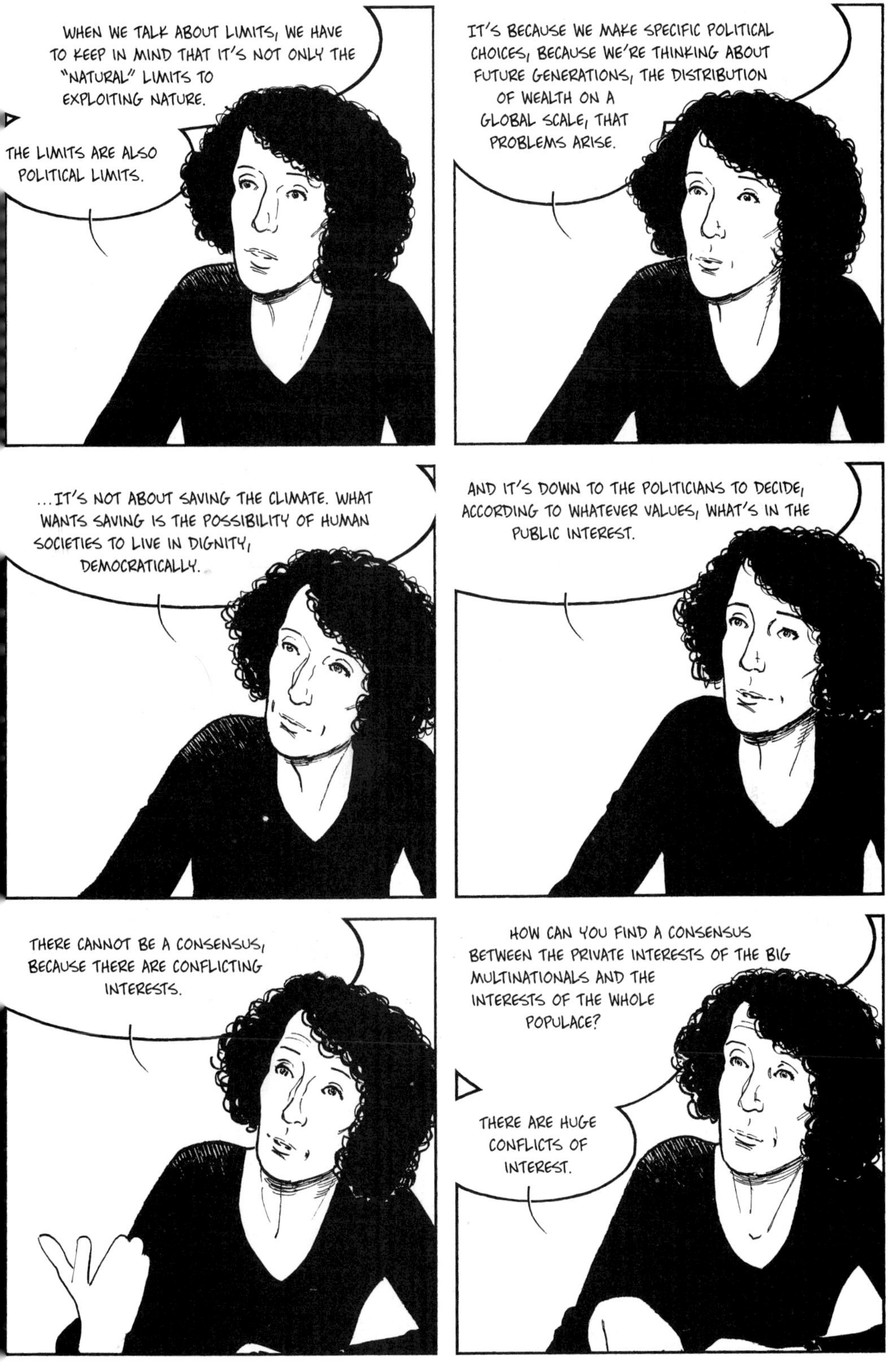
WHEN WE TALK ABOUT LIMITS, WE HAVE TO KEEP IN MIND THAT IT'S NOT ONLY THE "NATURAL" LIMITS TO EXPLOITING NATURE.
THE LIMITS ARE ALSO POLITICAL LIMITS.
IT'S BECAUSE WE MAKE SPECIFIC POLITICAL CHOICES, BECAUSE WE'RE THINKING ABOUT FUTURE GENERATIONS, THE DISTRIBUTION OF WEALTH ON A GLOBAL SCALE, THAT PROBLEMS ARISE.
...IT'S NOT ABOUT SAVING THE CLIMATE. WHAT WANTS SAVING IS THE POSSIBILITY OF HUMAN SOCIETIES TO LIVE IN DIGNITY, DEMOCRATICALLY.
AND IT'S DOWN TO THE POLITICIANS TO DECIDE, ACCORDING TO WHATEVER VALUES, WHAT'S IN THE PUBLIC INTEREST.
THERE CANNOT BE A CONSENSUS, BECAUSE THERE ARE CONFLICTING INTERESTS.
HOW CAN YOU FIND A CONSENSUS BETWEEN THE PRIVATE INTERESTS OF THE BIG MULTINATIONALS AND THE INTERESTS OF THE WHOLE POPULACE?
THERE ARE HUGE CONFLICTS OF INTEREST.

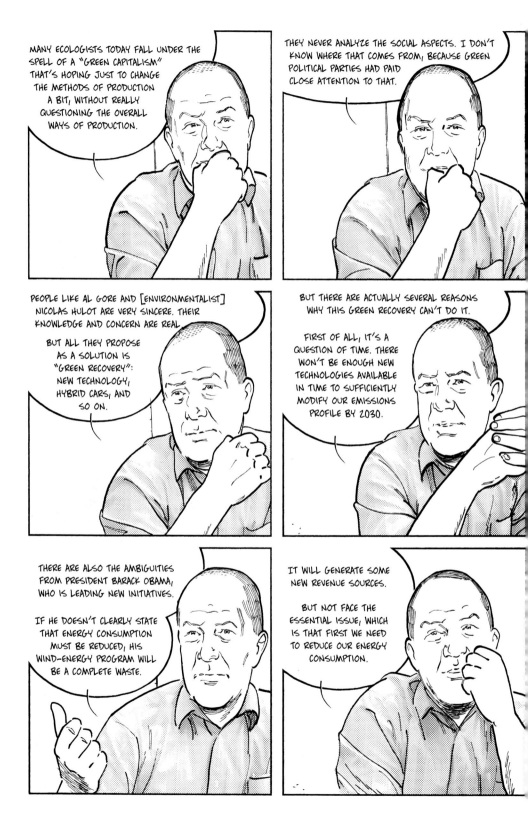

MANY ECOLOGISTS TODAY FALL UNDER THE SPELL OF A "GREEN CAPITALISM" THAT'S HOPING JUST TO CHANGE THE METHODS OF PRODUCTION A BIT, WITHOUT REALLY QUESTIONING THE OVERALL WAYS OF PRODUCTION.

THEY NEVER ANALYZE THE SOCIAL ASPECTS. I DON'T KNOW WHERE THAT COMES FROM, BECAUSE GREEN POLITICAL PARTIES HAD PAID CLOSE ATTENTION TO THAT.

PEOPLE LIKE AL GORE AND [ENVIRONMENTALIST] NICOLAS HULOT ARE VERY SINCERE. THEIR KNOWLEDGE AND CONCERN ARE REAL.

BUT ALL THEY PROPOSE AS A SOLUTION IS "GREEN RECOVERY": NEW TECHNOLOGY, HYBRID CARS, AND SO ON.

BUT THERE ARE ACTUALLY SEVERAL REASONS WHY THIS GREEN RECOVERY CAN'T DO IT.

FIRST OF ALL, IT'S A QUESTION OF TIME. THERE WON'T BE ENOUGH NEW TECHNOLOGIES AVAILABLE IN TIME TO SUFFICIENTLY MODIFY OUR EMISSIONS PROFILE BY 2030.

THERE ARE ALSO THE AMBIGUITIES FROM PRESIDENT BARACK OBAMA, WHO IS LEADING NEW INITIATIVES.

IF HE DOESN'T CLEARLY STATE THAT ENERGY CONSUMPTION MUST BE REDUCED, HIS WIND-ENERGY PROGRAM WILL BE A COMPLETE WASTE.

IT WILL GENERATE SOME NEW REVENUE SOURCES.

BUT NOT FACE THE ESSENTIAL ISSUE, WHICH IS THAT FIRST WE NEED TO REDUCE OUR ENERGY CONSUMPTION.

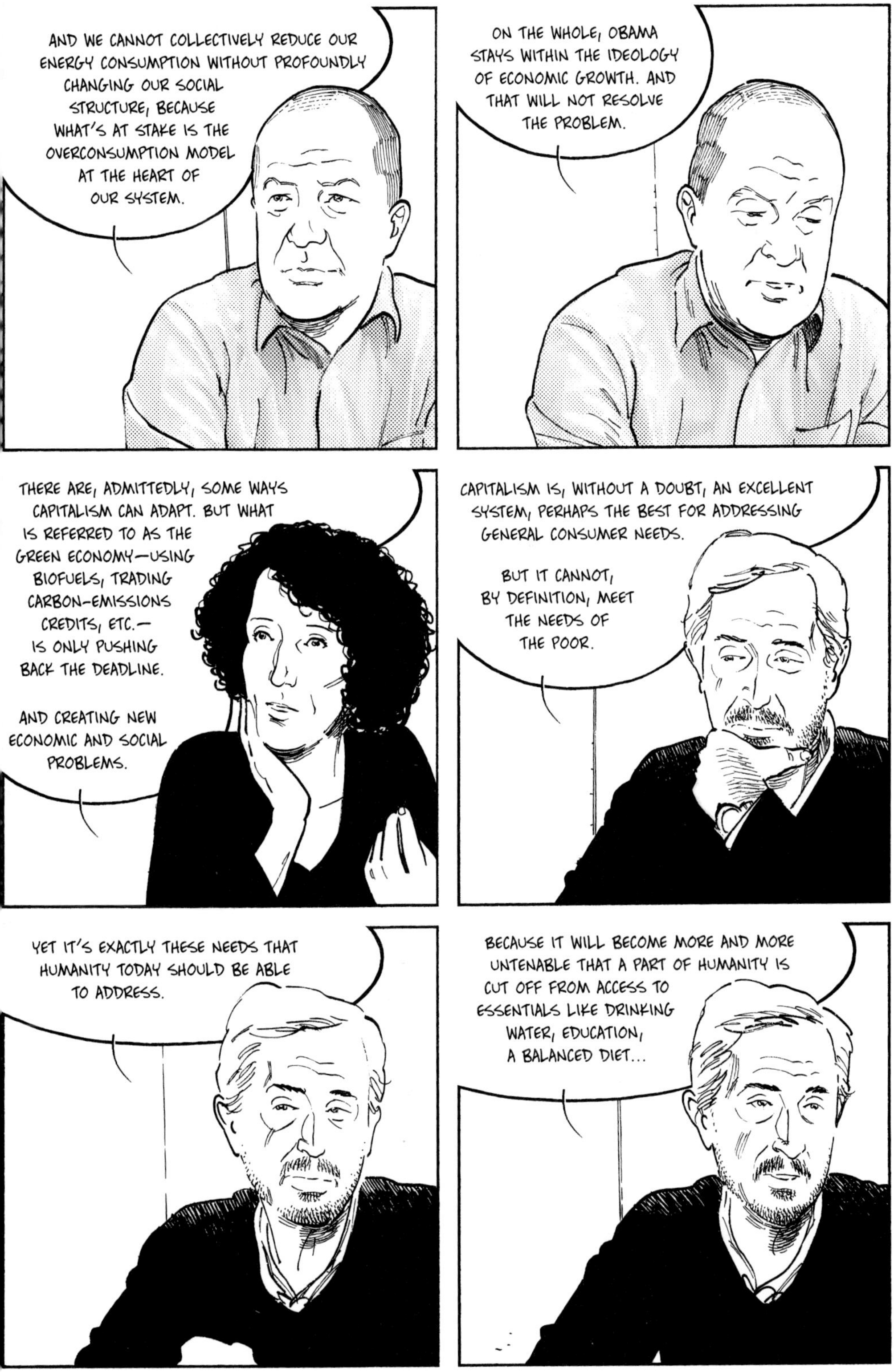
AND WE CANNOT COLLECTIVELY REDUCE OUR ENERGY CONSUMPTION WITHOUT PROFOUNDLY CHANGING OUR SOCIAL STRUCTURE, BECAUSE WHAT'S AT STAKE IS THE OVERCONSUMPTION MODEL AT THE HEART OF OUR SYSTEM.
ON THE WHOLE, OBAMA STAYS WITHIN THE IDEOLOGY OF ECONOMIC GROWTH. AND THAT WILL NOT RESOLVE THE PROBLEM.
THERE ARE, ADMITTEDLY, SOME WAYS CAPITALISM CAN ADAPT. BUT WHAT IS REFERRED TO AS THE GREEN ECONOMY—USING BIOFUELS, TRADING CARBON-EMISSIONS CREDITS, ETC.—IS ONLY PUSHING BACK THE DEADLINE.
AND CREATING NEW ECONOMIC AND SOCIAL PROBLEMS.
CAPITALISM IS, WITHOUT A DOUBT, AN EXCELLENT SYSTEM, PERHAPS THE BEST FOR ADDRESSING GENERAL CONSUMER NEEDS.
BUT IT CANNOT, BY DEFINITION, MEET THE NEEDS OF THE POOR.
YET IT'S EXACTLY THESE NEEDS THAT HUMANITY TODAY SHOULD BE ABLE TO ADDRESS.
BECAUSE IT WILL BECOME MORE AND MORE UNTENABLE THAT A PART OF HUMANITY IS CUT OFF FROM ACCESS TO ESSENTIALS LIKE DRINKING WATER, EDUCATION, A BALANCED DIET...

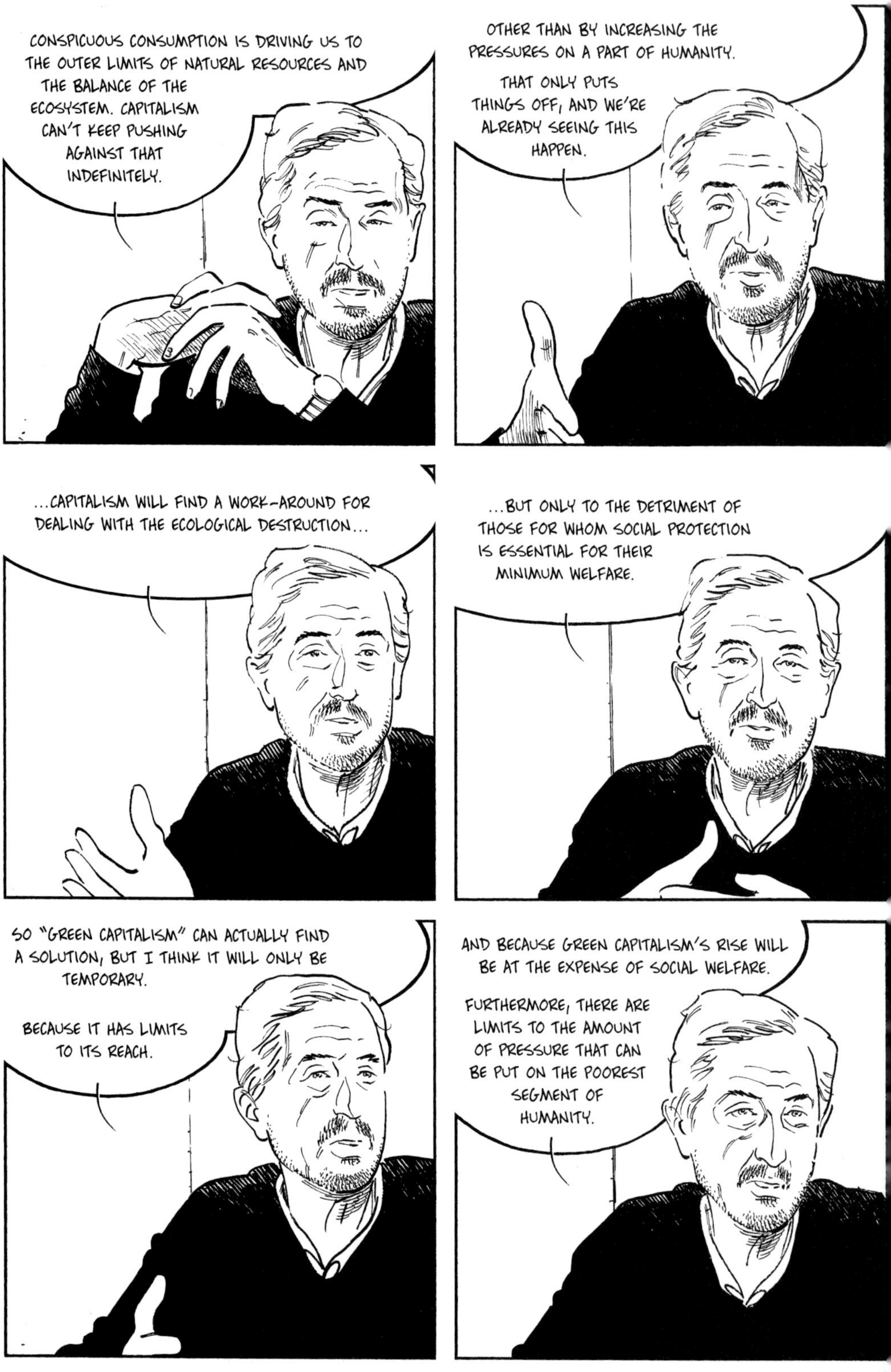
CONSPICUOUS CONSUMPTION IS DRIVING US TO THE OUTER LIMITS OF NATURAL RESOURCES AND THE BALANCE OF THE ECOSYSTEM. CAPITALISM CAN'T KEEP PUSHING AGAINST THAT INDEFINITELY.
OTHER THAN BY INCREASING THE PRESSURES ON A PART OF HUMANITY.
THAT ONLY PUTS THINGS OFF, AND WE'RE ALREADY SEEING THIS HAPPEN.
...CAPITALISM WILL FIND A WORK-AROUND FOR DEALING WITH THE ECOLOGICAL DESTRUCTION...
...BUT ONLY TO THE DETRIMENT OF THOSE FOR WHOM SOCIAL PROTECTION IS ESSENTIAL FOR THEIR MINIMUM WELFARE.
SO "GREEN CAPITALISM" CAN ACTUALLY FIND A SOLUTION, BUT I THINK IT WILL ONLY BE TEMPORARY.
BECAUSE IT HAS LIMITS TO ITS REACH.
AND BECAUSE GREEN CAPITALISM'S RISE WILL BE AT THE EXPENSE OF SOCIAL WELFARE.
FURTHERMORE, THERE ARE LIMITS TO THE AMOUNT OF PRESSURE THAT CAN BE PUT ON THE POOREST SEGMENT OF HUMANITY.

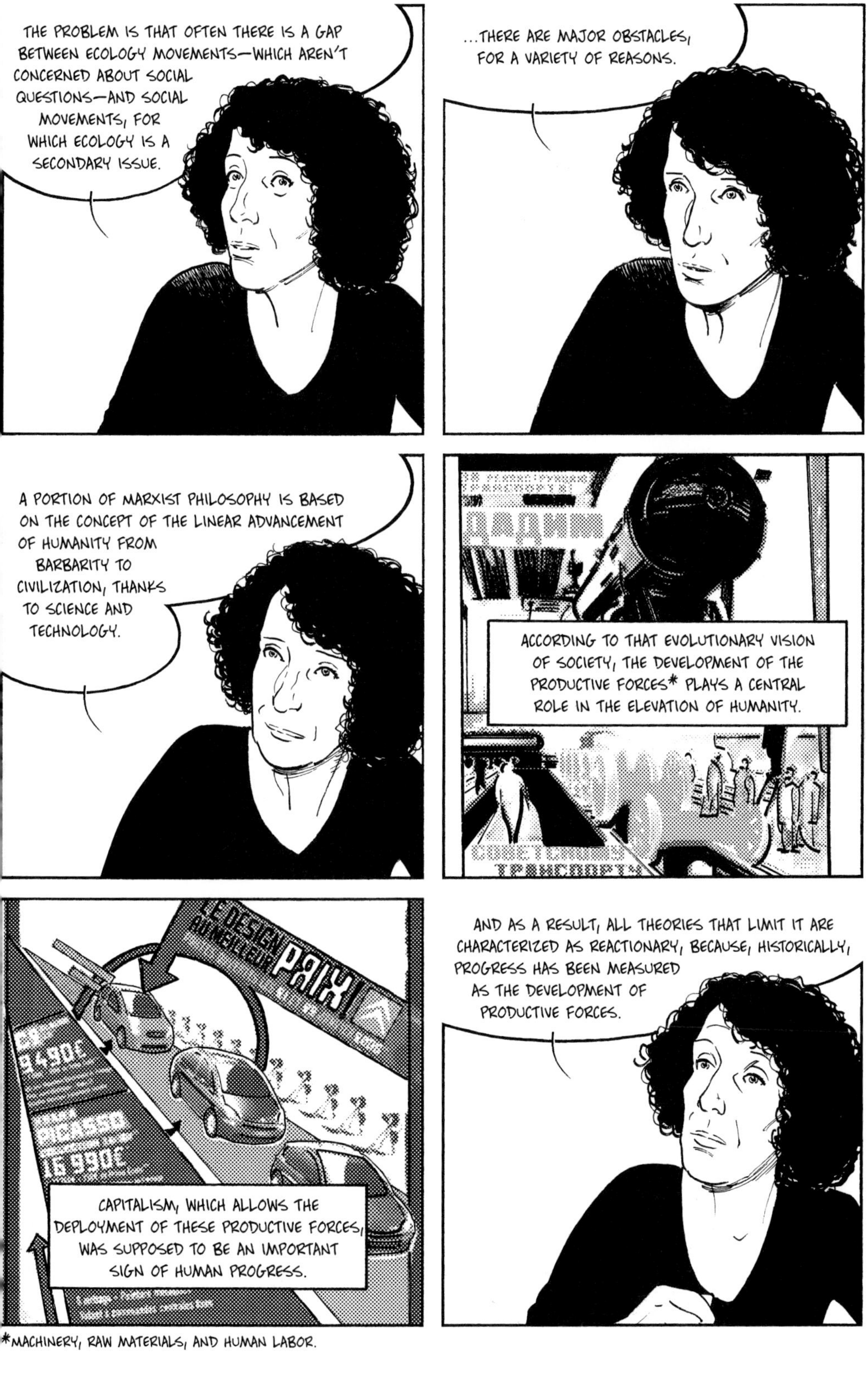
THE PROBLEM IS THAT OFTEN THERE IS A GAP BETWEEN ECOLOGY MOVEMENTS—WHICH AREN'T CONCERNED ABOUT SOCIAL QUESTIONS—AND SOCIAL MOVEMENTS, FOR WHICH ECOLOGY IS A SECONDARY ISSUE.
...THERE ARE MAJOR OBSTACLES, FOR A VARIETY OF REASONS.
A PORTION OF MARXIST PHILOSOPHY IS BASED ON THE CONCEPT OF THE LINEAR ADVANCEMENT OF HUMANITY FROM BARBARITY TO CIVILIZATION, THANKS TO SCIENCE AND TECHNOLOGY.
ACCORDING TO THAT EVOLUTIONARY VISION OF SOCIETY, THE DEVELOPMENT OF THE PRODUCTIVE FORCES* PLAYS A CENTRAL ROLE IN THE ELEVATION OF HUMANITY.
LE DESIGN
PRIX
PICASSO
15 990€
CAPITALISM, WHICH ALLOWS THE DEPLOYMENT OF THESE PRODUCTIVE FORCES, WAS SUPPOSED TO BE AN IMPORTANT SIGN OF HUMAN PROGRESS.
AND AS A RESULT, ALL THEORIES THAT LIMIT IT ARE CHARACTERIZED AS REACTIONARY, BECAUSE, HISTORICALLY, PROGRESS HAS BEEN MEASURED AS THE DEVELOPMENT OF PRODUCTIVE FORCES.

*MACHINERY, RAW MATERIALS, AND HUMAN LABOR.

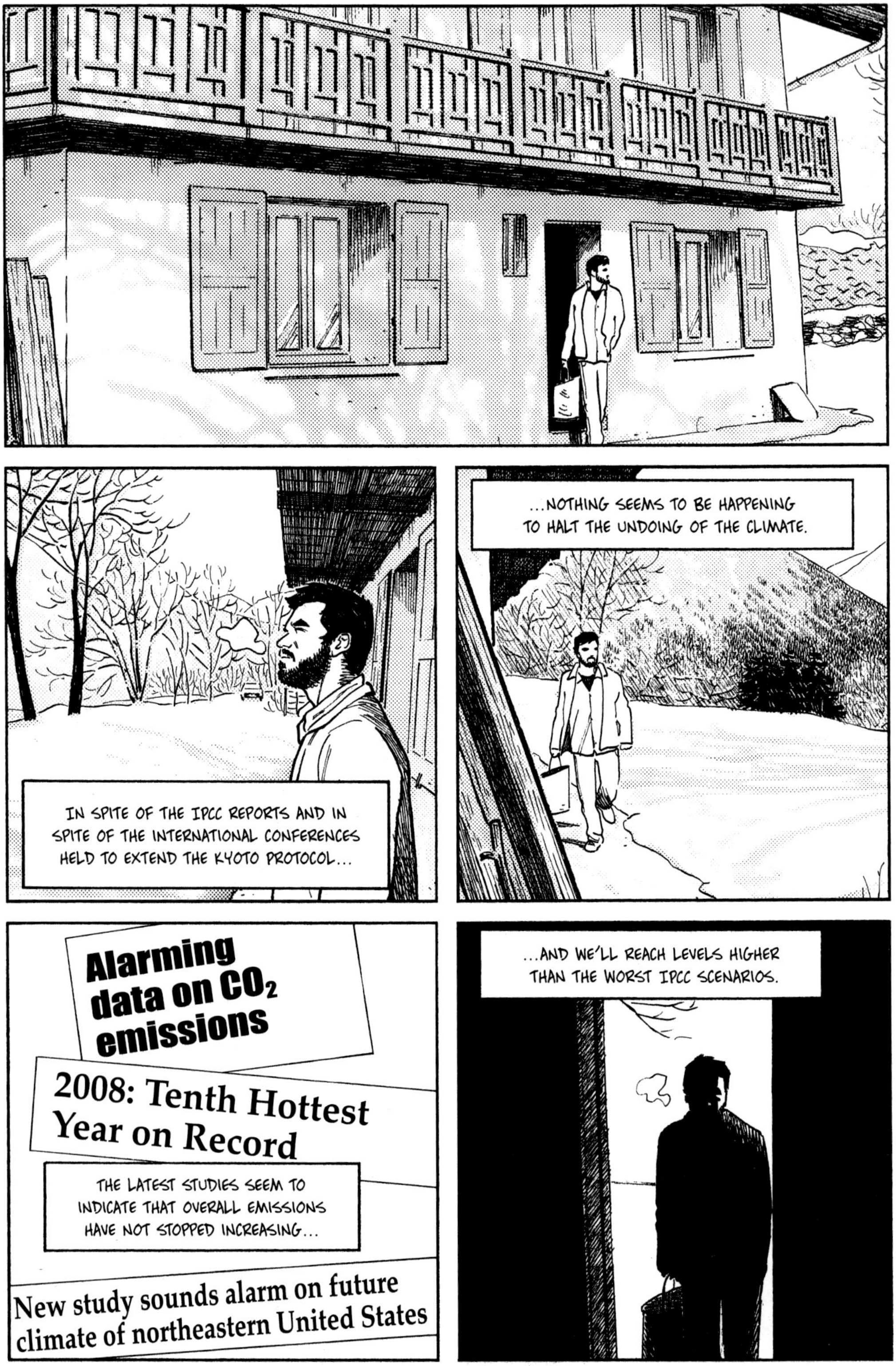
IN SPITE OF THE IPCC REPORTS AND IN SPITE OF THE INTERNATIONAL CONFERENCES HELD TO EXTEND THE KYOTO PROTOCOL...
...NOTHING SEEMS TO BE HAPPENING TO HALT THE UNDOING OF THE CLIMATE.
Alarming data on CO_2 emissions
2008: Tenth Hottest Year on Record
THE LATEST STUDIES SEEM TO INDICATE THAT OVERALL EMISSIONS HAVE NOT STOPPED INCREASING...
New study sounds alarm on future climate of northeastern United States
...AND WE'LL REACH LEVELS HIGHER THAN THE WORST IPCC SCENARIOS.

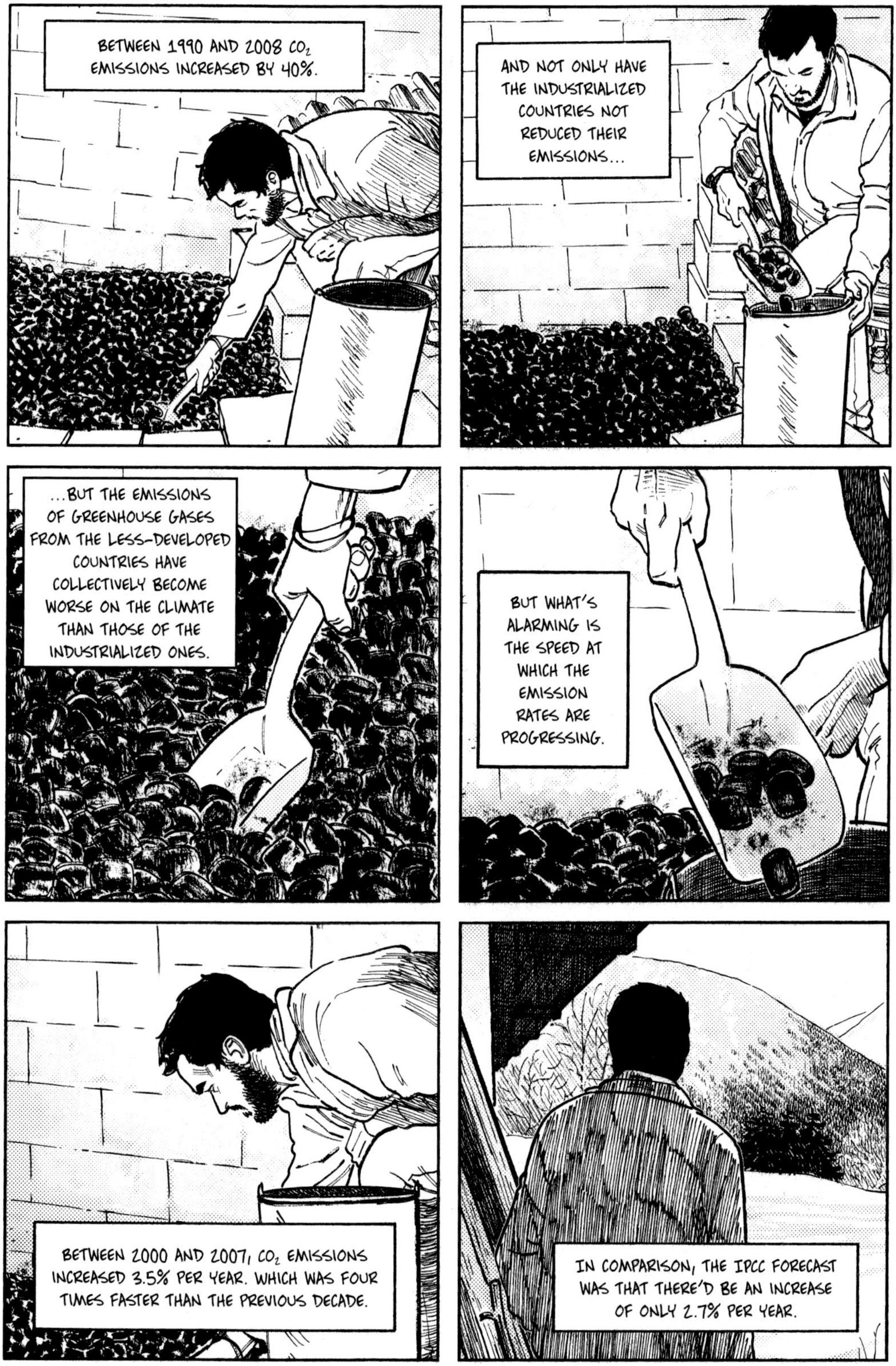
BETWEEN 1990 AND 2008 CO_2 EMISSIONS INCREASED BY 40%.
AND NOT ONLY HAVE THE INDUSTRIALIZED COUNTRIES NOT REDUCED THEIR EMISSIONS...
...BUT THE EMISSIONS OF GREENHOUSE GASES FROM THE LESS-DEVELOPED COUNTRIES HAVE COLLECTIVELY BECOME WORSE ON THE CLIMATE THAN THOSE OF THE INDUSTRIALIZED ONES.
BUT WHAT'S ALARMING IS THE SPEED AT WHICH THE EMISSION RATES ARE PROGRESSING.
BETWEEN 2000 AND 2007, CO_2 EMISSIONS INCREASED 3.5% PER YEAR. WHICH WAS FOUR TIMES FASTER THAN THE PREVIOUS DECADE.
IN COMPARISON, THE IPCC FORECAST WAS THAT THERE'D BE AN INCREASE OF ONLY 2.7% PER YEAR.

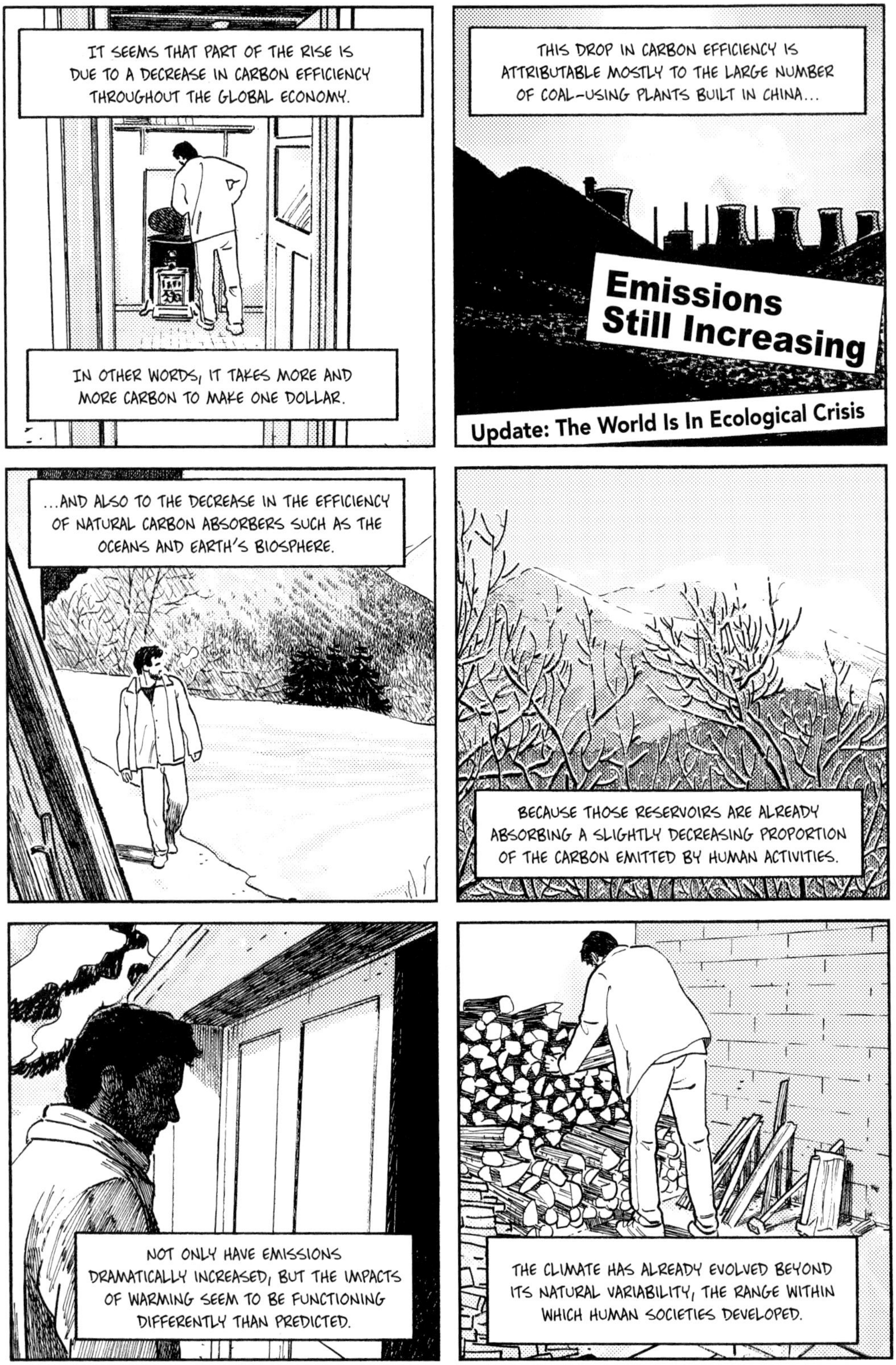
IT SEEMS THAT PART OF THE RISE IS DUE TO A DECREASE IN CARBON EFFICIENCY THROUGHOUT THE GLOBAL ECONOMY.
IN OTHER WORDS, IT TAKES MORE AND MORE CARBON TO MAKE ONE DOLLAR.
THIS DROP IN CARBON EFFICIENCY IS ATTRIBUTABLE MOSTLY TO THE LARGE NUMBER OF COAL-USING PLANTS BUILT IN CHINA...
Emissions Still Increasing
Update: The World Is In Ecological Crisis
...AND ALSO TO THE DECREASE IN THE EFFICIENCY OF NATURAL CARBON ABSORBERS SUCH AS THE OCEANS AND EARTH'S BIOSPHERE.
BECAUSE THOSE RESERVOIRS ARE ALREADY ABSORBING A SLIGHTLY DECREASING PROPORTION OF THE CARBON EMITTED BY HUMAN ACTIVITIES.
NOT ONLY HAVE EMISSIONS DRAMATICALLY INCREASED, BUT THE IMPACTS OF WARMING SEEM TO BE FUNCTIONING DIFFERENTLY THAN PREDICTED.
THE CLIMATE HAS ALREADY EVOLVED BEYOND ITS NATURAL VARIABILITY, THE RANGE WITHIN WHICH HUMAN SOCIETIES DEVELOPED.

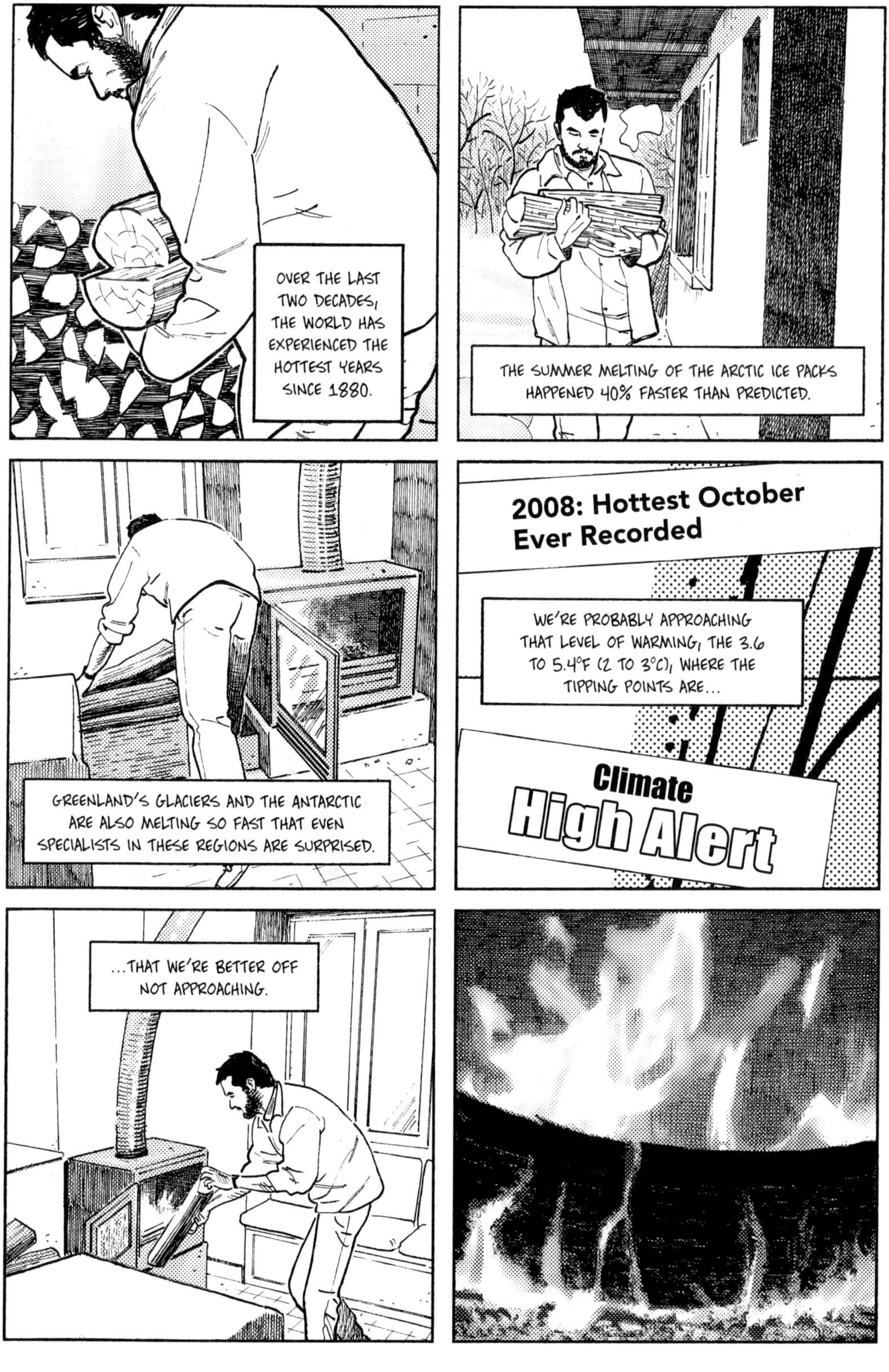
OVER THE LAST TWO DECADES, THE WORLD HAS EXPERIENCED THE HOTTEST YEARS SINCE 1880.
THE SUMMER MELTING OF THE ARCTIC ICE PACKS HAPPENED 40% FASTER THAN PREDICTED.
GREENLAND'S GLACIERS AND THE ANTARCTIC ARE ALSO MELTING SO FAST THAT EVEN SPECIALISTS IN THESE REGIONS ARE SURPRISED.
2008: Hottest October Ever Recorded
WE'RE PROBABLY APPROACHING THAT LEVEL OF WARMING, THE 3.6 TO 5.4°F (2 TO 3°C), WHERE THE TIPPING POINTS ARE...
Climate
High Alert
...THAT WE'RE BETTER OFF NOT APPROACHING.

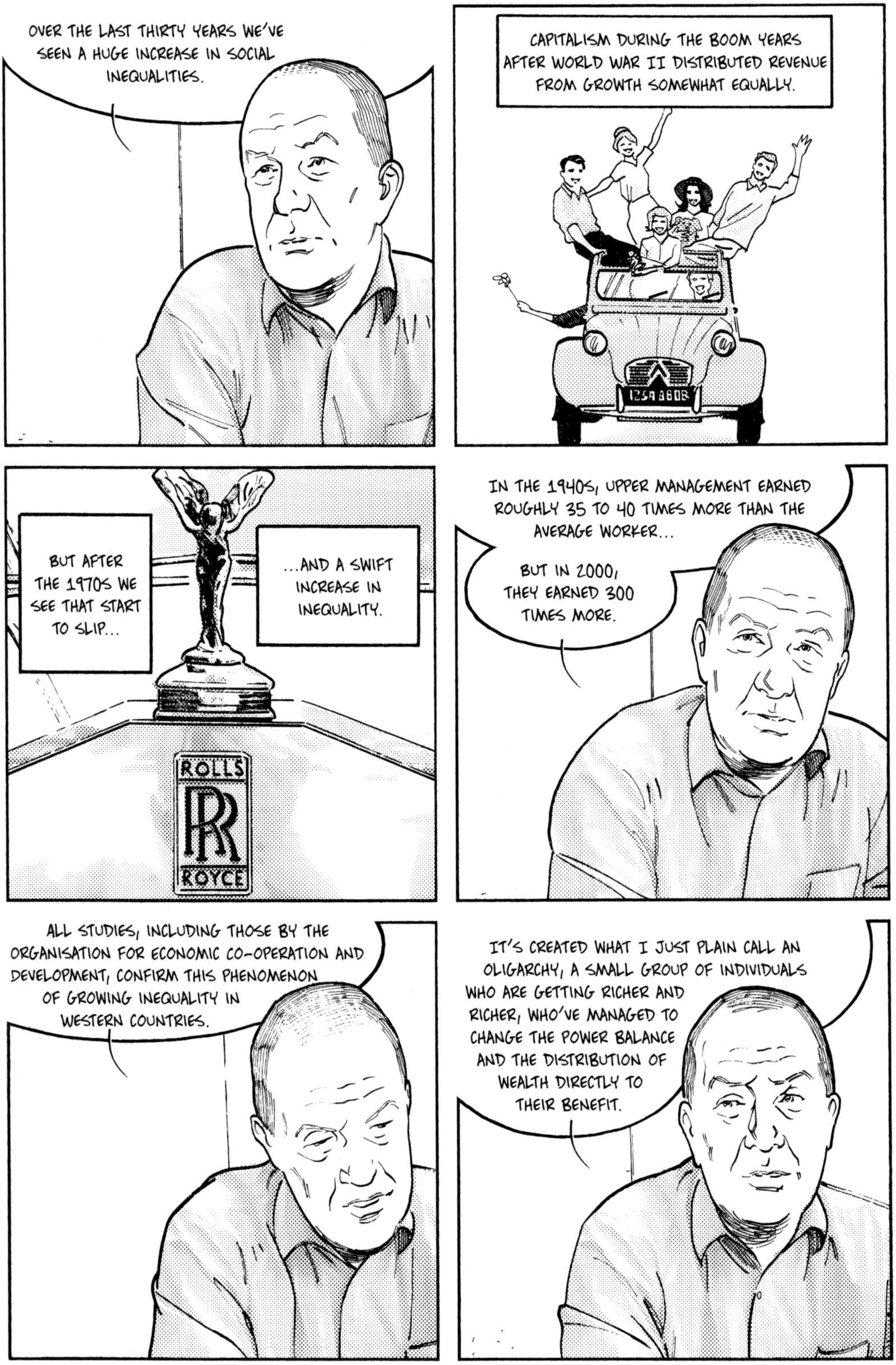
OVER THE LAST THIRTY YEARS WE'VE SEEN A HUGE INCREASE IN SOCIAL INEQUALITIES.
CAPITALISM DURING THE BOOM YEARS AFTER WORLD WAR II DISTRIBUTED REVENUE FROM GROWTH SOMEWHAT EQUALLY.
125A 8608
BUT AFTER THE 1970S WE SEE THAT START TO SLIP...
...AND A SWIFT INCREASE IN INEQUALITY.
ROLLS RR ROYCE
IN THE 1940S, UPPER MANAGEMENT EARNED ROUGHLY 35 TO 40 TIMES MORE THAN THE AVERAGE WORKER...
BUT IN 2000, THEY EARNED 300 TIMES MORE.
ALL STUDIES, INCLUDING THOSE BY THE ORGANISATION FOR ECONOMIC CO-OPERATION AND DEVELOPMENT, CONFIRM THIS PHENOMENON OF GROWING INEQUALITY IN WESTERN COUNTRIES.
IT'S CREATED WHAT I JUST PLAIN CALL AN OLIGARCHY, A SMALL GROUP OF INDIVIDUALS WHO ARE GETTING RICHER AND RICHER, WHO'VE MANAGED TO CHANGE THE POWER BALANCE AND THE DISTRIBUTION OF WEALTH DIRECTLY TO THEIR BENEFIT.

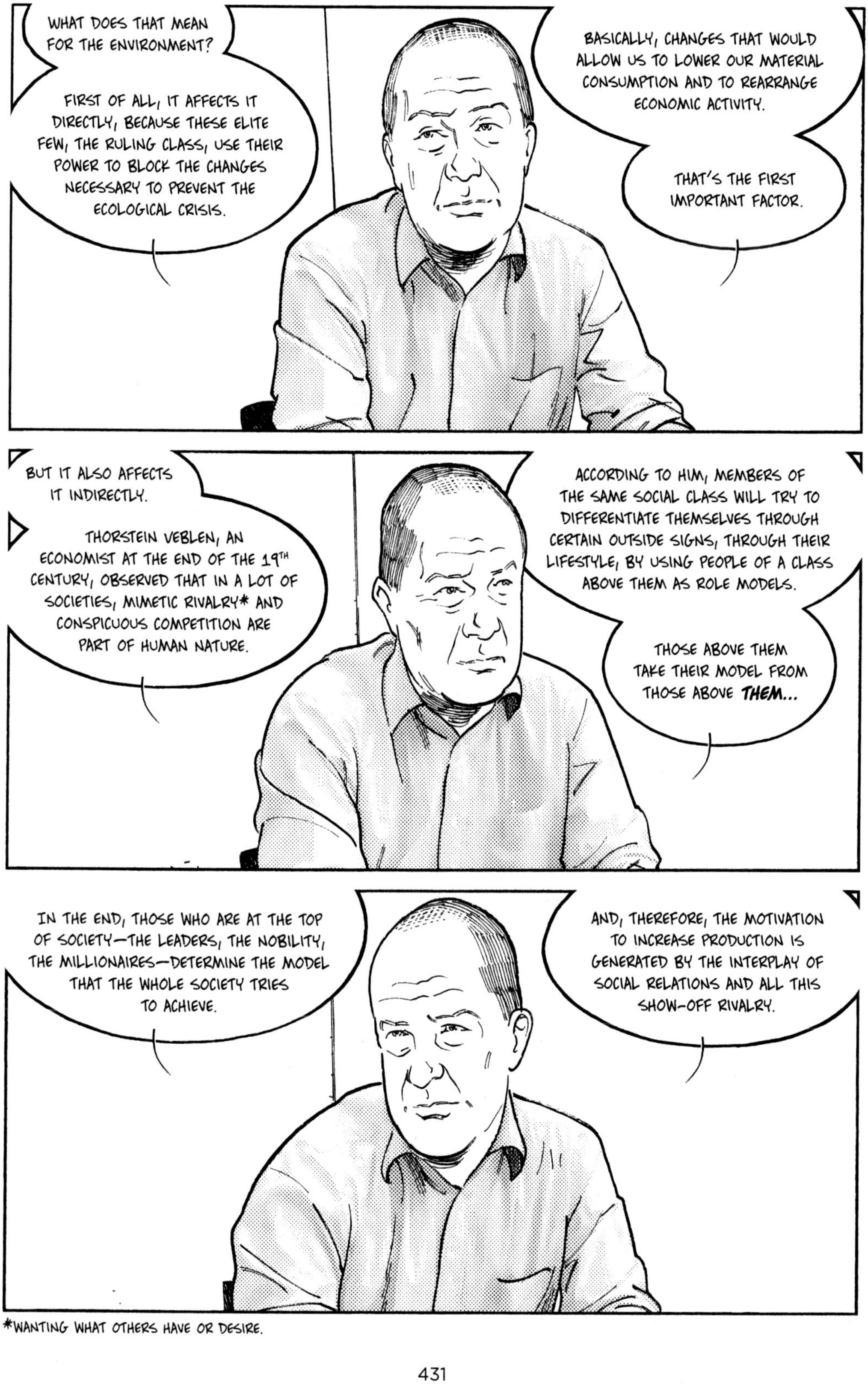
WHAT DOES THAT MEAN FOR THE ENVIRONMENT?
FIRST OF ALL, IT AFFECTS IT DIRECTLY, BECAUSE THESE ELITE FEW, THE RULING CLASS, USE THEIR POWER TO BLOCK THE CHANGES NECESSARY TO PREVENT THE ECOLOGICAL CRISIS.
BASICALLY, CHANGES THAT WOULD ALLOW US TO LOWER OUR MATERIAL CONSUMPTION AND TO REARRANGE ECONOMIC ACTIVITY.
THAT'S THE FIRST IMPORTANT FACTOR.
BUT IT ALSO AFFECTS IT INDIRECTLY.
THORSTEIN VEBLEN, AN ECONOMIST AT THE END OF THE 19TH CENTURY, OBSERVED THAT IN A LOT OF SOCIETIES, MIMETIC RIVALRY* AND CONSPICUOUS COMPETITION ARE PART OF HUMAN NATURE.
ACCORDING TO HIM, MEMBERS OF THE SAME SOCIAL CLASS WILL TRY TO DIFFERENTIATE THEMSELVES THROUGH CERTAIN OUTSIDE SIGNS, THROUGH THEIR LIFESTYLE, BY USING PEOPLE OF A CLASS ABOVE THEM AS ROLE MODELS.
THOSE ABOVE THEM TAKE THEIR MODEL FROM THOSE ABOVE ***THEM...***
IN THE END, THOSE WHO ARE AT THE TOP OF SOCIETY—THE LEADERS, THE NOBILITY, THE MILLIONAIRES—DETERMINE THE MODEL THAT THE WHOLE SOCIETY TRIES TO ACHIEVE.
AND, THEREFORE, THE MOTIVATION TO INCREASE PRODUCTION IS GENERATED BY THE INTERPLAY OF SOCIAL RELATIONS AND ALL THIS SHOW-OFF RIVALRY.

*WANTING WHAT OTHERS HAVE OR DESIRE.

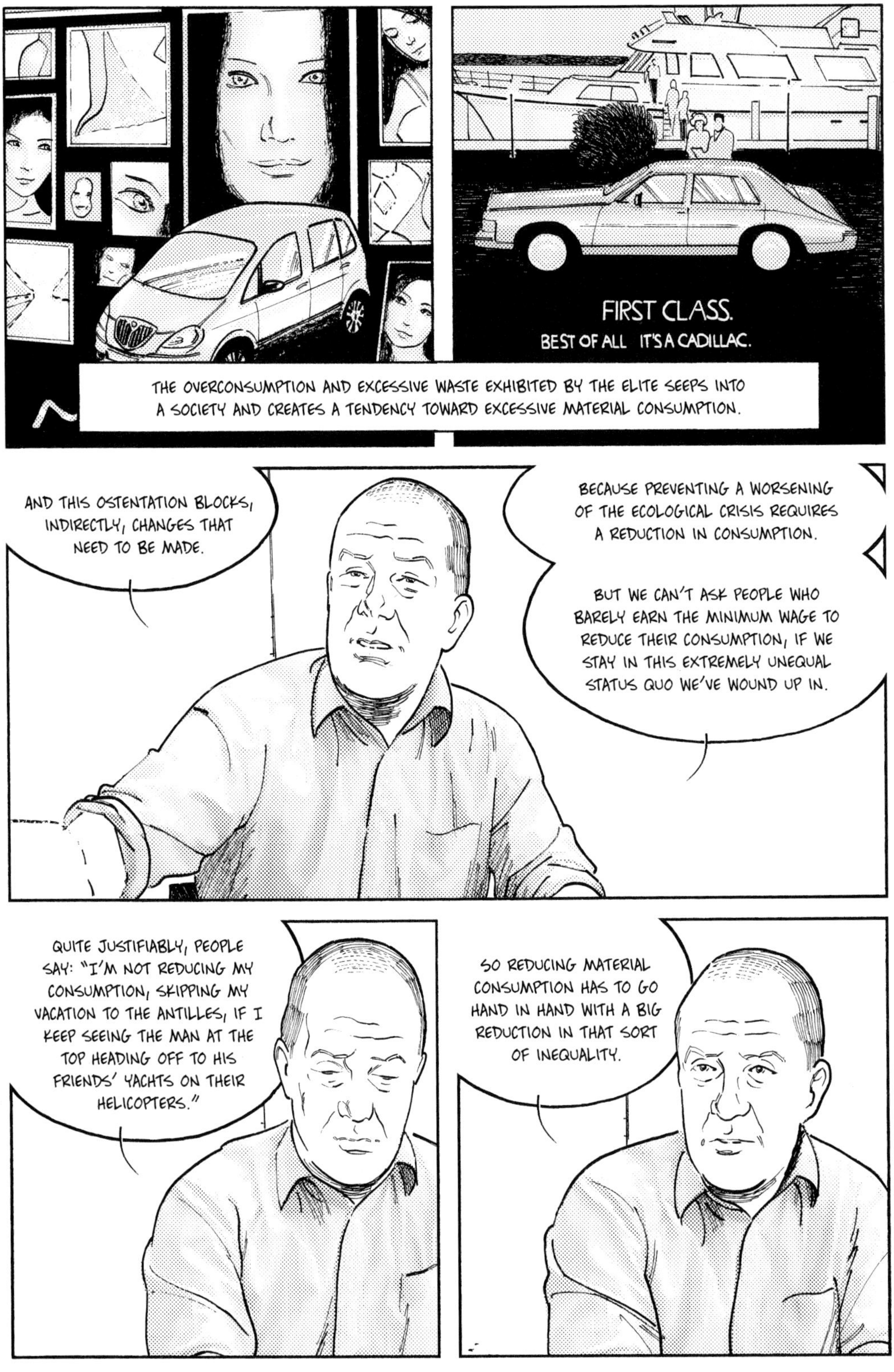
FIRST CLASS.
BEST OF ALL IT'S A CADILLAC.
THE OVERCONSUMPTION AND EXCESSIVE WASTE EXHIBITED BY THE ELITE SEEPS INTO A SOCIETY AND CREATES A TENDENCY TOWARD EXCESSIVE MATERIAL CONSUMPTION.
AND THIS OSTENTATION BLOCKS, INDIRECTLY, CHANGES THAT NEED TO BE MADE.
BECAUSE PREVENTING A WORSENING OF THE ECOLOGICAL CRISIS REQUIRES A REDUCTION IN CONSUMPTION.
BUT WE CAN'T ASK PEOPLE WHO BARELY EARN THE MINIMUM WAGE TO REDUCE THEIR CONSUMPTION, IF WE STAY IN THIS EXTREMELY UNEQUAL STATUS QUO WE'VE WOUND UP IN.
QUITE JUSTIFIABLY, PEOPLE SAY: "I'M NOT REDUCING MY CONSUMPTION, SKIPPING MY VACATION TO THE ANTILLES, IF I KEEP SEEING THE MAN AT THE TOP HEADING OFF TO HIS FRIENDS' YACHTS ON THEIR HELICOPTERS."
SO REDUCING MATERIAL CONSUMPTION HAS TO GO HAND IN HAND WITH A BIG REDUCTION IN THAT SORT OF INEQUALITY.

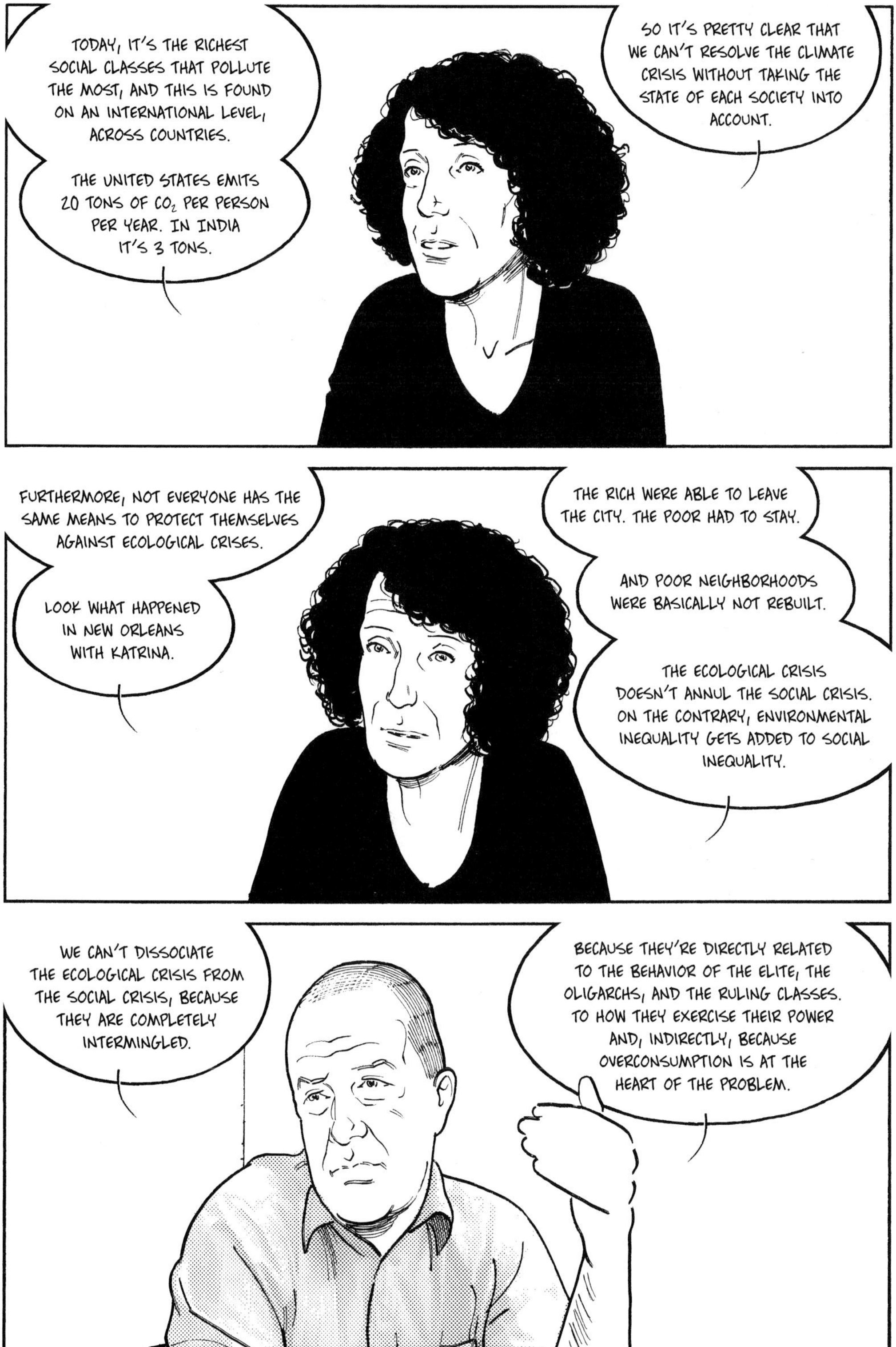
TODAY, IT'S THE RICHEST SOCIAL CLASSES THAT POLLUTE THE MOST, AND THIS IS FOUND ON AN INTERNATIONAL LEVEL, ACROSS COUNTRIES.
THE UNITED STATES EMITS 20 TONS OF CO_2 PER PERSON PER YEAR. IN INDIA IT'S 3 TONS.
SO IT'S PRETTY CLEAR THAT WE CAN'T RESOLVE THE CLIMATE CRISIS WITHOUT TAKING THE STATE OF EACH SOCIETY INTO ACCOUNT.
FURTHERMORE, NOT EVERYONE HAS THE SAME MEANS TO PROTECT THEMSELVES AGAINST ECOLOGICAL CRISES.
LOOK WHAT HAPPENED IN NEW ORLEANS WITH KATRINA.
THE RICH WERE ABLE TO LEAVE THE CITY. THE POOR HAD TO STAY.
AND POOR NEIGHBORHOODS WERE BASICALLY NOT REBUILT.
THE ECOLOGICAL CRISIS DOESN'T ANNUL THE SOCIAL CRISIS. ON THE CONTRARY, ENVIRONMENTAL INEQUALITY GETS ADDED TO SOCIAL INEQUALITY.
WE CAN'T DISSOCIATE THE ECOLOGICAL CRISIS FROM THE SOCIAL CRISIS, BECAUSE THEY ARE COMPLETELY INTERMINGLED.
BECAUSE THEY'RE DIRECTLY RELATED TO THE BEHAVIOR OF THE ELITE, THE OLIGARCHS, AND THE RULING CLASSES. TO HOW THEY EXERCISE THEIR POWER AND, INDIRECTLY, BECAUSE OVERCONSUMPTION IS AT THE HEART OF THE PROBLEM.

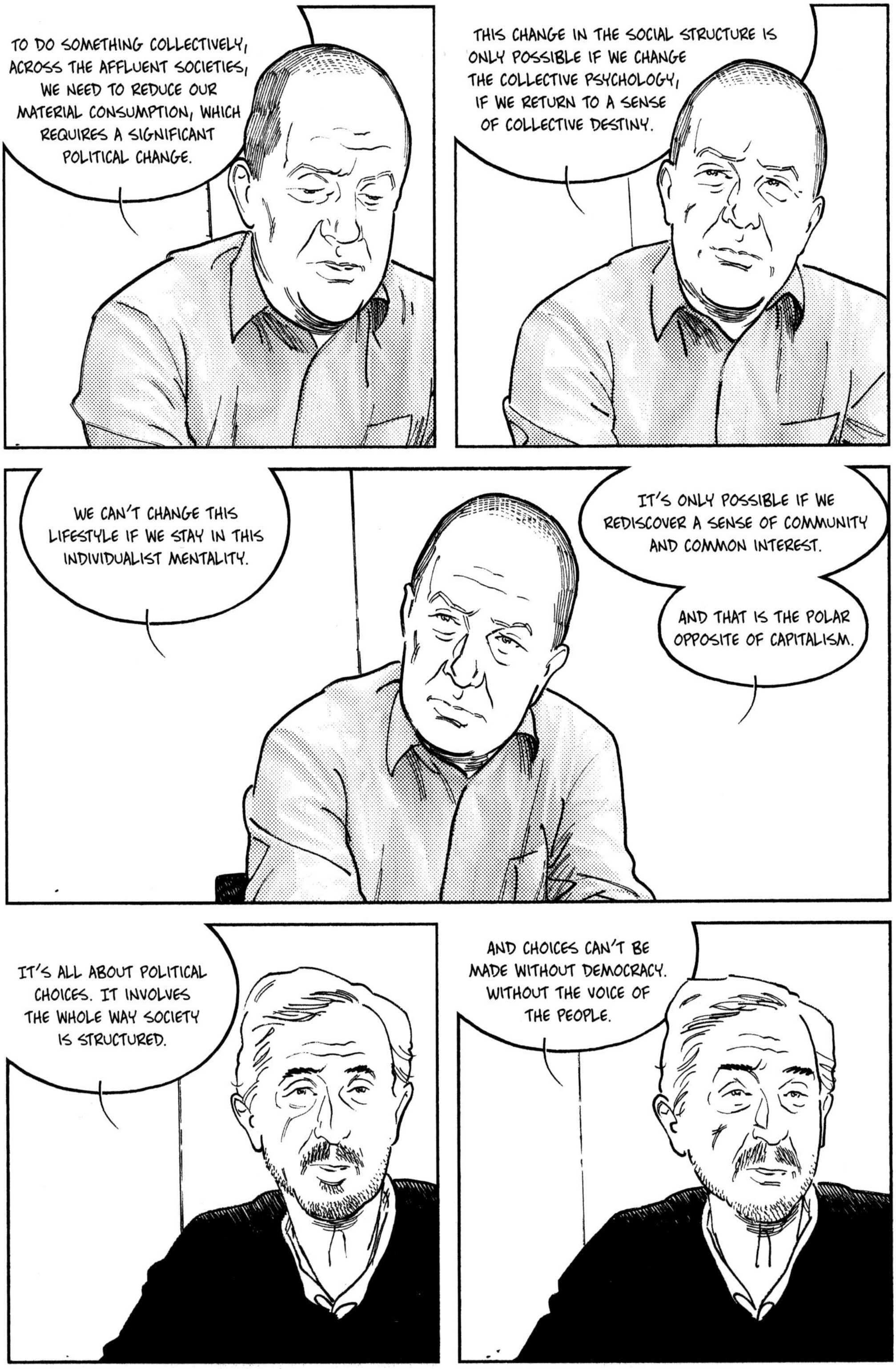
TO DO SOMETHING COLLECTIVELY, ACROSS THE AFFLUENT SOCIETIES, WE NEED TO REDUCE OUR MATERIAL CONSUMPTION, WHICH REQUIRES A SIGNIFICANT POLITICAL CHANGE.
THIS CHANGE IN THE SOCIAL STRUCTURE IS ONLY POSSIBLE IF WE CHANGE THE COLLECTIVE PSYCHOLOGY, IF WE RETURN TO A SENSE OF COLLECTIVE DESTINY.
WE CAN'T CHANGE THIS LIFESTYLE IF WE STAY IN THIS INDIVIDUALIST MENTALITY.
IT'S ONLY POSSIBLE IF WE REDISCOVER A SENSE OF COMMUNITY AND COMMON INTEREST.
AND THAT IS THE POLAR OPPOSITE OF CAPITALISM.
IT'S ALL ABOUT POLITICAL CHOICES. IT INVOLVES THE WHOLE WAY SOCIETY IS STRUCTURED.
AND CHOICES CAN'T BE MADE WITHOUT DEMOCRACY. WITHOUT THE VOICE OF THE PEOPLE.

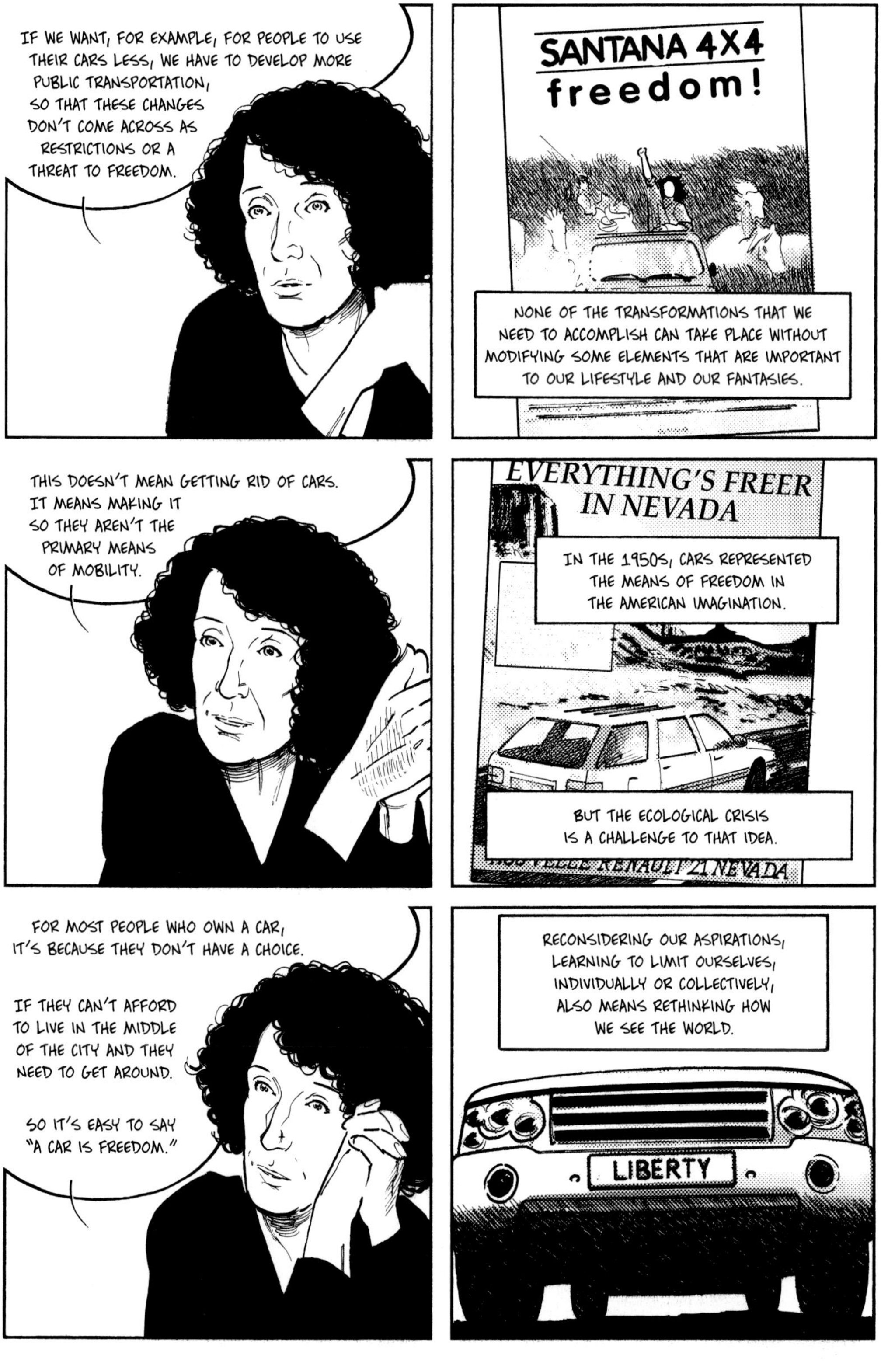
IF WE WANT, FOR EXAMPLE, FOR PEOPLE TO USE THEIR CARS LESS, WE HAVE TO DEVELOP MORE PUBLIC TRANSPORTATION, SO THAT THESE CHANGES DON'T COME ACROSS AS RESTRICTIONS OR A THREAT TO FREEDOM.
SANTANA 4X4
freedom!
NONE OF THE TRANSFORMATIONS THAT WE NEED TO ACCOMPLISH CAN TAKE PLACE WITHOUT MODIFYING SOME ELEMENTS THAT ARE IMPORTANT TO OUR LIFESTYLE AND OUR FANTASIES.
THIS DOESN'T MEAN GETTING RID OF CARS. IT MEANS MAKING IT SO THEY AREN'T THE PRIMARY MEANS OF MOBILITY.
EVERYTHING'S FREER IN NEVADA
IN THE 1950S, CARS REPRESENTED THE MEANS OF FREEDOM IN THE AMERICAN IMAGINATION.
BUT THE ECOLOGICAL CRISIS IS A CHALLENGE TO THAT IDEA.
RENAULT 21 NEVADA
FOR MOST PEOPLE WHO OWN A CAR, IT'S BECAUSE THEY DON'T HAVE A CHOICE.
IF THEY CAN'T AFFORD TO LIVE IN THE MIDDLE OF THE CITY AND THEY NEED TO GET AROUND.
SO IT'S EASY TO SAY "A CAR IS FREEDOM."
RECONSIDERING OUR ASPIRATIONS, LEARNING TO LIMIT OURSELVES, INDIVIDUALLY OR COLLECTIVELY, ALSO MEANS RETHINKING HOW WE SEE THE WORLD.
LIBERTY

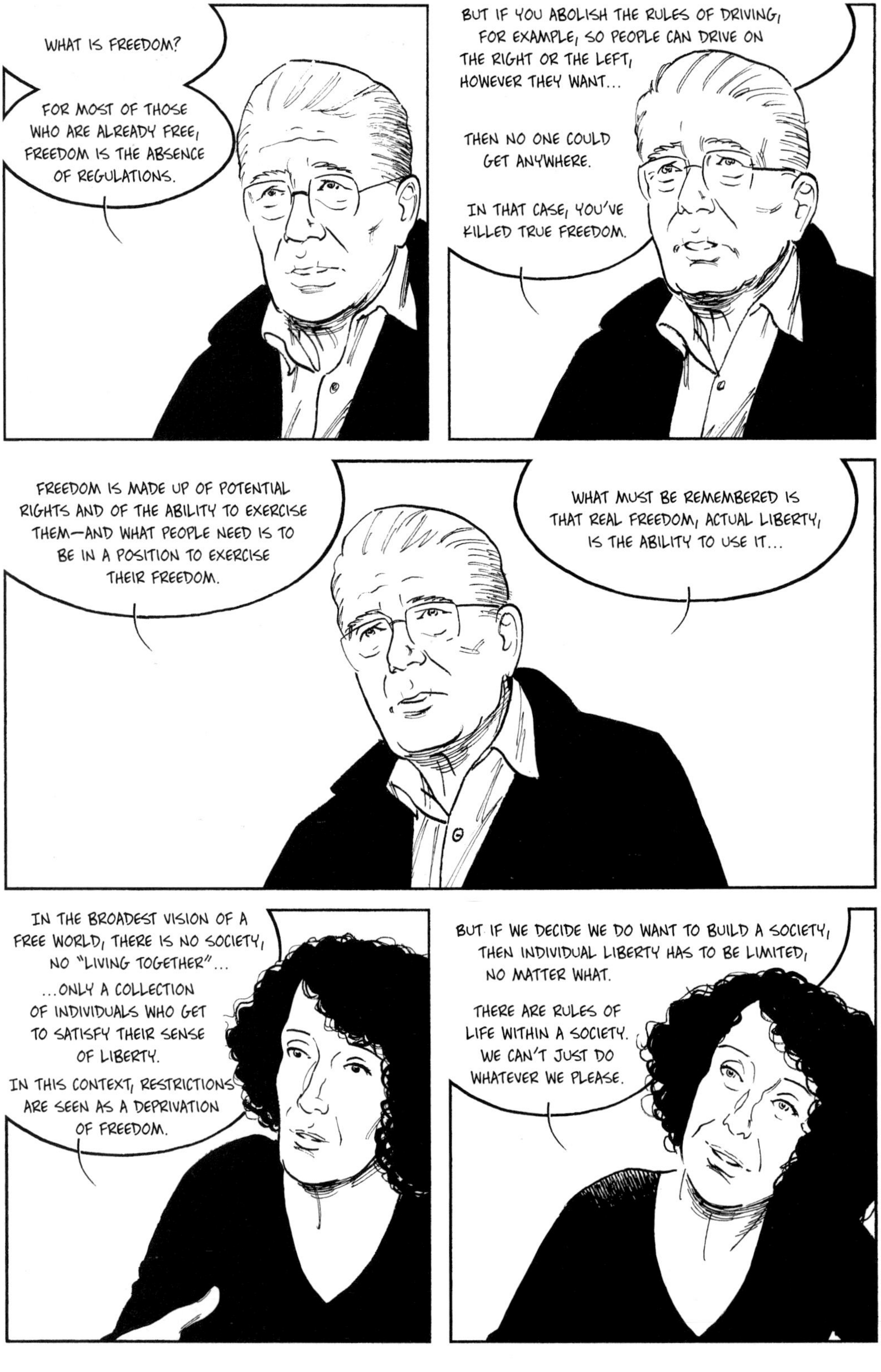
WHAT IS FREEDOM?
FOR MOST OF THOSE WHO ARE ALREADY FREE, FREEDOM IS THE ABSENCE OF REGULATIONS.
BUT IF YOU ABOLISH THE RULES OF DRIVING, FOR EXAMPLE, SO PEOPLE CAN DRIVE ON THE RIGHT OR THE LEFT, HOWEVER THEY WANT...
THEN NO ONE COULD GET ANYWHERE.
IN THAT CASE, YOU'VE KILLED TRUE FREEDOM.
FREEDOM IS MADE UP OF POTENTIAL RIGHTS AND OF THE ABILITY TO EXERCISE THEM—AND WHAT PEOPLE NEED IS TO BE IN A POSITION TO EXERCISE THEIR FREEDOM.
WHAT MUST BE REMEMBERED IS THAT REAL FREEDOM, ACTUAL LIBERTY, IS THE ABILITY TO USE IT...
IN THE BROADEST VISION OF A FREE WORLD, THERE IS NO SOCIETY, NO "LIVING TOGETHER"...
...ONLY A COLLECTION OF INDIVIDUALS WHO GET TO SATISFY THEIR SENSE OF LIBERTY.
IN THIS CONTEXT, RESTRICTIONS ARE SEEN AS A DEPRIVATION OF FREEDOM.
BUT IF WE DECIDE WE DO WANT TO BUILD A SOCIETY, THEN INDIVIDUAL LIBERTY HAS TO BE LIMITED, NO MATTER WHAT.
THERE ARE RULES OF LIFE WITHIN A SOCIETY. WE CAN'T JUST DO WHATEVER WE PLEASE.

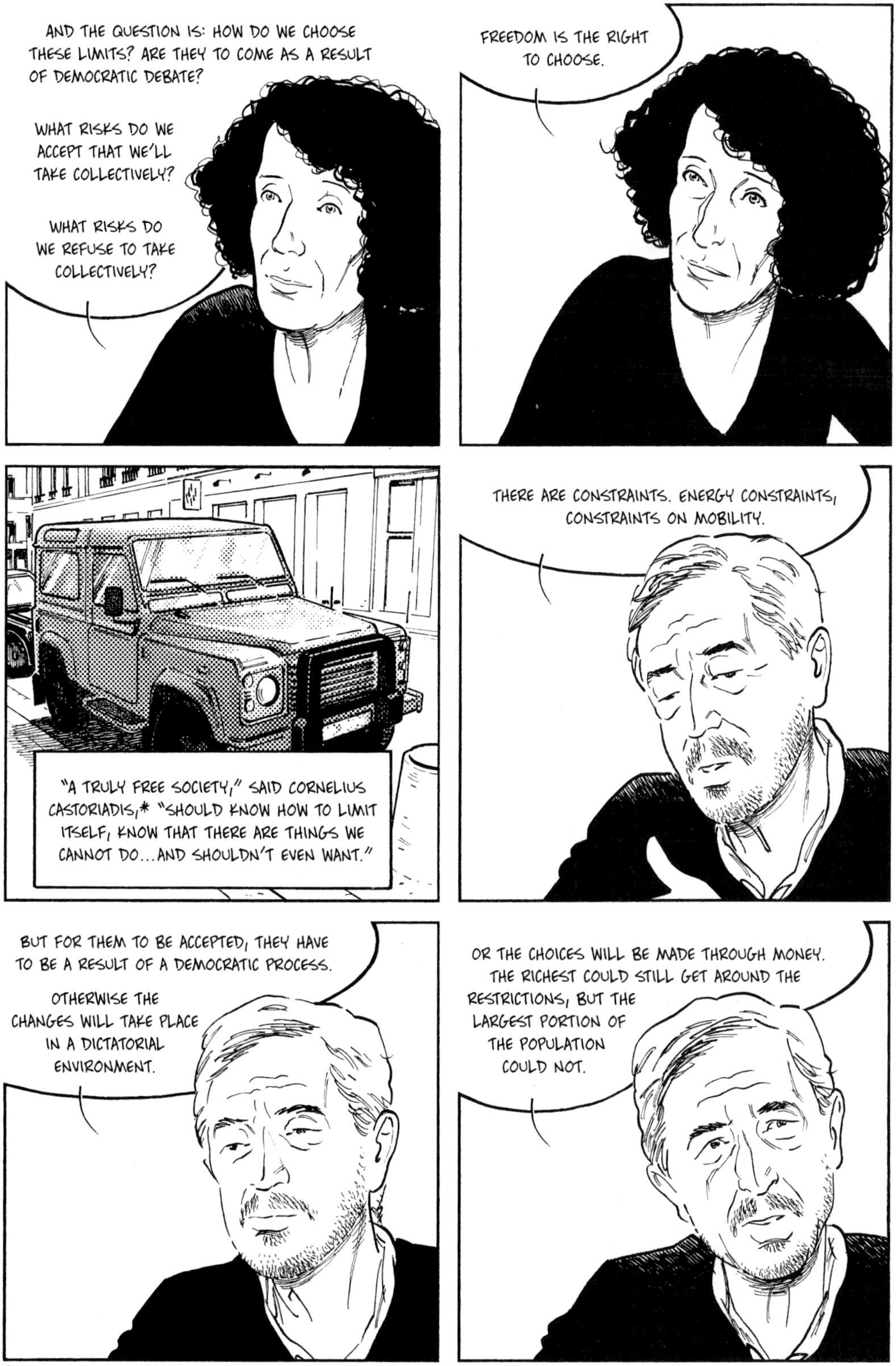
AND THE QUESTION IS: HOW DO WE CHOOSE THESE LIMITS? ARE THEY TO COME AS A RESULT OF DEMOCRATIC DEBATE?
WHAT RISKS DO WE ACCEPT THAT WE'LL TAKE COLLECTIVELY?
WHAT RISKS DO WE REFUSE TO TAKE COLLECTIVELY?
FREEDOM IS THE RIGHT TO CHOOSE.
"A TRULY FREE SOCIETY," SAID CORNELIUS CASTORIADIS,* "SHOULD KNOW HOW TO LIMIT ITSELF, KNOW THAT THERE ARE THINGS WE CANNOT DO... AND SHOULDN'T EVEN WANT."
THERE ARE CONSTRAINTS. ENERGY CONSTRAINTS, CONSTRAINTS ON MOBILITY.
BUT FOR THEM TO BE ACCEPTED, THEY HAVE TO BE A RESULT OF A DEMOCRATIC PROCESS.
OTHERWISE THE CHANGES WILL TAKE PLACE IN A DICTATORIAL ENVIRONMENT.
OR THE CHOICES WILL BE MADE THROUGH MONEY. THE RICHEST COULD STILL GET AROUND THE RESTRICTIONS, BUT THE LARGEST PORTION OF THE POPULATION COULD NOT.

*20TH-CENTURY GREEK-FRENCH PHILOSOPHER AND ECONOMIST.

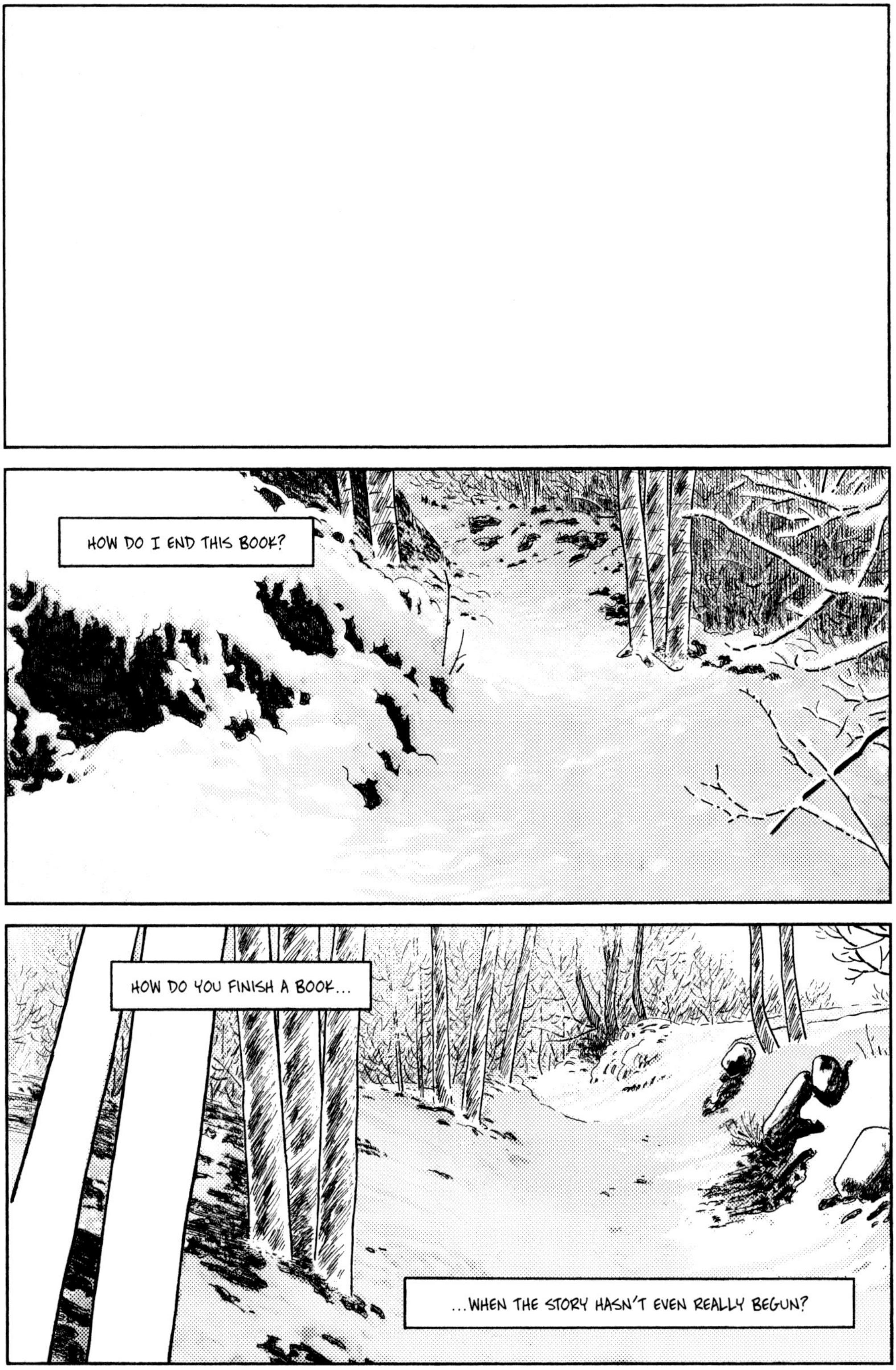
HOW DO I END THIS BOOK?
HOW DO YOU FINISH A BOOK...
...WHEN THE STORY HASN'T EVEN REALLY BEGUN?

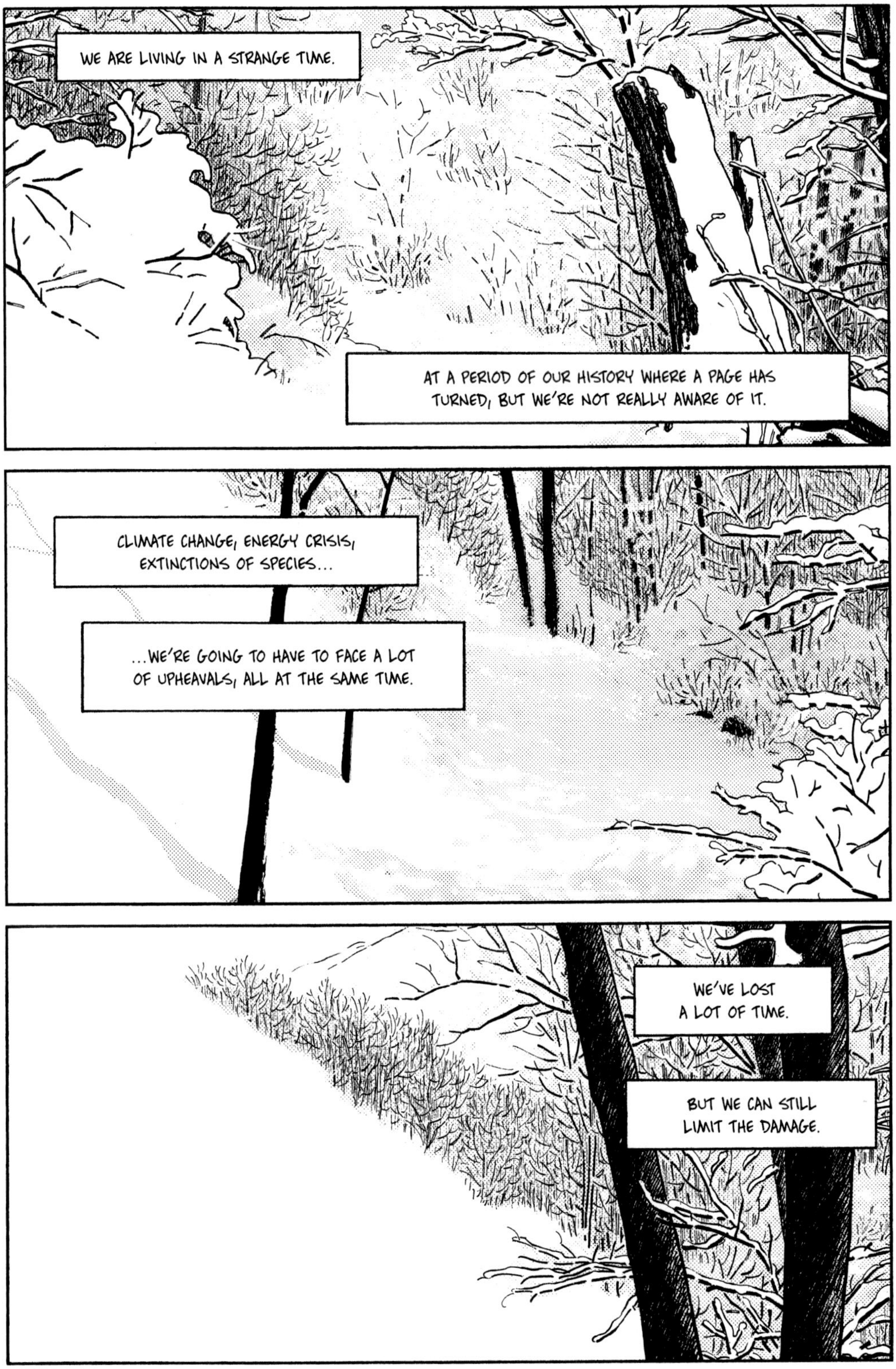
WE ARE LIVING IN A STRANGE TIME.
AT A PERIOD OF OUR HISTORY WHERE A PAGE HAS TURNED, BUT WE'RE NOT REALLY AWARE OF IT.
CLIMATE CHANGE, ENERGY CRISIS, EXTINCTIONS OF SPECIES...
...WE'RE GOING TO HAVE TO FACE A LOT OF UPHEAVALS, ALL AT THE SAME TIME.
WE'VE LOST A LOT OF TIME.
BUT WE CAN STILL LIMIT THE DAMAGE.

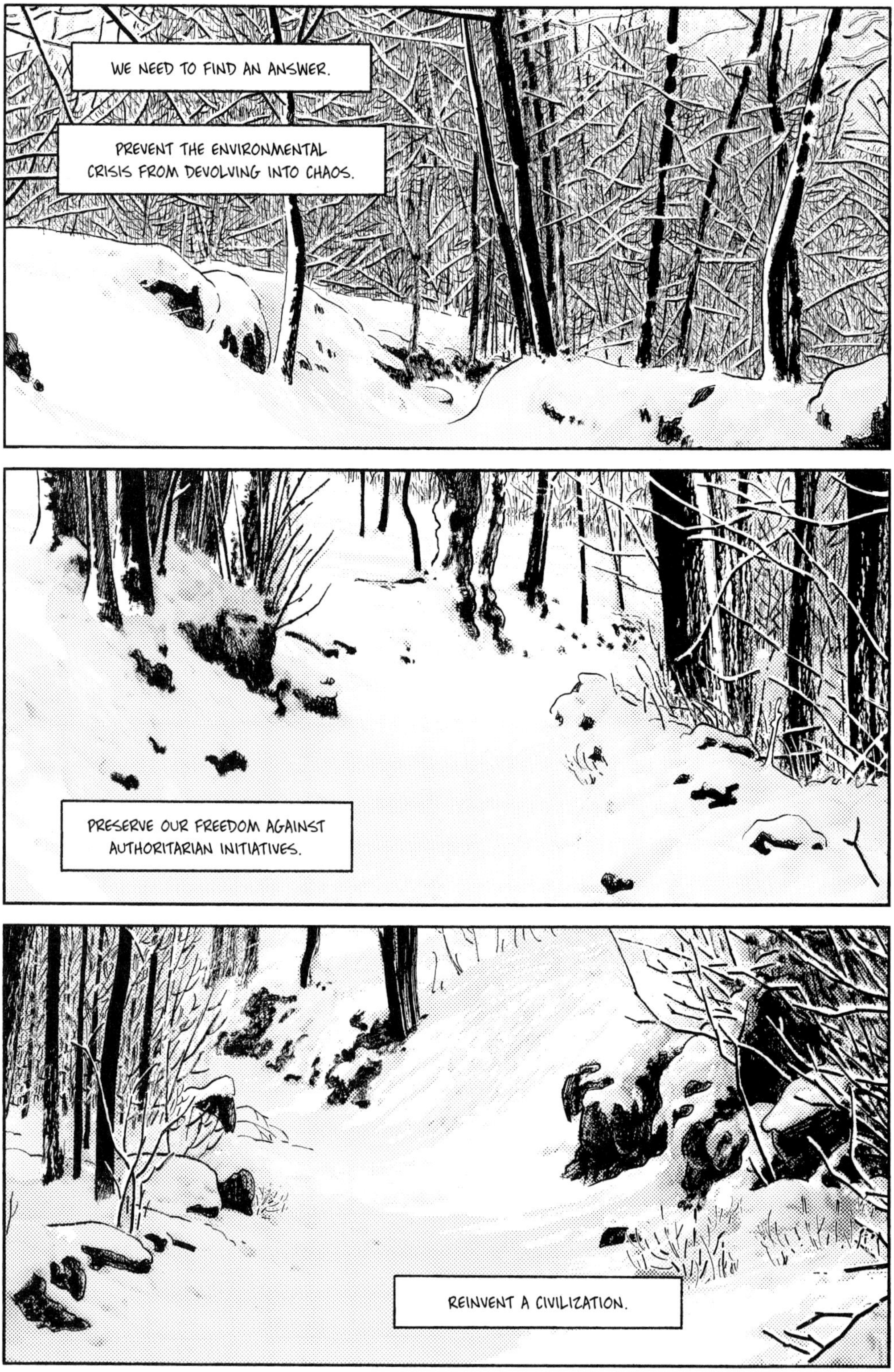
WE NEED TO FIND AN ANSWER.
PREVENT THE ENVIRONMENTAL CRISIS FROM DEVOLVING INTO CHAOS.
PRESERVE OUR FREEDOM AGAINST AUTHORITARIAN INITIATIVES.
REINVENT A CIVILIZATION.

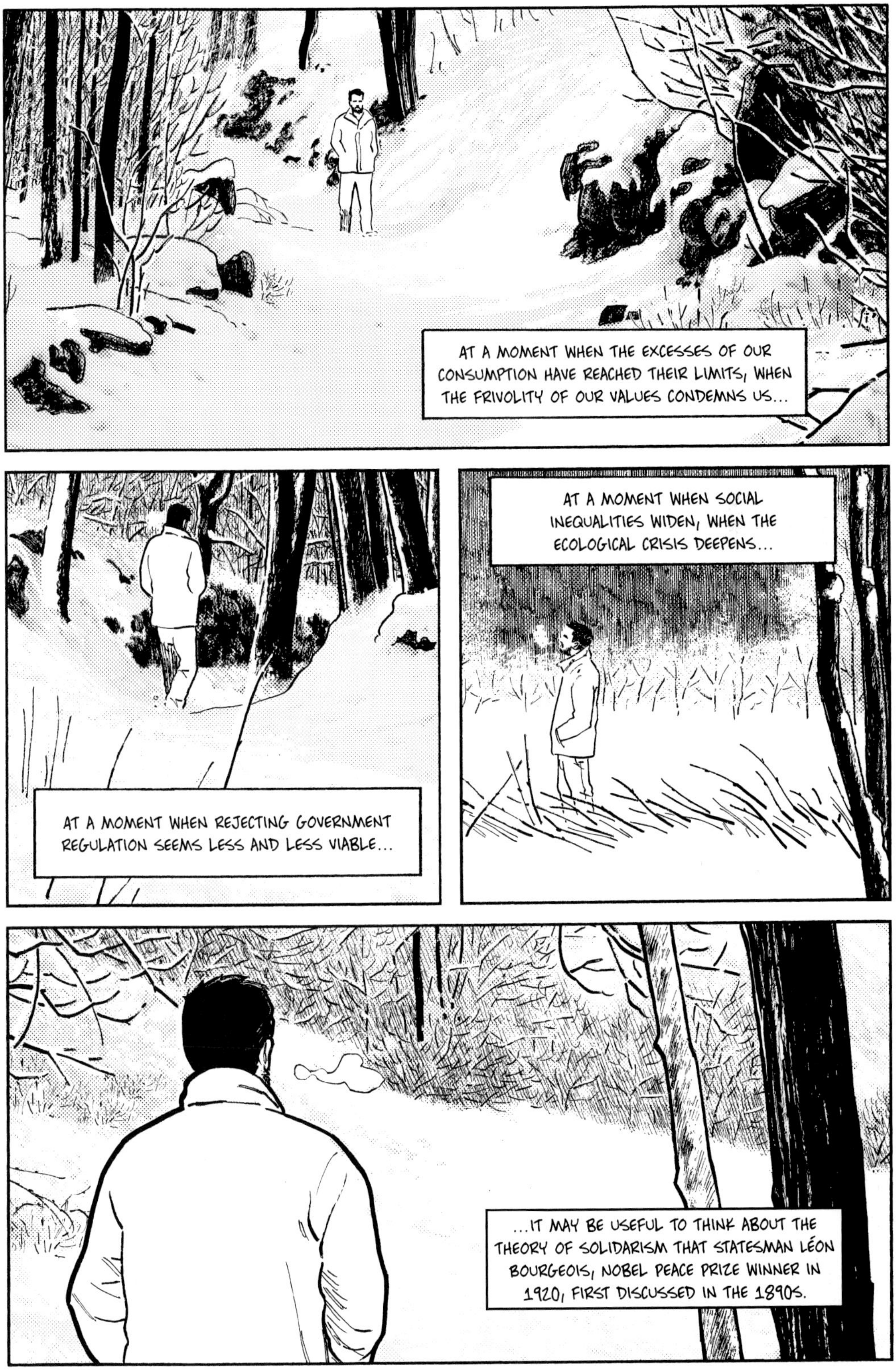
AT A MOMENT WHEN THE EXCESSES OF OUR CONSUMPTION HAVE REACHED THEIR LIMITS, WHEN THE FRIVOLITY OF OUR VALUES CONDEMNS US...
AT A MOMENT WHEN REJECTING GOVERNMENT REGULATION SEEMS LESS AND LESS VIABLE...
AT A MOMENT WHEN SOCIAL INEQUALITIES WIDEN, WHEN THE ECOLOGICAL CRISIS DEEPENS...
...IT MAY BE USEFUL TO THINK ABOUT THE THEORY OF SOLIDARISM THAT STATESMAN LÉON BOURGEOIS, NOBEL PEACE PRIZE WINNER IN 1920, FIRST DISCUSSED IN THE 1890S.

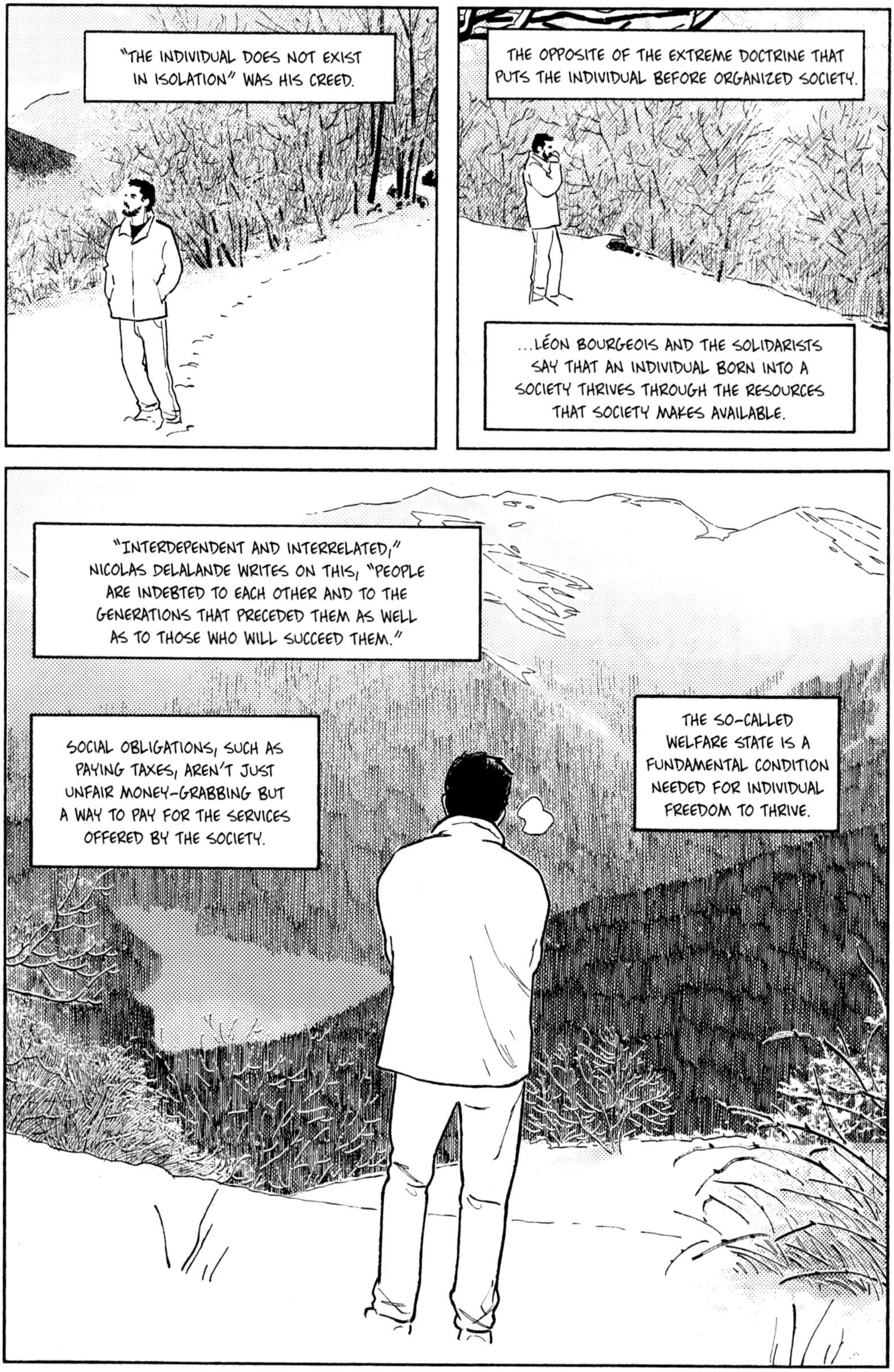
"THE INDIVIDUAL DOES NOT EXIST IN ISOLATION" WAS HIS CREED.
THE OPPOSITE OF THE EXTREME DOCTRINE THAT PUTS THE INDIVIDUAL BEFORE ORGANIZED SOCIETY.
...LÉON BOURGEOIS AND THE SOLIDARISTS SAY THAT AN INDIVIDUAL BORN INTO A SOCIETY THRIVES THROUGH THE RESOURCES THAT SOCIETY MAKES AVAILABLE.
"INTERDEPENDENT AND INTERRELATED," NICOLAS DELALANDE WRITES ON THIS, "PEOPLE ARE INDEBTED TO EACH OTHER AND TO THE GENERATIONS THAT PRECEDED THEM AS WELL AS TO THOSE WHO WILL SUCCEED THEM."
SOCIAL OBLIGATIONS, SUCH AS PAYING TAXES, AREN'T JUST UNFAIR MONEY-GRABBING BUT A WAY TO PAY FOR THE SERVICES OFFERED BY THE SOCIETY.
THE SO-CALLED WELFARE STATE IS A FUNDAMENTAL CONDITION NEEDED FOR INDIVIDUAL FREEDOM TO THRIVE.

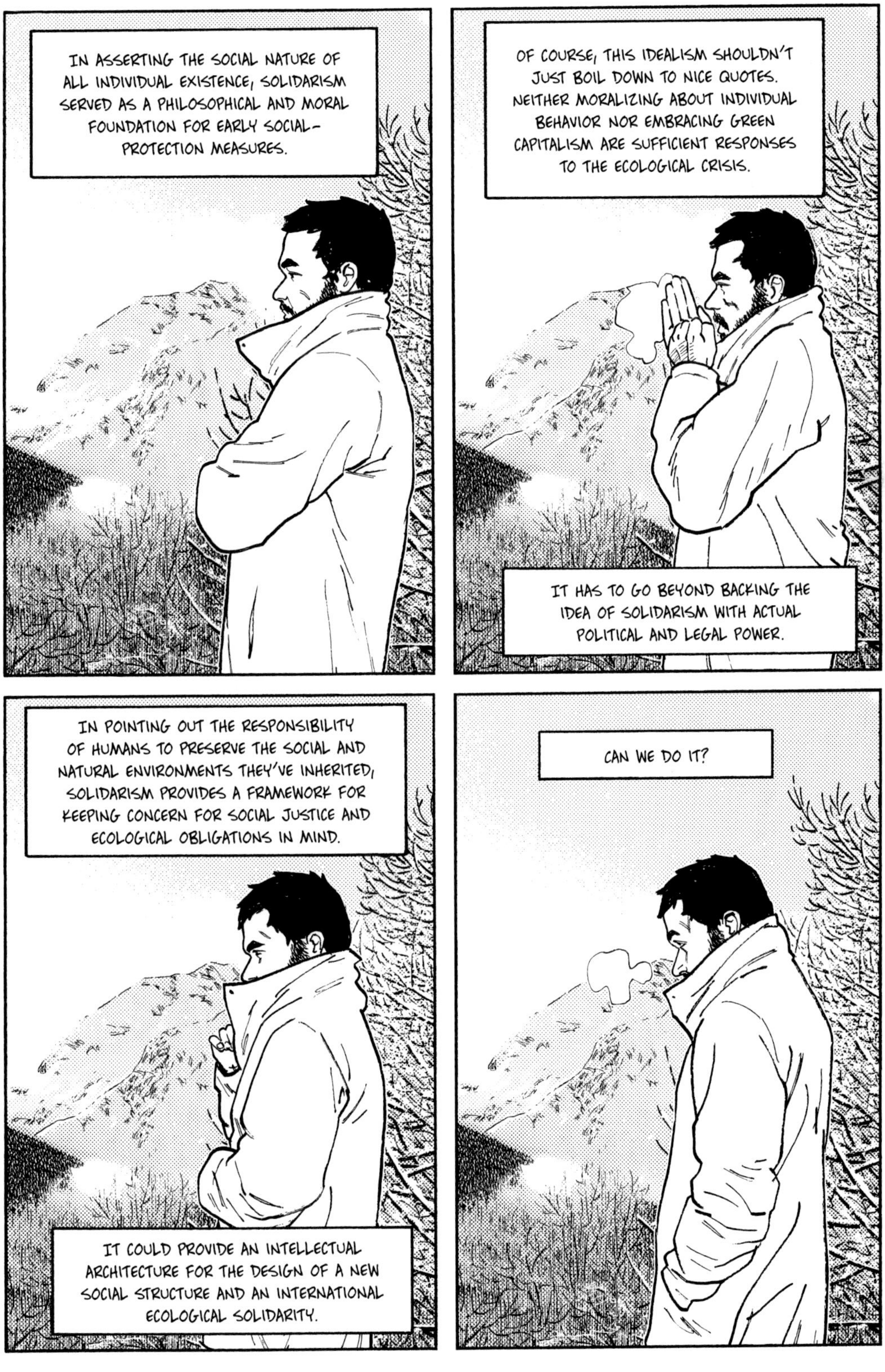
IN ASSERTING THE SOCIAL NATURE OF ALL INDIVIDUAL EXISTENCE, SOLIDARISM SERVED AS A PHILOSOPHICAL AND MORAL FOUNDATION FOR EARLY SOCIAL-PROTECTION MEASURES.
OF COURSE, THIS IDEALISM SHOULDN'T JUST BOIL DOWN TO NICE QUOTES. NEITHER MORALIZING ABOUT INDIVIDUAL BEHAVIOR NOR EMBRACING GREEN CAPITALISM ARE SUFFICIENT RESPONSES TO THE ECOLOGICAL CRISIS.
IT HAS TO GO BEYOND BACKING THE IDEA OF SOLIDARISM WITH ACTUAL POLITICAL AND LEGAL POWER.
IN POINTING OUT THE RESPONSIBILITY OF HUMANS TO PRESERVE THE SOCIAL AND NATURAL ENVIRONMENTS THEY'VE INHERITED, SOLIDARISM PROVIDES A FRAMEWORK FOR KEEPING CONCERN FOR SOCIAL JUSTICE AND ECOLOGICAL OBLIGATIONS IN MIND.
IT COULD PROVIDE AN INTELLECTUAL ARCHITECTURE FOR THE DESIGN OF A NEW SOCIAL STRUCTURE AND AN INTERNATIONAL ECOLOGICAL SOLIDARITY.
CAN WE DO IT?

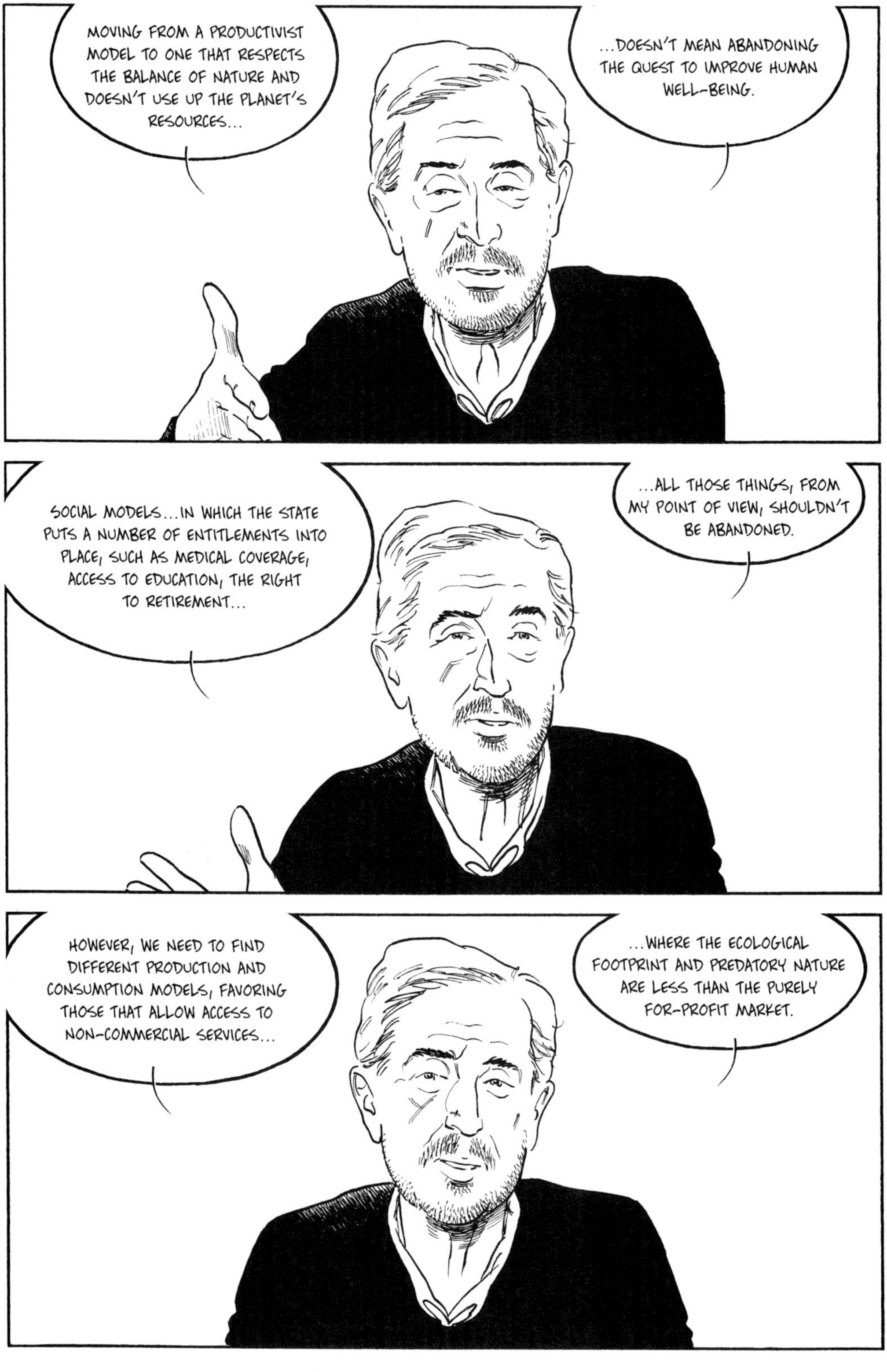
MOVING FROM A PRODUCTIVIST MODEL TO ONE THAT RESPECTS THE BALANCE OF NATURE AND DOESN'T USE UP THE PLANET'S RESOURCES...
...DOESN'T MEAN ABANDONING THE QUEST TO IMPROVE HUMAN WELL-BEING.
SOCIAL MODELS...IN WHICH THE STATE PUTS A NUMBER OF ENTITLEMENTS INTO PLACE, SUCH AS MEDICAL COVERAGE, ACCESS TO EDUCATION, THE RIGHT TO RETIREMENT...
...ALL THOSE THINGS, FROM MY POINT OF VIEW, SHOULDN'T BE ABANDONED.
HOWEVER, WE NEED TO FIND DIFFERENT PRODUCTION AND CONSUMPTION MODELS, FAVORING THOSE THAT ALLOW ACCESS TO NON-COMMERCIAL SERVICES...
...WHERE THE ECOLOGICAL FOOTPRINT AND PREDATORY NATURE ARE LESS THAN THE PURELY FOR-PROFIT MARKET.

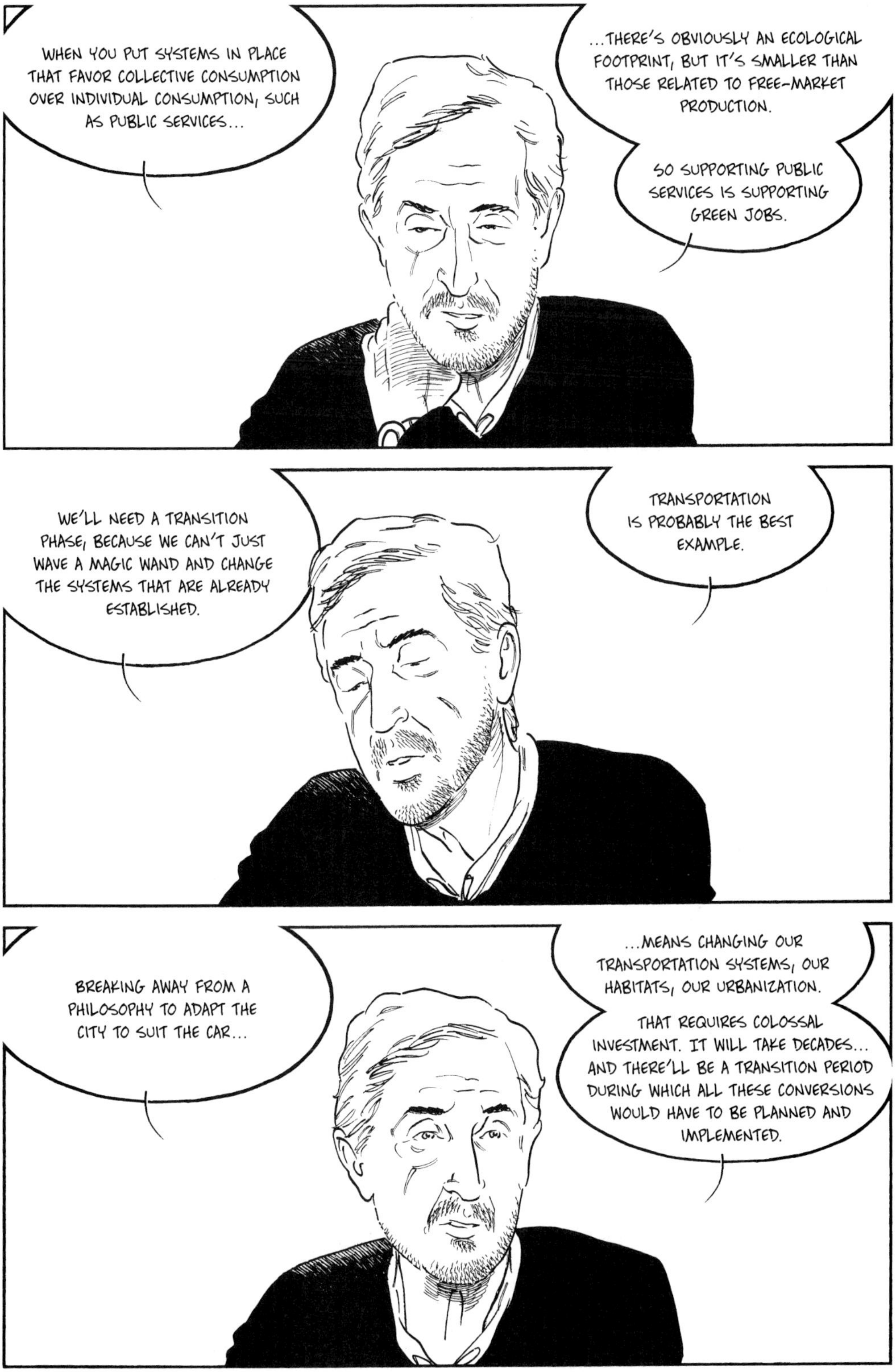
WHEN YOU PUT SYSTEMS IN PLACE THAT FAVOR COLLECTIVE CONSUMPTION OVER INDIVIDUAL CONSUMPTION, SUCH AS PUBLIC SERVICES...
...THERE'S OBVIOUSLY AN ECOLOGICAL FOOTPRINT, BUT IT'S SMALLER THAN THOSE RELATED TO FREE-MARKET PRODUCTION.
SO SUPPORTING PUBLIC SERVICES IS SUPPORTING GREEN JOBS.
WE'LL NEED A TRANSITION PHASE, BECAUSE WE CAN'T JUST WAVE A MAGIC WAND AND CHANGE THE SYSTEMS THAT ARE ALREADY ESTABLISHED.
TRANSPORTATION IS PROBABLY THE BEST EXAMPLE.
BREAKING AWAY FROM A PHILOSOPHY TO ADAPT THE CITY TO SUIT THE CAR...
...MEANS CHANGING OUR TRANSPORTATION SYSTEMS, OUR HABITATS, OUR URBANIZATION.
THAT REQUIRES COLOSSAL INVESTMENT. IT WILL TAKE DECADES... AND THERE'LL BE A TRANSITION PERIOD DURING WHICH ALL THESE CONVERSIONS WOULD HAVE TO BE PLANNED AND IMPLEMENTED.

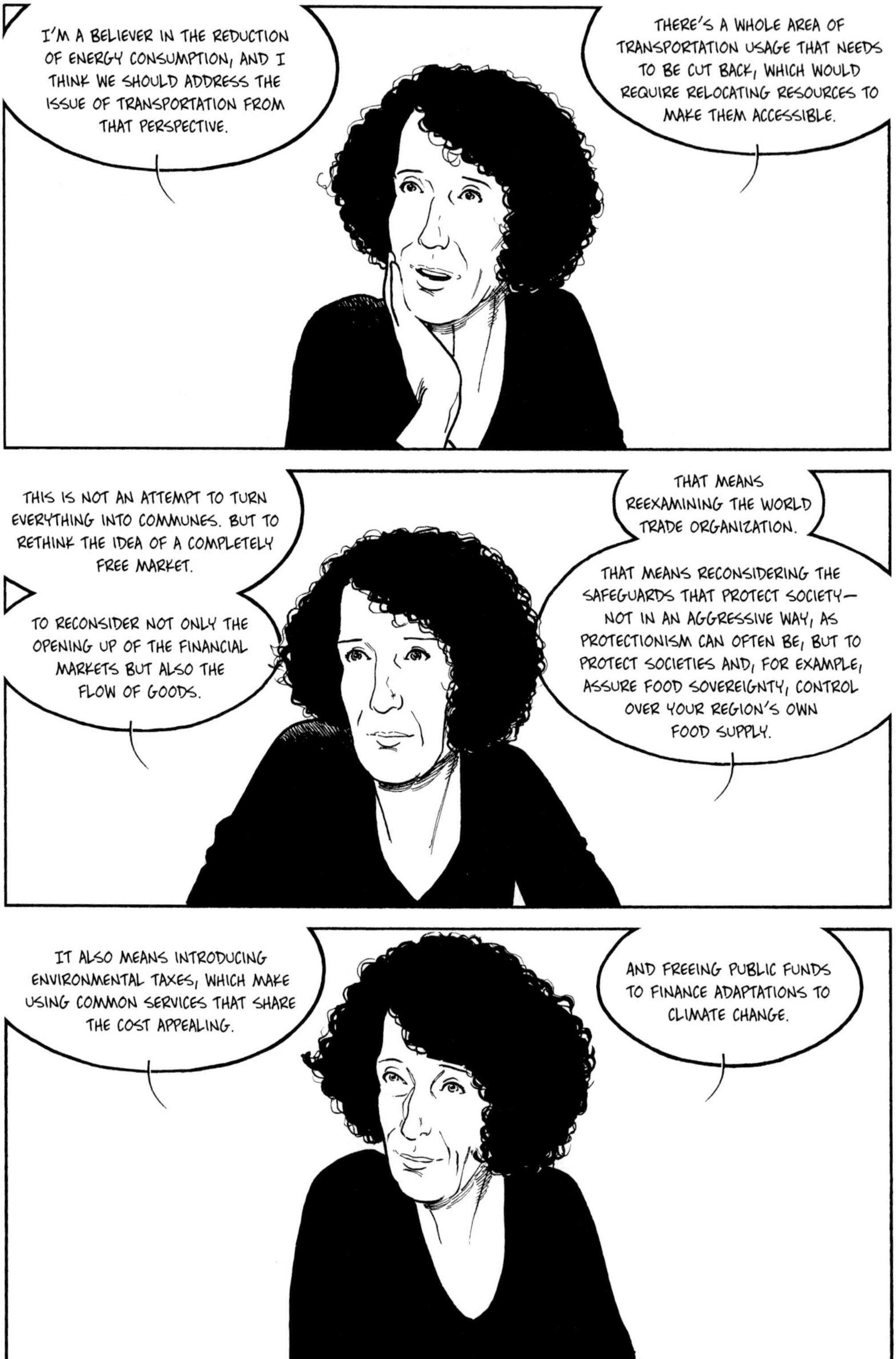
I'M A BELIEVER IN THE REDUCTION OF ENERGY CONSUMPTION, AND I THINK WE SHOULD ADDRESS THE ISSUE OF TRANSPORTATION FROM THAT PERSPECTIVE.
THERE'S A WHOLE AREA OF TRANSPORTATION USAGE THAT NEEDS TO BE CUT BACK, WHICH WOULD REQUIRE RELOCATING RESOURCES TO MAKE THEM ACCESSIBLE.
THIS IS NOT AN ATTEMPT TO TURN EVERYTHING INTO COMMUNES. BUT TO RETHINK THE IDEA OF A COMPLETELY FREE MARKET.
TO RECONSIDER NOT ONLY THE OPENING UP OF THE FINANCIAL MARKETS BUT ALSO THE FLOW OF GOODS.
THAT MEANS REEXAMINING THE WORLD TRADE ORGANIZATION.
THAT MEANS RECONSIDERING THE SAFEGUARDS THAT PROTECT SOCIETY— NOT IN AN AGGRESSIVE WAY, AS PROTECTIONISM CAN OFTEN BE, BUT TO PROTECT SOCIETIES AND, FOR EXAMPLE, ASSURE FOOD SOVEREIGNTY, CONTROL OVER YOUR REGION'S OWN FOOD SUPPLY.
IT ALSO MEANS INTRODUCING ENVIRONMENTAL TAXES, WHICH MAKE USING COMMON SERVICES THAT SHARE THE COST APPEALING.
AND FREEING PUBLIC FUNDS TO FINANCE ADAPTATIONS TO CLIMATE CHANGE.

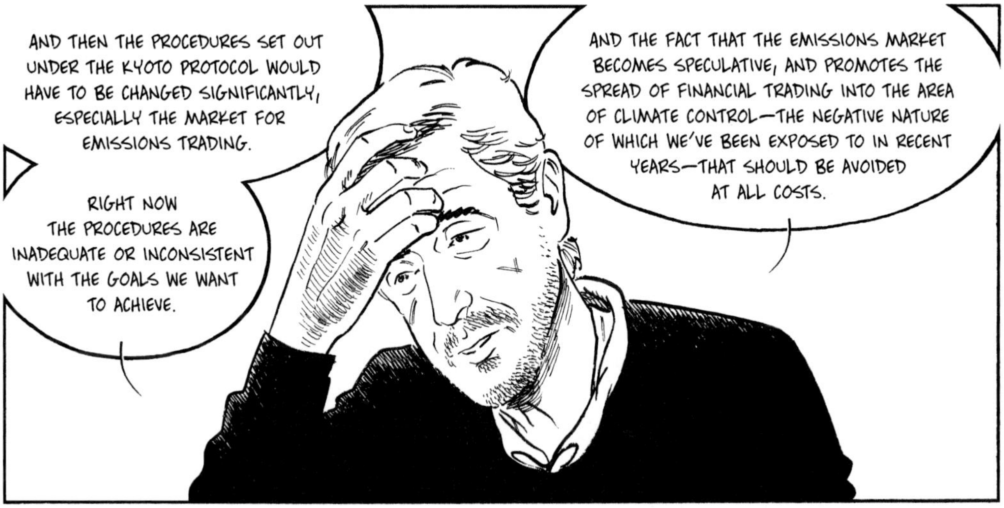
AND THEN THE PROCEDURES SET OUT UNDER THE KYOTO PROTOCOL WOULD HAVE TO BE CHANGED SIGNIFICANTLY, ESPECIALLY THE MARKET FOR EMISSIONS TRADING.
RIGHT NOW THE PROCEDURES ARE INADEQUATE OR INCONSISTENT WITH THE GOALS WE WANT TO ACHIEVE.
AND THE FACT THAT THE EMISSIONS MARKET BECOMES SPECULATIVE, AND PROMOTES THE SPREAD OF FINANCIAL TRADING INTO THE AREA OF CLIMATE CONTROL—THE NEGATIVE NATURE OF WHICH WE'VE BEEN EXPOSED TO IN RECENT YEARS—THAT SHOULD BE AVOIDED AT ALL COSTS.

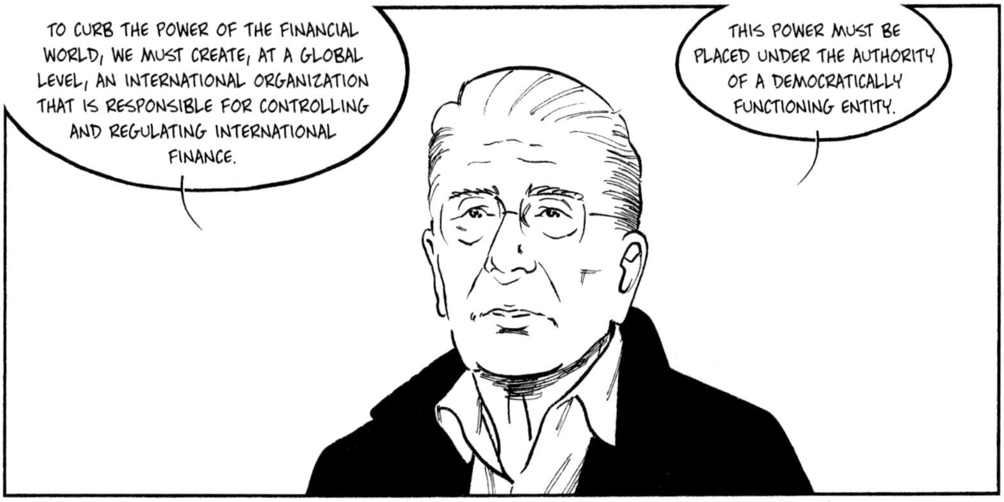
TO CURB THE POWER OF THE FINANCIAL WORLD, WE MUST CREATE, AT A GLOBAL LEVEL, AN INTERNATIONAL ORGANIZATION THAT IS RESPONSIBLE FOR CONTROLLING AND REGULATING INTERNATIONAL FINANCE.
THIS POWER MUST BE PLACED UNDER THE AUTHORITY OF A DEMOCRATICALLY FUNCTIONING ENTITY.

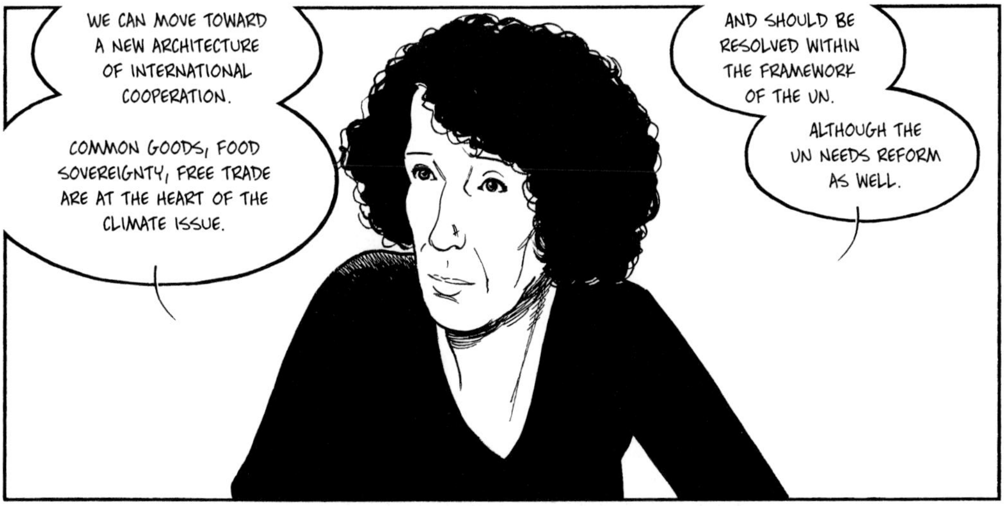
WE CAN MOVE TOWARD A NEW ARCHITECTURE OF INTERNATIONAL COOPERATION.
COMMON GOODS, FOOD SOVEREIGNTY, FREE TRADE ARE AT THE HEART OF THE CLIMATE ISSUE.
AND SHOULD BE RESOLVED WITHIN THE FRAMEWORK OF THE UN.
ALTHOUGH THE UN NEEDS REFORM AS WELL.

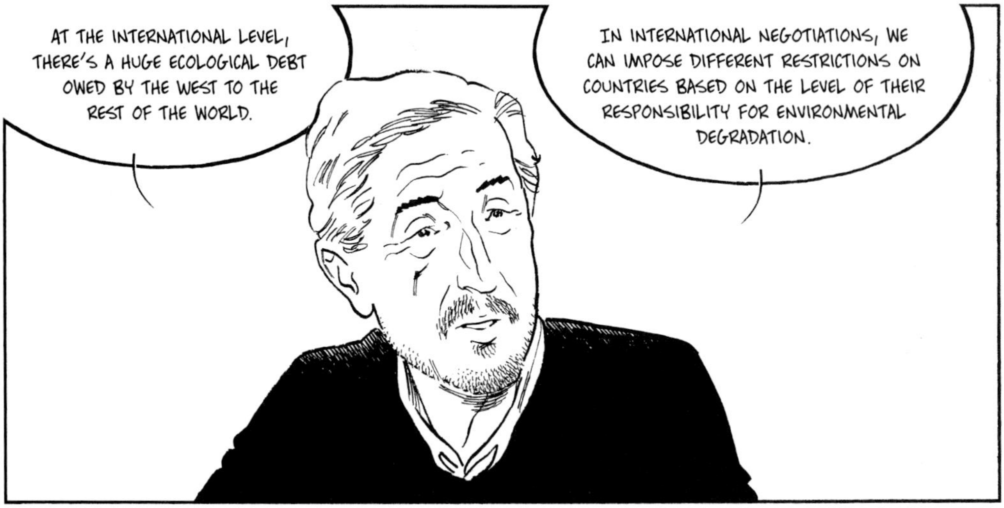
AT THE INTERNATIONAL LEVEL, THERE'S A HUGE ECOLOGICAL DEBT OWED BY THE WEST TO THE REST OF THE WORLD.
IN INTERNATIONAL NEGOTIATIONS, WE CAN IMPOSE DIFFERENT RESTRICTIONS ON COUNTRIES BASED ON THE LEVEL OF THEIR RESPONSIBILITY FOR ENVIRONMENTAL DEGRADATION.

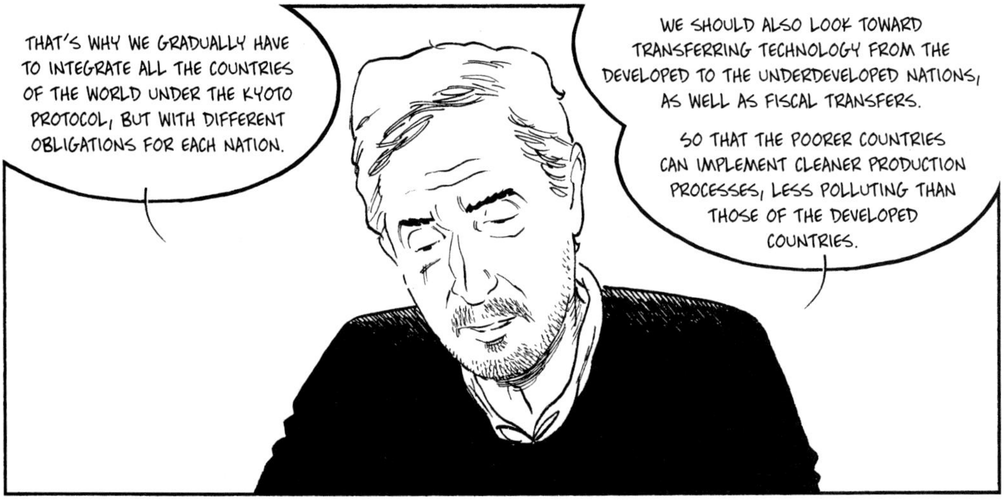
THAT'S WHY WE GRADUALLY HAVE TO INTEGRATE ALL THE COUNTRIES OF THE WORLD UNDER THE KYOTO PROTOCOL, BUT WITH DIFFERENT OBLIGATIONS FOR EACH NATION.
WE SHOULD ALSO LOOK TOWARD TRANSFERRING TECHNOLOGY FROM THE DEVELOPED TO THE UNDERDEVELOPED NATIONS, AS WELL AS FISCAL TRANSFERS.
SO THAT THE POORER COUNTRIES CAN IMPLEMENT CLEANER PRODUCTION PROCESSES, LESS POLLUTING THAN THOSE OF THE DEVELOPED COUNTRIES.

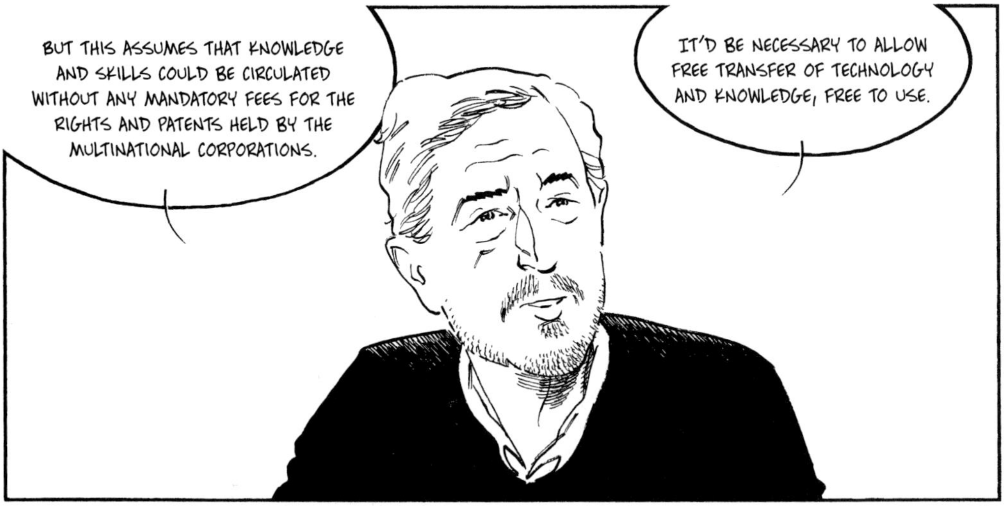
BUT THIS ASSUMES THAT KNOWLEDGE AND SKILLS COULD BE CIRCULATED WITHOUT ANY MANDATORY FEES FOR THE RIGHTS AND PATENTS HELD BY THE MULTINATIONAL CORPORATIONS.
IT'D BE NECESSARY TO ALLOW FREE TRANSFER OF TECHNOLOGY AND KNOWLEDGE, FREE TO USE.

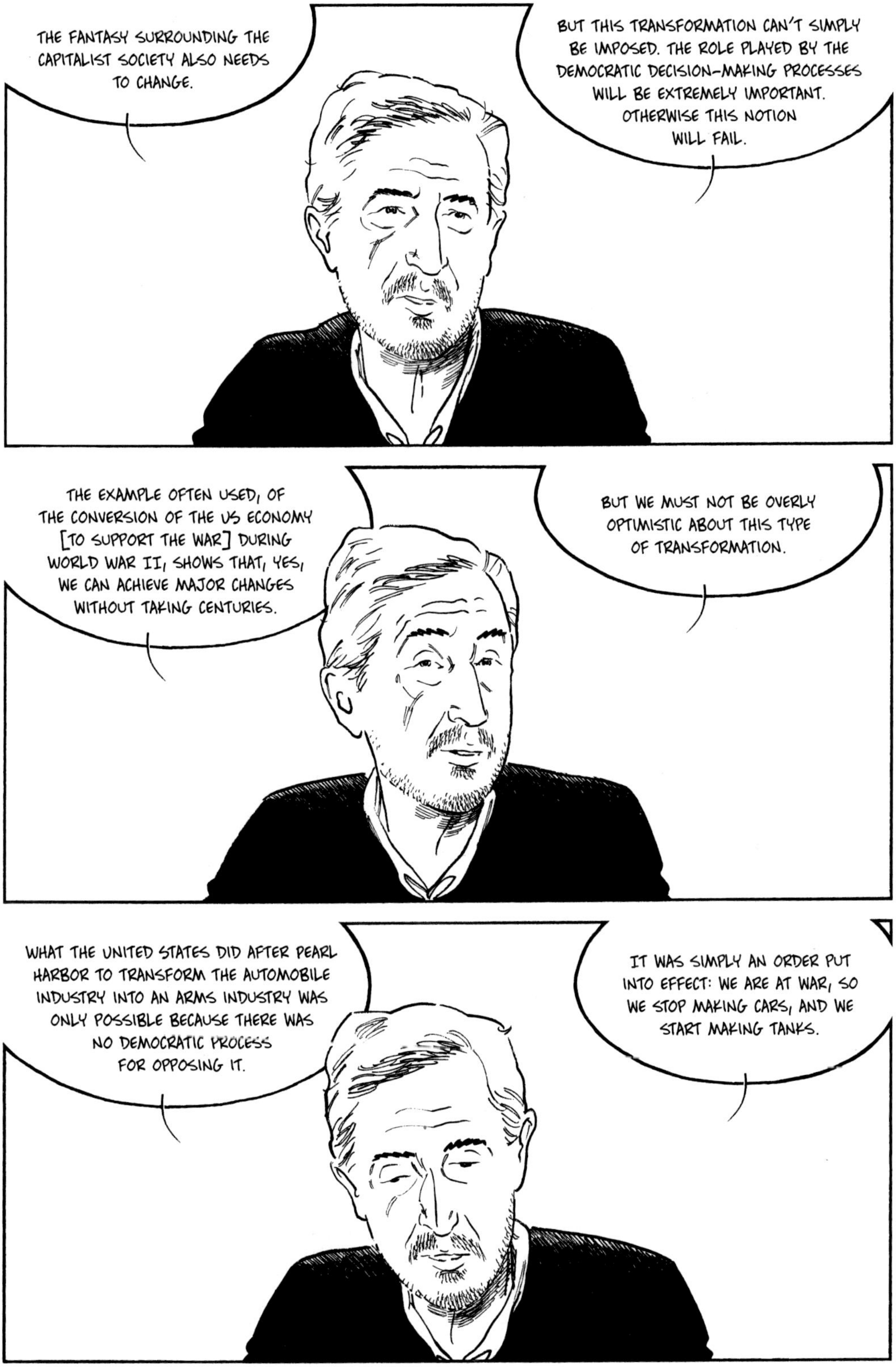
THE FANTASY SURROUNDING THE CAPITALIST SOCIETY ALSO NEEDS TO CHANGE.
BUT THIS TRANSFORMATION CAN'T SIMPLY BE IMPOSED. THE ROLE PLAYED BY THE DEMOCRATIC DECISION-MAKING PROCESSES WILL BE EXTREMELY IMPORTANT. OTHERWISE THIS NOTION WILL FAIL.
THE EXAMPLE OFTEN USED, OF THE CONVERSION OF THE US ECONOMY [TO SUPPORT THE WAR] DURING WORLD WAR II, SHOWS THAT, YES, WE CAN ACHIEVE MAJOR CHANGES WITHOUT TAKING CENTURIES.
BUT WE MUST NOT BE OVERLY OPTIMISTIC ABOUT THIS TYPE OF TRANSFORMATION.
WHAT THE UNITED STATES DID AFTER PEARL HARBOR TO TRANSFORM THE AUTOMOBILE INDUSTRY INTO AN ARMS INDUSTRY WAS ONLY POSSIBLE BECAUSE THERE WAS NO DEMOCRATIC PROCESS FOR OPPOSING IT.
IT WAS SIMPLY AN ORDER PUT INTO EFFECT: WE ARE AT WAR, SO WE STOP MAKING CARS, AND WE START MAKING TANKS.

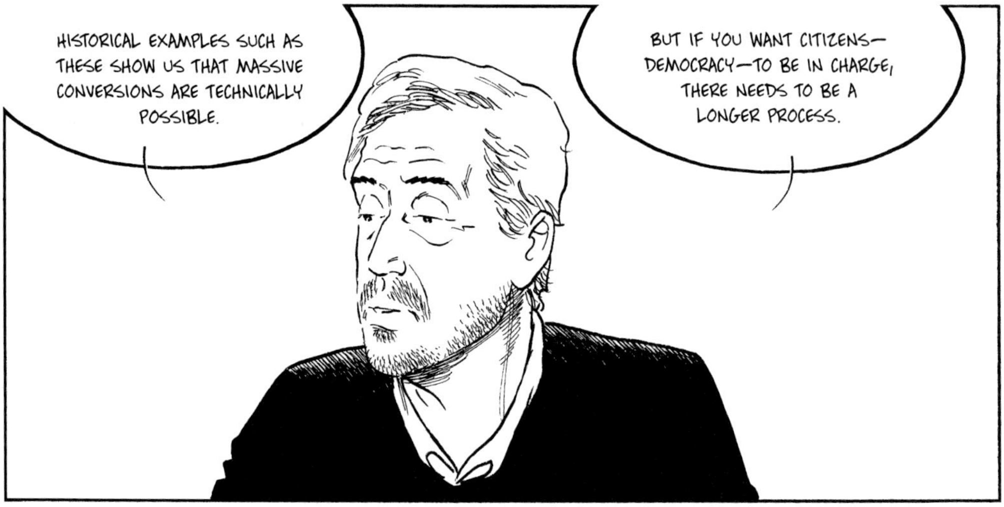
HISTORICAL EXAMPLES SUCH AS THESE SHOW US THAT MASSIVE CONVERSIONS ARE TECHNICALLY POSSIBLE.
BUT IF YOU WANT CITIZENS—DEMOCRACY—TO BE IN CHARGE, THERE NEEDS TO BE A LONGER PROCESS.

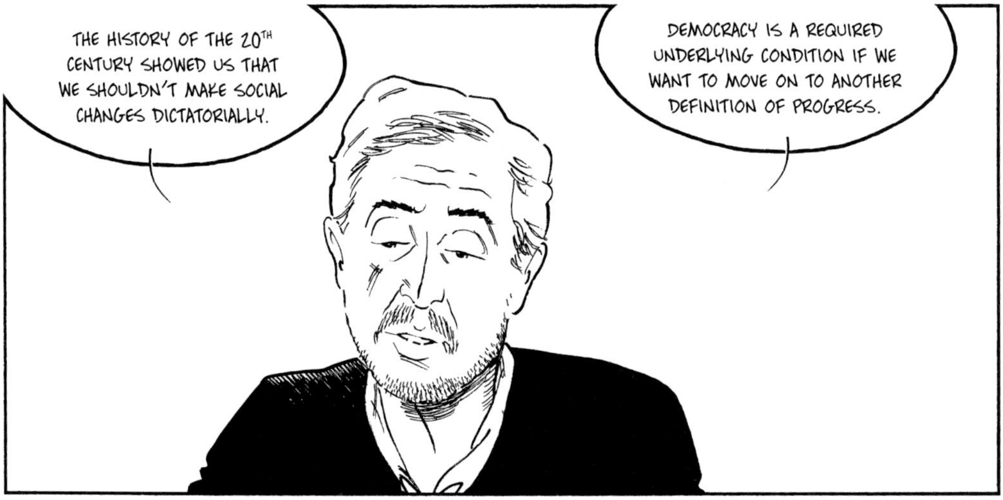
THE HISTORY OF THE 20TH CENTURY SHOWED US THAT WE SHOULDN'T MAKE SOCIAL CHANGES DICTATORIALLY.
DEMOCRACY IS A REQUIRED UNDERLYING CONDITION IF WE WANT TO MOVE ON TO ANOTHER DEFINITION OF PROGRESS.

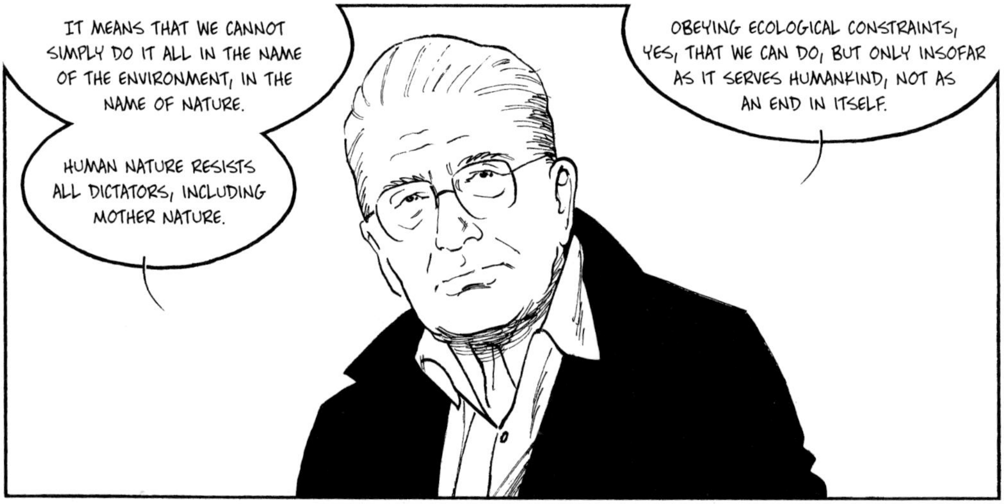
IT MEANS THAT WE CANNOT SIMPLY DO IT ALL IN THE NAME OF THE ENVIRONMENT, IN THE NAME OF NATURE.
HUMAN NATURE RESISTS ALL DICTATORS, INCLUDING MOTHER NATURE.
OBEYING ECOLOGICAL CONSTRAINTS, YES, THAT WE CAN DO, BUT ONLY INSOFAR AS IT SERVES HUMANKIND, NOT AS AN END IN ITSELF.

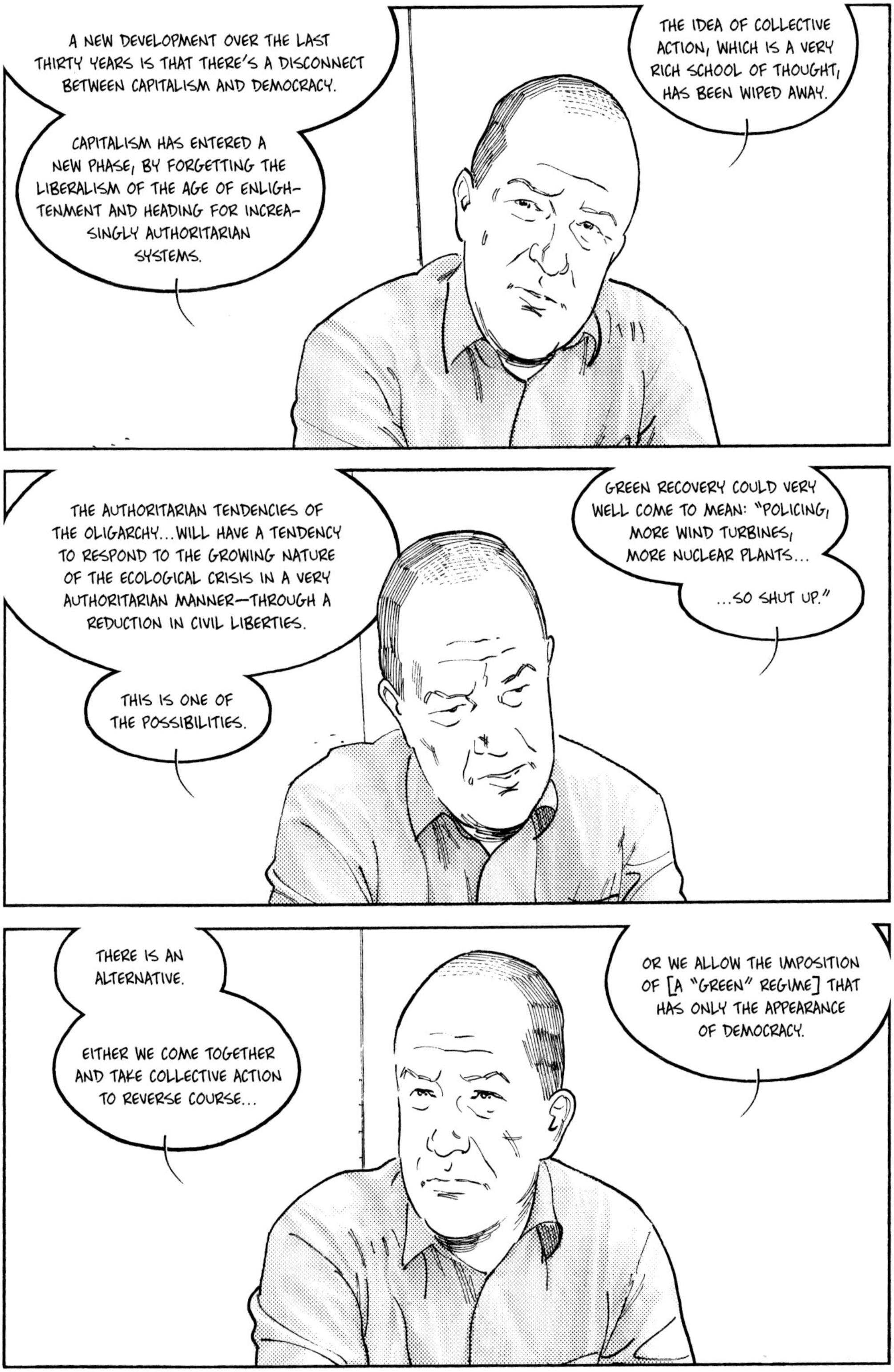
A NEW DEVELOPMENT OVER THE LAST THIRTY YEARS IS THAT THERE'S A DISCONNECT BETWEEN CAPITALISM AND DEMOCRACY.
CAPITALISM HAS ENTERED A NEW PHASE, BY FORGETTING THE LIBERALISM OF THE AGE OF ENLIGHTENMENT AND HEADING FOR INCREASINGLY AUTHORITARIAN SYSTEMS.
THE IDEA OF COLLECTIVE ACTION, WHICH IS A VERY RICH SCHOOL OF THOUGHT, HAS BEEN WIPED AWAY.
THE AUTHORITARIAN TENDENCIES OF THE OLIGARCHY...WILL HAVE A TENDENCY TO RESPOND TO THE GROWING NATURE OF THE ECOLOGICAL CRISIS IN A VERY AUTHORITARIAN MANNER—THROUGH A REDUCTION IN CIVIL LIBERTIES.
THIS IS ONE OF THE POSSIBILITIES.
GREEN RECOVERY COULD VERY WELL COME TO MEAN: "POLICING, MORE WIND TURBINES, MORE NUCLEAR PLANTS...
...SO SHUT UP."
THERE IS AN ALTERNATIVE.
EITHER WE COME TOGETHER AND TAKE COLLECTIVE ACTION TO REVERSE COURSE...
OR WE ALLOW THE IMPOSITION OF [A "GREEN" REGIME] THAT HAS ONLY THE APPEARANCE OF DEMOCRACY.

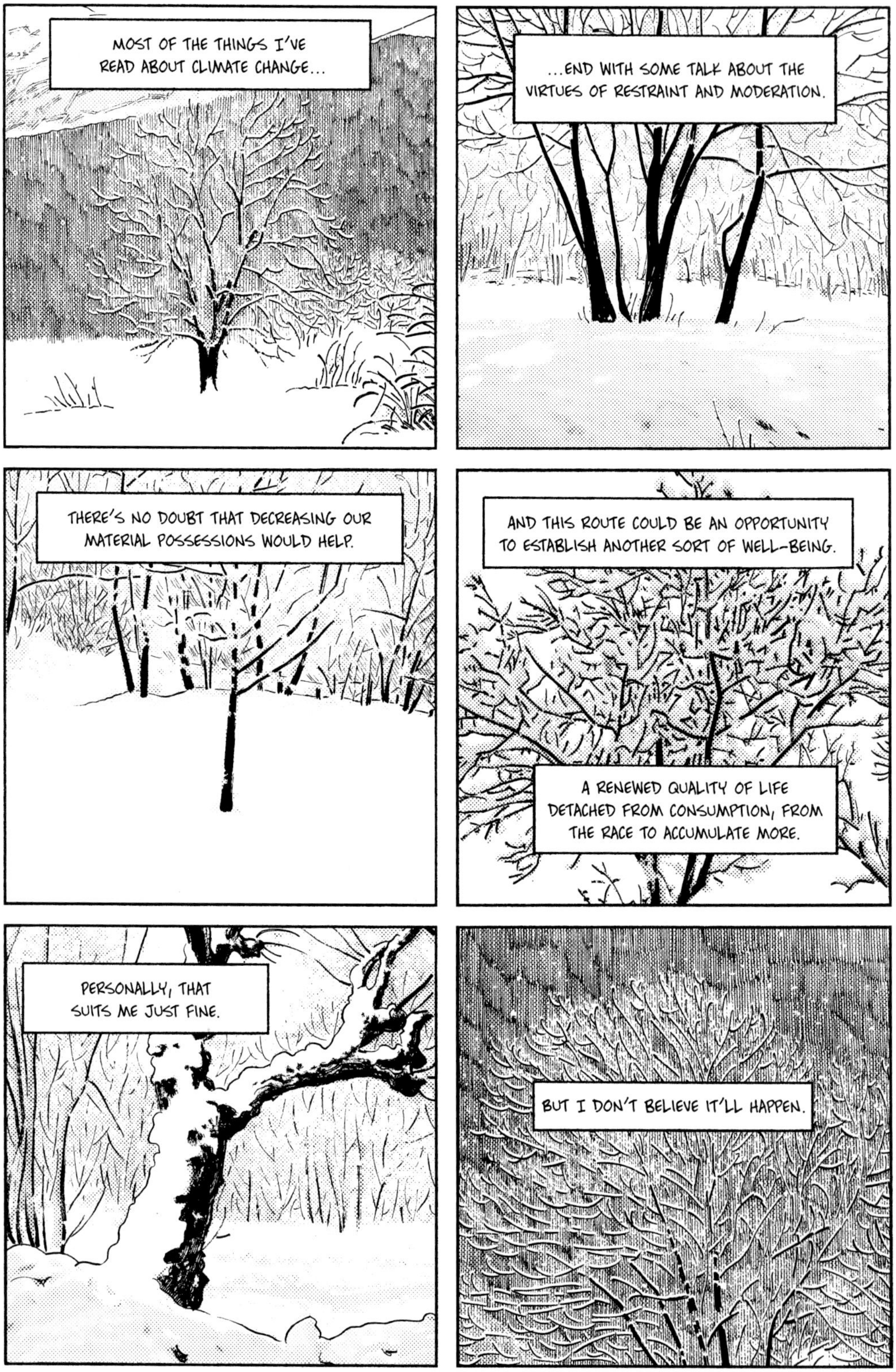
MOST OF THE THINGS I'VE READ ABOUT CLIMATE CHANGE...
...END WITH SOME TALK ABOUT THE VIRTUES OF RESTRAINT AND MODERATION.
THERE'S NO DOUBT THAT DECREASING OUR MATERIAL POSSESSIONS WOULD HELP.
AND THIS ROUTE COULD BE AN OPPORTUNITY TO ESTABLISH ANOTHER SORT OF WELL-BEING.
A RENEWED QUALITY OF LIFE DETACHED FROM CONSUMPTION, FROM THE RACE TO ACCUMULATE MORE.
PERSONALLY, THAT SUITS ME JUST FINE.
BUT I DON'T BELIEVE IT'LL HAPPEN.

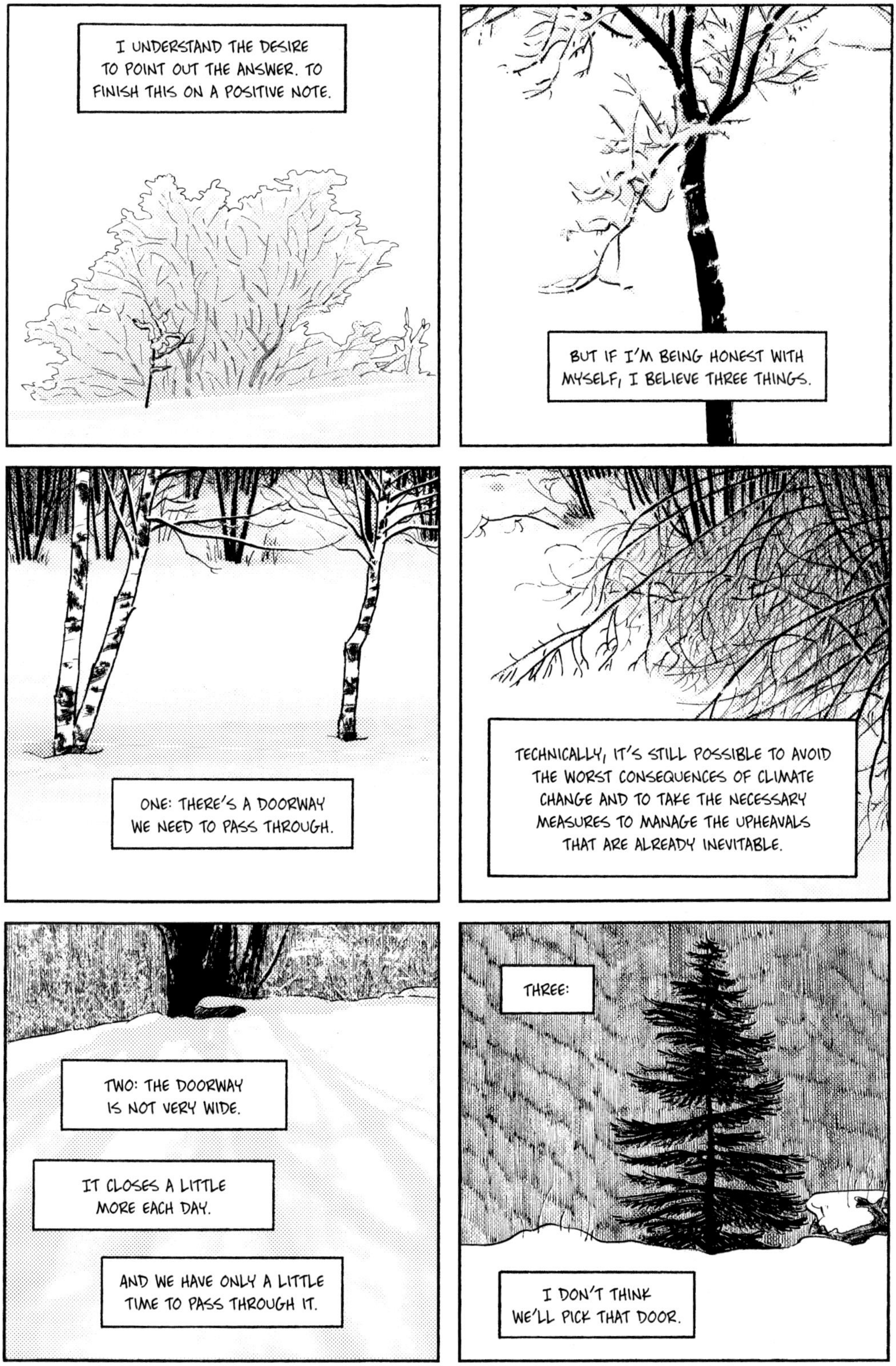
I UNDERSTAND THE DESIRE TO POINT OUT THE ANSWER. TO FINISH THIS ON A POSITIVE NOTE.
BUT IF I'M BEING HONEST WITH MYSELF, I BELIEVE THREE THINGS.
ONE: THERE'S A DOORWAY WE NEED TO PASS THROUGH.
TECHNICALLY, IT'S STILL POSSIBLE TO AVOID THE WORST CONSEQUENCES OF CLIMATE CHANGE AND TO TAKE THE NECESSARY MEASURES TO MANAGE THE UPHEAVALS THAT ARE ALREADY INEVITABLE.
TWO: THE DOORWAY IS NOT VERY WIDE.
IT CLOSES A LITTLE MORE EACH DAY.
AND WE HAVE ONLY A LITTLE TIME TO PASS THROUGH IT.
THREE:
I DON'T THINK WE'LL PICK THAT DOOR.

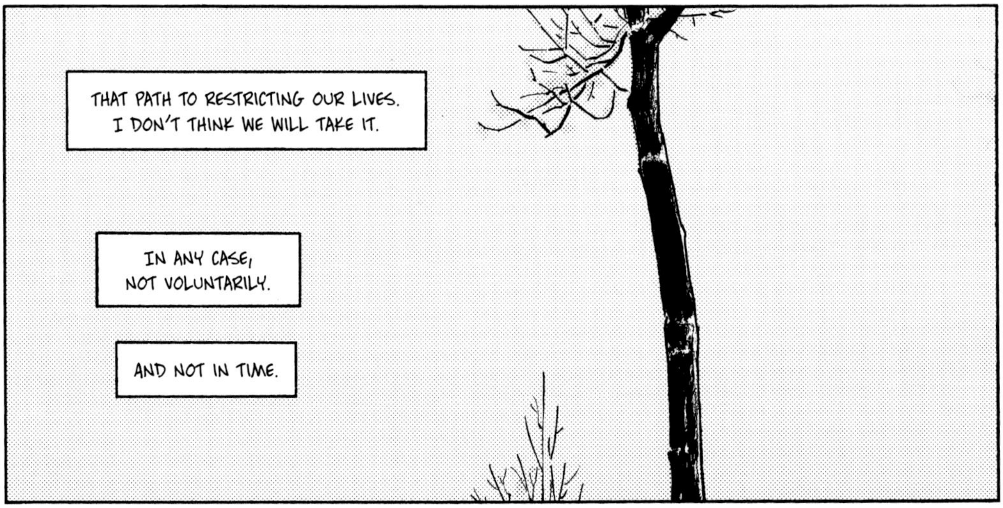
THAT PATH TO RESTRICTING OUR LIVES. I DON'T THINK WE WILL TAKE IT.
IN ANY CASE, NOT VOLUNTARILY.
AND NOT IN TIME.

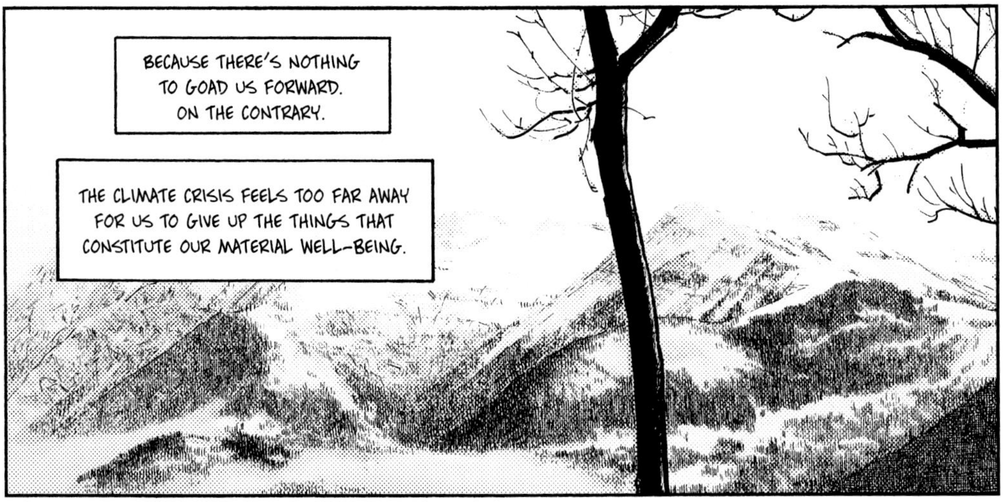
BECAUSE THERE'S NOTHING TO GOAD US FORWARD. ON THE CONTRARY.
THE CLIMATE CRISIS FEELS TOO FAR AWAY FOR US TO GIVE UP THE THINGS THAT CONSTITUTE OUR MATERIAL WELL-BEING.

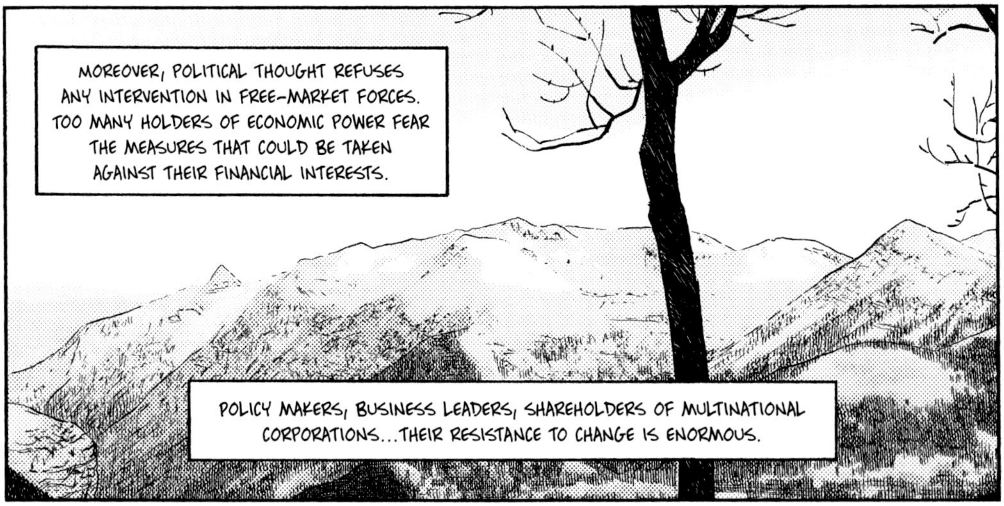
MOREOVER, POLITICAL THOUGHT REFUSES ANY INTERVENTION IN FREE-MARKET FORCES. TOO MANY HOLDERS OF ECONOMIC POWER FEAR THE MEASURES THAT COULD BE TAKEN AGAINST THEIR FINANCIAL INTERESTS.
POLICY MAKERS, BUSINESS LEADERS, SHAREHOLDERS OF MULTINATIONAL CORPORATIONS...THEIR RESISTANCE TO CHANGE IS ENORMOUS.

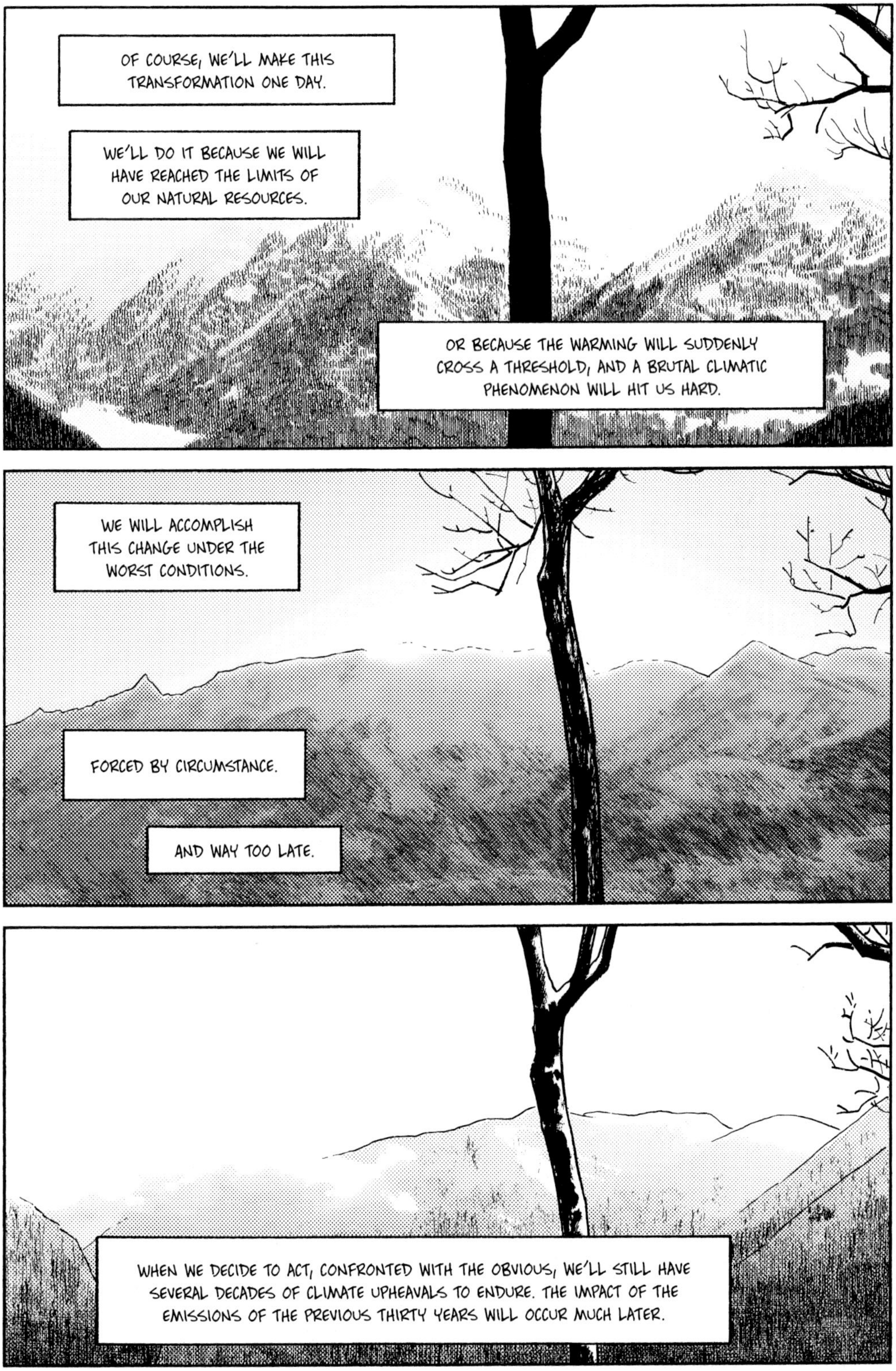
OF COURSE, WE'LL MAKE THIS TRANSFORMATION ONE DAY.
WE'LL DO IT BECAUSE WE WILL HAVE REACHED THE LIMITS OF OUR NATURAL RESOURCES.
OR BECAUSE THE WARMING WILL SUDDENLY CROSS A THRESHOLD, AND A BRUTAL CLIMATIC PHENOMENON WILL HIT US HARD.
WE WILL ACCOMPLISH THIS CHANGE UNDER THE WORST CONDITIONS.
FORCED BY CIRCUMSTANCE.
AND WAY TOO LATE.
WHEN WE DECIDE TO ACT, CONFRONTED WITH THE OBVIOUS, WE'LL STILL HAVE SEVERAL DECADES OF CLIMATE UPHEAVALS TO ENDURE. THE IMPACT OF THE EMISSIONS OF THE PREVIOUS THIRTY YEARS WILL OCCUR MUCH LATER.

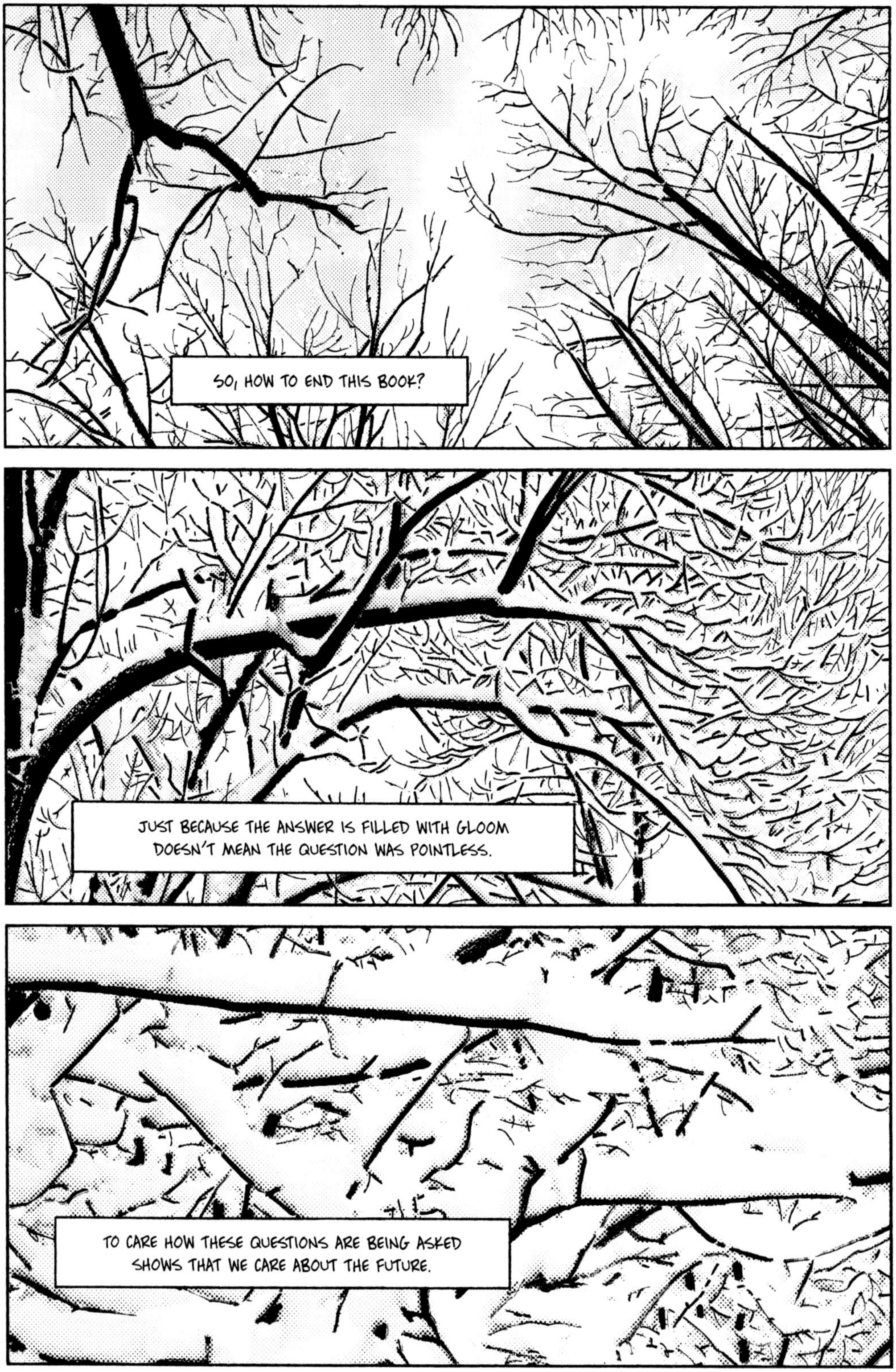
SO, HOW TO END THIS BOOK?
JUST BECAUSE THE ANSWER IS FILLED WITH GLOOM DOESN'T MEAN THE QUESTION WAS POINTLESS.
TO CARE HOW THESE QUESTIONS ARE BEING ASKED SHOWS THAT WE CARE ABOUT THE FUTURE.

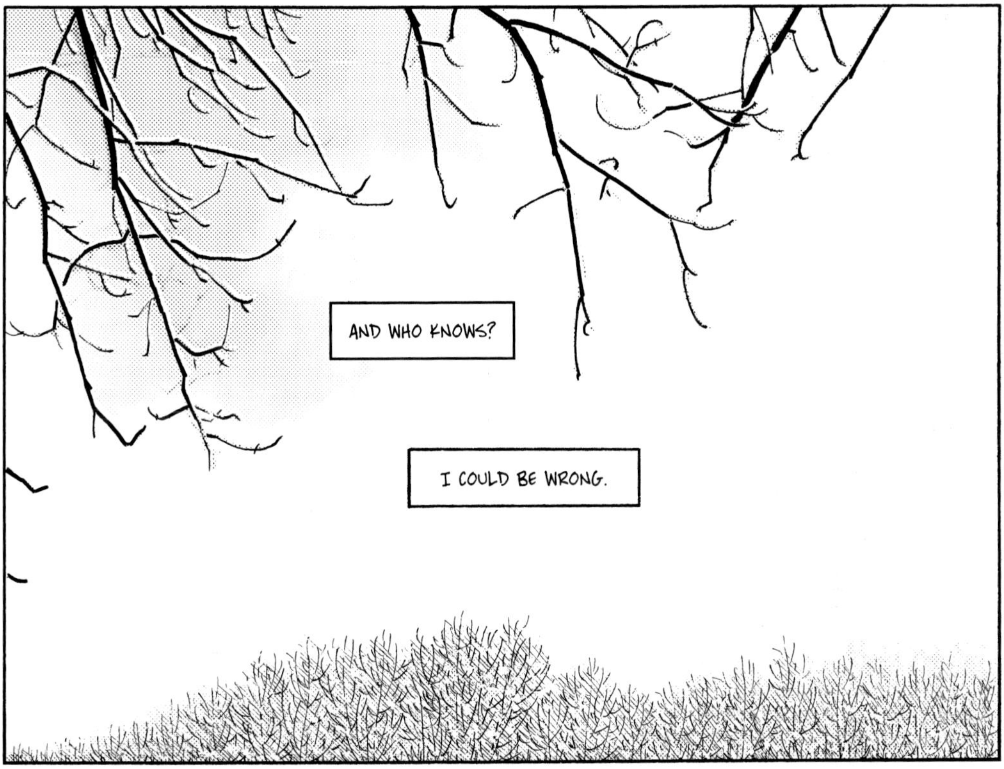
AND WHO KNOWS?
I COULD BE WRONG.
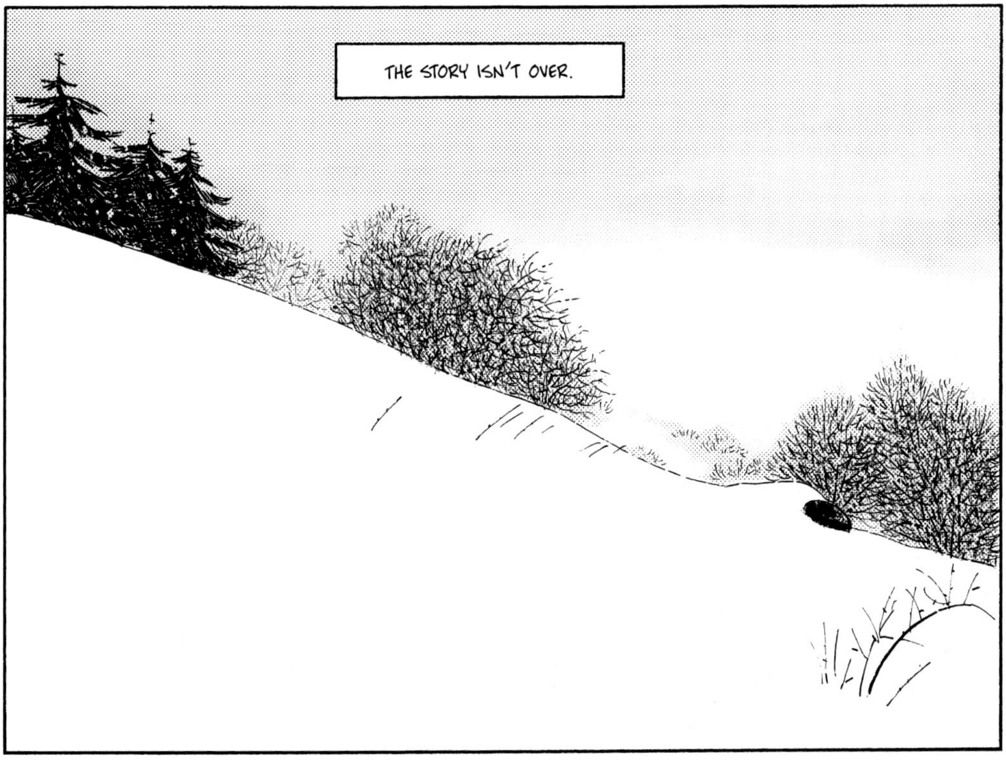
THE STORY ISN'T OVER.

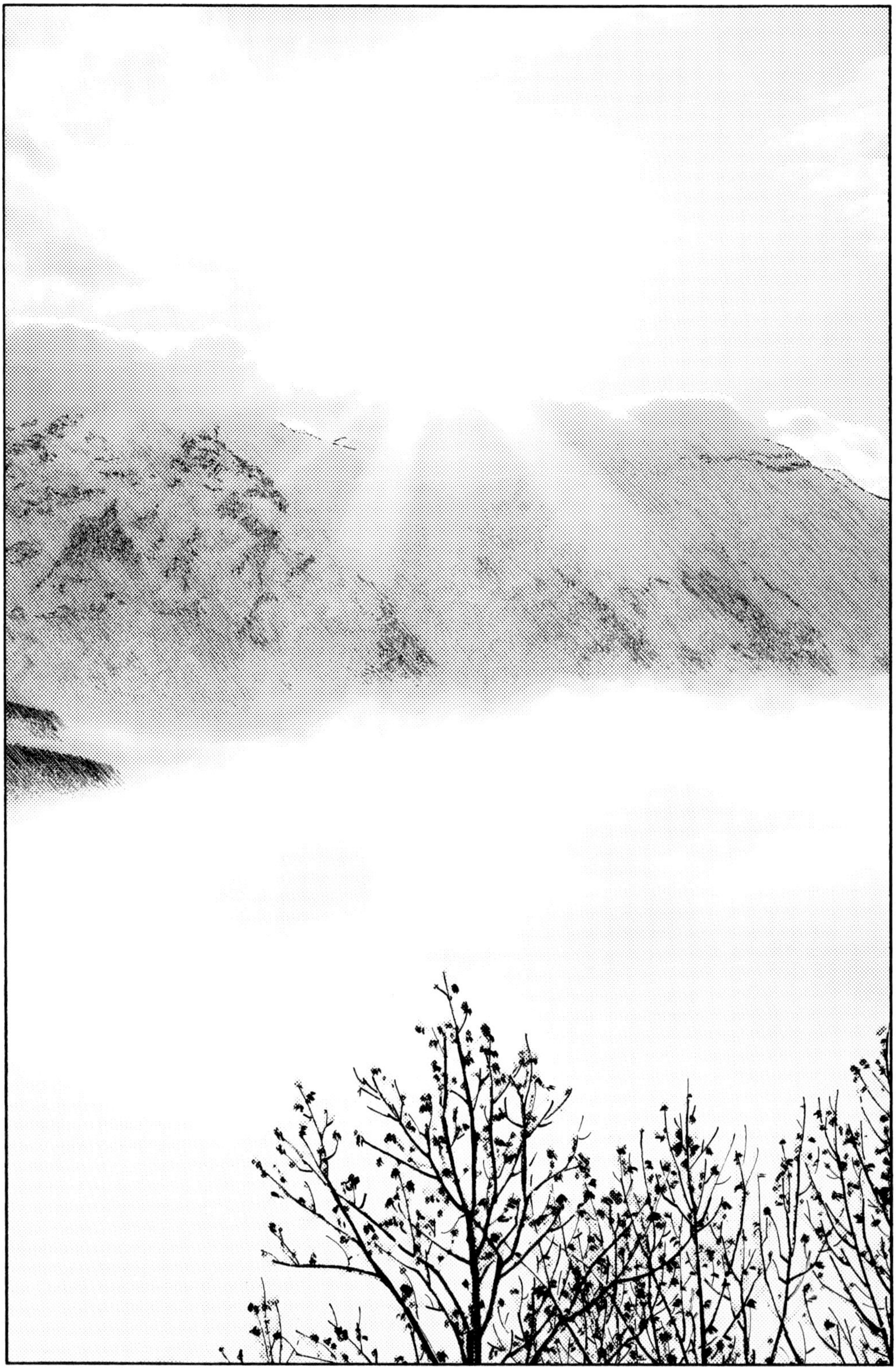

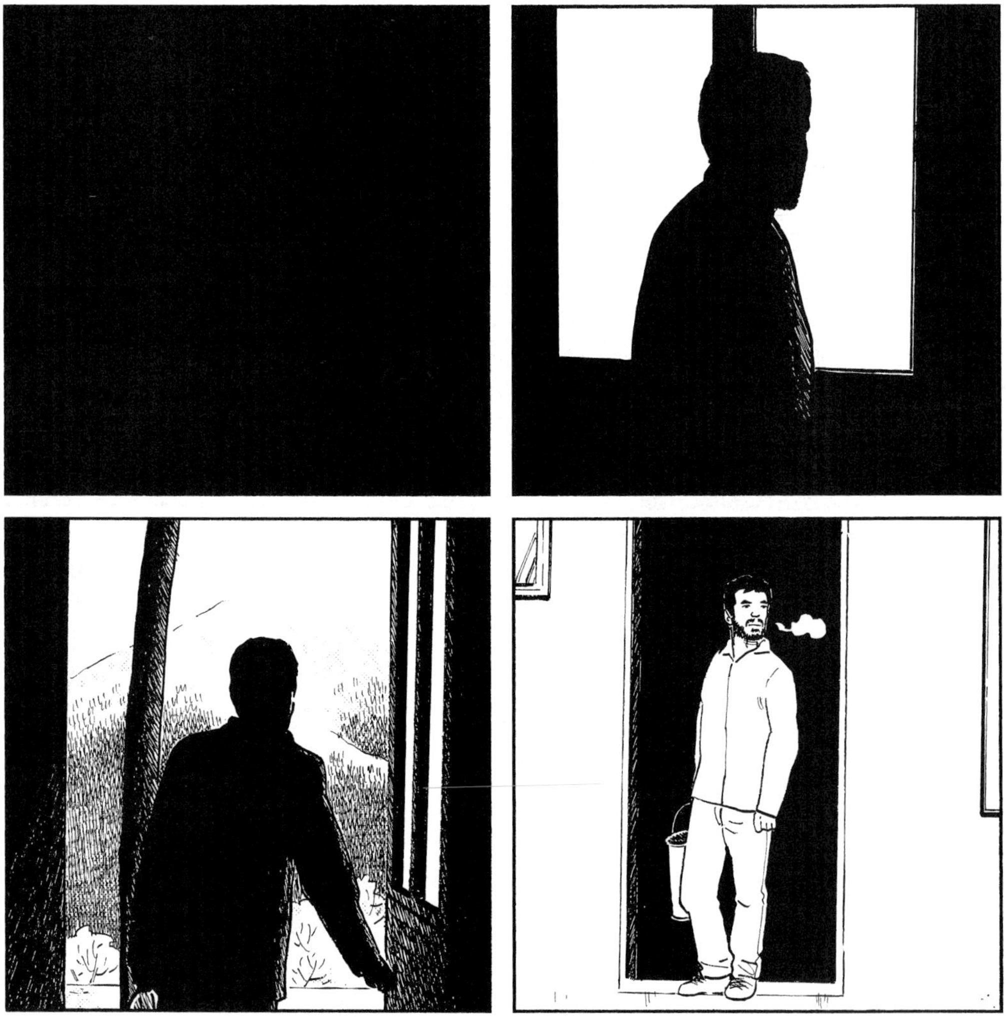

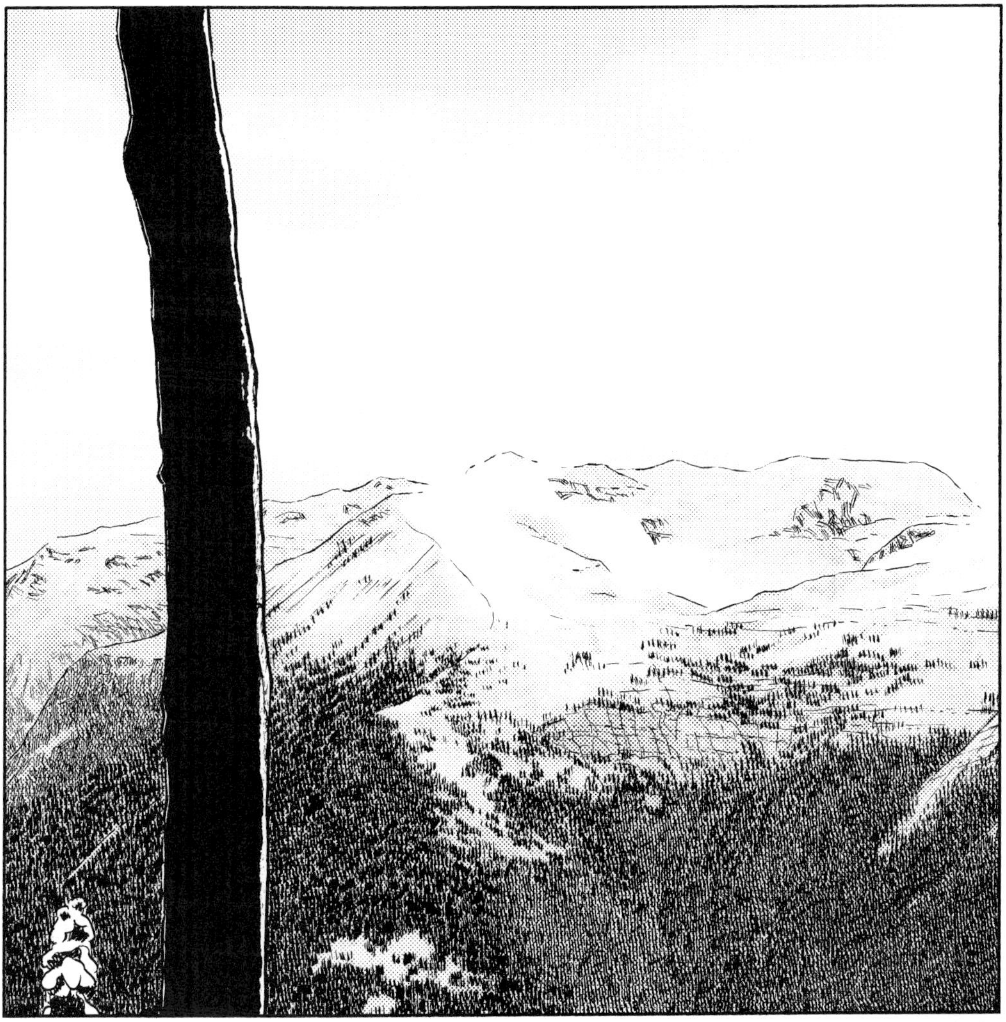

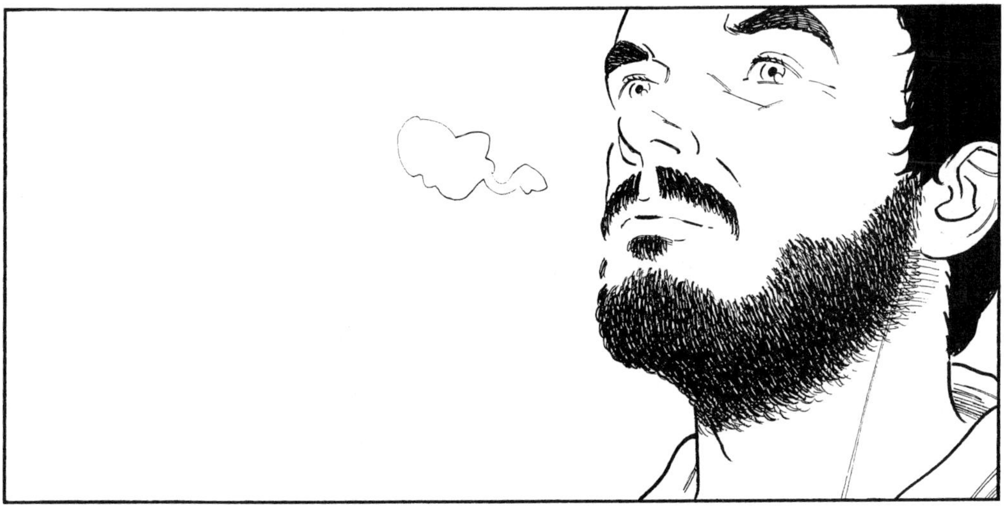

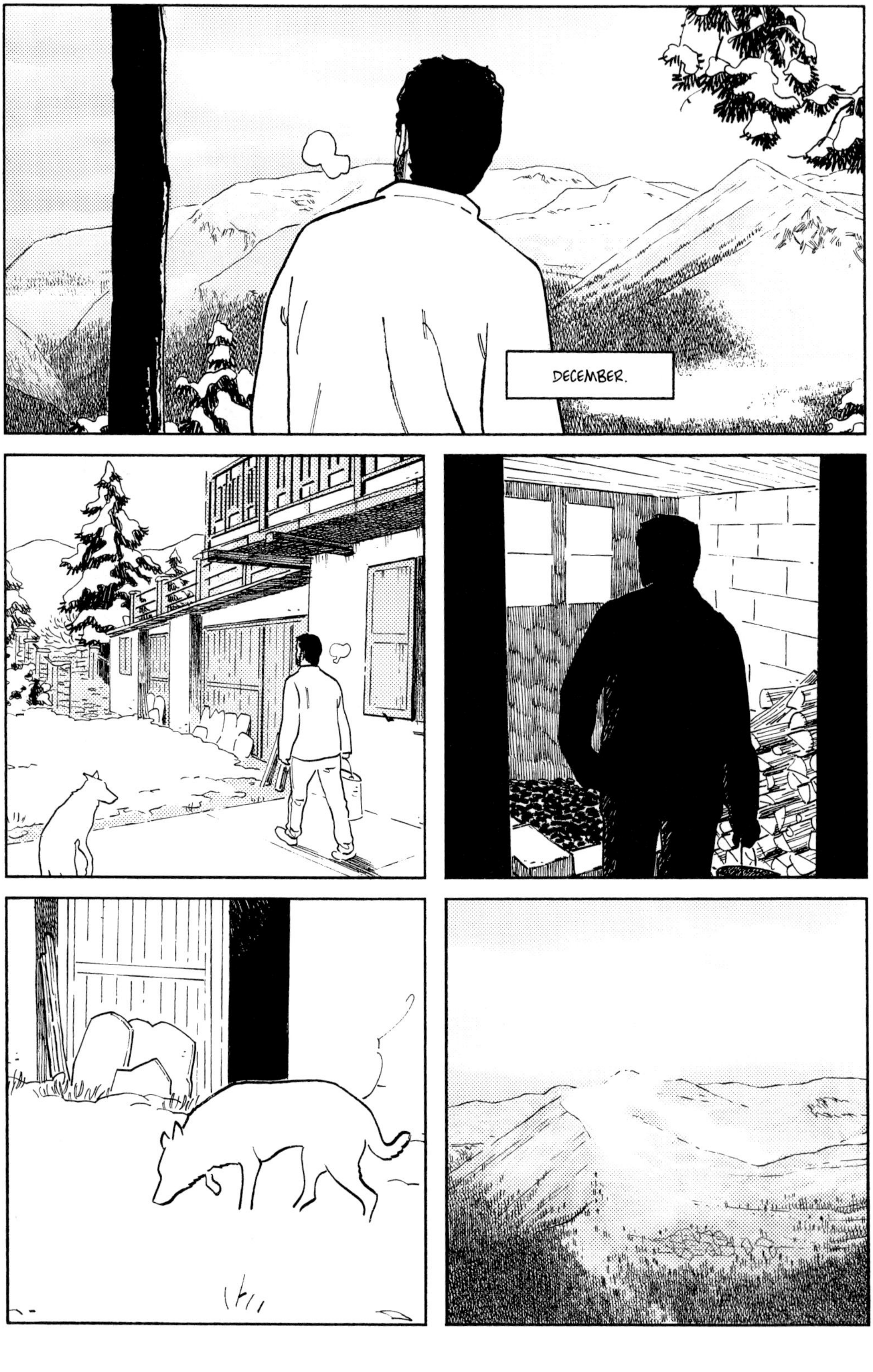
DECEMBER.

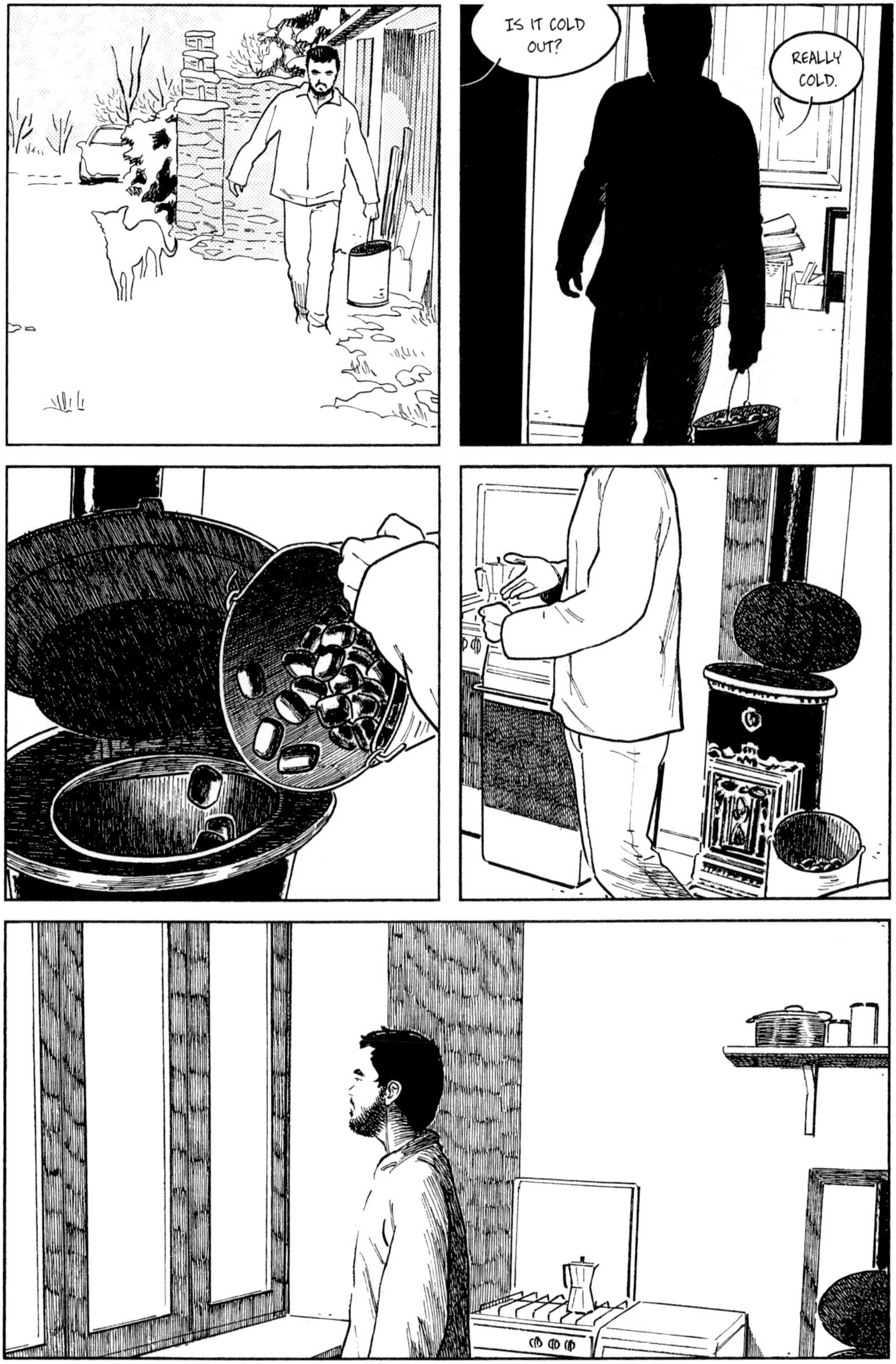
IS IT COLD OUT?
REALLY COLD.

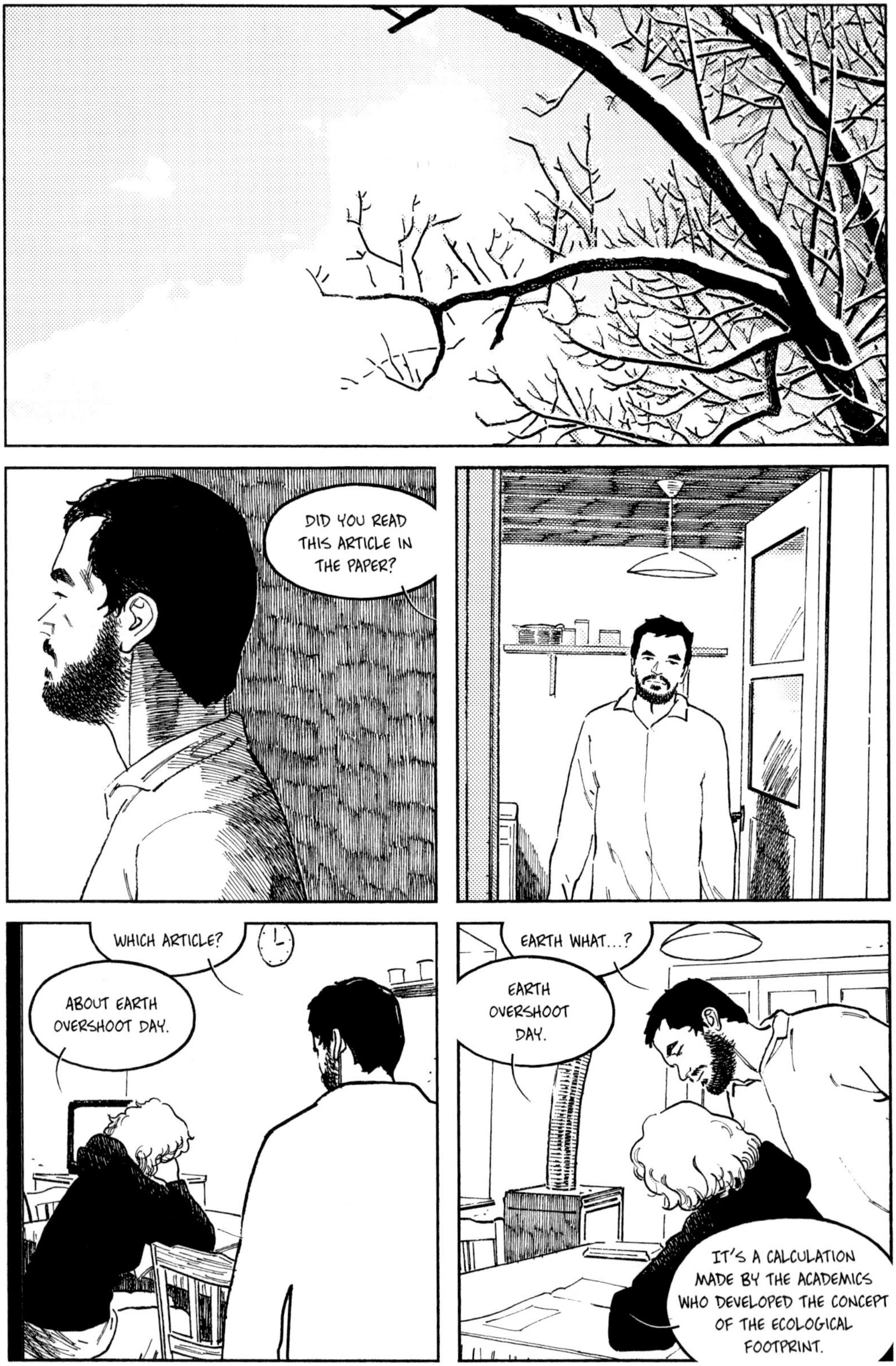
DID YOU READ THIS ARTICLE IN THE PAPER?
WHICH ARTICLE?
ABOUT EARTH OVERSHOOT DAY.
EARTH WHAT...?
EARTH OVERSHOOT DAY.
IT'S A CALCULATION MADE BY THE ACADEMICS WHO DEVELOPED THE CONCEPT OF THE ECOLOGICAL FOOTPRINT.

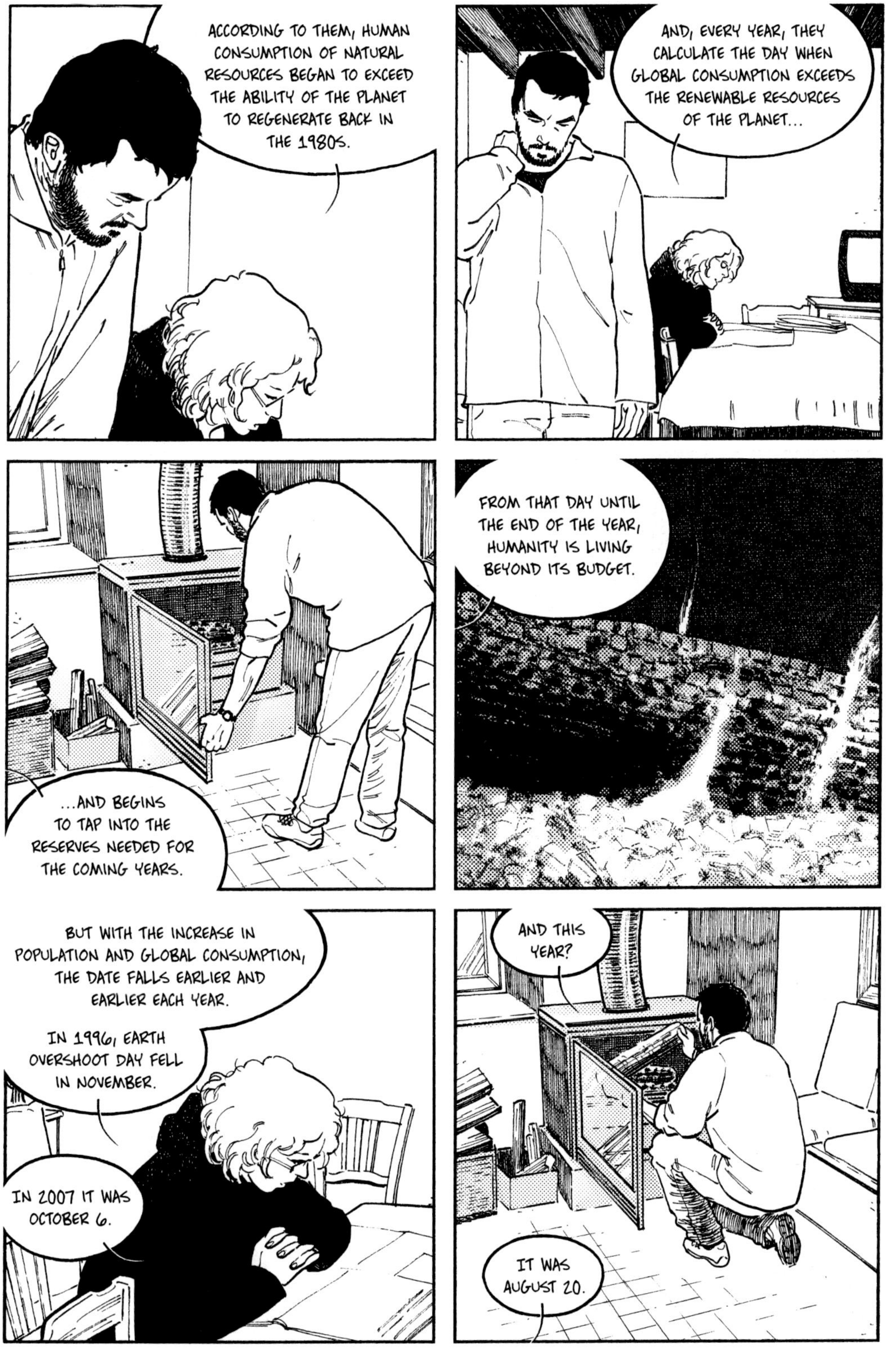
ACCORDING TO THEM, HUMAN CONSUMPTION OF NATURAL RESOURCES BEGAN TO EXCEED THE ABILITY OF THE PLANET TO REGENERATE BACK IN THE 1980S.
AND, EVERY YEAR, THEY CALCULATE THE DAY WHEN GLOBAL CONSUMPTION EXCEEDS THE RENEWABLE RESOURCES OF THE PLANET...
FROM THAT DAY UNTIL THE END OF THE YEAR, HUMANITY IS LIVING BEYOND ITS BUDGET.
...AND BEGINS TO TAP INTO THE RESERVES NEEDED FOR THE COMING YEARS.
BUT WITH THE INCREASE IN POPULATION AND GLOBAL CONSUMPTION, THE DATE FALLS EARLIER AND EARLIER EACH YEAR.
IN 1996, EARTH OVERSHOOT DAY FELL IN NOVEMBER.
IN 2007 IT WAS OCTOBER 6.
AND THIS YEAR?
IT WAS AUGUST 20.

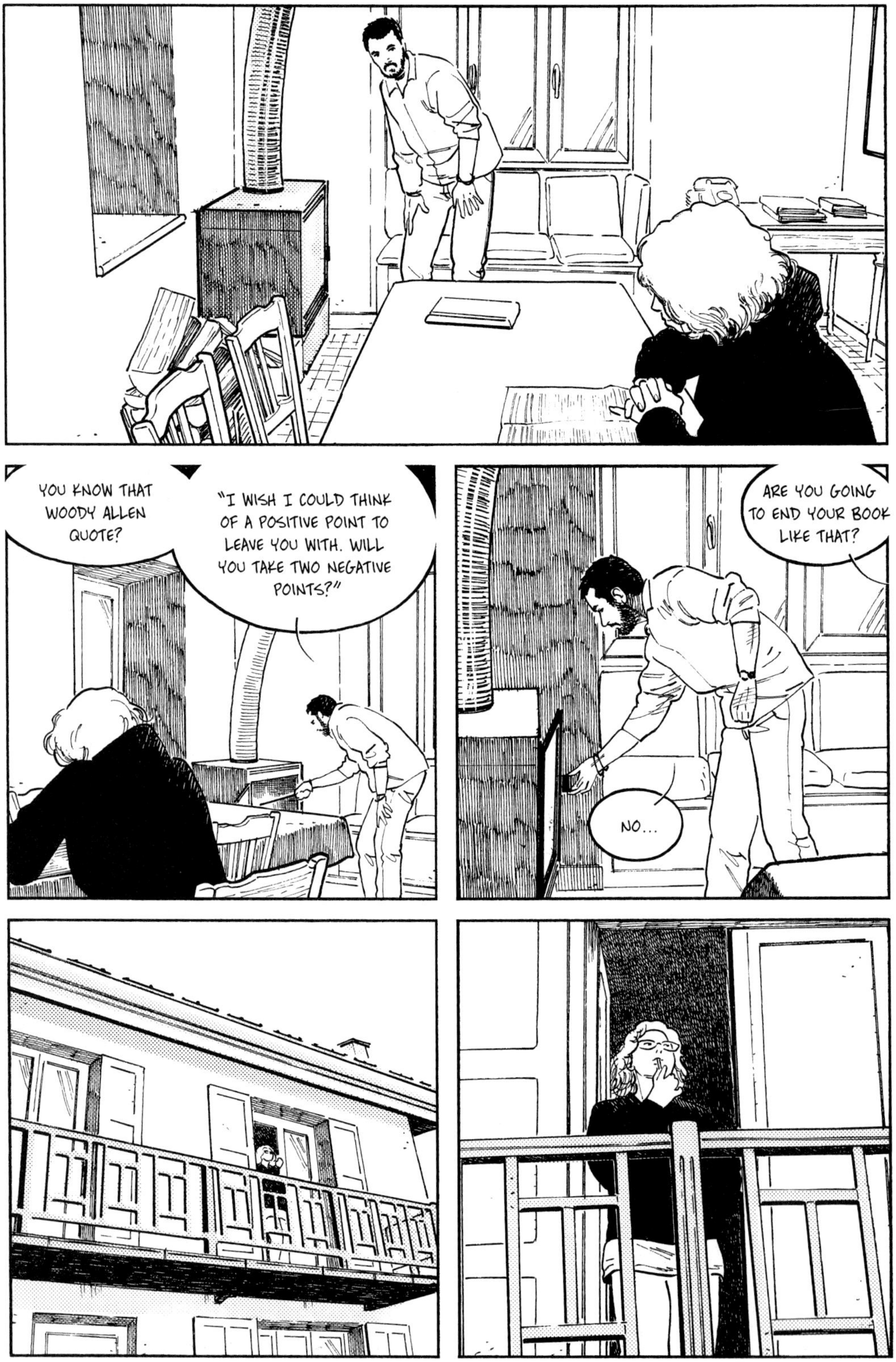
YOU KNOW THAT WOODY ALLEN QUOTE?
"I WISH I COULD THINK OF A POSITIVE POINT TO LEAVE YOU WITH. WILL YOU TAKE TWO NEGATIVE POINTS?"
ARE YOU GOING TO END YOUR BOOK LIKE THAT?
NO...

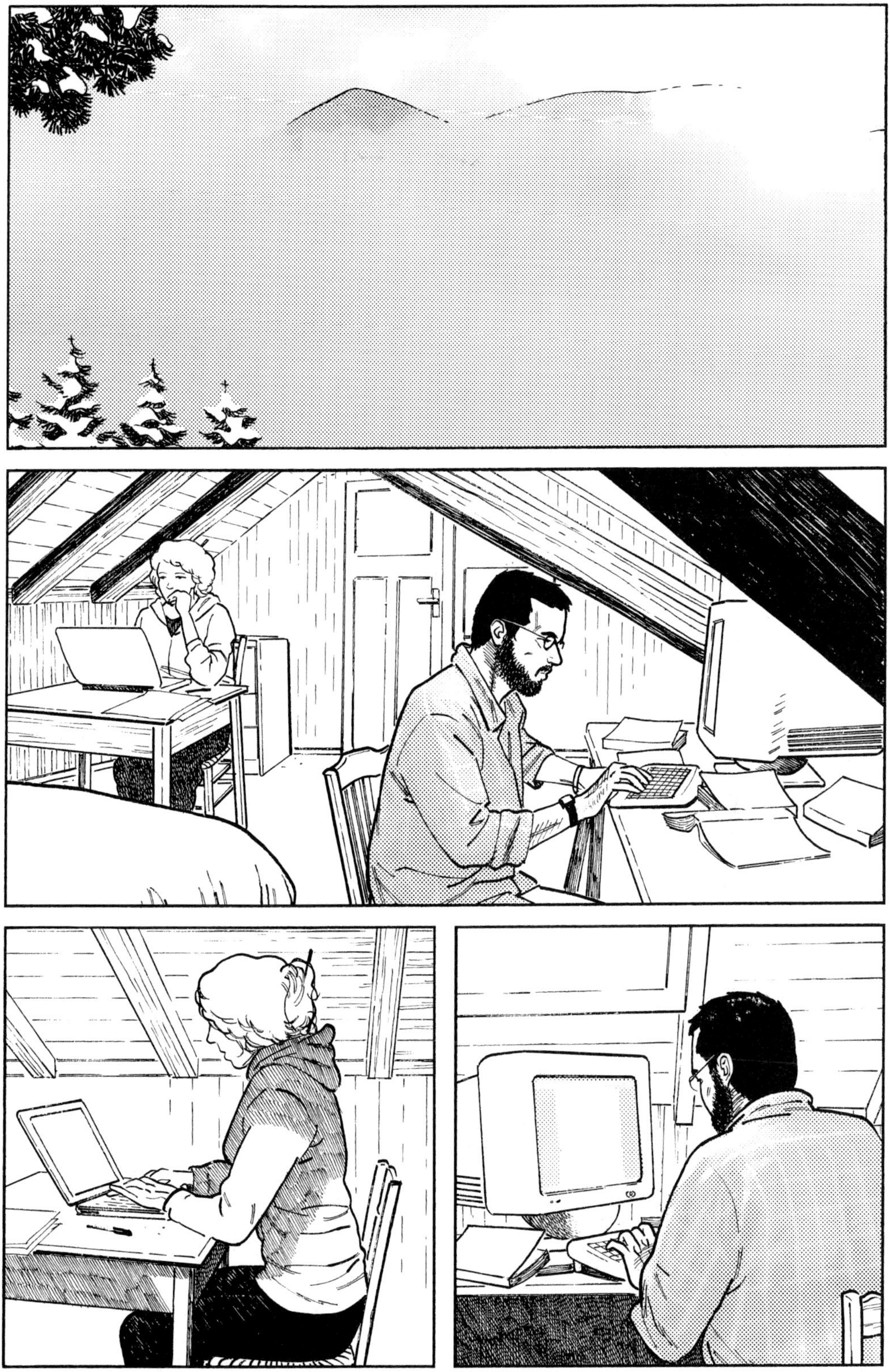

SOURCES

Philippe Squarzoni interviewed nine experts on climate science and the socioeconomic issues surrounding climate change.

The IPCC

The Intergovernmental Panel on Climate Change (IPCC) was created in 1988 at the initiative of two United Nations organizations, the World Meteorological Organization (WMO) and the UN Environment Programme (UNEP), to provide a clear overview of the most recent scientific and socioeconomic information on climate science. Thousands of scientists worldwide volunteer their efforts to the IPCC, and currently 195 countries are members.

Jean Jouzel is a renowned climatologist and the director of research at the Laboratory for Climate Sciences and the Environment (LSCE). He is a member of the American Geophysical Union and the International Glaciological Society and has been director of research at France's Atomic Energy Commission (CEA) since 1995. He has been involved with the IPCC for over twenty years and was a lead author on the second and third IPCC reports. Starting with his early work in 1968 studying polar ice core samples, he has been in the forefront of disseminating information to the public about the impact of human activities on the climate. As vice-chair of the IPCC Working Group on the scientific evidence of climate change, he received the Nobel Peace Prize in 2007 with Al Gore and his fellow IPCC members. Jean Jouzel has authored or co-authored over 250 publications and might very well be one of the most-cited authors in climate science. His book *The White Planet: The Evolution and Future of Our Frozen World* was released in English in 2013 from Princeton University Press.

Hervé Le Treut is a climatologist and director of the Dynamic Meteorology Laboratory of the Pierre Simon Laplace Institute (IPSL) for research in environmental sciences. IPSL gathers six laboratories in developing a common strategy for the study of Earth as a whole and for the study of other objects in the solar system. Hervé Le Treut participated in the fourth and fifth IPCC reports as a lead author and coordinating lead author, respectively, and is a member of the Joint Scientific Committee of the World Climate Research Programme.

Stéphane Hallegatte is a climatologist and engineer and a senior economist with the World Bank Sustainable Development Network. His expertise includes macroeconomic dynamics and green growth strategies, urban environmental policies, climate change vulnerability and adaptation, and disaster risk management. He was a lead author for the 2012 IPCC special report on managing the risks of extreme events and disasters and for its fifth assessment report, to be published in 2014.

Other Experts

Geneviève Azam is an economist whose specialty is exploring the links among ecology, economy, and society. She is a member of the Scientific Council of ATTAC (Association for the Taxation of Financial Transactions and Aid to Citizens), an organization dedicated to developing sustainable globalization and ecological alternatives.

Hélène Gassin is a specialist in environmental management, coauthor of *So watt? L'énergie: Une affaire de citoyens* (So watt? Energy: A civic issue), and commissioner of the French Energy Regulatory Authority (CRE). Previously she monitored international negotiations for Greenpeace.

Jean-Marie Harribey is an economist specializing in social protection and sustainable development. He was co-president of ATTAC from 2006 to 2009 and is a member of its Scientific Council.

Hervé Kempf is an international journalist specializing in environmental issues and is a daily contributor to the influential French newspaper *Le Monde*.

Bernard Laponche is a nuclear physicist. He was an engineer at the CEA and general director of the French Environment and Energy Management Agency (ADEME), a public agency under the joint authority of the Ministry for Higher Education and Research and the Ministry for Ecology, Sustainable Development, and Energy.

René Passet is an economist, a development specialist, and professor emeritus at the Sorbonne.

BIBLIOGRAPHY

Ambroise-Rendu, Marc. *Des cancres à l'Élysée: 5 présidents face à la crise écologique* [Dunces in the Élysée: five presidents face the ecological crisis]. Paris: Éditions Jacob Duvernet, 2007.

Brautigan, Richard. *A Confederate General From Big Sur.* New York: Ballantine Books, 1977.

———. *In Watermelon Sugar.* New York: Vintage/Ebury/Random House, 2002.

Céline, Louis-Ferdinand. *Journey to the End of the Night.* Translated by Ralph Manheim. New York: New Directions, 1983.

Collectif Argos. *Climate Refugees.* Preface by Jean Jouzel. First edition. Cambridge: The MIT Press, 2010.

Delalande, Nicolas. "Le solidarisme de Léon Bourgeois, un socialisme libéral?" ["The solidarism of Léon Bourgeois: A social liberal?"] *La Vie des Idées* (January 30, 2008). http://www.laviedesidees.fr/Le-solidarisme-de-Leon-Bourgeois.html.

Flaubert, Gustave. *Salammbô.* Translated by Mary French Sheldon. London and New York: Saxon & Co., 1885.

Foucart, Stéphane. *Le populisme climatique: Claude Allègre et Cie, enquête sur les ennemis de la science* [Climate populism: Claude Allègre and co. investigate the enemies of science]. Paris: Éditions Denoël, 2010.

Gassin, Hélène and Benjamin Dessus. *So watt? L'énergie: Une affaire de citoyens* [So watt? Energy: A civic issue]. La Tour-d'Aigues, France: Éditions l'Aube, 2006

IPCC, 2007: *Climate Change 2007: Impacts, Adaptation and Vulnerability. Contribution of Working Group II to the Fourth Assessment Report of the Intergovernmental Panel on Climate Change,* M.L. Parry, O.F. Canziani, J.P. Palutikof, P.J. van der Linden and C.E. Hanson, Eds., Cambridge University Press, Cambridge, UK, 7-22.

IPCC, 2007: *Climate Change 2007: Mitigation. Contribution of Working Group III to the Fourth Assessment Report of the Intergovernmental Panel on Climate Change* [B. Metz, O.R. Davidson, P.R. Bosch, R. Dave, L.A. Meyer (eds)], Cambridge University Press, Cambridge, United Kingdom and New York, NY, USA.

IPCC, 2007: *Climate Change 2007: The Physical Science Basis. Contribution of Working Group I to the Fourth Assessment Report of the Intergovernmental Panel on Climate Change* [Solomon, S., D. Qin, M. Manning, Z. Chen, M. Marquis, K.B. Averyt, M.Tignor and H.L. Miller (eds.)]. Cambridge University Press, Cambridge, United Kingdom and New York, NY, USA.

IPCC, 2007: *Climate Change 2007: Synthesis Report. Contribution of Working Groups I, II and III to the Fourth Assessment Report of the Intergovernmental Panel on Climate Change* [Core Writing Team, Pachauri, R.K and Reisinger, A. (eds.)]. IPCC, Geneva, Switzerland, 104 pp.

IPCC, 2011: *IPCC Special Report on Renewable Energy Sources and Climate Change Mitigation* [O. Edenhofer, R. Pichs-Madruga, Y. Sokona, K. Seyboth, P. Matschoss, S. Kadner, T. Zwickel, P. Eickemeier, G. Hansen, S. Schlömer, C. von Stechow (eds)], Cambridge University Press, Cambridge, United Kingdom and New York, NY, USA.

IPCC, 2012: *Managing the Risks of Extreme Events and Disasters to Advance Climate Change Adaptation. A Special Report of Working Groups I and II of the Intergovernmental Panel on Climate Change* [Field, C.B., V. Barros, T.F. Stocker, D. Qin, D.J. Dokken, K.L. Ebi, M.D. Mastrandrea, K.J. Mach, G.-K. Plattner, S.K. Allen, M. Tignor, and P.M. Midgley (eds.)]. Cambridge University Press, Cambridge, UK, and New York, NY, USA, 582 pp.

IPCC, 2013: Summary for Policymakers. In: *Climate Change 2013: The Physical Science Basis. Contribution of Working Group I to the Fifth Assessment Report of the Intergovernmental Panel on Climate Change* [Stocker, T.F., D. Qin, G.-K. Plattner, M. Tignor, S. K. Allen, J. Boschung, A. Nauels, Y. Xia, V. Bex and P.M. Midgley (eds.)]. Cambridge University Press, Cambridge, United Kingdom and New York, NY, USA.

Jancovici, Jean-Marc. *L'avenir climatique: Quel temps ferons-nous?* [The future climate: How much time do we have?]. Paris: Éditions du Seuil, 2005.

Jancovici, Jean-Marc and Alain Grandjean. *Le plein s'il vous plaît! La solution au problème de l'énergie* [Fill it up, please! The solution to the energy problem]. Paris: Points, 2007.

Jouzel, Jean and Anne Debroise, *Le climat, jeu dangereux* [The climate: A dangerous game]. Paris: Éditions Dunod, 2007.

Jouzel, Jean, Claude Lorius, and Dominque Raynaud. *The White Planet: The Evolution and Future of Our Frozen World.* Princeton, NJ: Princeton University Press, 2013.

Kempf, Hervé. *Pour sauver la planète, sortez du capitalism* [To save the planet, get out of capitalism]. Paris: Éditions du Seuil, 2009.

——. *How the Rich Are Destroying the Earth.* White River Junction, VT: Chelsea Green Publishing, 2008.

Lawrence, T. E. *Seven Pillars of Wisdom: A Triumph.* New York: Anchor Books/Random House, 1991.

Le Treut, Hervé and Jean-Marc Jancovici. *L'effet de serre: Allons-nous changer le climat?* [The greenhouse effect: Are we changing the climate?]. Paris: Flammarion, 2009.

Meadows, Donella H., Dennis L. Meadows, and Jorgen Randers. *The Limits to Growth.* White River Junction, VT: Chelsea Green Publishing, 2012.

Nicolino, Fabrice. *Biocarburants, la fausse solution* [Biofuels, the false solution]. Paris: Hachette, 2010.

Pascal, Blaise. *Pensées sur la religion et sur quelques autres sujets* [Thoughts on religion and on some other subjects]. eBooksLib, 2011, Kindle edition.

Proust, Marcel. *À la recherche du temps perdu* [*In Search of Lost Time*]. Paris: Gallimard, 1919.

Rimbaud, Arthur. *A Season in Hell.* Translated by Louise Varèse. Norfolk: New Directions, 1945.

Valéry, Paul. *Regards sur le monde actuel* [*Reflections on the World Today*]. Paris: Librairie Stock, Delamain et Boutelleau, 1931.

ENDNOTES TO THE INTRODUCTION

page 7   *most important onshore denning habitat for polar bears*   George M. Durner et al., "Catalogue of Polar Bear (*Ursus maritimus*) Maternal Den Locations in the Beaufort Sea and Neighboring Regions, Alaska, 1910–2010," U.S. Geological Services Data Series 568, 2010.

page 7   *largest wetlands complex in all of the circumpolar Arctic*   CAVM Team, 2003, Circumpolar Arctic Vegetation Map [Scale 1:7,500,000], Conservation of Arctic Flora and Fauna (CAFF) Map No. 1, Anchorage, Alaska: U.S. Fish and Wildlife Service. www.geobotany.uaf.edu/cavm (accessed Jan. 22, 2014).

page 7   *greatest known densities of nesting shorebirds and waterfowl*   Brad A. Andres, James A. Johnson, Stephen C. Brown, and Richard B. Lanctot, "Shorebirds Breeding in Unusually High Densities in the Teshekpuk Lake Special Area, Alaska," Lakewood, Colorado: U.S. Fish and Wildlife Service.

page 7   *oil and gas in the Arctic Ocean*   The U.S. Arctic Ocean has 23.6 billion barrels of technically recoverable oil and 104.41 trillion cubic feet of technically recoverable gas. *See* U.S. Dep't of the Interior Bureau of Ocean Energy Mgmt., "Proposed Final Outer Continental Shelf Oil & Gas Leasing Program 2012–2017" (June 2012), http://boem.gov/uploaded Files/BOEM/Oil_and_Gas_Energy_Program/Leasing/Five_Year_Program/2012-2017_ Five_Year_Program/PFP%2012-17.pdf (accessed Jan. 22, 2014). One barrel of oil = 0.43 metric tons of CO_2. *See* U.S. Environmental Protection Agency, "Green Power Equivalency Calculator Methodologies," http://www.epa.gov/greenpower/pubs/calcmeth.htm (accessed Nov. 18, 2013). One cubic foot of gas = 0.0545 kg CO_2. *See* U.S. Envtl. Prot. Agency, "2012 Climate Registry Default Emission Factors" (Jan. 6, 2012), http://www.theclimateregistry.org/downloads/2012/01/2012-Climate-Registry-Default-Emissions-Factors.pdf (accessed Jan. 22, 2014). Producing and burning all reserves would result in 15.8 billion metric tons CO_2 (10.1 billion from oil and 5.7 billion from gas).

page 7   *more than nine years of emissions*   According to the U.S. Environmental Protection Agency, in 2011 the CO_2 equivalent for all transportation emissions in the U.S. was 1.745 billion metric tons. *See* U.S. Envtl. Prot. Agency, "Inventory of U.S. Greenhouse Gas Emissions and Sinks: 1990–2011" (Apr. 2013), Executive Summary, http://www.epa.gov/climatechange/Downloads/ghgemissions/US-GHG-Inventory-2013-ES.pdf (accessed Jan. 22, 2014).

ACKNOWLEDGMENTS

I especially want to thank Geneviève Azam, Hélène Gassin, Stéphane Halle-gate, Jean-Marie Harribey, Jean Jouzel, Hervé Kempf, Bernard Laponche, Hervé Le Treut, and René Passet. Thank you also to Mary Benilde, Gus Massiah, Sandrine Mathy, and Susan George.

This book also owes a great debt to the research and work of Marc Ambroise-Rendu, Édouard Bard, Marie-Odile Delacour, Bertrand Barré, Philippe Bovet, Lester Brown, Loïc Chauveau, Laure Chémery, Anne De-broise, Nicolas Delalande, Olivier Delbard Frédéric Denhez, Benjamin Dessus, Jean-Louis Fellous, Geneviève Férone, Stéphane Foucart, Al Gore, Alain Granjean, Sylvestre Huet, Nicolas Hulot, Jean-Marc Jancovici, Spike Lee, Patrick Le Tréhondat, Fabrice Nicolino, Bernard Perret, Yves Sciama, Patrick Silberstein, Jacques Testart, and Christine Tréguier.

A tribute is made in these pages to the works of Charlie Chaplin, Francis Ford Coppola, Michael Curtiz, Walt Disney, John Ford, Alfred Hitchcock, Irvin Kershner, Stanley Kubrick, Akira Kurosawa, David Lean, Sidney Lumet, Alan Parker, Arthur Penn, Sydney Pollack, Martin Scorsese, and Steven Spielberg.

Thank you to Émilie Huard for the invitation. Thanks to Fabrice Neaud, Guillaume Bouzard, and Étienne Davodeau for the helping hand at the start of this project.

Thank you to Grégoire Seguin for his support and confidence. Bérénice, Violaine, Prune, Angèle, and Caroline were kind enough to read parts of the manuscript.

And thank you to Camille, above all.

ABOUT THE AUTHOR

Philippe Squarzoni spent his childhood in Ardèche and Reunion Island and attended college in Lyon. Between 1994 and 1996, he visited Croatia and the former Yugoslavia several times as a volunteer in a conflict-resolution project. Over the next few years he traveled through Mexico, Palestine, and Israel as a human-rights observer. His first political graphic novels—*Garduno, en temps de paix* (Garduno in peacetime) and *Zapata, en temps de guerre* (Zapata in wartime)—were published in 2002 and 2003. He has written on the subject of infanticide in *Crash-Text*, the Holocaust in *Drancy-Berlin-Oswiecim*, and about mental disability in *Les mots de Louise,* and is the illustrator of a work by Richard Brautigan.

INDEX